STP 1007

Aquatic Toxicology and Environmental Fate: Eleventh Volume

Glenn W. Suter II and Michael A. Lewis, editors

ASTM
1916 Race Street
Philadelphia, PA 19103

ASTM Publication Code Number (PCN): 04-010070-48
ISBN: 0-8031-1180-0
ISSN: 1040-306X

NOTE

The Society is not responsible, as a body,
for the statements and opinions
advanced in this publication.

Peer Review Policy

Each paper published in this volume was evaluated by three peer reviewers. The authors addressed all of the reviewers' comments to the satisfaction of both the technical editor(s) and the ASTM Committee on Publications.

The quality of the papers in this publication reflects not only the obvious efforts of the authors and the technical editor(s), but also the work of these peer reviewers. The ASTM Committee on Publications acknowledges with appreciation their dedication and contribution of time and effort on behalf of ASTM.

Printed in Ann Arbor, MI
May 1989

Foreword

This publication, *Aquatic Toxicology and Environmental Fate: Eleventh Volume,* contains papers presented at the Eleventh Symposium on Aquatic Toxicology and Hazard Assessment, which was held 10–12 May 1987, in Cincinnati, Ohio. This eleventh symposium was part of a series of annual symposia on Aquatic Toxicology sponsored by ASTM, through its Committee E-47 on Biological Effects and Environmental Fate. The chairmen of this symposium were Glenn W. Suter II, Oak Ridge National Laboratory, and Michael A. Lewis, Procter and Gamble Co. Both men also served as editors of this publication.

Contents

Introduction

This volume marks not only the beginning of the second decade of ASTM symposia on aquatic toxicology, but also the beginning of the second decade of aquatic toxicology as a mature science. The science of aquatic toxicology matured in 1977 in the sense that it adopted a paradigm for its practice: hazard assessment by tiered testing and iterative comparison of expected ambient and toxic concentrations. That paradigm was first described in the first ASTM aquatic toxicology symposium *(ASTM STP 634, Aquatic Toxicology and Hazard Evaluation: First Symposium).* The past decade was largely devoted to implementing this paradigm.

As the second decade begins, an alternate paradigm, risk assessment, is gaining popularity with regulatory agencies. Risk assessment is concerned with estimating the probability of specific undesirable effects. It is more demanding than hazard assessment in that it must deal explicitly with the nature and magnitude of effects and with uncertainty about those effects. It will require development of new methods and refinement of old methods of testing and analysis.

In addition, a number of problems are apparent that have not been adequately addressed. These include (1) large-scale cumulative effects of multiple pollution sources on regional communities and populations, such as the Chesapeake Bay and striped bass; (2) effects of spills, storm runoff, and other nonroutine exposures with complex temporal dynamics; (3) effects of variation in organism condition on susceptibility; (4) predicting effects of new chemicals whose mode of action is unknown; (5) cumulative effects of pollution, harvesting, modification of the physical environment, and natural stochastic processes; and (6) balancing risks and benefits of different levels of testing, monitoring, and waste treatment.

These and other problems constitute a significant challenge for the coming decade. The editors believe that the papers presented in this volume demonstrate a capability and a willingness to tackle these difficult problems.

Glenn W. Suter II

Oak Ridge National Laboratory, Oak Ridge, TN 37831; symposium chairman and editor.

Michael A. Lewis

Procter and Gamble Co., Cincinnati, OH 45217; symposium chairman and editor.

Fate of Chemicals in Water and Sediment

James G. Sanders[1] *and George R. Abbe*[1]

Silver Transport and Impact in Estuarine and Marine Systems

REFERENCE: Sanders, J. G. and Abbe, G. R., "**Silver Transport and Impact in Estuarine and Marine Systems,**" *Aquatic Toxicology and Environmental Fate: Eleventh Volume, ASTM STP 1007,* G. W. Suter II and M. A. Lewis, Eds., American Society for Testing and Materials, Philadelphia, 1989, pp. 5–18.

ABSTRACT: The transport of silver through estuarine and coastal marine systems is dependent upon biological uptake and incorporation. Uptake by phytoplankton is rapid, in proportion to silver concentration, and inversely proportional to salinity. In contrast to studies performed with other toxic metals, silver availability appears to be controlled by both the free silver ion concentration and the concentrations of other silver complexes, perhaps the ion pair AgCl. Silver incorporated by phytoplankton is not lost as salinity increases; as a result, silver associated with cellular material is largely retained within the estuary. Phytoplankton exhibit a variable sensitivity to silver. Sensitive species exhibit a marked delay in the onset of growth in response to silver at low concentrations, even though maximum growth rates are similar to controls. Sublethal effects, such as delayed growth, may not be observed during classical toxicity testing of a population; however, the effects to the community may be quite significant. A delay in the onset of growth reduces the ability of a population to respond to short-term favorable conditions and to succeed within the community, ensuring the dominance of resistant species with high initial rates of growth. Reliable assessment of the vulnerability of dynamic ecosystems requires that studies of potential pollutant impact consider geochemical and biogeochemical transformations and sublethal pressures upon community structure as well as the toxicity of the pollutant to keystone species.

KEY WORDS: algae, biogeochemistry, estuaries, silver, sublethal effects, toxicity testing, aquatic toxicology

The transport of trace metals through coastal zones depends on physical processes which control water flow and the mixing of fresh and salt water and on geochemical processes that occur as salinity increases as the element passes from the river into the estuary. These two subjects have received a great deal of study [*1–6*] and will continue to receive attention in years to come. Trace metal transport is also dependent upon biological processes. Trace metals can be actively or passively taken up by organisms, altering their partitioning, chemical reactivity, and biological availability; sometimes such biological alterations are quite large. Such biologically mediated reactions are the subject of this article.

Silver is an example of a biologically active trace metal. A natural product of geological weathering, it is found normally in rivers at concentrations of 0.3 $\mu g\ L^{-1}$ [*6*] and in oceans at 0.1 to 5.0 ng L^{-1} [*7,8*]. Concentrations in estuaries fall somewhere in between. It is quite toxic to aquatic organisms, although the level of silver necessary to impair organism function varies greatly. Toxicity data are difficult to assess, however, because of the large dif-

[1] Director and senior scientist, respectively, The Academy of Natural Sciences, Division of Environmental Research, Benedict Estuarine Research Laboratory, Benedict, MD 20612.

ferences in experimental design and the amount of attention paid to ensuring that the species tested are appropriate and the design approximates natural conditions. Too often, bioassays are performed under geochemical conditions that are not at all similar to natural aquatic systems, and the geochemical controls of trace metal speciation and reactivity are ignored.

Over the past two years, the authors of this paper have been studying the interactions of silver with estuarine and coastal marine phytoplankton. We are interested in the role of phytoplankton in the transport and fate of silver, and in biological and geochemical controls of silver toxicity. This article presents the results of our work and attempts to place these results into an overall perspective.

Methods

Phytoplankton

A number of species of phytoplankton were maintained in the laboratory and used for experiments, including *Thalassiosira pseudonana,* Clone SWAN1; *T. pseudonana,* Clone W; *Skeletonema costatum,* Clone SCOST; *Isochrysis* sp., Clone TISO; *Prorocentrum mariae-lebouriae,* Clone PROML; and *Chroomonas* sp., Clone LB2000. SWAN1, W, and TISO were obtained from the Provasoli-Guillard Center for Culture of Marine Phytoplankton; LB2000 was obtained from the Algal Culture Collection of Texas; PROML was obtained from Dr. Richard Rivkin; and SCOST was isolated by Stephen Cibik. These species cover a wide range of different algal taxa, including centric diatoms, chrysophytes, cryptophytes, dinoflagellates, and chlorophytes, and represent the major dominant species within the Chesapeake Bay, other temperate estuaries, and neritic waters. These clones and others are maintained on a regular basis under conditions of 25°C and 15 to 20 parts per thousand salinity in f/2 medium [*9*].

Silver Uptake

Three of the above species (SWAN1, TISO, and PROML) were maintained at three salinities (10, 15, and 20 parts per thousand) at 25°C. Culture medium varied in that concentrations of nitrogen, phosphorus, silicon, and vitamins were adjusted to f/5 [*9*]; no trace metals or chelators were added. Each species was inoculated into fresh medium at each salinity at cell densities of 10^3 cells mL^{-1} and allowed to grow to stationary phase (5 to 7 days). Silver was then added at three concentrations (0, 2, and 5 $\mu g\ L^{-1}$), and cellular uptake was determined after 24 and 48 h. Ambient silver concentrations were 0.05 $\mu g\ L^{-1}$. Cells were removed from the culture medium by gentle filtration through acid-washed glass fiber filters. The filters were washed and the cell material and filtrate were retained for silver analysis.

In a parallel experiment, phytoplankton were loaded with silver as above, then transferred to a higher salinity. After 24 h, the cells were filtered and the cell material and filtrate analyzed to determine the extent of silver desorption.

Silver Toxicity

Sensitivity of the above phytoplankton species was measured using essentially the technique of Murphy and Belastock [*10*], as modified by Sanders [*11*]. Phytoplankton were placed into 25-mm glass test tubes containing medium with concentrations of nitrogen, phosphorus, silicon, and vitamins equivalent to f/5 (no trace metal or chelator additions) at initial cell densities of 10^3 cells mL^{-1}. Culture tubes were prepared in triplicate at 5 different silver concentrations (0.05, 2.0, 5.0, 10.0, and 20.0 $\mu g\ L^{-1}$) and 4 different salinities (7.5, 15, 22.5, and 30 parts per thousand.) Culture tubes were maintained at 65 μeinsteins $m^{-2}\ s^{-1}$ on a 12:12 light/dark cycle at 25°C. The biomass of each culture tube was estimated daily at the same hour by direct measurement of *in vivo* fluorescence in a Turner-Designs fluorometer, a rapid and efficient method of estimating chlorophyll concentrations [*12*]. Growth, in fluorescence doublings per day [*11*], was calculated for each species under the various culture conditions.

Silver Analysis

Water—The silver concentration of low salinity water was determined by direct injection graphite furnace atomic absorption spectrophotometry (GFAAS), with standards prepared in the same salinity. Water samples of salinities >15 parts per thousand were analyzed after chelation and solvent extraction with standard addition [*13*]. Water samples with >15 parts per thousand salinity and with silver concentrations >0.5 $\mu g\ L^{-1}$ were also measured by direct injection. Limits of detection were 0.1 $\mu g\ L^{-1}$ for the direct injection technique and 0.02 $\mu g\ L^{-1}$ for the chelation and solvent extraction technique.

Tissues—Algal material was digested in teflon vials in redistilled concentrated nitric acid (HNO_3) using the technique of Sanders and Windom [*14*]. After 18 h of cold digestion, the vials were heated to 90°C until nearly dry. Residues were dissolved in 1% redistilled HNO_3 and analyzed by GFAAS with standards prepared in the appropriate matrix.

Modeling of Chemical Speciation

The changes in silver's chemical form (speciation) and complexation with varying silver concentrations and salinities were modeled using a chemical speciation model, MICROQL [*15*], modified by Riedel.[2] The model addressed the complexation reactions of 6 cations and 6 anions and the formation of 43 chemical species at constant pH (8.0) and calculated the predicted concentrations of 15 silver species. The total concentrations of all ions were set using realistic values for their concentrations at the salinities under study; silver concentrations were adjusted as necessary to match the additions above.

Results

Silver Uptake

All species of phytoplankton took up large quanities of silver; however, the uptake varied between species. The chrysophyte, *Isochrysis* sp., took up more silver on a weight basis than the other species, with the diatom *Thalassiosira pseudonana* incorporating almost as much silver (Fig. 1). *Prorocentrum mariae-lebouriae* took up much less silver. On average, algae grown in control media contained 0.33 ± 0.08 $\mu g\ g^{-1}$ (mean ± standard error); cells exposed to 2 $\mu g\ L^{-1}$ silver contained an average of 18.73 ± 1.62 $\mu g\ g^{-1}$, and cells exposed to 5 $\mu g\ L^{-1}$ contained an average of 42.81 ± 3.90 $\mu g\ g^{-1}$. The total amount of silver removed from solution by cellular uptake ranged between 17 to 68% of the silver available.

Salinity had a significant effect on silver uptake by all species at each concentration of silver [two-way analysis of variance (ANOVA); $P < 0.05$], as illustrated in Fig. 1. The highest uptake of silver occurred at the lowest salinity. Overall, doubling the salinity from

[2] Riedel, G. F., personal communication.

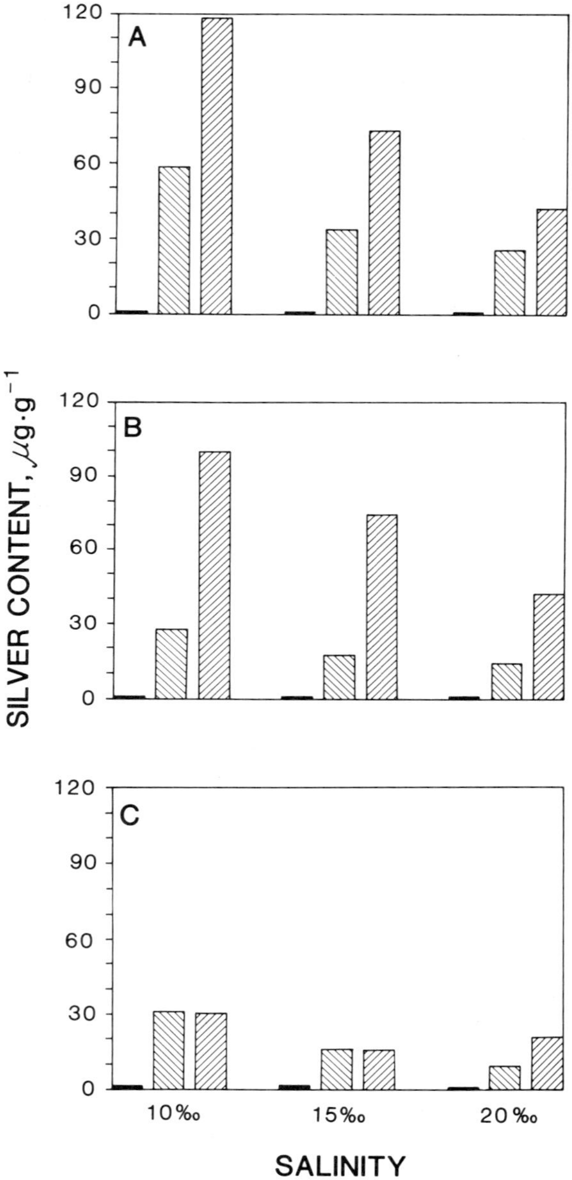

FIG. 1—*Accumulation of silver, in micrograms per gram, by three phytoplankton species at varying silver concentrations (0.05, 2.0, and 5.0 μg L^{-1}) and salinities (10, 15, and 20 parts per thousand):* ■ = *0.05 μg L^{-1};* ▧ = *2.0 μg L^{-1};* ▨ = *5.0 μg L^{-1};* (a) Isochrysis *sp.;* (b) Thalassiosira pseudonana; *and* (c) Prorocentrum mariae-lebouriae.

10 to 20 parts per thousand resulted in a reduction in silver uptake of 50%. Uptake of silver was intermediate at the intermediate salinity.

The duration of exposure to silver affected the cellular content of two algal species, *Thalassiosira pseudonana* and *Isochrysis* sp. These species exhibited decreases in total cell content of silver when exposure increased from 24 to 48 h. These decreases averaged 34% for *T. pseudonana* and 69% for *Isochrysis* sp. The third species, *Prorocentrum mariae-lebouriae,* did not exhibit a similar decrease over time.

When silver-laden phytoplankton were resuspended into culture media containing a higher salinity, no significant release of silver was noted. Phytoplankton exposed to a salinity increase of 5 parts per thousand exhibited an average loss of 1.4 $\mu g\ g^{-1}$, or 6% of total silver associated with the cell. Cells exposed to a 10 parts per thousand increase released 0.1 $\mu g\ g^{-1}$, or 0.5%. Both losses were well within the analytical error of the measurement (9%) and, as such, were considered negligible.

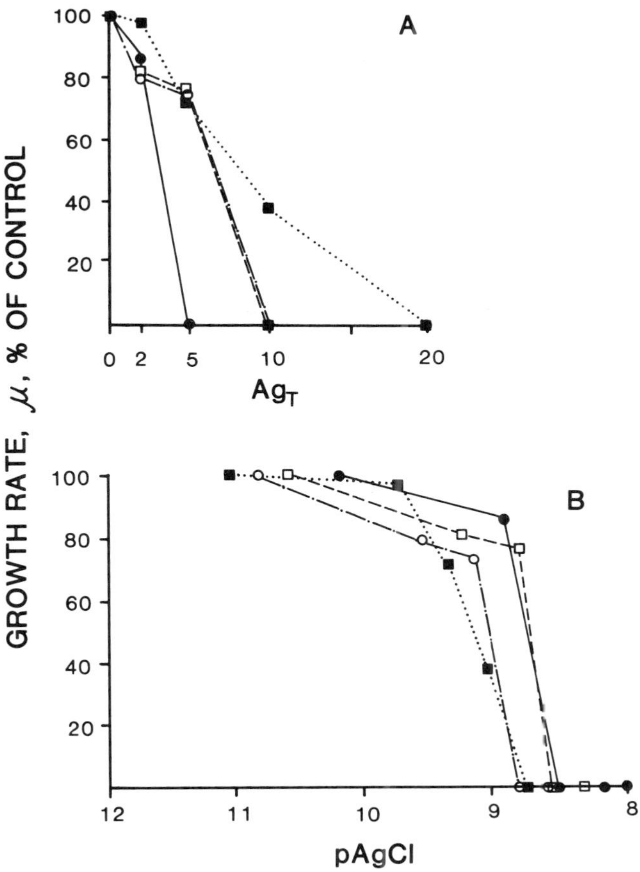

FIG. 2—*Inhibition of* Prorocentrum mariae-lebouriae *growth, as a percentage of control, with increasing silver concentrations at four salinities:* ●———● = *7.5 parts per thousand;* □ – □ = *15 parts per thousand;* ○— · —○ = *22.5 parts per thousand;* ■ · · · ■ = *30 parts per thousand;* (a) *expressed as a function of total dissolved silver concentration, in micrograms per litre;* (b) *expressed as a function of pAgCl (—log AgCl concentration, in moles per litre).*

Silver Toxicity

There were marked differences between the sensitivity of algal species to silver. The two flagellates tested, *Prorocentrum mariae-labouriae* and *Chroomonas* sp., exhibited the highest sensitivity to silver at all salinities (Fig. 2). The diatoms, *Skeletonema costatum* and *Thalassiosira pseudonana*, were considerably less sensitive (Fig. 3). The algal species exhibited three different responses to silver: cessation of growth, delayed onset of growth (increased lag phase), or unaffected growth. All three responses are visible in the illustration of *T. pseudonana* growth at 7.5 parts per thousand (Fig. 4). Cultures that exhibited delayed onset of growth (for example, cultures exposed to 5 and 10 $\mu g\ L^{-1}$ in Fig. 4) often eventually grew at a rate equal to controls; however, because of the initial lag period, the yield (cell biomass) with time of these cultures remained lower throughout the experiment (Table 1).

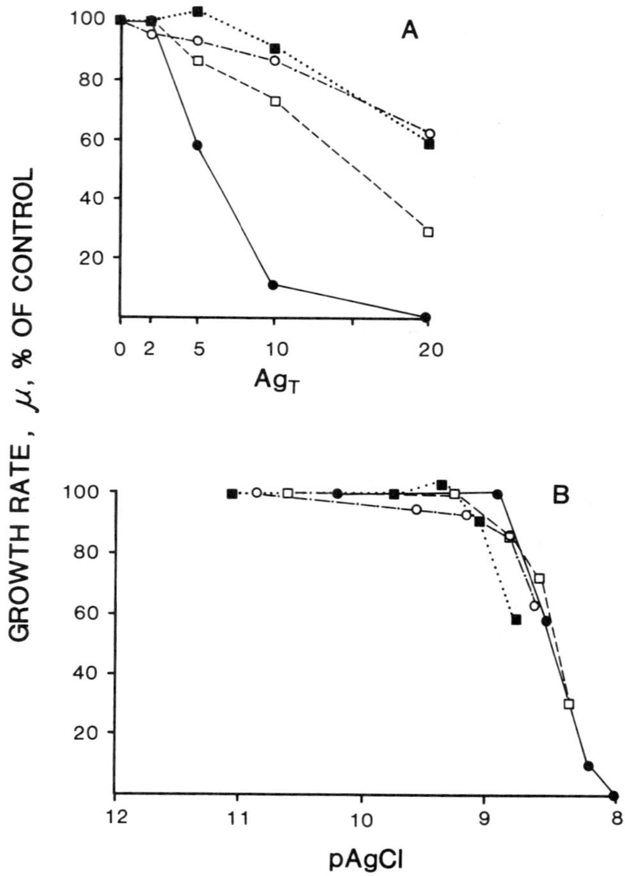

FIG. 3—*Inhibition of* Skeletonema costatum *growth, as a percentage of control, with increasing silver concentrations at four salinities:* ●——● *= 7.5 parts per thousand;* □— —□ *= 15 parts per thousand;* ○—·—○ *= 22.5 parts per thousand;* ■ · · · ■ *= 30 parts per thousand;* (a) *expressed as a function of total dissolved silver concentration, in micrograms per litre;* (b) *expressed as a function of pAgCl (*−log *AgCl concentration, in moles per litre.)*

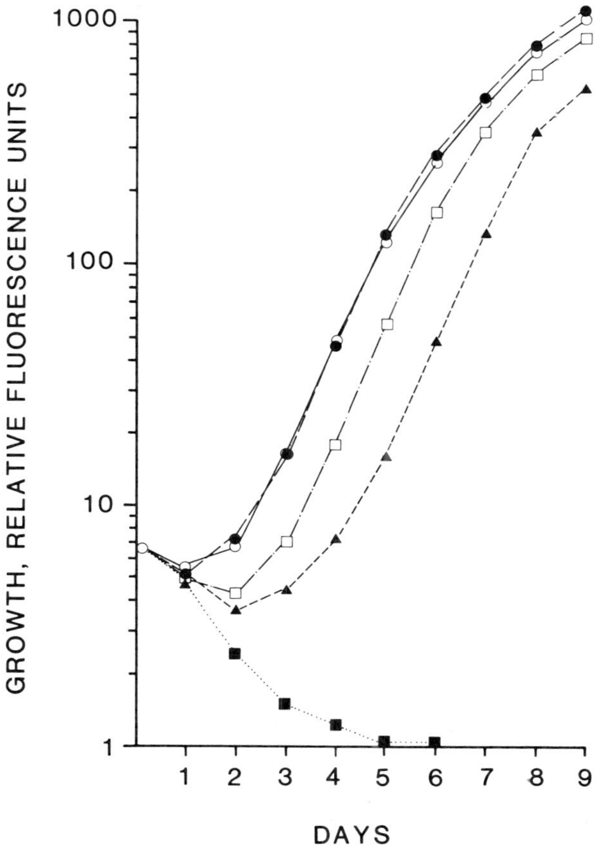

FIG. 4—*Growth of* Thalassiosira pseudonana *at a salinity of 7.5 parts per thousand and at various silver concentrations:* ○——○ = *0.05 μg L^{-1};* ●— — ● = *2.0 μg L^{-1};* □— · —□ = *5.0 μg L^{-1};* ▲——▲ = *10 μg L^{-1};* ■ · · · ■ = *20 μg L^{-1}.*

Often, the reduction in yield was a more sensitive indicator of algal stress by silver than was the growth rate measurement (Tables 1 and 2).

Regardless of species, silver toxicity was inversely proportional to salinity. Species increased their tolerance to silver at higher salinities (Table 2). Response surfaces for two species, *Prorocentrum mariae-labouriae* and *Skeletonema costatum,* are shown in Figs. 5 and 6.

Discussion

Silver uptake by phytoplankton is rapid, in proportion to silver concentration, and in general accordance with Freundlich adsorption isotherms [*16*], suggesting that the initial sorption is a function of surface chemistry. In addition, salinity plays an overriding role in the interactions of silver with phytoplankton. Silver uptake, although quite variable among species, is strongly controlled by salinity, as is the sorption of silver to abiotic particulates [*17*]. The strong degree of control that salinity exerts on silver uptake is caused by changes

TABLE 1—*Algal growth rate [in divisions per day (percentage of control)], length of lag phase (in days), and yield [in relative biomass units (percentage of control)] after 5 days of growth for species grown at different salinities under varying silver concentrations.*

Species	Ag, μg L⁻¹	7.5 Parts per Thousand			15 Parts per Thousand			22.5 Parts per Thousand			30 Parts per Thousand		
		Growth Rate, divisions/day (% of control)	Lag Phase, days	Yield, relative biomass units (% of control)	Growth Rate, divisions/day (% of control)	Lag Phase, days	Yield, relative biomass units (% of control)	Growth Rate, divisions/day (% of control)	Lag Phase, days	Yield, relative biomass units (% of control)	Growth Rate, divisions/day (% of control)	Lag Phase, days	Yield, relative biomass units (% of control)
Skeletonema costatum	0.05	1.57 (100)	1.0	81 (100)	1.58 (100)	0.8	85 (100)	1.41 (100)	0.9	50 (100)	1.69 (100)	1.0	96 (100)
	2.0	1.58 (101)	1.0	78 (96)	1.58 (100)	0.8	85 (100)	1.34 (95)	0.9	40 (80)	1.69 (100)	1.0	106 (110)
	5.0	0.92 (59)	1.3	12 (15)	1.37 (87)	1.0	48 (56)	1.32 (94)	0.9	41 (82)	1.73 (102)	1.1	96 (100)
	10.0	0.18 (11)	1.9	1.2 (1.5)	1.15 (73)	1.2	24 (28)	1.23 (87)	0.9	31 (62)	1.53 (91)	1.4	58 (60)
	20.0	0 (0)	...	0 (0)	0.47 (30)	1.1	3.5 (4.1)	0.89 (63)	1.0	12 (24)	0.99 (59)	1.4	13 (14)
Thalassiosira pseudonana	0.05	1.41 (100)	2.0	19 (100)	1.96 (100)	1.9	61 (100)	1.94 (100)	1.9	64 (100)	1.97 (100)	2.1	49 (100)
	2.0	1.42 (101)	1.9	20 (105)	1.96 (100)	1.9	61 (100)	2.19 (113)	1.9	107 (167)	2.00 (102)	2.2	49 (100)
	5.0	1.26 (89)	2.9	8.7 (46)	1.66 (85)	2.2	28 (46)	1.91 (98)	2.0	55 (86)	1.83 (93)	2.3	34 (69)
	10.0	0.72 (51)	3.9	2.4 (13)	1.36 (69)	2.9	11 (18)	1.74 (90)	2.2	32 (50)	1.46 (74)	2.6	16 (33)
	20.0	0 (0)	...	0 (0)	0.54 (28)	4.7	1.6 (2.6)	1.06 (55)	3.1	6.7 (10)	0.98 (50)	3.2	5.2 (11)
Chroomonas sp.	0.05	0.84 (100)	2.4	8.6 (100)	1.17 (100)	1.8	30 (100)	1.03 (100)	2.2	15 (100)	0.85 (100)	2.4	9.2 (100)
	2.0	0.37 (44)	4.5	1.9 (22)	1.22 (104)	1.9	31 (103)	1.20 (117)	2.2	24 (160)	0.91 (107)	2.7	11 (120)
	5.0	0 (0)	...	0 (0)	0 (0)	...	0 (0)	0 (0)	...	0 (0)	0.65 (76)	3.0	4.9 (53)
	10.0	0 (0)	...	0 (0)	0 (0)	...	0 (0)	0 (0)	...	0 (0)	0 (0)	...	0 (0)
	20.0	0 (0)	...	0 (0)	0 (0)	...	0 (0)	0 (0)	...	0 (0)	0 (0)	...	0 (0)
Prorocentrum mariae-lebouriae	0.05	0.79 (100)	1.9	7.3 (100)	0.85 (100)	1.5	11 (100)	0.82 (100)	2.3	7.9 (100)	0.69 (100)	2.7	5.5 (100)
	2.0	0.68 (86)	2.1	4.8 (66)	0.69 (81)	2.0	6.8 (62)	0.65 (79)	2.9	4.3 (54)	0.67 (97)	2.8	4.9 (89)
	5.0	0 (0)	...	0 (0)	0.65 (76)	3.0	4.4 (40)	0.60 (73)	3.2	3.8 (48)	0.49 (71)	3.2	2.8 (51)
	10.0	0 (0)	...	0 (0)	0 (0)	...	0 (0)	0 (0)	...	0 (0)	0.26 (38)	5.9	1.0 (18)
	20.0	0 (0)	...	0 (0)	0 (0)	...	0 (0)	0 (0)	...	0 (0)	0 (0)	...	0 (0)

TABLE 2—*Average silver toxicity (EC_{50}, in micrograms per litre, the concentration at which algal growth or culture yield is 50% of control) of alga species grown at different salinities.*

Species	EC_{50} for Growth, $\mu g\ L^{-1}$				EC_{50} for Yield, $\mu g\ L^{-1}$			
	7.5 parts per thousand salinity	15 parts per thousand salinity	22.5 parts per thousand salinity	30 parts per thousand salinity	7.5 parts per thousand salinty	15 parts per thousand salinity	22.5 parts per thousand salinity	30 parts per thousand salinity
Skeletonema costatum	5.9	15.4	>20	>20	3.8	6.2	13.2	12.4
Thalassiosira pseudonana	10.2	14.5	19.8	>20	4.8	4.8	10.0	8.1
Chroomonas sp.	1.8	3.6	3.6	6.7	1.4	3.6	4.1	5.2
Prorocentrum mariae-lebouriae	3.3	6.7	6.7	8.2	2.9	3.6	4.2	5.3

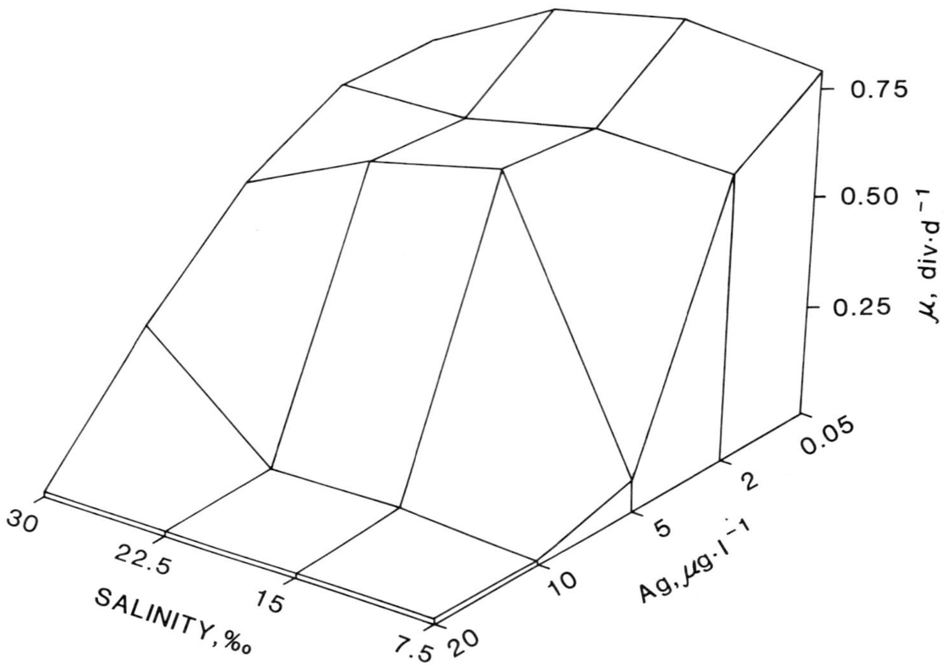

FIG. 5—*The effect of silver concentration and salinity on growth rate of* Prorocentrum mariae-lebouriae.

in the chemical speciation of silver brought about by the strong affinity of Ag^+ ions for Cl^- ions [*18*]. As salinity increases from 10 to 20 parts per thousand, the free silver ion concentration is decreased by a factor of 5, the concentration of the ion pair AgCl is decreased by a factor of 2, and other chloride complexes, such as $AgCl_3^{2-}$ and $AgCl_4^{3-}$ are increased in concentration (Fig. 7; see also Ref *19*). Such changes in chemical speciation dramatically affect silver behavior. Trace metal availability to biota consistently has been shown to be a function of free metal ion concentration [*20–22*]; thus, as the quantity of Ag^+ is decreased with increasing salinity, silver uptake is decreased similarly. However, silver availability may not be governed solely by Ag^+. Our results, and those of Engel et al. [*19*] indicate that other silver complexes, perhaps AgCl, are available and taken up by biota. The decrease in silver uptake as salinity increased from 10 to 20 parts per thousand (Fig. 1) is exactly the same as the decrease in AgCl concentration, strongly suggesting that the monochloro complex is available for uptake.

Silver accumulation, on a weight basis, in general, was similar between the several algal species and suspended particulates. For example, at a silver concentration of 2.0 μg L^{-1}, and a salinity of 10 to 15 parts per thousand, between 10 and 50 μg of silver were taken up per gram of suspended matter—either living or nonliving [*17*]. Variability in silver accumulation between algal species was probably due to cell size. Silver body burdens at 10 parts per thousand varied between 0.5 fg cell $^{-1}$ for the small *Thalassiosira pseudonana* (4 by 5 μm) to 22.6 fg cell $^{-1}$ for the much larger *Prorocentrum mariae-lebouriae* (14 by 17 μm). Cellular silver concentrations (at constant silver loading and salinity) were strongly correlated with cell surface area (r^2 = 0.92) and cell volume (r^2 = 0.97), as has been shown in other studies of metal/phytoplankton interactions [*16*].

Although silver accumulation by phytoplankton and suspended sediments was similar, the degree of silver desorption was not. Sediment particles, when resuspended into water of higher salinity, desorbed substantial quantities of silver. For example, a salinity increase from 0 to 15 parts per thousand caused release of 61% of the silver associated with sediments [*17*]. Similar experiments with phytoplankton exhibited negligible (<1%) release after a 10 parts per thousand salinity increase. Apparently, silver associated with cellular material either is held internally or is bound to cell surfaces more tightly than to sediment particles.

This difference in the degree of silver desorption between the two particle types has considerable implication for silver transport through estuaries. Additions of dissolved silver from anthropogenic sources to rivers and from the low salinity portions of estuaries will readily sorb to both living cells and sediment particles. Silver sorbed to suspended sediments will largely desorb as the particle moves through the estuary, complexing with chloride and remaining in a dissolved state. Thus, its transport through the estuary would be unimpeded. Silver incorporated by phytoplankton, however, will remain particulate until the cell lyses and the silver is remineralized. Because biogenic carbon is largely retained within the estuary [*23*], the transport of silver will be impeded, perhaps considerably, in productive estuaries where biogenic particles are an important component of suspended matter. Therefore, as with other trace elements, biological uptake can be an important modifier in the transport of an element through estuaries [*24*].

Silver toxicity to phytoplankton is also strongly controlled by salinity. Evidence from these experiments indicates that AgCl may be the primary toxic form of silver to phyto-

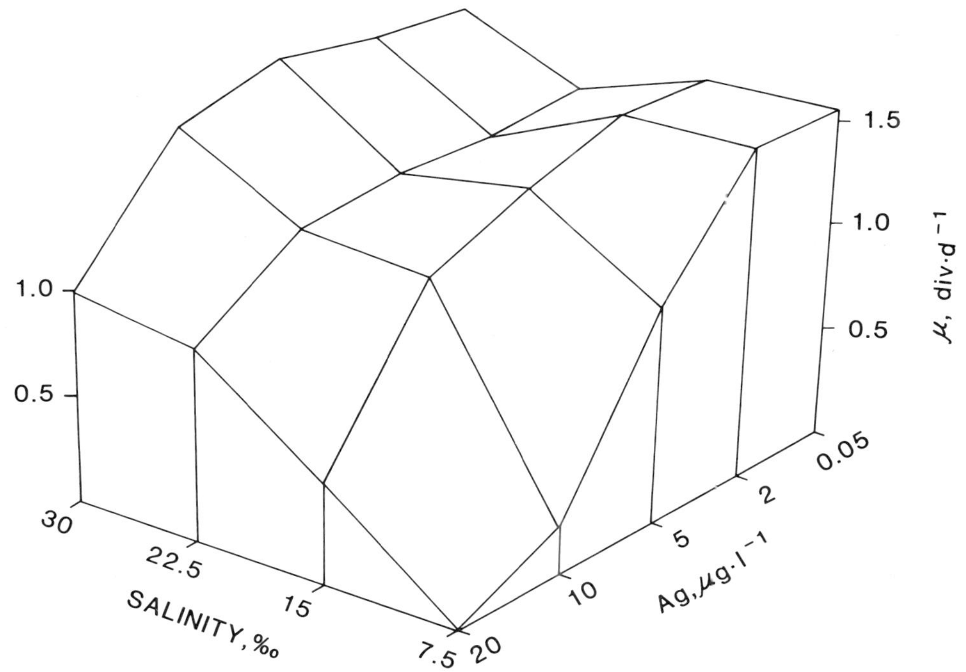

FIG. 6—*The effect of silver concentration and salinity on growth rate of* Skeletonema costatum.

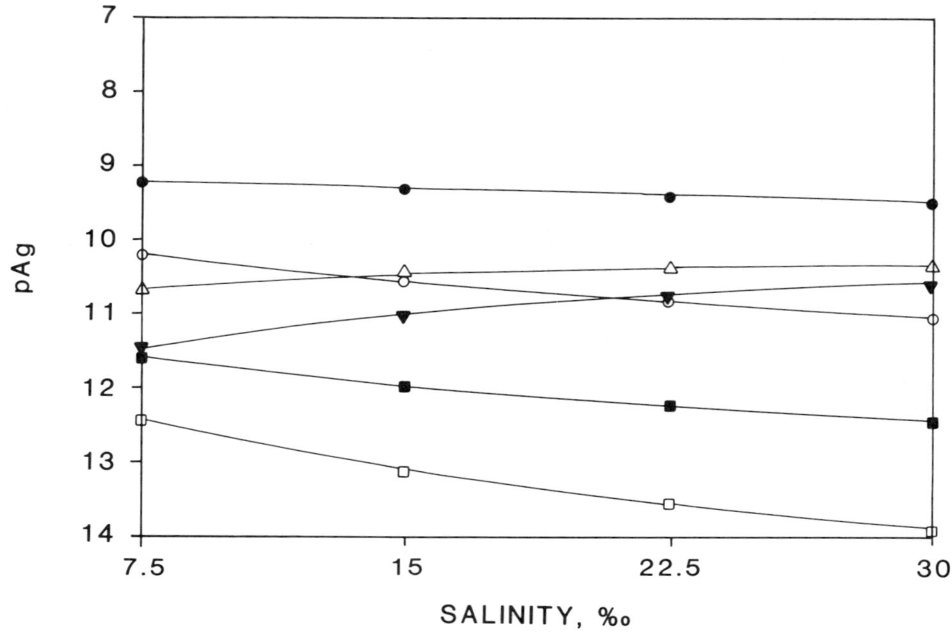

FIG. 7—*Effect of salinity on the concentration of various silver species, expressed as* −log *concentration, in moles per litre:* ● = $AgCl_2^-$; △ = $AgCl_3^{2-}$; ▼ = $AgCl_4^{3-}$; ○ = $AgCl$; ■ = $AgBr$; □ = Ag^+.

plankton. If the toxicity results are viewed in light of AgCl concentration, rather than total silver concentration (Figs. 2*b* and 3*b* relative to Figs. 2*a* and 3*a*), differences in silver toxicity between the different salinities disappear. Similar results have been presented for the free ions of copper [*20*] and cadmium [*21*] and may well be true for many cations.

Although all species studied showed the same general relationship between toxicity and AgCl concentration, there was considerable variability in sensitivity (Table 2). The flagellated species were considerably more sensitive to silver than the two diatoms. Differential sensitivity to a pollutant means that, under conditions of chronic loading of the pollutant, sensitive species of phytoplankton may be removed from the community while more resistant species continue to grow unimpaired. Thus, within the phytoplankton community, where rapid growth and species succession are the norm, the result of low levels of silver may be a shift in the dominant algal species toward silver-resistant forms, and not an overall decrease in primary productivity. Such a shift, seemingly of little import and difficult to detect after the fact, can be of great significance to the ecosystem if the species that have disappeared were important food items. In general, pollution-induced shifts in species dominance tend to favor forms that are undesirable as food to herbivores [*25,26*]. As many predators are quite selective in their prey, the disappearance of preferred food items can lead to shifts in the trophic structure of the community, impacting the success and survival of members of higher trophic levels [*24,27–29*]. Therefore, an apparently small impact at the lowest trophic level can lead to large, unpredicted changes to higher trophic levels and to the structure of an estuarine ecosystem. Too often, such indirect effects of a pollutant are not taken into consideration when an assessment of potential impact is made.

A common response to silver in these experiments was a marked delay in the onset of growth (for example, Fig. 4 and Tables 1 and 2) particularly at sublethal concentrations of silver. Assessments of contaminant toxicity to phytoplankton often fail to take into account this delay, but rather consider only the maximum growth rate, even though the effects to the community may be quite significant. As demonstrated in Tables 1 and 2, lengthening the lag phase dramatically reduced cell yields and resulted in much less biomass at any subsequent time. In a natural system, and particularly in an estuary, there is large variability in conditions such as light, nutrient concentrations, and physical stability. These factors largely control a cell's ability to succeed within a community. A delay in the onset of growth reduces the ability of a population to respond to short-term, favorable conditions, thus reducing its ability to succeed within the phytoplankton community and ensuring the dominance of resistant species with high initial rates of growth. Therefore, estimates of cell yield not only provide a more sensitive measurement of toxicity, but also provide a more realistic predictor of actual inhibition in a natural system. Such measurements should be included in all routine toxicity tests using algae.

Assessment of the health or vulnerability of dynamic, estuarine and marine ecosystems requires that studies of potential pollutant impact consider biogeochemical transformations, partitioning between dissolved and particulate phases, and sublethal pressures upon community structure, as well as the toxicity of the pollutant to keystone species. Only through such integrated studies can reliable predictions concerning pollutant impacts be generated.

Acknowledgments

We thank G. Riedel for comments and support, S. Cibik for assistance with algal cultures, and J. Bianchi for technical assistance. All three significantly improved this study through their interest and effort. This work was supported by the Baltimore Gas and Electric Co.

References

[1] Burton, J. D. in *Estuarine Chemistry,* J. D. Burton and P. S. Liss, Eds., Academic Press, London, 1976, Chapter 1, pp. 1–36.

[2] Boyle, E. A., Edmond, J. M., and Sholkovitz, E. R., *Geochimica et Cosmochimica Acta,* Vol. 41, 1977, pp. 1313–1324.

[3] Sholkovitz, E. R., *Earth and Planetary Science Letters,* Vol. 41, 1978, pp. 77–86.

[4] Officer, C. B., *Estuarine and Coastal Marine Science,* Vol. 9, 1979, pp. 91–94.

[5] Bourg, A. C. M. in *Trace Metals in Sea Water,* C. S. Wong, E. Boyle, K. W. Bruland, J. D. Burton, and E. D. Goldberg, Eds., Plenum Press, New York, 1983, pp. 195–208.

[6] Martin, J-M. and Whitfield, M. in *Trace Metals in Sea Water,* C. S. Wong, E. Boyle, K. W. Bruland, J. D. Burton, and E. D. Goldberg, Eds., Plenum Press, New York, 1983, pp. 265–296.

[7] Martin, J. H., Knauer, G. A., and Gordon, R. M., *Nature,* Vol. 305, 1983, pp. 306–309.

[8] Bloom, N. S. and Crecelius, E. A., *Analytica Chimica Acta,* Vol. 156, 1984, pp. 139–145.

[9] Guillard, R. R. L. and Ryther, J. H., *Canadian Journal of Microbiology,* Vol. 8, 1962, pp. 229–239.

[10] Murphy, L. S. and Belastock, R. A., *Limnology and Oceanography,* Vol. 25, 1980, pp. 160–165.

[11] Sanders, J. G., *Environmental Science and Technology,* Vol. 18, 1984, pp. 383–385.

[12] D'Elia, C. F., Sanders, J. G., and Boynton, W. R., *Canadian Journal of Fisheries and Aquatic Sciences,* Vol. 43, 1986, pp. 397–406.

[13] Kinrade, J. D. and Van Loon, J. C., *Analytical Chemistry,* Vol. 46, 1974, pp. 1894–1898.

[14] Sanders, J. G. and Windom, H. L., *Estuarine and Coastal Marine Science,* Vol. 10, 1980, pp. 555–567.

[15] Westall, J., "MICROQL I. A Chemical Equilibrium Program in BASIC," Technical Report, Swiss Federal Institute of Technology EAWAG, Duebendorf, Switzerland, 1979.

[16] Fisher, N. S., Bohé, M., and Teyssié, J.-L., *Marine Ecology—Progress Series,* Vol. 18, 1984, pp. 201–213.

[17] Sanders, J. G. and Abbe, G. R., *Continental Shelf Research,* Vol. 7, 1987, pp. 1357–1361.
[18] Jenne, E. A., Girvin, D. C., Ball, J. W., and Buchard, J. M. in *Environmental Impacts of Artificial Ice Nucleating Agents,* D. A. Klein, Ed., Dowden, Hutchinson, & Ross, Stroudsburg, PA, 1978, pp. 41–61.
[19] Engel, D. W., Sunda, W. G., and Fowler, B. A. in *Biological Monitoring of Marine Pollutants,* W. B. Vernberg, F. P. Thurberg, A. Calabrese, and F. J. Vernberg, Eds., Academic Press, New York, 1981, pp. 127–144.
[20] Sunda, W. G., and Guillard, R. R. L., *Journal of Marine Research,* Vol. 34, 1976, pp. 511–529.
[21] Sunda, W. G., Engel, D. W., and Thuotte, R. M., *Environmental Science and Technology,* Vol. 12, 1978, pp. 409–413.
[22] Morel, F. M. M. and Morel-Laurens, N. M. L. in *Trace Metals in Sea Water,* C. S. Wong, E. Boyle, K. W. Bruland, J. D. Burton, and E. D. Goldberg, Eds., Plenum Press, New York, 1983, pp. 841–870.
[23] Smullen, J. T., Taft, J. L., and Macknis, J., "Nutrient and Sediment Loads to the Tidal Chesapeake System" in *Chesapeake Bay Program Technical Studies: A Synthesis,* E. G. Macalaster, D. A. Barker, and M. Kasper, Eds., Environmental Protection Agency, Washington, DC, 1982, pp. 150–251.
[24] Sanders, J. G., *Canadian Journal of Fisheries and Aquatic Science,* Vol. 43, 1986, pp. 694–699.
[25] Greve, W. and Parsons, T. R., *Helgolaender Wissenschaftliche Meeresuntersuchungen,* Vol. 30, 1977, pp. 662–672.
[26] O'Connors, H. B., Wurster, C. F., Powers, C. D., and Rowland, R. G., *Science,* Vol. 201, 1978, pp. 737–739.
[27] Ryther, J. H., *Science,* Vol. 166, 1969, pp. 72–76.
[28] Parsons, T. R. in *The Ecology of the Seas,* D. H. Cushing and J. J. Walsh, Eds., W. B. Saunders Co., Philadelphia, 1976, pp. 81–97.
[29] Landry, M. R., *Helgolaender Wissenschaftliche Meeresuntersuchungen,* Vol. 30, 1977, pp. 8–17.

William J. Adams,[1] Wilfred J. Renaudette,[1] John D. Doi,[1] Mitch G. Stepro,[1] Michael W. Tucker,[1] Richard A. Kimerle,[1] Brenda B. Franklin,[2] and J. Vincent Nabholz[3]

Experimental Freshwater Microcosm Biodegradability Study of Butyl Benzyl Phthalate

REFERENCE: Adams, W. J., Renaudette, W. J., Doi, J. D., Stepro, M. G., Tucker, M. W., Kimerle, R. A., Franklin, B. B., and Nabholz, J. V., **"Experimental Freshwater Microcosm Biodegradability Study of Butyl Benzyl Phthalate,"** *Aquatic Toxicology and Environmental Fate: Eleventh Volume, ASTM STP 1007,* G. W. Suter II and M. A. Lewis, Eds., American Society for Testing and Materials, Philadelphia, 1989, pp. 19–40.

ABSTRACT: A 30-day freshwater microcosm study was conducted to measure the biodegradability of butyl benzyl phthalate (BBP) in water and sediment. Microcosms were designed to simulate the flow rate, water to sediment surface ratio, and environmental conditions of the Illinois River. Water and undisturbed sediment cores were brought into the laboratory from the Illinois River and placed in 53-L all-glass microcosms. BBP was tested at 10 and 100 μg/L and at 100 μg/L sterilized with five replicates at each concentration. BBP was added to each microcosm three times per week at one half the initial dose. The microcosms were sealed to trap radiolabeled carbon dioxide (^{14}C-CO_2) for purposes of assessing mineralization and obtaining a mass balance on each microcosm. Uniformly ring-labeled ^{14}C-BBP was used as the test material and analyses were performed for parent BBP as well as for carbon-14 (^{14}C) intermediates.

The results of this study indicated that BBP was readily degraded in both water and sediment. The estimated half-life in water for primary degradation was typically two days or less. Mineralization of BBP did occur, with an average of 10.4% of the total BBP dosed in each microcosm trapped as $^{14}CO_2$. Concentrations of BBP on the sediments were lower than predicted from sediment partition coefficients. (200 to 600 μg/kg) and ranged from 7 to 23 μg/kg. Both the sediment and water analyses showed that the majority of the ^{14}C activity detected was not BBP. The transient appearance of primary metabolites was observed, however, the preponderance of ^{14}C remaining in solution was not BBP.

A mass balance of each component (water, sediment, carbon dioxide, and so forth) in each microcosm indicated that 78.2% of the total amount of ^{14}C-BBP added to the microcosms could be accounted for. Approximately 50% of the mass remained as ^{14}C metabolites in the water, and only 1.8% and <0.1% were found as ^{14}C and parent BBP, respectively, on the sediment. The remainder of the ^{14}C-BBP was found on the glass, in silicone joints and stoppers, in evolved CO_2, or was removed by water sampling.

The collective data indicate that BBP should not be a persistent chemical in freshwater aquatic environments.

KEY WORDS: microcosm, butyl benzyl phthalate, sediment biodegradation and metabolites, aquatic toxicology

[1] Science fellow, research chemist, research specialist, research technicians, and senior fellow, respectively, Monsanto Co., Environmental Sciences Center, St. Louis, MO 63167.

[2] Group leader—Aquatic Toxicology, ABC Laboratories, Columbia, MO 65205.

[3] Ecologist, U.S. Environmental Protection Agency, Office of Pesticides and Toxic Substances, Health and Environmental Review Division, Washington, DC 20460.

To evaluate the potential for butyl benzyl phthalate (BBP) to degrade in surface waters and sediments, Monsanto Co. conducted a microcosm environmental fate study with BBP. The microcosm was designed to simulate certain aspects of a freshwater river system and relied upon the environmental fate portions of previously published reports by Parez [1,2].

The following parameters were considered a necessary part of the microcosm test design: (1) undisturbed sediment core samples, (2) mass balance of BBP in the test system, (3) carbon-14 (^{14}C)-labeled BBP (^{14}C-BBP) for measuring degradation and resulting carbon dioxide (CO_2) evolution, (4) confirmation of sediment and water ^{14}C-BBP measurements by gas chromatography or high-performance liquid chromatography or both, and (5) river ecosystem simulation with emphasis on sediment degradation and not on ecological effects. A river ecosystem was chosen because both manufacturing and major end-use customers for BBP are located on freshwater rivers rather than lakes or estuaries. The Illinois River was chosen in this simulation because it is a large river, receives multiple waste discharges with both industrial and municipal sources, and is close enough to St. Louis to obtain and transport sediment and water to Monsanto's laboratory for the microcosm test. Numerous studies have been conducted with BBP to determine its toxicity to aquatic organisms [3], therefore, ecological effects testing was not considered necessary. Additional effects testing was performed at the request of the U.S. Environmental Protection Agency to derive a water quality criterion. This will be reported at a later date.

Methods

Experimental Design

A 30-day microcosm study was designed. The study began 7 April 1986 and ended 7 May 1986. Two test concentrations, 10 and 100 μg/L BBP, plus a sterile control containing 100 μg/L BBP were used. The control microcosms received 222 mg/L of mercuric chloride at the initiation of the test and again on Day 11. Each concentration was replicated five times. The total number of microcosms used was 15 (3 treatments $\times$ 5 replicates). Each microcosm was assigned a number and randomly assigned to a specific location in an environmental room. Each microcosm contained 45 L of Illinois River water and one undisturbed sediment core collected from the Illinois River (Fig. 1). The microcosms consisted of 37.5 by 37.5 by 37.5 cm (inside diameter) glass containers which rested on wooden platforms above magnetic stirrers in a temperature-controlled room. The sediment core tubes were thick-walled (6-mm) glass cylinders open at both ends and were 10-cm tall and 10 cm in diameter. The sediment cores were collected in the littoral zone of the river by pushing the core 10 cm into the sediment, covering the top with a petri dish, and lifting the core from the river with a shovel. A petri dish was then snugly fit under each core. The core was placed into a 1-L container filled with river water and was transported back to the laboratory. The water volume to core surface area ratio in the microcosm was equal to that in the Illinois River. The microcosm design was specifically chosen to be large enough to accommodate a 10-cm core and maintain the appropriate water volume to sediment surface area ratio. Data were obtained from the U.S. Army Corps of Engineers to properly design the water to sediment ratio, which was approximately 0.5097m^3 to 0.0929m^2 (18 ft^3 to 1.0 ft^2). Water and sediment were collected three miles north of Grafton, IL (that is, approximately 30 miles north of St. Louis, MO).

All testing was done in an environmental chamber in which temperature and lighting were controlled at 20 $\pm$ 1°C and 13:11 h of light and darkness. Fluorescent natural spectrum lighting was used with the intensity adjusted to simulate incident light on the Illinois River on a cloudy day (400 $\pm$ 40 foot candles).

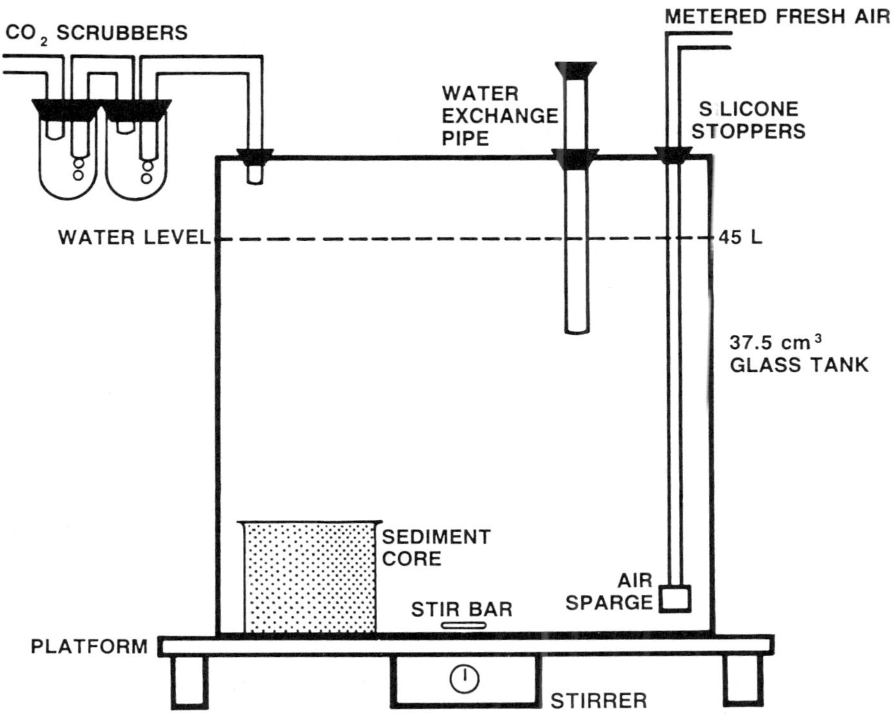

FIG. 1—*Experimental freshwater microcosm with an undisturbed sediment core.*

Turbulence and flux in the microcosms were provided by magnetic stirrers adjusted to the river flow rate. The cores was placed in the front left corner section of each microcosm and adjusted so that the required flow rate was attained across the core-soil surface to simulate the natural system mixing and enhance chemical mass transfer. Each microcosm was adjusted to contain a flow velocity of 6 cm/s across the core surface, using a strobe light to calibrate the stir bar revolutions. A flow meter was previously used to calibrate the stir bar revolutions with flow rate. Use of the strobe light eliminated the need to insert the flow meter into each microcosm.

Air was sparged through the microcosms at 20 mL/min to ensure aeration at levels approximately the Illinois River. Aeration was also used to collect carbon-14-labeled CO_2 (^{14}C-CO_2) in CO_2 scrubber traps formed from the degradation of BBP. Before starting the 30-day study, each microcosm was sealed with a glass plate top using silicone glue. Carbon dioxide was collected in two 0.5 *N* potassium hydroxide traps attached to the top of the microcosms.

One litre of water was removed and replaced with one litre of fresh Illinois River water every Monday, Wednesday, and Friday in each microcosm. Water exchanges in the microcosms were designed to simulate intermittent discharges of BBP into a river system and to develop a steady-state BBP concentration. Illinois River water was collected and brought to the laboratory once each week and held at 4°C until used. When the litre of water was added to each microcosm, BBP was spiked at an amount equivalent to 50% of the initial dose. The 100 control (c), 100, and 10 μg/L microcosms initially received 4.5, 4.5, and 0.36

mg total, respectively. The amount of dimethylformamide (DMF) solvent initially added was 0.2 mL/L or less in all microcosms and 0.01 mL/L in subsequent additions.

Water was added and removed from each microcosm by means of a 1.5-cm hole in the top of the microcosm. Air was added and gases vented by means of similar holes. Each hole was fitted with a silicone stopper to insure that the entire system was water and gas tight. All microcosms were checked for water leaks before the test and for air and water leaks after sealing the microcosms. No air was provided to the microcosms until the top seal was dry, which was on Day 2. Microcosm Numbers 9 and 11 did develop water leaks which necessitated lowering the water levels on Day 2 to below the leak level. Water removed was held for 48 h in a separate container and placed back into the respective microcosm after the leak was repaired. After adding the water back to both microcosms, the concentration of BBP was brought up to the starting level. Both of these microcosms were sacrificed on Day 11.

The microcosm study was designed to provide sufficient replication, and so that one microcosm chamber at each of the two treatment levels (10 and 100 μg/L) and the control could be sacrificed during the study on Days 11 and 22. The selection of microcosms to be sacrificed was not random. On Day 31, the remaining three microcosms at each treatment level and control were terminated. Sediment samples, as well as other appropriate samples, were collected for analyses each time a microcosm was sacrificed to ensure that a mass balance could be calculated for each microcosm. Additional details on the experimental design are available upon request [*4*].

BBP Chemical Purity Data

The chemical purity and identity of the unlabeled BBP neat chemical (sold commercially as Santicizer 160, Lot DD243) used in this study were determined by high-performance liquid chromatography (HPLC) and gas chromatography/mass spectroscopy (GC-MS) analysis, respectively. The purity of Lot DD243 was 98.6% BBP.

Isotopic purity of the ^{14}C-BBP (uniformly ring labeled on the phthalate ring) was determined before starting the study. The isotopic purity was determined by reversed-phase HPLC analysis and fractionation in combination with radiometric measurement of the specific peaks. The data indicate that 95% or more of the ^{14}C activity is contained in the BBP peak.

Characterization of monobutyl phthalate, monobenzyl phthalate, and phthalic acid standards was performed by combined high-performance liquid chromatography/mass spectroscopy (HPLC-MS). These data confirmed the presence of the parent ion in each sample and indicated sample purity in the range of 80 to 90%.

Verification of Stock Solution Concentrations

A separate stock solution was prepared for the 10 μg/L and 100 μg/L microcosm exposure levels. Each stock solution was a mixture of labeled and unlabeled BBP. The ^{14}C fraction of the two stocks used to spike the 10 and 100 μg/L microcosms were 45.6 and 5.60%, respectively. The specific activity of the two stock solutions was 37 024 and 3489 disintegrations per minute (dpm)/μg, respectively.

Stock solutions were also made in order to prepare blind-spiked samples for quality assurance/quality control (QA/QC) purposes. The concentration of BBP in each stock solution was determined by HPLC. The radioactivity was determined by liquid scintillation counting (LSC) of 4 μL of each stock solution. Nominal and measured concentrations were in good agreement.

Methods of Analysis for Water

Water Quality Characterization—Alkalinity, pH, hardness, dissolved oxygen, and conductivity were measured on Days 1, 8, 17, and 31 of the study. Methods of analysis followed the standard procedures described in Monsanto's Standard Operating Procedures [*5–9*]. Temperature was monitored each workday by checking the temperature of the environmental room and measuring the temperature of the water in a 2-L graduated cylinder placed under the fluorescent lights. The temperature inside each microcosm was monitored on Day 17 with a Fisher Scientific calibrated mercury thermometer.

Total suspended solids were measured on Days 1 and 31 by taking 100 mL of river water to dryness and measuring the total solids remaining in the drying dish. Dissolved and total organic carbon were measured on Days 1, 24, and 31. Fifty-millilitre samples from each microcosm were sent to the Industrial Testing Laboratory (St. Louis, MO) for analysis. Dissolved versus total organic carbon was distinguished by centrifuging the samples at 12 000 rpm for 10 min prior to analysis. Total organic carbon (TOC) was determined by a TOC analyzer.

Bacterial colony counts for the water phase were made on Days 1, 19, and 31 using prepared agar Millipore plates (plastic cassettes) designed for enumerating total bacteria (Fisher Scientific). Colonies on each plate were counted under a microscope, and the total number of colonies observed were reported.

Method Validation for BBP Analysis in Water—Water was analyzed for parent BBP by HPLC via hexane extraction. Parent BBP was also determined by hexane extraction and liquid scintillation counting of the ^{14}C activity in the hexane extracts. Total ^{14}C activity (that is, parent BBP and metabolites) was determined in the water by direct counting of the water samples by LSC. Each analytical method used to measure BBP was validated by spiking BBP or ^{14}C-BBP or both into Illinois River water at three or four levels, each in triplicate. The spiking levels spanned the range of interest, typically 1.2 to 2400 μg/L. At least three sample blanks were analyzed along with the spiked samples. Blank and fortified samples were analyzed along with the microcosm water samples for quality control purposes.

The percent recoveries for the various method validation procedures ranged from 85.5 to 100.9% as indicated: (1) LSC water analysis, 87.5%; (2) LSC hexane extracts, 100.9%; and (3) HPLC hexane extracts, 85.5%. Results of the analyses were not corrected for each percent recovery since the recoveries were high. The limit of detection (LOD) for each method was determined to be the lowest validated concentration which gave a response greater than river water blanks by plus or minus two standard deviations. The LOD for the three methods of analysis were (1) HPLC—hexane extraction, 2.74 μg/L; (2) LSC—^{14}C aqueous measurement, 1.2 μg/L; and (3) LSC—^{14}C hexane extraction, 1.2 μg/L.

The extraction procedure for BBP from river water samples gave recoveries in excess of 100%, when not corrected for the amount found in the blanks. River water blanks provided a positive response equivalent to an average of 2.74 μg/L BBP. This amount was subtracted from the results for each hexane-extracted sample.

BBP Metabolite Extraction and Identification—The metabolites of BBP are thought to be polar water soluble organic acids [*10,11*]. Direct aqueous HPLC injection was used to identify the metabolites.

Standards obtained included phthalic acid, monobutyl phthalate (one-half ester), and monobenzyl phthalate (one-half ester). The latter chemical was custom synthesized for this study. These three standards were chosen because they were thought to be the primary intermediate metabolites of BBP and were detectable by direct aqueous HPLC injection.

During the 30-day microcosm study, water samples (20 mL) were collected, stored at

4°C, and saved for direct aqueous HPLC analysis. Microcosm water samples collected at several time points were analyzed by HPLC direct aqueous injection.

Surface Film Concentrations—Measurement of the total ^{14}C activity contained in the surface water/film of each microcosm was determined when each microcosm was sacrificed. Surface water concentrations were made by placing three Whatman paper cellulose filter pads (36.5 cm^2) on the surface of the water of each microcosm at sacrifice. After saturation, the filter was placed into 15 mL of scintillation cocktail and counted (on Day 31) by LSC. Subsurface ^{14}C concentrations were measured by saturating three pads in water drawn from beneath the surface of each microcosm and analyzing the pads by LSC. Subsurface concentrations were also determined by LSC counts of the ^{14}C activity in 5 mL of water removed from beneath the surface and counted by LSC.

Carbon Dioxide Analysis

Carbon dioxide resulting from the metabolism of ^{14}C-BBP was collected by bubbling air into each microcosm and forcing all existing gas to pass through two 100-mL potassium hydroxide (KOH) (25% volume/volume) scrubbers connected in a series. The scrubbers were removed and the KOH replaced at least twice a week for all treatments except the controls which were changed once a week. The KOH was counted by LSC for ^{14}C-CO_2. Potassium hydroxide blanks were analyzed and demonstrated that no additional correction for background was necessary. The amount of ^{14}C-CO_2 produced was directly proportional to the amount of radioactivity associated with the carbon atoms of the uniformly labeled phthalate ring. Therefore, the amount of ^{14}C-BBP metabolized was obtained by determining the radioactivity of the KOH and dividing by the specific activity of the ^{14}C-BBP.

A trap for volatile BBP was not used in combination with the CO_2 traps because a preliminary microcosm study using a Tenax column in front of two KOH scrubbers indicated this was not necessary.

Sediment Analysis

Sediment cores were removed from the microcosms on specified days when the microcosms were sacrificed, that is, Days 11, 22, and 31. Each core was subsampled by forcing a 10-mm (inside diameter) glass tubing through the entire length of the core. Three subsamples were taken from each core and kept frozen until analyzed. Prior to analysis, each subcore was divided into 2-cm sections, that is 0 to 2 cm, 2 to 4 cm and so forth. These sections were dried at 60°C, mixed, subsampled in duplicate, and analyzed by oxidation/combustion and collection of ^{14}C-CO_2. This provided a measure of the total radioactivity in the sediment. Calculation of ^{14}C-BBP concentrations was performed by using the specific activity of the BBP in each microcosm and expressing the ^{14}C as ^{14}C-BBP. Each subcore was analyzed only to the depth where the ^{14}C was less than or equal to the LOD. The amount of ^{14}C measured in the three sediment subcores was used to calculate the total amount of ^{14}C, expressed as ^{14}C-BBP, in each 10-cm microcosm core. This was necessary to obtain a mass balance on the ^{14}C-BBP spiked in each microcosm.

The oxidation-combustion procedure was validated by spiking dried Illinois River sediment with ^{14}C-BBP and determining the LOD and the average percent recovery (117 $\pm$ 17%).

All core segments containing ^{14}C at greater than or equal to the LOD were extracted with hexane to recover parent BBP. Extracts were analyzed by LSC and electron capture gas chromatography (GC-EC). The extraction procedure was validated by adding ^{14}C-BBP to Illinois River sediments at three levels in each of six replicates, extracting the sediments with hexane, measuring the BBP concentration, and calculating the average recovery (86.4 ± 5.7%). Eighteen sediment samples fortified with BBP used for quality control were also analyzed by GC-EC and LSC.

Subsequent to the hexane extraction, each section that had been previously analyzed was reanalyzed (in duplicate) by oxidation/combustion and ^{14}C LSC measurement, thus providing a ^{14}C measurement before and after extraction. All sediment sample extractions, oxidation/combustions, and gas chromatography (GC) analyses were performed at ABC Laboratories (Columbia, MO). The limit of detection via LSC and GC ranged from 0.003 to 0.04 and 0.001 to 0.023 mg/kg, respectively.

Mass Balance of BBP (Radioactive Budget)

Each time a microcosm was sacrificed, ^{14}C measurements were made on the water, sediment, walls of the container, KOH scrubbers, and silicone glue in the joints in order to complete a mass balance for each microcosm. Each time a microcosm was sacrificed, the sediment core and water were removed and the microcosms were washed twice with 500 mL of methanol. Methanol rinses were combined and analyzed in duplicate via LSC.

The small amount of silicone sealant in the joints of the microcosm walls and bottoms of the three microcosms was totally removed with a razor blade. The average weight of the silicone was 53.2 g/microcosm and only varied among individual microcosms by 2 to 4 g. This value (53.2 g) was then assigned to each microcosm, and the mean concentration of ^{14}C-BBP in the silicone sealant for each microcosm was determined by assaying four representative samples (0.2 to 0.4 g) of silicone from each microcosm. Each sample was placed in scintillation cocktail for 18 h prior to analysis. The cocktail was 50% xylene and served as an excellent solvent for extracting the ^{14}C-BBP and partially dissolving the silicone. The samples were measured for ^{14}C by LSC. In order to validate the procedure, the extracted silicone samples were combusted and counted by LSC. They showed no detectable amount of ^{14}C, indicating the extraction was complete.

Silicone stoppers in the top of the microcosms were analyzed, as described above, for the sealant. A sample (0.1 to 0.3 g) was removed from the top and bottom of 18 out of the 48 stoppers. This was done to calculate the average amount of ^{14}C-BBP in the stoppers, since they were not numbered and at the time of analysis could no longer be identified with a particular microcosm. These data were not included in the mass balance calculations because of the inability to assign a particular stopper to a specific microcosm.

The ^{14}C-BBP data obtained from each microcosm compartment were used to calculate a total radioactivity budget for each microcosm. The sum of the components was compared with the total amount spiked into each microcosm to calculate the percent recovery. The ^{14}C budget was calculated as micrograms of ^{14}C-BBP rather than disintegrations per minute per microcosm. The values are interchangeable by multiplying the number of micrograms of ^{14}C-BBP by the appropriate specific activity, that is, 3489 and 37 024 dpm/μg for the 10 and 100 μg/L treatment levels, respectively. The term "parent" BBP is used throughout this report to mean intact (nondegraded) BBP. In contrast to this, the term "^{14}C-BBP" is frequently used to denote total ^{14}C activity or percent BBP plus metabolites.

Statistical Analysis of ^{14}C-BBP Data

Statistical analyses were performed to compare the mean concentration of ^{14}C-BBP in microcosm sediments or water among the three exposure conditions, that is, 10, 100, and 100 (control) μg/L. Results were statistically analyzed on Days 10, 22, 29, and 31.

Typically, there was an order of magnitude difference between the values measured at the lowest exposure concentration and those obtained at the higher exposure concentration. Homogeneity of variance could not be demonstrated through the use of nontransformed or log-transformed data. A likelihood ratio test available in SAS's DISCRIM procedure was used to test for equality of variance [*12*]. Since sample variances were not homogeneous, nonparametric techniques were used to test whether the ^{14}C-BBP water and sediment concentrations were different between treatments and between treatments and controls.

A Kruskal-Wallis one-way analysis of variance was conducted. If the overall test results were significant at the 0.05 level, then a nonparametric multiple comparison to determine the pattern of differences among the exposure concentrations was done in a manner similar to a Neuman-Keuls test [*13*].

Results

Water Quality Parameters

Water quality parameters for the Illinois River and the microcosms on Day 1 and subsequent days showed reasonable agreement. The following measurements were made on the Illinois River water at the beginning of the study: temperature = 18°C, dissolved oxygen = 9.3 mg/L, pH = 8.4, conductivity = 580 umhos, alkalinity = 138 mg/L, hardness = 292 mg/L, and river flow rate near the bottom = 6.0 cm/s. The dissolved oxygen, conductivity, and hardness in the microcosms remained relatively constant during the entire study and averaged 9.5 ± 0.5 mg/L, 648 ± 48 umhos, and 289 ± 58 mg/L, respectively. The pH ranged from 7.6 to 8.6. Alkalinity ranged from 154 to 236 mg/L and tended to increase with time. This increase in alkalinity probably reflects a shift in the carbonate-bicarbonate cycle as a function of algal productivity. The temperature was relatively constant during the 30-day study and averaged 20.5°C (range 19.4 to 21.9°C).

Suspended solids measurements at the beginning and end of the study did not show major changes, although there was some variability. The average concentration was 514 mg/L, a typical concentration for the Illinois River.

Total and dissolved organic carbon (TOC and DOC) measured in the microcosms on Days 1 and 31 showed reasonably low values with one or two exceptions. The TOC and DOC values were similar and ranged from 7.9 to 60 mg/L. The river values for DOC and TOC on Day 1 were 11.4 and 7.9 mg/L, respectively. Day 24 TOC and DOC values were much higher; analytical error is suspected, and the data are not reported here. Sediment core TOC values were typical for most rivers. The sediment TOC values averaged 1.04% and ranged from 0.74 to 1.52%.

Bacterial plate counts indicated the control microcosms were in fact sterile. No colonies were observed on the plate agar. Numerous colonies were observed on the treatment microcosm plates.

BBP—Water Analyses

The ^{14}C water measurements, expressed as ^{14}C-BBP for both total and dissolved ^{14}C-BBP, increased with time in all microcosms. This was expected since each microcosm was spiked with ^{14}C-BBP three times per week. The values for total activity varied by a factor of 10 between the two treatment levels as expected, and good agreement between replicates was obtained. Differences between the treatment levels are significant at the 0.05 level. The difference between dissolved and total activity, expressed as ^{14}C-BBP, is quite small, indicating that very little of the radioactivity was bound to the suspended particles (Figs. 2 to 4). This was true for the control and treatment microcosms.

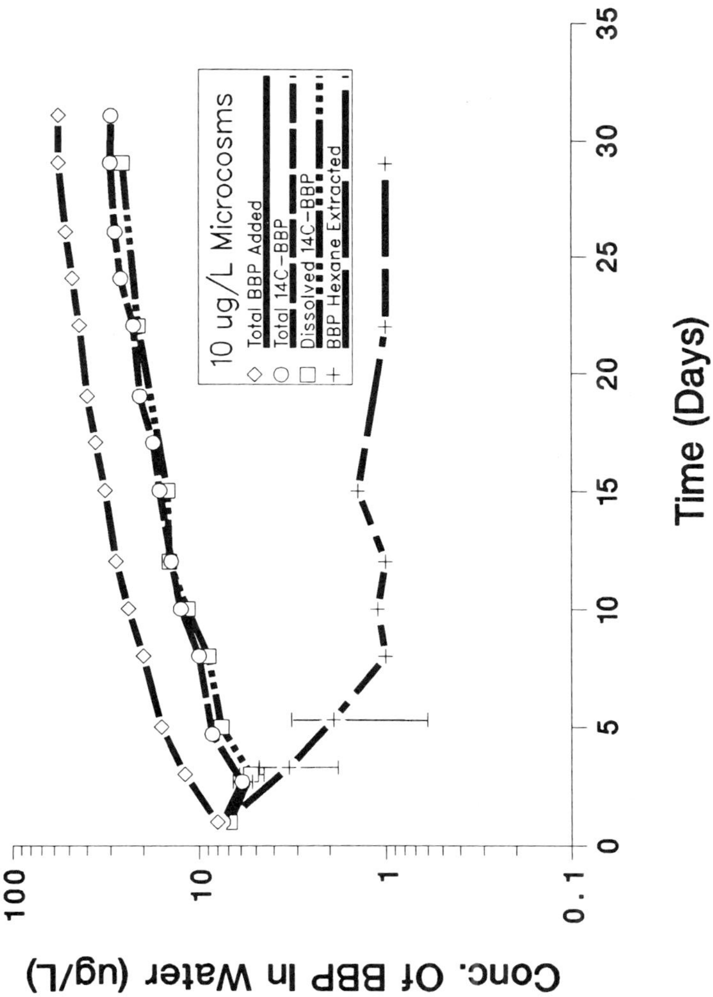

FIG. 2—*A comparison of parent BBP (hexane extracted) in water with the total and dissolved water concentrations (^{14}C-BBP) and with the total amount in micrograms of ^{14}C-BBP added over time in the 10 μg/L microcosms.*

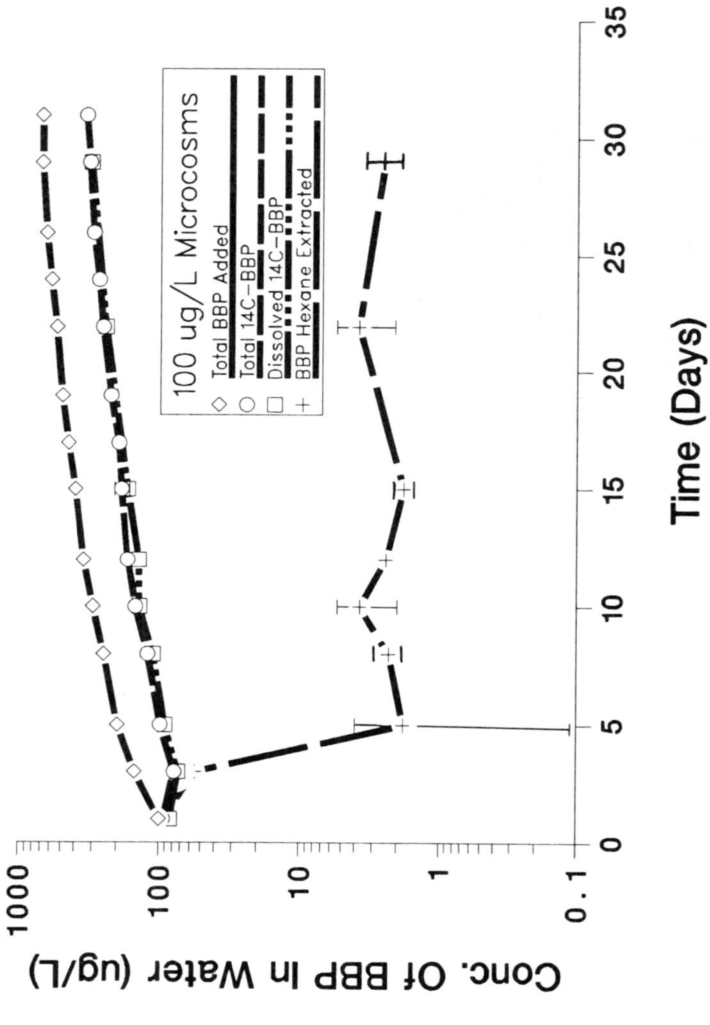

FIG. 3—*A comparison of parent BBP (hexane extracted) in water with the total and dissolved water concentrations (^{14}C-BBP) and with the total amount in micrograms of ^{14}C-BBP added over time in the 100 μg/L microcosms.*

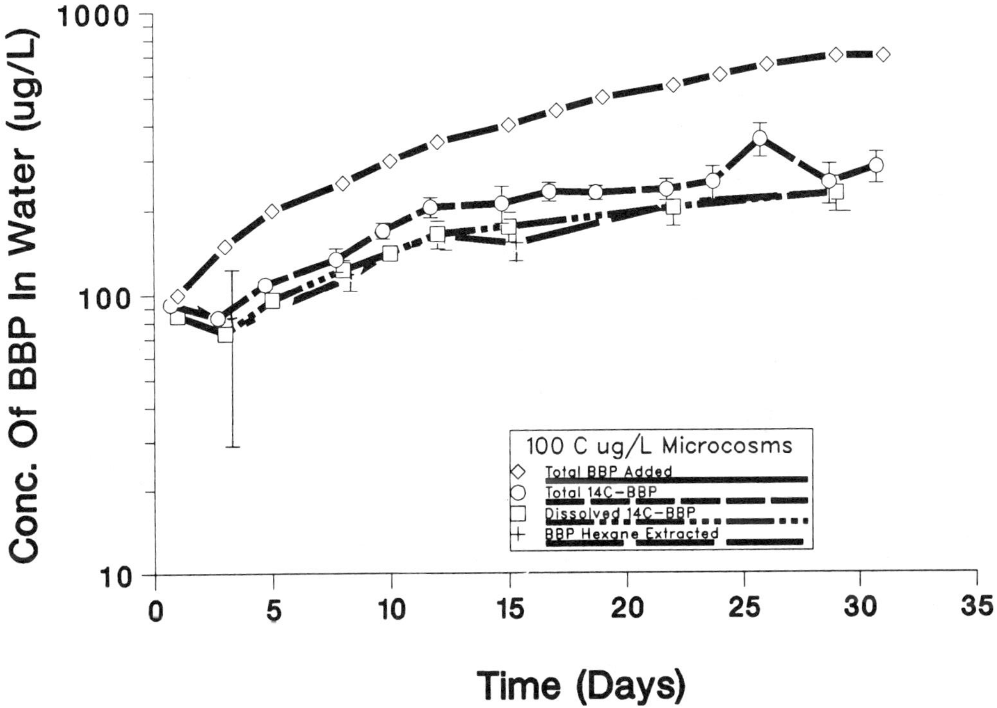

FIG. 4—*A comparison of parent BBP (hexane extracted) in water with the total and dissolved water concentrations (^{14}C-BBP) and with the total amount in micrograms of ^{14}C-BBP added over time in the 100 μg/L control microcosms.*

Analysis of hexane extracts of water by LSC indicate that the ^{14}C activity in the water phase is primarily not parent BBP. The data show good agreement between replicates and a statistical difference ($P < 0.05$) between the two treatments and the control. Parent BBP concentrations in the treatments rapidly declined to 50% by Day 3, to 1 to 20% by Day 5, and remained low (<1.0 to 4.3 μg/L) during the remainder of the study (Figs. 2 and 3). In contrast to this, the control microcosms, which contained mercuric chloride as a sterilant, consistently contained 80% or more of the total ^{14}C activity as parent BBP (Fig. 4). These data clearly show that the degradation mechanism is by microbial action, and that BBP in the water phase is rapidly converted to a more polar water soluble metabolite. A comparison of the two treatments with the control demonstrates this point (Fig. 5).

The disappearance of parent BBP in the treatment microcosms determined by LSC was confirmed by HPLC analysis of selected water samples. Samples were periodically collected from one 100-μg/L microcosm and one control microcosm throughout the experiment and from most of the microcosms on Days 11 and 31 for HPLC analysis. Parent BBP concentrations in the treatment microcosms were typically less than the limits of detection as determined by the three methods of analysis. For the sterile controls, parent BBP values varied by a factor of 2 among the three methods and were typically in the range of 100 to 300 μg/L. The HPLC data served to confirm the validity of measuring parent BBP by liquid scintillation counts of hexane extracts (Figs. 2 to 4).

Surface Film Measurement for BBP

Surface and subsurface concentrations of ^{14}C-BBP were similar in the 100 and 10 μg/L microcosms. Surface concentrations in the control microcosms were higher by an order of magnitude than the subsurface concentrations. This difference was most likely due to the

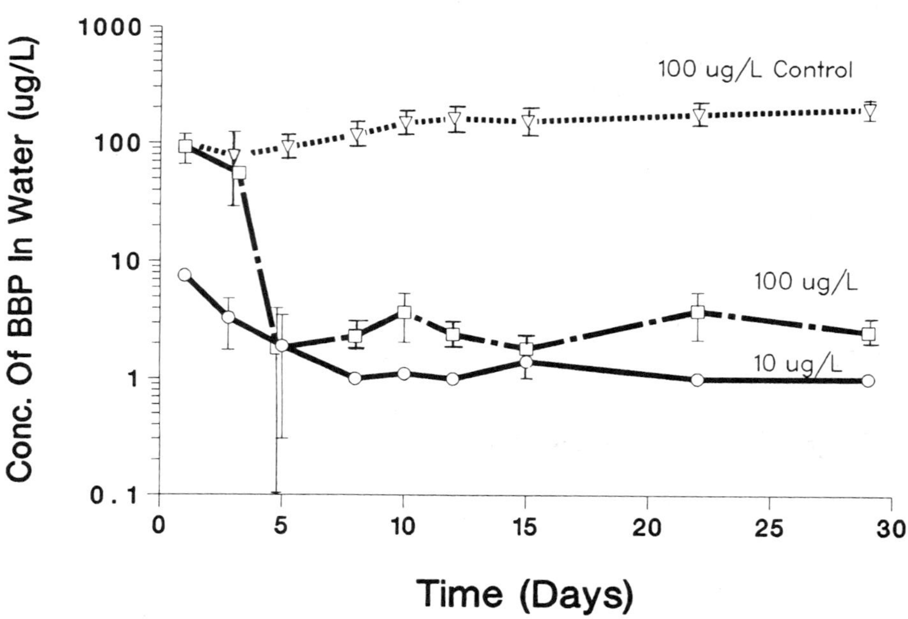

FIG. 5—*A comparison of the hexane extractable ^{14}C-BBP between the sterilized control and nonsterilized microcosms.*

fact that control microcosms contained primarily parent BBP which has a relatively low solubility in water (2.9 mg/L) and a density of <1.0 and, therefore, migrated to the water surface. The fact that control microcosms were sterilized indicates that higher surface concentrations were not due to adsorption to surface microbial populations. In the microcosms which contained viable bacteria, there was no indication of an increase of ^{14}C activity at the surface. The amount present at the surface of the controls was insignificant in terms of the average percent recovery in each control microcosm, and no correction in the mass balance was made.

BBP-CO_2 Analyses

Results of the LSC measurements on the KOH scrubbers indicate that ^{14}C-CO_2 was being produced as early as Day 5, the first sample day, and that BBP was being mineralized. The average percent mineralization of the total BBP added to each noncontrol microcosm sacrificed on Days 22 and 31 was 10.4 (±1.8%). There was no statistical difference between the 10 and the 100 μg/L treatment levels. In comparison, the controls showed an average percent mineralization of only 0.12%. These data indicate that the ultimate fate of BBP in the nonsterilized microcosms is biological degradation to carbon dioxide.

TABLE 1—*Measured concentration of ^{14}C-BBP in the microcosm sediments as a function of the sediment depth and length of exposure to ^{14}C-BBP before extraction.*

			Mean Concentration at Depth of Sediment Core		
Day of Sacrifice	Microcosm Number	Exposure Concentration, μg/L	0 to 2 cm, mg/kg	2 to 4 cm, mg/kg	4 to 6 cm, mg/kg
11	9	10	0.24 (0.03)[a]	<LOD[b]	NA[c]
11	13	100	2.30 (0.47)	<LOD	NA
11	11	100C[d]	0.59 (0.21)	<LOD	NA
22	3	10	0.14 (0.10)	0.003 (0.0004)	<LOD
22	1	100	0.76 (0.36)	<LOD	NA
22	10	100C	1.80 (0.49)	0.034 (0.001)	NA
31	4	10	0.25 (0.11)	<LOD	NA
31	8	10	0.18 (0.05)	0.005 (0.002)	0.005[e]
31	15	10	0.27 (0.11)	<LOD	NA
Mean			0.23 (0.05)		
31	6	100	0.85 (0.15)	<LOD	NA
31	7	100	4.54 (1.47)	0.09 (0.05)	0.12 (0.11)
31	16	100	1.77 (0.23)	0.08 (0.07)	0.16[e]
Mean			2.39 (1.92)		
31	2	100C	1.20 (0.25)	<LOD	NA
31	5	100C	2.45 (0.10)	0.16 (0.11)	0.36 (0.22)
31	14	100C	1.72 (0.44)	<LOD	NA
Mean			1.79 (0.63)		

[a] Numbers in parentheses are one standard deviation.
[b] LOD = limit of detection as defined in the methods section: 0.03 to 0.04, 0.03 to 0.04, and 0.003 mg/kg of ^{14}C-BBP for the 100C (control) 100, and 10 μg/L microcosms, respectively.
[c] NA = not analyzed.
[d] 100C denotes the control microcosms spiked initially at 100 ppb of ^{14}C-BBP.
[e] Insufficient numbers above LOD were available to calculate a standard deviation.

BBP—Sediment Analysis

Results of the sediment oxidation-combustion $^{14}C-CO_2$ analyses, expressed as ^{14}C-BBP, indicate that the vast majority of the ^{14}C sediment activity is located in the top 2 cm (Table 1). Depth of ^{14}C penetration into the sediments did not differ between the treatments or between the treatments and the controls. There was an approximate factor of 10 difference between the total ^{14}C activity on the sediments for the 10 and 100 μg/L treatment levels. The means were not statistically different at the 0.05 level due to the variability in the 100 μg/L and 100 (control) μg/L sediment values, but they were significantly different at the 0.1 level. These data indicate that the concentration of total activity on the sediments was a function of water exposure concentration as expected. There was no significant difference between the average concentration of total ^{14}C activity on the sediments of the 100 μg/L controls and the 100 μg/L treatment microcosms. This indicated that the mercuric chloride sterilization did not affect the partitioning of ^{14}C activity.

A comparison of the amount of ^{14}C activity on the sediments as a function of time shows no increase with time after Day 10 for the treatment microcosms, indicating equilibrium was reached very rapidly. This is indicative of a water soluble chemical. In contrast to this, a gradual increase in ^{14}C activity on the sediments of the controls was observed (Fig. 6). The difference between the treatment and control sediments suggests that parent BBP may be partitioning to the sediments in the sterile contols, whereas this is not occurring in the treatment samples because BBP is rapidly degraded in the water phase. Confirmation of this difference was obtained by LCS analysis of hexane extracts of the sediment cores. These data indicate that parent BBP was not prevalent in the treatment sediment cores. Only one treatment core contained a detectable amount of parent BBP. In contrast to this, the controls on Day 31 all contained approximately 0.05 mg/kg BBP in the top 2 cm and

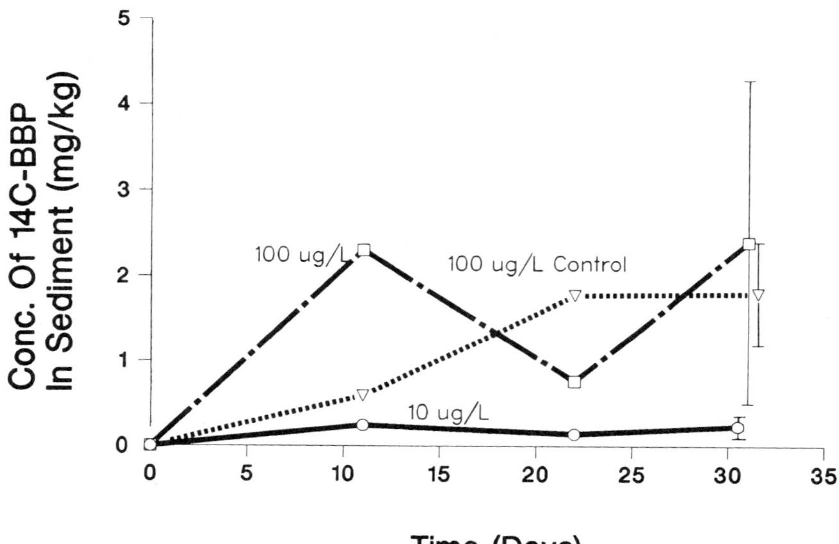

FIG. 6—*A comparison of the concentration of* ^{14}C*-BBP (total activity) in the top 2.0 cm of each microcosm sediment core.*

less than the LOD at lower depths. None of the sediment cores contained levels of parent BBP that approached the concentration of total ^{14}C activity on the sediment. Even the control sediment contained an order of magnitude less parent BBP than total activity. This indicates that the mercuric chloride sterilant placed in the water phase probably did not totally sterilize the sediments and that BBP was biologically degraded on the sediment. It may also indicate that chemical degradation occurred on the sediment. Photodegradation is unlikely since BBP has been shown to be stable in the presence of light [*3*].

Each sediment sample was reanalyzed by oxidation-combustion and LSC analysis following hexane extraction. These results agree with measurements made prior to the hexane extraction, demonstrating that very little parent BBP was present on the sediment (Table 2) and that no detectable loss of activity due to the hexane extraction occurred.

The results for parent BBP obtained by extracting sediment with hexane and analyzing the hexane by LSC were confirmed by electron capture GC (Table 3). The GC data are consistent with the LSC data, indicating little parent BBP was present on the sediments. Only one positive finding (0.007 mg/kg) was observed in the top 2.0 cm of all the treatment sediment cores. The values in Table 3 are almost entirely reported as less than a given limit of detection. These additional analyses demonstrate the general absence of parent BBP on the sediment of the nonsterile microcosms at a detection limit of 0.001 to 0.023 mg/kg. Concentrations of parent BBP on the sediment from sterile microcosms were typically 0.07 to 0.08 mg/kg.

TABLE 2—*A comparison of the ^{14}C activity, expressed as ^{14}C-BBP, in the 0 to 2-cm layer of each microcosm core before and after hexane extraction and in the hexane extract.*

Microcosm Number	Exposure Concentration, μg/L	Day of Study	^{14}C-BBP Concentration, mg/kg: Before Extraction	Hexane Extract	After Extraction
9	10	11	0.24	<LOD[a]	0.27
13	100	11	2.30	<LOD	3.03
11	100C[b]	11	0.59	<LOD	0.61
3	10	22	0.14	<LOD	0.16
1	100	22	0.76	<LOD	0.85
10	100C	22	1.80	0.06	1.59
4	10	31	0.25	<LOD	0.22
8	10	31	0.18	<LOD	0.24
15	10	31	0.27	<LOD	0.29
Mean			0.23	...	0.25
6	100	31	0.85	<LOD	0.81
7	100	31	4.54	<LOD	4.84
16	100	31	1.77	0.04	1.85
Mean			2.39	...	2.50
2	100C	31	1.20	0.04	1.28
5	100C	31	2.45	0.05	2.56
14	100C	31	1.72	0.05	2.06
Mean			1.79	0.05	1.97

[a] <LOD indicates the concentration was less than the limit of detection as described in the methods.

[b] 100C denotes the control microcosm.

TABLE 3—*Measured concentrations of parent BBP in the microcosm sediments in the top 0 to 2 cm; the samples were extracted with hexane and analyzed by electron capture gas chromatography.*

Microcosm Number	Exposure Concentration, μg/L	Day of Study	Average Concentrations of BBP, mg/kg, at Various Sediment Depths		
			0 to 2 cm	2 to 4 cm	4 to 6 cm
9	10	11	<0.015[a]	...[b]	...[b]
13	100	11	0.007	...[b]	...[b]
11	100C[c]	11	<0.022	...[b]	...[b]
3	10	22	<0.017	...[b]	...[b]
1	100	22	<0.020	...[b]	...[b]
10	100C	22	0.069	...[b]	...[b]
4	10	31	<0.020	...[b]	...[b]
8	10	31	<0.023	...[b]	...[b]
15	10	31	<0.019	...[b]	...[b]
6	100	31	<0.022	...[b]	...[b]
7	100	31	<0.013	...[b]	...[b]
16	100	31	<0.020	...[b]	...[b]
2	100C	31	0.079	...[b]	...[b]
5	100C	31	0.068	...[b]	...[b]
14	100C	31	0.079	...[b]	...[b]

[a] The symbol < indicates the limit of detection for a specific sample. It is variable because the amount of dilution used varied with each sample prior to injection in the GC.

[b] Sample not analyzed by GC because the concentration of BBP in the overlying sediment approached the limit of detection, indicating BBP would not be present at greater sediment depth or because LSC analysis indicated the absence of parent BBP.

[c] 100C denotes the control microcosm.

BBP—Mass Balance (Radioactivity Budget)

After the study termination, a measure of the total radioactivity of each microcosm compartment (water, sediment, glass, silicone, and so forth) was calculated in order to obtain a mass balance for each microcosm (Table 4). In each microcosm, the greatest amount of ^{14}C activity was in the water phase at sacrifice. The mean value was 50.8% of the total added. The average amount of ^{14}C in the sediment was only 1.8% (≤0.1% as parent BBP). Average amounts of ^{14}C in other components were 5.6% in the microcosm solvent rinse and 7.7% in the silicone sealant, and 6.6% of the total added was removed by the water replacement during the study. The amount in the scrubbers ranged from 0.1% for the controls to 12.0% for one of the 100 μg/L treatments. The mass balance results indicate that the average percent recovery of the total ^{14}C activity added to each microcosm was 78.2 ± 11.2% (Table 4). Considering the large size of the microcosms, the number of compartments to account for, the number of analytical procedures involved, the extremely large number of samples analyzed, and the fact that the samples were not corrected for analytical recovery losses, the microcosm recoveries are considered acceptable.

An additional element contributing to the loss of ^{14}C activity was found to be the silicone stoppers. Analyses indicated that the stoppers contained significant quantities of ^{14}C-BBP. The quantity of ^{14}C-BBP in each silicone stopper could not be added to the mass recovery total for each microcosm because the stoppers were not maintained with their respective microcosms after the study was completed.

The average amount of ^{14}C-BBP found in the stoppers was approximately 2 to 3% of the total number of micrograms added to each aquaria. One could legitimately add this

TABLE 4—*Mass balance and percent recovery for all* ^{14}*C-BBP added to each microcosm; all calculations are based on* ^{14}*C measurements and not analysis for parent BBP.*

Microcosm Number	Exposure Concentration, μg/L	Day of Sacrifice	Total Micrograms BBP Added	Total Micrograms Removed by Water Exchange	Total Micrograms in Water at Sacrifice	Total Micrograms Trapped in CO_2 Scrubbers	Total Micrograms in Microcosm Wash	Total Micrograms in Sediment	Total Micrograms in Silicone Sealant	Total Micrograms Recovered	Percent Recovery
9	10	11	1 074	36.1 (3.4)[a]	661 (61.5)	10.2 (0.9)	133 (12.4)	48.0 (4.5)	68.0 (6.3)	956	89.0
13	100	11	13 431	452 (3.4)	7 769 (57.8)	39.9 (0.3)	815 (6.1)	460 (3.4)	3 043 (22.7)	12 579	93.7
11	100C[b]	11	13 431	495 (3.7)	8 623 (64.2)	10.2 (0.1)	1 173 (8.7)	118 (0.9)	4 217 (31.4)	14 636	109
3	10	22	1 790	107 (6.0)	1 040 (58.1)	142 (7.9)	63.4 (3.5)	28.6 (1.6)	44.1 (2.5)	1 425	79.6
1	100	22	22 385	1 208 (5.4)	11 025 (49.3)	2 388 (10.7)	2 099 (9.4)	152 (0.7)	777 (3.5)	17 644	78.8
10	100C	22	22 385	1 485 (6.6)	9 450 (42.2)	27.2 (0.1)	1 082 (4.8)	367 (1.6)	2 387 (10.7)	14 798	66.1
4	10	31	2 506	203 (8.1)	1 287 (51.4)	278 (11.1)	104 (4.2)	50.0 (2.0)	58.0 (2.3)	1 980	79.0
8	10	31	2 506	203 (8.1)	1 373 (54.8)	294 (11.7)	90.9 (3.6)	38.0 (1.5)	22.7 (0.9)	2 022	80.7
15	10	31	2 506	212 (8.5)	1 377 (54.9)	185 (7.4)	67.8 (2.7)	54.0 (2.2)	60.1 (2.4)	1 956	78.1
6	100	31	31 339	2 400 (7.7)	15 255 (48.7)	3381 (10.8)	1 443 (4.6)	170 (0.5)	24.8 (0.1)	22 674	72.4
7	100	31	31 339	2 380 (7.6)	15 255 (48.7)	3755 (12.0)	1 067 (3.4)	950 (3.0)	205 (0.7)	23 612	75.3
16	100	31	31 339	2 352 (7.5)	15 075 (48.1)	3733 (11.9)	982 (3.1)	402 (1.3)	1 058 (3.4)	23 602	75.3
2	100C	31	31 339	2 247 (7.2)	13 050 (41.6)	55.5 (0.2)	1 891 (6.0)	240 (0.8)	2 285 (7.3)	19 769	63.1
5	100C	31	31 339	2 428 (7.7)	10 980 (35.0)	30.1 (0.1)	944 (3.0)	594 (1.9)	1 918 (6.1)	16 894	53.9
14	100C	31	31 339	2 599 (8.3)	14 175 (45.2)	22.5 (0.1)	2 616 (8.3)	344 (1.1)	4 917 (15.7)	24 674	78.8
Mean percent recovery				6.6	50.8	...[c]	5.6	1.8	7.7		78.2
One standard deviation				1.8	8.0	...[c]	2.9	1.1	9.0		12.9
Precision, % RSD				27.3	15.7	...[c]	51.8	61.1	116.9		16.5

[a] Values in parenthesis represent the percentage of the total recovered.
[b] 100C denotes the control microcosm.
[c] The mean percent recovery of carbon dioxide for the 10, 100, and 100C Day 31 microcosms was 10.1, 11.5, and 0.13%, respectively.

amount to the average mass balance percent recovery, which would bring it above 80%. On a worst case basis, one could assume that the three stoppers with the largest amount of ^{14}C-BBP all came from the one microcosm with the lowest percent recovery (#5-100C). The amount contained in these three stoppers would bring the recovery of this microcosm up to 84.1% from 53.9%.

Metabolite Identification in Water and Sediment

The results of the analyses performed by reversed-phase HPLC direct aqueous injection of microcosm water indicated that only one peak was observed by this method of analysis. Based on retention time and spectral analysis of this peak, it has been identified as monobutyl phthalate or monobenzyl phthalate (Fig. 7). The peaks for these two chemicals were not separable by the method used. Numerous other GC and GC-MS methods of analysis were tried, but were not capable of identifying polar water-soluble acid compounds.

Water samples from microcosm numbers 13 and 16 (100 μg/L) were analyzed immediately (1 to 2 days) after collection, and they both contained only one identifiable peak, that is, monobutyl or monobenzyl phthalate (Fig. 7). An additional peak was observed in both the river water blank and the microcosm samples. This peak is attributed to an unknown background ultraviolet (UV)-absorbing component of the river water. None of the samples contained peaks which would indicate the presence of the other metabolites that might be expected (that is, phthalic acid).

HPLC analyses of stored samples (60 days at 4°C) indicated there were no major peaks observed in any of the samples, including the controls. This disappearance is attributed to hydrolysis and biodegradation. Ester linkages are known to hydrolyze, particularly at a pH above 8. Confirmation of the degradation of the metabolites was obtained by counting (LSC) two samples from Day 22 which showed no metabolites by HPLC. LSC indicated the ^{14}C was still in the samples after 60 days. Both samples were then analyzed by reversed-phase fractionation and LSC counts of 30 s fractions. It was determined that nearly all ^{14}C activity was associated with the solvent peak. This substantiates the degradation of ^{14}C-BBP to very short-chain water-soluble acids or to a mixture of ^{14}C microbial biomass or both. In either case it is clear that the phthalate ring has been cleaved and that the metabolites of BBP are not stable even at 4°C.

The HPLC standards of the metabolites made up in acetonitrile and water were also observed to be unstable from one day to the next, and it was very difficult to obtain a linear response across a range of concentrations or volumes of injections over time.

Results of the sediment extraction and analysis for BBP metabolites indicate that phthalic acid as well as monobutyl and monobenzyl phthalate could be identified in the sediments, but not necessarily in the same samples. Quantitation of samples was not possible because the peaks were very close to the solvent front and because sediment blanks provided positive results, indicating that the rigor of the extraction procedure was removing polar substances from the Illinois River sediment which gave positive interferences. Six samples spiked at 100 μg/kg for validation purposes all yielded results much higher (greater than ten times) than the nominal level.

Discussion

Water

BBP spiked into Illinois River water rapidly degraded following a 1 to 2 day acclimation period. The half-life of the parent BBP immediately after the study began (Days 1 to 3) was 1.5 days in the 10 μg/L microcosms and 2.2 days in the 100 μg/L microcosms (assum-

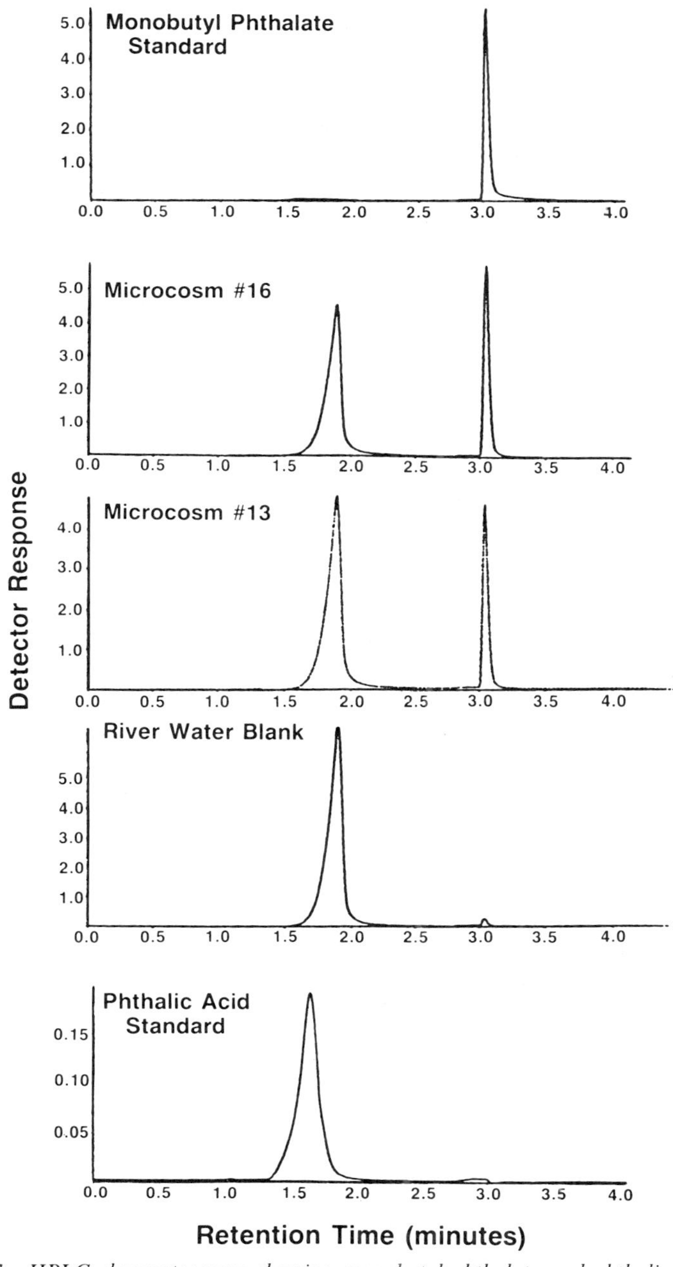

FIG. 7—*HPLC chromatograms showing monobutyl phthalate and phthalic acid standards as well as water samples from Microcosms 13 and 16 and from the Illinois River.*

ing first order decay). The average amount of CO_2 produced in the 10 and 100 μg/L microcosms by Day 22 and 31 was 10.4% of the total amount (in micrograms) of BBP added. Calculation of a precise half-life for ultimate degradation (CO_2 production) of BBP is difficult because of the continued addition of BBP three times per week. The 10.4% recovery of BBP in 30 days as CO_2 does not properly reflect the actual time to convert BBP to CO_2 because of repetitive additions of BBP. If one assumes biomass in the microcosms is constant, and the biodegradation rate is independent of concentration, approximate half-lives of 27.5 and 20.5 days can be obtained from a plot of the number of micrograms of BBP collected as CO_2 over time (days) for the 10 and 100 μg/L microcosms, respectively.

Half-lives for primary and ultimate degradation obtained in the present study are similar to those reported by Saeger et al. [*14*]. They reported half-lives of <2.0 days and 13 days, respectively, for primary and ultimate BBP degradation in microcosms (ecocores) simulating a freshwater lake-sediment ecosystem. Saeger et al. [*15*] also conducted a freshwater river die-away study (Mississippi River water, no sediment) and obtained a primary degradation half-life of <3.0 days. These data are all consistent and point to a rapid degradation of parent BBP in water and a somewhat slower, but continuous degradation of metabolites to CO_2. These data are supported by the reports by Perez et al. [*16*] for di-(2-ethylhexyl) phthalate (DEHP) and by Walker et al. [*17*] for di-*n*-butyl phthalate.

Sediment

Concentrations of parent BBP measured on sediments in all microcosms were less than would be expected on the basis of the sediment-water partition coefficient (K_p = 68 to 350; K_{oc} = 2.0 × 10^4) measured for BBP [*14*]. The average sediment organic carbon (OC) content of the microcosm cores was approximately 1.0%. The K_p for a 1.0% OC sediment would be 200, that is, $K_p = K_{oc} \times F_{oc}$ where F_{oc} equals the fraction organic carbon. Therefore, $2.0 \times 10^4 \times 1.0/100 = 200$. Using a sediment-water K_p of 200, a concentration of 2.0 and 20 mg/kg would be expected on the sediments in the 10 and 100 μg/L microcosms, respectively. If the estimated concentration of parent BBP on the sediments is made using the mean concentration of parent BBP in the water phase of the 10 and 100 μg/L microcosms, values of 200 μg/kg (200 × 1.0 μg/L) and 600 μg/kg (200 × 3.0 μg/L) would be expected, respectively. This assumes no degradation of parent BBP in the water column or sediments. None of the sediments approached their respective predicted concentrations (Table 3). Differences between predicted and observed values ranged from a factor of 7.6 to 294.

These data indicate that BBP is undergoing biodegradation on the sediments in the microcosms. This conclusion is consistent with the fact that BBP concentrations did not increase in the sediments after Day 10, the initial measurement (Fig. 6). Lack of parent BBP and ^{14}C activity in sediments below the 0 to 2 cm segment is further evidence of degradation on the sediments. In contrast to this, Perez et al. [*16*] reported DEHP concentrations of 6160 μg/kg in sediments from microcosms dosed at 100 μg/L. They also detected DEHP down to 7 cm in sediment cores. Data from the present study are clearly different from DEHP even though both chemicals were dosed at the same concentration, and both underwent rapid degradation in the water column. Biodegradation of BBP on sediments is not unique, dibutyl phthalate has also been shown to degrade rapidly in the presence of sediment [*17*].

The above BBP data are consistent with the environmental monitoring data of Gledhill et al. [*3*] who reported a mean BBP sediment concentration of ≤0.136 mg/kg. The limit of detection in the study by Gledhill et al. was only 0.1 mg/kg, which biased the data on the high side since most of the values were reported as less than the LOD.

A clear rationale for the difference between BBP and DEHP degradation on sediments is obtained from the studies of Johnson et al. [*18*] and Shelton et al. [*19*]. They showed that chain length and branching of the alkyl chain on the phthalate moiety reduces the potential for aerobic biodegradation on sediments. Linear dibutyl phthalate (DBP) degraded 25% compared with 2% or less for DEHP, di-isooctyl phthalate, and di-isononyl phthalate. BBP is much more structurally similar to DBP than to DEHP. Shelton et al. [*19*] showed similar results under anaerobic conditions with BBP undergoing complete degradation and DEHP showing little or no degradation. From these two studies, it can be concluded that two factors limit the degradation of DEHP but not BBP: (1) an alkyl chain length of greater than six carbons plus branching on the alkyl chain, and (2) anaerobic conditions. DEHP appears to need aerobic conditions for rapid degradation and BBP does not. Most sediments tend to be oxygen limited within a few millimetres of the surface, even if they are aerobic at the surface.

There is a wealth of evidence from previous studies [*11,19,20*], as well as the present study, indicating that the metabolic pathway is from parent BBP to monobutyl phthalate (and perhaps monobenzyl phthalate) to more polar water-soluble metabolites including phthalic acid, 4.5 dihydroxy phthalic acid, oxalic acid, formic acid (HCOOH), and eventually CO_2 and water. It is very probable that other intermediate chemicals are involved such as the aldehydes, ortho phthaldehyde, and phthaledhydic acid, both of which precede the splitting of the benzene ring as well as oxomalonic, malonic, and benzoic acid subsequent to ring splitting.

For purposes of environmental assessment of BBP metabolites, 48-h static acute toxicity tests with *Daphnia magna* and monobutyl phthalate, monobenzyl phthalate, and phthalic acid were conducted [*21*]. The results of these studies provided the following 48-h median effective concentration (EC_{50}) values: phthalic acid >640 mg/L, monobutyl phthalate >329 mg/L, and monobenzyl phthalate >160 mg/L. In each case, no toxicity was observed at the highest concentration tested. Concentrations up to and slightly exceeding the solubility limits of the two half esters were tested. In comparison, the 48-h EC_{50} value reported for BBP is 3.7 mg/L [*4*]. These data indicate the metabolites of BBP are much less toxic than the parent chemical and should be less of an aquatic environmental concern.

Conclusions

In conclusion, BBP undergoes rapid degradation in the water and on sediments. The half-life in water and sediment were $\leq$2 and 10 days, respectively. Data obtained in the present study with undisturbed river sediment cores did not differ from previous results [*14*] with disturbed lake sediments. These data indicate BBP is not a persistent chemical in freshwater aquatic environments.

References

[*1*] "Experimental Marine Microcosm Test Protocol and Support Document: Measurement of the Ecological Effects, Fate, and Transport of Chemicals In Site-Specific Marine Ecosystems," *EPA Ecological Research Series,* No. 60013-83-055, U.S. Environmental Protection Agency, Washington, DC, 1983.

[*2*] Adams, W. J., "Experimental Freshwater Microcosm Test Protocols," Environmental Sciences Center Special Study Protocol MSL-5218 (ESC-EAG-85-61), Monsanto Co., St. Louis, MO, 1985.

[*3*] Gledhill, W. E., Kaley, R. G., Adams, W. J., Hicks, O., Micahel, P. R., Saeger, V. W., and LeBlanc, G. A., "An Environmental Safety Assessment of Butylbenzyl Phthalate," *Environmental Sciences and Technology,* Vol. 14, 1980, pp. 301–305.

[*4*] Adams, W. J., Renaudette, W. J., Doi, J. D., Stepro, M. G., and Tucker, M. W., "Experimental Freshwater Microcosm Biodegradation Study of Butyl Benzyl Phthalate," Environmental Sci-

ences Center Report No. ESC-EAG-86-101, MSL-6045, Monsanto Co., St. Louis, MO, 1986, p. 107.

[5] "Standard Operating Procedure for pH Determination," Environmental Science Center Document No. ESC-86-SOP-006, Monsanto Co., St. Louis, MO, 1986.

[6] "Standard Operating Procedure for Alkalinity Measurement," Environmental Sciences Center Document No. ESC-86-SOP-007, Monsanto Co., St. Louis, MO, 1986.

[7] "Standard Operating Procedure for Hardness Determination," Environmental Sciences Center Document No. ESC-86-SOP-008, Monsanto Co., St. Louis, MO, 1986.

[8] "Standard Operating Procedure for Dissolved Oxygen Determination—YSI Meter," Environmental Sciences Center Document No. ESC-86-SOP-010, Monsanto Co., St. Louis, MO, 1986.

[9] "Standard Operating Procedure for Conductivity Measurement," Environmental Sciences Center Document No. ESC-86-SOP-011, Monsanto Co., St. Louis, MO, 1986.

[10] Saeger, V. W. and Tucker, E. S., "Biodegradation of Phthalic Acid Esters in River Water and Activated Sludge," *Applied Environmental Microbiology,* Vol. 31, 1976, pp. 29–34.

[11] Legube, B., Guyon, S., Sugimitsu, H., and Dore, M., "Ozonation of Naphthalene in Aqueous Solution—I. Ozone Consumption and Ozonation Products," *Water Research,* Vol. 20, No. 2, 1986, pp. 197–208.

[12] *SAS Users Guide: Statistics,* version 5 ed., SAS Institute Inc., Cary, NC, 1985.

[13] Zar, J. H., *Biostatistical Analysis,* 2nd ed., Prentice-Hall, Englewood Cliffs, NJ, 1984.

[14] Saeger, V. W., Kuehnel, R. G., Linck, C., and Adams, W. J., "The Environmental Fate of Santicizer® 160 (Butylbenzyl Phthalate) in Lake Water-Sediment Microcosms," Environmental Sciences Report No. ES-82-SS-53, Monsanto Co., St. Louis, MO, 1982, p. 26.

[15] Saeger, V. W., Davis, J. L., Kuehnel, R. G., Lewis, M. A., Linck, C., and Adams, W. J., "Santicizer® 160 River Die-Away Biodegradation Rate Study," Environmental Sciences Report No. ES-82-SS-6, Monsanto Co., St. Louis, MO, 1982, p. 20.

[16] Perez, K. T., Davey, E. W., Lackie, N. F., Morrison, G. E., Murphy, P. G., Soper, A. E., and Winslow, D. L., "Environmental Assessment of a Phthalate Ester, Di (2-Ethylhexyl) Phthalate (DEHP), Derived From a Marine Microcosm," *Aquatic Toxicology and Hazard Assessment: Sixth Symposium, ASTM STP 802,* W. E. Bishop, R. D. Cardwell, and B. B. Heidolph, Eds., American Society for Testing and Materials, Philadelphia, 1983, pp. 180–191.

[17] Walker, W. W., Cripe, C. R., Pritchard, P. H., and Bourquin, A. W., "Dibutyl Phthalate Degradation in Estuarine and Freshwater Sites," *Chemosphere,* Vol. 13, No. 12, 1984, pp. 1283–1294.

[18] Johnson, B. T., Heitkamp, M. A., and Jones, J. R., "Environmental and Chemical Factors Influencing the Biodegradation of Phthalic Acid Esters in Freshwater Sediments," *Environmental Pollution (Series B),* Vol. 8, 1984, pp. 101–118.

[19] Shelton, D. R., Boyd, S. A., and Tiedje, J. M., *Environmental Science and Technology,* Vol. 18, No. 2, 1984, pp. 93–97.

[20] Keyser, P., Bosayya, P. G., Eaton, R. W., and Ribbons, D. W., "Biodegradation of the Phthalates and Their Esters by Bacteria," *Environmental Health Perspectives,* Vol. 18, 1976, pp. 154–166.

[21] Tucker, M. W. and Adams, W. J., "A Measure of the Acute Toxicity of Butyl Benzyl Phthalate Metabolites to *Daphnia magna,*" Environmental Sciences Report No. ESC-EAG-86-100, MSL-6037, Monsanto Co., St. Louis, MO, 1986.

Roy M. Ventullo,[1] *Michael A. Lewis,*[2] *and Robert J. Larson*[2]

Response of Aquatic Microbial Communities to Surfactants

REFERENCE: Ventullo, R. M., Lewis, M. A., and Larson, R. J., "**Response of Aquatic Microbial Communities to Surfactants,**" *Aquatic Toxicology and Environmental Fate: Eleventh Volume, ASTM STP 1007,* G. W. Suter II and M. A. Lewis, Eds., American Society for Testing and Materials, Philadelphia, 1989, pp. 41–58.

ABSTRACT: The response of microbial communities to three classes of surfactants was studied in a stream and lake ecosystem. Dodecyltrimethylammonium chloride (TMAC, cationic) had no effect on bacterial biomass or carbon turnover in the sediment, water column, or epilithon in a first-order stream that was continuously dosed with TMAC for 23 days. Biodegradation of TMAC was evident at all sites and showed a time-dependent increase only at the highest (0.25 mg/L) dose site. The effects of long-chain nonionic (linear primary alcohol ethoxylate: Neodol 45-7) and anionic (linear alkylbenzene sulfonate) surfactants were studied *in situ* in Acton Lake, Ohio. The chemicals were tested over a concentration range of 0.001 to 10 mg/L in both short-term (3 h) and long-term (21 day) exposures. Neither of the chemicals had averse effects on the microbial biomass as measured by acridine orange direct counts. In the acute test of these chemicals, decreased turnover of glucose was noted in the concentration range of 0.1 to 10 mg/L surfactant. However, in chronic tests (21 day) the communities recovered, and carbon turnover returned to preexposure levels. Chronic exposure also resulted in an adaptive response as indicated by increased degradation of the chemicals after exposure. Our studies indicate that exposure of lake and stream microbial communities to surfactants had no long-term detrimental effects on bacterial heterotrophy and resulted in increased biodegradation of these materials.

KEY WORDS: microbial response, heterotrophic activity, biodegradation, adaptation, surfactants, aquatic toxicology

Surfactants represent a commercially important class of chemicals in the United States. More than 2×10^6 metric tons of surfactant were produced during 1982 [*1*]. These materials have a variety of consumer uses, including use in sanitizers, disinfectants, and detergents. While most of the surfactants are physically removed or degraded during sewage treatment, there is the potential for some release into aquatic environments as a component of municipal sewage effluent. Previous work on the biological effect of surfactants has been largely concerned with algae [*2–4*] and other macroorganisms [*5–7*]. The impact of surfactants on the bacteria of receiving waters has not been extensively studied [*8,9*].

Microbial communities play a quantitatively important role in biogeochemical cycling and carbon and energy flow in aquatic systems. In addition, their ability to degrade xenobiotic chemicals (biodegradation) is recognized as an important removal mechanism for chemicals that reach the aquatic environment. Knowledge of the response of the micro-

[1] Associate professor, Department of Biology, University of Dayton, Dayton, OH 45469.

[2] Toxicologist and section head, respectively, Environmental Safety Department, Procter & Gamble Co., Ivorydale Technical Center, Cincinnati, OH 45217.

biota to chemicals in aquatic environments can enhance our ability to predict the ecological effects of chemicals.

With few exceptions [*10–13*] the assays used to determine the effect of organic chemicals on microorganisms are carried out using specific bacteria or groups of bacteria in the laboratory [*14–16*]. Many of the methods have been developed for use as a screen for toxic chemicals in aquatic systems. Similarly, estimates of biodegradation of chemicals are often generated in the laboratory in biodegradability screening tests [*17,18*]. However, the relationship between these laboratory-derived data and information derived in field studies is not well defined.

Generally, the laboratory experimental conditions do not accurately simulate natural aquatic environments, where both physical and chemical parameters (pH, oxygen, and nutrient levels) as well as biological factors such as disease, predation, competition, and community structure and function are varied. Thus, the ability of laboratory testing to reflect a realistic assessment of the effect of chemicals on natural microbial communities is not known.

Our approach in this study was to measure the response of natural microbial communities to surfactants exposed *in situ*. We measured the fate and effects of three classes of detergent-related surfactants in natural water systems. These studies were conducted in both a lotic and lentic system.

Materials

Microbial communities were exposed to three surfactants in this study, each representing a major class of surfactant (Fig. 1). Dodecyltrimethylammonium chloride (TMAC) was obtained from the Sherex Chemical Co. (Dublin, OH). Linear alkylbenzene sulfonate (LAS) and Neodol 45-7 (LAE) were obtained from the Shell Chemical Co. (Houston, TX). Specific carbon-14 radiolabeled [1-^{14}C] dodecyltrimethylammonium chloride (64.6 μCi/mg; Amersham Corp., Arlington Heights, IL), [U-^{14}C-ring] tridecylbenzene sulfonate (24 μCi/mg; New England Nuclear, Boston, MA), and [U-^{14}C-ethoxylate] dodecylnonylethoxylate (5.4 μCi/mg; Procter and Gamble Co., Cincinnati, OH) were used in biodegradation studies. An amino acid mixture (L-[U-^{14}C]; 2200 μCi/mg), D-[U-^{14}C] glucose (1600 μCi/mg), and glutamic acid (1550 μCi/mg) were obtained from Amersham Corp.; these were used in effect studies.

Methods

Stream Dosing Study

Study Site—The Howesville Creek is a small stream located southeast of the town of Jasonville in western Indiana (Fig. 2). The Howesville Creek is a highly productive stream which supports a prolific algal community as well as a macroinvertebrate population typical of an organically enriched stream. From its source in a small lake just southeast of Jasonville to its mouth where it joins the Eel River, the Howesville Creek is approximately 12 miles long. The study area began 1 mile below the source and extended for another 2.5 miles of the stream. The Howesville Creek in the study area consists of pools and riffles and has a sandy sediment. The stream varies approximately 0.5 to 3 m in width and 1 to 100 cm in depth. Travel times and distances between test sites are presented in Fig. 3. The study area was located below the outfall of the Jasonville sewage treatment plant (Fig. 2). Although effluent from the treatment plant constituted most of the stream flow during the study period, the wastewater was effectively treated by the aerated lagoons and sand filters,

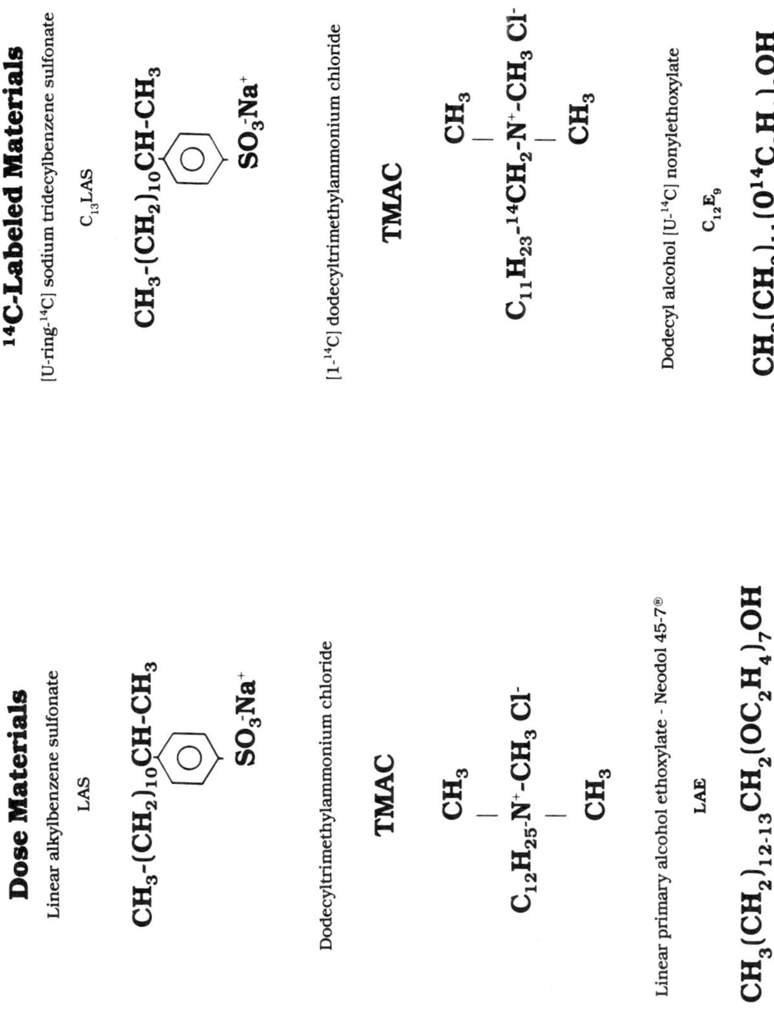

FIG. 1—*Test compounds used in the stream and enclosure studies.*

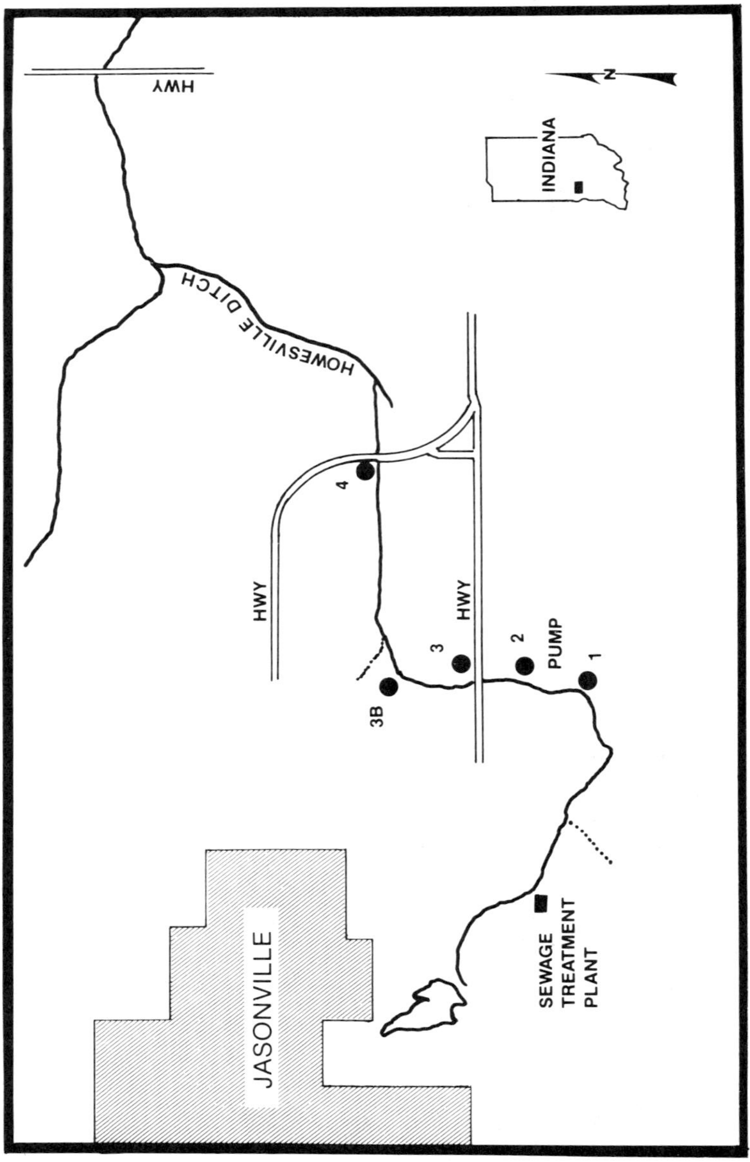

FIG. 2—*Howesville Creek study site. The numbers represent the sample sites described in Fig. 3.*

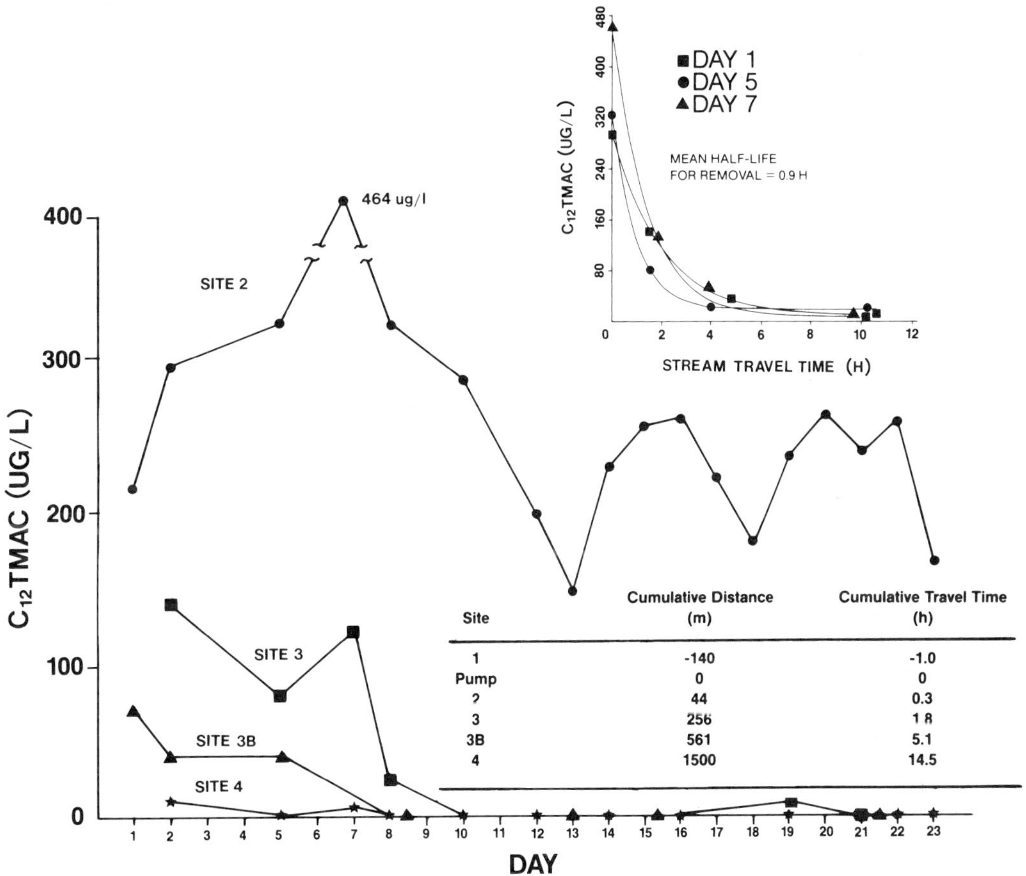

Site	Cumulative Distance (m)	Cumulative Travel Time (h)
1	-140	-1.0
Pump	0	0
2	44	0.3
3	256	1.8
3B	561	5.1
4	1500	14.5

FIG. 3—*Location of sampling stations including distance, traveling time, and measured TMAC concentrations along Howesville Creek.*

resulting in a relatively clean effluent [2]. The water quality of Howesville Creek during the period of this study is described in detail elsewhere [2].

Stream Flow—Flumes constructed of cinder blocks were placed in the stream at Sites 1, 3, and 4. Stream linear velocity was measured at the flumes with a Model 2030 flowmeter (General Oceanics, Miami, FL). Stream flow was calculated by multiplying the cross-sectional area by the linear velocity of the stream at the flume. Stream travel time between sites was measured by adding a dye to the stream. A liquid concentrate of rhodamine (Formulabs, Escondido, CA) was added as a bolus dose. Stream water was then sampled downstream and the absorbance at 553 nm measured. The peak absorbance downstream constituted the travel time to that point.

Stream Dosing—TMAC was continuously added to the Howesville Creek at a target in-stream concentration of 0.2 mg/L. The dosing apparatus consisted of a Brinkman 1P-12 peristaltic pump (Ismatec, Zurich, Switzerland) powered by a 12-V marine battery. The pump was modified with a voltage regulator to convert it from ac to dc.

TMAC was delivered to the stream through a single line of silicone tubing which was fastened to a steel fence post in the middle of the stream. TMAC dripped into the stream at approximately 4 mL/min. Stainless-steel shields placed in the stream below the TMAC outfall facilitated the dispersion of TMAC. To maintain a constant in-stream TMAC concentration, the pump rate was adjusted at least twice daily in proportion to stream flow, which was also measured at least twice daily.

TMAC Analysis—Stream water (500 mL) was collected 4 to 5 times from each site during the study and preserved with 1% formalin. Samples were analyzed by the method of Wee and Kennedy [*19*].

Study Duration—During a three-week baseline period, no dosing of the stream occurred. TMAC was then dosed to the stream for approximately three weeks during which the response of sediment, water column, and epilithic microbial communities was assessed.

Sample Collection and Preparation—Water, sediment, and epilithon used in the biological studies were collected at locations above (Site 1, −140 m) and below (Site 2, 44 m and Site 4, 1500 m) the TMAC dosing site. Epilithon was allowed to colonize unglazed ceramic tiles at Site 1, 2, and 4 for three weeks before dosing started. Incubation racks suspended the plates vertically in the water column to reduce sedimentation on the plates. Each rack contained eleven 7.7-cm by 15.2-cm tiles. Natural sediment from the stream botton and grab samples of the water column were also collected. Sediment was collected by placing an open-ended cylinder into the sediments, removing the overlying water, and collecting the upper 8 cm of sediment in sterile mason jars. All samples collected in the field were chilled and transported to the laboratory for testing within 4 h of collection. Water samples were collected in 4-L (1-gal) plastic containers and tested as collected with no further treatment.

Bottom sediment (~100 mL) was diluted tenfold with 900 mL of filter-sterilized (0.2 μm) creek water to make sediment-water suspensions (97 to 197 g/L). Epilithon suspensions were made by scraping a 220-cm^2 area from the tiles, mixing in a waring blender, and diluting with 1 L of filter-sterilized creek water.

Microbial Effects Studies—The effect of TMAC on the number and activity of heterotrophic bacteria in Howesville Creek was determined in water, sediment, and epilithon compartments. Total microbial numbers were determined by the acridine orange direct count (AODC) procedure [*20*], using the modification of Kirchman and Mitchell [*21*] to count both particle-bound and unattached cells. Bacterial heterotrophic activity was determined by the turnover time-tracer approach [*22*] in 100 mL serum bottles containing 10 mL of water or 10 mL of sediment or epilithon suspensions. The test substrates (~10 μg/L) included ^{14}C-glucose, ^{14}C-glutamic acid, or ^{14}C-mixed amino acids (MAA). Turnover time, T_t, values, or the time required for the complete metabolism [uptake + radiolabeled carbon dioxide ($^{14}CO_2$)] of the substrate at their natural or *in situ* concentrations, were calculated by the method of Gocke [*23*]. For sediment suspensions, T_t values were based on respired $^{14}CO_2$ only.

Biodegradation Kinetic Studies—The kinetics of biodegradation of TMAC in Howesville Creek were measured using both the single concentration turnover time method [*23*] and time-course techniques. Turnover times for TMAC were determined at 12°C as described for the natural substrates above. The ^{14}C-TMAC was added at initial concentrations ranging from 1 to 20 μg/L, and in all cases represented <10% of the *in situ* pool of TMAC present.

Time-course kinetic studies were conducted, as previously described [*22,24*], in triplicate 2-L Erlenmeyer flasks containing 1 L of creek water alone (44 m site) or water containing 1 and 10 g/L (dry weight) of added bottom sediments collected from the same site. Samples were incubated on a rotary shaker (120 rpm) at 25°C. The head space of the test flasks was continuously aerated with CO_2-free air, and $^{14}CO_2$ was trapped in 10 mL of 1.5 *N* potassium hydroxide. At various intervals, 10-mL aliquots of water were taken by syringe from the flasks and filtered through 0.2-μm filters to remove particulate matter. The filters were washed and the filtrate was added to 125-ml biometer flasks and acidified with ~1 mL of concentrated hydrochloric acid. The $^{14}Co_2$ released was trapped in 2 ml of 1.5 *N* potassium hydroxide in the side arm of the biometer. Samples from the biometer side arms and the external base traps were counted to determine the total amount of $^{14}CO_2$ produced. Data for percent $^{14}CO_2$ production were analyzed by various nonlinear product models to generate parameter estimates for the first order rate constant for degradation (k_1), normalized for the extent of $^{14}CO_2$ production. Parameter estimates were obtained by least squares analysis using the NONLIN computer program, as previously described [*24*].

Lake Enclosure Study

Study Area—*In situ* lake studies were conducted in Acton Lake, located in Preble and Butler Counties, Ohio (Fig. 4). Acton Lake is a 250 hectare reservoir located approximately 60 km northwest of Cincinnati, OH. The lake receives no direct wastewater discharge, but is exposed to potential contaminants via septic tank seepage and agricultural runoff. The limnology and plankton of Acton Lake have been previously described [*25*].

Surfactant Analysis—Surfactant concentrations were based on the mean-measured levels of material in the vessels. The anionic surfactant was analyzed by the standard methylene blue active substances (MBAS) method [*26*]. Nonionic surfactant was measured by the cobalt thiocyanate active substances (CTAS) method [*26*].

Bacterial Heterotrophic Activity—The effect of LAS and LAE on the density (numbers) and heterotrophic activity of bacterial communities in Acton Lake was determined in short-term acute (3-h) studies as well as long-term chronic (21-day) studies. Total microbial numbers were determined by the AODC procedure discussed above. Bacterial heterotrophic activity was determined by the single concentration turnover time-tracer approach [*23*]. Serum bottles contained 10 to 20 mL of lake water and 0.01 mg/L ^{14}C-glucose as the test substrate. Short-term acute studies were conducted by collecting lake water from the photic zone and distributing it to triplicate 300-mL biological oxygen demand (BOD) bottles containing various concentrations of surfactants (0.001 to 50 mg/L). Triplicate controls received only water and no test material. The bottles were incubated in the lake at a depth of ~1 m for 3 h (Fig. 5) and then removed and assayed for AODC and glucose turnover. Water samples for AODC determinations were preserved in 10% formalin buffered with 100 m*M* sodium phosphate (pH 7.0). Significant differences in AODC and glucose turn-

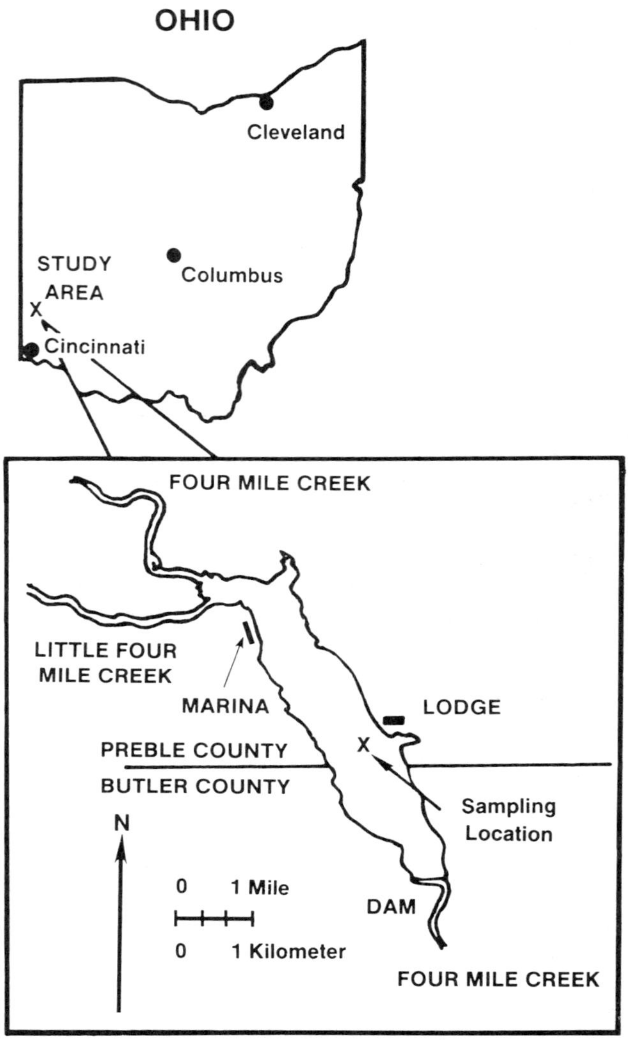

FIG. 4—*Acton Lake study site.*

over between control and test treatments were determined at the 95% confidence level by analysis of variance (Model I single classification) techniques.

Long-term chronic studies were conducted for a LAS and LAE, also using water collected from the photic zone. Lake water was pumped into large polyethylene tubs and thoroughly mixed by mechanical stirrers. The water was then distributed to duplicate sterile 26-L polycarbonate carboys containing various concentrations of LAS and LAE. A paired control carboy containing no test material was included with each set of duplicate test carboys. Sets of three carboys were mixed, capped, and suspended in the lake (1-m depth) for a period of 21 days. At the end of the incubation period, water samples were collected from control and test carboys and assayed for AODC and glucose T_t relative to unexposed controls. Oxygen levels did not drop below 4 mg/L throughout the incubation (data not

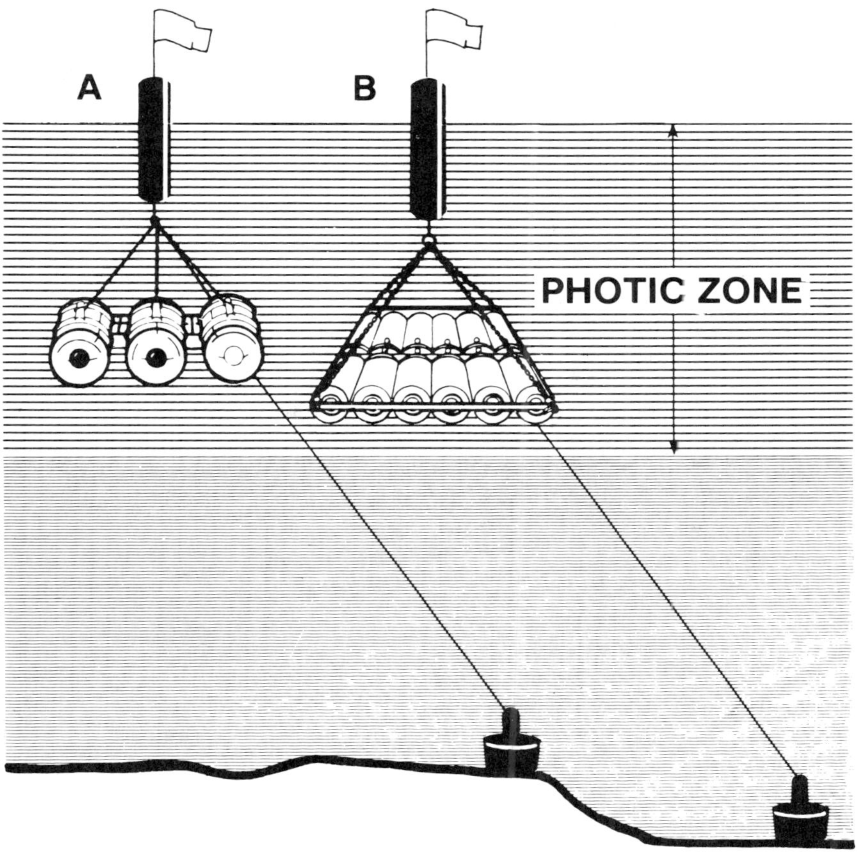

FIG. 5—*Enclosures used in the Acton Lake study:* (a) *Translucent chambers (26 L), two of which contained surfactant and one served as an unamended control in the chronic studies;* (b) *Bottles (300-mL) that were used for short-term acute tests. All the vessels were suspended at a depth of 1 m in Acton Lake.*

shown). Significant differences in AODC and glucose turnover between control and test treatments were determined at the 95% confidence level by analysis of variance (Model I single classification) techniques. Concentrations of anionic and nonionic surfactants were measured at intervals during the test period [*26*].

Biodegradability Studies—The kinetics of biodegradation of ^{14}C-LAS and ^{14}C-LAE in Acton Lake water were measured using the biodegradation potential technique. Briefly, various concentrations of LAS and LAE (~2 to 1500 μg/L) were incubated at 19°C in 10 to 20 mL of water for short periods, and the fraction, f, of radiolabel metabolized to $^{14}CO_2$ was determined.

Data for the rate, v, of degradation in nanograms per litre per hour at various added substrate concentrations, A, were analyzed by Eq 1, which is a hyperbolic saturation curve of the Michaelis-Menten type.

$$v = \frac{f}{t} \cdot A = \frac{V_{max} \cdot A}{K_a + A} \tag{1}$$

where f = $^{14}CO_2/^{14}C$-added, t = incubation time in hours, A = added substrate concentration in micrograms per litre, V_{max} = maximum degradation rate in nanograms per litre per hour, K_a = half saturation constant where $v = 0.5\ V_{max}$ in micrograms per litre. Parameter estimates for V_{max} and K_a were obtained by least squares analysis using a nonlinear computer program, as previously described [*8*].

Specific activity indices (SAI) were calculated by dividing V_{max} by the AODC [*27*] or the most probable number (MPN) of degraders [*9*].

Enumeration of Surfactant Degraders—The method described by Ventullo and Larson [*9*] was used to determine the most probable number (MPN) of bacteria capable of metabolizing LAS and LAE. Filter-sterilized lake water supplemented with the following constituents served as the growth media: ammonium nitrate (0.1%), sodium phosphate monobasic (0.08%) and dibasic (0.1%), sodium chloride (0.1%), calcium chloride dihydrate (0.002%), and sodium bicarbonate (0.05%). Radiolabeled LAS or LAE (10 to 40 μg/L) was added to five replicate serum bottles containing serial (1:10) dilutions of lake water. The bottles were incubated quiescently in the dark for 5 to 6 weeks at 15°C. Individual bottles were acidified and the $^{14}CO_2$ produced was collected on fluted wicks soaked with 1.5 *N* potassium hydroxide. Replicates evolving at least twice the dpm of the killed control were counted as positive. The MPN of degraders and associated 95% confidence intervals were calculated by a computer program [*28*].

Results

Stream Dosing Study

Microbial Effects Studies—Bacterial abundance in the water, sediment, and epilithon compartments of Howesville Creek is shown in Table 1. TMAC appeared to have little effect on bacterial numbers in the water, sediment, or epilithon compartments over the range of concentrations dosed to Howesville Creek (~150 to 460 μg/L). Little disruption of gross cell structure was evident when the cells were viewed after staining with acridine orange for the direct counting procedure. In addition, there appears to be little effect on heterotrophic activity of the bacterial communities exposed to TMAC (Tables 2 and 3).

TABLE 1—*Bacterial abundance in water, sediment, and epilithon from Howesville Creek.*

		Total Number of Bacteria, ×10^6		
Day[a]	Compartment	Site 1	Site 2	Site 4
2	water[b]	0.38	0.53	ND[c]
	sediment[d]	47.0	110.0	ND
	epilithon[e]	12.0	58.0	28.0
10	water	0.39	0.32	0.06
	sediment	24.0	20.0	ND
16	water	0.34	0.41	0.22
	sediment	180.0	2600.0	ND
23	water	0.12	0.40	ND
	sediment	75.0	86.1	290.3
	epilithon	22.0	9.0	17.0

[a] Time relative to the start of dosing.
[b] Cells per millilitre.
[c] Not determined.
[d] Cells per gram dry weight.
[e] Cells per square centimetre.

TABLE 2—*Turnover times for natural compounds in Howesville Creek sediments.*

Day[a]	Compound	Turnover Time, h[b]	
		Control (Site 1)	Dose (Site 2)
−6[c]	glucose	...	1
	glutamate	...	1
2	glucose	6	3
	glutamate	2	1
9	glucose	4	4
	glutamate	0.6	0.7
16	glucose	4	2
	glutamate	0.3	0.1
23	glucose	5	5
	glutamate	0.3	0.2

[a] Time relative to start of dosing.
[b] Turnover time, T_t, per gram weight sediment. Sediment concentrations ranged from 97 to 197 g/L.
[c] Six days before dosing.

TABLE 3—*Turnover times for natural compounds in Howesville Creek water column.*

Day[a]	Compound	Turnover Time, h		
		Control (Site 1)	Dose (Site 2)	Dose (Site 4)
−6[b]	glucose	...	17	12
	MAA[c]	...	19	7
2	glucose	7	11	...
	MAA	9	10	...
9	glucose	8	5	4
	MAA	5	5	5
16	glucose	9	8	17
	MAA	9	8	10
23	glucose	21	25	18
	MAA	4	11	5

[a] Time relative to start of dosing.
[b] Six days before dosing.
[c] Amino acid mixture.

Turnover time, T_t, values for glucose, glutamate, and MAA varied little between the control and dose sites for specific sampling periods. T_t values for the sediments also remained relatively constant during the course of the 23 day study, varying in many cases by less than a factor of 2. Overall, measured values for bacterial biomass and aerobic heterotrophic activity in Howesville Creek were fairly typical for a small fresh water stream and are unaffected by the level of TMAC being tested.

Removal of TMAC in Howesville Creek—Figure 3 shows the measured water column concentrations of TMAC in Howesville Creek at various sites as a function of the length of dosing period. From the figure, it is apparent that TMAC concentration decreased sharply with distance downstream and was essentially nondetectable after Day 10 of dosing at a distance of 256 m (10 h) downstream from the dose site. When the measured TMAC concentrations were plotted as a function of stream travel time, the rate of removal followed a typical first order relationship in all three cases (Fig. 3, inset). The estimated half-life for removal, based on the mean first order rate constant after 2, 5, and 7 days of dosing was ~1 h (0.96 ± 0.23 h). This half-life reflects removal of TMAC due to sorption to suspended solids and bottom sediments as well as biodegradation by suspended and attached microbial communities.

Heterotrophic Activity—The results of the single concentration turnover time studies to determine biodegradation rates (half-lives) for TMAC in water, sediment, and epilithon suspensions are summarized in Table 4. Degradation rates in sediments exhibited more variability and were significantly slower (approximately fivefold to twentyfold) and than the rates in either the water or epilithon compartments (Table 4). Slow TMAC degradation in sediment was not due to the absence of TMAC degraders, since MPN analyses indicated significant levels of bacteria capable of degrading TMAC (10^4 to 10^5 cells/g dry sediment). A significant adaptive response was noted for Site 2 sediment communities as compared with those at Site 1 and 4. The half-lives for TMAC were 2 to 20 times greater in sediment exposed to the quaternary ammonium compound for as little as 2 days (Table 4).

For the water compartment, TMAC degradation was significantly faster at Site 2, below the dosing area, where the highest TMAC exposures occurred (Fig. 1) than at either the control or postdose sites. In general, the half-live values for TMAC in water from below the dosing site was relatively constant over the 21-day study period, averaging ~2 days.

The limited data for the epilithon compartment indicates degradation rates were comparable for the control and dose sites, and somewhat slower at Site 4. T_t values for TMAC in epilithon collected from both above and below the dosing area also showed relatively little variation, averaging ~2.5 days.

Time-Course Biodegradation Kinetic Studies—Ultimate biodegradation of TMAC in Howesville Creek water and water-containing bottom sediments (Site 2) followed approximate first order kinetics over an initial concentration range of 10 to 100 μg/L with the rate constants and half-lives given in Table 5. The amount of $^{14}CO_2$ produced during biodegradation was extensive (80 to 90%), and no lag phase was observed prior to the onset of degradation. Both $^{14}CO_2$ production and removal of ^{14}C activity from solution (data not shown) followed comparable kinetic patterns, with the average half-lives for mineralization

TABLE 4—*Half-life for TMAC estimated from biodegradation by water, epilithon, and sediments based on the turnover time method.*

		Half-Life, days		
Day[a]	Compartment	Control (Site 1)	Dose (Site 2)	Dose (Site 4)
−6[b]	water	...	...	...
	sediment	...	7.9	...
2	water	4.3	2.3	...
	sediment	8.7	4.8	...
	epilithon	2.8	3.3	7.3
9	water	6.9	1.8	3.7
	sediment	17.1	4.5	...
16	water	8.04	1.8	15.7
	sediment	88	4.6	...
23	water	6.7	1.5	5.6
	sediment	38	2.3	...
	epilithon	2.0	2.0	8.3

[a] Time relative to start of dosing.
[b] Six days before dosing.

TABLE 5—*Effect of sediment on ultimate degradation of TMAC by Howesville Creek Water.*[a]

TMAC Concentration, μg/L	Sediment Concentration, g/L	$^{14}CO_2$ Produced, %[a]	Rate Constant, days^{-1}	Half-Life, h
10	...	87.5	2.51	...
100	...	79.3	2.63	6
100	1	70.5	0.56	30
100	10	64.3	0.20	83

[a] Water from Site 2.

and removal, based on the mean first order rate constants at 10 and 100 μg/L, of ~6 and 7 h, respectively. In the presence of sediments, the total $^{14}CO_2$ evolved was decreased, and the half-lives were increased twofold to fivefold over those in the water. The greater degradation noted in this experiment, when compared with the single concentration method, is probably due to the temperature difference and the fact that the water in this experiment was two to three-days old when we used it.

Lake Enclosure Study

Microbial Effects Studies—The results of the short-term dose response studies to determine the effects of LAS and LAE on the heterotrophic microbial communities in Acton Lake are summarized in Table 6. Both chemicals were tested several times during a four-month period, and the data shown are the means for all experiments.

Bacterial densities in Acton Lake ranged from 0.5 to 2.2 × 10^6 cells/mL during the test period and exhibited no significant change as a function of dose material or time of exposure (ANOVA). However, for one of the surfactants, LAS, there was a dose-dependent

TABLE 6—*Effect of acute and chronic exposure to surfactants on glucose turnover by heterotrophic bacteria in Acton Lake.*

	Mean Glucose Turnover Time, h[a]			
	C_{13}LAS		$C_{14-15}E_7$	
Concentration, mg/L	Acute Exposure	Chronic Exposure	Acute Exposure	Chronic Exposure
Control	10 (*)	6	20	9
0.001	15 (*)	ND[b]	ND	ND
0.01	21 (*)	ND	ND	5
0.1	17 (*)	7	22	4
0.5	46 (**)	ND	28	ND
1.0	96 (**)	3	29	6
5.0	556	2	ND	2
10.0	712	ND	27	2
50.0	ND	3	17	5
100	ND	3	ND	ND

[a] Within the first column, values with similar symbols are not significantly different. There are no differences among values in any other column ($P \leq 0.05$, n = 16 for each mean; a Model I single-classification ANOVA was used).

[b] Not determined.

increase in the turnover time for glucose, exhibiting a first-effect concentration (FEC) of 0.5 mg/L. At and above the FEC, longer turnover times, indicating decreased heterotrophic activity, was evident. The FEC for the LAE was not reached even at concentrations as high as 50 mg/L, indicating it was significantly less toxic than the LAS.

The results of the long-term dose-response studies are summarized in Table 6. These chronic studies were carried out to determine if there was an amelioration of the acute effects or a decrease in the FEC due to extended exposure. Bacterial abundance in the long-term studies were again unaffected by the wide range of surfactants concentrations (0.001 to 100 mg/L) used in this study (data not shown). Overall, the heterotrophic activity of the communities incubated in the enclosures was greater than that measured in the acute test (Table 6) or outside the chambers during the incubation period (mean T_t, 14 ± 8 h; range 4 to 21 h). In contrast to the acute studies above, a FEC for LAS was not detected. The highest concentration of LAS used in the chronic study was 100 times greater than the acute FEC. The response of Acton Lake microbial communities to chronic exposure to LAS resulted in a selection of more tolerant bacterial species and a recovery of glucose metabolism. A FEC for LAE was, as during the acute test, not evident.

Biodegradation Studies—The results of studies to examine the effect of exposure time and concentration on microbial biodegradative ability are shown in Table 7. Chronic (21 day) exposure of bacterial communities to LAS resulted in a significant increase in the biodegradation potential of the microbiota (*t*-test). While degradation rates remained unchanged in the unexposed control, there was a significant time- and concentration-dependent adaptation to LAS in the dosed enclosures. V_{max} values can be normalized for the number of active bacteria in a community [*9,27*] by dividing the V_{max} value by the mean MPN of degraders to obtain a specific activity index (10^{-6} ng/cell/h). For LAS, the specific activity index increased from an average 0.0001 ng/cell/h for the unexposed communities to 0.0018 ng/cell/h for the 5.0 mg/L dose level. Based on the limited data available, it appears that increased LAS degradation is due to both the greater number of degraders (Table 7) and the greater activity per cell.

Similar results were obtained in the LAE studies. The V_{max} for $C_{12}E_9$ is initially twenty-fold higher than for LAS and exhibits a concentration-dependent adaptation (Table 7). In contrast to the LAS study, the increased activity is primarily due only to the increased number of degraders since the specific activity index value for LAE degradation decreased from 116 to 73 × 10^{-6} ng/cell/h.

Table 8 shows the nominal and measured concentrations of surfactants in the carboys

TABLE 7—*Development of surfactant biodegradation capability in Acton Lake as a function of surfactant exposure.*

Nominal Surfactant Concentration, mg/L	V_{max}, ng/L/h		MPN of Degraders, 10^5 cells/L	
	C_{13}LAS	$C_{12}E_9$	C_{13} LAS	$C_{12}E_9$
0 (Control, Day 0)	2.6 (0.8 to 4.2)[a]	52 (21 to 70)	0.4 (0.1 to 0.5)[b]	4.5 (1.5 to 15.3)
0 (Control, Day 21)	3.1 (1.0 to 5.1)	ND[c]	0.2 (0.1 to 0.6)	3.7 (1.6 to 8.5)
0.01	7.6 (2.8 to 18)	ND	ND	2.2 (0.8 to 5.7)
5.0	586 (123 to 905)	236 (139 to 413)	3.3 (1.2 to 9.3)	32.5 (11.4 to 92.1)

[a] Value in parentheses are 95% confidence interval of parameter estimate.
[b] Values in parentheses are 95% confidence intervals of mean value.
[c] Not determined.

TABLE 8—*Nominal and measured concentrations for anionic and nonionic surfactants in Acton Lake microbial effects study.*

Surfactant	Nominal Concentration, mg/L	Measured Concentration, mg/L		
	Day 0	Day 0	Day 10	Day 21
$C_{13}LAS$	0.05	0.06	0.053	ND[a]
	0.1	0.08	0.13	ND
	1.0	0.9	1.03	<0.05
	5.0	5.3	3.8	1.2
	25	30.3	25.9	26.4
	100	122	116	116
$C_{14-15}E_7$	0.01	0.009	<0.001	<0.001
	0.1	0.09	0.002	<0.001
	1.0	0.81	0.02	0.05
	5.0	4.73	0.36	0.16
	10	8.45	1.94	0.34
	20	17.8	11.0	9.3

[a] Not detected.

during the chronic study. Both a time- and concentration-dependent decrease in surfactants was detected. Only at the two highest concentrations of LAS and the highest dose level of LAE did the measured concentration remain at or near the initial dose level.

Discussion

In developing realistic approaches for the exposure assessment of chemicals in aquatic environments, the use of field studies provides a great degree of realism [*2,29–32*]. We have, in this study, attempted to assess the response of natural microbial communities *in situ,* to environmentally relevant concentrations of three surfactants. Since lotic systems are the most difficult to model in the laboratory, we chose to dose a small stream and assess microbial communities from the water column, sediment, and epilithon to obtain field-derived data. All of the time-dependent variables such as flow, mixing, turbulence, and nutrient concentrations were allowed to influence the response of the microbiota. For lentic systems, enclosing small portions of the natural plankton communities has generally met with success [*2,25,33,34*], and this approach was used in the Acton Lake study.

The response of the microbiota to chemicals in the environment can be separated into a toxicity and biodegradative component. With respect to microbial biomass, there appeared to be little dose- or time-related change in cell number in either system. Damage to cell membranes or walls or both, resulting in lysis, was not evident. A lack of cell lysis by TMAC has also been reported by Ventullo and Larson [*9*] for Acton Lake microbial communities. This lack of cell disruption was not an unexpected result, since the maximum exposure concentration used in either study was less than 1% of the concentration generally reported to cause disruption of bacterial membranes [*35,36*]. Microbial effects assays which measure the functioning of microbiota in nutrient cycling certainly measures a process that, if affected, could have a significant ecological impact. We used the single concentration turnover time approach to measure the effect of the surfactants on stream and lake glucose and amino acid metabolism. This method has been used extensively to monitor heterotrophic activity in aquatic environments [*9,23,27,37*]. TMAC had no significant effect on the microbial activity in Howesville Creek (Tables 2 and 3). The lack of effect at Sites 1 and 4 is not unexpected, since TMAC concentrations were below detection at these sites (Fig. 3). However, even after exposure for 21 days at Site 2, no effect on the heterotrophic activity of Howesville Creek microbiota was noted.

Based on the results of the study in Acton Lake, there is a significant acute response of glucose metabolism to LAS but not to LAE (Table 6). These data follow the typical pattern of toxicity noted for surfactants; that is, nonionics, in general, are less active against bacteria than anionics or cationics [*35,36*]. It is important to note that chronic exposure of the microbiota to LAS led to an adaptive response, a recovery of heterotrophy at the 0.5-mg/L dose level, and a significant (hundredfold) increase in the FEC (Table 6). Some of the recovery was no doubt due to loss of surfactant from the enclosures by biodegradation, sorption, or loss thru the caps (Table 8). However, chemical analysis for LAS indicated that after day 21, enclosures dosed with ≥5 mg/L still had concentrations greater than the acute FEC. The exposure time, for which the concentration was greater than the acute FEC for the 1 mg/L dosage, was at least ten days. Nonetheless, both the concentration and length of exposure were sufficient to produce a recovery response to LAS at concentrations as high as 100 mg/L.

The aquatic microbial communities also exhibited a biodegradative response to the surfactants. This response is commonly described as adaptation or acclimation [*9,38*], which refers to a process where the biodegradation of the chemical is significantly increased as a result of prior exposure to the chemical. While an increase in degradation rate was not evident for the epilithon, both the water column and sediment microbiota from Site 2 in Howesville Creek exhibited adaptation to TMAC (Table 4). Increases in biodegradation capability (as indicated by decreased half-lives) were evident at Site 2 relative to the unexposed sites (1 and 4). Although a greater number of bacteria exists in the sediment (Table 1), the rates in the water column were always higher. In the experiment, where sediment was added at varying concentrations to the water from Site 2, the presence of sediment sharply decreased the rate and extent of mineralization of TMAC (Table 5). Since quaternary ammonium compounds have large K_d values [*36,39*], sorption of TMAC to sediments may have decreased the bioavailability of TMAC. It is possible, with the exception of the high-dose site (Site 2), that only the top few millimetres of sediment were exposed *in situ.* The exposure of the sediment communities to TMAC would be limited at Site 4. The rates measured in the time-course study, where the particulate matter was constantly suspended by shaking, were greater than in the short-term studies, where static incubations were used. This indicates that dispersion, as might occur in lotic systems, can be an important factor controlling the kinetics of TMAC degradation in sediment-water systems.

Chronic exposure of lake microbial communities to LAE and LAS also resulted in a biodegradation adaptive response. At the 5 mg/L dose level, a substantial increase in biodegradation and number of degraders occurred relative to the control and other dose levels (Table 6). Because enclosures dosed with intermediate concentrations of surfactants were not assayed, we cannot readily determine the threshold level for acclimation for either chemical. For LAE, the actual exposure concentration was <0.3 mg/L after ten days (Table 8), indicating that adaptation was occurring at a much lower concentration than the 5 mg/L nominal level. Similarly, a decreased level for LAS was noted (3.8 mg/L) at Day 10, but it was not as great a loss. In regards to the microbiota in these systems, while the MPN of degraders increased, no significant increase in the total number of bacteria was observed. Therefore, there must have been a shift in the community structure to account for the increased number of degraders (Table 7).

We have assessed the response of microbial communities in a lotic and lentic environment to three detergent related surfactants. TMAC and LAE had no acute or chronic effect on the heterotrophic activity as measured by glucose metabolism. Acute exposure to LAS did have an adverse effect on heterotrophy at concentrations >0.5 mg/L. An adaptation process, which resulted in selection of a more tolerant community and recovery of heterotrophy, occurred in the chronic test. No significant adverse effects were observed at concentrations nominally 100-fold higher than the FEC measured in the acute tests. Chronic exposure (2 to 21 day) of water column (Acton Lake and Howesville Creek) and sediment (Howesville Creek) microbial communities also resulted in an adaptive response in the fate studies.

As often happens with field studies, we were not always able to accumulate enough data to allow statistical comparisons. However, certain conclusions can be drawn from the data. *In situ* exposure of microbial communities to parts per billion levels of surfactants resulted in no long-lasting adverse affect on heterotrophic activity in either lentic or lotic environments. In addition, biodegradation is stimulated by exposure to the chemical, resulting in an effective removal of the chemicals.

Acknowledgments

We would like to thank personnel of North Texas State University for field support at Howesville Creek. We were assisted at Howesville Creek by M. J. Taylor, D. Davidson, and T. Williams. We thank J. A. Staubach, B. S. Schwab, B. G. Hamm, and T. Williams for their technical assistance. A special thanks to the reviewers.

References

[*1*] *Chemical Economics Handbook,* SRI International, Menlo Park, CA, 1981.

[*2*] Lewis, M. A., Taylor, M. J., and Larson, R. J. "Structural and Functional Response of Natural Phytoplankton and Periphyton to a Cationic Surfactant with Considerations on Environmental Fate," *Community Toxicity Testing, ASTM STP 920,* J. Cairns, Jr., Ed., American Society for Testing and Materials, Philadelphia, 1986, pp. 241–268.

[*3*] Payne, A. G. and Hall, R. H. *Mitteilungen der Internationalen Vereinigung für Theoretische und Angewandte Limnologie,* Vol. 21, 1978, pp. 507–520.

[*4*] Yamane, A., Okada, M., and Sudo, R., *Water Research,* Vol. 18, 1984, pp. 1101–1105.

[*5*] Bishop, W. E. and Perry, R. L., "Development and Evaluation of a Flow Thru Inhibition Test with Duckweed, *Lemna minor,*" *Aquatic Toxicology and Hazard Assessment* (Fourth Conference), *ASTM STP* 737, D. R. Branson and K. L. Dickson, Eds., American Society for Testing and Materials, Philadelphia, 1981, pp. 421–435.

[*6*] Lee, C. M., Fullard, J. F., and Huntington, E., *Journal of Testing and Evaluation,* Vol. 8, 1980, pp. 282–287.

[*7*] Adams, W. J., Kimerle, R. A., and Mosher, R. G. "Aquatic Safety Assessments of Chemicals Sorbed to Sediments," *Aquatic Toxicology and Hazard Assessment (Seventh Symposim), ASTM STP 854,* R. D. Cardwell, R. Purdy, and R. C. Bahner, Eds., American Society for Testing and Materials, Philadelphia, 1985, pp. 429–453.

[*8*] Larson, R. J. and Maki, A. W., "Effect of Linear Alkylbenzene Sulfonate on the Structure and Function of Microbial Communities in Model Ecosystems," *Aquatic Toxicology and Hazard Assessment (Fifth Symposium), ASTM STP 766,* J. G. Pearson, R. B. Foster, and W. E. Bishop, Eds. American Society for Testing and Materials, Philadelphia, 1982, pp. 120–136.

[*9*] Ventullo, R. M. and Larson, R. J., *Applied Environmental Microbiology,* Vol. 51, 1986, pp. 356–361.

[*10*] Albright, L. J., Wentworth, J. W., and Wilson, E. M., *Water Research,* 1972, Vol. 6, pp. 1589–1594.

[*11*] Barnhart, C. L. and Vestal, J. R., *Applied Environmental Microbiology,* Vol. 46, 1983, pp. 970–977.

[*12*] Bauer, J. E. and Capone, D. G., *Applied Environmental Microbiology,* Vol. 49, 1985, pp. 828–835.

[13] Burton, G. A. and Lanza, G. R., "Sediment Microbial Activity Tests for the Detection of Toxicant Impacts," *Aquatic Toxicology and Hazard Assessment (Seventh Symposium), ASTM STP 854,* R. D. Cardwell, R. Purdy, and R. C. Bahner, Eds., American Society for Testing and Materials, Philadelphia, 1985, pp. 214–228.
[14] Bitton, G., "Bacterial and Biochemical Tests for Assessing Chemical Toxicity in the Aquatic Environment: A Review," *CRC Critical Reviews in Environmental Control,* Vol. 13, 1983, pp. 51–67.
[15] Dutka, B. J., Nyholm, N., and Petersen, J., *Water Research,* Vol. 17, 1983, pp. 1363–1368.
[16] Greene, J. C., Miller, W. E., Debacon, M. K., Long, M. A., and Bartels, C. L., *Archives of Environmental Contamination and Toxicology,* Vol. 14, 1985, pp. 659–667.
[17] Larson, R. J., *Residue Review,* Vol. 85, 1983, pp. 159–171.
[18] Gerike, P. and Fischer, W. K., *Ecotoxicology and Environmental Safety,* Vol. 3, 1979, pp. 159–173.
[19] Wee, V. T. and Kennedy, J. M., *Analytical Chemistry,* Vol. 54, 1982, pp. 1631–1633.
[20] Hobbie, J. E., Daley, R. J., and Jasper, S., *Applied Environmental Microbiology,* Vol. 33, 1977, pp. 1225–1228.
[21] Kirchman, D. and Mitchell, R., *Applied Environmental Microbiology,* Vol. 43, 1982, pp. 200–209.
[22] Larson, R. J. and Payne, A. G., *Applied Environmental Microbiology,* Vol. 41, 1981, pp. 621–627.
[23] Gocke, K., *Marine Biology,* Vol. 42, 1977, pp. 131–141.
[24] Larson, R. J., "Kinetic and Ecological Approaches for Predicting Biodegradation Rates of Xenobiotic Organic Chemicals in Natural Ecosystems," *Current Perspectives in Microbial Ecology,* M. J. Klug and C. A. Reddy, Eds., Washington, DC, 1984, pp. 677–686.
[25] Lewis, M. A. and Hamm, B. G., *Water Research,* Vol. 20, 1986, pp. 1575–1582.
[26] *Standard Methods for the Examination of Water and Wastewater,* 16th Ed., American Public Health Association, Washington, DC, 1985.
[27] Wright, R. T. *Applied Environmental Microbiology,* Vol. 36, 1978, pp. 297–305.
[28] Clarke, K. R. and Owens, N. J. P., *Journal of Microbiological Methods,* Vol. 1, 1983, pp. 133–137.
[29] Cairns, J., *Water Research,* Vol. 16, 1982, pp. 941–952.
[30] Buikema, A. L., Niederlehner, B. R., Cairns, J., *Water Research,* Vol. 16, 1982, pp. 239–262.
[31] Matthews, R. A., Buikema, A. L., Cairns, J., and Rodgers, J. H., *Water Research,* Vol. 16, 1982, pp. 129–139.
[32] National Research Council, *Testing for the Effects of Chemicals on Ecosystems,* National Academy Press, Washington, DC, 1981.
[33] Marshall, J. S. and Mellinger, D. L. "An *In Situ* Experimental Method for Toxicological Studies on Natural Plankton Communities," *Aquatic Toxicology (Third Conference,) ASTM STP 707,* J. G. Eaton, P. R. Parrish, and A.C. Hendricks, Eds. American Society for Testing and Materials, Philadelphia, 1980, pp. 27–39.
[34] Sanders, J., Ryther, J., and Batchelder, J., *Journal of Experimental Marine Biology and Ecology,* Vol. 49, 1981, pp. 81–102.
[35] Boetheling, R. S., *Water Research,* Vol. 18, 1984, pp. 1061–1076.
[36] Swisher, R. D., *Surfactant Biodegradation,* second ed., Marcel Dekker, Inc., New York, NY, 1987.
[37] Kirchman, D. and Mitchell, R., *Applied Environmental Microbiology,* Vol. 43, 1982, pp. 200–209.
[38] Spain, J. C., Pritchard, P. H., and Bourquin, A. W., *Applied Environmental Microbiology,* Vol. 40, 1980, pp. 726–734.
[39] Larson, R. J. and Vashon, R. D., *Developments in Industrial Microbiology,* Vol. 24, 1983, pp. 425–434.

Roberta R. Fulthorpe,[1] *Neil A. Straus,*[2] *and R. Campbell Wyndham*[1]

Bacterial Adaptation to Chlorobenzoate Contamination in the Niagara Region Investigated by DNA:DNA Colony Hybridization

REFERENCE: Fulthorpe, R. R., Straus, N. A., and Wyndham, R. C., **"Bacterial Adaptation to Chlorobenzoate Contamination in the Niagara Region Investigated by DNA:DNA Colony Hybridization,"** *Aquatic Toxicology and Environmental Fate: Eleventh Volume, ASTM STP 1007,* G. W. Suter II and M. A. Lewis, Eds., American Society for Testing and Materials, Philadelphia, 1989, pp. 59–71.

ABSTRACT: Chloroaromatic biodegradation was studied in samples of water and the sediment-water interface of Bloody Run Creek, a tributary of the Niagara River with a source adjacent to the Hyde Park Industrial Landfill in New York State. *Alcaligenes* spp. were found which metabolized chlorobenzoates by virtue of an 85 kilobase plasmid, pBR60. These isolates were obtained following continuous culture enrichment of populations from the sediment-water interface, selecting isolates for their metabolism of 3-chlorobenzoate as a carbon and energy source. To determine whether plasmid deoxyribonucleic acid (DNA) sequences were present in a significant number of organisms in the water and surface sediments of the creek, samples were diluted and spread on nitrocellulose filters overlying solid media containing chlorobenzoates and traces of yeast extract. A control creek with a similar superficial geology but lacking contaminants was also sampled. Colony hybridization with phosphorus-32 (^{32}P)-labeled pBR60 revealed differences between samples from the two creeks in both the number of positive signals and the signal strength. Isolates, obtained from the creek water and surface sediments, which were capable of growth on 3-chlorobenzoate were screened by dot-blot hybridization using a specific probe, derived from pBR60, which detected sequences involved in chlorobenzoate catabolism. Evidence for the mobility of the plasmid between *Alcaligenes* and *Pseudomonas* species was also obtained. Technical problems with this approach, alternative methods, and applications are discussed.

KEY WORDS: chlorobenzoates, industrial pollution, bacteria, plasmids, catabolism, gene probe, DNA:DNA colony hybridization, aquatic toxicology

Adaptation of bacteria in environments exposed to organic chemical contaminants may involve enrichment of metabolically competent populations which are initially present at low numbers [*1,2,3,4*]. The rate at which populations increase to levels sufficient for contaminant removal depends on prior exposure [*2*], the presence or absence of particulates and sediments [*5*], nutrient status [*6,3*], the presence of predators, and even the age of the sample in a laboratory test [*3*]. Adaptation may also depend on the selection of phenotypic and genotypic variants of indigenous populations [*1*]. In a study of these variants [*1*], the effective size of a population was determined by the frequency of high-activity mutants within a parent population of a contaminant degrading *Acinetobacter* species. The relative

[1] NSERC scholar and associate professor, respectively, Department of Biology, Carleton University, Ottawa, K1S 5B6, Canada.

[2] Professor, Department of Botany, University of Toronto, Toronto, Ontario, M5S 1A1 Canada.

influence of these factors on rates of adaptation will necessarily be dependent on the chemical structure and environment.

Laboratory studies with recalcitrant organics have shown that the metabolic activity and substrate range of bacteria can be enhanced by plasmid conjugation and by mutation [*7,8*]. Resolving whether these molecular events occur in contaminated environments, and whether they are important in bacterial adaptation, is a major challenge. The basic problem is the detection of low-frequency molecular events in populations which are themselves a small fraction of the microbial community. Recently, specific gene probes for colony hybridization and total deoxyribonucleic acid (DNA) hybridization have been applied to samples of microbial communities contaminated with aromatic hydrocarbons [*9,10*], chlorobiphenyls [*11*], and mercury [*12*]. This approach, despite its limitations in terms of organism culturability and probe specificity [*10*], promises to close the gap between our knowledge of the molecular adaptability of bacteria in the laboratory and our knowledge of community changes as a result of exposure of natural populations to organic contaminants.

To address some of these problems, our laboratory has been investigating bacterial adaptation and contaminant degradation in surface runoff from the Hyde Park industrial landfill in Niagara Falls, New York. The landfill occupies 6.1 hectares (ha) and contains approximately 6.8×10^7 kg of chemical wastes deposited between 1953 and 1975. The major chemical contaminants from the site travel in nonaqueous and aqueous plumes through surface waters and groundwater. Soluble contaminants of these waters include copper (1 ppm), lead (1 ppm), zinc (31 ppm), chloroform (1 ppm), toluene (2 ppm), phenols (94 ppm), and mixed 2-, 3-, and 4-chlorobenzoic acids (189 ppm). In addition, there are parts per billion levels of other metals, chloroethanes, and chlorinated aromatics [*13*]. Although soils from the Hyde Park landfill have been the subject of some microbiological work [*14,15*], there is no information on the bacteria present in leachate waters or contaminated fresh water from the site. Occidental Chemical Corp. maintains a batch reactor facility for decontamination of leachate from the site. The facility uses a combination of activated carbon and biological treatments [*16*]. The bacterial community involved in the treatment process has not been characterized. In 1986 the authors of this paper began a study of indigenous sediment and water bacterial populations of Bloody Run Creek, near the Hyde Park landfill, reasoning that the prolonged contamination of these waters would have provided strong selection for contaminant biodegradation. The characterization of *Alcaligenes* and *Pseudomonas* species from the creek is reported elsewhere [*17,18*]. The following report describes the application of a novel catabolic plasmid isolated from one of these strains in field studies of the distribution of the catabolic genotype in freshwater streams in the Niagara region.

Materials and Methods

Site Description and Sampling

Bloody Run Creek (43°, 08′, 10″N, 79°, 02′, 30″W), a small tributary of the Niagara River, collects surface and near-surface runoff from land adjacent to the Hyde Park industrial landfill in Niagara Falls, New York. Bloody Run Creek has been the subject of remedial cleanup activities and has been isolated from the landfill, although sediments remain contaminated. Theatre Creek is a small tributary of the Niagara River which flows from a woodlot at the base of the Niagara escarpment at Lewiston, New York, through dolomite and shale formations, to the Niagara River. There is no industrial or landfill input to this control creek. Both sites are indicated on the map in Fig. 1. Samples of water and disturbed material from the sediment-water interface were taken in February 1987 from the Bloody

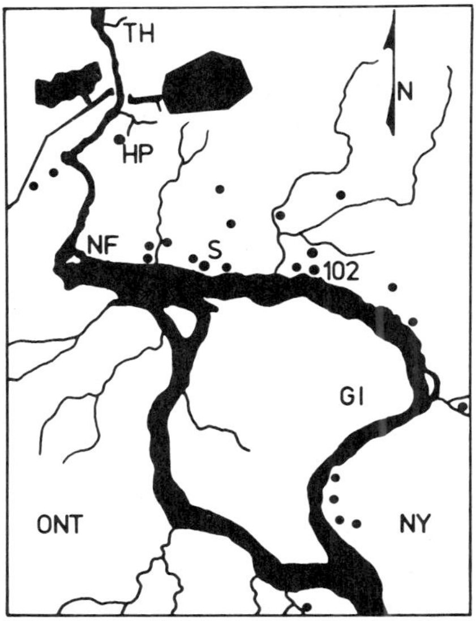

FIG. 1—*The Niagara River in the vicinity of Niagara Falls* (NF), *Ontario, and Grand Island* (GI), *New York. Sampling sites were on Bloody Run Creek adjacent to Hyde Park* (HP), *the Niagara River at the mouth of Bloody Run Creek, and on Theatre Creek* (TH). *Other industrial waste sites in the area, including the S area* (S) *and 102nd Street/Love Canal* (102), *are indicated by closed circles. The map is adapted from Ref* 13.

Run Creek and Theatre Creek, placed in sterile tubes, and taken to the laboratory. Water temperature at the time of sampling was 2 to 4°C. Dilutions were made in a minimal Medium A [*1*] and spread on 82.5-mm circular nitrocellulose filters (Biorad Laboratories), overlying media containing minimal Medium A, 2.0-m*M* 3-chlorobenzoate (Aldrich Chemical Co., 99^{+}%), and either 0.05 or 0.005% yeast extract (Difco Laboratories). The plates were incubated at 4 or 22°C for two to three weeks.

Reference Strains, Probe Preparation, and Hybridization

Alcaligenes sp. BR60 isolation and characterization have been described elsewhere [*17,18*]. Isolates from Bloody Run Creek were identified using morphological and biochemical criteria as specified [*19,20*].

Plasmids were isolated from *Alcaligenes* and *Pseudomonas* species using a scaled-up method of Crosa and Falkow [*21*]. Overnight cultures in 50 mL of TYE [Tryptone (Difco) (1.0%), yeast extract (Difco) (1.0%), and sodium chloride (NaCl) (0.5%)] were harvested (6000 × *g*), washed in 50-m*M* tris(hydroxymethyl)aminomethane (Tris)/hydrochloric acid (HCl) (pH 8.0), 10-m*M* ethylenediaminetetraacetic acid (EDTA) (TE), and lysed in 15 mL of 4% sodium dodecylsulfate (SDS) in TE at pH 12.5. Subsequent neutralization with 1.0 mL of 2-*M* Tris/HCl (pH 7.0) and chromosomal precipitation with 6 mL of 5-*M* NaCl were carried out on a scale 25 times the volumes previously described [*21*]. We included two phenol/chloroform extractions prior to precipitating with one volume of isopropanol. Plasmid preparations were then purified by density gradient centrifugation in cesium chloride (CsCl)/ethidium bromide [*22*].

Plasmid pBR60 was prepared for radiolabeling by digesting 0.4 μg with four units of

EcoRI restriction enzyme [Bethesda Research Laboratories (BRL)] as described in Ref *22*. The restriction fragments were denatured by boiling for 5 min and then labeled with phosphorus-32 (^{32}P)-deoxycytidine triphosphate (CTP) (Amersham) using the oligonucleotide primer/Klenow polymerase method [*23*]. The labeled probe was separated from unincorporated ^{32}P-CTP by column chromatography through Sephadex G-50, to yield a probe with a specific activity of 5×10^8 disintegrations per minute (dpm) per microgram of DNA.

A probe specific for the deletion region of pBR60 involved in chlorobenzoate metabolism was prepared from a clone of an EcoRI fragment. The construction of an EcoRI library of plasmid pBR60 in *Escherichia coli* JM101, and the localization of metabolic determinants to the deletion region of the plasmid, are described in another report [*17*]. Cloned DNA in plasmid pUC18 vector was purified from *E. coli* overnight cultures on TYE plus 50 μg mL^{-1} ampicillin by lysozyme lysis and alkaline SDS treatment, as described previously [*22,24*]. Plasmids were further purified from chromosomal DNA and ribonucleic acid (RNA) by column chromatography [*25*] on Sepharose 4B (Pharmacia). The EcoRI, 2.9-kilobase fragment E11 from this library was resolved from the pUC18 vector by EcoRI restriction digestion [*22*] followed by electrophoresis in low-melt agarose (Seaplaque, FMC Corp.). The excised fragment was radiolabeled in agarose following three-fold dilution using the oligonucleotide primer method described above.

Colony hybridizations were carried out on nitrocellulose filters following colony replication and treatment for colony lysis on Whatman 3MM paper, neutralization, and baking [*22*]. Hybridization was carried out as described [*17*] for 16 h at 42°C in 5× saline sodium citrate (SSC) [*22*], 50% formamide, 0.1% bovine serum albumin, 0.1% ficoll 400, 0.1% sodium dodecyl sulfate, 0.1% polyvinylpyrollidone 40 000, 50-m*M* sodium hydrogen phosphate ($NaHPO_4$), at pH 6.5. These hybridization conditions gave little nonspecific hybridization of plasmid pBR60 probes to nonhomologous DNAs. Autoradiograms on Kodak X-Omat film were exposed with a Du Pont Cronex intensifying screen, developed according to manufacturers' specifications, and reproduced on Kodak Technical Pan 2415 film for printing.

The E11 probe was used in dot-blot hybridizations with crude DNA extracts of isolates from Bloody Run Creek which were capable of growth on 3-chlorobenzoate. Overnight cultures of isolates in 1 mL of TYE were centrifuged in microcentrifuge tubes, and the pellet was resuspended in 100 μL of 0.5-*M* sodium hydroxide (NaOH), 0.05% SDS. Cell suspensions were heated to 37°C for 10 min for lysis, and then DNA was sheared by rapid pipetting through a 100-μL pipette tip. Crude DNA preparations were then screened for an estimate of yield by agarose gel electrophoresis, loading 5 μL in each lane, and were blotted in 5-μL drops in a grid pattern on nitrocellulose filters for hybridization. Following drying of the crude DNA, filters were floated on successive washes of 1-*M* Tris/HCl (pH 8.0), and the same buffer containing 1.5-*M* NaCl, blotting the filters dry between washes on 3MM paper. The filters were then air dried and baked at 80°C for 2 h. Hybridization using the E11 probe was carried out as described above.

Results and Discussion

Colony Hybridizations in Creek Water Samples

Colony development was rapid on the selection plates containing 2.0-m*M* 3-chlorobenzoate plus 0.05% or 0.005% yeast extract, and there was substantial accumulation of chlorocatechol metabolites in the medium surrounding many of these colonies. This was apparent from the formation of the distinctive purple-black autooxidation products of these metabolites [*26*]. Colony formation in the presence of these concentrations of yeast extract is not indicative of the use of chlorobenzoate as a source of carbon or energy.

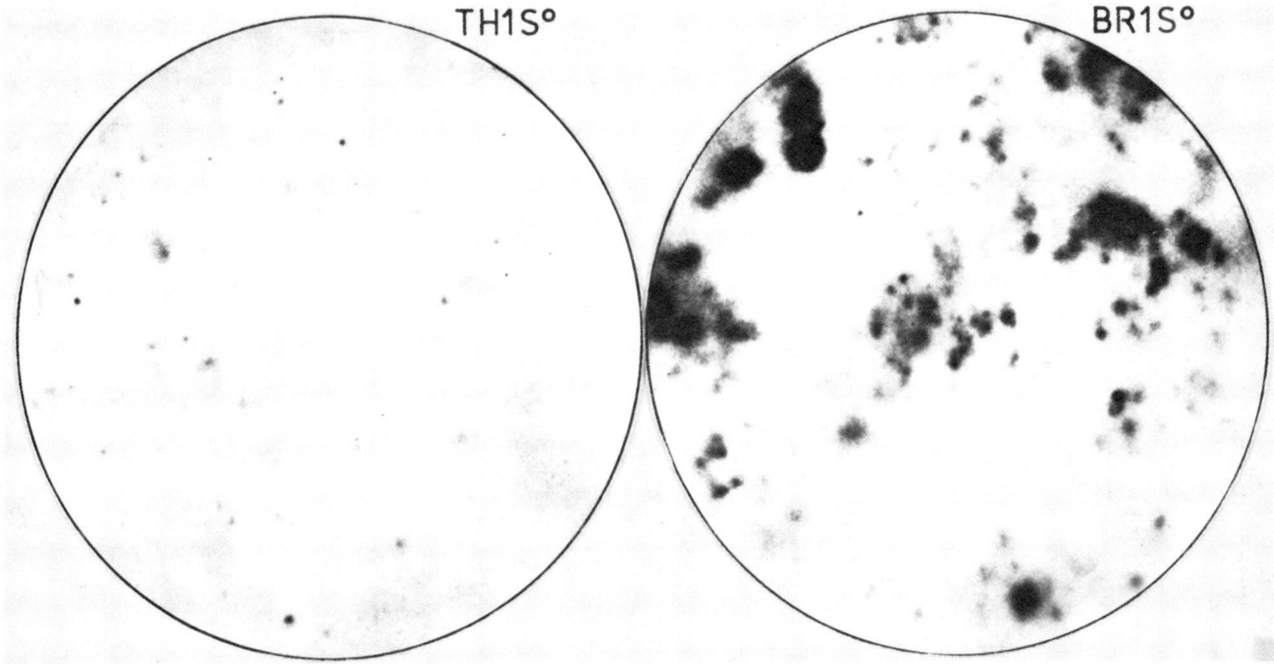

FIG. 2—*DNA:DNA colony hybridization to nitrocellulose filters inoculated with 50 μL of undiluted sediment from Theatre Creek Site 1* (TH1S^0) *and from Bloody Run Creek Site 1* (BR1S^0). *The probe used was whole-plasmid pBR60 from the 3-chlorobenzoate degrading isolate* Alcaligenes *sp. BR60. Plates were incubated for identical times. Total numbers of colonies on these plates, determined from lower dilutions on identical medium without nitrocellulose filters, were* 1.7×10^3 *for* TH1S° *and* 0.8×10^3 *for* BR1S°.

Hybridization of the chlorobenzoate degradative plasmid pBR60 to the colonies from these plates gave results like those of Fig. 2 for the Bloody Run and Theatre creeks. Both filters carried colonies from undiluted samples of sediment, with total counts [colony forming units (CFU) per plate], as determined from lower dilutions, indicated in the legend of Fig. 2. Further dilutions of these samples resulted in larger hybridization spots, presumably due to the relaxation of colony/colony competition for nutrients or other interactions. There was a background of small colonies on all filters which exhibited homology to pBR60, and these were included as a part of the total hybridization count. Colonies larger than 2 mm and yielding a positive hybridization signal, like the 14 formed on the BR1S° (Bloody Run Creek Site 1) filter of Fig. 2, were counted separately.

Table 1 summarizes the data obtained for colony-forming units and colony hybridization counts from samples of water and sediment taken at three sites. When small colonies were included in the total hybridization count, there were relatively uniform numbers obtained from water taken at the three sites. The exception was for the control (Theatre Creek) on media containing only 0.005% yeast extract, for which no positive signals were observed. In Niagara River water samples, the total hybridizations observed were greater than the total CFU per millilitre counted. This was attributed to the sensitivity of the hybridization technique, which detected small amounts of homologous DNA on nitrocellulose filters, where a colony may not have been counted. Total hybridizations from samples of sediments were about one order of magnitude higher than those for water. These counts were slightly higher for Bloody Run sediments than for the control sediments of Theatre Creek.

When positive hybridization signals larger than 2 mm were counted separately, a clearer pattern emerged. These colonies were found only in Bloody Run water and sediment and in the Niagara River water samples (Table 1). The counts obtained from water samples were only ten times higher than the practical detection limit of the enrichment method, or about 20 cells per millilitre. In the future we hope to increase the sensitivity of counting in aqueous samples by using a combination of most probable number (MPN) dilutions in conjunction with dot-blot hybridization [*27*]. This approach should improve sensitivity to as little as one cell per millilitre. Counts in sediment samples from Bloody Run Creek were well above our detection limit, with strong hybridization signals observed in dilutions of sediment down to 10^{-4}.

TABLE 1—*Colony forming units and colony hybridization counts.*

Sample	Yeast Extract, %[a]	CFU, mL^{-1}	Hybridization, mL^{-1} Total	Hybridization, mL^{-1} >2 mm
Water				
Bloody Run Creek	0.005	0.28×10^3	0.10×10^3	0
	0.05	1.5×10^3	0.20×10^3	0.20×10^3
Theatre Creek	0.005	0.60×10^3	0	0
	0.05	2.4×10^3	0.40×10^3	0
Niagara River	0.005	0.20×10^3	3.0×10^3	0.20×10^3
	0.05	0.40×10^3	1.0×10^3	0
Sediment				
Bloody Run Creek	0.005	2.4×10^4	1.7×10^4	0.50×10^4
	0.05	38.0×10^4	4.7×10^4	1.1×10^4
Theatre Creek	0.005	2.7×10^4	0.40×10^4	0
	0.05	440.0×10^4	1.2×10^4	0

[a] Concentration of yeast extract supplement in 2.0 m*M* 3-chlorobenzoate plates.

Studies of the rate of mineralization of ρ-nitrophenol [*3*], and other contaminants in fresh water [*4,2*], have established that an effective population size of contaminant-degrading organisms must be reached before significant removal occurs. This effective population size depends on the specific affinity of the organism for the contaminant, but generally lies in the range of 10^3 to 10^5 bacteria per millilitre [*3,4*]. The numbers of positive hybridization signals >2 mm in water samples of Bloody Run Creek and the Niagara River were below this range. If these are the only populations capable of chlorobenzoate degradation in the water, then we would expect limited removal of contaminants. Sediment surface samples from the contaminated creek contained on the order of 10^4 large hybridizing CFU per millilitre, a large enough community for degradation to begin almost immediately. These hypotheses are currently being tested using carbon-14 (^{14}C)-labeled chlorobenzoates and samples of water and sediments from the contaminated and control sites. In another report [*18*] we describe laboratory studies which show that phosphate limits contaminant removal in Bloody Run Creek water and that competition for phosphate by indigenous *Pseudomonas* species can effectively block chlorobenzoate degradation. Nutrient availability, and hence effective population size, may be factors controlling contaminant degradation at these sites.

Isolation and Characterization of Hybridizing Populations

Replication of colonies which formed strong hybridization signals resulted in the isolation of a number of strains with the ability to degrade 3-chlorobenzoate as the sole source of carbon and energy. Isolations were also made directly from samples of the sediment-water interface of Bloody Run Creek on 3-chlorobenzoate in liquid culture [*18*]. Our concern at this point was with the specificity of the whole pBR60 plasmid probe, and whether or not a strong hybridization signal was indicative of 3-chlorobenzoate degradative genes with sequence homology to those of pBR60. A probe specific to the region of pBR60 involved in chlorobenzoate degradation was therefore developed. This step was accomplished relatively quickly because of the unstable nature of plasmid pBR60. Figure 3 shows an EcoRI restriction digest of pBR60 and a deletion derivative of this plasmid, pBR40. A more complete account of the localization of the chlorobenzoate degradative genes to the deletion region of pBR60 is given elsewhere [*17*]. A specific probe was developed using a pUC18 clone of the 2.9-kilobase E11 fragment shown in Fig. 3. This fragment is not present in the deletion plasmid pBR40, which has lost the degradative determinants for 3-chlorobenzoate. The E11 probe was applied in hybridization studies on isolates from Bloody Run Creek.

Dot-blot hybridization results for a series of 24 isolates are shown in Fig. 4. Isolates 1, 9, 10, 11, and 13 from this experiment were taxonomically similar to the archetypal isolate *Alcaligenes* sp. BR60. Other isolates were sufficiently different to warrant characterization. As an example, the organism which generated hybridization Signal 4 of Fig. 4 was tentatively identified as a *Pseudomonas acidovorans* (isolate NR4), on the basis of biochemical and morphological characteristics. A partial list of these characteristics, which clearly distinguish this isolate from *Alcaligenes* sp. BR60, is presented in Table 2. Variations in plasmid content, discussed below, and morphology, also distinguish isolate NR4 from *Alcaligenes* sp. BR60.

The plasmid content of isolate NR4 was determined by agarose gel electrophoresis, along with crude and CsCl-purified pBR60 plasmid DNA (Fig. 5). *Pseudomonas acidovorans* NR4 carries a plasmid of the same molecular size as pBR60 (85 kb). This plasmid exhibits identical deletion frequencies for the loss of 14 kb of DNA to yield a plasmid of 71 kb (pBR40). The deletion derivative can be seen as a plasmid band just below the 85 kb plas-

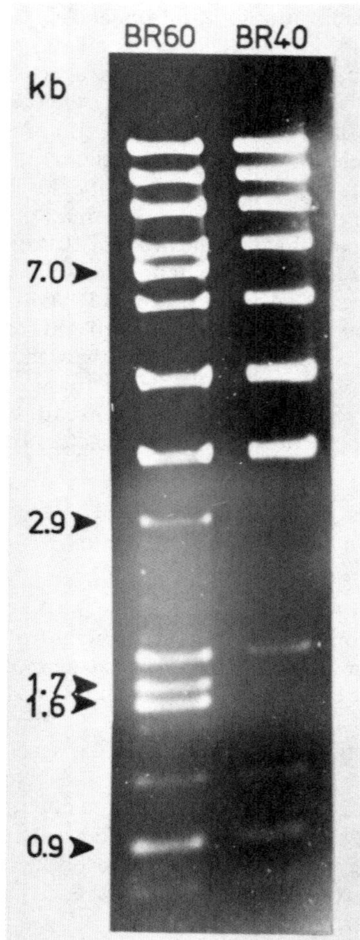

FIG. 3—*EcoRI restriction digests of plasmid pBR60 and a deletion derivative of this plasmid, pBR40, resolved by agarose gel electrophoresis. The size in kilobases (kb) of fragments lost from pBR60 during the deletion is indicated. Fragment E11 is a 2.9 kb fragment of pBR60 which is lost on deletion of a total of 14 kb of DNA to give pBR40. Three-Chlorobenzoate degradative activity is encoded at least partially by the deletion region of pBR60* [17].

mids in Lanes 2 and 3 of Fig. 5. *Pseudomonas acidovorans* NR4 also contains plasmids in the 100 to 120 kb molecular size range (Fig. 5). Work on restriction analyses and hybridization of probe E11 to the 85 kb plasmid of isolate NR4 indicates that the homology lies in the plasmid and not in the chromosomal DNA of this isolate. EcoRI restriction digestion of the 85 kb plasmid of isolate NR4 (data not shown) gave a fragment pattern identical to that of pBR60 (Fig. 3). The evidence suggests that pBR60 exists in at least two different indigenous populations of bacteria in Bloody Run Creek. Other isolates from enrichment cultures and from the pBR60 hybridization replica plates of sediment and water samples are currently being screened to establish the diversity and genetic background of organisms carrying chlorobenzoate degradative genes.

One of many questions which remain to be answered in freshwater adaptation studies concerns the importance of "ecotypes" in microbial communities. We have very little information on the geographical distribution of degradative strains, or the degradative plasmids they carry. Plasmid pBR60 from Bloody Run Creek appears to be unique in its restriction digest pattern (Fig. 3) [*17*] when compared with other 3-chlorobenzoate catabolic plasmids [*28–30*]. We have demonstrated [*17*] that there is no homology of pBR60 to the 3-chlorobenzoate catabolic genes of pAC27, the 110 kb plasmid of *Pseudomonas putida* isolated from sewage in Niskayuna, New York [*28*]. However, there is homology between pAC27 and 3-chlorobenzoate catabolic plasmids isolated from soils in Australia (pJP4) and sewage in Göttingen, West Germany (pB13) [*7,31*]. When pJP4 was used as a probe to screen 2,4-dichlorophenoxyacetic acid degrading, aquatic *Alcaligenes* species from Oregon, strong homology was detected in 1 of 30 isolates [*32*].

Taken collectively these observations suggest that the number of catabolic genotypes for chlorobenzoates may be small, but that there is room for independent evolution of eco-

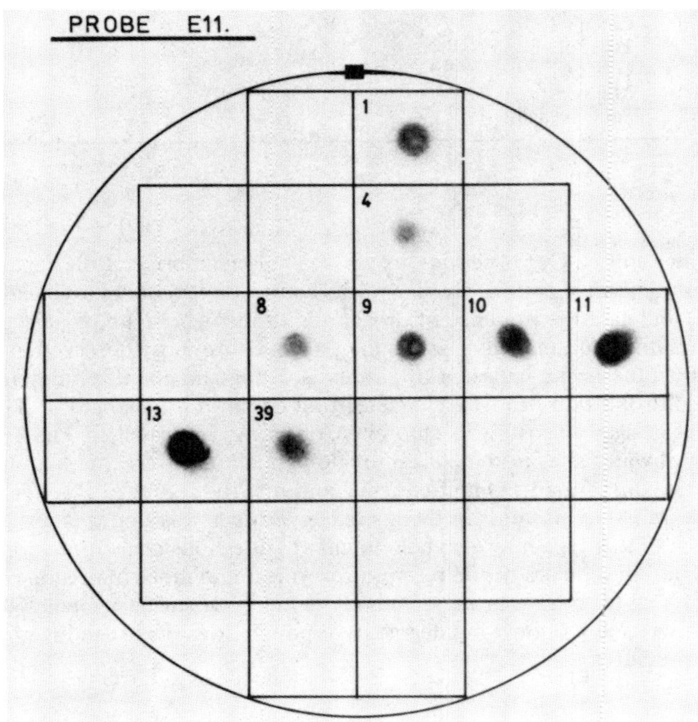

FIG. 4—*Dot-blot hybridization of radiolabeled fragment E11 (2.9 kb) to total DNA extracts of isolates from Bloody Run Creek. Approximately 0.2 μg of DNA from each of 24 isolates was blotted to the grid squares. Probing was carried out at high stringency, as described in the materials and methods section.*

TABLE 2—*Taxonomic characteristics of* Alcaligenes *sp. BR60 and* P. acidovorans *NR4.*

Characteristic	BR60	NR4
Cell length, μm	0.8 to 1.2	1.0 to 1.4
Motility	+	+
Flagella	>1	>1
Catalase	weak +	+
Oxidase	+	+
Oxidation/fermentation (fructose)	−	0 (acid)
Gelatin hydrolysis	−	−
40°C Growth	−	−
Growth substrates		
Succinate	+	+
3-Chlorobenzoic acid	+	+
Cis,cis-muconic acid	+	+
ρ-Hydroxybenzoic acid	+	+
Acetamide	−	+
Propionamide	−	+
2,3-Butanediol	−	+
D-mannitol	−	+
D-gluconate	+	+
D-fructose	−	+
Catechol	−	+
Ethanol	−	+
Glucose	−	−
Maltose	−	−
Salicylic acid	−	−

types within this subset. Our observations of weak colony hybridization signals in samples from an uncontaminated control creek, and the cometabolism of chlorobenzoates in these samples, may indicate the presence of progenitor strains which, under selection pressure from contaminant input, may give rise to the type of community observed in Bloody Run Creek. This hypothesis can be tested by simply isolating and characterizing the positively hybridizing control strains and by characterizing community changes in Theatre Creek samples on prolonged exposure to chlorobenzoates. As indicated in Fig. 1, there is no shortage of freshwater sites in the Niagara region which have received loadings of chlorinated aromatic contaminants. The frequency and distribution of pBR60 and other chloroaromatic degradative plasmids in these streams are currently under investigation.

An approach of this type may also help to answer questions concerning horizontal plasmid mobility and the constraints on plasmid-carrying populations in natural communities. And finally, a long-term monitoring of Bloody Run Creek during remedial cleanup work at Hyde Park will tell us how long degradative populations persist following removal of the selection pressure.

Acknowledgments

We greatly appreciate the assistance of Mary Barbara and Ellen Schwartzel in this research, which was supported by the Natural Sciences and Engineering Research Council of Canada through Operating and Strategic Grants to authors Wyndham and Straus.

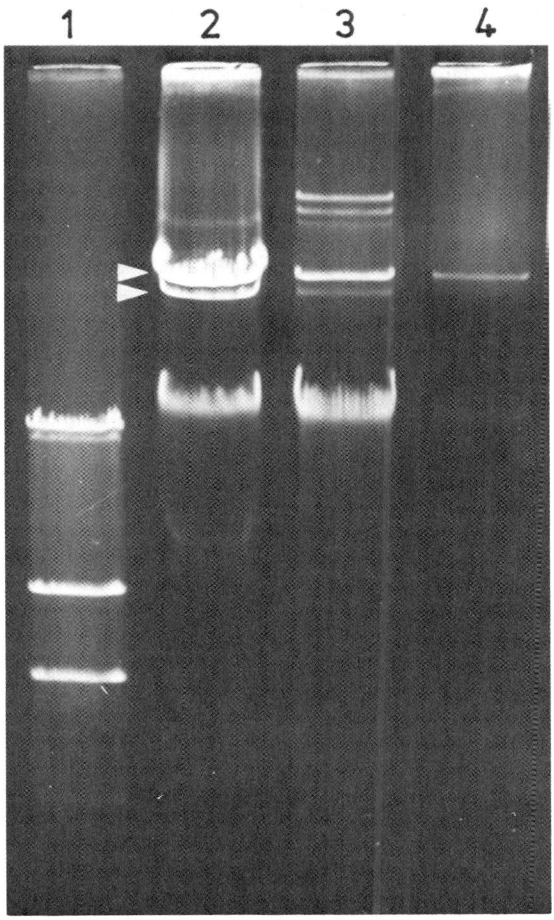

FIG. 5—*Agarose gel electrophoresis of plasmid extracts of Niagara region isolates: Lane 1, λ DNA/HindIII digest marker fragments; Lane 2,* Alcaligenes *sp. BR60 crude plasmid DNA showing pBR60 (85 kb, top arrow) and pBR40 (71 kb, bottom arrow); Lane 3,* Pseudomonas acidovorans *NR4 crude plasmid DNA showing plasmids corresponding to pBR60 and pBR40, as well as larger plasmids (>100 kb); and Lane 4, CsCl-purified pBR60. The broad lower band in Lanes 2 and 3 is chromosomal DNA.*

References

[*1*] Wyndham, R. C., "Evolved Aniline Catabolism in *Acinetobacter calcoaceticus* During Continuous Culture of River Water," *Applied Environmental Microbiology,* Vol. 51, 1986, pp. 781–789.

[*2*] Spain, J. C., Pritchard, P. H., and Bourquin, A. W., "Effects of Adaptation on Biodegradation Rates in Sediment-Water Cores from Estuarine and Freshwater Environments," *Applied Environmental Microbiology,* Vol. 40, 1980, pp. 726–734.

[*3*] Wiggins, B. A., Jones, S. H., and Alexander, M., "Explanations for the Acclimation Period Preceding the Mineralization of Organic Chemicals in Aquatic Environments," *Applied Environmental Microbiology,* Vol. 53, 1987, pp. 791–796.

[*4*] Ventullo, R. M. and Larson, R. J., "Adaptation of Aquatic Microbial Communities to Quater-

nary Ammonium Compounds," *Applied Environmental Microbiology,* Vol. 51, 1986, pp. 356–361.

[*5*] Spain, J. C., van Veld, P. A., Monti, C. A., Pritchard, P. H., and Cripe, C. R., "Comparison of ρ-Nitrophenol Biodegradation in Field and Laboratory Test Systems," *Applied Environmental Microbiology,* Vol. 48, 1984, pp. 944–950.

[*6*] Lewis, D. L., Kollig, H. P., and Hodson, R. E., "Nutrient Limitation and Adaptation of Microbial Populations to Chemical Transformations," *Applied Environmental Microbiology,* Vol. 51, 1986, pp. 598–603.

[*7*] Ghosal, D., You, I.-S., Chatterjee, D. K., and Chakrabarty, A. M., "Microbial Degradation of Halogenated Compounds," *Science,* Vol. 228, 1985, pp. 135–142.

[*8*] Latorre, J., Reineke, W., and Knackmuss, H.-J., "Microbial Metabolism of Chloroanilines: Enhanced Evolution by Natural Genetic Exchange," *Archives of Microbiology,* Vol. 140, 1984, pp. 159–165.

[*9*] Sayler, G. S., Shields, M. S., Tedford, E. T., Breen, A., Hooper, S. W., Sirotkin, K. M., and Davis, J. W., "Application of DNA-DNA Colony Hybridization to the Detection of Catabolic Genotypes in Environmental Samples," *Applied Environmental Microbiology,* Vol. 49, 1985, pp. 1295–1303.

[*10*] Sayler, G. S., Jain, R. K., Ogram, A., Pettigrew, C. A., Houston, L., Blackburn, J., and Riggsby, W. S., "Applications for DNA Probes in Biodegradation Research," *Proceedings,* Fourth International Symposium on Microbial Ecology, Ljubljana, Yugoslavia, 1986.

[*11*] Pettigrew, C. A. and Sayler, G. S., "The Use of DNA:DNA Colony Hybridization in the Rapid Isolation of 4-Chlorobiphenyl Degradative Bacterial Phenotypes," *Journal of Microbiological Methods,* Vol. 5, 1986, pp. 205–213.

[*12*] Barkay, T., Fouts, D. L., and Olson, B. H., "The Preparation of a DNA Gene Probe for the Detection of Mercury Resistance Genes in Gram-Negative Bacterial Communities," *Applied Environmental Microbiology,* Vol. 49, 1985, pp. 686–692.

[*13*] Wagner, E. T. and Collin, R. L., "Report of the Niagara River Toxics Committee," Environment Canada, Communications Branch, Regional Director General's Office, Toronto, Canada, 1984.

[*14*] Vandenbergh, P. A., Olsen, R. H., and Colaruotolo, J. F., "Isolation and Genetic Characterization of Bacteria that Degrade Chloroaromatic Compounds," *Applied Environmental Microbiology,* Vol. 42, 1981, pp. 737–739.

[*15*] Pierce, G. E., Robinson, J. B., and Colaruotolo, J. R., "Substrate Diversity of *Pseudomonas* Species Containing Chlorotoluene Degradative Plasmids," *Developments in Industrial Microbiology,* Vol. 24, 1983, pp. 499–507.

[*16*] Ying, W-C., Bonk, R. R., Lloyd, V. J., and Sojka, S. A., "Biological Treatment of a Landfill Leachate in Sequencing Batch Reactors," *Environmental Progress,* Vol. 5, 1986, pp. 41–50.

[*17*] Wyndham, R. C., Singh, R. K., and Straus, N. A., "Catabolic Instability, Plasmid Gene Deletion and Recombination in *Alcaligenes* sp. BR60," *Archives of Microbiology,* Vol. 150, 1988, pp. 237–243.

[*18*] Wyndham, R. C. and Straus, N. A., "Chlorobenzoate Catabolism and Interactions Between *Alcaligenes* and *Pseudomonas* Species from Bloody Run Creek," *Archives of Microbiology,* Vol. 150, 1988, pp. 230–236.

[*19*] Starr, M. P., Stolp, H., Truper, H. G., Balows, A., and Schlegel, H. G., Eds., *The Prokaryotes,* Vol. 1, Springer Verlag, New York, 1981, pp. 719–741.

[*20*] Krieg, N. R. and Holt, J. G., Eds. *Bergey's Manual of Determinative Bacteriology,* Vol. 1, Williams and Wilkins, Baltimore, 1984, pp. 141–199.

[*21*] Crosa, J. H. and Falkow, S. in *Manual of Methods for General Bacteriology,* P. Gerhardt, R. G. E. Murray, R. N. Costilow, E. W. Nester, W. A. Wood, N. R. Krieg, and G. B. Phillips, Eds., American Society for Microbiology, Washington, DC, Chapter 15.2, pp. 269–270.

[*22*] Maniatis, T., Fritsch, E. F., and Sambrook, J., *Molecular Cloning: A Laboratory Manual,* Cold Spring Harbor Laboratory, Cold Spring Harbor, NY, 1982.

[*23*] Feinberg, A. P. and Vogelstein, B., "Addendum: A Technique for Radiolabeling DNA Restriction Fragments to High Specific Activity," *Analytical Biochemistry.* Vol. 137, 1984, pp. 266–267.

[*24*] Birnboim, H. C. and Doly, J., "A Rapid Alkaline Extraction Procedure for Screening Recombinant DNA," *Nucleic Acids Research,* Vol. 7, 1979, pp. 1513–1523.

[*25*] Micard, D., Sobrier, M. L., Couderc, J. L., and Dastugue, B., "Purification of RNA-Free Plasmid DNA Using Alkaline Extraction Followed by Ultrogel A2 Column Chromatography, *Analytical Biochemistry,* Vol. 148, 1985, pp. 121–126.

[*26*] Haller, H. D. and Finn, R. K., "Biodegradation of 3-Chlorobenzoate and Formation of Black Color in the Presence and Absence of Benzoate," *European Journal of Applied Microbiology and Biotechnology,* Vol. 8, 1979, pp. 191–205.

[27] Fredrickson, J. K., Bezdicek, D. F., Brockman, F. E., and Li, S. W., "Enumeration of Tn5 Mutant Bacteria in Soil by MPN-DNA Hybridization and Antibiotic Resistance," *Applied and Environmental Microbiology*, Vol. 54, 1988, pp. 446–449.
[28] Chatterjee, D. K. and Chakrabarty, A. M., "Restriction Mapping of a Chlorobenzoate Degradative Plasmid and Molecular Cloning of the Degradative Genes," *Gene*, Vol. 27, 1984, pp. 173–181.
[29] Don, R. H. and Pemberton, J. M., "Genetic and Physical Map of the 2,4-Dichlorophenoxyacetic Acid Degradative Plasmid pJP4," *Journal of Bacteriology*, Vol. 161, 1985, pp. 466–468.
[30] Weisshaar, M.-M., Franklin, F. C. H., and Reineke, W., "Molecular Cloning and Expression of the 3-Chlorobenzoate Degrading Genes from *Pseudomonas* sp. B13," *Journal of Bacteriology*, Vol. 169, 1987, pp. 394–402.
[31] Chatterjee, D. K. and Chakrabarty, A. M., "Genetic Homology Between Independently Isolated Chlorobenzoate Degradative Plasmids," *Journal of Bacteriology*, Vol. 153, 1983, pp. 532–534.
[32] Amy, P. S., Schulke, J. W., Frazier, L. M., and Seidler, R. J., "Characterization of Aquatic Bacteria and Cloning of Genes Specifying Partial Degradation of 2,4-Dichlorophenoxyacetic Acid," *Applied Environmental Microbiology*, Vol. 49, 1985, pp. 1237–1245.

DISCUSSION

F. B. Taub[1] *(written discussion)*—How successful are chlorobenzoate-degrading strains in the creek water, and what effect do alternative organic carbon sources have on this success?

R. R. Fulthorpe, N. A. Straus, and R. C. Wyndham (authors' closure)—Sequences homologous to the chlorobenzoate degradative plasmid (pBR60) are present in 10^4 culturable bacteria per millilitre of the surface sediments. Screening of these bacteria for the ability to grow on chlorobenzoates indicates that many will do so under laboratory conditions. However, until we can determine the rates of assimilation of chlorobenzoates by the indigenous bacteria using ^{14}C-labeled substrates, we really have no idea how effective these populations are *in situ*. We have detected unmetabolized chlorobenzoates in creek water samples in the summer when phosphate limitation is severe. Nutrient limitations and alternative organic carbon sources in the creek may reduce the capacity of chlorobenzoate-degrading bacteria to metabolize the contaminant. This may also apply to other contaminants and other degradative populations in the Niagara watershed.

[1] School of Fisheries, University of Washington, Seattle, WA 98195.

U. M. Cowgill[1]

Effect of the Mud-Water Interface on the Chemical Life of Lakes

REFERENCE: Cowgill, U. M., **"Effect of the Mud-Water Interface on the Chemical Life of Lakes,"** *Aquatic Toxicology and Environmental Fate: Eleventh Volume, ASTM STP 1007*, G. W. Suter II and M. A. Lewis, Eds., American Society for Testing and Materials, Philadelphia, 1989, pp. 72–81.

ABSTRACT: There is scant information concerning the chemical composition of the mud-water interface in shallow, aerobic, eutrophic bodies of water. A little more is known of this region in deep eutrophic lakes with periodically anaerobic hypolimnions. The object of this paper is to compare the composition of the mud-water interface in two such limnological systems and to demonstrate the effect this region exerts on the chemical life of lakes.

The shallow lake used as an example is the preserve of Lake Huleh, Israel, which is hypereutrophic and rarely contains water to a depth greater than 35 cm but has an area of 46.9 ha (116 acres). Linsley Pond, Connecticut, will be used as an example of a deeper (14.6 m) eutrophic lake which is thermally stratified except for the autumnal and vernal isothermal periods.

In a shallow aerobic pond such as the Huleh Preserve, elements that are sorbed by the mud fall into two groups: those that are sorbed and retained by the organic matter, and those that move in concert with the ferric hydroxide complex. Under anaerobic conditions such as those present in Linsley Pond much of the year, those same elements diffuse from the mud into the hypolimnion.

Those elements held most strongly in the surface waters, as contrasted to the mud-water interface in shallow aerobic systems such as the Huleh Preserve, diffuse from the reduced mud surface in anaerobic systems such as Linsley Pond.

Some elements diffuse from a reduced mud surface and under aerobic conditions attain equilibrium. These are rubidium, lithium, and aluminum.

Sodium, boron, silicon, chromium, nickel, and chlorine do not diffuse from a reduced mud surface. In aerobic systems chlorine attains equilibrium, while in the shallow Huleh Preserve the sodium concentrates in the surface water. Boron, silicon, chromium, and nickel in shallow systems become enriched in surface waters in contrast to the liquid or solid portion of the mud-water interface.

All elemental concentrations reported in this study represent the total amount detected. Speciation is not addressed. The cyclical nature of elemental concentrations noted in this study should provide a much-needed background for sediment quality criteria.

KEY WORDS: mud-water interface, elemental composition, eutrophication, ferric hydroxide complex, organic matter, aquatic toxicology

There is scant information concerning the chemical composition of the mud-water interface in shallow, aerobic, eutrophic bodies of water. A little more is known of this region in deep eutrophic lakes with periodically anaerobic hypolimnions. The object of this paper is to compare the chemical composition of the mud-water interface in two such limnological systems and to demonstrate the effect this region exerts on the chemical life of lakes.

[1] Associate environmental consultant, Dow Chemical Co., Midland, MI 48674.

The shallow lake used as an example is the preserve of Lake Huleh, Israel. This body of water was first described about the time of Christ [*1*] and at that time was referred to as Lake Semechonitis. This was the first lake formed by the Jordan River. It used to empty into the Sea of Galilee (Lake Tiberius). Lake Huleh was drained between 1951 and 1958. This former lake area now supports carp ponds (maintained with water diverted from the Jordan River) and cash crops such as corn, sorghum, potatoes, and cotton. A small preserve of the old lake is maintained. When the carp ponds are cleaned, their highly eutrophic water is drained into the Huleh Preserve. This produces a hypereutrophic pond which rarely contains water to a depth $>$35 cm. The size of the preserve at the time of the study was 46.9 ha (116 acres). Background information on this region as well as this study has been previously described [*2–6*].

Linsley Pond is located in North Branford, Connecticut. It will be used as the example of a deeper eutrophic lake. This pond is usually thermally stratified except for the autumnal and vernal isothermal periods [*7–8*]. The hypolimnion is anaerobic with the exception of the two homothermal periods. Linsley Pond is 14.6 m deep and has an area of 9.4 ha (23.3 acres). Background material on Linsley Pond may be found elsewhere [*7–16*].

Much work [*17–25*] has been done to elucidate the various mechanisms involved in the anoxic-oxygenated mud-water interface and how this redox situation affects the chemical composition of the water column.

This paper is devoted to the development of biogeochemical cycles of some 47 elements (all that could be detected) in the mud-water interface when it is oxygenated and when it is anoxic.

Materials and Methods

Field Procedures

The best way to take mud-water interface samples is with a Naumann Sampler [*26*], a device designed for this purpose. The liquid portion of the mud-water interface sample was removed with a syringe fitted with a glass needle, both fashioned from boron-free glass. Since this part of the sample is normally no more than a few millimetres thick, it was necessary to obtain many samples for analytical purposes and amalgamate these subsamples in polyethylene bottles. The mud subsamples were extruded from the Naumann Sampler into polyethylene beakers that were kept covered with parafilm.

Water samples were collected using a 2-L Van Doren polyvinyl chloride (PVC) bottle. In shallower areas a 0.5-L Van Doren (PVC) bottle was employed. In all cases 10 L were amalgamated in polyethylene carboys for the shallower areas and 20 L for the deeper regions of the water bodies.

Laboratory Procedures

All samples were prepared for analysis the day of collection. They were dried in a large plant drying oven at 60°C. The liquid portion of the mud-water interface samples was filtered through a thoroughly washed (boiling glass-distilled water) Whatman No. 40 filter paper. The small amount of material remaining on the filter paper was added to the appropriate mud sample.

The wet weight of the mud could not be accurately obtained, and therefore all chemical results for the mud are presented on a dry weight basis while water analyses are recorded as the composition of the water at the time of collection. All chemical data are expressed as milligrams per litre or micrograms per litre.

Laboratory techniques have been previously described [*8,9,13,27*]. Suffice it to state that lithium, silver, beryllium, cadmium, mercury, boron, tin, bismuth, vanadium, sodium, molybdenum, and cobalt were studied by optical emission spectrometry. Tellurium was measured by a colorimetric method described by Hubert [*28–29*] and then confirmed by atomic absorption-graphite furnace [*30*]. The results from the two methods were concordant within <1%. All other elements detected were studied by X-ray fluorescence spectrometry, employing a helium tunnel, using all X-ray tubes at maximum rating, and, whenever possible, counts were accumulated to 10 000. Thus, in total dissolved solids obtained from concentrating the water samples and in the dried mud, calcium, potassium, chlorine, copper, and strontium were determined with a standard counting error of <1%; bromine, silicon, aluminum, fluorine, iodine, iron, manganese, titanium, sulfur, and yttrium with a precision of <2%; phosphorus, gallium, lead, rubidium, selenium, niobium, scandium, and germanium with a precision of <3%; zirconium and cobalt with a precision of <5%; barium and zinc with a precision of <6%; chromium and arsenic with a precision of <7%; magnesium and nickel with a precision of <8%; lanthanum and cerium with a precision of <9%; and, finally, hafnium with a precision of <15%. Each of these elements was counted on three separate and nonconsecutive occasions, and the results were considered to be reproducible only when net counts agreed within <1%.

Table 1 shows the highest and lowest concentrations of elements (in milligrams per litre) found in concentrated water (total solids) and dried mud by optical emission spectrometry and the percent relative standard deviation for counts from 40 duplicates. These figures refer to elemental quantities from both the Huleh Preserve and Linsley Pond.

Results

The Huleh Preserve

Water samples were taken from the papyrus preserve at the north end of what remains of the old Lake Huleh. Such samples are referred to as Preserve-North. Water samples collected from the central open water portion are referred to as Preserve-Central. The liq-

TABLE 1—*Highest and lowest concentrations of elements (in milligrams per litre) found in total solids and dried mud by optical emission spectroscopy and the relative standard deviation (in percent) in counts from 40 duplicates.*[a]

Element	Concentration, mg L^{-1}		Relative Standard Deviation, %	
	Low	High	Low	High
Be	0.05	4.80	4.2	2.1
Hg	0.05	10.61	3.1	0.62
Bi	1.12×10^{-3}	1.30	2.5	1.5
Sn	2×10^{-7}	55.9	2.2	3.5
Co	0.11	12.51	2.9	1.3
B	3.95	9 100	5.3	1.4
Mo	0.044	7.50	1.4	7.1
V	8×10^{-5}	13.72	2.1	3.5
Ag	8×10^{-3}	0.80	1.9	3.12
Li	0.142	195	4.5	1.2
Na	210.9	36 351	1.7	2.2
Cd	0.42	12.96	6.6	2.5

[a] Figures refer to elemental quantities from both the Huleh Preserve and Linsley Pond.

uid portion of the mud-water interface samples are referred to in the text as mud-H_2O-north and mud-H_2O-central while the solid portions are entitled mud-north and mud-central.

A comparison of the chemical composition of the surface waters of the preserve with those of the mud-water interface provide some interesting results. Three types of elemental distribution were observed: those whose concentration attains equilibrium (Table 2), those that are enriched in the mud-water samples relative to the surface waters above (Table 3), and those that are enriched in the surface waters of the preserve (Tables 4 and 5).

When the ratio of the concentration of any given element in the mud/preserve is compared with that in the mud/liquid portion of the interface sample and the two resulting figures are not statistically significantly different from each other, it may be concluded that the concentration of such an element is essentially in equilibrium in this system. Elements that exhibit this sort of distribution in the Huleh Preserve are rubidium, chlorine, lithium and aluminum. The distribution of rubidium and chlorine is shown in Table 2 and will serve as an example of this type of behavior.

TABLE 2—*Distribution of rubidium and chlorine in the Huleh Preserve.*

	Rubidium, $\mu g\ L^{-1}$ or ratio	Chlorine, $\mu g\ L^{-1} \times 10^3$ or ratio
Substance		
Preserve-north	25.9	37.0
Preserve-central	19.4	34.2
Mud-H_2O-north	26.0	36.1
Mud-H_2O-central	18.7	34.8
Mud-north	27.4×10^3	280.0
Mud-central	27.7×10^3	310.0
Ratio		
Mud/preserve-north	1058	7.56
Mud/mud-H_2O-north	1054	7.76
Mud/preserve-central	1428	9.06
Mud/mud-H_2O-central	1481	8.90

TABLE 3—*Distribution of copper and arsenic in the Huleh Preserve.*

	Copper, $\mu g\ L^{-1}$ or ratio	Arsenic, $\mu g\ L^{-1}$ or ratio
Substance		
Preserve-north	47.5	0.24
Preserve-central	44.5	0.74
Mud-H_2O-north	127.6	1.19
Mud-H_2O-central	107.0	1.69
Mud-north	35.1×10^3	3.56×10^3
Mud-central	43.9×10^3	3.18×10^3
Ratio		
Mud/preserve-north	739	14 833
Mud/mud-H_2O-north	275	2 992
Mud/preserve-central	987	4 297
Mud/mud-H_2O-central	410	1 882

TABLE 4—*Distribution of cadmium and bromine in the Huleh Preserve.*

	Cadmium, $\mu g\ L^{-1}$ or ratio	Bromine, $\mu g\ L^{-1}$ or ratio
Substance		
Preserve-north	3.99	229.1
Preserve-central	3.02	177.1
Mud-H_2O-north	0.41	145.6
Mud-H_2O-central	0.27	131.9
Mud-north	1.06×10^3	32×10^3
Mud-central	1.04×10^3	31×10^3
Ratio		
Mud/preserve-north	266	140
Mud/mud-H_2O-north	2585	220
Mud/preserve-central	344	175
Mud/mud-H_2O-central	3852	235

TABLE 5—*Distribution of iron, beryllium, and molybdenum in the Huleh Preserve.*

	Iron, $\mu g\ L^{-1}$ or ratio	Beryllium, $\mu g\ L^{-1}$ or ratio	Molybdenum, $\mu g\ L^{-1}$ or ratio
Substance			
Preserve-north	970	0.296	0.97
Preserve-central	590	0.09	1.88
Mud-H_2O-north	800	0.081	0.39
Mud-H_2O-central	430	0.076	0.12
Mud-north	$40\,378.7 \times 10^3$	85.4	0.57×10^3
Mud-central	$39\,035.7 \times 10^3$	85.6	0.54×10^3
Ratio			
Mud/preserve-north	41 628	289	588
Mud/mud-H_2O-north	50 473	1054	1462
Mud/preserve-central	66 162	951	287
Mud/mud-H_2O-central	90 781	1126	4500

When the ratio of the concentration of any given element in the mud/preserve is compared with that of the mud/interface sample, and the mud/preserve figure is larger than that of the mud/interface figure, it may be concluded that the concentration of such an element is enriched in the waters of the preserve. Such elements are not sorbed by the mud. The majority of the elements detected in this study behave in this fashion. These are sodium, potassium, copper, silver, magnesium, calcium, strontium, barium, zinc, boron, gallium, scandium, yttrium, lanthanum, cerium, silicon, germanium, lead, hafnium, phosphorus, arsenic, niobium, sulfur, selenium, chromium, tellurium, fluorine, manganese, cobalt, and nickel. The distribution of copper and arsenic will serve as an example of this type of behavior and is shown in Table 3.

There are 13 elements that do not behave in either of the ways just described. When the ratio of the concentration of any given element in the mud/preserve sample is smaller than that in the mud/interface sample, it may be concluded that the concentration of the particular element is accumulated by the mud from the interface liquid in contact with it. Elements that fall into this category are cadmium, bromine, titanium, zirconium, cesium, iodine, vanadium, mercury, tin, bismuth, iron, beryllium, and molybdenum. The distribution of cadmium and bromine, as well as that of iron, beryllium, and molybdenum, will serve as examples of this type of behavior and is shown in Tables 4 and 5, respectively.

A statistical examination of the distribution of the concentration of these 13 elements in the top 387 cm of a lake core taken 1.2 km northeast of the preserve provides some illuminating results. Cadmium and bromine are coherent with organic matter as measured by loss of weight on ignition. Linear correlation coefficients, *r*, calculated among organic matter, bromine, and cadmium are highly significant, but no such significance was observed with the 11 remaining elements. These 11 elements are envisioned as moving together as a kind of ferric hydroxide "colloidal soup," with the remaining 10 elements moving in concert with iron as iron slowly falls out of the interface region into the mud. These results are shown in Table 6.

To further examine these chemical relationships, statistical analyses were carried out comparing the chemical composition of various locales within the preserve with one another. All correlation coefficients were statistically significant beyond the 0.1% level. The concentrations of these 13 elements in the liquid portion of the mud-water interface samples were highly related to the mud in all regions of the preserve studied. When this statistical procedure was applied to all 47 elements, there was no statistically significant relationship between the liquid portion of the mud-water interface and the solid portion. This was reasonable since only 13 of the 47 elements were sorbed by the mud. The statistical relationships are shown in Tables 7 and 8.

In conclusion, it appears that elements preferentially sorbed by the mud from the mud-water interface in shallow, aerobic aquatic systems, such as the Huleh Preserve, do so by two processes: organic matter sorption (cadmium, bromine) and the scavenger action of ferric hydroxide-type complexes (cesium, beryllium, mercury, tin, titanium, zirconium, bismuth, vanadium, molybdenum, and iodine).

Linsley Pond

The concentration of 35 elements was measured in each of 5 layers (surface to 2.5 m, 2.5 to 5 m, 5 to 8 m, 8 to 11 m, and 11 to 14 m) of Linsley Pond for 53 consecutive weeks. The techniques, analytical procedures, and data have been previously described [*7–11,13–*

TABLE 6—*Pertinent linear correlation coefficients of various chemical components in the top 387 cm of Core III of Lake Huleh.*[a]

Element	Organic Matter	Fe
Cd	0.5739	...
Br	0.6578	...
I	...	0.5225
Cs	...	0.9334
Be	...	0.9993
Hg	...	0.6259
Sn	...	0.5270
Ti	...	0.9914
Zr	...	0.9019
Bi	...	0.9988
V	...	0.9996
Mo	...	0.9706
Fe	...	1.0000

[a] $N = 39$, $r = 0.51$, $P < 0.001$.

TABLE 7—*Linear correlation coefficients between elemental and mineralogical concentrations of various regions of the Huleh Preserve.*[a]

	Mud-H_2O-North	Mud-H_2O-Central	Mud-North	Mud-Central	Preserve-North	Preserve-Central
Mud-H_2O-north	1.000	0.9701	NS[b]	NS	0.8825	0.8737
Mud-H_2O-central	0.9701	1.000	NS	NS	0.7890	0.7828
Mud-north	NS	NS	1.000	0.9996	NS	NS
Mud-central	NS	NS	0.9996	1.000	NS	NS
Preserve-north	0.8825	0.7890	NS	NS	1.000	0.9910
Preserve-central	0.8737	0.7828	NS	NS	0.9910	1.000

[a] N = 65; all significant beyond 0.1% level.
[b] NS = not significant.

TABLE 8—*Linear correlation coefficients among elemental concentrations that are sorbed by the mud.*[a]

	Mud-H_2O-North	Mud-H_2O-Central	Mud-North	Mud-Central
Mud-H_2O-north	1.000	0.9925	0.9643	0.9616
Mud-H_2O-central	0.9925	1.000	0.9367	0.9341
Mud-north	0.9643	0.9367	1.000	0.9999
Mud-central	0.9616	0.9341	0.9999	1.0000

[a] N = 13; all are significant beyond the 0.1% level.

14,27,31]. Since only the results are needed to develop geochemical cycles, a brief description of how they were obtained will be presented.

Linear correlation coefficients were calculated between the concentration of any given element in a specific layer with that of the same element in all the other layers for the 53 weeks of the study. Only those coefficients with a statistical significance in excess of 0.1% were accepted. These significant relationships were further analyzed by use of multiple correlation techniques. For example, the concentration of a given element in the surface to a 2.5-m depth would be employed as the dependent variable, and the independent variables would be the concentration of that element in all the other layers. The next multiple correlation coefficient would be calculated employing the concentration of this element in the 2.5 to 5-m layer as the dependent variable; the remaining concentrations of this element in all the other layers would be the independent variables. This procedure was repeated until all the concentrations of this element in each of the zones of the lake were treated individually as dependent variables. This statistical approach provides an estimate of mixing of individual concentrations of a given element throughout the lake and may be used to indicate which elements diffuse from the anoxic mud. Further discussion of this technique may be found in Cowgill [*32*].

Employing beryllium as an example (Table 9), the results of the F test as applied to the multiple correlation R when considering the two hypolimnetic layers are of interest. When the dependent variable is the beryllium concentration in the 8 to 11-m zone, the F value of R is 58.8. When the dependent variable is the beryllium concentration in the 11 to 14-m zone, the F value of R is 29.3. If both F values were the same, it could be postulated that diffusion from the mud was not in process. However, since 58.8 is considerably larger than 29.3, it may be hypothesized that beryllium was diffusing from the mud, probably as an amino acid [*33*], and hence contributing to the concentration of the 8 to 11-m zone and

TABLE 9—*The significant relationships of beryllium in various portions of the 14-m water column in Linsley Pond.*[a]

t Value of Partial Correlation Coefficient[b]					
0 to 2.5 m	2.5 to 5 m	5 to 8 m	8 to 11 m	11 to 14 m	*F* Value of *R*[c]
X[d]	...	5.13	...	...	32.6
...	*X*	...	3.99	...	58.8
5.13	...	*X*	...	...	46.4
...	3.99	...	*X*	5.12	58.8
...	...	...	5.12	*X*	29.3

[a] All values listed are significant beyond the 0.1% level of significance.
[b] The partial correlation coefficient is measured by the *t* (48 degrees of freedom) test.
[c] Significance of the multiple correlation coefficient *R* is measured by the *F* (4,48 degrees of freedom) test.
[d] *X* denotes the dependent variable.

thus causing the *F* value of *R* to be greater. The relationship between the beryllium concentration in the 8 to 11-m zone and in the 11 to 14-m zone is determined by that in the 11 to 14-m zone, which obtains a contribution from the diffusion of beryllium from the reduced mud surface. When the dependent variable is the beryllium concentration in the 11 to 14-m zone, and the beryllium concentration in each of the other layers serves as the independent variable, the *F* value of *R* is the lowest in the beryllium series, suggesting that the contribution from the upper water is not significant.

Employing this technique of estimating elemental circulation, potassium, rubidium, lithium, copper, silver, beryllium, magnesium, calcium, strontium, barium, zinc, mercury, aluminum, gallium, lead, titanium, phosphorus, bismuth, vanadium, sulfur, molybdenum, fluorine, bromine, iodine, iron, manganese, and cobalt all diffuse from the reduced mud surface into the deep hypolimnion. Some of these elements move from the mud into the water as a complex of some amino acid with their sulfide, while titanium probably diffuses as the citrate since there are no known soluble inorganic titanium compounds in nature.

Discussion

In shallow, eutrophic, aerobic systems, such as the Huleh Preserve, the mud sorbs cadmium, bromine (organic matter), cobalt, beryllium, mercury, tin, titanium, zirconium, bismuth, vanadium, molybdenum, iodine, and iron. In deep, eutrophic, anaerobic systems, such as Linsley Pond, 27 elements diffuse from the mud surface into the hypolimnetic waters above. Of these 27 elements, beryllium, mercury, titanium, bismuth, vanadium, molybdenum, iodine, iron, and bromine are sorbed by the mud in shallow, eutrophic, aerobic systems. It may be suggested that when anaerobic mud surfaces become oxygenated, these elements are sorbed by the mud from the oxygenated liquid portion of the mud-water interface. Cesium and zirconium were not detected in the Linsley Pond water at any time during the year of study. Tin and cadmium did not diffuse from the reduced mud surface into the hypolimnion in Linsley Pond due, most likely, to the very small quantities present. It is also important to note that on an annual basis, the concentration of cadmium in the 8 to 14-m zone is nearly constant. Furthermore, the mud surface in Linsley Pond has not been oxygenated during the autumnal and vernal homothermal periods for quite some time. Thus, it was not possible to discover whether in fact these elements are sorbed by the mud in the presence of oxygen in deeper eutrophic systems.

These two studies suggest that a group of elements sorbed by the mud from the liquid portion of the mud-water interface under aerobic conditions is released under anaerobic conditions from the unprotected mud (anoxic) surface to the deep cold waters of the hypolimnion. These elements are beryllium, mercury, titanium, bismuth, vanadium, molybdenum, iodine, iron, and bromine.

Another group of elements, under aerobic conditions, appears to accumulate in the liquid portion of the mud-water interface from the surface waters above. When the mud becomes anaerobic and is no longer protected by oxygen, these elements leave this liquid portion of the mud-water interface and move into the hypolimnion. Elements that appear to be distributed in this fashion are potassium, copper, silver, magnesium, calcium, strontium, barium, zinc, gallium, lead, phosphorus, sulfur, fluorine, manganese, and cobalt. No comment can be offered on the behavior of scandium, yttrium, lanthanum, cerium, germanium, hafnium, arsenic, niobium, selenium, and tellurium since these elements were not detected in Linsley Pond water.

Rubidium, lithium, and aluminum appear to diffuse from the reduced mud but attain equilibrium under oxygenated conditions in shallow ponds. Chlorine does not diffuse from the reduced mud surface but appears to be in equilibrium in shallow aerobic systems.

It was observed in a study [*16*] comparing geochemical relationships between zones in Linsley Pond that sodium and chlorine move together throughout the lake, and statistically speaking these two elements are not separable, that is, the independent variable cannot be ascertained. Furthermore, it is noteworthy that mercury and beryllium, aluminum and titanium, iron and titanium, vanadium and tin were coherent throughout the year of study in the 8 to 14-m section of Linsley Pond. This would suggest that though vanadium diffuses from a reduced mud surface, it becomes coherent with tin which does not diffuse, implying that though the concentration of tin is quite small, at least two different compounds involving tin are present. Similarly, aluminum attains equilibrium in aerobic systems but

TABLE 10—*Summary of results.*

Aerobic Condition, Huleh Preserve (N = 47)	Anaerobic Condition, Linsley Pond (N = 35)
	SIMULTANEOUS PROCESSES
SORBED BY MUD	DIFFUSES FROM MUD
Br, Be, Hg, Ti, Bi, V	Br, Be, Hg, Ti, Bi, V
Mo, I, Fe	Mo, I, Fe
Cs, Zr	not detected
Cd, Sn	do not diffuse from mud
SURFACE WATER-INTERFACE	DIFFUSES FROM MUD-HYPOLIMNION
K, Cu, Ag, Mg, Ca, Sr, Ba,	K, Cu, Ag, Mg, Ca, Sr, Ba,
Zn, Ga, Pb, P, S, F, Mn, Co,	Zn, Ga, Pb, P, S, F, Mn, Co
Sc, Y, La, Ce, Ge, Hf, As	not detected
Nb, Se, Te	not detected
ATTAINS EQUILIBRIUM	DIFFUSES FROM MUD-HYPOLIMNION
Rb, Li, Al	Rb, Li, Al
ATTAINS EQUILIBRIUM	DOES NOT DIFFUSE FROM MUD
Cl	Cl
SURFACE WATER-INTERFACE	DO NOT DIFFUSE FROM MUD
Na, B, Si, Cr, Ni	Na, B, Si, Cr, Ni

clearly is coherent with titanium in anaerobic ones, both diffusing from the reduced mud surface.

A summary of the discussion is presented in Table 10.

Acknowledgment

The National Science Foundation is gratefully acknowledged for its support of the Linsley Pond and Lake Huleh projects.

References

[*1*] Traill, R., "The Jewish War of Flavius Josephus," Vol. II, Houlston and Wright, London, 1862, pp. 46 and 53.
[*2*] Cowgill, U. M., *Archiv für Hydrobiologie,* Vol. 66, 1969, pp. 249–272.
[*3*] Cowgill, U. M., *Archiv für Hydrobiologie,* Vol. 77, 1973, pp. 421–474.
[*4*] Cowgill, U. M., *Nature,* Vol. 256, 1975, pp. 476–478.
[*5*] Cowgill, U. M., *Internationale Revue der gesamten Hydrobiologie,* Vol. 65, 1980, pp. 379–409.
[*6*] Hutchinson, G. E. and Cowgill, U. M., *Archiv für Hydrobiologie,* Vol. 72, 1973, pp. 145–185.
[*7*] Cowgill, U. M., *Developments in Applied Spectroscopy,* Vol. 6, 1968, pp. 299–321.
[*8*] Cowgill, U. M., *Archiv für Hydrobiologie,* Vol. 68, 1970, pp. 1–95.
[*9*] Cowgill, U. M., *Developments in Applied Spectroscopy,* Vol. 10, 1972, pp. 331–352.
[*10*] Cowgill, U. M., *Archiv für Hydrobiologie, Supplements,* Vol. 45, 1974, pp. 1–119.
[*11*] Cowgill, U. M., *Archiv für Hydrobiologie,* Vol. 74, 1974, pp. 350–374.
[*12*] Cowgill, U. M., *Internationale Vereinigung für Theoretische und Angewandt Limnologie,* Vol. 19, 1975, pp. 2749–2757.
[*13*] Cowgill, U. M., *Archiv für Hydrobiologie,* Vol. 78, 1976, pp. 279–309.
[*14*] Cowgill, U. M., *Archiv für Hydrobiologie, Supplements,* Vol. 50, 1976, pp. 401–438.
[*15*] Cowgill, U. M., *Archiv für Hydrobiologie,* Vol. 79, 1977, pp. 36–61.
[*16*] Cowgill, U. M., *Archiv für Hydrobiologie, Supplements,* Vol. 53, 1977, pp. 1–47.
[*17*] Hutchinson, G. E., *Ecological Monographs,* Vol. 11, 1941, pp. 21–60.
[*18*] Mackereth, F. J. H., *Philosophical Transactions of the Royal Society of London,* Series B, Vol. 250, 1966, pp. 165–213.
[*19*] Mortimer, C. H., *Journal of Ecology,* Vol. 29, 1941, pp. 280–329.
[*20*] Mortimer, C. H., *Journal of Ecology,* Vol. 30, 1942, pp. 147–201.
[*21*] Mortimer, C. H., *Limnology and Oceanography,* Vol. 16, 1971, pp. 387–404.
[*22*] Gorham, E., *Limnology and Oceanography,* Vol. 3, 1958, pp. 291–298.
[*23*] Gorham, E., *Limnology and Oceanography,* Vol. 10, 1965, pp. 268–279.
[*24*] Berner, R. A., *Limnology and Oceanography,* Vol. 22, 1977, pp. 781–786.
[*25*] Berner, R. A., *American Journal of Science,* Vol. 267, 1969, pp. 19–42.
[*26*] Naumann, E., *Die Binnengewasser,* Vol. 9, 1930, pp. 1–126.
[*27*] Cowgill, U. M., *Developments in Applied Spectroscopy,* Vol. 5, 1966, pp. 3–23.
[*28*] Chao, T. T., Sanzolone, R. F., and Hubert, A. E. *Analytica Chimica Acta,* Vol. 96, 1978, pp. 251–257.
[*29*] Hubert, A. E., *U.S. Geological Survey Professional Paper,* Vol. 750-B, 1971, pp. B188–190.
[*30*] Hubert, A. E., *U.S. Geological Survey Professional Paper,* Vol. 750-D, 1971, pp. D162–164.
[*31*] Cowgill, U. M., *Proceedings of the National Academy,* Vol. 57, 1967, pp. 198–200.
[*32*] Cowgill, U. M. in *Environmental Biogeochemistry,* Vol. 2, Ann Arbor Science, Ann Arbor, MI, 1976, pp. 731–741.
[*33*] Neuberg, C. and Mandl, I., *Archives of Biochemistry,* Vol. 19, 1948, pp. 149–161.

Wayne R. Munns, Jr.,[1] *and John F. Paul*[2]

Dredged Material Plume Dispersal in Central Long Island Sound

REFERENCE: Munns, W. R., Jr., and Paul, J. F., **"Dredged Material Plume Dispersal in Central Long Island Sound,"** *Aquatic Toxicology and Environmental Fate: Eleventh Volume, ASTM STP 1007,* G. W. Suter II and M. A. Lewis, Eds., American Society for Testing and Materials, Philadelphia, 1989, pp. 82–93.

ABSTRACT: A simulation model based upon *in situ* current velocity data and records of disposal events was developed to predict the chemical exposure field resulting from dredged material disposal plumes in Central Long Island Sound (CLIS) during the spring of 1983. In the model, plumes are assumed to form as subthermocline suspended particulate clouds which are uniform vertically and Gaussian horizontally. Horizontal cloud dimensions are allowed to grow with a constant diffusion velocity as the plume is transported by tidal currents. The model permits estimation of increases or elevations in suspended solids concentration above background levels at individual sites resulting from multiple disposal operations. Model predictions of time-averaged elevations compare favorably with those estimated from tissue residues of *Mytilus edulis* deployed in CLIS as part of a joint Environmental Protection Agency (EPA)/Army Corps of Engineers program to investigate dredged material disposal options. Predicted elevations ranged between 0.0 and 1.6 mg/L. Based upon bulk chemistry measurements, a dredged material concentration of 1.5 mg/L translates to a water column polychlorinated biphenyl (PCB) concentration of approximately 10 ng/L, or about one third of the EPA chronic water quality criterion for that compound. The model indicates the contaminant exposure field at five biological effects field stations to be derived primarily from plume passage, rather than from postdisposal transport from a nearby disposal mound, during the period of active dredged material disposal. Both near (<1 km) and far (>1 km) field dispersal of dredged material in the form of plumes is indicated to occur within this shallow marine system.

KEY WORDS: plumes, dredged material, Long Island Sound, polychlorinated biphenyls (PCBs), particles, dispersion model, simulation, turbidity, currents, *Mytilus edulis,* tissue residues, contaminant exposure field, transport, diffusion equation, aquatic toxicology

The high rates of sedimentation found in most marine harbors necessitate periodic dredging to maintain water depth and prevent shoaling. In addition to recreational and commercial boating, industrial and manufacturing activities are often associated with these waterways. Sediments removed during maintenance dredging may therefore contain elevated levels of organic and inorganic contaminants. Disposal of dredged sediments thus presents a problem which has been receiving increased attention. Potentially viable disposal options include transport to upland landfills, aquatic disposal, and using the material to create or expand marine wetlands.

[1] Senior biologist, Science Applications International Corp., Environmental Research Laboratory—Narragansett, South Ferry Road, Narragansett, RI 02882.

[2] Chief, Exposure Assessment Branch, U.S. Environmental Protection Agency, Environmental Research Laboratory—Narragansett, South Ferry Road, Narragansett, RI 02882.

In 1982, the U.S. Environmental Protection Agency (EPA) and the U.S. Army Corps of Engineers (COE) initiated a multi-year program to investigate the comparative risks associated with these disposal options [1]. The Field Verification Program (FVP) was designed to assess and field validate methodologies for the prediction of environmental impacts resulting from disposal of dredged materials. The COE was responsible for investigating the upland and wetland disposal options, whereas EPA's Environmental Research Laboratory—Narragansett (ERLN) investigated the aquatic disposal option.

The ERLN applied its risk (formerly termed hazard) assessment research strategy for ocean disposal [2,3] to the investigation of the aquatic disposal option. Risk assessment includes both characterization of the environmental exposure field resulting from disposal of the material (exposure assessment), and examination of biological effects resulting from exposure to the material of interest (effects assessment). A detailed exposure assessment involves development of process models to describe contaminant transport, transformation, and fate in the environment. Calibration and verification of these models requires the collection of physical, chemical, and biological data.

Black Rock Harbor, Connecticut, (Fig. 1) was the source of dredged material for use in the FVP. Between 26 April and 14 May 1983, approximately 55 000 m^3 of material dredged from this harbor were disposed at a predetermined location in the Central Long Island Sound (CLIS) disposal area (Fig. 1). The water depth in this area is a fairly uniform 20 m, with little variation in bottom topography. Greater than 88% of the variation in current velocity at the site is explained by the first four tidal harmonics [4], with primary flow in an east-west direction. Maximum current speeds approach 25 cm/s. The water column is stratified from April through October, displaying a sharp thermocline at a depth of approximately 8 m [5].

The FVP site (see Fig. 1) was chosen to be far removed from other active disposal sites in CLIS and to be unaffected by historic disposal activities in the area. Five sampling stations were established along the primary axis of current flow to examine biological effects resulting from the disposal operation: CTR was located at the center of the disposal mound; 200E, 400E, and 1000E were located 200, 400, and 1000 m east of CTR, respectively; and REFS, a reference station, was located approximately 3 km to the southwest of CTR. It was initially assumed that the disposal mound itself would be the sole source of environmental contamination occurring at these stations. As a corollary, it was additionally assumed that REFS would remain unaffected by the FVP disposal mound due to its distance from that site and the direction of prevailing water currents. However, evidence of temporally varying water column contamination at CTR and increased contaminant tissue residues of *Mytilus edulis* deployed at REFS as part of the study were detected prior to the disposal operation at the FVP site [5]. This information indicated potential passage of dredged material plumes created by concurrent disposal operations at other sites within the CLIS disposal area. Approximately 10^6 m^3 of dredged material were disposed in CLIS from March through June 1983. Detailed assessment of exposure conditions existing at the FVP biological effects stations therefore necessitated examination of the above mentioned assumptions and of the potential influence of concurrent disposal operations elsewhere in CLIS.

Here we present the results of simulations of the transport of dredged material disposed at the CLIS disposal area during the spring of 1983. The intent of this work was to determine whether disposal operations occurring elsewhere in CLIS could affect chemical exposure fields at the biological effects stations described above. Information gathered during the FVP concerning the physical characteristics of the dredged material and published estimates of initial plume composition and dimensions were used to simulate plume formation and dispersion. Current velocity data obtained from *in situ* instrumentation were used

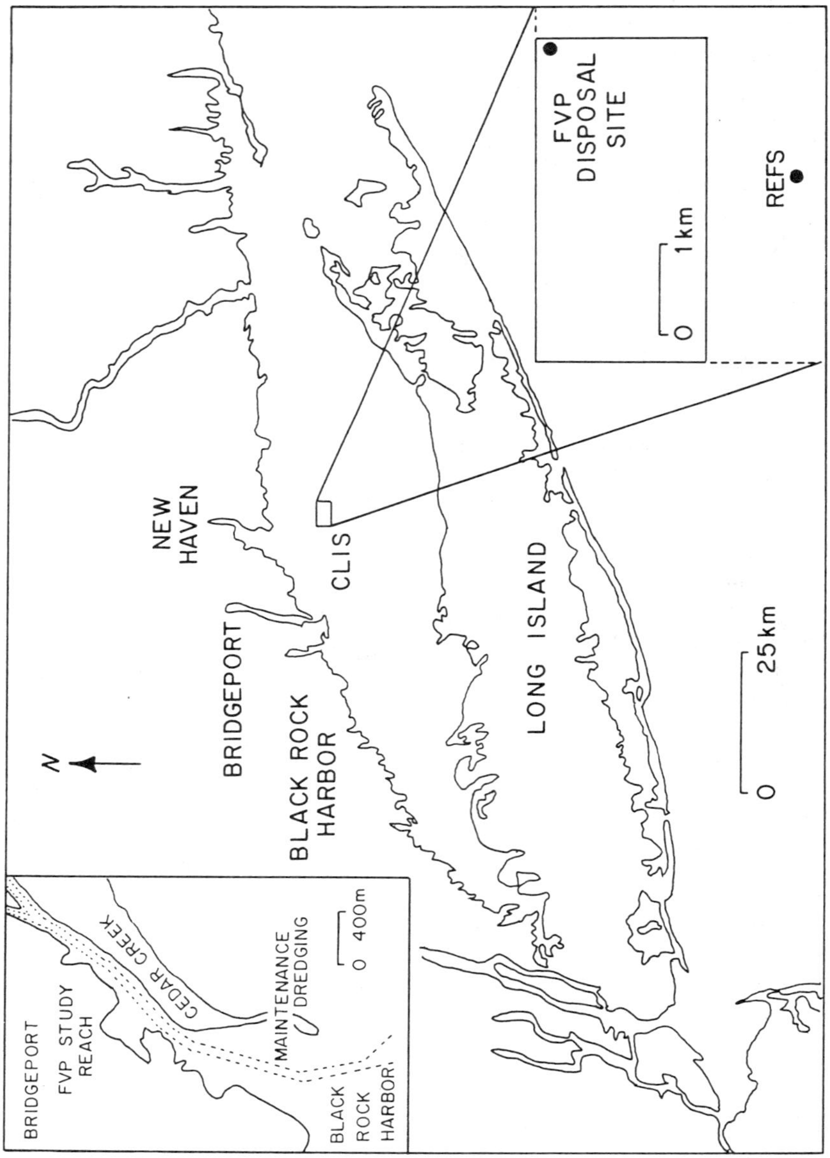

FIG. 1—*Long Island Sound and the locations of Black Rock Harbor, the CLIS disposal area, and the FVP study site.*

to transport each plume. A comparison of model predictions with another estimate of suspended dredged material concentrations is presented and implications of the results obtained through use of the model are discussed.

Procedure

Model Development

A model was developed to simulate the transport and dispersion of dredged material plumes within CLIS during the spring of 1983. Plumes were assumed to form below the seasonal thermocline (at 8 m from the surface [*5*]) as a result of retention of fine particulates and dissolved components within the water column following gravitational descent of the material released from scows. Material within each plume was assumed to be distributed uniformly in the vertical dimension and normally in the horizontal dimensions. This material was treated as being completely conserved; no losses occurred due to settling or other physical and chemical processes. Predictions of suspended solids elevations above background levels resulting from the passage of disposal plumes are therefore to be considered as upper bound estimates for transport.

The initial mass of suspended particulates within each plume was determined by the volume of dredged material released from the scow. Dumping logs containing load volumes were maintained by inspectors on the disposal vessels and were obtained from the New England Div., COE. Concentrations were established by multiplying the volume of material by its density (2.46 g/mL) and adjusting for its water content (72.6%). Previous studies have reported estimates of the amount of disposed material remaining in suspension as ranging from 1 to 5% [*6,7,8*]; we chose a conservative value of 0.25% to account for the particulate material that would rapidly settle out and not be transported. For a typical scow load of 2675 m^3, approximately 4500 kg of material was calculated to remain in the water column.

The initial distribution of mass within the plume, and subsequent dispersion of this mass, was controlled by the solution to the diffusion equation [*9*]

$$C(x,t) = \frac{M}{\sqrt{2\pi}\,(\sigma + \omega t)} \exp\left\{\frac{-x^2}{2(\sigma + \omega t)^2}\right\}$$

where $C(x,t)$ is the mass concentration at time t at a distance x from the plume's center, M is the concentration of mass in the y and z dimensions, σ is the initial width of the plume, and ω is the constant diffusion velocity or plume growth rate. We used 50 m as our estimate of initial plume width based upon studies of a similar dredged material disposed of in CLIS [*6*], and a diffusion velocity of 0.3 cm/s [*10*]. Distribution of a suspended mass of 4500 kg resulted in a peak suspended dredged material concentration of 60 mg/L at the plume's center.

The dumping logs also provided information concerning the approximate timing and location of individual scow releases. In addition to the FVP site, four other dump sites (denoted MQR, CS-1, CS-2, and SP) were actively used during the spring of 1983. Their locations relative to the FVP site and the biological effects stations are shown in Fig. 2. Individual scow releases at a specific disposal site generally occurred within a 200-m radius of that site's marker buoy. This uncertainty in disposal location was not incorporated into the model; all releases at an individual site were assumed to have taken place at the buoy.

Current velocity data used to transport plumes in the simulation were obtained from an electromagnetic current meter deployed in near-continuous fashion at the 1000E station

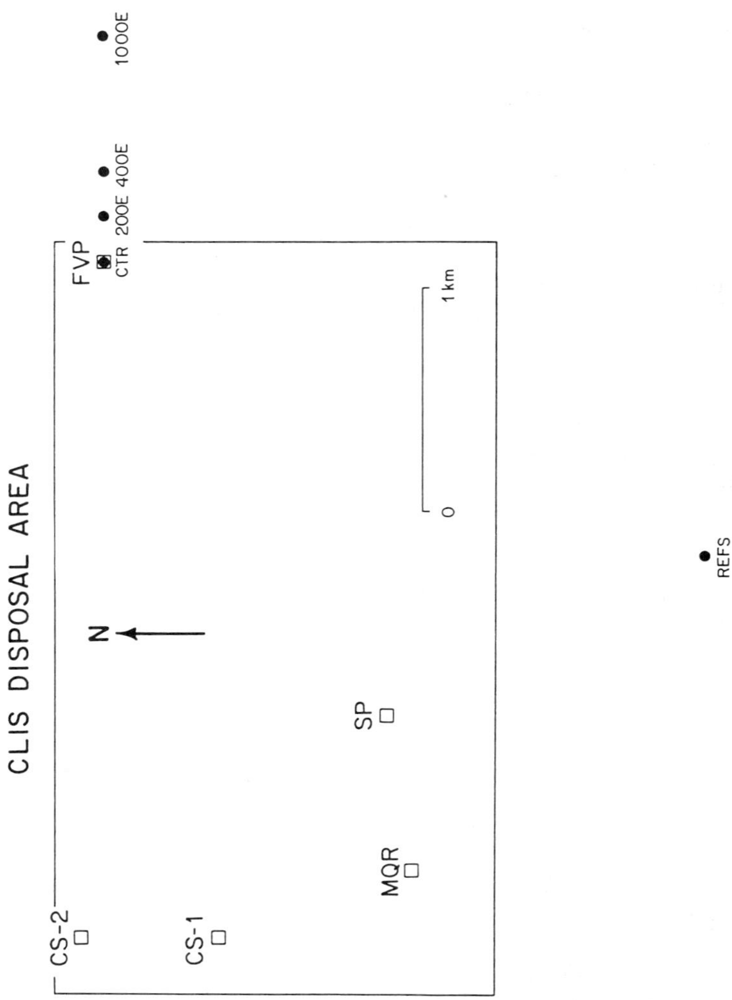

FIG. 2—*Locations of active disposal sites and biological effects stations in CLIS.*

during spring 1983 [*11*]. Horizontal velocities at 1 m above the bottom were burst sampled at 15 min intervals. For the purposes of this simulation, currents were assumed to be coherent over the entire study area and were represented by those measured at 1000E. The uniform topography and water depth of the study area suggest this assumption to be valid. Data from three deployments (NHDB09, NHDB10, and NHDB11) of the *in situ* current meter were used (see Table 1). These deployments bracketed the disposal operations at the FVP site (26 April to 14 May) and corresponded to periods of high disposal activity and dynamic changes in water column chemical exposures [*5*].

Model Behavior

Because of the *a posteriori* nature of this analysis, the model was not validated through field studies. Model behavior was therefore examined through a comparison with data collected independently as part of the FVP. Predictions of time-averaged exposures to the dredged material were compared with estimates derived for tissue residues of *Mytilus edulis* (blue mussels) deployed at 1 m above the bottom at each of the biological effects stations [*12*]. *Mytilus* deployment periods are given in Table 1. A laboratory-derived regression (see below) was used as a calibration line [*13*] to predict the concentrations of Black Rock Harbor (BRH) material sufficient to produce the tissue concentrations observed in the field over similar exposure times. It was assumed that *Mytilus* provided an integrated measure of water column exposure conditions. Laboratory studies [*12*] support this assumption.

Much of the material disposed in CLIS had been dredged from harbors other than Black Rock Harbor. Because we lacked information concerning the chemistry of this material, it was necessary to assume the bulk chemistry of BRH to be representative of all material disposed in CLIS during the study. Additionally, it was necessary to restrict the comparison to contaminants whose behavior would not be greatly affected by physical, chemical, and biological differences between the laboratory and field exposures. We chose to examine polychlorinated biphenyls (PCBs), conservative contaminants found in high levels (6.5 ± 0.4 μg/g dry weight [mean ± 1 standard error (SE)]) in the BRH-dredged material [*5*], for this comparison. Total PCBs were quantified as Arochlor 1254. The calibration line for PCBs was PCB residue = $1036.3 \times$ BRH concentration $+ 282.0$ ($R^2 = 0.999$; see Ref *12*).

Results

Model Behavior

The behavior of the simulation model can best be visualized by examining the results of a single disposal event. Figure 3 shows the first 12.5 h (one tidal cycle) of the predicted trajectory of the center of a plume created at MQR after a scow release on 2 May at 13:35.

TABLE 1—*Designations and periods of coverage of* in situ *instrumentation and* Mytilus edulis *deployments.*

	Designation	Period of Coverage, 1983
Instrument deployments	NHDB09	18 April to 26 April
	NHDB10	26 April to 23 May
	NHDB11	23 May to 16 June
Mytilus edulis deployments	T+0	22 April to 24 May
	T+2	23 April to 7 June

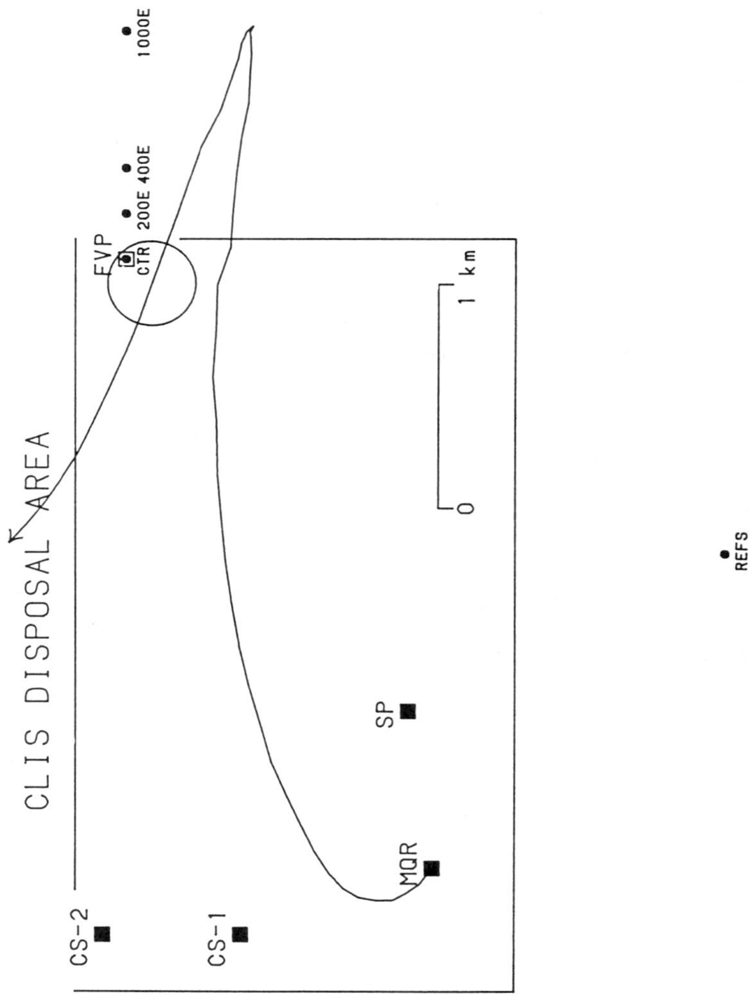

FIG. 3—Predicted trajectory of the center of a disposal plume created at MQR on 2 May. The disposal occurred at 13:35, shortly before high tide. The circle centered about the trajectory near CTR corresponds to the 10 mg/L isoconcentration predicted at that time.

This disposal occurred shortly before high tide. After a brief excursion to the northwest, the plume was transported to the east on an ebb tide, resulting in its passage near the biological effects stations. The following flood tide brought the plume to within 150 m of CTR before being transported out of the area. In Fig. 3, the circle centered about the trajectory near CTR corresponds to the predicted 10 mg/L isoconcentration of the plume at that time. The plume's passage resulted in a sharp elevation in suspended solids of approximately 12 mg/L as shown in Fig. 4 (*top*). The temporal pattern of suspended solids elevations resulting from all disposal operations during the NHDB10 deployment of the *in situ* current meter is shown in Fig. 4 (*bottom*). Fifty-nine strong (>5 mg/L) peaks were predicted to occur as a result of the 149 dumping events during this period.

Time-averaged suspended solids elevations predicted to occur at 400E, 1000E, and REFS during the two *Mytilus* deployments are given in Table 2. BRH concentrations required to produce the PCB tissue residues observed in *Mytilus* deployed during these periods are also given in this table (residue data were not collected at CTR or 200E during either deployment). Predicted dredged material concentrations are generally similar to those derived from the residue data.

Model Results

Model predictions of time-averaged dredged material concentrations resulting from operations at all five active disposal sites are given in Table 3. In general, average elevations fell between 0 and 1.5 mg of dredged material per litre. A graded exposure from CTR (high) to 1000E (low) was predicted for deployments NHDB09 and NHDB10, whereas the opposite trend was predicted for NHDB11. Relative to background suspended solids concentrations of 15 mg/L [*14*], these elevations appear trivial and would probably be undetectable on a time-averaged basis. However, a BRH concentration of 1.0 mg/L translates to a water column PCB concentration of approximately 6.5 ng/L. As background levels of PCBs in CLIS were determined to be on the order of 0.3 to 0.5 ng/L during spring 1983 [*5*], the predicted increase in water column contamination resulting from plume transport is significant. In addition, *Mytilus* displayed negative physiological responses to BRH concentrations as low as 1.5 mg/L (the lowest level tested) in related laboratory tests [*12*]. It should be noted that a BRH concentration of 1.5 mg/L translates to a water column PCB concentration of about 10 ng/L, or about one third of the EPA chronic water quality criterion [*15*].

The relative contribution of disposal at the FVP site to overall exposure conditions at the biological effects stations is also indicated in Table 3. When model conditions were restricted to scow releases at that site only (during deployment NHDB10), predicted time-averaged concentrations of suspended dredged materials were only 12 to 50% of those obtained when operations at all active sites were considered.

Relatively large elevations were predicted to occur at the REFS station during deployment NHDB09, prior to the FVP disposal operation. Additionally, plume passage was predicted to influence water column chemistry at CTR during this deployment. These results agree with the pre-FVP contamination discussed earlier, providing further support of model predictions.

Discussion

The results of these simulations indicate that disposal operations unassociated with the FVP study may have influenced the chemical exposure fields experienced by biota at the FVP biological effects stations. The relative contribution of these distant disposal opera-

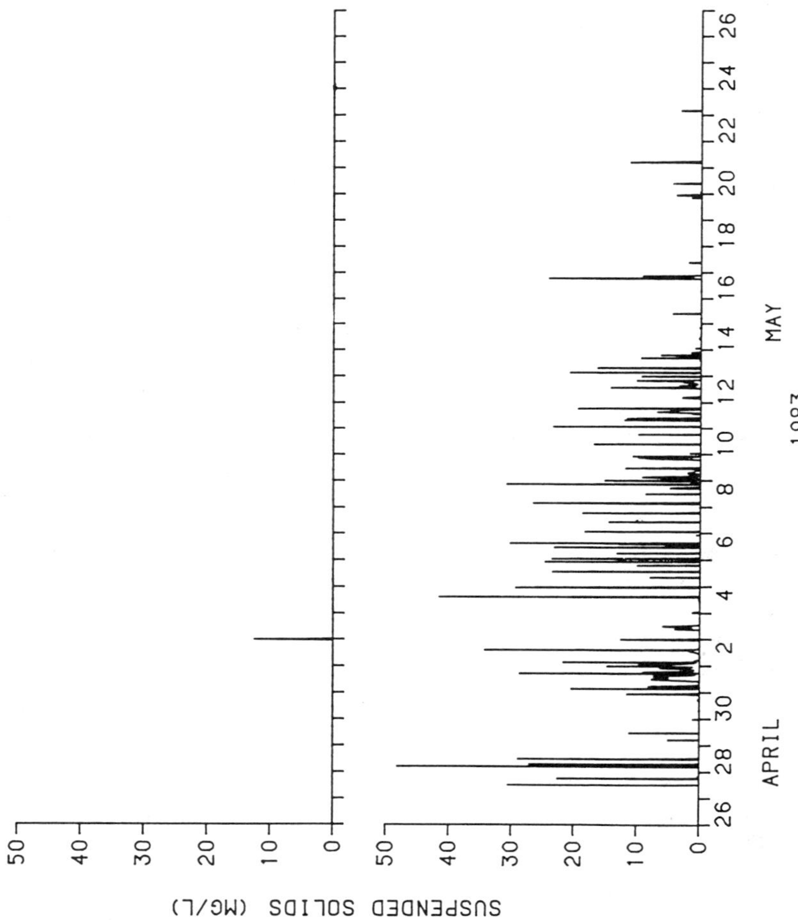

FIG. 4—*Temporal record at CTR of predicted suspended solids elevations above background resulting from a single scow release at MQR on 2 May* (top), *and from all disposal activities during the NHDB10 deployment of the* in situ *current meter* (bottom).

TABLE 2—*Comparison of predicted suspended dredged material elevations (in milligrams per litre) with dredged material exposures estimated from* Mytilus *tissue residues (in milligrams per litre).*

Mytilus Deployment		Biological Effects Station		
		400E	1000E	REFS
T+0	model prediction	...	0.42 ± 0.08[a]	0.0
	residue estimate	NC[b]	0.50 ± 0.07	0.0
T+2	model prediction	0.97 ± 0.02	1.01 ± 0.02	0.01 ± 0.00
	residue estimate	0.85 ± 0.06	0.44 ± 0.07	0.0

[a] Mean ±95% confidence interval.
[b] Data not collected.

TABLE 3—*Predicted time-averaged suspended dredged material elevations (in milligrams per litre) at biological effects stations.*

Deployment Period	Biological Effects Station				
	CTR	200E	400E	1000E	REFS
NHDB09 (all sites)	0.29 ± 0.14[a]	0.22 ± 0.12	0.13 ± 0.08	0.02 ± 0.02	0.61 ± 0.12
NHDB10 (all sites)	0.81 ± 0.14	0.68 ± 0.12	0.55 ± 0.10	0.42 ± 0.08	0.0
NHDB10 (FVP site)	0.37 ± 0.12	0.27 ± 0.10	0.17 ± 0.08	0.05 ± 0.04	0.0
NHDB11 (all sites)	1.06 ± 0.08	1.19 ± 0.08	1.33 ± 0.08	1.60 ± 0.10	0.02 ± 0.00

[a] Mean ±95% confidence interval.

tions was potentially large, as indicated by model results when scow releases were restricted solely to the FVP site during deployment period NHDB10 (Table 3). The initial assumption in the project that the FVP disposal mound would be the sole source of contamination is therefore suspect. Indeed, elevations in contaminant levels (as estimated by *Mytilus* tissue residues) during the period of active disposal in CLIS can be attributed for the most part to plume passage, rather than contaminant transport from the disposal mound itself. This general result has implications in regard to the risk associated with the disposal of contaminated materials in shallow marine systems. Further theoretical and empirical work is needed for evaluation of the relative contributions of plume passage and postdisposal transport of contaminants to the chemical exposure field experienced by organisms around disposal areas.

In addition to near-field (<1 km) influences, these results also suggest that disposal operations in shallow, high-energy marine systems such as CLIS may result in far-field (>1 km) transport of environmental contaminants. For instance, significant suspended solids elevations were predicted to occur at REFS during the NHDB09 instrument deployment period, when the nearest scow releases were approximately 1.5 km to the northeast (at SP, see Fig. 2). At this time, a more detailed assessment of the extent of lateral transport of dredged material cannot be determined by our model, as we assume complete conservation of plume constituents (that is, no gravitational settling by particles or loss due to other physical and chemical processes).

Some additional comments regarding model refinement and validation are cogent. The model currently assumes a fixed percentage of the original mass released from the scow to remain in the water column as a plume. No regard is given to particle retention within the plume as related to particle size. Small particles generally settle more slowly than large particles and presumably display higher mass-specific contaminant concentrations due to an inverse surface area (with associated number of chemical "attachment sites") to volume relationship. Contaminant concentrations associated with the plume material may therefore be underestimated by the bulk chemistry of the material released. This error is confounded by the assumption of conservation of material within the plume during transport and dispersion. A more accurate estimate of chemical exposures at the biological effects stations would be achieved by incorporating information concerning both differential particle retention and settling and particle size specific chemistry into the plume formation and transport phases of the simulation. Ideally, aspects of contaminant phase partitioning could also be incorporated. With the collection of this type of information through empirical studies of the dredged materials, the model could also be used to predict off-mound benthic contamination associated with dredged material disposal.

Acknowledgments

We wish to acknowledge F. Bohlen for providing the *in situ* current velocity data and W. Nelson for calculating the levels of dredged material sufficient to produce the *Mytilus* PCB tissue residues observed in the field. We also thank G. Pesch and J. Gentile for providing the impetus to complete this work, and J. Scott, J. Gentile, G. Pesch, C. J. Meise, J. Nocito, T. O'Connor, and an anonymous reviewer for providing helpful and timely comments on the report. Wayne R. Munns, Jr. was supported under Contract No. 68-03-3236 to Science Applications International Corp., Allen D. Beck, Project Officer. This report has been reviewed by the U.S. Environmental Protection Agency, Environmental Research Laboratory at Narragansett, Rhode Island, and approved for publication. Approval does not signify that the contents necessarily reflect the views and policies of EPA, nor does the mention of trade-names or commercial products constitute endorsement or recommendation for use. This is contribution No. 847 of the Environmental Research Laboratory—Narragansett.

References

[*1*] Gentile, J. H., Pesch, G., and Dillon, T. M. in *Oceanic Processes in Marine Pollution,* Vol. 5, D. A. Wolfe and T. P. O'Connor, Eds., Kreiger, Melborne, FL, 1988, pp. 115–122.

[*2*] Gentile, J. H., Bierman, V. J., Jr., Paul, J. F., Walker, H. A., and Miller, D. C. in *Oceanic Processes in Marine Pollution,* Vol. 3, M. A. Champ and P. K. Park, Eds., Kreiger, Melborne, FL, in press.

[*3*] Bierman, V. J., Jr., Gentile, J. H., Paul, J. F., Miller, D. C., and Brungs, W. A., "Research Strategy for Ocean Disposal: Conceptual Strategy and Case Study," *Environmental Hazard Assessment of Effluents,* H. L. Bergman, R. A. Kimerle, and A. W. Maki, Eds., Pergamon Press, New York, pp. 313–329.

[*4*] Paul, J. F., Bierman, V. J., Jr., Davis, W. R., Hoffman, G. L., Munns, W. R., Jr., Pesch, C. E., Rogerson, P. F., and Schimmel, S. C., in *Oceanic Processes in Marine Pollution,* Vol. 5, D. A. Wolfe and T. P. O'Connor, Eds., Kreiger, Melborne, FL, in press.

[*5*] Munns, W. R., Jr., Paul, J. F., Bierman, V. J., Jr., Davis, W. R., Galloway, W. B., Hoffman, G. L., Rogerson, P. F., and Pruell, R. J., "Exposure Assessment Component of the Field Verification Program: Overview and Data Presentation," EPA 600/3-88-XXX, U.S. Environmental Protection Agency Environmental Research Laboratory, Narragansett, RI, 1988, pp. 123–135.

[*6*] Gordon, R. B., *Estuarine and Coastal Marine Science,* Vol. 2, 1974, pp. 349–358.

[*7*] Custar, C. and Wakeman, T., "Dredge Disposal Study San Francisco Bay and Estuary," Main Report, U.S. Army Corps of Engineers, San Francisco, CA, 1977.

[8] DeLoch, S. R. and Waring, E. G., *Dredging and Dredged Material Disposal,* Vol. 1, American Society of Civil Engineers, New York, NY, 1984, pp. 569–578.

[9] Carslaw, H. S. and Jaeger, J. C., *Conduction of Heat in Solids,* Oxford University Press, London, 1959.

[10] Csanady, G. T. in *Ocean Dumping of Industrial Wastes,* B. H. Ketchum, D. R. Kester, and P. K. Park, Eds., Plenum Press, New York, 1981, pp. 109–129.

[11] Bohlen, W. F. and Winnick, K. B., "Time Series Observations of the Near-Bottom Suspended Material Field Adjacent to the Field Verification Program Site, Central Long Island Sound Disposal Area," project report prepared for the U.S. Environmental Protection Agency, University of Connecticut, Groton, CT, 1986.

[12] Nelson, W. G., Phelps, D. K., Galloway, W. B., Rogerson, P. F., and Pruell, R. J., "Effects of Black Rock Harbor Dredged Material on the Scope for Growth of the Blue Mussel *Mytilus edulis* after Laboratory and Field Exposures," Technical Report D-87-7, U.S. Army Engineer Waterways Experiment Station, Vicksburg, MI, 1987.

[13] Snedecor, G. W. and Cochran, W. G., *Statistical Methods,* The Iowa State University Press, Ames, IA, 1980, p. 169.

[14] Bohlen, W. F. and Winnick, K. B., "Observations on Near-Bottom Suspended Material Concentrations at the FVP Site, Central Long Island Sound Dredged Materials Disposal Area: Immediate Pre-Disposal, Disposal, and Immediate Post-Disposal Period, 18 April 1983–29 June 1983," Project report prepared for Science Applications Inc., University of Connecticut, Groton, CT, 1984.

[15] "Ambient Water Quality Criteria for Polychlorinated Biphenyls," EPA 440/5-80-068, U.S. Environmental Protection Agency, Office of Water Regulations and Standards Div., Washington, DC, 1980.

David A. Graves,[1] Jeffrey W. Foltz,[2] James J. Colcolough,[3] and Paul M. Horton[4]

Toxicity of Percolate from Decontaminated Pesticide-Contaminated Soils

REFERENCE: Graves, D. A., Foltz, J. W., Colcolough, J. J., and Horton, P. M., "**Toxicity of Percolate from Decontaminated Pesticide-Contaminated Soils,**" *Aquatic Toxicology and Environmental Fate: Eleventh Volume, ASTM STP 1007,* G. W. Suter II and M. A. Lewis, Eds., American Society for Testing and Materials, Philadelphia, 1989, pp. 94–100.

ABSTRACT: Pesticide storage sites and aerial application landing sites all represent potential sources of surface and groundwater contamination. Cleansing operations associated with spraying equipment and leakage of containment areas during torrential rains both represent vehicles for pesticide contamination of adjacent soils. The objectives of this research were to measure toxicity of percolate from treated contaminated soils and to determine what level of treatment would produce a nontoxic percolate. Two treatments were examined, activated carbon and agricultural limestone. Treatments were applied to soils contaminated with either toxaphene (1167 mg/kg), chlorpyrifos (96 mg/kg), or carbofuran (876 mg/kg). The toxicity of control water and percolates from uncontaminated soils (control), contaminated soil, and treated soils were measured with fathead minnows. Agricultural limestone was not effective in reducing the toxicity of contaminated soils. Percolate from contaminated soils resulted in 100% mortality and percolate from uncontaminated soils resulted in no mortality. Levels of activated carbon required for 50% survival of fathead minnows at 96 h were 62.7, 0.3, and 46.3 g of activated carbon per kilogram of soil for toxaphene, chlorpyrifos, and carbofuran contaminated soils, respectively. Expressed relative to soil contamination levels, the application rates which would produce 50% survival were 54, 3, and 57 mg of activated carbon per milligram of soil pesticide for toxaphene, chlorpyrifos, and carbofuran, respectively.

KEY WORDS: toxicity, decontamination, activated carbon, agricultural lime, fathead minnow, toxaphene, chlorpyrifos, carbofuran, aquatic toxicology

The manufacture, transportation, and application of agricultural pesticides provides many opportunities for soils to become contaminated with pesticides. Contamination in excess of recommended application rates may result from accidental spills, misapplication, or mishandling of equipment.

Past research efforts have investigated soil decontamination as a means of reducing phytotoxicity and uptake of herbicides by agricultural crops. Activated carbon has been extensively researched as an adsorbent [*1*], and it may provide protection for several years [*2*]. Lime and its influence on soil pH have also been investigated as a means of reducing herbicide toxicity [*3,4*].

[1] Graduate student, Department of Health and Environmental Control, Columbia, SC 29202.

[2] Associate professor of fisheries, Department of Aquaculture, Fisheries, and Wildlife, Clemson University, Clemson, SC 29634-0362.

[3] Chemist, Agricultural Chemical Services, Clemson University, Clemson, SC 29634.

[4] Extension specialist, Department of Entomology, Clemson University, Clemson, SC 29634-0365.

Soil contamination also poses a threat to aquatic environments by way of surface water runoff and percolation through saturated soils. Contamination of soils by stored toxaphene at a locality near Bishopville, South Carolina, dictated the need to develop a decontamination procedure due to the possibility that the toxaphene could percolate through soils during heavy rainfalls and enter adjacent watersheds. Also selected for study as soil contaminants were chlorpyrifos and carbofuran. Chlorpyrifos has been used extensively as a mosquito larvicide. Carbofuran is used on corn and other crops, and both pesticides exhibit anticholinesterase activity [*5,6,7*]. However, toxicity of chlorpyrifos to fish is approximately one order of magnitude greater than the toxicity of carbofuran [*8*]. Toxicity of toxaphene to fish is comparable to chlorpyrifos. Johnson and Finley [*8*] reported the median lethal concentration (LC_{50}) at 96 h for rainbow trout to be 0.011 and 0.007 mg/L (11 and 7 ppb) for toxaphene and chlorpyrifos, respectively.

The objectives of this study were (1) to examine toxicity to fathead minnows *(Pimephales promelas)* of percolates produced by percolating reconstituted water through contaminated, treated contaminated, and uncontaminated soils; (2) to ascertain the treatment level in contaminated soil that resulted in the 50% survival of test organisms; and (3) to compare the effectiveness of activated carbon and lime as treatments of contaminated soils.

Procedures

Source and Preparation of Soils

Toxaphene-contaminated soil was collected from the contamination site. Uncontaminated soil, for comparison with toxaphene-contaminated soil, was collected from a wooded area near the site of contamination. Uncontaminated soil, for contamination with chlorpyrifos and carbofuran, was collected from a wooded area with similar soil characteristics. These samples were passed through an 18-mesh U.S. sieve to produce samples of uniform particle size.

Chlorpyrifos-contaminated soil was produced by spraying 10 kg of uncontaminated soil with 1.0 g of technical grade pesticide dissolved in 1.0 L of acetone. Carbofuran-contaminated soil was produced similarly with 9.0 g of technical grade pesticide dissolved in 1.0 L of acetone. A 0.6 by 0.6-m wooden frame with a plywood bottom was constructed to hold the weighed soil at an average depth of 1.5 cm. Pesticide was applied with a 7.6-L stainless steel spray applicator. Newly contaminated soil was allowed to air dry and was then held in a 20-L glass jar. The wooden frame was lined with aluminum foil with each contamination procedure to prevent cross contamination.

Treatment and Percolation of Soils

Treatment levels were based upon range-finding tests at 0, 1, 10, and 100-g treatment per kilogram of soil. Procedures for range-finding tests were identical to procedures for definitive tests described below. Activated carbon and lime treatments (Tables 1, 2, and 3) were introduced into contaminated soil in a 20-L glass jar and thoroughly mixed by agitation. Duplicate samples were produced by introducing doubled treatments into 2.0 kg of soil, which were then divided into 1.0-kg samples after mixing. Soil samples were placed in 1.0-L glass beakers for water saturation. Very soft reconstituted water was prepared according to the U.S. Environmental Protection Agency (USEPA) [*9*]. Duplicate samples of treated-contaminated soils, untreated contaminated soils, and uncontaminated soils were saturated with 400 mL of reconstituted water. Samples were covered with parafilm

TABLE 1—*Mortality of fathead minnows exposed to percolate from toxaphene-contaminated (1167 mg/kg) soils treated with lime and activated carbon.*

Treatment, g/kg		Percent Mortality at Designated Hour[a]							
Lime[b]	Activated[c] Carbon	1	3	6	12	24	48	72	96
0	0	100, 100	...	...	...	...	...	...	...
22	0	100, 100	...	...	...	...	...	...	...
44	0	100, 100	...	...	...	...	...	...	...
88	0	100, 100	...	...	...	...	...	...	...
176	0	100, 100	...	...	...	...	...	...	...
0	11	10, 0	70, 100	...	...	100, 100	...	...	...
0	22	0, 0	30, 30	...	...	100, 100	...	...	...
0	44	0, 0	0, 0	...	...	60, 80	80, 90	80, 100	80, 100
0	88	0, 0	0, 0	...	...	0, 0	0, 10	0, 10	0, 10

[a] Replicate mortalities are separated by commas.
[b] Mortalities at 96 h for duplicate exposures to uncontaminated soil and control water were 0, 0 and 0, 0%.
[c] Mortalities at 96 h for duplicate exposures to uncontaminated soil and control water were 0, 0 and 0, 10%.

TABLE 2—*Mortality of fathead minnows exposed to percolate from chlorpyrifos-contaminated (97 mg/kg) soils treated with lime and activated carbon.*

Treatment, g/kg		Percent Mortality at Designated Hour[a]							
Lime[b,c]	Activated[d] Carbon	1	3	6	12	24	48	72	96
0	0	0, 0	0, 0	0, 0	30, 40	50, 70	70, 90	80, 90	90, 100
15	0	0, 0	0, 0	0, 0	40, 30	80, 60	80, 60	80, 60	90, 80
116	0	0, 0	0, 0	0, 0	0, 30	30, 70	50, 70	70, 90	90, 100
0	0.025	0, 0	0, 0	0, 0	0, 0	70, 40	70, 50	70, 50	80, 50
0	0.050	0, 0	0, 0	0, 0	0, 0	50, 60	60, 60	60, 60	60, 60
0	0.100	0, 0	0, 0	0, 0	0, 0	40, 30	50, 40	50, 50	50, 50
0	0.200	0, 0	0, 0	0, 0	0, 0	60, 40	60, 40	60, 40	60, 40
0	0.400	0, 0	0, 0	0, 0	0, 0	30, 50	30, 60	30, 60	30, 60
0	2.000	0, 0	0, 0	0, 0	0, 0	0, 0	0, 0	0, 0	0, 0

[a] Replicate mortalities are separated by commas.
[b] Mortalities at 96 h for duplicate exposures to uncontaminated soil and control water were 0, 0 and 0, 0%.
[c] Intermediate and higher levels of lime were not tested because intermediate levels would be ineffective and higher levels impractical for field application.
[d] Mortalities at 96 h for duplicate exposures to uncontaminated soil and control water were 0, 0 and 0, 0%.

and allowed to stand for 16 h. Parafilm covers were not allowed to come in contact with the soil, water, or percolate.

Next, standing water from each soil sample was decanted through a paper coffee filter in an individual 20-cm diameter glass funnel into a separate 5.0-L wide-mouth glass bottle. Soil samples were then flushed with 1000 mL of reconstituted water as follows. Approxi-

TABLE 3—*Mortality of fathead minnows exposed to percolate from carbofuran-contaminated (876 mg/kg) soils treated with lime and activated carbon.*

Treatment, g/kg		Percent Mortality at Designated Hour[a]							
Lime[b,c]	Activated[d] Carbon	1	3	6	12	24	48	72	96
0	0	100, 100	...	...	...	...	...	...	...
1	0	100, 100	...	...	...	...	...	...	...
100	0	100, 100	...	...	...	...	...	...	...
0	4	100, 100	...	...	...	...	...	...	...
0	8	100, 100	...	...	...	...	...	...	...
0	16	100, 100	...	...	...	...	...	...	...
0	32	0, 10	0, 10	0, 10	0, 10	0, 10	10, 10	30, 20	50, 50
0	64	0, 0	0, 0	0, 0	10, 0	20, 0	20, 20	40, 30	40, 30
0	100	0, 0	0, 0	0, 0	0, 0	0, 0	0, 0	0, 0	0, 10

[a] Replicate mortalities are separated by commas.
[b] Mortalities at 96 h for duplicate exposures to uncontaminated soil and control water were 0, 0 and 0, 0%.
[c] Intermediate and higher levels of lime were not tested because intermediate levels would be ineffective and higher levels impractical for field application.
[d] Mortalities at 96 h for duplicate exposures to uncontaminated soil and control water were 0, 0 and 0, 0%.

mately 25% of each saturated soil sample was added to its corresponding funnel, and 250 mL of reconstituted water was slowly passed through the soil. Filters were changed and the process repeated until 1000 mL of water had been percolated through each sample.

Resulting percolates were then filtered through No. 54 hardened Whatman filters in a Buchner funnel by vacuum filtration. When percolates produced were <1000 mL, reconstituted water was added to achieve a total volume of 1000 mL. Each percolated soil sample produced between 970 mL and 1190 mL of percolate. Reconstituted water subjected to identical filtration procedures served as controls.

Toxicity Test Procedures

Duplicate 1.0-L samples of percolates and controls in separate 5.0-L wide-mouth glass bottles (that is, test vessels) were randomly assigned horizontal positions in the experiment room. Aeration through a disposable glass pipette was initiated 10 min prior to the addition of fish to the test vessels and was continued throughout the experiment. Fish were added two at a time from left to right until all vessels contained ten individuals. Mortalities were recorded at 1, 3, 6, 12, 24, 48, 72, and 96 h. Dissolved oxygen and pH were measured at 1, 24, 48, 72, and 96 h. Dissolved oxygen and pH probes were rinsed extensively between readings, and the order of readings was controls first, followed by progressively lower levels of activated carbon. After 100% mortality or 96 h, whichever came first, the entire percolate sample was frozen for later determination of pesticide concentration. Fish were frozen as mortality occurred and their lengths and weights later measured to the nearest millimetre and 0.01 g, respectively.

Fathead minnows were selected as test organisms due to the extent to which they have been used in toxicity testing and the ease with which they are cultured. Young-of-the-year fathead minnows averaging 29-mm total length and 0.23-g wet weight were used as test organisms. Water temperature was maintained at 24 ± 1°C. Photoperiod, pH, and dissolved oxygen were maintained according to USEPA methods [*9*].

Analysis of Soils and Percolates

Dry soils and percolates were analyzed for pesticide residues. Five-gram soil subsamples were extracted in 5 mL of acetone. Percolate samples were extracted with 15 mL of methylene chloride in 500-mL of percolate. Pesticide concentrations were measured by gas chromatography (electron-capture mode). Methods of extraction and analysis were according to USEPA methods [*10*]. Soils were analyzed for pH, cation exchange capacity, total organic carbon, and percentage silt and sand.

Statistical Analysis

Definitive test data were analyzed by a standard probit analysis using 96-h mortality data to calculate the level of treatment necessary to provide 50% survival after 96-h exposure.

Results

Characteristics of Soils

Pesticide concentrations in contaminated soils were 1167 [standard error (SE) = 154, n = 2], 97 (SE = 41, n = 2), and 876 (SE = 72, n = 2) mg pesticide per kilogram dry soil for toxaphene, chlorpyrifos, and carbofuran, respectively. The soil pH averaged 5.1, and the cation exchange capacity was 8 meq/100 g of soil. Total organic carbon, silt, and sand averaged 0.8, 7, and 85%, respectively.

Treatment with Lime

Lime was ineffective in producing tolerable percolates from any of the three pesticides. One-hundred percent mortality of minnows occurred within 1 h at all levels of lime treatment of toxaphene-contaminated soils (Table 1). Mortality of minnows in percolates from lime-treated chlorpyrifos-contaminated soils showed some survival to 24 h. However, by 96 h, mortality was identical to untreated contaminated soil (Table 2). Lime treatment of carbofuran-contaminated soil resulted in 100% mortality by 1 h (Table 3). Additional testing of lime was not pursued because field applications of rates in excess of 100 g/kg of soil (that is, 10%) would be impractical.

Treatment with Activated Carbon

Toxaphene—One-hundred percent mortality of minnows placed in percolate from untreated toxaphene-contaminated soils occurred within 1 h (Table 1). Percolates of soils treated with 11 and 22 g/kg of soil resulted in longer survival, but 100% of the fish had died by 24-h elapsed time (Table 1). Toxaphene concentrations were reduced from 0.026 mg/L (SE = 0.004, n = 4) in percolate from untreated soils to 0.025 mg/L (SE = 0.006, n = 4), and 0.007 mg/L (SE = 0.002, n = 4) in percolates from the 11 and 22-g/kg soil treatment levels, respectively. Percolate from soils treated with 44 g/kg resulted in 10% survival (that is, 90% mortality), with most individuals surviving 48 h (Table 1). Toxaphene was reduced to 0.002 mg/L (SE $<$ 0.001, n = 4). Survival in percolate treated with 88 g/kg of soil was nearly equivalent to controls, and toxaphene was reduced to 0.001 mg/L (SE $<$ 0.001, n = 4). The level of activated carbon required for 50% survival at 96 h

TABLE 4—*Levels of activated carbon required for 50% survival of fathead minnows in percolates from soils contaminated with three pesticides.*

Pesticide	Contamination Level, mg/kg soil	LC_{50}, mg activated carbon/kg soil	95% Confidence Limits, mg activated carbon/kg soil		Activated Carbon Level, mg activated carbon/mg soil pesticide
			Lower	Upper	
Toxaphene	1167	62.7	54.1	72.6	54
Chlorpyrifos	97	0.3	0.0	0.9	3
Carbofuran	876	46.3	35.0	52.4	53

was 62.7 g of activated carbon per kilogram soil which was equivalent to 54 mg of activated carbon per milligram of soil toxaphene (Table 4).

Chlorpyrifos—No mortalities were observed in the first 12 h of exposure to percolates from activated carbon-treated chlorpyrifos-contaminated soils (Table 2). Ninety-six hour mortality ranged from 95% in the untreated soil to 65 and 45% in the 0.025 and 0.4 g/kg treatment levels (Table 2). Percolate from contaminated soil treated with 2.0 g/kg resulted in no mortality. Percolate from soil treated with 0.025 g of activated carbon per kilogram averaged 0.001 mg/L (SE = 0). All other treatment levels produced percolates with undetectable levels of chlorpyrifos. The level of activated carbon required for 50% survival at 96 h was 0.34 g/kg of soil which was equivalent to 3 mg of activated carbon per milligram of soil chlorpyrifos (Table 4).

Carbofuran—Percolates from untreated carbofuran-contaminated soils and soils treated with 4, 8, and 16 g of activated carbon per kilogram soil resulted in 100% mortality by 1 h (Table 3). Treatment with 32, 64, and 100 g/kg of soil resulted in 50, 35, and 0% mortality at 96 h, respectively. No mortalities were observed in controls (Table 3). Carbofuran in percolates from untreated carbofuran-contaminated soils averaged 7.320-mg/L (SE = 0.380, n = 2) carbofuran, respectively. Treatment with 32 and 64 g/kg averaged 0.240 (SE = 0.020, n = 2) and 0.120 (SE = 0.020, n = 2) mg/L. Carbofuran was undetectable in controls and soils treated with 100 g of activated carbon per kilogram soil. The level of activated carbon required for 50% survival was 46.34 g/kg of carbofuran-contaminated soil (Table 4). This was equivalent to 53 mg of activated carbon per milligram of soil carbofuran.

Discussion

Lime treatment of soils contaminated with the pesticides investigated in this study was ineffective in reducing potential hazard to aquatic environments. Treatment of test soils with activated carbon resulted in the direct reduction of percolate toxicity to fish. Toxicity was reduced as the treatment level was increased. Pesticide residues were adsorbed to the activated carbon. Chlorpyrifos was strongly adsorbed in comparison to toxaphene and carbofuran. Percolates from chlorpyrifos-contaminated soils treated with activated carbon contained <0.001 mg/L. Percolate concentrations of toxaphene and carbofuran ranged from 0.001 to 0.026 mg/L for toxaphene and 0.0 to 0.240 mg/kg for carbofuran. Pesticide concentrations in percolates of soils treated with activated carbon at rates resulting in 50%

survival were near LC_{50} concentrations. According to Johnson and Finley [*8*], LC_{50} values for carbofuran, toxaphene, and chlorpyrifos are 872, 18, and 7 mg/L of water (that is, parts per billion). In this study, pesticide concentrations at the two treatment levels which bracketed each LC_{50} for activated carbon were 240 and 120, 2 and 1, and <1 mg/L of water (that is, parts per billion) for carbofuran, toxaphene, and chlorpyrifos, respectively. The actual pesticide concentrations are close to those reported by Johnson and Finley [*8*], and the relative toxicities are identical.

Summary

Toxicity testing can be used to evaluate the effectiveness of a soil treatment in reducing the toxicity of percolate from pesticide-contaminated soil in the aquatic environment. Data generated through toxicity testing can be used to calculate treatment levels necessary to reduce percolate toxicity to environmentally safe levels. This information may reveal an economically feasible temporary solution to a contamination problem. Temporary alternatives such as use of activated carbon could prevent contamination of ground and surface water, while permanent solutions are evaluated and implemented.

Acknowledgment

Technical Contribution No. 2873, South Carolina Agricultural Experiment Station, Clemson University, Clemson, SC.

References

[*1*] Foy, C. L. and S. W. Bingham, "Some Research Approaches Toward Minimizing Herbicidal Residues in the Environment," *Residue Reviews,* Vol. 29, 1969, pp. 105–135.

[*2*] Lichtenstein, E. P., Schulz, K. R., and Furemann, T. W., "Long-Term Effects of Carbon in Reducing Uptake of Insecticidal Soil Residues by Crops," *Journal of Economic Entomology,* Vol. 64, 1971, pp. 585–588.

[*3*] Weber, J. B., Perry, D. W., and Ibaraki, K., "Effect of pH on the Phytotoxicity of Prometryne Applied to Synthetic Soil Media," *Weed Science,* Vol. 16, 1968, pp. 134–136.

[*4*] Upchurch, R. P., "Behavior of Herbicide in Soil," *Residue Reviews,* Vol. 16, 1966, pp. 134–136.

[*5*] Macek, K. K., Walsh, D. F., Hogan, J. W., and Holz, D. D., "Toxicity of the Insecticide Dursban to Fish and Aquatic Invertebrates in Ponds," *Transactions of the American Fisheries Society,* Vol. 101, 1972, pp. 420–427.

[*6*] Thirugnanam, M. and Forgash, A. J., "Environmental Impact of Mosquito Pesticides: Toxicity and Anticholinesterase Activity of Chlorpyrifos to Fish in a Salt Marsh Habitat, *Archives of Environmental Contamination and Toxicology,* Vol. 5, 1977, pp. 415–425.

[*7*] Eisler, R., "Carbofuran Hazards to Fish, Wildlife, and Invertebrates: A Synoptic Review," U.S. Fish and Wildlife Service Biological Report 85(1.3) U.S. Fish and Wildlife Service, Washington, DC, 1985.

[*8*] Johnson, W. W. and Finley, M. T., "Handbook of Acute Toxicity of Chemicals to Fish and Aquatic Invertebrates," U.S. Fish and Wildlife Service Research Publication, No. 137, U.S. Fish and Wildlife Service, Washington, DC, 1980.

[*9*] "Methods for Acute Toxicity Tests with Fish, Macroinvertebrates, and Amphibians," EPA/600/3-75/009, U.S. Environmental Protection Agency, Washington, DC, 1975.

[*10*] "Analysis of Pesticide Residues in Human and Environmental Samples," EPA/600/8-80/038, U.S. Environmental Protection Agency, Washington, DC, 1980.

Jean A. Nocito,[1] Henry A. Walker,[2] John F. Paul,[2] and Charles A. Menzie[3]

Application of a Risk Assessment Framework for Marine Disposal of Sewage Sludge at Midshelf and Offshelf Sites

REFERENCE: Nocito, J. A., Walker, H. A., Paul, J. F., and Menzie, C. A., "**Application of a Risk Assessment Framework for Marine Disposal of Sewage Sludge at Midshelf and Offshelf Sites,**" *Aquatic Toxicology and Environmental Fate: Eleventh Volume, ASTM STP 1007,* G. W. Suter II and M. A. Lewis, Eds., American Society for Testing and Materials, Philadelphia, 1989, pp. 101–120.

ABSTRACT: A risk assessment framework was developed for predicting impacts from midshelf and offshelf disposal of sewage sludge. In the application presented, only the impacts resulting from bioaccumulation of the contaminants were considered. Disposal sites were assumed to be characteristic of sites in the Middle Atlantic Bight. The contaminant transport model used assumed all contaminants were associated with particulates and calculated the flux through the water column and into the sediments. Since degradation losses and resuspension were not explicitly included in the model, results obtained represented upper-bound estimates of time-averaged contaminant flux to the sediments. Cross-shelf profiles of maximum concentrations in the sediment surface layer were calculated by using an expression that accounted for mixing in the bioturbated zone. By assuming that organisms reach equilibrium with the surrounding sediment, concentrations of neutral hydrophobic organic contaminants in infaunal organisms were computed by using a preference factor (PF) based upon the organic carbon content of the sediments and the lipid content of the organisms. Results indicated that body burdens may be at least a half-order of magnitude higher in organisms living in the vicinity of the midshelf site than those living offshelf. Predicted tissue residues were compared with the Food and Drug Administration (FDA) tolerance level for polychlorinated biphenyls (PCBs) and the action level for chlordane to determine potential impacts from bioaccumulation of these contaminants in the tissues of organisms. By using the U.S. Environmental Protection Agency (EPA) Integrated Risk Information System (IRIS), the reference dose (RfD) for chlordane was obtained and used to evaluate the risk to humans based on a dose to man calculation. Carcinogenic risk estimates were made using estimated potency factors. Projected results indicated that tissue concentrations resulting from sludge disposal would be at least an order of magnitude lower than the FDA tolerance level for PCBs at both sites, even after 100 years of disposal. The simulated midshelf disposal for a 100-year period resulted in body burdens of 0.28 ppm chlordane compared to the FDA action level of 0.3 ppm. However, the calculated RfD for chlordane was exceeded after 70 years of midshelf disposal. Although higher noncarcinogenic risks were associated with midshelf disposal of chlordane after 100 years, higher carcinogenic risks were associated with midshelf disposal of PCBs. Because of large uncertainties in modeling the various processes affecting transport,

[1] Physical scientist, Science Applications International Corp., U.S. Environmental Protection Agency, Environmental Research Laboratory—Narragansett, Narragansett, RI 02882; currently PhD candidate, University of Massachusetts, Amherst, MA 07003.

[2] Environmental scientist and branch chief, Exposure Assessment Group, respectively, U.S. Environmental Protection Agency, Environmental Research Laboratory—Narragansett, Narragansett, RI 02882.

[3] Senior scientist, Charles Menzie Associates, Westford, MA 01886.

fate, and contaminant uptake in the benthic food chain, the results are only intended to provide estimates of relative risks related to midshelf versus offshelf disposal options. Based on relative comparisons made in this analysis, higher human health risks would be associated with midshelf disposal than with offshelf disposal.

KEY WORDS: sewage sludge, benthic flux, risk assessment framework, preference factor, contaminant mass loading, Middle Atlantic Bight, ocean disposal, polychlorinated biphenyls, (PCBs), chlordane, aquatic toxicology

With the enactment of the Marine Protection, Research, and Sanctuaries Act (MPRSA) in 1972, Congress authorized the U.S. Environmental Protection Agency (EPA) to designate suitable sites and grant permits for ocean dumping of various materials, including sewage sludge. Preference is for sites located beyond the edge of the continental shelf whenever feasible, as stated in Section 102 of MPRSA [*1*]. EPA has denied petitions from New York and New Jersey municipalities to redesignate the 12-mile site for sewage sludge disposal [*2*]. This site has been in use since 1924. In 1984 the EPA designated an alternate site, a section of the 106-mile ocean disposal site (referred to as deep-water municipal disposal site, DWMDS) for the disposal of municipal sewage sludge on a five-year interim basis [*2*]. The phase-out of sludge disposal operations at the 12-mile site commenced in March of 1986, and disposal there ended by December of 1987 [*3*].

As more restrictions are placed on land-based waste disposal sites, municipalities other than New York are beginning to consider application for ocean disposal permits.[4] The suitability of using sites on the continental shelf rather than deep-ocean sites is being considered. To address this issue, EPA needs to develop a method for evaluating the relative impacts of midshelf versus offshelf disposal.

A risk assessment framework based on projected mass loadings of sewage sludge is one method for determining whether there are dangers to man or the ecosystem associated with disposal at the sites. A first step in assessing potential impacts involves upper-bound calculations that often are based on limited data. Calculations of this type can rank the potential degrees of impact at different sites. Upper-bound assessments will aid in evaluating potential impacts to the marine environment and can be useful in site designation and the review of dumping permit applications.

A risk assessment approach has been used by the EPA Environmental Research Laboratory—Narragansett (ERLN) for evaluating potential upper water column impacts associated with sewage sludge disposal at the 106-mile site [*4*]. The model predicts long-term, time-averaged elevations in waste constituent concentrations that depend on mass loading rates to the site [*4,5*]. It was assumed that uniform mixing occurs over the upper water column, defined by a seasonal thermocline in the summer and a permanent pycnocline in winter. Settling of particulates from this upper mixed layer was not included, and thus benthic exposures were not determined using this model.

The benthic component of a marine ecosystem is important because of (1) its role in recycling nutrients and contaminants from the ocean floor, and (2) certain fishery resources depend on the benthic food chain. Risk assessment, therefore, must address potential benthic impacts related to ocean disposal of sewage sludge. Such assessments are particularly important because contaminants accumulate in many marine sediments over a long time. Thus for long-term disposal, the time scales appropriate for assessment of benthic

[4] O'Connor, T., National Ocean Survey, National Oceanic and Atmospheric Administration, Rockville, MD, personal communication, 1986.

impacts should be much longer (10 to 100 years) than those used in upper water column assessments (seasons).

A few offshelf benthic flux calculations have been made related to offshelf waste disposal practices. O'Connor et al. [*6*] performed scoping calculations to project the benthic flux from sludge disposal at the 106-mile site, assuming that 20% of the solids would settle at a rate of 10^{-2} cm/s or faster. Preliminary estimates of the settled sludge flux were defined simply as mass of particles per unit time spread over the area of horizontal dispersion achieved in a 200-day period. Partch [*7*] refined this estimate by determining fallout patterns of particulates based on various settling velocity distributions. However, his results were not scaled to benthic flux rates based on projected sewage sludge loading rates.

A comparative analysis of potential impacts resulting from midshelf and offshelf marine disposal of sewage sludge is presented in a risk assessment framework. Risk assessment involves utilization of available physical, chemical, and biological data, along with information pertaining to transport pathways, to determine potential ecological and human health risks associated with exposure to a contaminant. In our application of the risk assessment framework, residues in benthic species are used as contaminant vectors to demersal fisheries and then to man to assess human health risks associated with sewage sludge disposal. The primary objectives of this paper are (1) to present a framework which could be used for performing assessments of potential benthic risks from the disposal of sewage sludge, and (2) to compare the relative impacts of disposing sewage sludge at midshelf versus offshelf sites. After presenting a brief description of the risk assessment framework, the relative risks associated with midshelf versus offshelf disposal of sewage sludge containing two persistent hydrophobic organic compounds, PCBs and chlordane, will be shown.

Risk Assessment Framework

The risk assessment framework used for ocean disposal is based on the work of Paul et al. [*4*]. To evaluate the hazard associated with exposure to toxic compounds, risk assessment involves the following: two information gathering steps (site and waste characterization), followed by three evaluation steps (exposure assessment, hazard assessment, and risk characterization). Note that this terminology is consistent with that used in environmental risk assessment [*4,8*]. Information concerning site and waste characteristics is used in all three assessment steps. Exposure assessment relates the transport and fate of source input of contaminants to resultant concentrations in the environment. Hazard assessment is used to relate environmental concentrations to biological effects, such as toxicity to organisms and bioaccumulation. The risk characterization step combines the exposure and effects results to provide a scientific basis for estimating risks. When this procedure is applied to multiple sites, relative comparisons can then be made pertaining to impacts from waste disposal. The framework for exposure and hazard assessments and risk characterization is illustrated in Fig. 1.

A transport model was used to relate the waste input to benthic flux of a contaminant to the bottom sediment. Estimated exposure concentrations then can be used to evaluate two types of risk associated with the benthic component: (1) ecological and (2) human health. Ecological risks posed to benthic communities were not calculated due to the unavailability of sediment quality criteria and toxicological data on indigenous organisms. Exposure concentrations in sediment were used to predict bioaccumulation in tissues of benthic organisms. Human health risks were evaluated based on doses to man and FDA tolerance and action levels for contaminants in tissues (Fig. 1).

FDA tolerance and action levels have been set for certain toxic compounds based in part

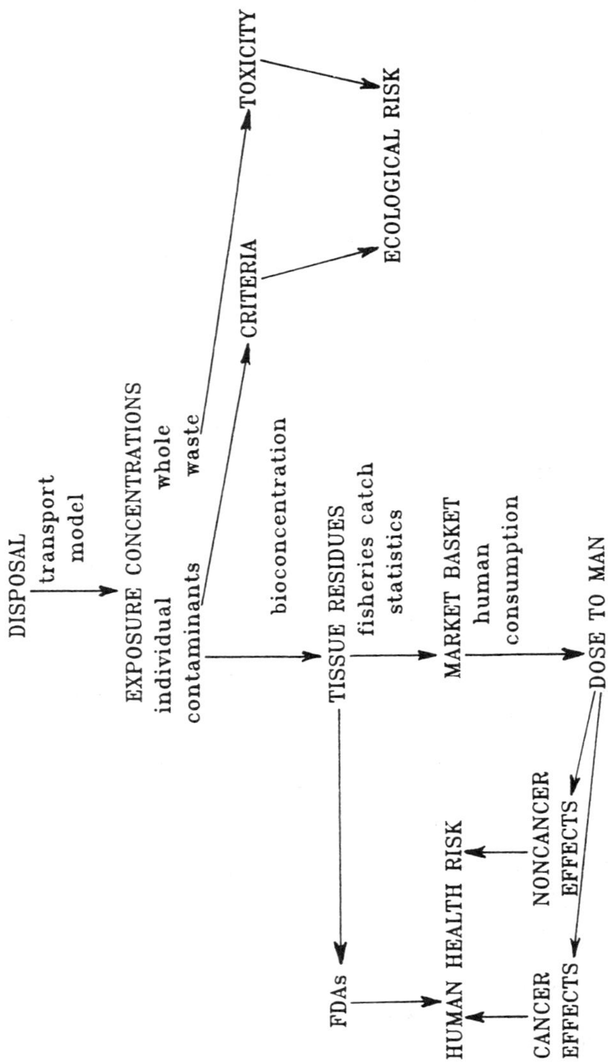

FIG. 1—*Risk assessment framework—Environmental Research Laboratory, Narragansett, RI.*

on consideration of the potential human health effects. Direct comparison is made between Food and Drug Administration (FDA) levels and predicted tissue residues in fish. Calculation of dose to man is more involved, relating fisheries catch statistics and human consumption data to body burdens in fish. This resultant dose is then compared with a reference dose for humans to determine potential noncarcinogenic risks. Carcinogenic risks are determined by multiplying the computed daily dose by an estimated potency factor for a particular toxicant.

In our application, calculations lead to upper-bound estimates of risks associated with bioaccumulation of specified contaminants via the benthic food chain. Because of large uncertainties in modeling the various processes affecting transport, fate, and contaminant uptake via the benthic food chain, the results only are intended to provide estimates of the relative risk related to midshelf versus offshelf disposal options. Although a formal analysis of error propagation has not been conducted, uncertainties that enter at various points in the calculation are discussed.

Site Characterization

Two hypothetical sites were used in the example calculations, one located on the outer continental shelf and the other on the continental slope and rise. Both sites were chosen to typify characteristics of the Middle Atlantic Bight (MAB) region of the northeast coast of the United States. The midshelf site is located approximately 70-km offshore, with a water depth of 60 m; whereas the offshelf site lies along the 2000-m isobath, approximately 150-km offshore. Figure 2 illustrates the cross-shelf profile of the continental shelf, slope, and rise regions. In this analysis, it was assumed that no variation occurs in the along-shelf direction. Transport characteristics were determined from current meter data from the Mid-Atlantic Slope and Rise Physical Oceanography Study (MASAR/POS). This study examined circulation features above the continental slope and rise of the Middle Atlantic Bight from 1983 to 1986 [*9*]. Along the southern array, located offshore of Virginia Beach, 40 and 2000-m deep mooring arrays were used. This was the only synoptic set of midshelf and offshelf current meter data available. The velocity field at the 40-m depth mooring represented midshelf site conditions. A summary of the physical oceanographic conditions of MAB is presented in Ingham et al. [*10*].

The MAB continental shelf is characterized by sandy (>75% sand) or gravelly sand sediments up to depths of 200 m, with an increasing percentage of silt and clay as depths increase down slope. In contrast, sediments on the MAB continental slope and rise are predominately silts and clays with <10% sand-sized particles [*11*]. Oceanographic conditions occurring in the shelf-slope region lend insight into the physical processes governing this sediment size distribution. Researchers have observed near-bottom currents to be sufficiently large under the influence of storm events to resuspend shelf sediments [*12,13*]. These periodic large-scale, storm-related transport events are separated by long periods of minimal transport. However, the near-bottom continental slope area is generally a tranquil environment devoid or nearly devoid of sufficient currents to transport sediments [*14*].

Sediment patterns that influence benthic community structure along the shelf are determined in part by ridge and swale topography. Swales serve as depositional zones for fine-grained sediment, whereas coarser sediments are deposited on ridges. The MAB outer continental shelf, particularly in swale habitats, is reported to have the highest number of infaunal species of all shelf bathymetric zones [*15*]. Swales on the outer continental shelf

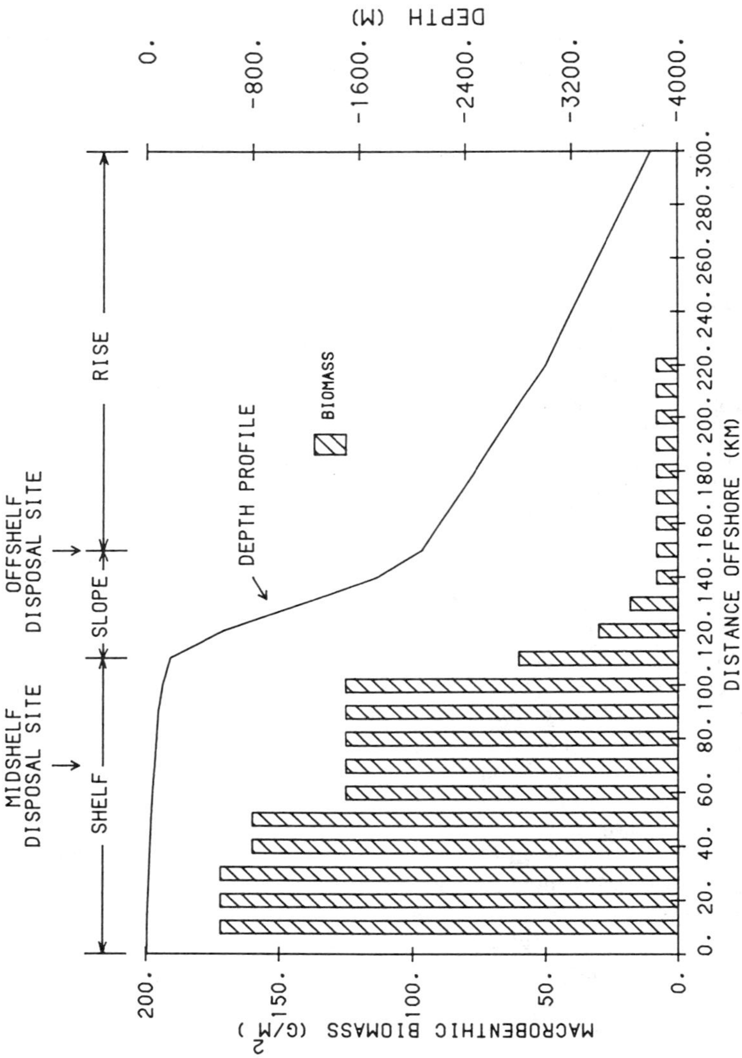

FIG. 2—*Depth distribution of macrobenthic biomass.*

may serve as depositional areas where detritus collects and is thus more available to benthic infauna. The potential for bioaccumulation in swale areas is greater relative to ridge areas due to the longer residence times and larger populations which inhabit these areas, although resuspension eventually may transport the sediment offshelf.

The mid-Atlantic shelf is a highly productive region, with benthic communities primarily influenced by sediment type, temperature, and availability of food [*15*]. Figure 2 depicts a high density of macrobenthic biomass (total weight per unit area) in this region [*16,17*]. The outer continental shelf is defined as that portion of the shelf with water depths ranging from approximately 50 to 200 m and is inhabited at least seasonally by many recreational and commercial species [*15*]. Dominant infaunal groups found on the outer continental shelf are amphipods and polychaetes. For benthos dwelling on the shelf and upper slope, detritus is the primary food source [*18*].

With an increase in water depth and distance from land, infauna generally decrease in numbers of species, abundance, and biomass per unit area. However, demersal fish species, such as tilefish *(Lopholatilus chamaeleonticeps)* and graysole *(Glyptocephalus cynoglossus),* and crustaceans, namely lobster *(Homarus americanus)* and red crab *(Cancer borealis),* support fisheries on the upper continental slope in the MAB. The primary food source for these species is benthic infauna. Dominant infaunal groups inhabiting the continental slope are polychaetes and bivalves, with polychaetes predominant on the rise. Benthic communities in the slope region also are influenced by food supply [*19*] and temperature [*16*], which may be the reason for the decline in biomass with distance from the more highly productive shelf (Fig. 2).

Waste Characterization

The sludge volume used in this analysis was assumed equivalent to the annual loading rate of approximately 7.3 million wet metric tons of sludge recently disposed at the 12-mile site [*3*]. Waste constituent concentrations were based on reported values from New York publicly owned treatment works (POTW) sludges [*20,21*]. The concentration of polychlorinated biphenyls (PCBs) was 2.57 mg/kg dry weight, and for chlordane it was 5.26 mg/kg dry weight [*20*]. Assuming that sludge is 5% solids, the corresponding daily loading rate was 2.57 kg/day and 5.26 kg/day for PCBs and chlordane, respectively.

Data on settling characteristics of New York and New Jersey sludges are generally not available. The settling velocity distribution used was based on Faisst's data for southern California sludges [*22*]. The settling velocity classes used and the corresponding proportion of total mass associated with each class are summarized in Table 1.

TABLE 1—*Settling velocity distribution assumed for sludge.*

Settling Velocity, cm/s	Fraction of Total Mass
0.0001	0.269
0.0003	0.118
0.001	0.137
0.003	0.135
0.01	0.137
0.03	0.090
0.1	0.114

Exposure Assessment

Description of the Layered Two-Dimensional Model

An extension of the modeling approach used to provide upper-bound estimates of contaminant distribution in the upper mixed layer resulting from sludge disposal at the DWMDS [*4,5*] was used to describe the transport and fate of sewage particles. This model utilizes an application of Taylor's theory [*23*] to estimate the dispersal of waste plumes originating from the disposal site. Described as discrete two-dimensional Gaussian plumes, these waste plumes are advected by the mean current and dispersed according to the variance-covariance structure of the currents. Based on the probabilistic nature of these currents, estimates were made of the long-term, time-averaged concentration distribution in the upper-mixed layer of the water column.

Whereas the previously developed model provides upper-bound estimates of contaminant distribution in the upper-mixed layer, the layered two-dimensional model used here estimates upper-bound contaminant flux through the water column and into bottom sediments. In the layered two-dimensional modeling approach, waste plumes were assumed to settle through the water column according to the specified settling velocity distribution (as given in Table 1). To obtain upper-bound estimates of contaminant flux to the sediment, it was assumed that all contaminants are associated with the particulate phase of the waste (no partitioning between dissolved and particulate phases) and are not degraded or transformed in the water column and bottom sediments. Model calculations are based on a unit daily loading rate of 1 kg/day of a conservative substance.

In addition to assumptions defined in Paul et al. [*4*] and Walker et al. [*5*], the layered two-dimensional model assumes that sewage sludge particulates will not be subject to further horizontal dispersion once deposited on the ocean bottom. Thus, the ocean bottom was treated as an absorbing boundary. Although particles deposited on the outer continental shelf may be resuspended by episodic storm events, this process was not simulated. This conservative assumption is consistent with providing upper-bound estimates of benthic impacts.

The midshelf site was modeled as one water column layer and the offshelf site as three layers, based on available current meter data. Low-pass filtered current meter statistics are summarized in Table 2. Simulations were run for a 200-day period for the offshelf site, by which time steady-state benthic flux rates were reached. Because of a lower mean velocity, the duration of simulations for the midshelf site was for a 300-day period. Independent simulations were run for each settling size fraction in Table 1 to determine the benthic flux

TABLE 2—*Low-pass filtered current meter statistics as obtained from MASAR/POS study.*[a]

	Along the Shelf		Across the Shelf	
	Mean, cm/s	Variance, $(cm/s)^2$	Mean, cm/s	Variance, $(cm/s)^2$
60-m isobath	−3.19	180.2	0.24	40.9
2000-m isobath				
(0 to 100 m)	−13.9	232.2	2.26	213.8
(100 to 500 m)	−9.0	97.2	1.87	38.1
(500 to 4000m)	−3.1	12.7	−1.29	6.1

[a] The orientation is defined as follows: for the along-shore component, the negative sign indicates a downstream flow; the cross-shelf component is negative on shore and positive offshore.

distribution that resulted from each settling class. Results then were rescaled according to the percent mass associated with each settling velocity category.

Predicted Exposure Field

Long-term, benthic accumulation results were based on a particulate loading of 10^6 kg of solids per day; that is, a volume of 20 000 m^3 of sludge with 5% solids. The predicted maximum flux of 100 g of solids/m^2/year caused by disposal at the midshelf site was approximately one and a half orders of magnitude larger than that at the offshelf site, 5.0 g of solids/m^2/year. However, material would be dispersed over a much larger area at the off-shelf site. The depositional area from the midshelf site would be more elongated than that for the offshelf simulation due to the relatively lower variance in the cross-shelf direction.

Contaminant flux distributions were obtained by rescaling results using contaminant specific loading rates; namely, 2.57 kg/day of PCBs and 5.26 kg/day of chlordane. Cross-shelf profiles of average contaminant flux to sediment then were calculated by averaging flux values in the along-shelf direction over a 500-km distance. Since this analysis was comparative, average cross-shelf contaminant profiles were used to portray the results from disposal at midshelf and offshelf locations.

Once the spatial distribution of solids and contaminant flux from water column to sediment surface has been estimated, the mixing due to biological activity was accounted for in the upper few centimetres of sediments. Dayal et al. [*24*] used the distribution of cesium-137 (^{137}Cs) with depth to calculate a bioturbation coefficient of 1.14 cm^2/year. However, other investigators have found sedimentary distribution of radionuclides to be consistent with lower coefficient values [*6*]. Since less biological activity is expected to occur in deep ocean sediments than on the shelf, we used a bioturbation coefficient of 1.0 cm^2/year for the midshelf site and 0.1 cm^2/year for the offshelf site, both of which are at the low end of reported values.

To calculate contaminant concentration in sediment resulting from bioturbation, an expression for distribution of a substance continuously supplied to and diffusing into a plane is used [*25*]. This expression is defined in O'Connor et al. [*6*] as follows

$$C(x,t) = Q\left[\left(\frac{t}{\pi D}\right)^{1/2} e^{-x^2/4Dt} - \frac{x}{2D}\, erfc\, \frac{x}{\sqrt{4Dt}}\right]$$

where

C = concentration of contaminant in sediment, $\mu g/cm^3$,
Q = deposition rate of contaminant, $\mu g/cm^2$/year,
t = time, year,
D = bioturbation coefficient, cm^2/year,
x = depth into sediment, cm,
$C(x,0)$ = 0, and
$erfc$ = complimentary error function [*26*].

To estimate the impacts caused by long-term disposal, we calculated both 10 and 100-year disposal periods. Contaminant concentrations were averaged over the top 2 cm of sediment for determining exposures to the benthos. Summarized in Table 3 are relative contaminant concentrations based on a 1 kg/day loading rate for midshelf and offshelf disposal after 10 and 100 years. Contaminant concentrations are rescaled accordingly, using specified loading rates. Resultant maximum sediment concentration at the midshelf site was approximately half an order of magnitude larger than that at the offshelf site. Expo-

TABLE 3—*Relative contaminant concentration based on 1 kg/day loading rate.*

	Sediment Concentration, ng/g sediment (dry weight)			
	Midshelf Disposal		Offshelf Disposal	
Distance, km	10 Years	100 Years	10 Years	100 Years
5.0	0.0	0.0	2.0×10^{-2}	1.0×10^{-1}
15.0	2.5×10^{-3}	9.6×10^{-3}	3.0×10^{-2}	1.5×10^{-1}
25.0	1.3×10^{-2}	4.8×10^{-2}	4.0×10^{-2}	2.0×10^{-1}
35.0	6.0×10^{-2}	2.3×10^{-1}	5.5×10^{-2}	2.8×10^{-1}
45.0	2.3×10^{-1}	8.7×10^{-1}	7.5×10^{-2}	3.8×10^{-1}
55.0	6.6×10^{-1}	2.5	8.5×10^{-2}	4.3×10^{-1}
65.0	1.7	6.4	8.0×10^{-2}	4.0×10^{-1}
75.0	3.5	13.4	8.5×10^{-2}	4.3×10^{-1}
85.0	3.5	13.2	1.2×10^{-1}	6.0×10^{-1}
95.0	1.6	6.2	1.4×10^{-1}	7.0×10^{-1}
105.0	7.9×10^{-1}	3.0	1.8×10^{-1}	8.8×10^{-1}
115.0	2.8×10^{-1}	1.1	2.1×10^{-1}	1.1
125.0	1.3×10^{-2}	4.8×10^{-2}	3.4×10^{-1}	1.7
135.0	2.5×10^{-3}	9.6×10^{-3}	1.0	5.2
145.0	0.0	0.0	1.1	5.5
155.0	2.5×10^{-3}	9.6×10^{-3}	9.2×10^{-1}	4.6
165.0	0.0	0.0	5.4×10^{-1}	2.7
175.0	0.0	0.0	3.9×10^{-1}	2.0
185.0	0.0	0.0	2.1×10^{-1}	1.1
195.0	0.0	0.0	1.2×10^{-1}	5.8×10^{-1}
205.0	0.0	0.0	4.5×10^{-2}	2.3×10^{-1}
215.0	0.0	0.0	2.0×10^{-2}	1.0×10^{-1}
225.0	0.0	0.0	5.0×10^{-3}	2.5×10^{-2}

sure levels for the 100-year period were approximately five times higher than for the corresponding 10-year period for each site.

Hazard Assessment

Bioaccumulation of persistent organic compounds such as PCBs and chlordane by benthic organisms is an important factor in assessing risks. An equilibrium partitioning model [*27*] currently undergoing tests at EPA/Environmental Research Laboratory—Narragansett (ERLN) estimates the potential accumulation of persistent organic compounds into tissues of infaunal organisms. An underlying assumption in this model is that equilibrium exists between lipids in the organism and organic carbon in the sediment for the organic contaminants. Thus, fugacity in organism and sediment are equal. It is assumed that no kinetic or steric barriers to the free exchange of contaminants are present. In this preference factor analysis, bioaccumulation is a redistribution of contaminants between sources, namely organic carbon in sediments and sinks, such as lipids in organisms. Thus for neutral hydrophobic organic compounds, the following equation expresses this relationship

$$\text{PF}\left(\frac{\dfrac{C^i}{\text{grams organism (dry weight)}}}{\dfrac{\text{grams lipid}}{\text{grams organism (dry weight)}}}\right) = \left(\frac{\dfrac{C^i}{\text{grams sediment (dry weight)}}}{\dfrac{\text{grams (OC)}}{\text{grams sediment (dry weight)}}}\right)$$

where PF is the preference factor, superscript i refers to contaminant, the lipid content is expressed as percentage of organism dry weight, OC is the total organic carbon in sediment expressed as percent, and C^i is the concentration in respective components.

PF describes the relative affinity of a compound for organic carbon versus lipid pool, and theoretically has a value of approximately 0.5 [*27*]. Based on the work of Lake et al. [*27*], PF values for chlorinated compounds representing both field and laboratory exposures agree well with this theoretical value and provide evidence to justify this approach for certain chlorinated organics, organisms, and exposures. The two persistent hydrophobic organic compounds, PCBs and chlordane, were chosen because of their suitability in the PF analysis and their projected mass loadings that resulted from sewage sludge disposal. This model is not applicable to contaminants that are metabolized, such as polyaromatic hydrocarbons (PAHs). The PF model is useful as a first-level screening tool for assessing advisability of ocean disposal of a given waste [*27*].

Because of their relative abundance along the continental shelf, slope, and rise, polychaetes were used in the example calculations. The percentage of lipids in polychaetes ranges from 3 to 4%, depending on spawning cycle.[5] A preference factor of 0.5 was used for both PCBs and chlordane, based on preliminary results from Lake et al. [*27*]. Sediment concentrations were converted to dry weight by assuming a sediment solids density of 2.7 g/cm^3 and a porosity of 80% [*6*]. Estimates of ambient organic carbon content at each site were based on reported data from New York City special permit application documents for ocean disposal of municipal wastes at 106-mile and 60-mile sites [*28,29*]. The organic carbon content of marine sediments often decreases with increasing grain size. For this analysis, 1% organic carbon was used for the offshelf site and 0.5% organic carbon for the midshelf site [*28,29*]. The amount of organic carbon associated with particulate flux to bottom sediments was estimated by assuming that organic carbon is conserved in the water column. It is unrealistic to assume organic carbon would be conserved in marine sediments on time scales of 100 years.[6] Therefore, the organic carbon content of sediment in the example calculations is assumed equivalent to ambient values at each site. This assumption is consistent with providing an upper-bound estimate of benthic tissue residues since in PF analysis, the higher the sediment organic carbon content, the lower the tissue residue in the organism, assuming all other values are equal.

Resultant body burdens in polychaetes versus distance offshore for PCBs and chlordane are depicted in Fig. 3. Tissue residue values in Fig. 3 were converted to wet-weight basis by assuming a wet-to-dry weight ratio of 4. Thomann suggested a wet-to-dry weight ratio of 5 for zooplankton and 4 for upper levels of the food chain [*30*]. After 100 years, a maximum chlordane concentration of 0.28 ppm was predicted in polychaetes from midshelf disposal, compared to 0.055 ppm from offshelf disposal. These values were approximately half an order of magnitude larger than the amount that the organism would accumulate in only 10 years. Note that concentration peaks occur approximately 80 km from shore for midshelf disposal, whereas peaks for offshelf disposal occur 140-km offshore. For a 10-year disposal period, tissue residues were 6.5 times higher in polychaetes impacted by midshelf disposal than those exposed by offshelf disposal. This difference drops slightly, to a factor of 5, for a 100-year disposal period.

---

[5] Rubinstein, N., U.S. Environmental Protection Agency, Environmental Research Laboratory—Narragansett, Narragansett, RI, personal communication, 1987.

[6] Bender, M., University of Rhode Island, Graduate School of Oceanography, Kingston, RI, personal communication, 1986.

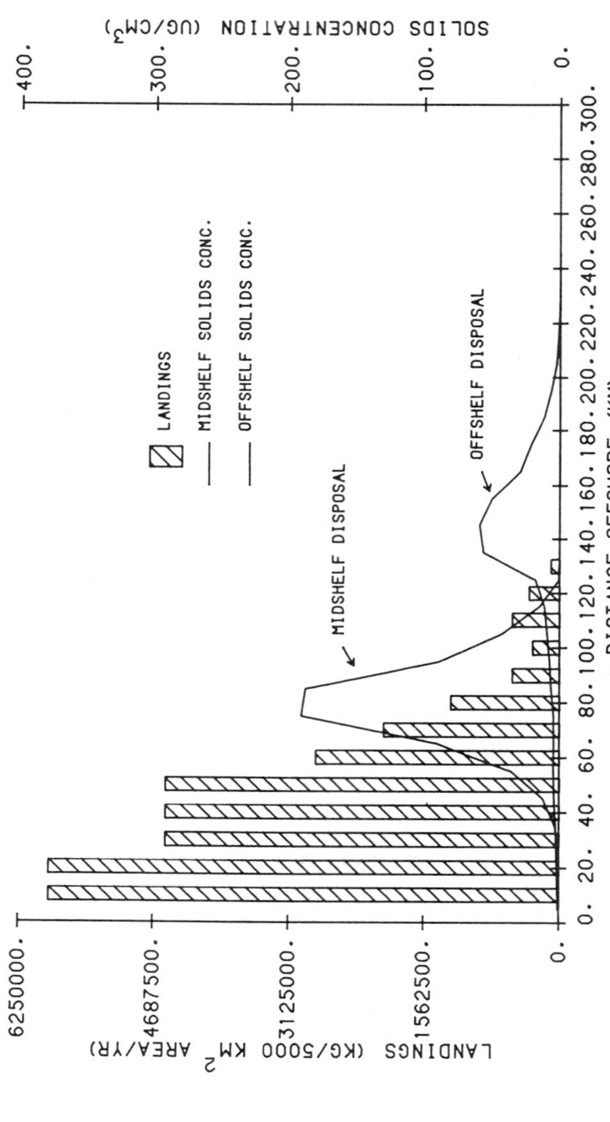

FIG. 3—*Comparison of cross-shelf solids concentration and demersal seafood landings.*

Risk Characterization: Human Health

As noted earlier, benthic infauna are vital components of the food supply for several commercially important demersal fish species. To evaluate the potential for bioaccumulation of contaminants at the two sites, a comparison of the potential impacts associated with midshelf versus offshelf disposal were made in two ways. The first is a qualitative comparison between the relative abundance of macrobenthic organisms and the projected sediment contaminant fluxes. Then, projected body burdens for each site were compared with human health end points to assess the impacts from sewage sludge disposal quantitatively.

Profiles of the onshore-offshore gradient of macrobenthic biomass and demersal seafood landings have been provided by Abt Associates Inc. in conjunction with this work [*17*]. Biomass measurements of macrobenthic organisms were used in a comparison of the two sites, since biomass is generally a good indicator of the relative amount of food available to support higher trophic levels. The relationship between biomass and depth was constructed based on data from Wigley and Theroux [*16*]. Profiles of commercial fisheries landings have been developed by Abt Associates for the mid-Atlantic continental shelf, slope, and rise, based on commercial catch data maintained by the National Marine Fisheries Service (NMFS) [*17*].

Illustrated in Fig. 4 is the onshore-offshore gradient in demersal seafood landings overlaid by the average concentration of sewage sludge solids during a ten-year period. Note that profiles for individual contaminants would have the same shape as this particulate flux profile, but would be rescaled according to the particular contaminant loading rate. As depicted in Fig. 4, particulate flux from disposal at the midshelf site would impact a higher density of demersal seafood landings than that at the offshelf site. Similar results would occur when macrobenthic biomass data in Fig. 2 are overlaid with solids concentration profiles, although landings data decline more rapidly with distance from the shore than that portrayed for macrobenthic biomass. Due to the greater number of fishes exposed to elevated solids concentrations at the midshelf site, it is inferred that there is qualitatively a higher risk associated with the disposal of sewage sludge at the midshelf site than at the offshelf site.

To determine human health risks associated with the projected level of organism contamination as illustrated in Fig. 3, the following assumptions were made: (1) polychaetes are the primary food source for commercially fished demersal species, (2) fishes from this area eat contaminated polychaetes in the cross-shelf zone of maximum exposure, and (3) fishes are assumed to equilibrate in their tissues the same concentration of PCBs and chlordane that are in the polychaetes. Three human health criteria were used in this analysis: (1) the FDA tolerance level for PCBs and the action level for chlordane, (2) the EPA reference dose (RfD) for chlordane, and (3) EPA-estimated potency factors to determine increased cancer risk levels for PCBs and chlordane.

The U.S. Food and Drug Administration set a tolerance level for PCB residues in edible portions of fishes and shellfish as 2.0 ppm on a wet weight basis. This was in response to results of studies showing that consumption of PCB-contaminated food may pose cancer risks to humans [*31*]. Based on our estimates for ten years of disposal, tissue residue levels in fish caught 80 km from shore (midshelf site) would be 0.036 ppm (wet weight), compared to a level of 0.005 ppm in fish caught 135 km offshore (offshelf site). These values increase by a factor of 5 for 100 years of disposal. Although increased levels of PCBs are associated with fishes in contact with sediment contaminated by midshelf disposal of sewage sludge, these values are still at least an order of magnitude lower than the FDA tolerance level, even after 100 years of disposal.

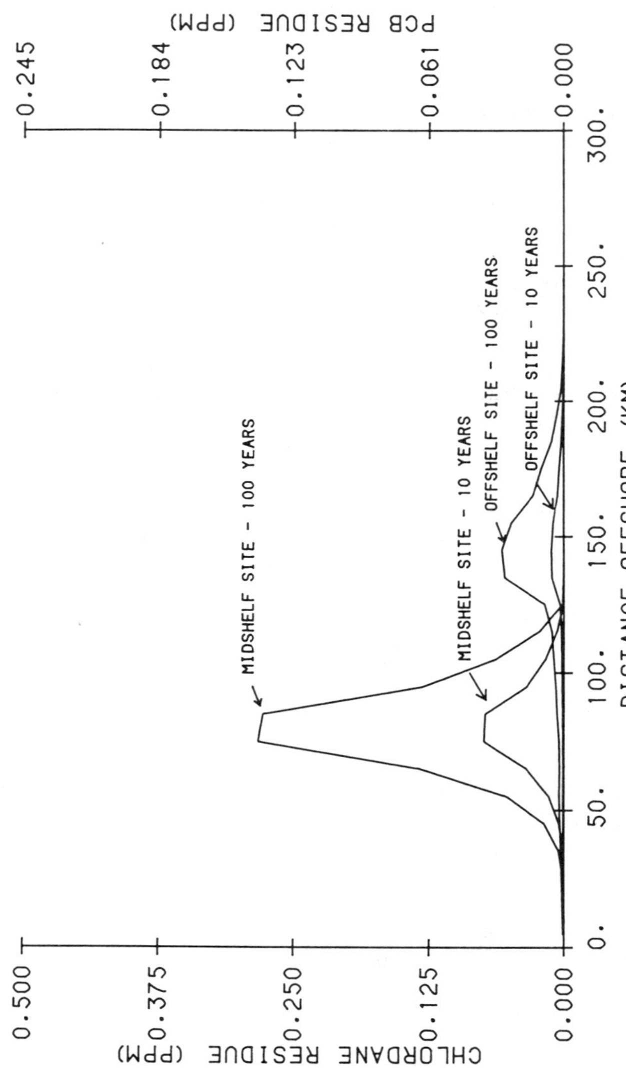

FIG. 4—*Cross-shelf profile of project tissue residues.*

Based on marketability for human consumption, the FDA action level for chlordane was set at 0.3 ppm wet weight for fishes [*32*]. Projected tissue residue levels from sewage sludge disposal at both sites were at least half an order of magnitude lower than this FDA action level, although midshelf disposal resulted in a higher value than offshelf disposal, that is 0.074 ppm versus 0.012 ppm. However, after 100 years of disposal, tissue residue levels associated with midshelf disposal approach the FDA action level; namely 0.28 ppm (midshelf). Note that offshelf disposal after 100 years would result in body burdens lower than that at the midshelf site after only 10 years. Considering that chlordane residues in fishes after 100 years of midshelf disposal would be within a calculated 6% of the FDA action level, much higher risks to human health are associated with the midshelf disposal option.

To deal with the increased need to provide risk assessment and risk management information in conjunction with ongoing EPA statutory activities, EPA has implemented the Integrated Risk Information System (IRIS) [*33*]. This system contains both toxicity and regulatory information on various chemicals. Those chemicals listed in IRIS, selected on the basis of both their toxicological properties and occurrence in the environment, are included only after a comprehensive review of chronic toxicity data by various work groups. PCBs are not yet included in IRIS because reviews are in progress.

EPA quantifies the toxicological effects of chemicals by assessing both noncarcinogenic and carcinogenic health effects. For noncarcinogenic effects, EPA uses the reference dose to determine whether an individual exposed to a particular compound is at risk. Reference dose is defined as an estimate, with uncertainty spanning approximately an order of magnitude, of the daily exposure to human population that is likely to be without an appreciable risk of deleterious effects during a lifetime [*33*]. Operationally it is derived from a no-observed-adverse-effects level (NOAEL) and an uncertainty factor which accounts for extrapolation from animals to humans. Note that reference dose is based on the assumption that thresholds may exist for certain toxic effects such as cellular necrosis, but may not exist for other toxic effects such as carcinogenicity [*33*]. The reference dose for chlordane is 5×10^{-5} mg/kg/day [*34*]. The daily dose to humans is calculated by multiplying the projected maximum body burdens in fishes by the ratio of fish consumption to human body weight. It is estimated that an individual living in the Middle Atlantic region consumes on average 16.2 g/day of fish [*35*]. Assuming that the seafood diet of a 70-kg individual consists entirely of fish from the zone of maximum exposure, the dose of chlordane to this individual would be below the RfD level at both sites after 10 years of disposal; specifically, 1.7×10^{-5} mg/kg/day (midshelf) and 2.5×10^{-6} mg/kg/day (offshelf). However, if disposal were to continue for an additional 90 years, peak concentrations at the midshelf site, 6.5×10^{-5} mg/kg/day, would exceed the RfD. Predicted exceedance would occur after 70 years of disposal at the loading rate of 5.26 kg/day of chlordane. Even after 100 years, the dose resulting from offshelf disposal, 1.27×10^{-5} mg/kg/day, would not even exceed the dose reached after only 10 years of disposal at the midshelf site (1.7×10^{-5}). Based on these results, there is a significantly increased risk associated with disposing of sewage sludge containing chlordane at midshelf as compared with the offshelf site.

For potential carcinogens, chemical levels are correlated with carcinogenic risk estimates by using a unit risk value together with the assumption of lifetime consumption of fish. The unit risk in fish tissue is defined as the increased cancer risk to a person who throughout life eats fish contaminated with 1 μg/g of a toxicant. It is calculated by multiplying a computed daily dose to man by the estimated potency factor for the contaminant. Values for fish consumption rate and human body weight used in the previous calculation are used here. The potency factor for chlordane is 1.3 per mg/kg/day [*36*]; for PCBs it is 7.7 per mg/kg/day [*37*]. Based on these values, the unit risk in fish tissue for chlordane is computed as 3.01×10^{-4}/ppm; for PCBs it is 1.78×10^{-3}/ppm. Unit risk is then multiplied by a contaminant residue level to determine the increased cancer risk level. Midshelf disposal of chlordane after 100 years resulted in an increased cancer risk level of 8.0×10^{-5} as compared with an offshelf level of 1.6×10^{-5}. For PCBs, midshelf disposal after 100 years resulted in an increased cancer risk level of 2.5×10^{-4} versus 5.0×10^{-5} for the offshelf site. Although midshelf disposal of chlordane was associated with higher human health risks compared with the FDA action level and the RfD for noncarcinogenic effects, there was a higher carcinogenic risk associated with midshelf disposal of PCBs.

Discussion

An application of the risk assessment framework presented in this paper has utility when comparing the potential risks associated with ocean disposal of sewage sludge on a relative basis. Because of uncertainties associated with the physical, chemical, and biological processes involved and with limited data, it is difficult to determine the absolute risks associated with disposing of wastes at midshelf and offshelf locations. To predict absolute human health risks, all sources of contamination, such as contaminant uptake via drinking water, air, and other food sources must be taken into account. Since this has not been done, the risk projections in this analysis should only be compared on a relative basis and may provide a low estimate of absolute risk if used alone.

It was assumed that the only source of contaminant residues in fishes was the result of feeding on benthic infauna. The potential importance of this pathway has been described in experimental systems where dietary accumulation of PCBs from a contaminated sediment by a demersal fish was the dominant exposure route [*38*]. Neither contaminant uptake from water nor biomagnification in the food chain was considered. Recent studies suggest that biomagnification may be the dominant process in accumulation of hydrophobic organic compounds in higher trophic levels [*39*]. Although the probability is unknown that a fish would obtain its food supply entirely from the highly contaminated area, it is conceivable that a fish could accumulate the estimated concentration in its tissues by eating prey with lower body burdens.

Although higher concentrations in sediment were associated with midshelf disposal operations, much of the fine-grained material deposited on the continental shelf is expected to be resuspended by episodic storms. An exception to this is in swale areas characterized by lower energy bottom environments. Elevated concentrations of contaminants may accumulate temporarily in these depositional areas. Even though swales can erode over time, residence time in these areas may be sufficient for bioaccumulation to occur. If long-term transport resulted in the redistribution of material along similar isobaths, the predicted cross-shelf distribution would be reasonable. If resuspension resulted in eventual transport of fine-grained material off the shelf, impacts may not be as localized as predicted.

Lack of site-specific data prevented us from making direct comparisons of the model results with sediment and tissue residue values from the field. The only comparable data from this region are those from the New York Bight, although loadings and site characteristics are different than the sites described in this analysis. However, to discern how these upper-bound estimates may compare with field data, values reported [*40,41*] from the New York Bight were compared with midshelf site results. Maximum cross-shelf sediment concentration for PCBs was computed as 34.4 ppb for the 100-year disposal period compared with reported [*40*] PCB levels in sediments of 50 to 150 ppb in the New York Bight apex. Chlordane concentrations in sediments of Christiaensen Basin, a known depositional zone in the New York Bight, are reported [*41*] as <10 ppb compared with the maximum cross-shelf concentration of 70 ppb for 100-years of midshelf disposal. Thus, projected upper-bound sediment PCB and chlordane concentrations are within an order of magnitude of those measured in the New York Bight. In order to put these values in perspective, one should also realize that PCBs and chlordane only have been manufactured and thus disposed of during the last few decades.

Summary

A comparative risk assessment determined potential impacts associated with marine disposal of sewage sludge at midshelf and offshelf sites. Upper-bound estimates of particulate flux to bottom sediments were determined from a layered two-dimensional model. Results showed that offshelf disposal would disperse and subsequently deposit material over a much greater area than would midshelf disposal. Cross-shelf profiles of sediment concentrations for PCBs and chlordane were at least half an order of magnitude larger in the vicinity of the midshelf site compared with the offshelf site. Although relatively low concentrations of contaminants were estimated to accumulate in tissues of benthic organisms, potential impacts from midshelf disposal may be higher because much more biomass would be exposed due to midshelf disposal. Calculations showed that concentrations of PCBs at both sites would be below FDA tolerance levels; however, chlordane residues were within a calculated 6% of the FDA action level after 100 years of midshelf disposal. After 70 years of midshelf disposal, the predicted chlordane dose would exceed the RfD. Higher carcinogenic risk levels were associated with midshelf disposal of PCBs. Thus on a comparative basis, human health impacts were predicted to be higher as the result of midshelf disposal of sludges compared with offshelf disposal.

This analysis provides a conservative upper-bound estimate of potential benthic impacts as measured by bioaccumulation resulting from the disposal of municipal wastes at midshelf and offshelf sites. Methods presented in this analysis can be used in a comparative risk assessment framework. However, these methods are still in their developmental phase and must undergo further testing and validation before they can become quantitative tools to predict risks associated with the marine disposal of wastes. To test and validate these tools, a comprehensive biomonitoring program on a regional scale should be implemented to determine actual contaminant concentrations that are accumulated in tissues of both demersal fish species and benthic organisms.

Acknowledgments

We would like to thank Peter Hamilton at Science Applications International Corp. for providing us with current meter statistics data from the MASAR/POS study, and James Lake and Norman Rubinstein for their assistance with preference factor analysis work. Special thanks to John Gentile, John Scott, and Jan Prager for their critical review of the manuscript. J. Nocito was supported under Contract No. 68-03-3236 to Science Applications International Corp., Allen D. Beck, Project Officer, and C. Menzie under Contract No. 86-60 to Abt Associates Inc., Richard Wells, Project Director. This report has been reviewed by the U.S. Environmental Protection Agency, Environmental Research Laboratory at Narragansett, Rhode Island, and approved for publication. Approval does not signify that the contents necessarily reflect the views and policies of EPA, nor does the mention of trade names or commercial products constitute endorsement or recommendation for use. This is Contribution No. 848 of the Environmental Research Laboratory—Narragansett.

References

[*1*] Code of Federal Regulations, Title 40, *Protection of Environment,* Environmental Protection Agency, Washington, DC, chapter 1, subchapter H—Ocean Dumping.

[*2*] U.S. Environmental Protection Agency, "Ocean Dumping: Final Designation of Site," *Federal Register,* Vol. 49, 1984, pp. 19005–19012.

[*3*] Santoro, E. D. and Suszkowski, D. J., "Current Status—Phase Out of Ocean Dumping of Sewage Sludge in the New York Bight Apex," International Ocean Disposal Symposium, Asilomar Conference Center, Pacific Grove, CA, 18–21 April 1986.

[*4*] Paul, J. F., Bierman, V. J., Jr., Walker, H. A., and Gentile, J. H., "Application of a Hazard Assessment Research Strategy for Waste Disposal at the 106-Mile Ocean Disposal Site," *Oceanic Processes in Marine Pollution,* D. Hood and A. Schoener, Eds., Vol. 4, Scientific Monitoring Strategy for Ocean Waste Disposal, Kreiger, Malabar, FL, 1987.

[*5*] Walker, H. A., Paul, J. F., and Bierman, V. J., Jr., "Methods for Waste Load Allocation of Municipal Sewage Sludge at the 106-Mile Ocean Disposal Site," *Environmental Toxicology and Chemistry,* Vol. 6. No. 6, 1987.

[*6*] O'Connor, T. P., Okubo, A., Champ, M. A., and Park, P. K., "Projected Consequences of Dumping Sewage Sludge at Deep Ocean Sites Near New York Bight," *Canadian Journal of Fisheries and Aquatic Sciences,* Vol. 40, Supplement 2, 1983, pp. 228–241.

[*7*] Partch, E. N., "Estimation of the Dispersal of Sinking Particles from Deep Water Dumpsite 106 Using Site D Current Meter Records," Marine Sciences Research Center, State University of New York at Stony Brook, Stony Brook, NY, 1985.

[*8*] Bierman, V. J., Jr., Gentile, J. H., Paul, J. F., Miller, D. C., and Brungs, W. A., "Research Strategy for Ocean Disposal: Conceptual Framework and Case Study," *Environmental Hazard Assessment of Effluents,* H. Bergman, R. A. Kimerle, and A. W. Maki, Eds., 1986; *Proceedings,* Pellston Environmental Workshop, Cody, WY, 22–27 Aug. 1982. Pergamon Press Inc.

[*9*] Science Applications International Corp., "Mid-Atlantic Slope and Rise Physical Oceanography Study (MASAR/POS)," Semiannual Report No. 3, MMS Contract No. 14-12-0001-30066, Minerals Management Service, Vienna, VA, 1986.

[*10*] Ingham, M. C., Ed., "Summary of the Physical Oceanographic Processes and Features Pertinent to Pollution Distribution in the Coastal and Offshore Waters of the Northeast United States, Virginia to Maine," NOAA Technical Memorandum NMFS-F/NEC-17, National Oceanic and Atmospheric Administration, Washington, DC, 1982.

[*11*] O'Connor, T. P., "106-Mile Site Sediment Characteristics," *106-Mile Site Characterization Update,* J. B. Pearce, D. C. Miller, and C. Berman, Eds., NMFS-F/NEC-26, NOAA Technical Memorandum, National Oceanic and Atmospheric Administration, Washington, DC, 1983, Chapter 3.

[*12*] Butman, B., Noble, M., and Folger, D. W., "Long-Term Observations of Bottom Current and Bottom Sediment Movement on the Middle Atlantic Continental Shelf," *Middle Atlantic Outer Continental Shelf Environmental Studies,* D. W. Folger, Ed., Vol. 3, Geological Studies, Virginia Institute of Marine Science, Gloucester Point, VA, 1977, pp. 2–1–2–61.

[*13*] Swift, D. J. P., Greeland, F. L., Gadd, P. E., Han, G., Lavelle, J. W., and Stubblefield, W. L., "Morphologic Evolution and Coastal Sand Transport, New York–New Jersey Shelf," Special Symposium on the Middle Atlantic Continental Shelf and the New York Bight, M. G. Gross, Ed., American Society of Limnology and Oceanography, Lawrence, KS, 1976, pp. 69–89.

[*14*] Heezen, D. C,. "Photographic Reconnaissance of Continental Slope and Upper Continental Rise," May 1974 Baseline Investigation of Deepwater Dumpsite 106, NOAA Dumpsite Evaluation Report 75-1, U.S. Department of Commerce, National Oceanic and Atmospheric Administration, Rockville, MD, 1975, pp. 27–103.

[*15*] Steimle, F. W. and McNulty, J. K., "Benthic Fauna," *106-Mile Site Characterization Update,* NMFS-F/NEC-26, J. B. Pearce, D. C. Miller, and C. Berman, Eds., National Oceanic and Atmospheric Administration, Washington, DC, 1983, Chapter 8.

[*16*] Wigley, R. L. and Theroux, R. B., "Atlantic Continental Shelf and Slope of the United States—Macrobenthic Invertebrate Fauna of the Middle Atlantic Bight Region—Faunal Composition and Quantitative Distribution," Geological Survey Professional Paper 529-N, United States Geological Survey, Denver, CO, 1981.

[*17*] "Ocean Disposal of Municipal Sewage Sludge: A Comparative Analysis of Mid-Shelf and Deep Ocean Dumpsites," AAI 86-60, Abt Associates Inc., prepared for the U.S. Environmental Protection Agency, Office of Policy Analysis, Washington, DC, 1986.

[18] Rowe, G. T. and Staresinic, N., "Sources of Organic Matter to the Deep-Sea Benthos," Ambio Special Report, No. 6, Royal Swedish Academy of Science, Pergamon Press, Stockholm, Sweden, 1979, pp. 19–23.

[19] Thiel, H, "Structural Aspects of the Deep-Sea Benthos," Ambio Special Report No. 6, Royal Swedish Academy of Science, Pergamon Press, Stockholm, Sweden, 1979. pp. 19–23.

[20] "A Comparison of Studies of Toxic Substances in POTW Sludges," U.S. Environmental Protection Agency, Office of Water Regulations and Standards, Criteria and Standards Division, Washington, DC, 1984.

[21] "A Special Permit Application for the Disposal of Sewage Sludge from Twelve New York City Water Pollution Control Plants at the 12-Mile Site," New York City Department of Environmental Protection, prepared by Ecological Analysts, Inc., Sparks, MD, and SEAMOcean, Inc., Wheaton, MD, 1983.

[22] Faisst, W. K., "Characterization of Particles in Digested Sewage Sludge," *Particles in Water,* M. C. Kavanaugh and J. O. Leckie, Ed., ACS Advances in Chemistry, Series No. 189, American Chemical Society, Washington, DC, 1981, pp. 259–282.

[23] Taylor, G. I., "Diffusion by Continuous Movements," *Proceedings of the London Mathematical Society,* Series A, Vol. 20, 1921, pp. 196–211.

[24] Dayal, R., Okubo, A., Duedall, I. W., and Ramamoorthy, A., "Radionuclide Redistribution Mechanisms at the 2800-m Atlantic Nuclear Waste Disposal Site," *Deep-Sea Research,* Vol. 26A, 1979, pp. 1329–1345.

[25] Carslaw, M. S. and Jaeger, J. C., *Conduction of Heat in Solids,* 2nd ed., Oxford University Press, London, 1959.

[26] Fischer, H. B., List, E. J., Koh, R. C. Y., Imberger, J., and Brooks, N. H., *Mixing in Inland and Coastal Waters,* Academic Press, New York, 1979, p. 40.

[27] Lake, J. L., Rubinstein, N., and Pavignano, S., "Predicting Bioaccumulation: Development of a Simple Partitioning Model for Use as a Screening Tool for Regulating Ocean Disposal of Wastes," *Proceedings,* Sixth Pellston Workshop, Florissant, CO, 12–17 Aug. 1984.

[28] "Technical Information to Support the Redesignation of the 60-Mile Site for the Ocean Disposal of Municipal Sewage Sludge," New York City Department of Environmental Protection, prepared by Ecological Analysts, Inc., Sparks, MD and SEAMOcean, Inc., Wheaton, MD, 1983.

[29] "Technical Information to Support the Redesignation of the 106-mile Site for the Ocean Disposal of Municipal Sewage Sludge," New York City Department of Environmental Protection, prepared by Ecological Analysts, Inc., Sparks, MD and SEAMOcean, Inc., Wheaton, MD, 1983.

[30] Thomann, R. V., "Equilibrium Model of Fate of Microcontaminants in Diverse Aquatic Food Chains, *Canadian Journal of Fisheries and Aquatic Sciences,* Vol. 38, 1980, pp. 280–296.

[31] U.S. Food and Drug Administration, "Polychlorinated Biphenyls Decision," *Federal Register,* Vol. 49, 1984, pp. 21511–21520.

[32] "Compliance Policy Guide 7141.01," Attachment B3, U.S. Food and Drug Administration, Washington, DC, 31 July 1979.

[33] "Integrated Risk Information System Supportive Documentation," EPA/600/8-86/032a, Office of Health and Environmental Assessment, U.S. Environmental Protection Agency, Washington, DC, 1987.

[34] "Integrated Risk Information System (IRIS): Reference Dose (RfD) for Oral Exposure for Chlordane," Office of Health and Environmental Assessment, U.S. Environmental Protection Agency, Environmental Criteria and Assessment Office, Cincinnati, OH, 1985.

[35] "Seafood Consumption Data Analysis: Final Report," prepared by SRI International, Menlo Park, CA under Contract No. 68-01-3887, U.S. Environmental Protection Agency, Washington, DC, 1980.

[36] "Integrated Risk Information System (IRIS): Risk Estimate for Carcinogenicity for Chlordane," Office of Health and Environmental Assessment, U.S. Environmental Protection Agency, Environmental Criteria and Assessment Office, Cincinnati, OH, 1987.

[37] "Integrated Risk Information System (IRIS): Risk Estimate for Carcinogenicity for PCBs," Office of Health and Environmental Assessment, U.S. Environmental Protection Agency, Environmental Criteria and Assessment Office, Cincinnati, OH, 1987.

[38] Rubinstein, N. I., Gilliam, W. T., and Gregory, N. R., "Dietary Accumulation of PCBs from a Contaminated Sediment Source by a Demersal Fish *(Leiostomus xanthurus),*" *Aquatic Toxicology,* Vol. 5, 1984, pp. 331–342.

[39] Connolly, J. P. and Pederson, C. J., "Thermodynamic Based Evaluation of Organic Chemical Accumulation," *Environmental Science and Technology,* Vol. 22, 1988, pp. 99–103.

[*40*] "Contaminants in New York Bight and Long Island Sound Sediments and Demersal Species, and Contaminant Effects on Benthos, Summer 1980," R. Reid, J. O'Reilly, and V. Zdanowicz, Eds, NOAA Technical Memorandum NMFS-F/NEC-16, National Oceanic and Atmospheric Administration, Washington, DC, 1982.

[*41*] MacLeod, W. D., Jr., Ramos, L. S., Friedman, A. J., Burrows, D. G., Prohaska, P. G., Fisher, D. L., and Brown, D. W., "Analysis of Residual Chlorinated Hydrocarbons, Aromatic Hydrocarbons and Related Compounds in Selected Sources, Sinks, and Biota of the New York Bight," NOAA Technical Memorandum OMPA-6, National Oceanic and Atmospheric Administration, Washington, DC, 1981.

Effluent Toxicity

Jon Doi[1] *and Donald R. Grothe*[1]

Use of Fractionation and Chemical Analysis Schemes for Plant Effluent Toxicity Evaluations

REFERENCE: Doi, J. and Grothe, D. R., "**Use of Fractionation and Chemical Analysis Schemes for Plant Effluent Toxicity Evaluations,**" *Aquatic Toxicology and Environmental Fate: Eleventh Volume, ASTM STP 1007,* G. W. Suter II and M. A. Lewis, Eds., American Society for Testing and Materials, Philadelphia, 1989, pp. 123–138.

ABSTRACT: A new field of analysis has arisen for the evaluation of wastewater samples in response to forthcoming regulations. Instead of using current methodology to determine if target compounds are present in environmental water samples (for example, priority pollutant analysis), Monsanto's Environmental Sciences Center (ESC) has developed a scheme to identify compounds that are either toxic to aquatic organisms or major sources of total organic carbon in wastewaters. This procedure may be necessary with the upcoming toxicity reduction evaluation (TRE) program sponsored by the United States Environmental Protection Agency (USEPA). The ESC scheme uses simple column chromatography techniques to initially fractionate the wastewater into separate inorganic and organic constituents. Each fraction is evaluated for toxicity to a common test organism, *Daphnia magna.* The fraction containing the toxicity is then analyzed to identify the sources of toxicity. Several case studies are presented to show how the ESC fractionation and chemical analysis scheme has been used to isolate and identify the factors responsible for wastewater toxicity.

KEY WORDS: chemical fractionation, toxicity identification, effluent toxicity, toxicity reduction evaluation, aquatic toxicology

In 1984, the United States Environmental Protection Agency (EPA) issued its Policy for the Development of Water Quality-Based Toxics Control. The policy, which is being implemented through the National Pollutant Discharge Elimination System (NPDES) permit program, integrates traditional chemical analyses and whole effluent toxicity test methods to assess and control the release of toxic chemicals into the nation's waters. Procedural documents [*1,2*] have also been issued which provide recommendations for implementing EPA's strategy for toxic substances.

An important aspect of EPA's policy is the use of effluent toxicity tests to determine if a discharge is likely to have any adverse effect on representative species in the aquatic environment. The EPA has recommended the concept of tier testing, that is, collecting data in sequential stages, as a cost-effective method for generating the data necessary to assess impact. Based upon the relationship between toxicological effects data and environmental exposure, decisions can be made regarding a discharge's likely impact on the aquatic environment. In situations where a discharge is likely to be environmentally unacceptable, field confirmation studies, toxicity identification or reduction procedures or a combination of these may have to be implemented to better understand or improve effluent quality.

[1] Senior research chemist and senior environmental specialist, respectively, Monsanto Co., Environmental Sciences Center, St. Louis, MO 63167.

Within recent years, several protocols have been developed for isolating and identifying toxic components in complex waste mixtures [*3–6*]. Walsh and Garnas [*3*] have developed a procedure which utilizes various exchange resins to separate an effluent into organic and inorganic fractions. If either fraction exhibits toxicity, subfractionation procedures are implemented using ion exchange resins for inorganics and extraction under basic and acidic conditions for organics. Chemical analyses and toxicity tests are conducted on each fraction or subfraction to determine if the toxic component can be identified.

The purpose of this paper is to present an approach for isolating and identifying the factors responsible for effluent toxicity. The approach, which is based upon the concepts of Walsh and Garnas, centers on the use of column chromatographic techniques (that is, cation and anion exchange resins, activated carbon, and silica gel) to separate a complex mixture into organic and inorganic fractions. Each fraction is then evaluated for its acute toxicity to an aquatic organism and analyzed by the most appropriate analytical technique. If the toxicity lies in the organic fraction, another organic separation scheme is used to isolate the toxic components. An additional parameter that is monitored is the amount of total organic carbon (TOC) in the various fractions. In this way, the quantity of organics as well as the toxicity of each fraction is being tracked through the fractionation scheme. The goal of these protocols is to identify the toxic components by answering the question of what the toxic compounds in the water are rather than whether there are particular compounds in the water. Three case studies are presented to illustrate how the protocols, coupled with a knowledge of aquatic toxicology and chemistry, can be used to successfully identify the toxic components in complex waste streams.

Materials and Methods

Phase I Toxicity Evaluation Scheme

The chemical fractionation scheme used to isolate fractions containing compounds responsible for the toxicity of environmental water samples is shown in Fig. 1. The treatments can be used sequentially, as shown in Fig. 1, or individually. By running the treatments in series, different components are taken out sequentially, and the wastewater should not be toxic at the end of the fractionation procedure. By using the different treatments individually, a better understanding of what kinds of compounds are removed can be obtained.

Cation (Bio-Rad AG 50W-X8, 20 to 50 mesh, hydrogen form, strongly acidic) and anion (Bio-Rad AG 1-X8, 20 to 50 mesh, hydroxide form, strongly basic) exchange resins are used to separate out the inorganic fraction of the wastewater sample. *Daphnia magna* acute toxicity tests are performed after each ion exchange resin to determine which inorganic fraction, if either, is causing the toxicity.

The original, "natural" effluent is also put through a granular activated carbon (GAC) (Nuchar WV-G, 12 to 40 mesh) column, followed by a silica gel (Fisher chromatographic grade, 60 to 200 mesh) column. In general, GAC removes the less polar organics, while silica gel removes the more polar organic compounds. The ion exchange columns are put at the end of the organic removal side of the protocol (right side of Fig. 1) to remove the rest of the toxicity if it is due to both organic and inorganic compounds.

Toxicity tests are performed after each resin treatment. TOC and ICP (inductively coupled plasma) measurements are taken for the evaluation of organic and inorganic materials, respectively, when the bioassay tests indicate that the toxicity likely lies in either fraction. From the analytical results, a decision is made regarding whether the toxicity lies in the organic, inorganic, or both fractions. If the toxicity lies in the organic fraction, another protocol is used to further fractionate the sample and attempt to identify the source or

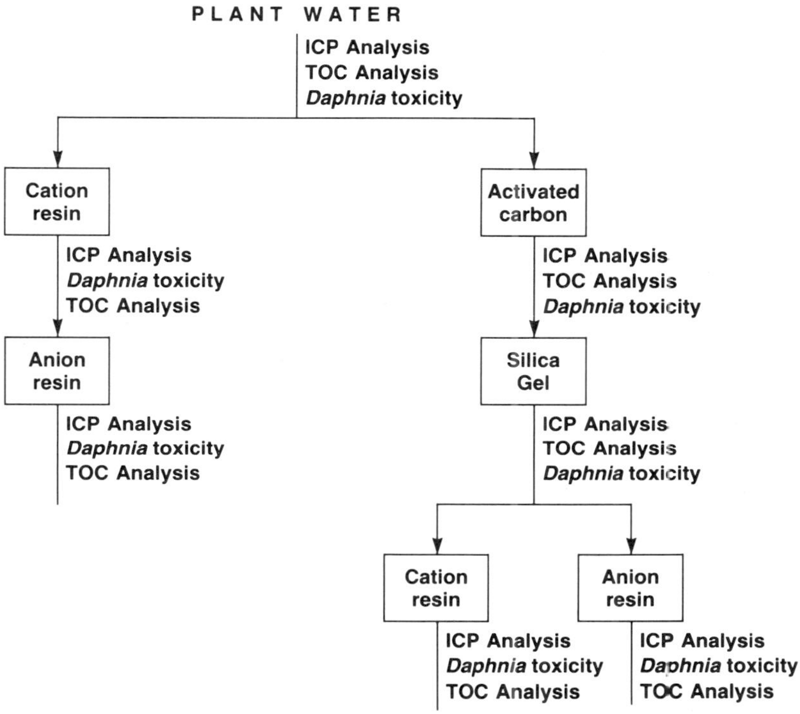

FIG. 1—*Phase I testing strategy for the evaluation of toxicity in wastewater samples.*

sources of toxicity (see Figs. 2 and 3). If the toxicity lies in the inorganic fraction, a determination can usually be made on what is causing the toxicity.

Additional insight on the sources of toxicity is obtained from synthetic effluent tests. The synthetic effluent is composed solely of the suspected compounds in a matrix as similar to the natural effluent as possible (for example, effluent with the toxicity removed, plant intake water, and so forth) and at the same concentration as in the natural undiluted effluent. Acute toxicity tests are performed on the synthetic wastewater and compared with the toxicity of the natural effluent. While the synthetic effluent does not contain all the chemical constituents that could be in the original wastewater, if both toxicity tests show comparable results, this is strong evidence that the suspected compounds are indeed the actual toxicants. However, when utilizing synthetic effluents, it should be recognized that differences in the water quality characteristics of the natural versus the synthetic effluent could effect correlations between the test results of the two wastewaters. The definitive test would be to remove the suspected compounds from the plant process and retest the natural effluent. If the wastewater shows no toxicity, then the suspected compounds are the actual toxicants. This has been done in one case, which will be described later (see Case Study 2).

Organic Compound Fractionation and Characterization Scheme

If the Phase I testing procedure indicates that the toxic compounds are organic, then the wastewater is put through the ESC (Environmental Sciences Center) scheme, shown in

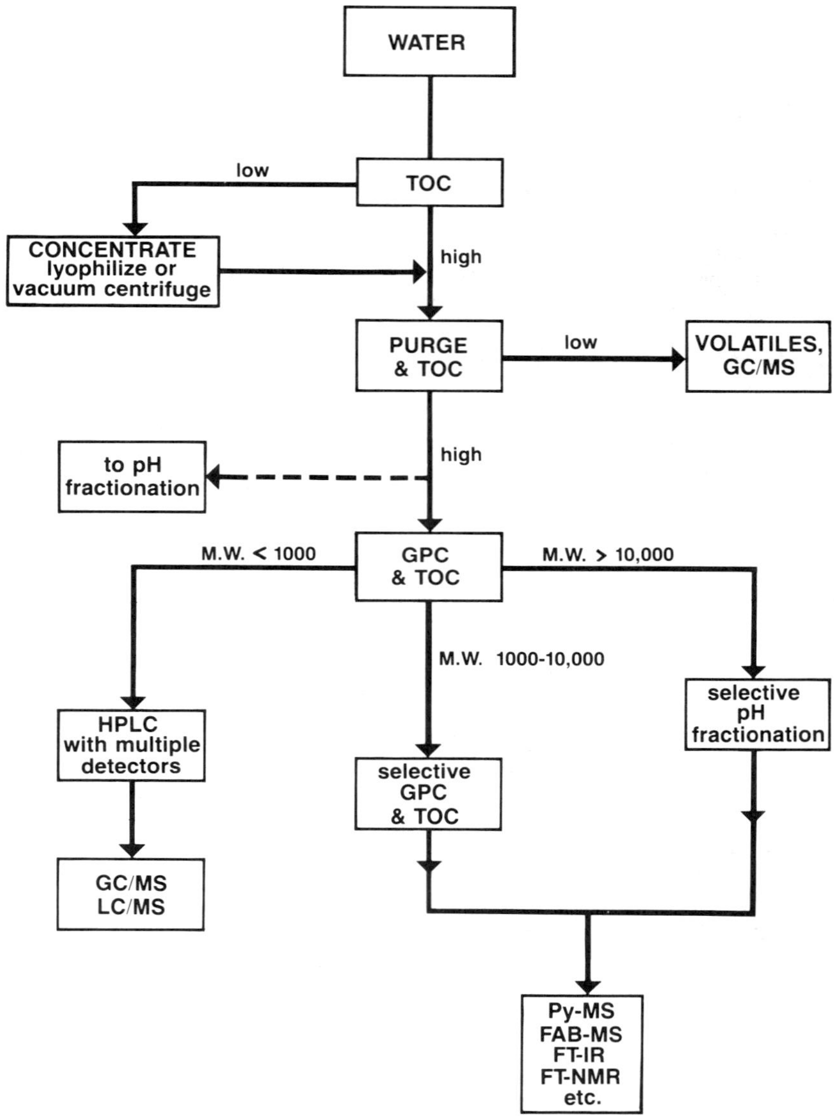

FIG. 2—*General scheme for the fractionation and analysis of organic compounds in environmental water samples.*

Figs. 2 and 3. This procedure can also be used if the wastewater needs to be chemically characterized, independent of toxicity testing. This situation could arise if one wanted to know the major organic components of the effluent. In either case, the ESC protocol is based on toxicity and TOC measurements, which allows the investigator to mass balance both the toxicity (in toxicity units) and the organic material (in ppm carbon) as the effluent is taken through the fractionation procedure. It is important to note that aquatic toxicity

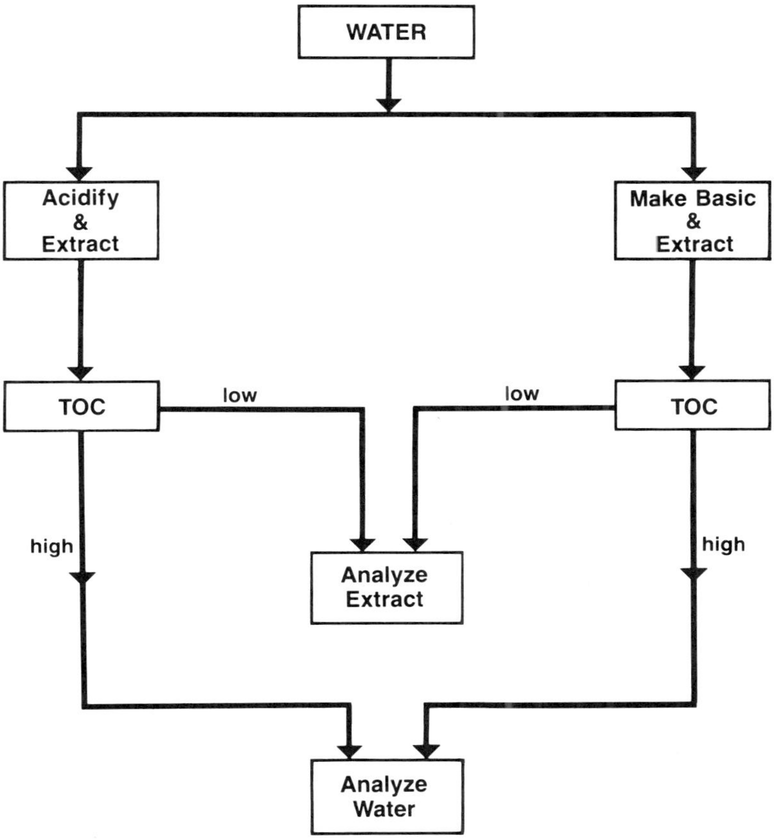

FIG. 3—*Scheme for the pH fractionation of extractable organic compounds in environmental water samples.*

and TOC measurements do not necessarily track together even though the toxicants are organic *if* the organic compounds are toxic at very low levels [that is, low ppb to high parts per trillion (ppt)]. Since TOC can only be routinely measured down to 100 ppb, trace toxic compounds will not be tracked by TOC measurements. The TOC measurements are taken on a Dohrmann DC-80 or DC-180 TOC analyzer. The toxicity of each fraction can be tracked using aquatic organisms such as *D. magna.* These analytical and toxicological measurements provide not only information on treatability, but also the extraction and elution efficiency of the columns. The fractionation procedure is described below:

1. See Fig. 2. A TOC measurement is taken of the natural undiluted wastewater sample. If the TOC reading is low (that is, less than a few parts per million carbon), the organics are concentrated by removing the water. It is important to note that any volatile organic compounds will be lost in this concentration step. Also, it is possible that some compounds may chemically change from the concentration procedure. It is important to carry out proper quality assurance/quality control (QA/QC) procedures, with blanks and spikes during the fractionation procedure.

2. After the concentration step (if necessary), the sample is purged with helium, and the TOC is again measured. If organic carbon has been purged from the water, then the purgeable organic carbon (POC) is measured by purge-and-trap gas chromatography/mass spectrometry (GC/MS).

3. The purged water sample is then subjected to pH fractionation (see Fig. 3). In this step, the sample is acidified with sulfuric acid (1:1 concentrated $H_2SO_4 : H_2O$) [American Chemical Society (ACS) grade] to a pH of 2 and extracted with an organic solvent that can be easily purged from water, such as dichloromethane or hexane [Burdick and Jackson, ultraviolet (UV) grade]. Then the water is made basic with sodium hydroxide (10 *N* NaOH in Milli-Q water) (ACS grade) to a pH of 11 and extracted again with an organic solvent. After each extraction, the water is sparged with helium to remove the residual organic solvent, and TOC measurements are taken. If a significant amount of the organic carbon lies in the acid/neutral or base extractables fraction, then wide-scan GC/MS analysis is used to characterize the extractables.

4. The organics that remain are considered nonvolatile organics (NVO), that is, compounds that cannot be purged or extracted. At this point (see Fig. 2), gel permeation chromatography (GPC) is used to fractionate the NVOs by separating the compounds according to molecular size [Perkin Elmer Series 4 high-performance liquid chromatography (HPLC) system, Perkin Elmer Chromatographics 4 Data System, Kratos 783 variable wavelength UV-visible detector or the Varian 9060 photodiode UV array detector and Waters Zorbax PSM trimodal size exclusion GPC columns]. Three broad molecular weight (MW) ranges are separated.

5. The low-MW fraction is further fractionated and characterized by HPLC, using a variety of analytical detectors (for example, UV photodiode array, refractive index, fluorescence, electrochemical, and so forth). More sophisticated characterization techniques are available if needed [for example, liquid chromatography (LC)/MS, LC/FT-IR (Fourier transform infrared), supercritical fluid chromatography (SFC)/MS, and so forth].

6. The mid-MW fraction can be further separated into more defined MW ranges by GPC columns of higher selectivity (Waters Zorbax PSM 60 or 100 GPC columns). Characterization techniques such as pyrolysis MS (Py-MS), fast atom bombardment MS (FAB-MS), FT-IR, SFC/FT-IR/MS, FT nuclear magnetic resonance (FT-NMR) can be used in an attempt to characterize these organic compounds.

7. Little can be done with high-MW compounds, although gross separations into acidic, neutral, and basic fractions can be performed. One can then use the same analytical techniques as described above in steps 5 and 6.

8. Blank and control experiments (using Milli-Q and well water) are carried out at each step of the fractionation procedure to ensure that sample manipulation does not introduce either artifactual TOC or toxicity into the protocol.

It should be noted that TOC measurements are taken at each step in the fractionation and characterization scheme. When this procedure is used to identify the toxic organic components in a wastewater sample, then acute toxicity tests are also done at each step of the scheme. Only the fractions of interest are analyzed further. If the toxicity does not lie in the purgeables or extractables fraction, then only the nonvolatile organics are investigated further.

Procedure

The following case studies illustrate how ESCs Phase I fractionation scheme was used to identify the factors responsible for toxicity in several complex effluents.

Case Studies

Effluents from three chemical manufacturing plant sites were used to evaluate the utility of the toxicity evaluation scheme for isolating toxic components in complex waste mixtures. The effluents were selected for the study based upon the results of *D. magna* acute toxicity tests which indicated that the median effective concentration (EC_{50}) values for the effluents were typically <100%. Ten litres of effluent were collected from each plant site over a 24-h period and shipped on ice via an overnight courier to Monsanto's Environmental Sciences Center in St. Louis. All samples were maintained in the laboratory at 4°C prior to fractionation and analysis. Existing chemical and biological effects data for each wastewater or a knowledge of plant processes or both were utilized to provide information on potential toxicants in the wastewater. A brief description of each effluent as well as the procedures that were utilized to isolate and identify the toxic components are presented below.

Case Study 1—The effluent evaluated in this study was a relatively complex waste mixture, containing a variety of inorganic and organic constituents. One of the characteristic features of the wastewater was its relatively high ammonia content (100 to 300 mg/L). Since correlations typically existed between toxicity and ammonia levels throughout the year, it was speculated that ammonia was a factor responsible for the wastewater's toxicity to *D. magna*. However, studies by various investigators failed to confirm ammonia as the toxic component.

In order to determine if the toxic components in the effluent could be isolated, chemical fractionation tests were initiated based upon the Phase I testing scheme. Total organic carbon, inorganics, total ammonia, and *D. magna* acute toxicity were determined for each effluent fraction. A Dohrmann Model DC-80 TOC Analyzer, an inductively coupled argon plasma atomic emission spectrometer (ICP-AES) and an ammonia-selective electrode were utilized to measure total organic carbon, metals, and total ammonia, respectively. *Daphnia* tests were conducted in duplicate based upon procedures presented in EPA's "Methods for Measuring the Acute Toxicity of Effluents to Freshwater and Marine Aquatic Organisms" [*7*]. The undiluted effluent was set up in duplicate for each fractionation test. Ten first instar daphnids (<24 h old) and 100 mL of test solution were utilized for each replicate sample.

Due to the fact that little information could be found in the published literature regarding the acute toxicity of ammonia to *D. magna*, an acute toxicity test was conducted with ammonium chloride to define the acute effect level. The results of the test were utilized to determine if a correlation existed between effluent toxicity and unionized ammonia levels. A 300 mg/L ammonium chloride stock solution was utilized as the stock solution for the ammonia test. Six effluent concentrations and a control were set up using well water obtained from eastern Missouri as dilution water. The pH of the stock solution and dilution water were 8.0 and 8.6, respectively. The water had a total hardness of 160 mg/L. Test procedures for the acute tests were based upon the previously mentioned acute toxicity procedures. The test temperature was 21°C. The 48-h EC_{50} and 95% confidence limits were calculated by a computer program for calculating median lethal concentration (LC_{50}) values.[2] Unionized ammonia levels were computed from measured total ammonia concentrations using procedures described by Emerson et al. [*8*].

---

[2] Stephan, C. E., "Computer Programs for Calculating the LC_{50} and 95% Confidence Interval," EPA-600/4-85-013, Appendix E, U.S. Environmental Protection Agency, Environmental Research Laboratory—Duluth, Duluth, MN, March 1985.

Case Study 2—The wastewater evaluated in this case study theoretically had few chemical constituents since it was comprised of condenser/cooling water blowdown and storm water runoff. The wastewater received no treatment prior to discharge to the receiving stream. Major chemical constituents known to be present in the waste stream are hexavalent chromium, chlorine, and a biocide, which were utilized by the manufacturing facility as inhibitors of microbial growth, algae, and corrosion. The concentrations of these compounds in the wastewater are relatively constant throughout the year.

During on-site toxicological testing of the waste stream in 1985, the undiluted effluent was found to be acutely toxic to *D. magna.* Chemical analysis of the wastewater indicated that hexavalent chromium (100 to 200 ppb) was at or near 48-h EC_{50} values reported in the literature for *D. magna* (20 to 212 ppb). The acute toxicity of the biocide to *Daphnia* was unknown. In order to determine if metals could be contributing to effluent toxicity, a chelator [1 mg/L sodium nitrilotriacetic acid (NTA)] was added to a subsample of the effluent to bind with the metals and make them unavailable for uptake. Sodium thiosulfate (1.0 mg/L) and sodium sulfite (4.2 mg/L) were also added to subsamples of the effluent to neutralize any chlorine and biocide that may have been present in the effluent. The concentration of chlorine and biocide were determined before and after chemical neutralization via amperometric titration and liquid chromatography, respectively. Each effluent sample was evaluated for its acute toxicity to *D. magna* to determine if chelation or chemical neutralization had any effect in reducing the toxicity of the effluent.

Due to the fact that the waste stream being investigated was comprised of four storm water and cooling water inputs, *Daphnia* toxicity tests were also conducted on water samples collected from various locations along the wastewater drainage system. The results of the *D. magna* tests were used to determine if toxicity could be traced to a particular water source. The *Daphnia* acute tests were conducted as previously described for Case Study 1. ICP, TOC, and hexavalent chromium analyses were conducted on each of the water samples to determine if a correlation existed between toxicity and a specific chemical constituent.

Chemical fractionation tests were also initiated with the wastewater. Three resins were utilized in the fractionation procedure: granular activated carbon, a cation exchange resin, and an anion exchange resin. Separate 200 mL subsamples of the effluent were treated with each resin and evaluated for acute toxicity to *Daphnia.* Hexavalent chromium concentrations were measured in the effluent before and after each treatment to determine if there was a correlation between toxicity and chromium levels.

Case Study 3—The effluent evaluated in this study is chemically complex, containing a variety of organic and inorganic constituents. One of the characteristic features of the effluent is its very high conductivity, which ranges from approximately 10 000 to 25 000 micromhos/cm. The major components contributing to the high conductivity are sodium (1020 mg/L), calcium (3000 mg/L), and chloride (7310 mg/L) ions. Since the mid-1970s, effluent toxicity tests have been periodically conducted on the effluent to evaluate its effect on the fathead minnow. The tests, which were conducted for both in-house and regulatory compliance monitoring purposes, indicated that the 96-h LC_{50} value typically fell within the range of approximately 45 to 80%. The fathead minnow has generally been used as the test species due to regional regulatory test requirements. Because the effluent is very saline, it has been speculated by a number of investigators that salinity could either be directly responsible for effluent toxicity or indirectly responsible by stressing the organism during the test.

In 1981, a study was initiated to determine if salinity could be responsible for effluent toxicity. The study focused on evaluating the acute toxicity of two effluent samples. The first effluent, which was designated as "natural effluent," was obtained from the plant's discharge sampling point. The second effluent, designated as "synthetic effluent," was prepared by adding sodium chloride and calcium chloride to well water obtained from eastern Missouri so that the concentration of sodium (1020 mg/L), calcium (3000 mg/L), and chloride (7310 mg/L) were identical to the concentrations in the natural effluent. Each effluent sample was then evaluated for its acute toxicity to the fathead minnow. Similar tests were also conducted in 1986.

Although the initial investigation of the factors responsible for effluent toxicity focused on an evaluation of sodium, calcium, and chloride levels, research by Rodgers et al.[3] indicated that high calcium levels could be responsible for the toxicity of some plant discharges. These researchers found that effluents having a calcium/sodium ratio of 15:1 exhibited high test organism mortality, while effluents with a ratio of 1:20 showed low test organism mortality levels. Calcium acute effect levels were also established by the researchers for several aquatic species (*Daphnia pulex* 48-h EC_{50} = 499 mg/L; Mysidopsis 96-h LC_{50} = 927 mg/L; fathead minnow 96-h LC_{50} = 2766 mg/L). Because calcium levels in the natural effluent were above those reported to be toxic to the fathead minnow, calcium levels in the natural effluent fathead minnow test solutions were evaluated to determine if a correlation existed between test organism mortality and calcium concentrations. In addition, screening acute toxicity tests were conducted on individual solutions of sodium chloride and calcium chloride to determine the approximate acute effect concentration for *D. magna,* which is commonly used for effluent toxicity investigations.

Chemical fractionation tests were performed with the effluent to further determine if the factors responsible for effluent toxicity could be isolated. Procedures utilized for the fractionation tests were similar to those used in the first two case studies. The effluent was passed through four separate resin treatments (granular activated carbon, cation exchange resin, anion exchange resin, and cation followed by anion exchange resin) and evaluated for acute toxicity to *D. magna.* Total organic carbon and ICP analyses were also conducted on the effluent before and after each treatment to see if there was a correlation between TOC or any inorganic constituents or both and the toxicity results.

Results

Case Study 1

The results of chemical fractionation of the effluent are presented in Table 1. Of the six treatments evaluated, only sequential treatment with activated carbon, cation exchange resin, anion exchange resin, and zeolite were effective in eliminating the toxicity of the effluent. Zeolite, a form of molecular sieves composed of hydrous silicates, was used for removal of ammonia because it is used in fish tanks for the same purpose. All other treatments had no effect in reducing effluent toxicity. Analysis of TOC, ICP, and ammonia analyses for each treatment revealed no apparent correlation between toxicity and organic or inorganic constituents.

The results of the static acute toxicity test conducted with ammonium chloride indicated that the *D. magna* 48-h EC_{50} value (95% confidence limits) for total ammonia was 32 (18.1 to 36.2) mg/L. Calculation of unionized ammonia levels from total ammonia concentration, pH, and temperature of the effluent indicated that the 48-h EC_{50} value (95% confi-

[3] Dorn, P. B. and Rodgers, J. H., "Variability Associated with Identification of Toxics in NPEDES Effluent Toxicity Tests," *Environmental Toxicology and Chemistry*, in press.

TABLE 1—*Ammonia concentrations and acute toxicity of Case Study 1 effluent before and after chemical fractionation treatment.*

Treatment	48-h Mortality, %	Concentration, mg/L	
		Total NH_3	Unionized NH_3
None (original effluent)	100	315	1.9 to 5.9
Activated carbon	100	144	2.2
Anion resin	100	195	3.0
Cation resin	100	<5	<0.08
Cation and anion resins	100	6.4	0.4 to 0.19
Activated carbon, cation and anion resins	0	<5	<0.08
Zeolite	0	<5	<0.08

dence limits) for unionized ammonia was 0.35 (0.2 to 0.4) mg/L. Based upon these acute effect levels, the concentration of unionized ammonia in the untreated effluent, activated carbon, and anion treatments were well above the acute effect concentration for ammonia. However, unionized ammonia concentrations in all other treatments were below the acute effect concentration.

It was later found that the cation exchange resin itself introduced toxicity, but only when a significant portion of the exchange process occurred. A cation exchange resin control, established by passage of well water through the cation column and evaluated for acute toxicity to *Daphnia,* indicated the cation resin treatment was nontoxic. While the cation exchange resin significantly reduced the levels of inorganic cations (including ammonium ion) in all three studies, biological testing consistently showed toxicity to *D. magna.* This illustrated why, in this case study, the cation and the combination of cation and anion exchange resin treatments reduced the ammonia concentrations below toxic levels, but the effluent, after these treatments, was still toxic. The hypothesis that an unknown toxic component was coming off the cation exchange resin was confirmed when a 300 mg/L ammonium chloride stock solution in well water was passed through the cation column. The solution after treatment was still toxic even though total ammonia levels had been reduced to <1 mg/L, well below the 48-h EC_{50} level of 32 mg/L. No other toxic constituents had been present in the stock solution. It appears that a minimum amount of ion exchange is necessary to release enough toxic components to be acutely toxic to *D. magna.* That is why passing well water through the cation exchange column does not introduce acute toxicity even though well water contains a small amount of cations.

The granular activated carbon, cation and anion resins combination treatment, and the zeolite treatment removed the ammonia from the effluents and also removed the toxicity. Comparison of the natural undiluted effluent with a synthetic effluent containing solely ammonia (as ammonium chloride) at the same concentration, pH, and temperature, supported the hypothesis that unionized ammonia was the source of toxicity in this wastewater.

Case Study 2

The results of static acute toxicity tests conducted on this effluent indicated that treatment with sodium NTA, sodium thiosulfate, and sodium sulfite (for metals, chlorine, and biocide removal, respectively) had no effect in reducing the toxicity of the effluent.

Based on the results of acute toxicity, TOC, and ICP analyses, no correlation was found between effluent toxicity and the mentioned parameters. However, there was a correlation between toxicity and hexavalent chromium levels. Sampling location A, the most upstream sampling point, had the lowest acute toxicity of all sampling points with a 48-h EC_{50} value of 80%. This station also had the lowest concentration of hexavalent chromium (29 ppb), present as CrO_4^{2-} or chromium ion. Toxicity significantly increased at station B (EC_{50} = 40%) and fluctuated along the course of the wastewater canal. Increases in effluent toxicity accompanied corresponding increases in hexavalent chromium concentrations. Note that chelating agents such as NTA and EDTA (ethylenediaminetetraacetic acid) will not remove complex metal anions such as chromate.

A summary of chemical fractionation tests conducted with the effluent are presented in Table 2. Both activated carbon and the anion exchange resin eliminated the acute toxicity of the effluent. While no correlation was seen between toxicity and TOC or ICP analyses, there was a correlation between toxicity and hexavalent chromium. Hexavalent chromium concentrations in the untreated effluent (241 ppb) and cation resin treatment (210 ppb) were at or near acute effect levels reported in the literature for *D. magna* (20 to 212 ppb). All the daphnids exposed to these waters died within 48 h. Hexavalent chromium levels in the treatments showing no acute toxicity (activated carbon and anion exchange resin treatment) were either at or below the lowest literature reported acute effect concentrations (that is, 20 ppb).

Figure 4 shows a dose response curve for the effluent and associated hexavalent chromium concentrations. The measured 48-h EC_{50} of 41 ppb Cr^{6+} is based on the acute toxicity study of hexavalent chromium to *D. magna.* If one interpolates a 48-h EC_{50} for the natural effluent, one finds a value of 35 ppb Cr^{6+}. The similarities between these two values presents strong evidence that hexavalent chromium is the source of this wastewater's toxicity. The best evidence for toxicity source confirmation was obtained when hexavalent chromium alone was removed from the actual plant process. The effluent completely lost its toxicity.

Case Study 3

A summary of fathead minnow acute toxicity results for the natural and synthetic effluent tests are presented in Table 3. The 96-h LC_{50} values for the two effluents during the first, second, and sixth test periods were essentially identical, indicating a correlation between toxicity and concentrations of sodium, calcium, or chloride or a combination of these. Test results for the natural and synthetic effluents during the third, fourth, and fifth periods were significantly different, with the natural effluent exhibiting the higher toxicity in each case. While such results indicate that salinity may not always be solely responsible for effluent toxicity, salinity still cannot be ruled out as a contributing factor to the toxicity of the effluent. It is possible that the high salinity of the effluent could be stressing the test organisms, making them more susceptible to other chemicals in the wastewater.

TABLE 2—*Hexavalent chromium concentration and acute toxicity of Case Study 2 effluent before and after chemical fractionation treatment.*

Treatment	48-h Mortality, %	Cr^{6+}, ppb
None (original effluent)	100	241
Activated carbon	0	10
Cation resin	100	210
Anion resin	0	20

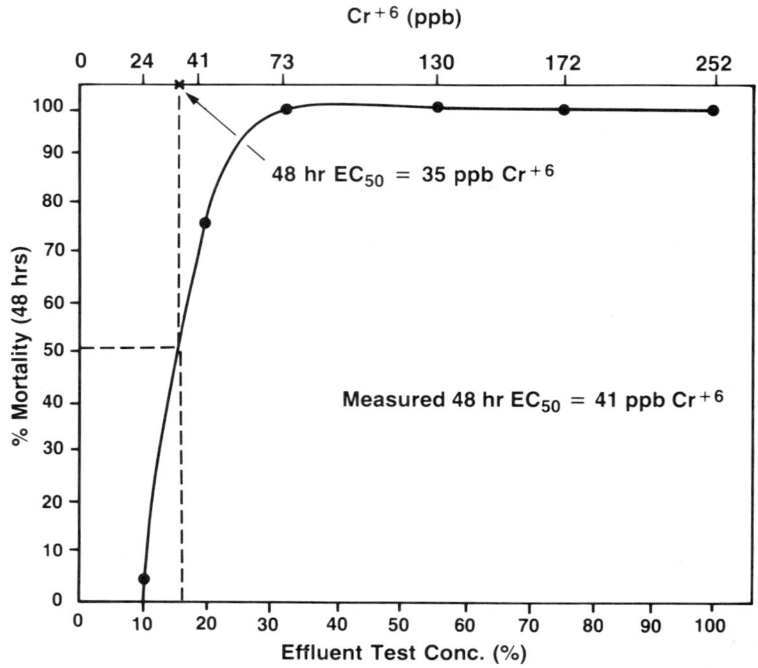

FIG. 4—*Dose response curve and associated hexavalent chromium concentrations for the Case Study 2 effluent.*

TABLE 3—*Summary of fathead minnow acute toxicity results for "natural" and synthetic effluents.*

	96-h LC_{50}, %	
Test Period	Natural Effluent	Synthetic Effluent
1	79	70
2	60	60
3	49	89.5
4	77	>100
5	64	80
6	80	81

Table 4 summarizes the percent mortality observed at each natural effluent test concentration and the corresponding calcium concentrations for the first test period. Based upon the test data, the 96-h LC_{50} for the natural effluent was 79%. The calcium concentration in the 100% effluent concentration was 3000 mg/L, with a predicted nominal calcium content in the 79% effluent of 2400 mg/L. This value is very similar to the 2766 mg/L effect level for calcium reported by Rodgers et al.,[3] indicating that calcium could be responsible for the effluent toxicity.

Based upon the results of *Daphnia* screening acute toxicity tests, the 48-h EC_{50} values

TABLE 4—*Summary of fathead minnow "natural" effluent acute toxicity data and corresponding levels of calcium in the test solutions.*

Effluent Test Concentration, %	48-h Mortality, %	Calcium Concentration, mg/L[a]
100	85	3000
56	5	1680
32	0	960
18	0	540
10	0	300
5.6	0	168
Control	0	0

[a] Nominal test concentrations.

for sodium, calcium, and chloride were determined to be >2400 mg/L, >500 but <900 mg/L, and 3000 to 4000 mg/L, respectively. A comparison of these effect concentrations to measured concentrations of sodium (1020 mg/L), calcium (3000 mg/L), and chloride (7310 mg/L) in the natural effluent indicates that the concentration of calcium and chloride in the undiluted effluent was two to four times above the EC_{50} level.

Table 5 presents the results of chemical fractionation tests with the undiluted natural effluent. Neither activated carbon, the cation exchange resin, or the anion exchange resin had any effect in reducing the toxicity of the effluent. However, treatment of the effluent with cation and anion exchange resins in sequence eliminated effluent toxicity. No correlation was observed between the toxicity of the various effluent treatments and TOC concentration. However, as indicated in Table 5, a correlation was found between toxicity and calcium and chloride ion levels. The concentration of calcium and chloride in the natural effluent, activated carbon, and anion exchange resin treatments were well above acute effect levels for *D. magna*. Chloride levels were also significantly higher than the effect level in the cation exchange resin treatment. No mortality was observed in the sequential cation and anion exchange resins treatment which had both very low levels of calcium and chloride.

Confirmation of calcium and chloride as the sources of toxicity in this wastewater was carried out by preparing a synthetic effluent containing only calcium, sodium, and chloride ions at the same level as the natural effluent and comparing the dose response curves of the synthetic and natural effluents. These data strongly indicate that calcium and chloride ions are a major source of toxicity in this effluent.

TABLE 5—*Summary of chemical fractionation tests of the undiluted "natural" effluent with corresponding toxicity results and calcium and chloride concentrations.*

Treatment	48-h Mortality, %	Concentration in Effluent, mg/L	
		Calcium	Chloride
None (original effluent)	100	2122	5603
Activated carbon	100	1887	5603
Cation resin	100	2.7	5443
Anion resin	100	1070	2446
Cation and anion resin	0	29	477

Discussion

Analysis of three plant wastewaters using the Phase I Toxicity Evaluation Protocol indicated that calcium, chloride, hexavalent chromium, and unionized ammonia were the primary factors responsible for the toxicity of the wastewaters investigated. During the toxicity investigations, background knowledge on the chemical composition of the wastewater and plant processes contributing to the waste stream provide valuable information on potential factors responsible for the toxicity of the wastewater. TOC and ICP analyses also proved to be valuable, cost effective procedures for determining resin removal efficiencies and whether wastewater toxicity was due to organic or inorganic constituents. Although the synthetic effluents used in these studies did not contain all the chemical constituents that were in the natural effluents, they were useful in providing additional evidence on the likely source of effluent toxicity by developing correlations between wastewater dose response curves and individual constitutent effect levels. The potential impact of differences in the water quality characteristics of natural and synthetic wastewaters should be recognized when trying to establish effects correlations.

In Case Study 3, it was learned that two chemicals (calcium and chloride), which are not normally considered toxic, can exist in effluents at acutely toxic levels (that is, calcium $>$ 499 mg/L, chloride $>$ 3000 mg/L). The resulting salinity can exceed the tolerances of most freshwater organisms. Such data raises serious questions regarding the validity of using freshwater organisms to evaluate the aquatic safety of highly saline effluents that are discharged into estuarine waters.

Based upon the results of the case studies, a better understanding was obtained of the capabilities and limitations of the various resins used in the Phase I protocol. While the cation exchange resin significantly reduced the levels of inorganic cations in all three wastewaters, biological testing of the resin treatments consistently showed acute toxicity to *D. magna*. Case Study 1 revealed that an unknown toxic component was likely being released from the resin during the cation exchange process. The unknown toxicant was not removed from solution after sequential treatment of the cation and anion exchange resins, but was removed after these two treatments plus activated carbon. In Case Study 3, any toxicants coming off the cation column were removed after sequential treatment with the anion column, because the post-cation and post-anion water was nontoxic. It appears that other factors (that is, pH, conductivity, hardness, TOC, and so forth) affect the ability of a sequential treatment of only cation and anion (Study 3) or cation, anion, and activated carbon (Study 1) to remove the unknown toxicant coming off the cation exchange resin. The results of the cation control tests indicate the importance of appropriate quality control procedures for determining the suitability of resins for toxicity identification purposes. The strong anion exchange resin removes the majority of the anions from the tested plant effluents, but it also appears to remove a significant portion of the cations, that is, calcium and ammonium ions. Similarly, although activated carbon is generally reported to remove nonpolar organic compounds, it also removes appreciable amounts of certain cations and anions, that is, GAC removes most of the chromate anions, about half of the ammonium cations, but very little calcium and chloride ions. Because of these results, the importance of taking analytical measurements after each treatment is shown.

Identifying the source of toxicity is a very important aspect of the overall TRE (toxicity reduction evaluation) strategy. The Phase I protocol is a general approach for fractionating complex wastewater effluents into their inorganic and organic components. If the toxicity lies in the inorganic fraction, fairly simple and routine analytical methodologies are used to identify the source of toxicity (that is, ICP for metals; ion-selective electrodes for ammonium, chloride, and cyanide ions; colorimetric analysis for hexavalent chromium; and so forth). In many cases, one can stop the work after the Phase I scheme, review the analytical data, make a decision on what compounds are causing the toxicity, and prepare a synthetic effluent to support one's suspicions. However, if the toxicity lies in the organic fraction, identifying the toxic components will be much more difficult. Instead of a few tens of metals and several anions to be concerned about in the inorganic fraction, the toxicity could lie with any of the several million organic compounds on the organic side of the fractionation scheme. This is why another, more complex organic fractionation/characterization scheme is invoked when the source of toxicity is organic in nature. It may well be true that the organic source of toxicity may never be absolutely identified due to the following: (1) the sheer number of possible compounds and (2) the lack of appropriate analytical techniques to *identify* organic compounds (for example, GC/MS is only able to identify about 10% of all organic compounds [*9,10*]). However, with the ESC organic fractionation/characterization scheme, even if the absolute identity of the toxicants are not determined, a great deal of information will be known about the toxic organic compounds: for example, is it purgeable, extractable, or nonvolatile, polar or nonpolar; what functional groups are present; and what is its molecular size and weight? With this information about the source of toxicity, one may be able to make an informed and cost-effective decision about how to treat the wastewater.

In summary, a toxicity evaluation scheme has been developed to fractionate wastewater into inorganic and organic fractions and then identify the toxic components in the wastewater. Confirmation of toxicity is determined using *D. magna* or other suitable test organisms. The procedure has been successfully used on three plant wastewaters. Verification of the suspected toxic components is accomplished using synthetic effluents containing only the suspected compounds.

Acknowledgments

The authors wish to thank Dr. Fred Hileman and Dr. Richard Kimerle for their ideas and creative thinking in this new field of toxicity testing and chemical analysis of wastewater. We also would like to thank Michael Tucker, Sandra Becker, and David McKenzie for their help in carrying out the experiments. The above mentioned people are all members of the Environmental Sciences Center at Monsanto Company.

References

[*1*] "Technical Support Document for Water Quality-Based Toxics Control," U.S. Environmental Protection Agency, Office of Water Enforcement and Permits, Washington, DC, 1985.

[*2*] "Permit Writer's Guide to Water Quality-Based Permitting for Toxic Pollutants," U.S. Environmental Protection Agency, Office of Water Enforcement and Permits, Washington, DC, 1987.

[*3*] Walsh, G. E. and Garnas, R. L., "Determination of Bioactivity of Chemical Fractions of Liquid Wastes Using Freshwater and Saltwater Algae and Crustaceans," *Environmental Science and Technology*, Vol. 17, No. 3, 1983, pp. 180–182.

[*4*] Parkhurst, B. R., Gehrs, C. W., and Rubin, I. B., "Value of Chemical Fractionation for Identifying the Toxic Components of Complex Aqueous Effluents," *Aquatic Toxicology (Second Conference), ASTM STP 667*, American Society for Testing and Materials, Philadelphia, 1979, pp. 122–130.

[*5*] Samolloff, M. R., Bell, J., Birkholz, D. A., Webster, G. R., Arnott, E. G., Pulak, R., and Madrid, A., "Combined Bioassay-Chemical Fractionation Scheme for the Determination and Ranking of Toxic Chemicals in Sediments," *Environmental Science and Technology*, Vol. 17, No. 6, 1983, pp. 329–334.

[*6*] Mount, D. I. and Anderson-Carnahan, L., "Methods for Toxicity Reduction Evaluations—Phase I: Toxicity Characterization Procedures," EPA-600/3-88-034, U.S. Environmental Protection Agency, Washington, DC, 1988.

[*7*] Peltier, W. H. and Weber, C. I., "Methods for Measuring the Acute Toxicity of Effluents to Freshwater and Marine Organisms," EPA-600/4-85-013, U.S. Environmental Protection Agency, Cincinnati, OH, 1985.

[*8*] Emerson, K., Russo, R. C., Lund, R. E., Thurston, R. V., "Aqueous Ammonia Equilibrium Calculations: Effect of pH and Temperature," *Journal of the Fisheries Research Board of Canada*, Vol. 32, No. 12., 1975, pp. 2379–2383.

[*9*] *Drinking Water and Health*, National Academy of Science, Washington, DC, 1977, pp. 489–856.

[*10*] Bedding, N. D., McIntyre, A. E., Perry, R., and Lester, J. N., "Organic Contaminants in the Aquatic Environment: I. Sources and Occurrence," *The Science of the Total Environment*, Vol. 25, No. 2, 1982, pp. 143–167.

William L. Goodfellow, Jr.,[1] Wayne L. McCulloch,[1] John A. Botts,[2] Arthur G. McDearmon,[3] and Dolloff F. Bishop[4]

Long-Term Multispecies Toxicity and Effluent Fractionation Study at a Municipal Wastewater Treatment Plant

REFERENCE: Goodfellow, W. L., Jr., McCulloch, W. L., Botts, J. A., McDearmon, A. G., and Bishop, D. F., **"Long-Term Multispecies Toxicity and Effluent Fractionation Study at a Municipal Wastewater Treatment Plant,"** *Aquatic Toxicology and Environmental Fate: Eleventh Volume, ASTM STP 1007,* G. W. Suter II and M. A. Lewis, Eds., American Society for Testing and Materials, Philadelphia, 1989, pp. 139–158.

ABSTRACT: Characterization of the acute and chronic toxicity of effluent samples from the Patapsco Wastewater Treatment Plant (WWTP), Baltimore, Maryland, was performed approximately twice weekly from April 1986 to January 1987. Acute toxicity of both the primary and secondary effluents was evaluated using *Ceriodaphnia dubia, Mysidopsis bahia,* and Microtox [a luminescent marine bacteria *Photobacterium phosphoreum* (Beckman Instruments, Inc.)]; chronic toxicity was evaluated using the *C. dubia* seven-day chronic test. The data indicate that *C. dubia* was the most sensitive indicator of acute toxicity for both primary and secondary effluent wastewaters. The acute responses for *M. bahia* and Microtox were considerably less sensitive than those for *C. dubia.* The 48-h median lethal concentration (LC_{50}) values for *C. dubia* ranged from 0.5 to 23.4% secondary effluent and 0.1 to 6.5% primary effluent wastewaters. Microtox 5-min median effective concentration (EC_{50}) values ranged from 9.4 to 100% secondary effluent and 0.6 to 19.5% primary effluent, respectively. *M. bahia* 96-h LC_{50} values were 8.9 to 100% secondary effluent and 14.8 to 35.4% primary effluent, respectively. The seven-day *C. dubia* chronic values (ChV) for the primary and secondary effluents were only slightly lower than the acute responses, ranging from <0.1 to 9.4% secondary effluent and <0.1 to 6.5% primary effluent, respectively.

As an additional characterization of the plant's treatment process effectiveness, influent and effluent samples were fractionated in an attempt to identify specific components of acute toxicity entering and leaving the WWTP. As a result of the fractionation study, nonpolar organic compounds appeared to be the major toxic component of the samples fractionated. While identification of the specific compounds causing toxicity in the nonpolar organic fractions was not possible, the octanol/water partition coefficient (Log *P*) values for the most toxic fractions were estimated to be approximately 6 to 9 in the summer samples and approximately 3.6 to 8.3 in the winter samples. Cations, anions, and polar organic compounds do not appear to be major contributors to the toxicity of the effluent samples fractionated. The percentage of toxicity reduction across the plant was approximately 50 to 90% for the three species tested.

KEY WORDS: municipal effluent, complex effluent, acute toxicity, chronic toxicity, *Ceriodaphnia dubia, Mysidopsis bahia,* Microtox, toxicity reduction evaluation, effluent fractionation, freshwater discharge, estuarine receiving waters, aquatic toxicology

[1] Scientist and senior scientist, respectively, EA Engineering, Science, and Technology, Inc., Hunt Valley/Loveton Center, Sparks, MD 21152.

[2] Environmental scientist, Engineering–Science, Inc., Fairfax, VA 22030.

[3] Engineer, City of Baltimore, Patapsco Wastewater Treatment Plant, Baltimore, MD 21226.

[4] Chief, Treatment and Assessment Branch, Water and Hazardous Waste Treatment Division, Risk Reduction Engineering Laboratory, U.S. Environmental Protection Agency, Cincinnati, OH 45268.

State and regional Environmental Protection Agency (EPA) permit authorities are including more and more effluent toxicity testing requirements in National Pollutant Discharge Elimination System (NPDES) permits as a result of U.S. EPA's water quality-based approach to toxics control. However, limited data are available on long-term variability of the toxicity associated with municipal wastewater treatment plant effluent, and even less information exists on freshwater municipal effluents entering estuarine systems.

As part of a toxicity reduction evaluation (TRE) conducted at the Patapsco Wastewater Treatment Plant (Patapsco WWTP) in Baltimore, Maryland, a long-term characterization of the acute and chronic toxicity of the plant's primary and secondary effluents was performed. The City of Baltimore had in place a toxics monitoring program in 1980 using Microtox *(Photobacterium phosphoreum)* and a modified respirometry test [*1*]. The Patapsco WWTP was chosen for this study because there was evidence of influent wastewater toxicity and toxicity pass-through to the receiving waters [*1*].

The Patapsco WWTP has an approximate flow of 1.5 to 2.0 m^3/s (35 to 45 million gallons per day), of which approximately 30% is industrial wastewater. Historically, toxic spills of short duration (usually <6 h) have been noted regularly by Microtox in the industrial wastewater stream. High concentrations (1000 $\mu g/L$ or greater) of xylenes, benzenes, and toluenes and measurable amounts of other priority pollutants have been consistently measured in these influent wastewaters [*2*].

The purpose of this study was to investigate the acute and chronic toxicity of the primary and secondary effluent wastewaters from the Patapsco WWTP and to identify the toxic components of the complex municipal wastewater. Acute toxicity tests were conducted to evaluate the effluent variability and to estimate the potential toxicity pass-through associated with secondary effluent. Microtox, *Mysidopsis bahia,* and *Ceriodaphnia dubia* were used as test species. Chronic toxicity was determined on the basis of survival and reproductive potential using *C. dubia.* Toxics identification procedures were used to evaluate the toxic component or components of the effluent in order to potentially treat it for the associated toxicity, or to implement better pretreatment controls on industries contributing the toxic components.

Materials and Methods

Sample Collection and Preparation

Secondary effluent samples were collected weekly (approximately two samples per week) from the Patapsco WWTP between April 1986 and January 1987. The samples were 24-h composite samples, collected continuously from the main discharge channel after the secondary clarifier units, but prior to chlorination. The effluent was pumped from the discharge channel by submersible pump (Little Giant Pump Co., Oklahoma City, Oklahoma) into a 18.9-L polyethylene carboy rigged with an overflow. The approximate 95% replacement time in the carboy sump was <20 min [*3*]. From the carboy sump, the effluent was delivered by metering pump (FMI, Inc., Oyster Bay, NY) at a rate sufficient to collect 37.8 L of effluent sample per 24-h period. The sample was collected in either chemically cleaned glass aquaria or polyethylene carboys kept on wet ice. Within 4 h after termination of the composite sampling period, the sample was transported back to the Aquatic Toxicology Laboratory at EA Engineering, Science, and Technology, Inc.

Primary effluent samples were collected less frequently than secondary effluent samples—approximately two per month from April 1986 to January 1987. The primary samples, also 24-h composite samples consisting of 24 separate hourly grab samples (totaling

37.8 L per sample), were taken as equal volume portions after the wastewater exited each operational primary settling tank. Hourly grab samples were necessary because the facility's design did not allow for continuous sampling of a common channel from the operating primary settling tanks. In addition, the 24-h sampling window took place 8 h prior to the secondary effluent sampling window in order to reflect the hydraulic retention time within the treatment processes. The primary samples were collected hourly using a sampling pole and were stored as a composite sample in polyethylene carboys at 4°C. At the end of the 24-h period, the composite sample was transported back to the authors' laboratory (EA) within 10 h after termination of the composite sampling period.

The effluent samples were given a laboratory accession number upon receipt of the sample at the laboratory. The high residual carbon dioxide (CO_2) concentration (supersaturated) present in the secondary effluent, due to the use of oxygen in the plant's bioreactors, made it necessary to allow the CO_2 levels in the effluent sample to equilibrate. This was performed by gently aerating the sample (approximately 1 h) until the pH reached 7.5. After the CO_2 levels in the effluent reached equilibrium with the air, an aliquot of the sample was removed for toxicity testing with *C. dubia*. Synthetic sea salts (Forty Fathoms Marine Mix; Marine Enterprises, Towson, MD) were then added to the remaining effluent sample to achieve a salinity of 20 parts per thousand for toxicity testing with *M. bahia*. The primary effluent sample was handled the same way as the secondary effluent sample with the exception of the CO_2 equilibration step, which was not necessary. Both types of samples were stored in a walk-in refrigerator at 4°C until needed.

Toxicity Testing

The three species of test organisms utilized were *P. phosphoreum* (Microtox), *M. bahia* (opossum shrimp), and *C. dubia* (water flea). Microtox supplies and reagents were purchased from Microbics, Inc. (Carlsbad, CA). *M. bahia* and *C. dubia* were obtained from multigenerational cultures maintained at the authors' laboratory. The mysid cultures were in 18.9 and 37.8-L, all-glass aquaria, outfitted with undergravel filtration and a substrate of crushed coral. The temperature of the system was maintained at 23 ± 2°C and a salinity of 20 parts per thousand. When juvenile mysids were needed, gravid females were isolated into a brood chamber/trap. The females were isolated in a screened chamber of the chamber/trap, and the juveniles (after release) were captured in a siphon trap. The juveniles were harvested daily. Only mysids, one to five-days old were used for testing. Cultures were fed newly hatched *Artemia* sp. (<24-h old). *Ceriodaphnia dubia* cultures were held in 1-L beakers (as bulk cultures) and individual 30-mL beakers containing one female each. This individual culture system allowed organisms of known parentage and age to be used for testing. On the morning of testing, any neonates (young) were removed from the cultures. Only neonates released in the next 8 h from broods of six or more were used for testing. The cultures were maintained at 25 ± 2°C and were fed the yeast/Cerophyll/trout chow recipe, supplemented with the algae *Selenastrum capricornutum,* as outlined in Horning and Weber [5]. All cultures were held in environmentally controlled rooms with a 16-h light and 8-h dark photoperiod.

The mysids were cultured and tested in artificial seawater [formulated from a dry sea salts mixture (Forty Fathoms Marine Mix) and activated-carbon-filtered, dechlorinated, municipal water]. Artificial seawater was prepared in large-capacity polyethylene tanks equipped with recirculating pumps to facilitate dissolution and mixing, and then it was adjusted to a final salinity of 20 parts per thousand. The *C. dubia* were cultured and tested in Patapsco River water collected from Henryton Bridge, Howard County, Maryland. Both dilution waters have proven safe for testing of aquatic organisms toxicity with no evident loss of fecundity, as evidenced by maintenance of the multigenerational cultures. Microtox assays were performed in the reconstituted sea salts supplied by Microbics, Inc.

General toxicity test methods followed Peltier and Weber [*4*] for acute toxicity tests, and Horning and Weber [*5*] for chronic toxicity tests. Only acute toxicity tests with control mortality $\leq$10% and chronic toxicity tests with control mortality $\leq$20% were used in this study. In addition, a minimum of 15 neonates per females was used in the controls to evaluate the acceptability of the chronic *C. dubia* tests. Microtox assays followed procedures outlined by Microbics, Inc. [*6*]. Evaluation of the several months of daily 5-min and 15-min exposure durations determined that no statistical significance ($P = 0.05$) existed in Microtox median effective concentration (EC_{50}) values of effluent samples from Patapsco WWTP. The mysids were exposed in 20-cm, round glass culture bowls containing 1-L volumes of test solution. Each treatment group consisted of two replicate test chambers with ten organisms per replicate. The *C. dubia* were tested in 30-mL plastic beakers containing 15 mL of test solution. Each treatment group consisted of ten replicate test containers with one organism per test chamber. Test solutions were measured with Class A volumetric glassware. Solutions were dispensed equally to all chambers. A solution dispenser similar to the one described in Mount and Norberg [*7*] was used for *C. dubia* testing. Test solutions were renewed daily in both the mysid and *C. dubia* tests.

During the solution renewal of the mysid tests, uneaten and dead *Artemia* sp. and other debris were removed from the bottom of the test chambers using a siphon tube. The water level in each chamber was lowered, leaving approximately 25% of test solution. New test solution was prepared in 2-L volumetric flasks on the day of the renewal, and it was dispensed equally (1 L) into both replicate test chambers. Twice daily the juveniles in each test chamber were fed 0.5 mL of a concentrated (0.05 g/mL) suspension of newly hatched (<24-h old) *Artemia* sp. New test solutions were prepared daily for *C. dubia* on the day of renewal, and they were dispensed as described above. The test organisms were transferred from the old test vessels to the new test vessels using a wide-bore glass tube. Caution was taken not to stress the test organism during transfer. The *C. dubia* were fed yeast/Cerophyll/trout chow (0.1 mL per chamber) daily.

Mortality observations were made at 24-h intervals (during solution renewal). Dead organisms were removed as they were observed. For the *C. dubia* chronic test, neonates were also counted and discarded. Temperature, pH, dissolved oxygen, and salinity (for mysids) were measured daily from the control and the low, medium, and high-test concentrations. Routine water quality determinations were performed in accordance with Standard Methods [*8*].

Statistical Analyses

The median lethal concentration (LC_{50}) estimating methods included maximum likelihood estimates based on the probit method [*9*] and the binomial and moving average methods [*10*]. Microtox EC_{50} values were determined by linear regression of the gamma values (light response) and the exposure concentration [*6*]. Effluent LC_{50}/EC_{50} values were based on nominal effluent concentrations (volume to volume measurements). Exposure durations and the resulting LC_{50}/EC_{50} values were as follows: Microtox, 5-min EC_{50}; *M. bahia,* 96-h LC_{50}; and *C. dubia,* 48-h LC_{50}. The 5-min EC_{50} represented the percentage of effluent which caused a 50% reduction in light emitted by the luminescent bacteria following a 5-min exposure. The 48-h and 96-h LC_{50} values represented the percentage of effluent which caused 50% mortality in the test organism during either 48 or 96 h of exposure.

Chronic values (ChVs) were determined by the calculation of mean young production

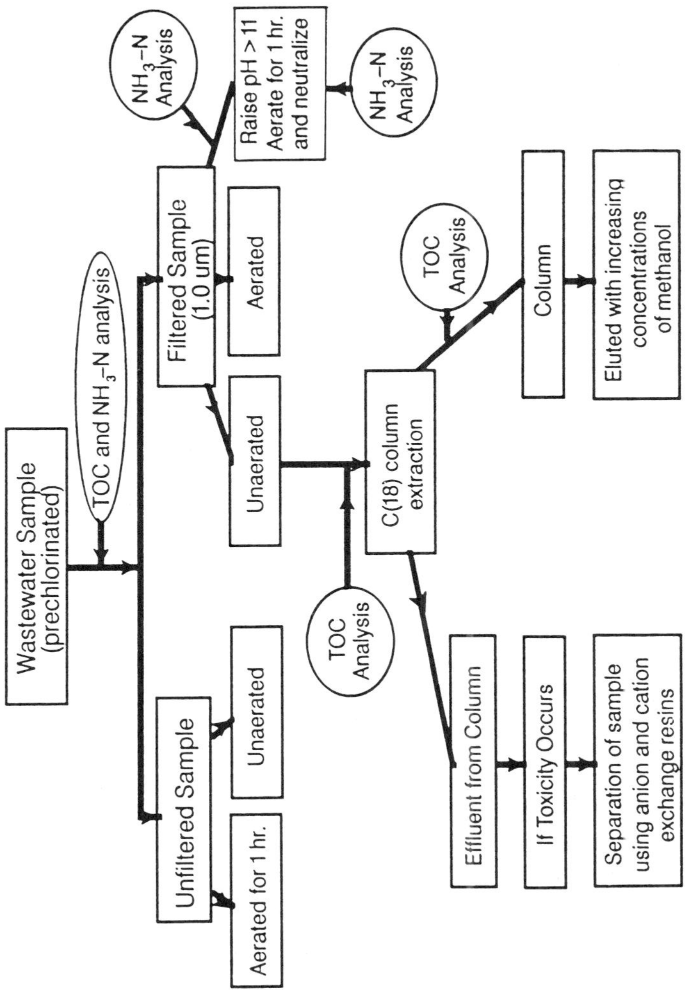

FIG. 1—*Toxicity identification procedure.*

of *C. dubia* for each exposure group using the total number of young produced by the original ten females during the seven days of exposure. In instances where there was a spilled replicate, or a male was noted, the original number exposed was corrected to N = 9 or N = 8 as appropriate. One-way analysis of variance (ANOVA) was performed on the data, based on the total females. This was done to detect statistically significant (P = 0.05) differences in young production between the exposure groups during a given test. If a statistically significant difference in treatment group means was detected by ANOVA, a Tukey's Studentized range test was performed to delineate which treatment group mean or means were significantly different (P = 0.05) from the control. Using this information, the no-observable-effect concentration (NOEC) and the lowest-observable-effect concentration (LOEC) were determined. The ChV was calculated as the lowest value of the geometric mean of the NOEC and the LOEC (reproductive), or the seven-day LC_{50} (survival).

Correlation of the acute Microtox, mysid, and *C. dubia* values and the *C. dubia* ChV was performed using Pearson's product-moment correlation [*11*]. Statistically significant correlations (P = 0.05) were further evaluated by plotting the two variables showing a significant correlation and by observing the data trends.

Effluent Fractionation

Five secondary effluent samples and one primary effluent sample were fractionated using the toxics identification approach shown in Fig. 1. This approach follows the toxics identification method developed by Anderson-Carnahan and Mount [*12*]. This procedure is a tiered approach and involves fractionation of the wastewater and testing of the fractions along with system blanks for acute toxicity. Initially, the whole wastewater and fractions of wastewaters treated by aeration, filtration, and pH adjustment were tested for toxicity using time lethality or definitive acute *C. dubia* tests or both. The toxicity tests on aerated wastewater samples were used to indicate if toxicity was associated with volatile compounds. The filtration step was designed to determine whether toxicity was present in the suspended particulate phase or in the soluble phase of the wastewater. Aeration, in conjunction with pH adjustment, was used to evaluate unionized ammonia toxicity. If the filtrated unaerated fraction was found to be toxic, further fractionation was conducted to determine whether toxicity was associated with nonpolar organic compounds or cation and anion compounds that may have been present in the wastewater. Fractionation of nonpolar organics was accomplished by passing the sample through a C_{18} solid-phase extraction column (J. T. Baker Chemical Co., Phillipsburg, NJ) and eluting the column with increasing concentrations of methanol (MeOH). Fractionation of cations and anions followed techniques outlined in Walsh and Garnas [*13*]. Dowex 2-X8, a strong base anion exchange resin, and Dowex 50W-X8, a strong acid cation exchange resin 20 to 50 mesh (J. T. Baker Chemical Co.), were used.

Time lethality tests [*12*] were used as the screening tier of each fractionation step. They consisted of three concentrations: 100, 50, and 25% fractionation samples. Exposure chambers were 30-mL plastic beakers containing 15 mL of test solution. Five *C. dubia* (<24-h old) were exposed to each concentration treatment. Mortality observations were made at 0.5, 1, 2, 4, 8, and 24 h. Definitive acute tests were also performed in 30-mL test chambers containing 10 mL of test solution. Two replicates of five organisms each were exposed to a concentration series of a minimum of five test concentrations and a control, using procedures outlined by Peltier and Weber [*4*] and by Anderson-Carnahan and Mount [*12*]. Exposure durations were 48 h, and the statistical analysis followed those previously discussed.

TABLE 1—*Acute and chronic toxicity data for three aquatic organisms exposed to secondary effluent wastewater from the Patapsco WWTP.*

Sample Date	Toxicity, % effluent			
	Microtox, 5-min EC_{50}	*Mysidopsis bahia*, 96-h LC_{50}	*Ceriodaphnia dubia* 48-h LC_{50}	*Ceriodaphnia dubia* 7-day CV[a]
4/9/86	61.2	34.8 (30 to 100)[b]	5.5 (3 to 10)	5.5
4/10/86	52.8	34.7 (26 to 47)	5.5 (3 to 10)	5.5
4/16/86	81.5	44.1 (36.1 to 67.2)	5.5 (3 to 10)	5.5
4/17/86	93.2	49.0 (35.3 to 77.4)	5.5 (3 to 10)	1.7
4/23/86	≥100	63.7 (47.9 to 77.4)	5.2 (3 to 10)	4.2
4/30/86	≥100	62.5 (39.3 to 80.4)	9.4 (6.5 to 13.5)	4.6
5/8/86	88.9	70.7 (50 to 100)	4.6 (3.2 to 6.5)	1.8
5/14/86	57.2	80.6 (50 to 100)	9.4 (6.5 to 13.5)	2.3
5/28/86	≥100	56.7 (25 to 100)	17.3 (6.5 to 40.5)	5.5
6/4/86	≥100	53.5 (25 to 100)	23.4 (13.5 to 40.5)	9.4
6/5/86	≥100	57.0 (50 to 100)	10.7 (6.5 to 40.5)	8.1
6/11/86	32.6	>100	18.1 (12.7 to 28.2)	4.6
6/12/86	82.1	66.2 (52.8 to 79.4)	8.7 (6.5 to 13.5)	1.2
6/18/86	. . .[c]	80.6 (68.1 to 100)	7.3 (5.4 to 10.0)	7.3
6/25/86	88.7	33.7 (26.2 to 44.1)	4.6 (3.2 to 6.5)	2.0
6/26/86	. . .[c]	19.6 (16.5 to 23.3)	2.2 (1.8 to 2.8)	1.6
7/9/86	. . .[c]	68.6 (60.7 to 76.9)	4.8 (3.8 to 6.6)	3.9
7/10/86	. . .[c]	35.7 (28.7 to 44.7)	9.4 (6.5 to 13.5)	3.4
7/16/86	. . .[c]	21.6 (16.6 to 28.0)	4.6 (3.2 to 6.5)	1.7
7/23/86	61.3	73.0 (61.5 to 88.0)	4.6 (3.2 to 6.5)	1.1

TABLE 1—*(Continued).*

Sample Date	Toxicity, % effluent			
	Microtox, 5-min EC_{50}	*Mysidopsis bahia,* 96-h LC_{50}	*Ceriodaphnia dubia*	
			48-h LC_{50}	7-day CV[a]
7/30/86	≥100	46.2 (35.7 to 100)	4.3 (3.1 to 6.4)	1.1
8/5/86	57.1	50.0 (27.8 to 62.2)	4.6 (3.2 to 6.5)	2.2
8/6/86	71.8	58.2 (44.6 to 68.3)	6.5 (3.2 to 13.5)	2.2
8/13/86	≥100	42.6 (33.6 to 52.0)	3.8 (1.6 to 6.5)	1.1
8/20/86	100	42.0 (35.0 to 57.7)	2.2 (1.6 to 3.2)	<0.2
8/21/86	>100	80.6 (68.1 to 100)	4.6 (3.2 to 6.5)	3.5
9/3/86	63.6	42.4 (32.2 to 92.5)	4.6 (3.2 to 6.5)	<0.2
9/10/86	77.8	38.5 (27.5 to 48.9)	9.4 (6.5 to 13.5)	4.6
9/17/86	≥100	84.7 (64.4 to 100)	9.4 (6.5 to 13.5)	2.1
9/18/86	≥100	29.5 (21.6 to 44.9)	9.4 (6.5 to 13.5)	2.3
9/24/86	71.8	44.1 (36.1 to 67.2)	9.4 (6.5 to 13.5)	2.3
9/25/86	67.8	56.7 (37.2 to 69.2)	8.7 (6.5 to 13.5)	3.8
10/1/86	9.4	22.5 (16.5 to 30.3)	2.9 (1.6 to 6.5)	1.1
10/2/86	≥100	35.4 (27.5 to 52.5)	7.7 (3.2 to 13.5)	3.4
10/8/86	100	100 (74.6 to >100)	4.6 (3.2 to 6.5)	1.1
10/15/86	43.8	25.9 (20.1 to 33.4)	1.0 (0.8 to 1.6)	<0.2
10/16/86	39.3	15.6 (11.8 to 20.3)	2.1 (1.6 to 3.2)	0.3
10/22/86	56.7	24.3 (18.0 to 32.8)	0.9 (0.3 to 1.8)	0.2
10/29/86	59.3	. . .[c]	6.5 (4.2 to 10.0)	1.8
11/12/86	94.3	11.7 (7.6 to 15.9)	4.7 (4.2 to 10.0)	2.3
11/13/86	≥100	46.4 (NC)[d]	6.5 (4.2 to 10.0)	2.5
12/10/86	≥100	16.0 (13.0 to 19.6)	2.7 (1.8 to 4.2)	0.5

TABLE 1—*(Continued).*

Sample Date	Toxicity, % effluent			
	Microtox, 5-min EC_{50}	*Mysidopsis bahia,* 96-h LC_{50}	*Ceriodaphnia dubia* 48-h LC_{50}	*Ceriodaphnia dubia* 7-day CV[a]
12/15/86	88.9	8.8 (5.9 to 11.5)	0.5 (0.3 to 0.8)	<0.1
1/6/87	70.6	21.8 (16.1 to 29.0)	0.7 (0.4 to 1.6)	<0.1
1/8/87	≥100	45.7 (31.4 to 78.9)	1.0 (0.3 to 1 8)	0.2

[a] CV is equivalent to the lowest of the seven-day LC_{50} values or the geometric mean of the no-observable-effect concentration and the lowest-observable-effect concentration.
[b] The numbers in parentheses indicate the 95% confidence interval.
[c] Acute test not performed on this sample.
[d] Confidence intervals were not able to be calculated.

Results and Discussion

Acute and Chronic Toxicity

The acute toxicity of the Patapsco WWTP's primary and secondary wastewater effluents was monitored approximately weekly from April 1986 to January 1987. Those samples evaluated included 45 secondary effluent samples and 13 primary effluent samples. When primary and secondary effluent samples were evaluated on the same day, the secondary sample was lagged 8 h so the sample would reflect the treatment plant's hydraulic retention time. Most of the samples were collected on a Thursday or Friday which historically have been two of the more toxic event days at the Patapsco WWTP, based on Microtox EC_{50} values [*1,2*]. Hence, the toxicity values presented in this study represent a bias toward the selection of sampling days for a higher potential of more toxic effluents. Tables 1 and 2 present the acute toxicity of Microtox, *Mysidopsis bahia,* and *Ceriodaphnia dubia* to both secondary and primary effluent samples from the Patapsco WWTP.

Of the three species tested, *C. dubia* was the most sensitive indicator of acute toxicity for both the seondary and primary effluent wastewaters. Mysids were generally more sensitive to secondary effluent wastewater than Microtox (Table 1), while Microtox was more sensitive to primary effluent wastewater than mysids (Table 2). The 48-h LC_{50} values for *C. dubia* ranged from 0.5 to 23.4% secondary effluent and from 0.1 to 6.5% primary effluent wastewaters. Microtox 5-min EC_{50} values ranged from 9.4 to 100% effluent and from 0.6 to 19.5% effluent, while *M. bahia* 96-h LC_{50} values ranged from 8.9 to 100% effluent and from 14.8 to 35.4% effluent for secondary and primary effluent wastewater samples, respectively. Generally, *C. dubia* was 7.5 and 12.5 times more sensitive than mysids and Microtox respectively, when exposed to secondary effluent, and 8.8 and 3 times more sensitive than mysids and Microtox, respectively, when exposed to primary effluent.

The correlation of the EC_{50}/LC_{50} values of the three species was evaluated using Pearson's product-moment correlation [*11*]. A significant correlation (P = 0.002) was observed for *C. dubia* and mysids exposed to secondary effluent; however, the r value demonstrates a weak correlation (r = 0.45). All other correlations of the acute values between the test organisms were not significant ($P < 0.05$). While the correlation between *C. dubia* and mysids was only 0.45, and the acute responses of one species may only be a weak predictor

TABLE 2—*Acute and chronic toxicity data for three aquatic organisms exposed to primary effluent wastewater from the Patapsco WWTP.*

Sample Date	Toxicity, % effluent			
	Microtox, 5-min EC_{50}	*Mysidopsis bahia,* 96-h LC_{50}	*Ceriodaphnia dubia* 48-h LC_{50}	*Ceriodaphnia dubia* 7-day CV[a]
4/23/86	19.5	21.1 (16.2 to 27.2)[b]	2.9 (1.6 to 3.2)	1.0
4/30/86	9.3	27.9 (20.0 to 33.2)	2.2 (1.6 to 3.2)	1.0
5/14/86	19.2	33.3 (25 to 50)	4.6 (3.2 to 6.5)	...[c]
5/28/86	13.6	20.6 (12.5 to 50)	2.2 (1.6 to 3.2)	1.0
6/18/86	8.7	35.4 (30.1 to 41.5)	1.4 (0.9 to 2.1)	0.4
7/16/86	3.7	18.2 (15.4 to 22.0)	1.1 (0.8 to 1.6)	0.3
7/23/86	4.4	26.9 (22.1 to 33.4)	1.1 (0.8 to 1.6)	0.3
8/6/86	4.7	15.1 (12.1 to 18.6)	1.1 (0.8 to 1.6)	0.3
8/13/86	12.4	14.8 (10.9 to 19.6)	4.6 (3.2 to 6.5)	1.2
9/10/86	0.6	23.7 (17.4 to 36.2)	4.6 (3.2 to 6.5)	1.5
10/8/86	3.2	30.1 (25.0 to 36.2)	1.2 (0.8 to 1.6)	0.3
10/22/86	3.3	15.3 (12.3 to 19.1)	0.1 (0.1 to 0.1)	<0.1
10/29/86	1.2	16.2 (11.4 to 21.7)	6.5 (0.2 to 10.0)	6.5

[a] CV is equivalent to the lowest of the seven-day LC_{50} values or the geometric mean of the no-observable-effect concentration and the lowest-observable-effect concentration.
[b] The numbers in parentheses indicate the 95% confidence interval.
[c] Chronic test not performed on this sample.

of the other species acute response, the LC_{50} values of mysids and *C. dubia* are somewhat in agreement, as demonstrated by the highly significant correlation ($P = 0.002$).

It is interesting to note that except for one month (October), the mean monthly 5-min EC_{50} values for Microtox exposed to secondary effluent were consistently 80.2 to 97.1% effluent (Table 3). However, the 5-min EC_{50} value for October was 58.4% effluent. This consistency of mean monthly acute toxicity responses is not evidenced with *M. bahia* and *C. dubia* as it was with the Microtox test (Table 3). From April to September, the *M. bahia* mean monthly 96-h LC_{50} ranged from 48.1 to 69.3% secondary effluent. However, starting in October an increase in toxicity was indicated by mean monthly 96-h LC_{50} values ranging from 12.4 to 37.2% secondary effluent. A similar increase in toxicity was observed for *C. dubia* starting in October; however, it was not as obvious an increase in toxicity as with

TABLE 3—*Mean (±standard deviation) monthly acute toxicity for three aquatic organisms exposed to secondary effluent wastewater from the Patapsco WWTP, April 1986–January 1987.*

	Acute Toxicity, % effluent					
	Microtox		*Mysidopsis bahia*		*Ceriodaphnia dubia*	
Month	5-min EC_{50}	*N*	96-h LC_{50}	*N*	48-h LC_{50}	*N*
April	84.4 (±20.3)	6	48.1 (±12.8)	6	6.1 (±1.6)	6
May	82.0 (±22.2)	3	69.3 (±12.0)	3	10.4 (±6.4)	3
June	80.7 (±27.9)	5	54.4 (±33.4)	7	10.7 (±7.5)	7
July	80.6 (±27.4)	2	49.0 (±21.7)	5	5.5 (±2.2)	5
August	85.8 (±20.2)	5	54.7 (±15.9)	5	4.3 (±1.6)	5
September	80.2 (±16.1)	6	49.3 (±19.4)	6	8.4 (±1.9)	6
October	58.4 (±32.8)	7	37.2 (±31.4)	6	3.7 (±2.7)	7
November	97.1 (±4.0)	2	29.1 (±24.5)	2	5.6 (±1.3)	2
December	94.4 (±7.8)	2	12.4 (±5.1)	2	1.6 (±1.5)	2
January	85.3 (±20.7)	2	33.8 (±16.9)	2	0.8 (±0.2)	2
Pooled means	79.3 (±23.4)	40	47.6 (±23.1)	44	6.3 (±4.6)	45

mysids. The mean monthly 48-h LC_{50} for *C. dubia* from April to September ranged from 4.3 to 10.7% secondary effluent, while from October to January the values ranged from 0.8 to 5.6% secondary effluent. The months where the effluent appeared to be most toxic to *M. bahia* and *C. dubia* were December and January, respectively. Figures 2 and 3 present the acute toxicity values for the three species when exposed to secondary and primary effluent.

Historical mean monthly averages of 5-min EC_{50} values for Microtox exposed to final effluent samples show a decrease of the effluent's toxicity to Microtox [*1*]. From November 1980 to September 1982, the mean monthly 5-min EC_{50} values ranged consistently from approximately 2.5 to 8% final effluent. In October 1982, the 5-min EC_{50} values for Microtox started increasing, with the exception of 5 months (October 1982 and March–June 1983), and the 5-min EC_{50} values were consistently above 20% final effluent [*1*]. This increase in the 5-min EC_{50} values was most likely the result of the Patapsco WWTP starting secondary treatment processes in September 1982 [*1*]. The resulting steady increase of 5-min EC_{50} values from September 1982 to present is probably due to the continual acclimation of the plant biomass to the influent wastewaters.

In another study performed at the Patapsco WWTP, 15-min EC_{50} values for Microtox were determined from postchlorinated secondary effluent samples in March 1984 [*14*]. The four 15-min EC_{50} values ranged from 1.5 to 10.2% effluent; however, the authors attribute this lower than normal toxicity as measured by Microtox to chlorine toxicity [*14*].

C. dubia exposed to secondary and primary effluent wastewater were evaluated using the short-term chronic method [*5*]. The ChVs from April 1986 to January 1987 for secondary and primary effluent samples are presented in Tables 1 and 2. The seven-day ChVs ranged from <0.1 to 9.4% secondary effluent (Table 1) and <0.1 to 6.5% primary effluent (Table 2). The mean [± standard deviation (SD)] ChVs for the 45 secondary and 13 primary effluent samples were 2.5 (±2.1) and 1.2 (±1.8)% effluent, respectively. A similar trend of increasing chronic toxicity starting in October was indicated by mean monthly seven-day ChVs. The mean monthly seven-day ChVs ranged from 0.2 to 4.9% secondary effluent.

The resulting acute/chronic ratios for the secondary and primary effluent samples ranged from 1.0 to 23.0 and 1.0 to 4.0, respectively. The mean (±SD) acute/chronic ratios were 4.0 (±3.6) and 2.9 (±1.1) for secondary and primary effluent samples, respectively.

In March 1984, *C. dubia* and *Pimephales promelas* (fathead minnow) short-term chronic

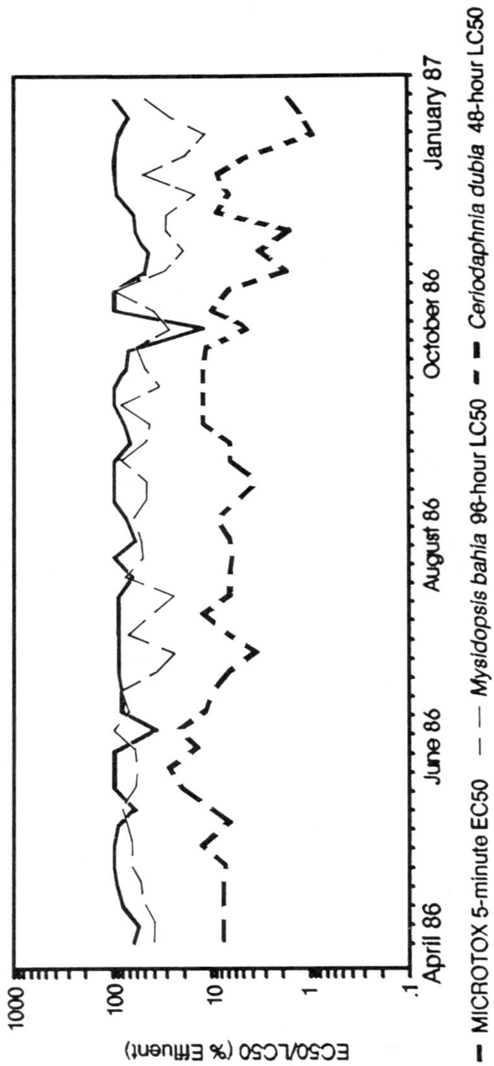

FIG. 2—*Acute toxicity of secondary effluent samples from the Patapsco WWTP using Microtox,* Mysidopsis bahia, *and* Ceriodaphnia dubia *as the test organisms.*

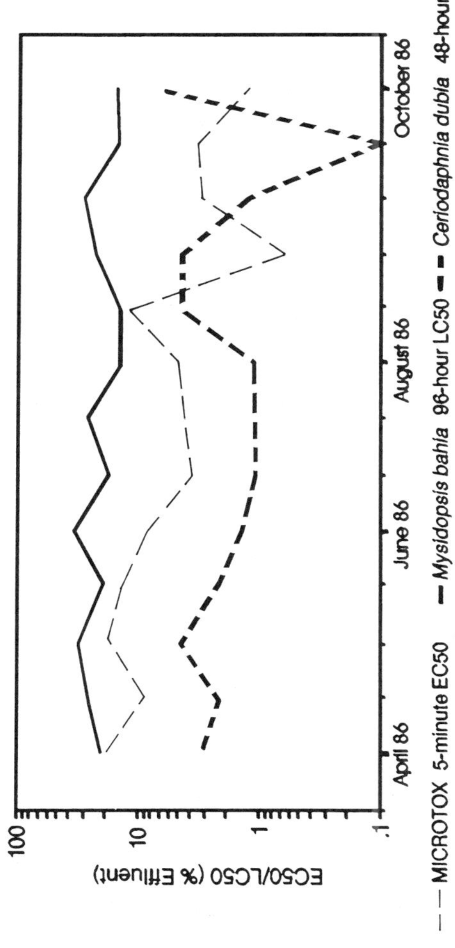

FIG. 3—*Acute toxicity of primary effluent samples from the Patapsco WWTP using Microtox,* Mysidopsis bahia, *and* Ceriodaphnia dubia *as the test organisms.*

tests were performed on postchlorinated secondary effluent from Patapsco WWTP [*14*]. The seven-day fathead minnow ChV was 54.7% effluent with an NOEC of 30% and an LOEC of 100% effluent. The *C. dubia* seven-day ChV was 0.53% effluent, with an NOEC of 0.37% and an LOEC of 0.75% effluent [*14*]. Both tests, performed using chlorinated effluent that was also supersaturated with CO_2, probably resulted in lower ChVs than would have been determined using unchlorinated effluent samples. Also, the ChV for the fathead minnow was subject to substantial error because of the exposure concentration interval between the NOEC and LOEC [*14*]. These two ChVs indicate that *C. dubia* are considerably more sensitive chronically to Patapsco secondary effluent than fathead minnows.

It is difficult to determine why the relative sensitivity of *C. dubia* and *M. bahia* differ in Patapsco wastewater because only a small data base exists in the literature for comparing these organisms. No studies could be found that compared *C. dubia* with other species; however, a few studies have compared *C. reticulata,* the same genus as *C. dubia,* to other organisms. Mount and Norberg [*7*] suggested that *C. reticulata, Daphnia pulex,* and *D. magna* have similar sensitivities to effluent and single chemicals, and generally did not differ in toxic response by more than a factor of 1 or 2 when compared with 13 substances tested. LeBlanc [*15*] compared the acute sensitivities of several species, including *D. magna* and *M. bahia,* to nonpesticide organics. Of the 19 different chemicals investigated, *M. bahia* was more sensitive to 13 compounds, *D. magna* was more sensitive to 4 compounds, and for 2 chemicals, comparisons could not be made [*15*].

Based on the few available studies, it can be concluded that *C. dubia* and *M. bahia* are similar in their acute responses to many toxic substances. However, it is difficult to determine the reason or reasons that *M. bahia* are less sensitive to the Patapsco WWTP wastewaters in this study. One possible reason may be that the toxicity of the effluent is physically or chemically modified, or both, by the addition of synthetic sea salts. The salts added to the test samples may coagulate or bind toxic organics and metals making them unavailable to the test organisms, and thereby reducing the toxicity to *M. bahia.* The possibility that toxicity is mitigated by the saltwater is important because the effluent from the Patapsco WWTP enters the Patapsco River estuary. Another possible reason for the lack of sensitivity of *M. bahia* may be that *C. dubia* is simply more sensitive to the toxic component or components of the Patapsco wastewater as evidenced by the differences in *C. dubia* and fathead minnow ChVs [*14*].

Effluent Fractionation

The chemical/physical fractionation was performed using a modified Anderson-Carnahan and Mount [*12*] procedure (Fig. 1). Five secondary effluent samples and one primary effluent sample was fractionated into chemical/physical fractions. Table 4 summarizes the results of the acute toxicity of each of the fractionated samples. In all cases, the most toxic fraction was found in the C_{18} fraction which suggests that of the samples evaluated, the toxic component or components were nonpolar compounds. The two methanol fractions from the C_{18} column that were consistently the most toxic were the 90% methanol fraction analyzed from the three samples fractionated in the summer and the 85% methanol samples fractionated during the winter. In all cases, the 25, 50, and 100% methanol fractions were nontoxic (48-h $LC_{50} > 400\%$ sample). Methanol fractions with 48-h LC_{50} values greater than 100% sample were achievable because the technique used concentrated the sample to a maximum test concentration of 500% sample.

The stepwise toxicity reduction can be defined as the percent of the toxicity removed in each step of the fractionation procedure. For example, the aeration step may remove 10%

the original samples.

Sample Type	SE,[a] 9 July 86[b]	PE, 23 July 86[c]	SE, 23 July 86	SE, 10 Dec. 86	SE, 6 Jan. 87	SE, 8 Jan. 87
Whole effluent	...[d]	...[d]	...[d]	1.7 (1.2 to 2.3)[e]	0.6 (0.5 to 1.0)	1.9 (1.1 to 3.4)
Whole effluent, aerated to constant pH	4.8 (3.8 to 6.6)	1.1 (0.8 to 1.6)	4.6 (3.2 to 6.5)	2.7 (1.8 to 4.2)	0.7 (0.4 to 1.6)	1.0 (0.3 to 1.8)
Filtered effluent (1.0 μm), unaerated	5.9 (3.2 to 13.5)	4.6 (3.2 to 6.5)	14.6 (6.5 to 40.5)	5.5 (3 to 10)	2.3 (1.5 to 3.4)	4.6 (3 to 7)
Filtered effluent, aerated to constant pH	9.4 (6.5 to 13.5)	2.9 (1.6 to 6.5)	9.4 (6.5 to 13.5)	5.5 (0.4 to 10)	4.6 (3 to 7)	2.5 (1.6 to 4.0)
Filtered effluent, raise pH >11, aerate 1 h, and neutralize	...[d]	...[d]	...[d]	5.5 (3 to 10)	19.4 (15 to 25)	7.5 (4.3 to 26.2)
C_{18} column pass-through	>100	>100	>100	51.5 (41.0 to 68.4)	>100	>100
C_{18} column eluted with						
25% MeOH	>500	>500	>500	>500	>500	>500
50% MeOH	>500	>500	>500	>500	>500	>500
75% MeOH	>500 ...	380.4 (250 to 500)	>500	42.7 (25 to 100)	133.4 (50 to 250)	24.0 (16.2 to 47.0)
80% MeOH	273.0 (100 to 500)	70.2 (50 to 100)	70.2 (50 to 100)	24.0 (21.2 to 34.8)	22.0 (19.4 to 25.0)	1.5 (1.0 to 2.8)
85% MeOH	76.1 (50 to 100)	22.6 (20.4 to 25.2)	35.4 (25 to 50)	15.6 (12.5 to 19.4)	10.8 (9.4 to 12.6)	1.3 (0.8 to 2.6)
90% MeOH	10.8 (9.5 to 12.6)	7.0 (5.0 to 8.8)	9.8 (6.1 to 14.3)	158.1 (100 to 250)	133.4 (50 to 250)	345.4 (25 to 50)
95% MeOH	38.0 (25 to 50)	32.9 (25 to 50)	32.9 (25 to 50)	>500	353.5 (250 to 500)	500 (250 to > 500)
100% MeOH	>500 ...	353.6 (250 to 500)	>500 ...	> 500 ...	426.8 (250 to 500)	426.8 (250 to 500)
Dowex 50W-X8 (cation exchange resin)	...[d] ...	...[d] ...	...[d] ...	50.0 (25 to 100)	...[d] ...	...[d] ...
Dowex 2-X8 (anion exchange resin)	...[d] ...	...[d] ...	...[d] ...	51.5 (41.0 to 68.4)	...[d] ...	...[d] ...

[a] SE = secondary effluent wastewater sample.
[b] Sample date.
[c] PE = primary effluent wastewater sample.
[d] Acute toxicity test not performed on this sample.
[e] The numbers in parentheses indicate the 95% confidence interval.

of the total toxicity of the effluent sample. The next step, filtration, may remove 50% of the toxicity remaining after aeration. This stepwise accounting of toxicity removal continues until all of the toxicity is accounted for (that is, 48-h $LC_{50} > 100\%$ sample) or if the toxicity was not removed with additional treatments (that is, cation or anion treatment for the sample on 10 December 1987). The stepwise reduction of toxicity for three of the fractionated samples is presented in Fig. 4.

Aeration of the effluent samples removed approximately 15 to 40% of the acute toxicity (Table 4, Fig. 4). Filtration of the effluent sample removed approximately 30 to 70% of the remaining toxicity after aeration. Ammonia stripping reduced the remaining toxicity of the effluent after filtration by 40 to 75% in only two of the three winter samples. The C_{18} column (nonpolar organics) removed >90% of the remaining acute toxicity after aeration, filtration, and ammonia stripping.

Further investigations of the acute toxicity of the fractionated effluent samples were performed on each percent methanol fraction. The relative toxicities of the nonpolar organic fractions for three of the effluent samples fractionated are presented in Fig. 5. In all cases, the most acutely toxic methanol fraction was the 85 and 90% fraction (Table 4, Fig. 5). An attempt was made to identify what the toxic component or components were in each of the toxic methanol fractions using gas chromatography/mass spectrometry (GC/MS) techniques; however, analytical determinations of the toxic components have not been successful to date in these samples. Approximate *n*-octanol-water partitioning coefficients (Log *P*) values were determined for some of the general percent methanol fractions. These preliminary Log *P* values for the methanol fractions are the following: 50% methanol fraction, Log $P = <1.5$; 75% methanol fraction, Log $P = 1.4$ to 2.4; 80% methanol fraction, Log $P = 2.7$ to 4.4; 85% methanol fraction, Log $P = 3.6$ to 8.3; and the 90% methanol fraction, Log $P = \sim 6$ to 9. While the analytical determination of the effluent sample is important, the knowledge of the approximate Log *P* of the toxic fraction is extremely helpful in the selection of treatment options (that is, powdered activated carbon, and so forth).

Fractionation of secondary effluent samples from the Patapsco WWTP was performed in March 1984 [*14*] using the Walsh and Garnas [*13*] procedure. In both samples that were fractionated, most of the toxicity was attributed to the base-neutral subfractions of the organic fraction. Chemical analyses on the base-neutral subfractions did not identify the toxic components among the pesticides, herbicides, polychlorinated biphenyls (PCBs), and priority pollutants tested [*14*].

Data gathered from the two fractionation efforts do agree that most of the toxicity is organic and more specifically a nonpolar organic compound. However, one sample that was fractionated using the Walsh and Garnas [*13*] procedure did indicate that some toxicity was coming from the cation fraction [*14*]. The sample that was fractionated on 10 December 1986 suggests indirectly that some of the toxicity may be from a polar organic compound. This indirect assumption is made because the C_{18} column and the cation and anion exchange resins treatments only removed enough toxicity to result in a 48-h LC_{50} of 51.5% effluent.

Summary and Conclusions

Of the three species utilized for acute toxicity testing, *Ceriodaphnia dubia* was the most sensitive species to Patapsco WWTP effluent wastewaters. *Mysidopsis bahia* was generally more sensitive to secondary effluent wastewaters than Microtox, while Microtox was more sensitive to primary effluent wastewaters than *M. bahia*. The 48-h LC_{50} values for *C. dubia* ranged from 0.5 to 23.4% secondary effluent and 0.1 to 6.5% primary effluent wastewaters. Microtox 5-min EC_{50} values ranged from 9.4 to 100% secondary effluent and 0.6 to 19.5%

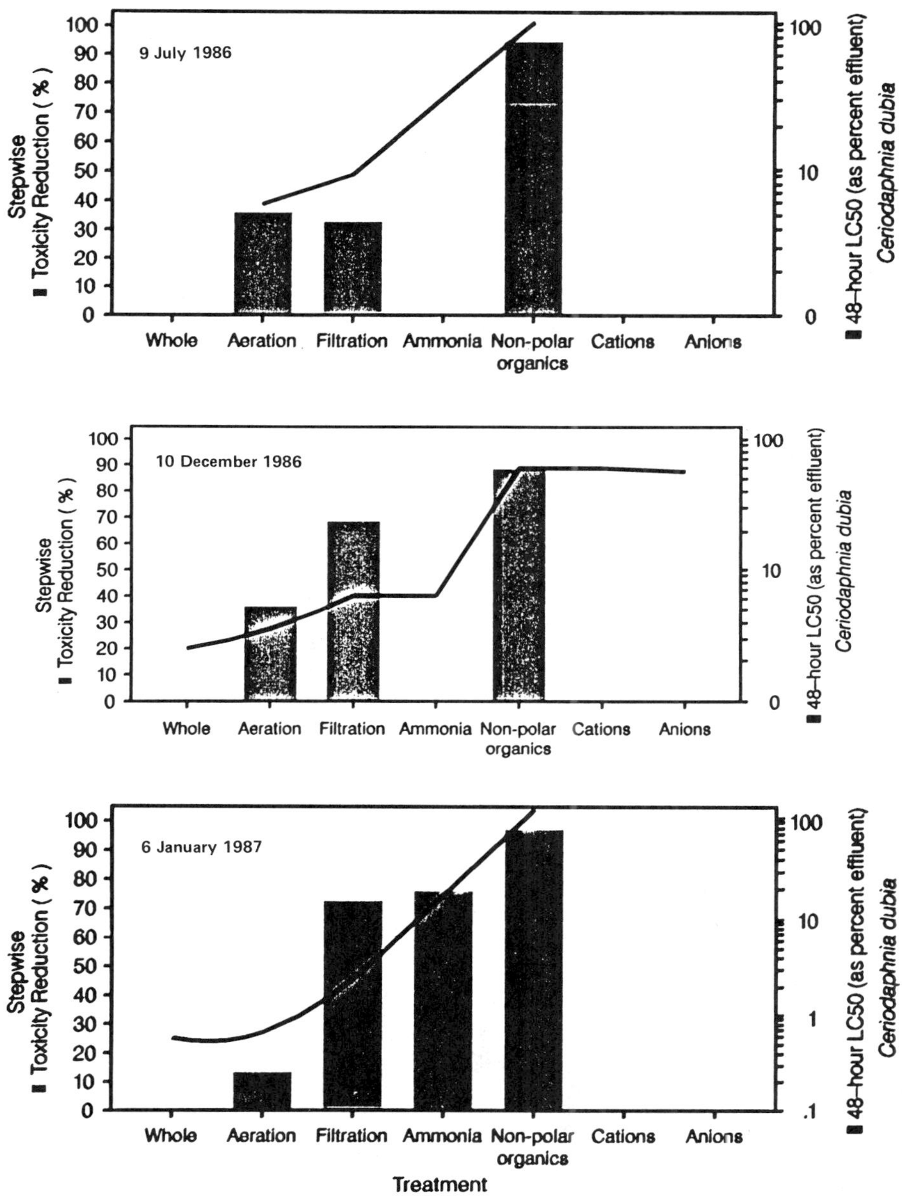

FIG. 4—*Stepwise toxicity reductions and acute toxicity of various fractions of a secondary effluent sample from the Patapsco WWTP.*

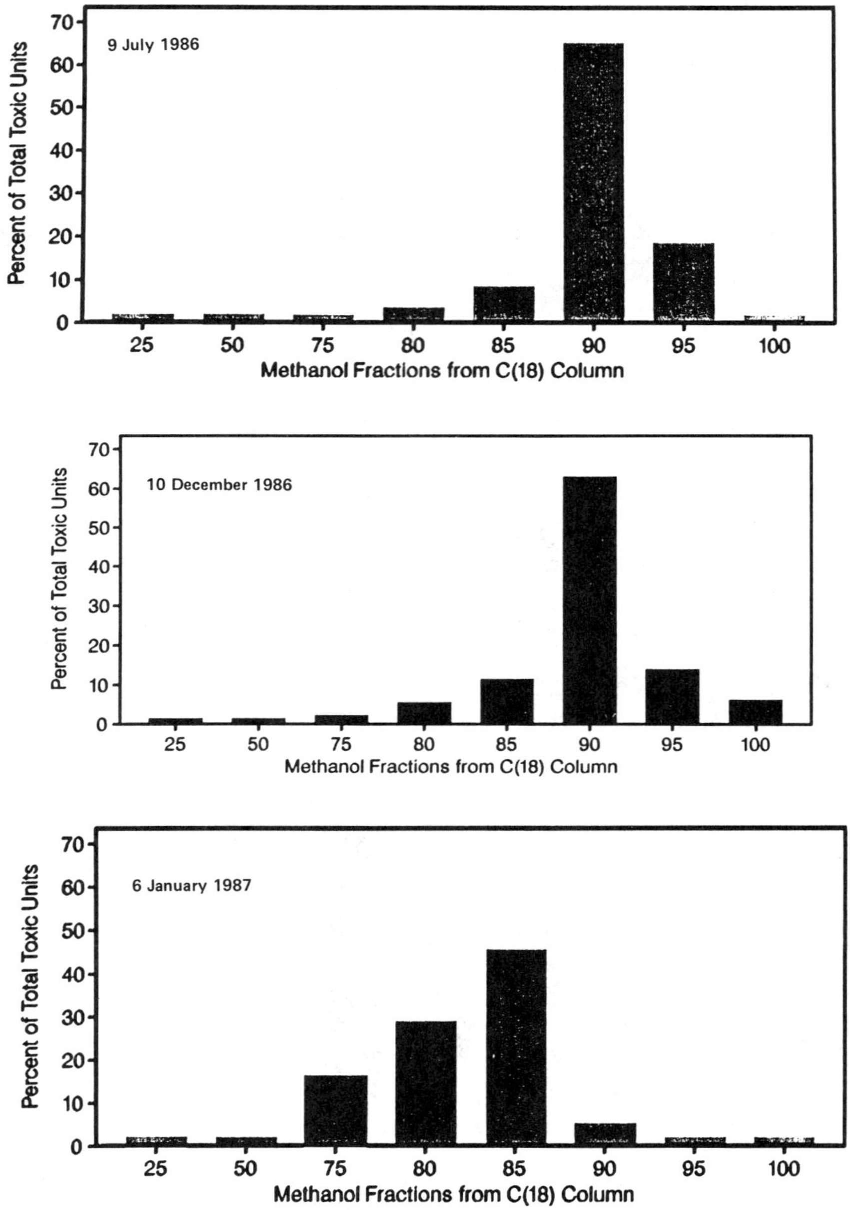

FIG. 5—*Relative toxicity of the nonpolar organic fractions of a secondary effluent sample from the Patapsco WWTP.*

primary effluent, respectively. *M. bahia* 96-h LC_{50} values were 8.9 to 100% secondary effluent and 14.8 to 35.4% primary effluent, respectively. A significant correlation of *C. dubia* and *M. bahia* acute values was observed; however, the *r* value was only 0.45. The percentage of toxicity reduction across the plant was approximately 50 to 90% for the three species tested. The observed acute/chronic ratios for *C. dubia* were low, approximately 3 to 4.

From the fractionation studies, aeration removed approximately 15 to 40% of the acute toxicity. The filtration step removed an additional 30 to 70% of the remaining toxicity after aeration, while ammonia stripping reduced the remaining toxicity of the effluent (after filtration) in two of the three winter samples by 40 to 75%. The nonpolar organic compounds appeared to be the major toxic component of the effluent samples fractionated, removing over 90% of the remaining toxicity. While identification of the specific compounds causing toxicity in the nonpolar organic fraction was not possible, the Log *P* values for the most toxic fractions were estimated to be approximately 6 to 9 in the three summer samples and approximately 3.6 to 8.3 in the three winter months. Cations, anions, and polar organic compounds do not appear to be major contributors to the toxicity of the effluent samples that were fractionated.

Acknowledgments

We would like to thank G. Slattery, P. Keenan, B. Sklar, R. Cox, E. Whitney, and D. Pietryka of the City of Baltimore for their valuable assistance in coordinating and performing the extensive analytical services required for this project. We would also like to thank D. Lemarie, J. Boraczek, N. Belinko, M. Kotulak, V. Sohn, K. Geckle, C. Pace, J. Ferris, C. Hill, B. Keisel, M. Fisher, B. Alianiello, R. Daniels, and M. Williams for their technical assistance in the toxicity and fractionation testing. We acknowledge W. Rue, D. Burris, and L. Cessar for their valuable assistance in the fractionation procedure. We especially thank D. Mount of U.S. EPA, Environmental Research Laboratory, Duluth, Minnesota, for allowing us to learn from his fractionation efforts. And finally, we wish to thank EA's Communications Department and Imagic for the preparation of this manuscript. The work being conducted for the Patapsco WWTP TRE is funded under a cooperative agreement between the U.S. EPA and the City of Baltimore.

References

[1] Slattery, G. H., "Effects of Toxic Influent on Patapsco Wastewater Treatment Plant Operations," *Proceedings,* Water Pollution Control Federation Annual Conference, New Orleans, LA, 1–4 Oct. 1984.

[2] "Evaluation of Alternatives for the Reduction of Toxic Inhibition at the Patapsco Wastewater Treatment Plant," prepared for the City of Baltimore, MD, Engineering-Science, Inc., 1985.

[3] Sprague, J. B., "Measurement of Pollutant Toxicity to Fish. I. Bioassay Methods for Acute Toxicity," *Water Resources,* Vol. 3, 1969, pp. 793–821.

[4] Peltier, W. H. and Weber, C. I., "Methods for Measuring the Acute Toxicity of Effluents to Freshwater and Marine Organisms," Third ed., EPA/600/4-85-013, U.S. Environmental Protection Agency, Environmental Monitoring and Support Laboratory, Cincinnati, OH, March 1985.

[5] Horning, W. B., II and Weber, C. I., "Short-Term Methods for Estimating the Chronic Toxicity of Effluents and Receiving Waters to Freshwater Organisms," EPA/600/4-85-014, U.S. Environmental Protection Agency, Environmental Monitoring and Support Laboratory, Cincinnati, OH, December 1985.

[6] *Microtox Operating System Manual,* Beckman Instruments, Inc., Carlsbad, CA, 1982.

[7] Mount, D. I. and Norberg, T. J., "A Seven-Day Life-Cycle Cladoceran Toxicity Test," *Environmental Toxicology and Chemistry,* Vol. 3, 1984, pp. 425–434.

[8] *Standard Methods for the Examination of Water and Wastewater,* 16th ed., American Public Health Association, Washington, DC, 1985.
[9] Finney, D. J., *Probit Analysis,* Cambridge University Press, London, 1971.
[10] Stephan, C. E., "Methods for Calculating an LC_{50}," *Aquatic Toxicology and Hazard Evaluation, ASTM STP 634,* F. L. Mayer and J. L. Hamelink, Eds., American Society for Testing and Materials, Philadelphia, 1977, pp. 65–84.
[11] *SAS User's Guide: Basics,* Version 5 ed., SAS Institute, Inc., Cary, NC, 1985.
[12] Anderson-Carnahan, L. and Mount, D. I., "Methods for Toxicity Reduction Evaluations, Phase I—Toxicity Characterization Procedures," U.S. Environmental Protection Agency, Washington, DC, 1987.
[13] Walsh, G. E. and Garnas, R. L., "Determination of Bioactivity of Chemical Fractions of Liquid Wastes Using Freshwater and Saltwater Algae and Crustaceans," *Environmental Science and Technology,* Vol. 17, No. 3, 1983, pp. 180–182.
[14] Mount, D. I., Steen, A. E., and Norberg-King, T., "The Validity of Effluent and Ambient Toxicity Tests for Predicting Biological Impact, Back River, Baltimore Harbor, Maryland," EPA/600/8-86-001, U.S. Environmental Protection Agency, Environmental Research Laboratory, Duluth, MN, July 1986.
[15] LeBlanc, G. A,. "Interspecies Relationship in Acute Toxicity of Chemicals to Aquatic Organisms," *Environmental Toxicology and Chemistry,* Vol. 3, 1984, pp. 47–60.

Steven C. Schimmel,[1] *Glen B. Thursby,*[2] *Margarete A. Heber,*[3] *and Melissa J. Chammas*[4]

Case Study of a Marine Discharge: Comparison of Effluent and Receiving Water Toxicity

REFERENCE: Schimmel, S. C., Thursby, G. B., Heber, M. A., and Chammas, M. J., **"Case Study of a Marine Discharge: Comparison of Effluent and Receiving Water Toxicity,"** *Aquatic Toxicology and Environmental Fate: Eleventh Volume, ASTM STP 1007,* G. W. Suter II and M. A. Lewis, Eds., American Society for Testing and Materials, Philadelphia, 1989, pp. 159–173.

ABSTRACT: An on-site investigation was conducted in northeast Florida to evaluate the toxicity of a pulp and paper mill discharge and to determine whether there was any receiving water toxicity associated with that discharge. The species tested included the macroalga *Champia parvula,* the mysid *Mysidopsis bahia,* and two fish species, *Cyprinodon variegatus* and *Menidia beryllina.* All methods evaluated the critical, sensitive life stages and estimated chronic toxicity. Test results indicated that effluent effect concentrations were approximately 1% for the two most sensitive species (mysids and the macroalga) and 32% for the least sensitive *(C. variegatus).* Receiving waters adjacent to the discharge were sampled and tested daily over a seven-day period. Water samples collected at stations closest to the discharge were toxic to the macroalga, and the effect concentrations (estimated by a previous dye study) were consistent with those determined in the effluent tests. In addition, the Environmental Protection Agency's (EPA's) Duluth Laboratory conducted on-site chemical fractionation studies using the freshwater crustacean, *Ceriodaphnia reticulata.* Results of these studies indicated that unionized ammonia was the primary toxic component of the effluent. Data in this case study, from the suite of methods developed (or modified) at EPA's Narragansett Laboratory, appear suitable for the derivation of toxicity-based National Pollutant Discharge Elimination System permits for the marine environment.

KEY WORDS: effluents, toxicity, marine, estuarine, pulp mill, fish, mysids, algae, ammonia, discharge permit, aquatic toxicology

Changes in the regulations within the National Pollutant Discharge Elimination System (NPDES) [*1*] of the 1977 Clean Water Act (PL 95-217) now allow for the incorporation of toxicity tests into a permit as regulatory tools. These changes have also been incorporated into the Water Quality Act of 1987. Specific effluent test methods have been developed to evaluate the toxicity of discharges to freshwater systems using the freshwater cladoceran, *Ceriodaphnia* sp. [*2*], and the fathead minnow, *Pimephales promelas* [*3*]. These methods have proven successful, and their use is now required in nearly 2000 of over 65 000 permits [*4*].

[1] Supervisory research aquatic biologist, Environmental Research Laboratory, U.S. Environmental Protection Agency, Narragansett, RI 02882.

[2] Research assistant professor, Botany Dept., University of Rhode Island, Kingston, RI 02881.

[3] Environmental protection specialist, Permits Div., U.S. Environmental Protection Agency, Washington, DC 20460.

[4] Associate biologist, Science Applications International Corp., Narragansett, RI 02882.

The development of effluent and ambient toxicity test methods for estuarine and marine organisms began at the Environmental Protection Agency (EPA) Environmental Research Laboratory at Narragansett, Rhode Island (ERLN) nearly two years after the freshwater tests were developed. Five species were selected on the basis of their ecological importance, successful use in the laboratory, and sensitivity to selected toxicants. They included a macroalga *(Champia parvula)*, a sea urchin *(Arbacia punctulata)*, a crustacean *(Mysidopsis bahia)* and two fish species *(Cyprinodon variegatus and Menidia beryllina)*. Toxicity test methodologies using these species were developed specifically for the evaluation of effluents entering marine and estuarine waters [5]. They were developed in the laboratory and then taken into the field and tested on-site in a mobile laboratory.

The purpose of the present study was (*a*) to evaluate the toxicity of an effluent from a pulp and paper mill that discharges into a marine receiving water using the suite of marine test organisms, (*b*) to determine the relationship of effluent toxicity and receiving water toxicity within and outside of a proposed mixing zone, and (*c*) to generate all the necessary toxicity data for deriving a discharge permit for the pulp mill plant. Presented here are the results of the effluent toxicity tests, receiving water test results, and data indicating the probable causes of the toxicity.

Discharge Site

The effluent selected for evaluation was from a pulp and paper mill located in Fernandina Beach, Florida on the Amelia River (Fig. 1). The plant discharges approximately 60 million litres each day (16 million gallons per day) to the river. The Amelia River at the point of discharge is a high-salinity tidal basin with a salinity averaging $\geq$30 parts per thousand. The existing permit allows the plant to discharge only during an outgoing tide, maximizing dilution in the receiving water. (Limited freshwater input provides little or no downstream transport other than during a receding tide.) The discharge pipe is in the river channel which is dredged and maintained as part of the Intercoastal Waterway system. The pipe is fitted with a multiple-port diffuser to enhance dilution. Channel depth at the discharge site is approximately 8 m, and the mean tidal amplitude is approximately 2 m.

A dye study was conducted by the EPA's Region IV Environmental Services Division (ESD) during 5–9 May 1986. Measurements of the dye cloud movement were used to develop wastewater dilution concentrations for periods of initial dilution, high slack tide, low slack tide, and average conditions. Results of the 5–9 May 1986 dye study were used in this report to estimate effluent concentrations in whole receiving water samples collected and tested one week later (14–21 May 1986). Most, but not all, dye sampling stations were used as receiving water stations. Receiving water sampling stations are shown in Fig. 1.

Effluent and Receiving Water Collections

Daily composite (24-h) effluent collections were made during the seven days of the study. Effluents were collected from a lagoon immediately adjacent to the discharge. Samples were collected and maintained in a chilled receptacle until they were used in the toxicity tests (within 18 h of collection). No physical or chemical alterations, other than temperature changes, were made on the effluent before they were diluted with seawater for testing.

Receiving waters were collected daily as grab samples from seven stations (1, 2, 3, 5, 8, 10, and 11), and all stations were sampled on 18–20 May 1986 (Fig. 1). Samples were collected using a nonmetallic pump that operated approximately 30-cm below the surface. All samples were collected during an ebbing tide that coincided with the time of effluent

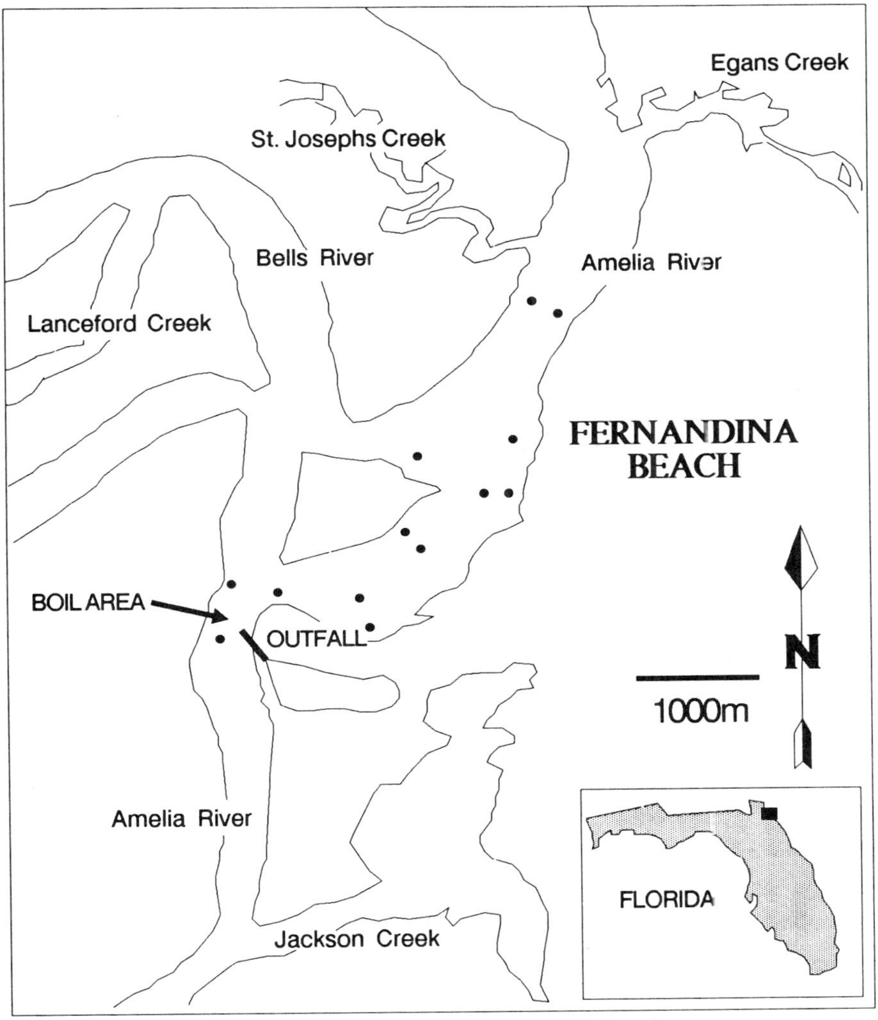

FIG. 1—*Amelia River receiving water sampling stations and location of pulp and paper mill effluent discharge.*

discharge. Receiving water samples were maintained on ice before they were used in toxicity tests. No physical or chemical alterations were made on the receiving water samples, other than temperature changes, before or during the toxicity tests.

Effluent and Receiving Water Toxicity Tests

Four marine species were used to evaluate the toxicity of the pulp mill effluent. The test exposure durations ranged from two days *(C. parvula)* to seven days (*M. Bahia, C. variegatus,* and *M. beryllina*). (*C. parvula* was exposed for two days, followed by a five-day-postexposure period.) Results from each test provided estimates of chronic toxicity in terms of either growth, survival, or reproduction. Abbreviated test methods for each species are discussed below.

Champia parvula *Reproduction Test*

The algal reproduction test uses the marine macroalga, *Champia parvula* [*6,7*]. This red marine macroalga is found throughout the East and Gulf Coasts of North America. In brief, the method consists of exposing both male and female algae to effluents or receiving waters for two days, during which time fertilization may occur. The female algae are then removed and placed in control medium for a five- to seven-day recovery period. The recovery period allows time for cystocarps (evidence of sexual reproduction) to mature. At the end of the recovery period, the number of cystocarps per plant are counted. The end point measured when testing effluents with *C. parvula* is the absence of sexual reproduction and the statistical differences from the controls. A $\geq$95% decrease in the numbers of cystocarps compared to the control response is considered "zero" reproduction. Statistical differences in the number of cystocarps formed are used to determine the toxic effects of receiving waters relative to control waters.

Mysidopsis bahia *Growth, Survival, and Fecundity Test*

The mysid, *Mysidopsis bahia,* is a sensitive test organism frequently used to measure the toxicity of substances in marine and estuarine waters [*8*]. The mysid test method consists of a daily renewal exposure of seven-day-old juveniles for seven consecutive days. The animals are fed newly hatched *Artemia* sp. daily [*9*]. During this period the animals mature and the females produce ova in their oviducts or developing embryos in their brood pouches. The end points for the test are reproduction (measured as the percentage of females with ova or developing embryos), growth (measured as dry weight), and survival.

Cyprinodon variegatus *Growth and Survival Test*

The seven-day-larval sheepshead minnow *(Cyprinodon variegatus)* growth and survival test [*10*] was modeled after the freshwater fathead minnow procedure [*3*]. It consists of a seven-day exposure of newly hatched larvae to effluents or receiving waters; test solutions are renewed daily during the test. The method derives two end points, growth (measured as dry weight) and survival. Larvae are fed *Artemia* sp. daily.

Menidia beryllina *Growth and Survival Test*

The seven-day-larval silverside *(Menidia beryllina)* growth and survival test [*11*] is nearly identical to that for the sheepshead minnow, except seven-day-old larvae (posthatch) are used. The silverside procedure was developed to evaluate lower salinity receiving waters ($\geq$5 parts per thousand) and is often more sensitive than *C. variegatus* to many pollutants.

Statistics

Results of the *C. parvula* and fish growth and survival tests were analyzed by one-way analysis of variance (ANOVA) followed by Dunnett's mean separation test to determine significant (α = 0.05) differences between experimental concentrations and the control.

Fish survival data were subjected to arc sine square root transformation before ANOVA. Mysid survival and reproduction data were analyzed by one-way ANOVA and by Duncan's multiple-range test after arc sine square root transformations were applied. Mysid growth data were not transformed. In addition to significant differences, the results of *C. parvula* reproductive tests are also presented in terms of no sexual reproduction. This is indicated when reproduction in a treatment is decreased ≥95% of that observed in the control [7].

Results

Effluent Toxicity Tests

Effluent from the Fernandina Beach pulp and paper mill was toxic to all species at concentrations ≥32% (see Tables 1 through 5). Table 5 also shows the effluent concentrations (range from 0.08 to 2.0%) at the Amelia River receiving water sampling stations. The red alga *(C. parvula)* and the mysid *(M. bahia)* were the tests most sensitive to the effluent; the estimated no-observed-effect concentration (NOEC) for both species was 1.0% effluent. Results of the two *C. parvula* tests (Table 1) indicate very severe effects on reproduction in the 2.5% effluent concentration (0 and 0.2 cystocarps per plant, respectively), whereas there were no statistically significant effects on reproduction in the 1.0% effluent concentration. Mysid egg production was affected in 3.2% effluent, but it was not significantly reduced in the 1.0% concentration (Table 2). Egg production in the control females was not as high as was expected (90%) because the test temperature was not maintained above 25°C throughout the study. Another result of the lower temperature was an increase in variability of the number of females with eggs; a 65% reduction from the control response was required to detect statistical significance. Nevertheless, significant differences in egg production were observed at all concentrations above 1.0% effluent.

Growth of surviving sheepshead minnow (Table 3) and silverside larvae (Table 4) was unaffected in all the effluent concentrations in which there were survivors. Survival of the sheepshead minnow was affected in the 32% effluent concentration, and survival for the silverside was significantly different from the control survival in 10% effluent. Although silverside larval survival in the 3.2% concentration was only 58%, high variability among the three replicates prohibited the detection of a significant difference in that concentration.

TABLE 1—*The effect of a pulp and paper mill effluent on the formation of cystocarps by* Champia parvula.[a]

Percent Effluent	Number of Cystocarps per Plant, ± SD $n = 3$	
	17 May	18 May
Control	11 ± 2	14 ± 5
0.5	12 ± 3	13 ± 3
1.0	13 ± 1	9 ± 3
2.5	0.2[b]	0[b]
5.0	0[b]	0[b]
10.0	0[b]	0[b]

[a] Temperature was 23 to 25°C, salinity was 30 parts per thousand, and light density was ~100 μeinstens m-2s-1 of daylight fluorescent light on a 16 h:8 h, light/dark cycle. No necrotic tissue was observed at any of the concentrations tested.

[b] Significantly different from the control ($\alpha = 0.05$).

TABLE 2—*The effects of a pulp and paper mill effluent on reproduction, growth, and survival of* Mysidopsis bahia *in a seven-day static renewal test.*[a]

Effluent Concentration, %	Survival, %	Females with Eggs, %	Mean Dry Weight per Female Mysid ± SD, mg	Mean Salinity ± SD, parts per thousand	Mean Temperature ± SD, °C	Mean Dissolved Oxygen ± SD, mg/L
Control (brine + DI)	100	47	0.291 ± 0.08	35 ± 2	23.4 ± 1.2	5.1 ± 0.9
1.0	100	20	0.239 ± 0.05	35 ± 2	24.3 ± 1.2	4.9 ± 1.1
3.2	88	4[b]	0.273 ± 0.09	35 ± 3	23.8 ± 1.0	4.8 ± 0.7
10.0	41[b]	6[b]	0.219 ± 0.06	35 ± 4	24.2 ± 0.7	4.3 ± 0.9
32.0	0[b]	...	...	33 ± 6	23.4 ± 0.7	4.3 ± 0.9
50.0	0[b]	...	...	32 ± 0	23.5 ± 0.0	4.2 ± 0.7

[a] Eight replicates of five mysids were used in each concentration.
[b] Significantly different from the control (α = 0.05).

TABLE 3—*The effects of a pulp and paper mill effluent on the growth and survival of sheepshead minnow* (Cyprinodon variegatus) *larvae during an on-site seven-day exposure.*[a]

Effluent Concentration, %	Survival, %	Mean Dry Weight per Larvae ± SD, mg	Mean Salinity ± SD, parts per thousand	Mean Temperature ± SD, °C	Mean Dissolved Oxygen + SD, mg/L
Control (brine + DI)	98	0.875 ± 0.14	33 ± 0	25.6 ± 0.4	5.1 ± 0.3
1.0	98	1.021 ± 0.12	34 ± 0	25.2 ± 0.4	5.0 ± 0.1
3.2	98	0.937 ± 0.13	33 ± 1	25.3 ± 0.3	4.6 ± 0.1
10.0	93	0.882 ± 0.05	33 ± 1	25.4 ± 0.0	3.6 ± 0.1
32.0	11[b]	0.889 ± 0.17	34 ± 0	24.9 ± 0.4	4.8 ± 0.8
50.0	0[b]	...	34 ± 0	25.1 ± 0.2	6.2 ± 0.2

[a] Three replicates of 15 larvae were used in each concentration.
[b] Significantly different from the control ($\alpha = 0.05$)

TABLE 4—*The effects of a pulp and paper mill effluent on the growth and survival of inland silverside* (Menidia beryllina) *larvae during an on-site seven-day exposure.*[a]

Effluent Concentration, %	Survival, %	Mean Dry Weight per Larvae ± SD, mg	Mean Salinity ± SD, parts per thousand	Mean Temperature ± SD, °C	Mean Dissolved Oxygen ± SD, mg/L
Control (brine + DI)	87	0.963 ± 0.02	34 ± 0	25.3 ± 0.5	4.6 ± 0.2
1.0	80	1.071 ± 0.08	34 ± 0	25.4 ± 0.3	4.4 ± 0.1
3.2	58[b]	0.988 ± 0.18	34 ± 0.4	25.3 ± 0.1	4.4 ± 0.2
10.0	33[c]	0.737 ± 0.34	34 ± 0	25.2 ± 0.4	3.7 ± 0.2
32.0	0[c]	...	35 ± 0	24.3 ± 0.4	6.8 ± 0.3
50.0	0[c]	...	34 ± 0	23.9 ± 0.4	4.9 ± 1.1

[a] Three replicates of 15 larvae were used in each concentration.
[b] High variability between three replicates; value not significantly different from the control.
[c] Significantly different from the control ($\alpha = 0.05$).

TABLE 5—*The effect of receiving waters from the Amelia River on the formation of cystocarps by* Champia parvula.[a]

		Number of Cystocarps per Plant						
		17 May Replicate[b]				18 May Replicate		
Station No.	Percent Effluent[a]	A	B	C	Mean	A	B	Mean
1	2.0	0.4	0.8	0.2	0.5[c]	0.0	1.6	0.8[c]
2	. . .[d]	17.2	12.2	14.6	14.7	20.6	17.4	19.0
3	. . .[e]	1.0	1.6	0.4	1.0[c]	2.0	4.0	3.0[c]
4	1.5	. . .	. . .	. . .	. . .	11.8	9.0	10.4
5	. . .	4.6	1.6	0.6	2.3[c]	0.0	1.2	0.6[c]
6	0.05	. . .	. . .	. . .	. . .	1.6	0.6	1.1[c]
7	1.4	. . .	. . .	. . .	. . .	2.4	4.4	3.4[c]
8	0.08	1.6	11.6	11.6	8.3	23.6	18.4	21.0
9	1.4	. . .	. . .	. . .	. . .	1.2	1.0	1.1[c]
10	0.5	14.4	15.4	15.8	15.2	10.4	16.6	13.5
11	0.2	10.8	13.2	14.4	12.8	26.6	13.3	19.9
12	0.9	. . .	. . .	. . .	. . .	. . .	. . .	. . .
13	0.7	. . .	. . .	. . .	. . .	8.4	6.0	7.2
14	0.5	. . .	. . .	. . .	. . .	2.0	8.6	5.3[c]
15	0.5	. . .	. . .	. . .	. . .	5.8	4.6	5.2[c]
16	. . .	. . .	. . .	. . .	. . .	6.0	9.8	7.9
17	0.6	. . .	. . .	. . .	. . .	12.0	11.4	11.7
18	0.5	. . .	. . .	. . .	. . .	7.8	11.6	9.7
19	0.4	. . .	. . .	. . .	. . .	15.0	5.8	10.4
20	0.4	. . .	. . .	. . .	. . .	10.8	9.6	10.2
21	0.5	. . .	. . .	. . .	. . .	9.2	10.0	9.6

[a] Effluent percents were calculated based on a dye study. No necrotic tissue was observed in any of the treatments.

[b] Mean of five plants per replicate.

[c] Statistically less than Station 2 (ANOVA followed by Dunnett's mean separation test).

[d] Control station, assume percent of effluent to be close to zero.

[e] No data available.

A 37.7% reduction from the *M. beryllina* control response was required to detect statistically significant differences.

Receiving Water Toxicity Tests

Toxicity was detected at several collection stations in the Amelia River by the red alga (Table 5) and the mysid (Table 6), but not by the sheepshead minnow (Table 7). Only the results of the red alga and the sheepshead minnow tests demonstrated a relationship or correlation between the results expected on the basis of the experimental effluent studies and those observed in the field. Significant reproductive effects were detected by *C. parvula* exposed to concentrations predicted to be toxic from the effluent toxicity tests and the dye concentrations (Tables 1 and 5, respectively). For example, on 18 May the calculated amount of effluent present at receiving water station 1 was 2%, and the mean number of cystocarps present was 0.8 (0.8 cystocarps, for the purpose of this test method, equals 0). The mean number of cystocarps present in the effluent toxicity test at 2.5% effluent on 18 May was 0. The same direct comparisons can be made for Stations 1, 6, 7, 9, 14, and 15 on 17 and 18 May (Tables 1 and 5 and Fig. 2). No significant differences were detected by

TABLE 6—*The effects of ambient receiving waters (Amelia River, FL) on reproduction, growth, and survival of* Mysidopsis bahia *in a seven-day static renewal test, conducted on-site in a mobile laboratory.*[a]

Receiving Water Station No.	Effluent Concentration, %	Survival, %	Females with Eggs ± SD, %	Mean Dry Weight per Female Mysid ± SD, mg	Mean Temperature ± SD, °C	Mean Dissolved Oxygen ± SD, mg/L
Control (brine + DI)	0.0	100	47	0.291 ± 0.08	23.4 ± 1.2	5.1 ± 0.9
2 (ambient control)	0.0	93	71	0.355 ± 0.115	24.5 ± 1.0	4.7 ± 0.9
1	2.0	90	64	0.314 ± 0.061	24.4 ± 1.0	4.0 ± 0.9
3	...	88	77	0.269 ± 0.071	24.3 ± 0.7	4.1 ± 0.6
5	...	98	47	0.369 ± 0.118	24.2 ± 0.8	4.0 ± 0.8
8	0.08	93	14[b]	0.271 ± 0.069	24.1 ± 0.7	4.4 ± 0.8
10	0.5	93	66	0.317 ± 0.087	24.3 ± 0.9	4.1 ± 0.8
11	0.2	93	31	0.274 ± 0.064	24.0 ± 0.6	4.0 ± 0.7

[a] Eight replicates of five mysids were used in each concentration. Mean salinity was 34.4 ± 2 parts per thousand for all treatments.
[b] Significantly different from the ambient control ($\alpha = 0.05$); value judged not to be valid.

TABLE 7—*The effects of ambient receiving waters (Amelia River, FL) on the growth and survival of sheepshead minnow* (Cyprinodon variegatus) *larvae during an on-site seven-day exposure.*[a]

Receiving Water Station No.	Effluent Concentration, %	Survival, %	Mean Dry Weight per Individual ± DS, mg	Mean Temperature ± DS, °C	Mean Dissolved Oxygen ± SD, mg/L
Control (brine + DI)	0.0	98	0.875 ± 0.14	25.6 ± 0.4	5.1 ± 0.3
2 (Ambient control)	0.0	98	1.046 ± 0.02	25.2 ± 0.1	4.5 ± 0.5
1	2.0	98	1.033 ± 0.04	25.1 ± 0.1	4.6 ± 0.2
3	...	100	0.972 ± 0.02	25.1 ± 0.1	4.5 ± 0.1
5	...	100	0.920 ± 0.03	25.2 ± 0.1	4.5 ± 0.1
8	0.08	95	1.069 ± 0.04	25.2 ± 0.1	4.9 ± 0.1
10	0.5	100	1.028 ± 0.05	25.1 ± 0.2	4.7 ± 0.1
11	0.2	100	0.993 ± 0.04	25.1 ± 0.1	4.7 ± 0.1

[a] Three replicates of 15 larvae were used in each treatment. Mean salinity was 33.8 ± 2 parts per thousand for all treatments.

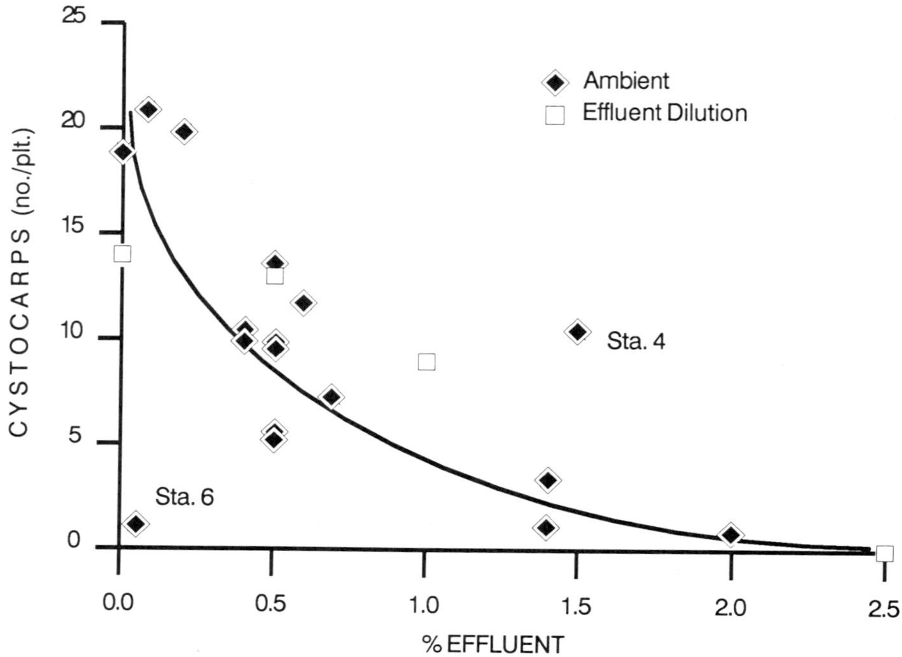

FIG. 2—*Number of* Champia parvula *cystocarps plotted against percent effluent for data from 18 May. The percent effluent for ambient samples was based on the dye study of the previous week (Table 5). The percent effluent for effluent dilution samples were derived from the effluent toxicity test (Table 1).*

the sheepshead minnow test, and none was anticipated because no toxicity was observed below 32% effluent (Table 3); no concentration of effluent over 2% was expected in the river based on the dye study.

The mysid receiving water toxicity study (Table 6) did not correlate well with the dye measurements conducted during the previous week (Table 5). Toxicity was detected only in Station 8 and was measured as the reduced number of females with eggs in their oviducts. Based on the previous dye study, the effluent concentration at Station 8 was expected to measure only 0.08% effluent. Other stations, where marginal toxicity may have been predicted, were not toxic (based on the effluent effect concentration of 3.2%; Table 2). Therefore, the reduction in mysid fecundity observed at Station 8 was statistically significant, but was not believed to be valid.

Probable Causes of Toxicity

Chemical fractionations and toxicity tests of the pulp and paper mill effluent were conducted on-site by Mount and Anderson-Carnahan (EPA, Duluth, MN, and EPA Region IV, Altanta, Georgia, respectively).[5] Their results indicated that ammonia appeared to be the primary cause of the observed toxicity. (The freshwater cladoceran, *Ceriodaphnia*

[5] Anderson-Carnahan, L., U.S. Environmental Protection Agency, Water Management Division, Region IV, Atlanta, GA, personal communication, December 1987.

TABLE 8—*Toxicity comparisons of calculated unionized ammonia in effluent from a pulp and paper mill and those from separate ammonia tests using* Champia parvula, Mysidopis bahia, *and* Menidia beryllina.[a]

Species	Effluent Tests: Lowest Effect Concentration, % effluent	Effluent Tests: Calculated Chronic Value for Unionized Ammonia, mg/L[c]	Ammonia Tests Chronic Value, mg/L
Champia parvula (red macroalga)	2.5	0.06	0.04
Mysidopsis bahia (mysid shrimp)	3.2	0.07	0.12
Menidia beryllina (inland silverside)	10.0	0.20	0.06

[a] The average measured total ammonia concentration in 100% effluent was 81 mg/L. The calculated unionized ammonia concentration in the effluent was 3.66 mg/L (pH = 8, 25°C, 32 parts per thousand salinity).

[b] Unpublished ERLN data, chronic numbers refer to calculated unionized ammonia concentrations from 2-day *C. parvula* tests, 32-day *M. bahia* tests, and 28-day *M. beryllina* tests.

[c] Estimated chronic value is the geometric mean of the highest no-effect concentration and the lowest effect concentration for 2-day tests *(C. parvula)* and 7-day tests (*M. bahia* and *M. beryllina*).

sp., was used in these evaluations.) The average measured concentration of total ammonia in 100% effluent was 81 mg/L.[5] The calculated unionized ammonia (NH_3) in the effluent at pH 8.0, 32 parts per thousand salinity is 3.66 mg/L.

The calculated unionized ammonia in the lowest whole effluent effect concentrations for *C. parvula, M. bahia,* and *M. beryllina* are shown in Table 8 along with data for unpublished chronic values from ERLN toxicity tests.[6] The two sets of values correlate well with one another. Results of the *C. parvula* two-day reproductive test were nearly identical. Similarly, comparisons of mysid and silverside chronic values differ by no more than a factor of 2 and 3, respectively. In addition, data from *C. parvula,* exposed only to ammonium chloride concentrations, closely matched that of its response to whole effluent (Fig. 3). Based on these data, it appears that unionized ammonia was the primary cause of toxicity to the three species, and it is likely to be the cause for *C. variegatus* as well.

Summary and Conclusions

Effluent from this pulp and paper mill was toxic to a variety of test organisms, and toxicity was detected in the receiving water. The most sensitive species tested was the red alga, *Champia parvula;* significant toxic effects with the algae were observed in effluent concentrations ≥2.5% effluent. The least sensitive species tested was the sheepshead minnow to which effluent concentrations of 32% and greater were required to elicit a toxic response. Receiving waters were toxic only to the red alga and these results correlated well with estimates of effluent concentrations (based on the previous dye study) and effluent effect concentrations. Sheepshead minnows were not sufficiently sensitive to the effluent to elicit an adverse response at the concentrations expected in the receiving water.

[6] Miller, D. C., U.S. Environmental Protection Agency, Environmental Research Laboratory, Narragansett, RI, personnel communication, December 1987.

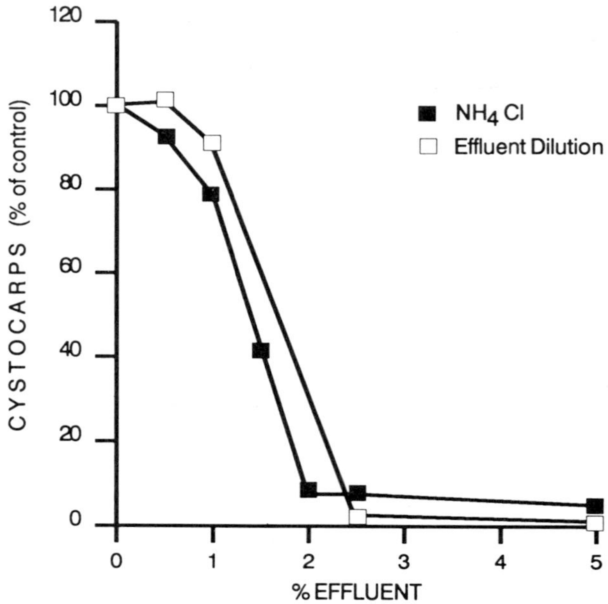

FIG. 3—*Number of* Champia parvula *cystocarps (as percent of control) plotted against percent effluent. The effluent data are averages from 17 May and 18 May. The ammonium chloride data are based on 81-mg ammonium per litre in the effluent. A stock solution of 26.7-mg ammonium chloride per 100 mL was used in the laboratory exposure.*

Chemical fractionation and subsequent testing of the effluent on-site using a freshwater cladoceran indicated that the primary cause of the toxicity was ammonia. This contention was corroborated by a comparison of unionized ammonia concentrations in the effluent with those reported toxic in separate laboratory studies of the marine species, *C. parvula, M. bahia,* and *M. beryllina.*

This case study illustrated: (1) the importance of using a wide taxonomic range of test species to elucidate the toxicity characteristics of an effluent, (2) the sensitivity of the suite of organisms selected to detect the effects of a discharge on a receiving water body, and (3) the ability of chemical fractionation procedures to detect the specific cause or causes of toxicity. The information provided here could serve as the basis of a toxicity-based discharge permit limit and would contribute substantially to an effective toxicity reduction evaluation.

Acknowledgments

We thank D. I. Mount of EPA's Environmental Research Laboratory (Duluth, MN) and L. Anderson-Carnahan of EPA's Region IV (Atlanta, Georgia) for their contributions in chemical fractionation data, and we also wish to thank D. C. Miller of EPA's Environmental Research Laboratory (Narragansett, RI) for the chronic ammonia toxicity data.

Although the research described in this article has been supported by the United States Environmental Protection Agency, it has not been subjected to Agency review and therefore does not necessarily reflect the views of the Agency, and no official endorsement should be inferred.

References

[*1*] U.S. Environmental Protection Agency, "Development of Water Quality-Based Permit Limitations for Toxic Pollutants: National Policy," Federal Register, Vol. 49, 1984, pp. 9016–9019.

[*2*] Mount, D. I. and Norberg, T. J., "A Seven-Day Life-Cycle Cladoceran Toxicity Test," *Environmental Toxicology and Chemistry,* Vol. 3, 1984, pp. 425–434.

[*3*] Norberg, T. J. and Mount, D. I., "A New Subchronic Fathead Minnow (*Pimephales promelas*) Toxicity Test," *Environmental Toxicology and Chemistry,* Vol. 4, 1985, pp. 711–718.

[*4*] Wall, T. M. and Hanmer, R. W., "Biological Testing to Control Toxic Water Pollutants," *Journal of the Water Pollution Control Federation,* Vol. 59, No. 1, 1987, pp. 7–12.

[*5*] *Short-Term Methods for Estimating the Chronic Toxicity of Effluents and Receiving Waters to Marine and Estuarine Organisms,* EPA 600/4-87/028, W. B. Horning, C. I. Weber, D. J. Klemm, T. W. Neiheisel, P. A. Lewis, E. L. Tobinson, J. Menbedick, and F. Kessler, Eds., U.S. Environmental Protection Agency, Washington, DC, 1988.

[*6*] Steele, R. L. and Thursby, G. B., "A Toxicity Test Using Life Stages of *Champia parvula* (Rhodophyta)," *Aquatic Toxicology and Hazard Assessment; Sixth Symposium, STP 802,* W. E. Bishop, R. D. Cardwell, and B. B. Heidolph, Eds., American Society for Testing and Materials, Philadelphia, 1983, pp. 73–89.

[*7*] Thursby, G. B. and Steele, R. L., "Comparison of Short- and Long-Term Sexual Reproduction Tests with the Marine Alga *Champia parvula,*" *Environmental Toxicology and Chemistry,* Vol. 5, 1986, pp. 1013–1018.

[*8*] Nimmo, D., Bahner, L. H., Rigby, R. A., Sheppard, J. M., and Wilson, A. J., "*Mysidopsis bahia:* An Estuarine Species Suitable for Life-Cycle Toxicity Tests to Determine the Effects of a Pollutant," *Aquatic Toxicology and Hazard Evaluation (First Conference), ASTM STP 634,* F. L. Mayer and J. L. Hamelink, Eds., American Society for Testing and Materials, Philadelphia, 1977, pp. 109–116.

[*9*] Lussier, S. M., Kuhn, A., and Sewall, J. "Guidance Manual for Conducting Seven-Day Mysid Survival/Growth/Reproduction Studies Using the Estuarine Mysid *Mysidopis bahia,*" *Users Guide to the Conduct and Interpretation of Complex Effluent Toxicity Tests at Estuarine/Marine Sites,* S. C. Schimmel, Ed., Contribution No. 796, Environmental Research Laboratory, U.S. Environmental Protection Agency, Narragansett, RI, 1986, pp. 86–124.

[*10*] Hughes, M. M., Heber, M. A., Schimmel, S. C., and Berry, W. J., Guidance Manual for Conducting Complex Effluent and Receiving Water Larval Fish Growth/Survival Studies with the Sheepshead Minnow *(Cyprinodon variegatus),*" *Users Guide to the Conduct and Interpretation of Complex Effluent Toxicity Tests at Estuarine/Marine Sites,* S. C. Schimmel, Eds., Contribution No. 796, Environmental Research Laboratory, U.S. Environmental Protection Agency, Narragansett, RI, 1986, pp. 125–179.

[*11*] Heber, M. A., Hughes, M. M., Schimmel, S. C., and Bengtson, D. A., "Guidance Manual for Rapid Chronic Toxicity Tests on Effluents and Receiving Waters with Larval Inland Silversides *(Menidia beryllina),*" *Users Guide to the Conduct and Interpretation of Complex Effluent Toxicity Tests at Estuarine/Marine Sites,* S. C. Schimmel, Ed., Contribution No. 796, Environmental Research Laboratory, U.S. Environmental Protection Agency, Narragansett, RI, 1986, pp. 180–250.

James R. Pratt,[1] Jody Mitchell,[2] Richard Ayers,[2] and John Cairns, Jr.[3]

Comparison of Estimates of Effects of a Complex Effluent at Differing Levels of Biological Organization

REFERENCE: Pratt, J. R., Mitchell, J., Ayers, R., and Cairns, J., Jr., **"Comparison of Estimates of Effects of a Complex Effluent at Differing Levels of Biological Organization,"** *Aquatic Toxicology and Environmental Fate: Eleventh Volume, ASTM STP 1007,* G. W. Suter II and M. A. Lewis, Eds., American Society for Testing and Materials, Philadelphia, 1989, pp. 174–188.

ABSTRACT: Impact of a toxic municipal sewage effluent containing textile dyes was evaluated using standard acute and chronic single-species tests and a microcosm test using indigenous microbiota from the receiving stream. Estimated effect levels were compared with the calculated in-stream waste concentration and measured impacts on stream microbiota and macrobenthos. Acute tests examined effluent effects on *Daphnia pulex,* and short-term chronic effluent dilution tests were conducted using *Ceriodaphnia dubia* and larval *Pimephales promelas.* Microcosm tests examined effluent effects on the structure and function of microbial communities on artificial substrates. Additional tests examined effects of receiving stream water on *C. dubia* and microcosms.

Effluent concentrations of 70% were acutely toxic in median effective concentrations) to *D. pulex.* Estimates of the no-observable-effect concentration (NOEC) were 10 and 7.5% for *C. dubia* and *P. promelas,* respectively. Responses in the microcosm systems were nonlinear and showed significant subsidy of both structure and function at concentrations of 30% and toxicity at concentrations >30%. Tests of receiving stream water below the effluent outfall showed a slight depression of reproduction in *C. dubia* and recovery to upstream levels for water sampled from farther downstream. Microcosms treated with ambient stream water did not show effects on taxonomic richness, but reduced midmorning oxygen levels were observed for microcosms dosed with water from stations below the effluent outfall. Surveys of in-stream microbiota and macrobenthos confirmed adverse impacts below the discharge and indicated recovery 11 km downstream. Microcosm tests were generally less sensitive than single-species tests in detecting toxicity of the effluent. None of the laboratory tests corresponded to observed in-stream effects. The sensitivity of the tests may have been improved by improving the test conditions, including continuous replacement of effluent, as was done in larval *P. promelas* tests. To be effective in predicting in-stream effects, the design of intensive toxicity surveys should make every effort to mimic local conditions.

KEY WORDS: *Ceriodaphnia, Pimephales promelas,* microcosm, artificial substrates, sewage effluent, effluent toxicity, aquatic toxicology

[1] Assistant professor, School of Forest Resources, The Pennsylvania State University, University Park, PA 16802.

[2] Biologists, State Water Control Board, Commonwealth of Virginia, Richmond, VA 23230.

[3] Professor and director, University Center for Environmental and Hazardous Materials Studies, Virginia Polytechnic Institute and State University, Blacksburg, VA 24061.

Little quantitative evidence is available for validating estimates of hazard for toxic discharges. Most hazard estimates are made from tests on sensitive single species, but compelling questions about the transferability of evidence from one level of biological organization to more complex interacting ecological systems [*1,2*] have summarized doubts that ecologists have in relying on single-species tests to protect ecosystems. To evaluate the effects of environmental stresses at the ecosystem level, a number of microcosms and mesocosms have been developed (for example, see Ref *3*), and some of these are presently in use in hazard evaluation schemes for new chemicals [*4–6*]. All of these microcosm systems are generic systems not necessarily representing a particular ecosystem, and none incorporate flow or continuous replacement of the test medium.

Few toxic chemicals are discharged singly in industrial or municipal discharges. Frequently, the particular combination of toxicants in an effluent is time varying and is hypothesized to have greater impact on ecosystems during low flow/low dilution periods when toxic materials are most concentrated in the water column. State regulatory agencies have limited resources and often only conduct a single toxicity assessment at a particualr site prior to recommending changes in limitations on a permitted discharge. The worst case of minimum receiving stream dilution is used for comparison to these single toxicity assessments. Recently, regulatory effort has been directed at managing toxicity from complex effluents rather than managing individual chemical concentrations [*7*]. The purpose of our research was to compare estimates of hazard based on single-species tests of complex effluents and a microcosm design using indigenous communities from the specific receiving system. These laboratory tests were followed by tests evaluating extant toxicity in the receiving system waters and by a survey of stream benthos to determine if the laboratory tests predicted in-stream effects.

Methods

Study Site

This study was conducted in the Smith River near Martinsville (Henry County), Virginia. The Smith River at Martinsville receives a complex toxic effluent from a publicly owned treatment works (POTW) at Martinsville, Virginia. In addition, two upstream effluent discharges have a potential contributing toxicity. Both upstream discharges are above a small dam and impoundment of the Smith River (Fig. 1). One input is from a nylon fiber and yarn manufacturer; the second input is from another publicly owned treatment works. Several kilometres upstream of these discharges is the Philpott Reservoir, which generates electricity on a peaking power-demand schedule. While all three potentially toxic discharges are discharged continuously, power generation at the Philpott Reservoir is reduced on weekends and decreases flows in the Smith River.

The City of Martinsville operates a secondary wastewater plant that discharges 0.25 m^3/s [5.7 MGD (million gallons per day)] to the Smith River. The plant influent is 70% industrial waste and includes waste from several textile mills, knitting plants, and other light industries. Inorganic dyes in the influent are not removed in the treatment process, and the effluent is highly colored. This color is maintained in the stream for several kilometres downstream. Flow in the Smith River is controlled by releases from the upstream Philpott Dam, resulting in fluctuating flow rates (0.52 to 3.1 m^3/s or 12 to 71 MGD). The estimated worst-case condition for a three-day low flow during a ten-year period (3Q10) is 2.35 m^3/s (54 MGD). The ratio of the average Martinsville POTW discharge to this estimated low flow gives an estimated in-stream waste concentration (IWC) of 9.5%. This value is used

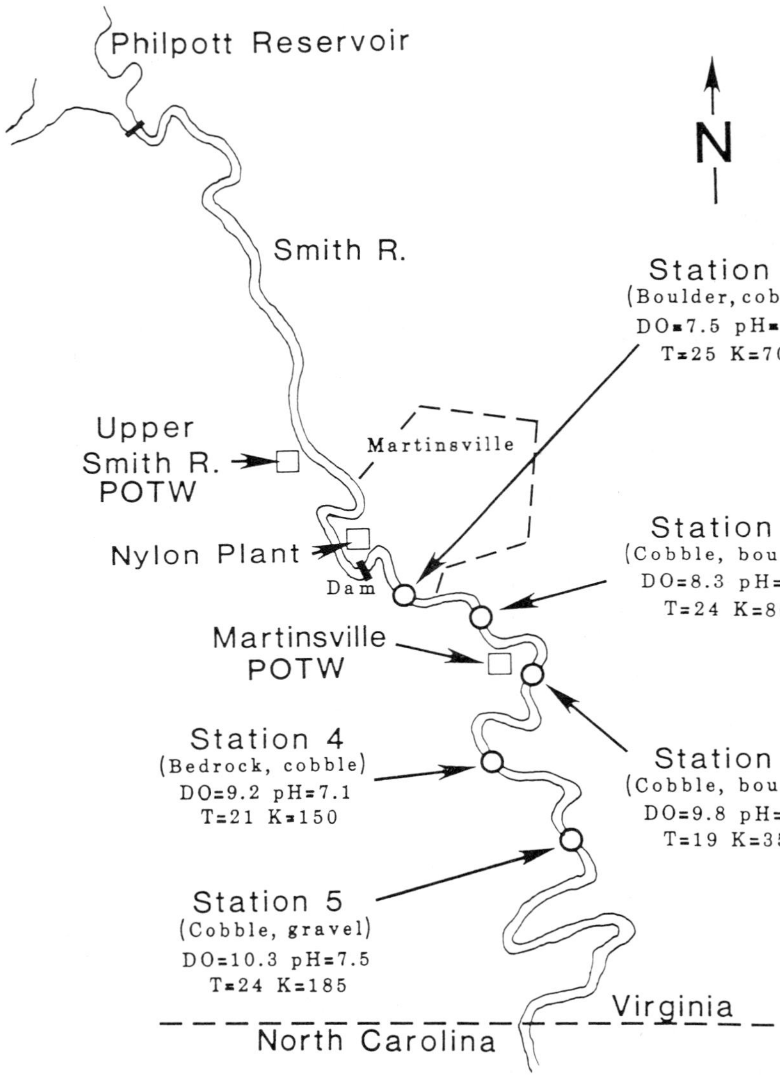

FIG. 1—*The Smith River (Virginia) study area. Major discharges and characteristics of the stream at the sampling stations are shown:* DO = *dissolved oxygen in milligram per litre;* T = *temperature in degrees centigrade;* K = *conductivity in μmhos/cm.*

for comparison with results of toxicity tests to determine if acute or chronic toxicity conditions may exist in the river downstream of the effluent discharge. Similar calculations for the nylon plant (the nearest potentially toxic discharge) upstream of the Martinsville POTW and the small dam in the City of Martinsville yield an IWC of 1.8%. The small dam between the nylon plant and the Martinsville POTW increases the detention time of the nylon plant waste and probably further reduces its potential toxicity.

Effluent Toxicity Tests

Standard effluent dilution toxicity tests were conducted on site to evaluate both acute and chronic toxicity of the Martinsville POTW effluent to sensitive single species according to methods outlined by the U.S. Environmental Protection Agency [*8,9*]. Acute 48-h static toxicity tests evaluated the mortality of *Daphnia pulex*. Tests of effects on light emission from luminescent bacteria (Microtox, Beckman Instruments) were done over 15-min periods using grab samples of effluent. Short-term chronic tests evaluated reproduction and growth using *Ceriodaphnia dubia* and larval fathead minnows *Pimephales promelas*, respectively. Activated charcoal dechlorinated tap water was used as dilution water in each test except *C. dubia* tests, in which filtered James River water collected above Richmond, Virginia, was used as the diluent.

Dilution waters differed among the tests but were chosen because of (*a*) the ease of experimental manipulation, (*b*) toxicity-free dilution water was needed, (*c*) the waters all had a hardness similar (~65 mg of calcium carbonate per litre) to that of the receiving water, and (*d*) the test organisms were expected to show adequate survival and reproduction in this water. James River water was used for *C. dubia* tests because the test organisms had been reared in this water and showed adequate growth and survival.

The seven-day *C. dubia* survival and reproduction test was conducted at effluent concentrations of 60, 30, 10, 3, and 1% plus controls. The seven-day larval *P. promelas* growth test was conducted using a proportional, flow-through mini-diluter system [*10*]. The effluent supply in the diluter was mixed at 30%. Posthatch larval *P. promelas* were exposed to concentrations of 30, 15, 7.5, 3.75, and 1.87% effluent. Temperature in the effluent supply fluctuated from 18 to 30°C during the test, but temperatures recorded in the actual test chambers showed that fish in the highest effluent dilution were exposed to a temperature range of 20 to 27°C.

A microcosm test was conducted at Virginia Tech laboratories using methods similar to those of Niederlehner et al. [*11*]. Polyurethane foam (PF) artificial substrates were placed in the Smith River immediately above the POTW outfall (Station 2, Fig. 1). The substrates were colonized for a period of seven days and then collected from the river *en masse* and returned to the laboratory in an insulated container partially filled with stream water. These substrates served as species sources representative of aerobic microbial communities in the Smith River. Effluent dilution toxicity tests were conducted in triplicate, using effluent concentrations of 100, 30, 10, 3, and 1% plus controls. Diluted effluent solutions were replaced on alternate days throughout the duration of the test. Activated charcoal dechlorinated tap water was used as the diluent for these tests. Microcosms were illuminated with daylight equivalent lights [Color Rendering Index (CRI) >90, Vita-Lites, Durotest, Inc.] at 5000 lx on a 16-h light/8-h dark schedule. A single colonized PF substrate was placed in the center of each 7-L test tank and surrounded by six initially barren, washed PF substrates (islands). The initially barren substrates served as new habitat to be colonized by species from the species source (epicenter). A single island substrate was removed after 2, 7, 14, and 21 days for examination.

The PF substrates removed from each test tank were squeezed into clean polystyrene collecting containers to remove as much of the contents of the substrate as possible. The contents were allowed to settle and then were examined for the numbers and kinds of protozoan species (ciliates, flagellates, amoebae) as described by Cairns et al. [*12*]. Standard protozoological keys were used to identify organisms encountered to the lowest practical taxon. Identification was usually possible to the genus level and frequently to the species (see Ref *12* for a list of taxonomic references). After 21 days, PF substrate samples from each test tank were subsampled for additional analyses, including the determination of total protein [*13,14*], ash-free dry weight [*15*], chlorophyll *a* [*15*], hexosamine (a measure of chitin [*16*]), bacterial numbers [*17*], and alkaline phosphatase activity [*18*].

Receiving Water Toxicity Tests

Grab samples of Smith River water were collected for seven successive days at Stations 1, 3, 4, and 5 (Fig. 1). Samples were filtered through a Wisconsin plankton net (50 μm mesh size) and used for daily renewals in seven-day *C. dubia* survival and reproduction tests and modified microcosm tests. Microcosm tests were conducted as described above, except that the test containers were 5-L glass aquaria that contained one naturally colonized species source from the Station 2 reference site and one initially barren PF substrate. Triplicate glass aquaria were filled with 3.5 L of Smith River water from each station. Test systems were illuminated using lights identical to those used in the effluent dilution toxicity tests, and the water was changed daily in each test tank by siphoning and carefully replacing the test volume. After seven days, both substrates were collected and sampled for the numbers of protozoan species. In addition, dissolved oxygen and pH were measured in each test container to provide an estimate of relative balance between photosynthetic oxygen production and respiratory oxygen consumption.

Stream Survey

A stream survey evaluating benthic macroinvertebrate and protozoan communities was conducted at the five survey stations shown in Fig. 1. The PF substrates identical to those used in the laboratory tests described previously were anchored on concrete blocks at each of the five sampling stations seven days prior to the benthic macroinvertebrate survey. After this colonization period, and at the same time as the benthic macroinvertebrate survey, replicate PF substrates were collected and examined for the numbers and kinds of protozoan species as described previously.

Benthic macroinvertebrate field methods were qualitative and involved the use of D-frame dip nets and hand sampling to sample all available habitats at each station to observe the maximum community diversity possible. Organisms were collected and identified to the family level and were assigned relative abundance values. Dissolved oxygen, temperature, pH, and conductivity were measured at each station, and the total residual chlorine was determined.

Data Analysis

Mortality of test organisms in standard single-species toxicity tests was analyzed by nonlinear interpolation or Fisher's exact test [*9,19*]. Species numbers, results of nontaxonomic analyses in microcosm tests, *C. dubia* reproduction, and *P. promelas* growth were analyzed using analysis of variance and multiple comparison test [*19,20*].

Results

Effluent discharged from the Martinsville POTW is colored by inorganic dyes from textile manufacturing. The plant influent has a high-salt (sodium chloride) content resulting in high effluent conductivity (>2000 μmho/cm). The relative concentration of the effluent plume can be traced by the relative elevation of conductivity in the river below the POTW discharge. Dilution of the effluent plume ranges from 8 to 47%, based on flows (0.52 to 3.1 m^3/s) from the Philpott Reservoir. The small impoundment at the City of Martinsville

TABLE 1—*Summary of laboratory toxicity test results for effluent from Martinsville POTW.*[a]

Test Species	Test Type and Duration	End Point	Effective Concentration, % (General Type of Response)
Daphnia pulex	static, 48 h	mortality	LC_{50} = 70 (I)[b]
Photobacterium (Microtox)	static, 15 min	light emission	EC_{50} = 24 to 27 (I)
Ceriodaphnia dubia		mortality	LC_{50} = 37 (I)
	static renewal, 7 days	reproduction	NOEC = 10, LOEC = 30 (I)
Pimephales promelas	continuous replacement, 7 days	growth	NOEC = 7.5, LOEC = 15.0 (I)
Microcosm	static renewal, 21 days	net production	
		protein	NOEC = 30, LOEC = 100 (S)[c]
		dry weight	NOEC = 30, LOEC = 100 (S)
		chlorophyll *a*	NOEC = 10, LOEC = 30 (S)
		hexosamine	NOEC = 30, LOEC = 100 (S)
		bacteria	NOEC = 10, LOEC = 30 (I)
		alkaline phosphatase activity	NOEC = 10, LOEC = 30 (I)
		protozoan species number	NOEC = 30, LOEC = 100 (I)

[a] The in-stream waste concentration was 9.5%.
[b] Response type: I = inhibition.
[c] Response type: S = stimulation.

uses the peak power demand water release to operate a run-of-the-river electrical generating plant. Conductivity above the POTW discharge ranged from 70 to 85 μmho/cm, but conductivity downstream from the discharge ranged from 150 to 455 μmho/cm, depending on the time of day (river discharge) and distance downstream (Fig. 1).

Effluent Toxicity Tests

The effluent was acutely toxic to all standard test species. The estimated median effective concentration (EC_{50}) and median lethal concentration (LC_{50}) ranged from 24 to 70% effluent (Table 1). The lowest effluent concentration producing an adverse response was in the Microtox short-term test (EC_{50} = 24%). No *C. dubia* survived at effluent concentrations >30% in chronic tests.

Chronic effect levels (LOEC) were 15 and 30% for *P. promelas* and *C. dubia*, respectively. The respective no-observed-effect concentrations (NOECs) were 7.5 and 10% (Table 1). The dose-response relationship for *C. dubia* was nonlinear (Fig. 2) and indicated some enhancement of reproduction at 1 and 3% effluent.

Microcosms showed enrichment responses for several variables (Tables 1 and 2). Estimates of net production (biomass) on artificial substrates showed stimulation at 10 to 30% effluent, probably due to phosphate enrichment (~5 mg/L in effluent). Nitrate concentrations were typically below detection limits (0.025 mg/L). Bacterial numbers and alkaline phosphatase activity were inhibited at ≥30% effluent. Protozoan species richness was reduced only at 100% effluent. Both single-species tests were more sensitive to effluent effects than the microcosm test (Fig. 2).

Receiving water toxicity tests showed low (60%) survival of *C. dubia* in water from Station 1. Survival may have been complicated by transfer of daphnids to this water without acclimation. Protozoan species numbers in the microcosms were similar to those found in the stream at this site. Surviving daphnids showed normal reproduction, and recovery was indicated by both daphnid survival and increased reproduction in water from Station 4 (Fig.3), although *C. dubia* reproduction decreased in water from the most downstream station. Reproduction in water from this site was still greater than in water from the upstream

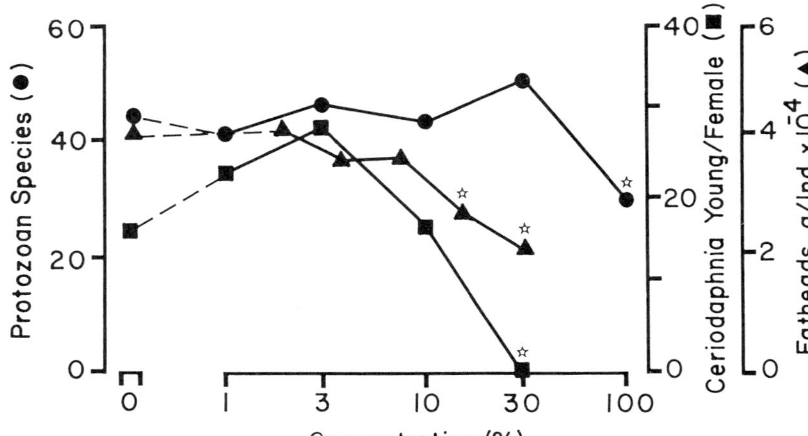

FIG. 2—*Effect of Martinsville POTW effluent (in percent) on* C. dubia *reproduction,* P. promelas *growth, and protozoan species numbers. Values marked with an asterisk are significantly different from controls.*

TABLE 2—*Summary of nontaxonomic measures from laboratory microcosms.*

Treatment	Total Protein, mg/L	Ash-Free Dry Weight, mg/L	Hexosamine, mg/L	Chlorophyll *a*, mg/L	Bacteria, cells/mL (ln)	Alkaline Phosphatase Activity[a]
Control	36.0	104.7	0.67	129.0	13.57	177.7
	(10.0)[b]	(37.1)	(0.44)	(55.6)	(0.05)	(12.7)
1%	27.3	69.0	0.30	102.3	13.56	202.7
	(3.1)	(3.6)	(0.09)	(38.5)	(0.27)	(41.8)
3%	34.7	63.0	0.23	111.2	13.50	214.0
	(5.0)	(11.8)	(0.23)	(38.6)	(0.09)	(101.2)
10%	44.7	83.0	0.50	422.7	13.35	111.3
	(9.0)	(20.2)	(0.13)	(129.7)	(0.26)	(2.5)
30%	58.7	88.3	1.27	952.3[c]	12.93[c]	82.0[c]
	(12.7)	(4.0)	(0.40)	(369.5)	(0.21)	(21.6)
100%	130.0[c]	173.7[c]	2.57[c]	1423.7[c]	11.72[c]	75.7[c]
	(73.7)	(53.9)	(1.48)	(952.1)	(0.14)	(7.5)
F	4.51	6.04	5.39	5.09	43.66	5.41
P	0.015	0.0051	0.0079	0.0098	0.0001	0.0078

[a] Alkaline phosphatase activity is expressed as nanomoles of *p*–nitrophenol per milligram of protein per hour.

[b] Values in parenthesis are the standard deviation.

[c] These values are significantly different from the controls. The F statistic for analysis of variance and the significance, P, are shown at the bottom of each column.

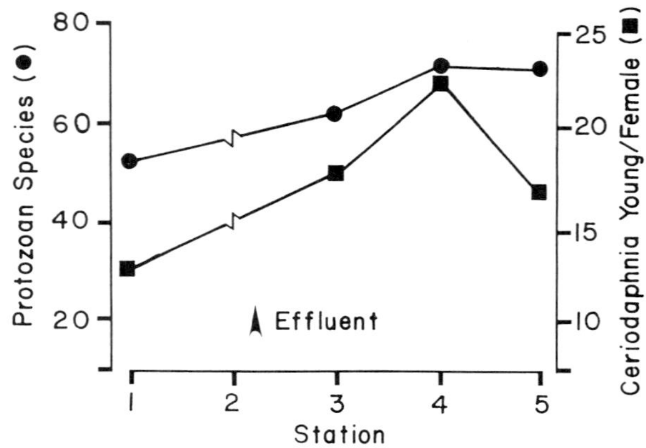

FIG. 3—*Effect of receiving water toxicity tests on* C. dubia *reproduction and protozoan species numbers.*

reference station. Elevated reproduction was paralleled by increases in the protozoan species number, but the decreased dissolved oxygen in the microcosm showed the influence of effluent loading at Station 3.

Organic loading or toxicity or both were evident in microcosm tests where midmorning dissolved oxygen was significantly depressed [$P < 0.001$; analysis of variance (ANOVA)] in water from all stations downstream of the POTW discharge in comparison with the oxygen levels in microcosms dosed with upstream water (Fig. 4). Oxygen depression was greatest for water from Station 3, and it decreased for water taken from stations at increasing distances downstream. Oxygen depression in microcosms corresponded to an observed reduction in taxonomic richness of the stream biota, but no corresponding reduction of protozoan species was observed in these microcosms.

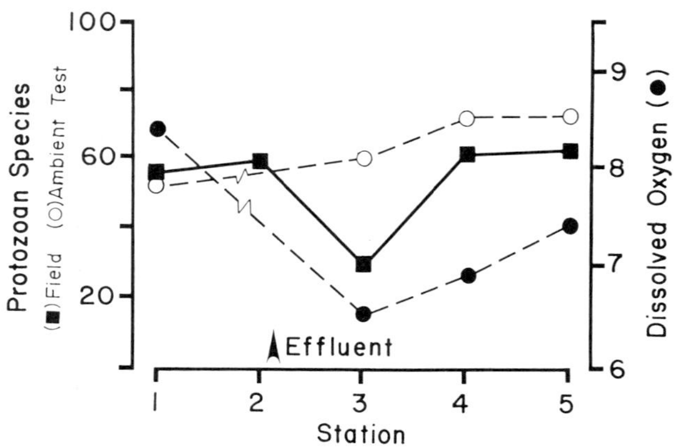

FIG. 4—*Numbers of protozoan species for stream survey and receiving water toxicity tests compared with the morning dissolved oxygen (in milligrams per litre) levels in receiving water toxicity tests.*

Stream Survey

Evaluations of stream biota (protozoa and macroinvertebrates) showed obvious adverse impact at the first stream station below the POTW discharge. Taxonomic richness and population sizes were reduced at this site, but recovered downstream (Figs. 4 and 5). Protozoan species richness decreased by nearly 50% at Station 3 compared with upstream stations. However, there were no significant differences in biomass (dry weight, $P = 0.5$) or *in vivo* fluorescence ($P = 0.11$) in the PF substrate samples. Macroinvertebrate family richness decreased similarly (Table 3), but invertebrate richness decreased from Station 1 to Station 2, a response not observed for protozoa. Specific observations for each station follow.

Station 1—The benthic habitat at this station was considered optimal, but macroinvertebrate population sizes showed that habitat was underutilized. Facultative organisms such as midges, blackflies, and ephemerellid mayflies were the most common forms observed. The benthic community was apparently affected by upstream pollution sources or changes in nutrients attributable to the upstream impoundment.

Station 2—The benthic community at this station was similar to, but slightly less diverse than, the community at Station 1. Facultative forms were the most common and indicated no improvement in water quality relative to upstream.

Station 3—This station showed obvious evidence of the POTW discharge. Water was highly colored (purple) and showed a large increase in conductivity. The density of macroinvertebrates was noticeably lower, and populations of both sensitive and tolerant forms were reduced compared to upstream, indicating mildly toxic pollution. Extreme toxicity was not indicated due to the presence of sensitive (heptageneid mayflies and hydroptilid caddis flies) and facultative taxa.

Station 4—The benthic community at this station was similar to communities at Stations 1 and 2, despite the persistence of discolored water. Mayfly populations were still low, indicating that recovery of stream benthos was incomplete.

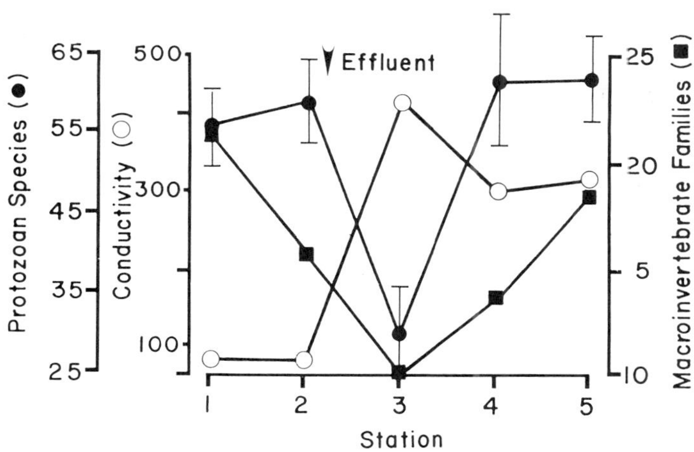

FIG. 5—*Comparison of protozoan and macroinvertebrate taxonomic richness in the stream survey and conductivity at each station.*

TABLE 3—*Summary of macroinvertebrate survey of Smith River, Virginia.*

Station	Distance from POTW Discharge	Macroinvertebrate Families, No.				Rating
		Sensitive	Facultative	Tolerant	Total	
1	4-km upstream	4	12	6	22	fair—low density in good habitat
2	100-m upstream	5	8	3	16	fair—similar to Station 1
3	800-m downstream	2	6	2	10	fair—obvious impact
4	5.5-km downstream	4	7	3	14	fair—mayfly density low
5	11-km downstream	6	8	5	19	fair to good

Station 5—This station was downstream of the Marrowbone and Letherwood Creeks, which provide additional dilution and species propagules to the Smith River. The benthic community at this station indicated more complete recovery than that at Station 4. A greater density of organisms was found here than at any other station surveyed.

Decreased biotic richness of stream communities was strongly related to the presence of the effluent when measured by elevated conductivity (Fig. 5). However, severe effects on stream communities were only observed at the first site below the POTW discharge. Recovery of protozoan species numbers was complete at Station 4, and macroinvertebrate richness returned to near the upstream range at Station 5. Decreased biotic richness was not attributable to low dissolved oxygen or residual chlorine.

Discussion

The purpose of biological testing of complex effluents is to estimate effects of complex (and varying) mixtures of potentially toxic chemicals discharged to aquatic ecosystems [*7*]. Test species and life stages are selected to provide sensitive biological indicators of toxic stress. In addition, tests have been developed to provide this sensitivity in a relatively short period compared with the life cycles of many species. Results of standardized tests of complex effluents have been compared with observed in-stream effects and found to be generally protective of ecosystem structure (for example, see Ref *21*). Many scientists have suggested, however, that tests incorporating more ecosystem components and processes will be necessary to provide adequate protection of complex natural systems (for example, see Refs *1* and *2*).

Effluent Toxicity Tests

Previous evaluations of acute toxicity at the Martinsville POTW had shown that fathead minnows were more sensitive to effluent effects than daphnids.[4] These experiments further

[4] Mitchell, J., Virginia State Water Control Board, Richmond, VA, unpublished data, 1985.

demonstrated this sensitivity and corresponded with other reports of the greater sensitivity of fathead minnow growth over survival in effluent tests [*21*]. Similar conclusions were reached by Benoit et al. [*10*] in tests of pure compounds, although literature reports suggest survival and growth may be equally sensitive [*22*].

Altered responses of *C. dubia* and microcosm biota occurred at similar effluent concentrations. Daphnid reproduction was slightly stimulated at low-effluent concentrations, and stimulatory effects on microcosm biomass estimators were also recorded (Table 2). The number of species of protozoa and summary microbial community measures in similar microcosms have been shown to be sensitive to low levels of pure compounds such as cadmium, copper, and chlorine [*11,23,24*]; however, no such sensitivity was observed in these experiments. The maintenance of protozoan species numbers in microcosms containing diluted effluent reflects nutrient addition, which is further substantiated by large increases in chlorophyll biomass with increasing effluent concentration. We have previously reported elevations of protozoan species numbers in response to enrichment [*25*]. Stimulation by phosphorus is consistent with decreased alkaline phosphatase activity; however, bacterial numbers were reduced at high effluent concentrations (>10%). Observed toxicity in 100% effluent may be a result of osmotic stress from the effluent rather than overt toxicity.

Receiving water toxicity tests showed effects on survival and, therefore, reproduction. However, surviving daphnids showed normal reproduction, and recovery was indicated by both daphnid survival and increased reproduction in water from Station 4. Averaging reproduction across the initial number of test organisms gave an ambiguous estimate of sublethal effects. While the causes of daphnid mortality can only be speculated upon, *surviving* daphnids exposed to water from Stations 1, 3, 4, and 5 averaged 20.5, 19.3, 24.1, and 18.4 young per female, respectively. This suggests there were only moderate effects on reproduction. There were no identified sources of toxicants to account for the decline in reproduction at Station 5. Similar confounding results of *C. dubia* survival and reproduction have been reported [*21*].

Elevated reproduction at Station 4 was paralleled by increases in the number of protozoan species in modified microcosms. Decreased dissolved oxygen in the microcosm showed the influence of effluent loading at Station 3. Enhancement of the number of species and *C. dubia* reproduction suggests organic enrichment. Changes in microcosm dissolved oxygen did not return to upstream levels even at the most downstream site.

Laboratory-Field Comparison

None of the laboratory tests suggested the severity of in-stream impacts. Laboratory tests indicated chronic toxicity of the Martinsville POTW near the IWC. Microcosm and *C. dubia* tests had NOECs of 10%. Microcosms showed stimulation for many variables. The NOEC for fathead minnow growth was 7.5%, and growth was reduced by 12 to 33% relative to the controls in test concentrations (7.5 to 15%) bracketing the IWC.

The single-species tests were better indicators of adverse in-stream effects than microcosms because adverse responses were observed at effluent dilutions near the IWC. Field studies showed that macroinvertebrate family diversity and protozoan species richness were reduced by approximately one half, below the effluent discharge. Taxonomic richness in microcosms did not show this effect except in 100% effluent. Other microcosm responses showed *stimulation* in response to effluent, and *C. dubia* reproduction was also stimulated at low effluent dilutions, although stimulation at effluent dilutions of <10% was followed by increasing mortality and decreased reproduction at dilutions of 10% and above.

Summary and Recommendations

Unlike studies of the toxicities of pure compounds, studies of effluents and in-stream effects include several uncontrolled factors. Effluent quality may vary with time. Stream discharge also varies with time as do the qualities of upstream discharges. Agencies regulating toxic discharges face fundamental problems in conducting evaluations of effluent toxicity. First, resources are limited; more than a single evaluation of a given discharger is unlikely. Second, technical support for such investigations is limited and some guidance is specifically lacking (for example, see Refs *7* through *9*). Differences in water quality are known to alter the measurement of toxicity, but guidelines [*8*] provide considerable latitude in selection of the dilution water.

We used dechlorinated tap water as diluent to avoid confounding the effects of possible upstream toxic sources and to maintain a diluent of a water hardness similar (also recommended) to that of the receiving stream. *Ceriodaphnia dubia* were tested in water in which they had been reared. However, such selections make comparisons with in-stream effects or receiving water (ambient) tests [*7*] problematic. Even intensive comparisons of laboratory and field results (see Ref *21*) have used inconsistent dilution water to maintain test organisms under optimal conditions. Efforts to create reconstituted water for *C. dubia* are recent and still beg the question to dilution water consistency. To some extent, investigators will be guided by the purpose of a specific study: if the objective is to estimate site-specific impacts, test organisms should be reared in or acclimated to reference water (a clean upstream source) from the receiving ecosystem; if the objective is to compare estimates of effects using several tests, all tests should use the same diluent. This will necessarily prolong studies or at least require longer lead times to prepare or acclimate test populations, and this may complicate continuous replacement tests in which large volumes of diluent are needed.

Microcosms may become useful tools for assessing impacts because they can simulate some ecosystem processes, but microcosms will also require environmental realism if moderate impacts are to be predicted. No standard microcosm methods are currently available for assessing effluent impacts. The microcosm exposure method used in this study (static replacement) failed to predict in-stream loss of taxonomic richness and incorrectly predicted enrichment when none was observed in the receiving stream. Similarly, modified microcosms showed limited ability to identify adverse conditions in receiving water (ambient) tests even though substantial impacts were observed in the receiving stream. A potentially useful microcosm design should incorporate continuous effluent replacement (preferably using reference site dilution water) to detect adverse effects. Daily replacement is unlikely to allow detection of effects except under highly toxic conditions. This observation is confirmed by the relatively greater sensitivity of *P. promelas* tests that utilized continuous effluent replacement.

Standard single-species chronic tests also failed to predict in-stream impacts at rational stream flows. The NOECs were near or above the IWC, but low flow conditions (3Q10) in the stream had not been approached during or for several months prior to this study. The *C. dubia* reproduction tests indicated enrichment at low-effluent concentrations and in receiving water (ambient) tests, but the recommended method of analysis of reproduction data [*9*] fails to account for mortality. In other words, surviving *C. dubia* may have elevated reproduction, but the average number of young per female is reduced by averaging across the dead adult females and thus is not a measure of the actual production per *surviving* female.

Receiving water (ambient) tests were among the poorest estimators of in-stream impacts. In this study, the daily collection of grab samples—as specified in technical support information [7]—were complicated by upstream regulation of flow so that dam releases could create high or low dilution depending on the timing of grab sample collection and on power generation and water release schedules. Such problems create serious difficulties in evaluating in-stream toxicity.

Acknowledgment

The advice and assistance of David Paylor, Virginia State Water Control Board, and John Dyches, City of Martinsville, are gratefully acknowledged. Portions of this research were supported by grants from the Air Force Office of Scientific Research (AFOSR-85-0324), the U.S. Environmental Protection Agency (R-812813-01-0), and Environmental and Water Resources Research (EWRR). Technical assistance was provided by Nancy Bowers, Carolyn Hax, and Barbara Niederlehner. Darla Donald provided editorial assistance and Betty Higginbotham assisted with the final manuscript.

References

[1] Cairns, J., Jr., *Hydrobiologia,* Vol. 100, 1983, pp. 47–57.
[2] Kimball, K. D. and Levin, S. A., *BioScience,* Vol. 35, 1985, pp. 165–171.
[3] Giesy, J. P., *Microcosms in Ecological Research,* National Technical Information Center, Springfield, VA, 1980.
[4] Leffler, J., "Tentative Protocol of an Aquatic Microcosm Screening Test for Ecosystem-Level Effects of Chemicals," Battelle Columbus Laboratories Subcontract T6411 (7197), Battelle Columbus Laboratories, Columbus, OH, 1981.
[5] Taub, F. B. in *Aquatic Microbial Communities,* J. Cairns, Jr., Ed., Garland Publishing Company, New York, 1977, pp. 105–138.
[6] Giddings, J. M., Franco, P. J., Cushman, R. M., Hook, L. A., Southworth, G. R., and Stewart, A. J., *Environmental Toxicology and Chemistry,* Vol. 3, 1984, pp. 465–488.
[7] *Technical Support Document for Water Quality-Based Toxics Control,* U.S. Environmental Protection Agency, Office of Water, Washington, DC, 1985.
[8] Peltier, W. and Weber, C. I., *Methods for Measuring the Acute Toxicity of Effluents to Freshwater and Marine Organisms,* third ed., EPA 600/4-85-013, Environmental Monitoring and Support Laboratory, U.S. Environmental Protection Agency, Cincinnati, OH, 1985.
[9] Horning, W. B. and Weber, C. I., *Short-term Methods for Estimating the Chronic Toxicity of Effluents and Receiving Waters to Freshwater Organisms,* EPA 600/4-85-014, Environmental Monitoring and Support Laboratory, U.S. Environmental Protection Agency, Cincinnati, OH, 1985.
[10] Benoit, D. A., Puglisi, F. A., and Olson, D. L., *Environmental Pollution* (Series A), Vol. 28, 1982, pp. 189–197.
[11] Niederlehner, B. R., Pratt, J. R., Buikema, A. L., Jr., and Cairns, J., Jr., *Environmental Toxicology and Chemistry,* Vol. 4, 1985, pp. 155–165.
[12] Cairns, J., Jr., Kuhn, D. L., and Plafkin, J. L. in *Methods and Measurements of Periphyton Communities: A Review, ASTM STP 690,* R. L. Weitzel, Ed., American Society for Testing and Materials, Philadelphia, 1979, pp. 34–57.
[13] Rausch, P., *Hydrobiologia,* Vol. 78, 1981, pp. 237–251.
[14] Bradford, M. M., *Analytical Biochemistry,* Vol. 72, 1976, pp. 248–254.
[15] *Standard Methods for the Examination of Water and Wastewater,* 16th ed., American Public Health Association, Washington, DC, 1985.
[16] Gatt, R. and Berman, E. R., *Analytical Biochemistry,* Vol. 15, 1966, pp. 167–171.
[17] Porter, K. G. and Feig, Y. S., *Limnology and Oceanography,* Vol. 25, 1980, pp. 943–948.
[18] Sayler, G. S., Puziss, M., and Silver, M., *Applied Environmental Microbiology,* Vol. 38, 1979, pp. 922–927.
[19] Sokal, R. R. and Rohlf, F. J., *Biometry,* W. H. Freeman and Company, New York, 1983.
[20] Dunnett, C. W., *Journal of the American Statistical Association,* Vol. 50, 1955, pp. 1906–1121.
[21] Mount, D. I., Thomas, N. A., Norberg, T. J., Barbour, M. T., Roush, T. H., and Brandes, W. F., *Effluent and Ambient Toxicity Testing and In-Stream Community Response on the Ottawa River,*

Lima, Ohio, EPA 600/3-84-080, Environmental Research Laboratory, U.S. Environmental Protection Agency, Duluth, MN, 1984.
[*22*] Woltering, D. M., *Aquatic Toxicology,* Vol. 5, 1984, pp. 1–21.
[*23*] Pratt, J. R., Niederlehner, B. R., Bowers, N., and Cairns, J., Jr., *Toxicity Assessment,* Vol. 2, 1987, pp. 417–436.
[*24*] Pratt, J. R., Bowers, N. J., Niederlehner, B. R., and Cairns, J., Jr., *Environmental Toxicology and Chemistry,* vol. 7, 1988, pp. 679–687.
[*25*] Pratt, J. R., Cairns, J., Jr., and Horowitz, R., *Hydrobiologia,* Vol. 148, 1987, pp. 159–174.

Brian J. Finlayson[1] *and Dennis C. Wilson*[2]

Evaluation of Lethal Levels, Release Criteria, and Water Quality Objectives for an Acid-Mine Waste

REFERENCE: Finlayson, B. J. and Wilson, D. C. **"Evaluation of Lethal Levels, Release Criteria, and Water Quality Objectives for an Acid-Mine Waste,"** *Aquatic Toxicology and Environmental Fate: Eleventh Volume, ASTM STP 1007,* G. W. Suter II and M. A. Lewis, Eds., American Society for Testing and Materials, Philadelphia, 1989, pp. 189–203.

ABSTRACT: Acid-mine discharges from the Spring Creek drainage contribute large quantities of toxic copper, zinc, and cadmium to anadromous chinook salmon *(Oncorhynchus tshawytscha)* and steelhead trout *(Salmo gairdneri)* fisheries below Keswick Dam after regulated release with the Sacramento River, California. Release criteria and water quality objectives for the protection of salmonids from the metals in the waste have been developed using laboratory toxicity tests and field monitoring studies. These metal criteria and objectives were evaluated by on-site toxicity tests and additional monitoring studies.

Chinook salmon and steelhead trout fry were exposed to Spring Creek acid-mine waste in on-site toxicity tests. Acutely lethal concentrations of copper, zinc, and cadmium from the waste were similar to those from laboratory tests on metal sulfate mixtures, indicating that these metals are the major toxic components of the waste. Monitoring the waste in the Sacramento River after controlled dilution using criteria for copper and zinc indicated the following: (1) mean concentrations of the metals approached or exceeded incipient lethal levels (96-h LC_{10} values); (2) short-term (<5 out of 24 h) peak concentrations of copper, zinc, and cadmium in the Sacramento River exceeded median lethal levels (96-h LC_{50} values) under normal operating conditions; and (3) no measurable loss of copper from adsorption onto particulate matter and subsequent precipitation and deposition occurred when the waste was diluted to levels tolerable to fish. New criteria for zinc and new release schedules for the copper criteria that will protect against acutely lethal levels should be developed.

KEY WORDS: toxicity, salmonids, copper, zinc, cadmium, water quality objectives, field tests, acid-mine waste, aquatic toxicology

The parallel developments of mining and hydroelectric resources in the upper Sacramento River basin of California have resulted in a compounding of adverse effects on anadromous chinook salmon *(Oncorhynchus tshawytscha)* and steelhead trout *(Salmo gairdneri)* resources. Acid-mine waste from Spring Creek containing toxic quantities of copper, zinc, and cadmium has been entering the Sacramento River near Redding since the inception of mining on Iron Mountain in the late 19th century (Fig. 1). With the construction of Shasta Dam in 1944, the toxic effects of the waste on anadromous fishes were no longer minimized by dilution flows from the upper Sacramento River [*1,2*]. Shasta Lake

[1] Environmental services supervisor, California Department of Fish and Game, Aquatic Toxicology Laboratory, Elk Grove, CA 95624.

[2] Environmental specialist, California Regional Water Quality Control Board, Central Valley Region, Redding, CA 96001.

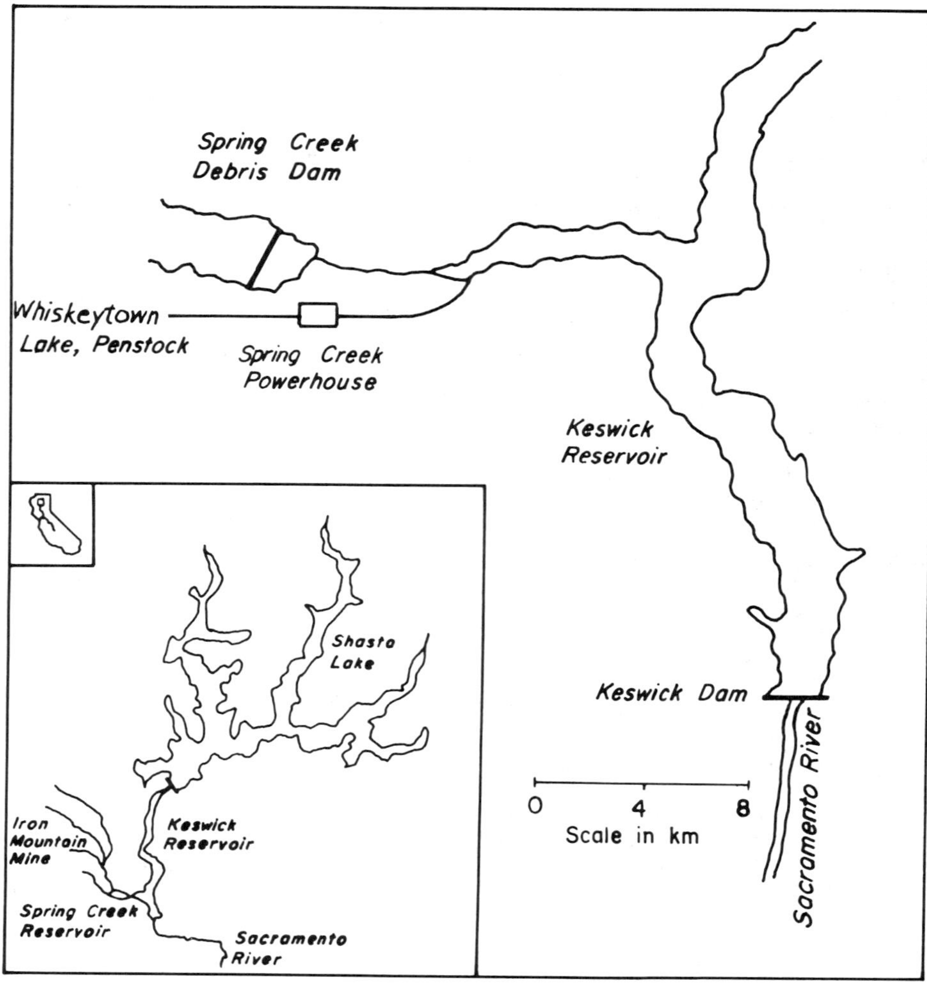

FIG. 1—*Upper Sacramento River basin, showing the location of Spring Creek and Keswick Reservoir and the sites for the toxicity tests and water samples.*

and Keswick Reservoir blocked both dilution flows and the passage of anadromous salmonids to upstream spawning areas. The resulting concentration of salmon below Keswick Dam, compounded with the increased proportion of acid-mine waste in the Sacramento River, resulted in extensive fish mortalities; for example, one third of the chinook salmon population died from metal poisoning before spawning in 1944.

The discharge of Spring Creek acid-mine waste into the Sacramento River basin was first regulated after the completion of the Spring Creek Debris Dam in 1963. From 1963 to 1978, controlled releases of the waste were made, based on estimated total copper concentrations at Keswick Dam thought not to be acutely toxic to salmonids [*3*]. Periodic fish kills continued during this 15-year period because these criteria did not consider the toxicity of "dissolved" copper, zinc, and cadmium concentrations [*4*]. Concentrations of dissolved metals refer to those fractions which pass through a membrane of 0.3 μm [*5*] or 0.45 μm [*6,7*] porosity. New copper and zinc criteria for Keswick Dam were developed by Wilson [*8*] in 1978 based on concentrations of dissolved copper and zinc found to be tolerable to salmonid eggs and fry by Finlayson and Ashuckian [*9*] and Finlayson and Verrue [*10*] and the known dilution flow capacity from the upper Sacramento River. The release schedules for the criteria assumed a 50% reduction in copper and a 10% reduction in zinc via adsorption onto particulate matter and subsequent precipitation and deposition as the waste was diluted in Keswick Reservoir [*8*]. Two sets of dissolved metal criteria were developed, one for normal discharge conditions and the other for emergencies when the Spring Creek Reservoir was close to overflowing and additional storage was needed during heavy storms. The normal and emergency criteria at Keswick Dam were 5 and 7.5 μg/L copper and 64 and 96 μg/L zinc, respectively [*8*].

Dissolved metal concentrations of 5.6 μg/L copper, 16 μg/L zinc, and 0.2 μg/L cadmium were instituted as water quality objectives for the Sacramento River below Keswick Dam by the California State Water Resources Control Board [*11*] in 1984, based on toxicity studies of Finlayson and Verrue [*5*]. The objectives reflect maximum metal concentrations tolerated for unimpaired aquatic life, and these are goals for water quality conditions in the Sacramento River drainage. The water quality objectives [*11*] for zinc and cadmium are not as yet reflected in the release criteria [*8*] for the Spring Creek acid-mine waste. Even if they were, these objectives cannot be met consistently because of the existing conditions of flow and metal concentrations from Spring Creek and the available dilution from the Sacramento River. Anticipated cleanup of the Iron Mountain mine area, increased holding facilities for the acid-mine waste in Spring Creek Reservoir, and increased dilution from the Sacramento River would allow for the attainment of the water quality objectives.

In this paper, we compare results of on-site toxicity tests on the Spring Creek acid-mine waste with the metal criteria and water quality objectives derived from laboratory tests on copper, zinc, and cadmium sulfate mixtures. We also present the results of additional monitoring studies to determine the loss of copper, zinc, and cadmium from the acid-mine waste in Keswick Reservoir.

Materials and Methods

Toxicity Tests

Test Organisms and Location—Juvenile steelhead trout were obtained from the California Department of Fish and Game (DFG) Nimbus Hatchery located on the American River, a tributary to the Sacramento River. Juvenile chinook salmon were obtained from the U.S. Fish and Wildlife Service Coleman National Hatchery located on Battle Creek, a tributary to the Sacramento River. Fish were subjected to normal hatchery cultural practices [*12*]. The fish were transported to the DFG mobile testing laboratory [*13*] located at the U.S. Bureau of Reclamation Spring Creek Powerhouse, near the Spring Creek arm of the Keswick Reservoir (Fig. 1). The fish were kept in 50-L Plexiglas troughs with an adequate supply of flowing water for 4 days prior to testing. The fish were not fed during acclimation and testing. Mean fork length ranged from 29 to 42 mm for salmon and 45 to 62 mm for steelhead trout.

Dilution Water and Acid-Mine Waste—Water quality from Whiskeytown Lake through the Spring Creek Powerhouse penstock used for dilution during the February through March 1979 and 1980 tests was very consistent: pH 7.5 to 7.6; 36 to 38 mg/L calcium carbonate ($CaCO_3$) alkalinity; 48 to 62 mg/L total dissolved solids; 39 to 40 mg/L $CaCO_3$ hardness; 8 to 10°C temperature; less than detectable concentrations (in micrograms per litre) of total (unfiltered) copper (<2), zinc (<2), cadmium (<0.2), aluminum (<10), iron (<10), and chromium (<1); and dissolved oxygen above 85% saturation. The water quality for Whiskeytown Lake is similar in chemical composition to Keswick Reservoir upstream of the confluence with Spring Creek. Tests were conducted in a room with constant temperature (12°C) control.

Fish were exposed in continuous-flow toxicity tests for 96 h, in 1979, and for 240 h, in 1980, with Spring Creek acid-mine waste pumped from 100 m downstream of the Spring Creek Debris Dam. The waste was checked analytically every 6 h during the tests for pH and for concentrations of total copper, zinc, cadmium (except 5–9 February 1979), aluminum, and iron. The results are shown in Table 1. The mean copper/zinc/cadmium ratios for total metals of the undiluted acid-mine waste during the tests ranged from 1:3.5:0.03 to 1:4.7:not analyzed. The total metal concentrations ranged from 1.59 to 4.76 mg/L copper, 6.49 to 18.9 mg/L zinc, and 0.05 to 0.14 mg/L cadmium. The pH ranged from 2.7 to 2.9.

Testing Protocol—Chinook salmon and steelhead trout were tested concurrently three times in 1979 and two times in 1980. Standard practices [ASTM Practice for Conducting Acute Toxicity Tests for Fishes, Macroinvertebrates, and Amphibians (E 729-80)] and procedures of Finlayson and Verrue [*5*] were followed for the acute toxicity tests. The tests were conducted in 12-L Plexiglas troughs with the water volume adjusted to 10 L: 50% water replacement occurred in 2.5 h. The troughs were dosed by proportional diluters and predilution systems [*9*]. Each test consisted of a control treatment in duplicate and a geometric series (dilution factor of 0.6) of five waste concentrations in duplicate, bracketing the LC_{50} value. The stocking density per trough varied with fish size from 15 (30 fish per concentration) for steelhead trout to 25 (50 fish per concentration) for salmon. Loading factors for the tests ranged from 0.05 to 0.15 g/L/day based on flow rates through the troughs.

The acid-mine waste was added to the troughs for 24 h prior to beginning the tests; metal concentrations approached nominal within 16 h. Fluorescent lighting was on between 0600 and 1800 h. Dead fish were removed and counted every 24 h.

Chemistry—Dissolved oxygen, temperature, and pH were measured in each trough daily, and hardness and alkalinity were measured twice during the tests.

Water samples (15 mL) from each trough were collected at 48-h intervals during the toxicity tests and analyzed for concentrations of dissolved copper, zinc, cadmium, aluminum, and iron by filtering the diluted waste through a 0.3 μm porosity glass fiber filter with a nylon syringe [*5*]. At the same time, water samples (30 mL) were also taken from the troughs and analyzed for concentrations of total metals. The samples were deposited in metal-free linear polyethylene narrow-mouth bottles and preserved with 0.3 mL of 6 *N* nitric acid (HNO_3).

Copper, zinc, cadmium, aluminum, and iron concentrations were determined by air-acetylene or graphite-furnace atomic absorption spectrophotometry, depending on the metal and concentration. The detection limits were 2 μg/L for copper and zinc, 0.2 μg/L for cadmium, 10 μg/L for aluminum, and 100 μg/L for iron. Cadmium was not measured in the test troughs for the 1979 tests. Analytical precision (coefficient of variation) for repeated analyses was ±4.4% for copper, ±6.7% for zinc, ±12% for cadmium, ±20% for aluminum, and ±10% for iron.

Statistics—Moving average analysis [*14*] was used to calculate LC_{50} values with angle (arcsin) transformation of mortality and $\log_{10}$ transformation of concentration, and 95% confidence intervals were calculated using Fieller's Theorem. Copper, zinc, and cadmium

TABLE 1—*Mean total metal concentrations, pH, and conductivity of Spring Creek acid-mine waste during toxicity tests.*

Test Duration	Copper, mg/L[a]	Zinc, mg/L	Cadmium, mg/L	Aluminum, mg/L	Iron, mg/L	pH	Conductivity, μS/cm at 25°C
5–9 Feb. 1979	4.01 (3.60 to 4.55)	18.9 (15.3 to 22.4)	. . .	25.8 (22.7 to 28.0)	94.8 (80.6 to 105)	2.8 (2.5 to 2.9)	920 (820 to 1000)
16–20 Feb. 1979	2.79 (1.92 to 3.17)	12.4 (7.8 to 14.4)	0.10 (0.08 to 0.12)	14.3 (8.8 to 18.0)	60.3 (36.5 to 77.5)	2.8 (2.6 to 3.0)	760 (550 to 870)
1–5 March 1979	1.59 (1.48 to 1.73)	6.49 (6.05 to 6.72)	0.05 (0.04 to 0.05)	11.7 (10.7 to 12.3)	45.7 (40.2 to 51.0)	2.7 (2.6 to 2.9)	520 (460 to 600)
4–14 Feb. 1980	4.76 (3.87 to 5.74)	18.4 (15.1 to 22.1)	0.14 (0.12 to 0.18)	34.9 (31.0 to 38.6)	138 (109 to 175)	2.8 (2.8 to 2.9)	1270 (1150 to 1460)
11–21 March 1980	3.45 (1.70 to 5.29)	12.2 (7.68 to 16.2)	0.09 (0.06 to 0.13)	20.4 (16.9 to 25.3)	113 (76.3 to 143)	2.9 (2.8 to 4.0)	950 (700 to 1100)

[a] Range of concentrations in parentheses.

concentrations at the LC_{50} level were interpolated from least-squares regression (waste concentration versus metal concentration). Mean 96-h and 240-h LC_{50} values for trout and chinook salmon were considered significantly different if 95% confidence limits did not overlap. Differences between any of the individual test LC_{50} values were not evaluated because of variability associated with testing different lots of fish and testing different copper/zinc/cadmium ratios for the dissolved metals in the diluted waste in concurrent tests.

Waste Dilution Studies

Twenty percent weight volume fluorescein or Rhodamine WB dye solutions were pumped into Spring Creek below Spring Creek Debris Dam for a 30-min period on 21 and 25 March and 3 April 1980 to determine the travel time of the waste between Spring Creek Debris Dam and Keswick Dam. Dye was added immediately before the Spring Creek Powerhouse began daily power generation. The studies were done under two operational schemes: (1) continual discharge from Spring Creek Debris Dam with daily, intermittent discharge from Spring Creek Powerhouse (21 and 25 March); and (2) coordination of Spring Creek Debris Dam and Spring Creek Powerhouse discharges (3 April). The emergency metal criteria were in effect on 21 March, and the normal metal criteria were in effect on 25 March and 3 April. Water was monitored for fluorescence for 24 h after the dye addition by analyzing water passing through one of the turbines inside Keswick Dam using a Turner Model III fluorometer. Fluorometer filters, light sources, and calibration methods were those recommended by the manufacturer. To determine the concentration of metals released to the Sacramento River, water samples were taken hourly from the exit of the fluorometer and analyzed for concentrations of total and dissolved metals. Surface water samples were collected from the Keswick Reservoir 1 km upstream of the Spring Creek arm on the three monitoring days and analyzed for concentrations of total and dissolved metals to determine the relative contribution of metals from the acid-mine waste.

Results

Toxicity Tests

Survival of chinook salmon and steelhead trout controls in the tests ranged from 99 to 100% and 92 to 100%, respectively. Concentrations of dissolved metals in the control troughs were ≤ 2 μg/L copper, ≤ 2 μg/L zinc, and ≤ 0.2 μg/L cadmium. Mean water temperatures in the troughs during the tests were 11.0 to 12.5°C and varied within 1°C among troughs in a test. Trough pH ranged from 5.2 to 7.5, alkalinity was 4 to 38 mg/L $CaCO_3$, and hardness was 52 to 35 mg/L $CaCO_3$ for the highest waste concentrations and controls, respectively.

Concentrations of dissolved copper, aluminum, and iron did not proportionately increase with increased waste concentrations because of precipitation; concentrations of total metals did proportionately increase. Concentrations of dissolved copper, zinc, cadmium, aluminum, and iron in the troughs averaged 40.5, 82.3, 74.4, <15, and <15% of total metal concentrations, respectively. The copper/zinc/cadmium ratio at the mean 96-h LC_{50} value for dissolved metals was 1:13:0.13.

The individual 96-h LC_{50} values of the waste to salmon and steelhead trout varied from 0.9 to 4.3% (Table 2). The individual 96-h LC_{50} values for salmon to dissolved metals ranged from 12 to 15 μg/L copper, 83 to 276 μg/L zinc, and 0.7 to 3.1 μg/L cadmium; for steelhead trout the values ranged from 12 to 16 μg/L copper, 113 to 224 μg/L zinc, and 1.4 to 1.9 μg/L cadmium. However, there were no significant differences between mean 96-h and 240-h LC_{50} values for either fish species and no significant differences between mean LC_{50} values for salmon and steelhead trout; the 95% confidence limits overlapped.

TABLE 2—*Concentrations of dissolved metals from acid-mine waste causing 50% (LC_{50}) mortality (95% confidence intervals in parentheses) to chinook salmon and steelhead trout in 96-h and 240-h.*

Species	Copper/Zinc/Cadmium Ratios[a]	96-h LC_{50} Values				240-h LC_{50} Values			
		Waste, %	Copper, μg/L	Zinc, μg/L	Cadmium, μg/L	Waste, %	Copper, μg/L	Zinc, μg/L	Cadmium, μg/L
Chinook salmon	1:4.7:. . .	1.0 (0.6 to 1.8)	<14[b]	160 (142 to 181)	. . .	. . .	. . .	. . .	. . .
	1:4.4:0.04	1.9 (1.3 to 2.8)	>7 < 25[b]	211 (175 to 256)	. . .	. . .	. . .	. . .	. . .
	1:4.1:0.03	4.3 (3.7 to 5.0)	15 (13 to 18)	276 (254 to 302)	. . .	. . .	. . .	. . .	. . .
	1:3.9:0.03	1.3 (1.2 to 1.4)	13 (12 to 13)	159 (137 to 179)	3.1 (3.0 to 3.2)	1.1 (1.0 to 1.4)	12 (11 to 13)	150 (130 to 179)	2.9 (2.8 to 3.2)
	1:3.5:0.03	1.0 (1.0 to 1.1)	12 (11 to 13)	83 (77 to 89)	0.7 (0.6 to 0.8)	0.5 (0.4 to 0.6)	10 (9 to 10)	50 (48 to 53)	0.4 (0.4 to 0.5)
$\bar{x}$[c]			13 (3)	178 (140)	1.9 (3.3)		11 (1)	100 (139)	1.6 (3.5)
Steelhead trout	1:4.7:. . .	0.9 (0.3 to 2.7)	<15[b]	176 (155 to 202)	. . .	. . .	. . .	. . .	. . .
	1:4.4:0.04	1.8 (1.3 to 2.7)	<15[b]	159 (127 to 205)	. . .	. . .	. . .	. . .	. . .
	1:4.1:0.03	1.9 (1.5 to 2.0)	12 (9 to 13)	113 (75 to 130)	. . .	. . .	. . .	. . .	. . .
	1:3.9:0.03	1.7 (1.5 to 1.8)	15 (14 to 16)	224 (197 to 256)	1.9 (1.7 to 2.1)	1.6 (1.5 to 1.8)	15 (14 to 16)	213 (185 to 244)	1.8 (1.6 to 2.0)
	1:3.5:0.03	0.9 (0.7 to 1.1)	16 (15 to 18)	162 (131 to 210)	1.4 (1.2 to 1.7)	0.8 (0.7 to 1.0)	15 (14 to 17)	150 (129 to 183)	1.3 (1.1 to 1.5)
$\bar{x}$[c]			14 (4)	167 (78)	1.6 (0.7)		15 (0)	181 (87)	1.6 (0.7)

[a] Based on mean concentration of total metals. Results presented in chronological order.
[b] Value not used in calculation of mean.
[c] Mean values with ±95% confidence limits (1.96 × standard deviation) in parentheses.

Waste Dilution Studies

The travel time between Spring Creek Debris Dam and Keswick Dam varied from 6.6 to 8.3 h and decreased with increased releases from Keswick Dam (Fig. 2). Patterns of dissolved copper, zinc, and cadmium concentrations over time at Keswick Dam were dependent on the operational scheme of Spring Creek Debris Dam and Spring Creek Powerhouse. When the release from Spring Creek Debris Dam was continuous and that from Spring Creek Powerhouse was intermittent, dissolved metal concentrations peaked at Keswick Dam and corresponded to the initial discharge from Spring Creek Powerhouse (Fig. 2). Total metal concentrations followed a similar pattern (not shown). Relatively constant concentrations of the three metals were present at Keswick Dam when the Spring Creek Debris Dam and Spring Creek Powerhouse releases were coordinated. Concentrations of iron and aluminum at Keswick Dam were relatively constant and unaffected by the discharge of acid-mine waste.

Maximum concentrations of dissolved metals at Keswick Dam on 21 and 25 March and 3 April were 27, 51, and 21 μg/L copper; 182, 214, and 68 μg/L zinc; and 1.4, 2.0, and 0.8 μg/L cadmium, respectively. Low concentrations of dissolved copper ($\leq$2.5 μg/L), zinc ($\leq$3.5 μg/L), and cadmium ($\leq$0.2 μg/L) were present in Keswick Reservoir upstream of the Spring Creek confluence. Mean concentrations of dissolved copper at Keswick Dam exceeded the metal release criteria in all three monitoring studies; the mean concentrations of dissolved zinc were either lower than or equal to the criteria (Table 3). The average copper/zinc/cadmium ratios for total and dissolved metals at Keswick Dam were similar to those at Spring Creek Debris Dam during normal release conditions, but the copper/zinc/cadmium ratio at Keswick Dam decreased during emergency release conditions (Table 3). Thus, when the waste was diluted to 0.37% or more (normal release criteria), no copper was lost, but approximately 40% of the copper was removed in Keswick Reservoir when the waste was diluted to 0.47% (emergency release criteria) (Table 4). The lack of copper removal during normal release conditions resulted in the criteria for copper being exceeded by up to 260% at Keswick Dam (Table 4).

Discussion

There were no significant differences between mean 96-h and mean 240-h LC_{50} values for either chinook salmon or steelhead trout to the dissolved metals, indicating that acutely lethal effects of the waste occur within 96-h (Table 2). Although the low 2.7 to 2.9 pH values contribute toxicity to the undiluted waste, the neutral 6.8 to 7.2 pH values do not contribute toxicity to the diluted waste at or below the LC_{50} values.

The results from our on-site tests on the waste were similar to those obtained from previous tests conducted at our laboratory, in water of similar quality (18 mg/L $CaCO_3$ hardness and alkalinity and 7.2 pH), on this waste [*9*], and, more importantly, on copper, zinc, and cadmium sulfate mixtures [*5*]. Finlayson and Ashuckian [*9*] estimated the dissolved metal 96-h LC_{50} values for steelhead fry to the waste at 10 μg/L copper and 180 μg/L zinc (cadmium was present but not measured) compared with the findings in this study of 14 μg/L copper and 187 μg/L zinc, respectively. Similarly, Finlayson and Verrue [*5*] estimated the dissolved metal 96-h LC_{50} values for chinook salmon fry to a copper, zinc, and cadmium sulfate mixture (copper/zinc/cadmium ratio of 1:12:0.08) at 18, 218, and 1.8 μg/L compared with the 96-h LC_{50} values from the acid-mine waste (copper/zinc/cadmium ratio of 1:14:0.15) in this study of 13, 179, and 1.9 μg/L, respectively. These comparisons indicate that the effects of the Spring Creek acid-mine waste on salmonids are reproducible and comparable to copper, zinc, and cadmium sulfate mixtures at the appropriate dissolved copper/zinc/cadmium ratio. Hence, copper, zinc, and cadmium in the dissolved forms are the major toxic components of the waste.

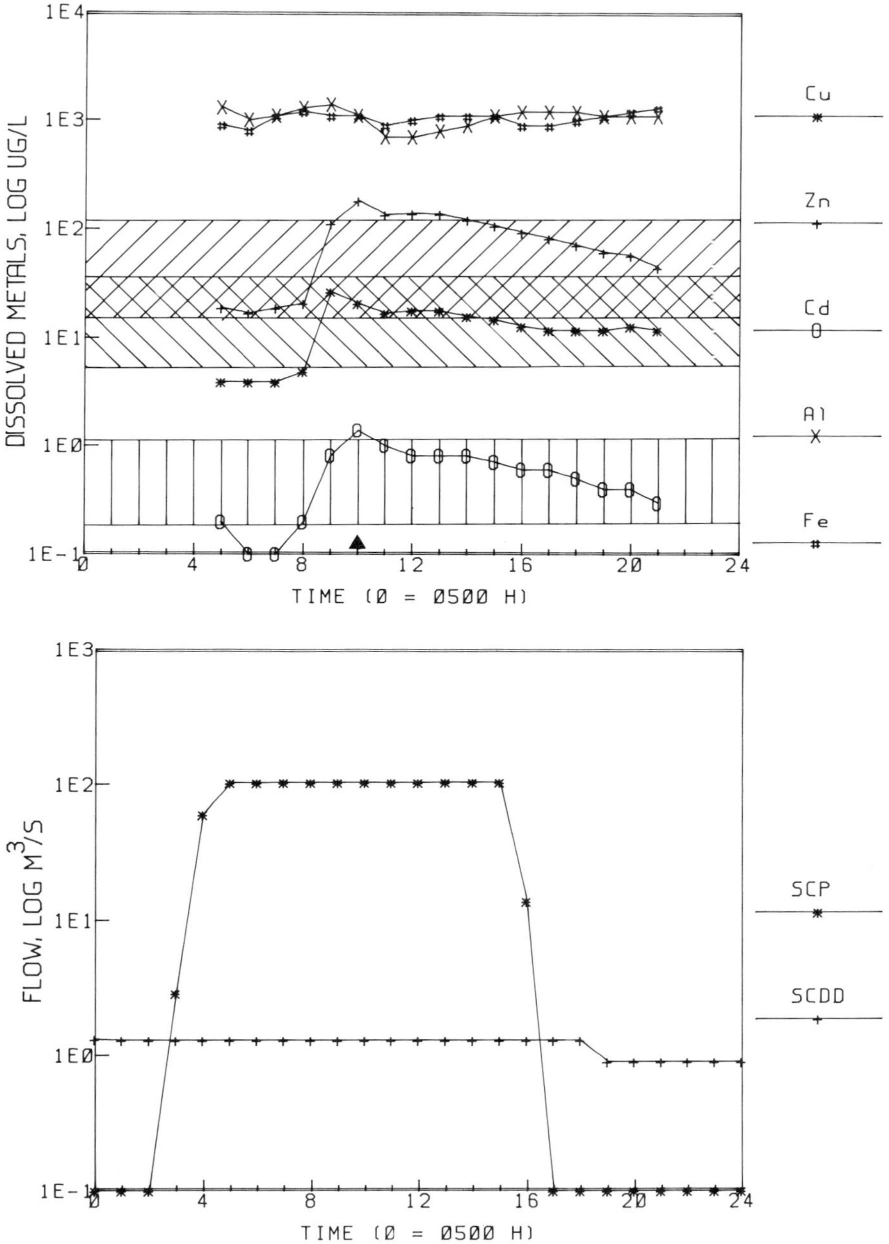

FIG. 2a—*Log_{10} concentrations (in micrograms per litre) of dissolved metals at Keswick Dam on 21–22 March 1980, with corresponding log_{10} releases (in cubic metres per second) from Spring Creek Debris Dam (SCDD) and Spring Creek Powerhouse (SCP). The mean release from Keswick Dam was 285.5 m^3/s on 21 March. Horizontal bars represent median lethal concentrations (upper limit) and water quality objectives (lower limit) for chinook salmon fry* [5,11] *to dissolved zinc (right to left line shading), copper (left to right shading), and cadmium (vertical shading). The arrival of the initial discharge from Spring Creek Powerhouse is represented by a solid triangle.*

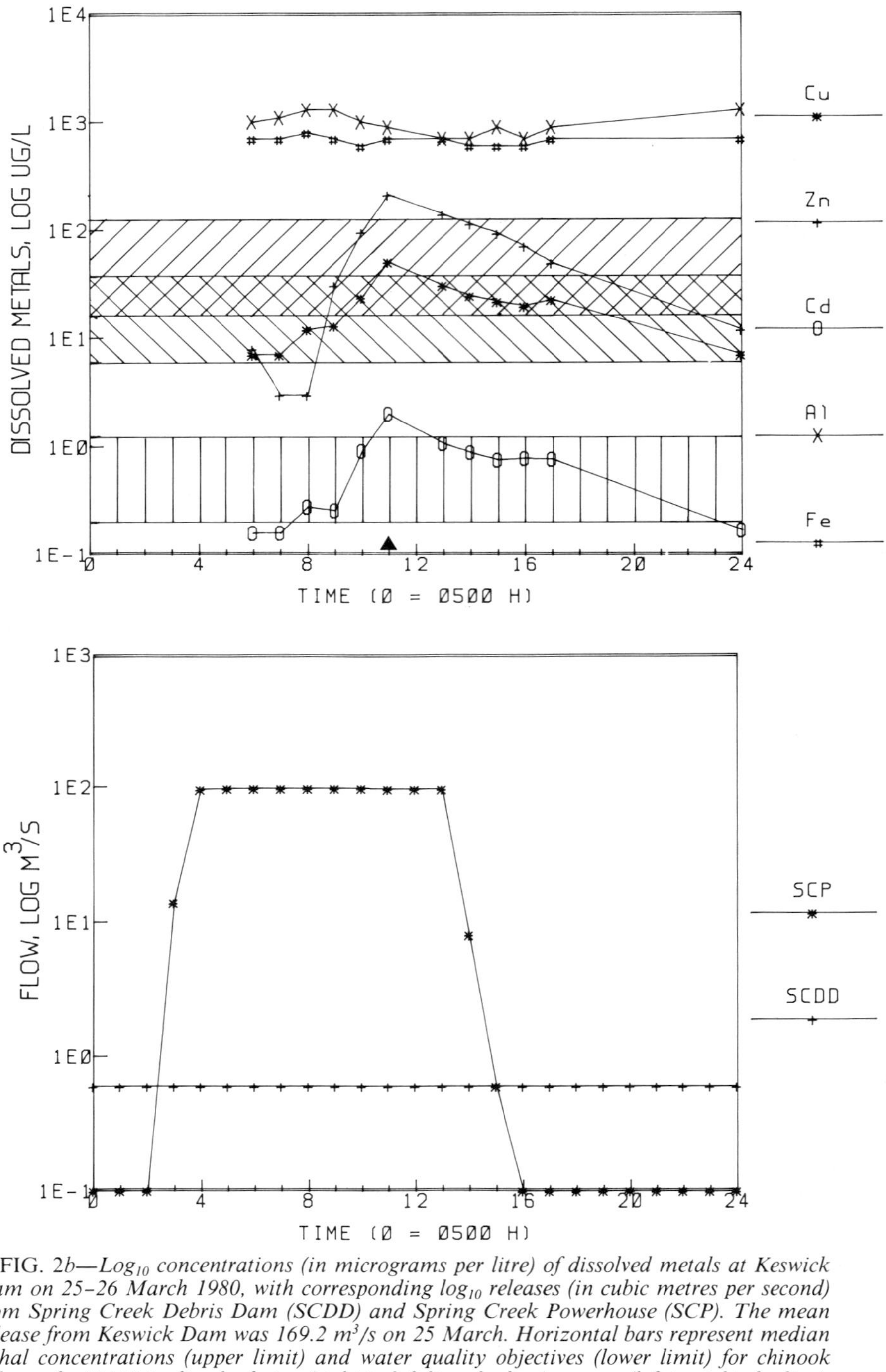

FIG. 2b—*Log*$_{10}$ *concentrations (in micrograms per litre) of dissolved metals at Keswick Dam on 25–26 March 1980, with corresponding log*$_{10}$ *releases (in cubic metres per second) from Spring Creek Debris Dam (SCDD) and Spring Creek Powerhouse (SCP). The mean release from Keswick Dam was 169.2 m*^{3}*/s on 25 March. Horizontal bars represent median lethal concentrations (upper limit) and water quality objectives (lower limit) for chinook salmon fry* [5,11] *to dissolved zinc (right to left line shading), copper (left to right shading), and cadmium (vertical shading). The arrival of the initial discharge from Spring Creek Powerhouse is represented by a solid triangle.*

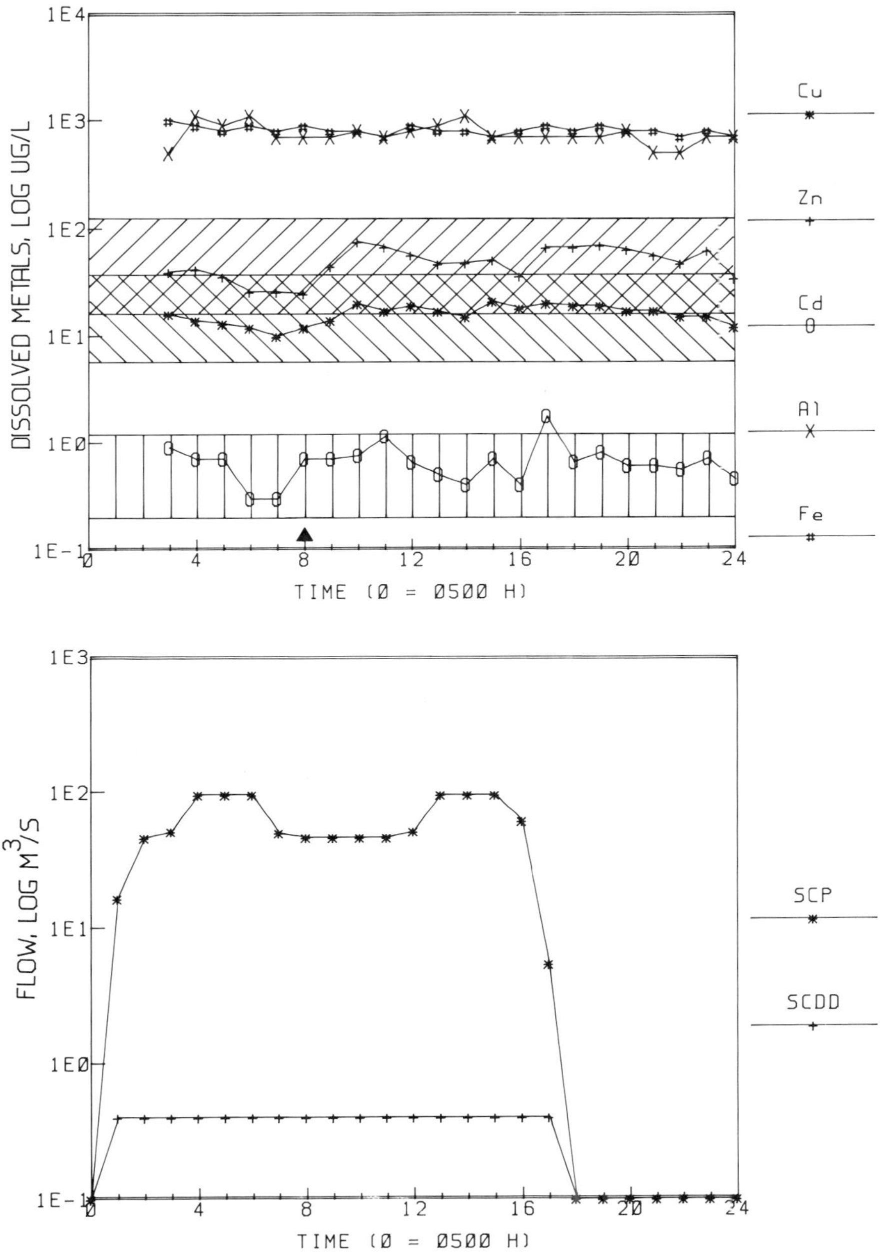

FIG. 2c—*Log_{10} concentrations (in micrograms per litre) of dissolved metals at Keswick Dam on 3–4 April 1980, with corresponding log_{10} releases (in cubic metres per second) from Spring Creek Debris Dam (SCDD) and Spring Creek Powerhouse (SCP). The mean release from Keswick Dam was 212.0 m^3/s on 3 April. Horizontal bars represent median lethal concentrations (upper limit) and water quality objectives (lower limit) for chinook salmon fry* [5,11] *to dissolved zinc (right to left line shading), copper (left to right shading), and cadmium (vertical shading). The arrival of the initial discharge from Spring Creek Powerhouse is represented by a solid triangle.*

TABLE 3—*Mean metal concentrations at Spring Creek Debris Dam, Keswick Reservoir (upstream of Spring Creek), and at Keswick Dam, and copper/zinc/cadmium ratios during emergency and normal release conditions in 1980.*

Date	Site	Copper, μg/L	Zinc, μg/L	Cadmium, μg/L	Copper/Zinc/Cadmium Ratios
21–22 March[a]	Spring Creek Debris Dam	5250	16 400	130	1:3.1:0.02
	Keswick Reservoir	4.0 (2.5)[b]	3.5 (3.0)	0.2 (0.1)	...
	Keswick Dam	19 (13)	92 (84)	0.7 (0.6)	1:4.8:0.04 (1:6.5:0.05)
25–26 March[c]	Spring Creek Debris Dam	6020	18 900	140	1:3.1:0.02
	Keswick Reservoir	3.0 (2.5)	3.0 (2.0)	0.2 (0.1)	...
	Keswick Dam	26 (20)	73 (64)	1.2 (0.6)	1:2.8:0.05 (1:3.1:0.03)
3–4 April[c]	Spring Creek Debris Dam	6250	20 500	160	1:3.1:0.02
	Keswick Reservoir	4.0 (2.3)	4.0 (3.0)	0.2 (0.1)	...
	Keswick Dam	20 (16)	64 (48)	1.0 (0.7)	1:3.2:0.05 (1:3.0:0.04)

[a] Emergency criteria of 15 (7.5) μg/L copper and 108 (96) μg/L zinc for Keswick Dam.
[b] Dissolved metal concentrations are in parentheses.
[c] Normal criteria of 10 (5) μg/L copper and 72 (64) μg/L zinc for Keswick Dam.

TABLE 4—*Release schedule dilution factors (Spring Creek Debris Dam/Keswick Dam) expressed as expected (based on dilution) and nominal (based on net total metal concentrations). Mean net metal concentrations also expressed as percent over release criteria.*

	Dilution Factor			Percentage over Release Criteria	
		Nominal			
Date	Expected	Copper	Zinc	Copper	Zinc
21–22 March[a]	0.0047	0.0029	0.0053	0 (33)[b]	−19 (−16)
25–26 March[c]	0.0037	0.0037	0.0036	130 (260)	−3 (−3)
3–4 April[c]	0.0021	0.0025	0.0029	60 (180)	−17 (−30)

[a] Emergency criteria of 15 (7.5) μg/L copper and 108 (96) μg/L zinc for Keswick Dam.
[b] Dissolved metal concentration values in parentheses.
[c] Normal criteria of 10 (5) μg/L copper and 72 (64) μg/L zinc for Keswick Dam.

Results for the metals tested individually by Finlayson and Verrue [*5*] were used as the basis for the Sacramento River basin plan water quality objectives for copper, zinc, and cadmium [*11*]. Tests on the individual metals were used because the three metal mixture tests indicated antagonistic (or independent) toxicity [*5*]. The objectives were developed by adjusting the U.S. Environmental Protection Agency equations [*6, 7*] for average allowable dissolved metal concentrations to equal one tenth the 96-h LC_{50} values [*15*] for chinook salmon fry [*5*] at 20 mg/L $CaCO_3$ hardness; the results are as follows:

$$\text{Copper objective} = e^{(0.905[\ln(\text{hardness})] - 1.613)}$$
$$\text{Zinc objective} = e^{(0.83[\ln(\text{hardness})] - 0.289)}$$
$$\text{Cadmium objective} = e^{(1.16[\ln(\text{hardness})] - 5.778)}$$

The concentrations estimated for salmon fry at 20 mg/L $CaCO_3$ hardness were adjusted upward for the 40 mg/L $CaCO_3$ hardness of the Sacramento River below Keswick Dam. For example, the three equations above at a hardness of 20 and 40 mg/L $CaCO_3$ result in average allowable concentrations of 3 and 5.6 μg/L copper, 9 and 16 μg/L zinc, and 0.1 an 0.2 μg/L cadmium, respectively. In comparison, the U.S. Environmental Protection Agency [*6, 7*] recommends at a hardness of 40 mg/L $CaCO_3$ average allowable concentrations of 5 μg/L copper, 47 μg/L zinc, and 1.6 μg/L cadmium for the protection of most aquatic species in a balanced community. The zinc and cadmium criteria recommended by the U.S. Environmental Protection Agency exceed the 96-h LC_{10} values reported by Finlayson and Verrue [*5*] of 40 μg/L zinc and 0.8 μg/L cadmium for chinook salmon fry exposed to metals individually. In recognition of the need for site-specific protective criteria, the U.S. Environmental Protection Agency [*6*] allows for the development of criteria for the protection of fish and aquatic life using either one species or a group of aquatic organisms (that is, salmonids) when it has been demonstrated that they are important (that is, recreationally or commercially) and more sensitive to the toxicant. The chinook salmon and steelhead trout tested here and by Finlayson and Ashuckian [*10*] and Finlayson and Verrue [*5*] appear to meet these specifications.

The emergency and normal criteria [*8*] for dissolved zinc of 96 and 64 μg/L, respectively, approach or exceed the 96-h LC_{10} value of 82 μg/L [*5*] for chinook salmon fry exposed to a three-metal mixture with a copper/zinc/cadmium ratio of 1:3:0.02, and these values exceed the water quality objective of 16 μg/L [*11*]. Even though the emergency and normal criteria [*8*] for dissolved copper of 7.5 and 5.0 μg/L, respectively, are in compliance with the water quality objective of 5.6 μg/L [*11*], these values were exceeded at Keswick Dam. The corresponding controlled release schedules for Spring Creek Debris Dam assume that adsorption onto particulate matter and subsequent precipitation and deposition of the metals in Keswick Reservoir will result in a 50% reduction of total and dissolved copper and a 10% reduction of total and dissolved zinc [*8*]. This assumption was based on monitoring data collected in 1978 when the waste was released into Keswick Reservoir at lower dilution rates during periods of high runoff, probably accompanied by higher suspended solids concentrations than were present in this study. Our monitoring data indicate, however, that little reduction of copper and zinc occurred when the waste was diluted to a strength of 0.37% or less; the copper/zinc/cadmium ratios of the waste remained unchanged between Spring Creek Debris Dam and Keswick Dam (Table 3). When the waste was diluted to a strength of 0.47% on 21 March, a substantial reduction (40%) of copper occurred in Keswick Reservoir based on a comparison of the expected and nominal dilution factors (Table 4). The reduction of copper in the Keswick Reservoir corresponded to the reduction of copper in troughs of the on-site toxicity tests. Precipitation and deposition can be expected to lower the concentrations of total and dissolved copper from the waste at lethal levels in Keswick Reservoir, but these phenomena appear to have no significant effect at sublethal levels reflected in the normal metal release criteria [*8*] and the Sacramento River basin water quality objectives [*11*].

The daily mean concentrations of dissolved metals during March and April 1980 (Table 3) at Keswick Dam approached or exceeded 96-h LC_{10} values reported by Finlayson and Verrue [*5*] of 26 μg/L copper, 82 μg/L zinc, and 0.8 μg/L cadmium for salmon fry exposed to a three metal mixture (copper/zinc/cadmium ratio of 1:3:0.02). Small and undetectable mortality of young salmon and steelhead trout present below Keswick Dam could have occurred during this study. Daily peaks (Fig. 2) of dissolved metals exceeded the 96-h LC_{50} values reported by Finlayson and Verrue [*5*] of 37 μg/L copper, 121 μg/L zinc, and 1.1 μg/L cadmium for salmon fry exposed to a three metal mixture (copper/zinc/cadmium ratio of 1:3:0.02). The daily 4 to 5-h peaks were the result of the discharge from Spring Creek Powerhouse scouring the accumulated acid-mine waste in the static Spring Creek arm of Keswick Reservoir when the discharges from Spring Creek Debris Dam and Spring Creek Powerhouse were not coordinated. The difference in toxicity between continuous and peak exposures at the same mean metal concentrations were not determined but can be assessed. The peaks probably did not enhance the acute lethal effects of the acid-mine waste; acute toxicity of copper can be accurately predicted from the mean metal concentration.[3] The effect on chronic toxicity is less clear. Seim et al. [*16*] found a two-fold increase in chronic toxicity (impaired growth) when steelhead trout eggs and fry were intermittently rather than continuously exposed to copper at the same mean concentration. However, the results from their study may not be applicable to the discharge of Spring Creek acid-mine waste. Seim et al. [*16*] used daily exposures in which copper concentrations quickly increased and decreased from stable 4.5-h exposure plateaus, four to five times the mean copper concentrations. Conversely, increasing and decreasing metal concentrations at Keswick Dam were less acute, and peak concentrations were approximately two to three times the mean metal concentrations.

Conclusions

The effects of the acid-mine waste can be estimated with toxicity tests on copper, zinc, and cadmium sulfate mixtures, and the water quality objectives for these three metals reflect site-specific conditions of the Sacramento River. New criteria for zinc and new release schedules for the copper criteria that will protect against acutely toxic levels should be developed soon for the Spring Creek acid-mine waste. The release criteria should reflect the Sacramento River water quality objectives following cleanup of the Iron Mountain area and increased holding facilities for the acid-mine waste in Spring Creek Reservoir.

Acknowledgments

J. Nelson, K. Verrue, and D. Knoff assisted in the study. This and other studies conducted on the Spring Creek acid-mine waste problem were supported by funds from the California Regional Water Quality Control Board-Central Valley Region.

References

[*1*] "Effects of Mine-Waste Pollution on Anadromous and Resident Fish in Upper Sacramento River, Shasta County, California," U.S. Fish and Wildlife Service, Washington, DC, 1959.

[3] Chapman, G., U.S. Environmental Protection Agency, personal communication, 1986.

[2] Fuller, R., Shay, J., Ferreria, R., and Hoffan, R., "An Evaluation of Problems Arising from Acid-Mine Drainage in the Vicinity of Shasta Lake, Shasta County, California," Water Resources Investigations Report 78-32, U.S. Geological Survey, Denver, CO, 1979.
[3] Lewis, R., "Recommended Flow Releases from Spring Creek Debris Dam for the Protection of Salmonid Fishes in the Sacramento River," California Department of Fish and Game, Region 1, Redding, CA, 1963.
[4] Finlayson, B. and Wilson, D., "Acid-Mine Waste: How it Affects King Salmon in the Upper Sacramento River," *Outdoor California,* November-December 1979, pp. 8–12.
[5] Finlayson, B. and Verrue, K., "Toxicities of Copper, Zinc, and Cadmium Mixtures to Juvenile Chinook Salmon," *Transactions of the American Fisheries Society,* Vol. 111, 1982, pp. 645–650.
[6] U.S. Environmental Protection Agency, "Water Quality Criteria Documents: Availability," *Federal Register,* Vol. 45, 1980, pp. 79318–79379.
[7] U.S. Environmental Protection Agency, "Water Quality Criteria; Requests for Documents," *Federal Register,* Vol. 49, 1984, pp. 4551–4554.
[8] Wilson, D, "Proposed Interim Release Flows from Spring Creek Debris Dam for the Protection of Salmonid Fishes in the Sacramento River," Memorandum to California Regional Water Quality Control Board-Central Valley Region, California Department of Fish and Game, Region 1, Redding, CA, 3 Aug. 1978.
[9] Finlayson, B. and Ashuckian, S., "Safe Zinc and Copper Levels from the Spring Creek Drainage for Steelhead Trout in the Upper Sacramento River, California," *California Fish and Game,* Vol. 65, 1979, pp. 80–99.
[10] Finlayson, B. and Verrue, K., "Estimated Safe Zinc and Copper Levels for Chinook Salmon, *Oncorhynchus tshawytscha,* in the Upper Sacramento River, California," *California Fish and Game,* Vol. 66, 1980, pp. 68–82.
[11] "Resolution 84-55, Amending the Water Quality Control Plan, Sacramento River Basin (5A)," California State Water Resources Control Board, Sacramento, CA, 1984.
[12] Leitritz, E. and Lewis, R., "Trout and Salmon Culture," *California Department of Fish and Game Fish Bulletin,* No. 164, 1979.
[13] Finlayson, B., Nelson, J., and Hansen, R., "Design Optimization of the Chlorination Process. Volume II: Comparison of Acute Toxicity of Chlorinated Effluents from Optimized and Existing Facilities," Research Report Series EPA-600/2-80-133, U.S. Environmental Protection Agency, Cincinnati, OH, 1980.
[14] Peltier, W. and Weber, C., "Methods for Measuring the Acute Toxicity of Effluents to Freshwater and Marine Organisms (3rd ed.)," Environmental Monitoring and Support Laboratory Report EPA/600/4-85/013, U.S. Environmental Protection Agency, Cincinnati, OH, 1985.
[15] Windom, H. C., Chakoumakos, C., Hodson, P. O., Nakatani, R. E., Reish, D. J., Sunda, W. G., and Swanson, D. L., "Copper," *Review of the EPA Red Book: Quality Criteria for Water,* R. V. Thurston, R. C. Russo, C. M. Fetterolf, T. A. Edsall, and Y. M. Baker, Eds., Water Quality Section, American Fisheries Society, Bethesda, MD, 1979, pp. 95–105.
[16] Seim, W., Curtis, L., Glenn, S., and Chapman, G., "Growth and Survival of Developing Steelhead Trout *(Salmo gairdneri)* Continuously or Intermittently Exposed to Copper," *Canadian Journal of Fisheries and Aquatic Sciences,* Vol. 41, 1984, pp. 433–438.

Albert Westerman[1]

Yearly Chronic Toxicity Comparisons of Stream Sediments and Waters

REFERENCE: Westerman, A., **"Yearly Chronic Toxicity Comparisons of Stream Sediments and Waters,"** *Aquatic Toxicology and Environmental Fate: Eleventh Volume, ASTM STP 1007,* G. W. Suter II and M. A. Lewis, Eds., American Society for Testing and Materials, Philadelphia, 1989, pp. 204–214.

ABSTRACT: Three years of fathead minnow *(Pimephales promelas)* embryo-larval chronic toxicity tests were conducted on stream sediments and waters collected from eight localities in the Yellow Creek drainage, in Kentucky. Test sites were chosen to assess the impact from the following point and nonpoint sources: (1) a wastewater treatment plant (WWTP), (2) urban runoff, (3) coal mining and washing activities, (4) septic tank seepage, and (5) dumping of trash into the creek. All tests were conducted in replicate, and sediments and water were analyzed for heavy metals (high concentrations of metals have historically been measured in the area). The toxicity of stream sediments and water decreased downstream from the WWTP with dilution and distance from the outfall. Results of the solid-phase sediment tests were very similar over the three years. However, considerable differences were noted in the yearly water testing. Sediment samples collected immediately below the WWTP demonstrated 82% toxicity in 1984, 81% toxicity in 1985, and 78% toxicity in 1987. Water samples from that location produced 21% lethality in 1984, 100% lethality in 1985, and 21% lethality in 1986. Control sediments and waters from the basin demonstrated very high survival over the three years (range 95 to 98%). From these data, solid-phase sediment toxicity tests are useful indicators of long-term impacts (over three years) while water-column toxicity tests are more representative of short-term effects. Under the constraints of time and resources that are placed on federal and state agencies and in areas that are monitored infrequently, sediment toxicity testing would be preferable to water-column testing to assess overall in-stream toxic conditions.

KEY WORDS: chronic toxicity, stream sediments, solid-phase sediment tests, water column tests, point and nonpoint pollution, embryo-larval toxicity tests, aquatic toxicology

Stream sediments have been demonstrated to be the principal "sink" for metallic and organic contaminants discharged into our waters [*1–11*]. Such stream sediments may contain 10 000 times or greater concentrations of contaminants than are found in the overlying water [*12*]. Chromium, cadmium, lead, iron, zinc, and various other heavy metals have been reported at concentrations up to 100 mg/g or more in bottom sediments [*13,14*]. As bottom sediments serve as spawning substrates for most species of fish and aquatic invertebrates, contaminants released from the sediments can kill the eggs and larvae of many fish and aquatic insect species. Killing the eggs and larvae can eliminate the reproductive potential of aquatic life populations and have long-term degrading effects on a stream's fauna.

[1] Aquatic toxicologist, Kentucky Department for Environmental Protection, Division of Water, Frankfort Office Park, Frankfort, KY 40601.

Yellow Creek's fauna (Bell County, Kentucky) have been under severe jeopardy because of a long history of heavy metal pollution in the basin [*15–19*]. The potential toxicity of any contaminant has been observed to be largely dependent upon its concentration, the chemical form in which the contaminant occurs, and the particular physicochemical variables of the medium in which it occurs during its exposure to aquatic life [*20*]. In an effort to bypass exhaustive chemical speciation studies, toxicity (bioavailability) can be measured directly through chronic toxicity tests of the stream sediments and water. The objectives of this investigation were to (1) determine the extent to which sediments taken from eight sites along a stream (Yellow Creek, Bell County, Kentucky) were chronically toxic to the early life stages of an endemic species to the stream, fathead minnow *(Pimephales promelas);* (2) assess the detrimental effects of stream water from those sites on early life stages of fathead minnows; (3) compare water-column toxicity data with solid-phase sediment toxicity data collected from the eight sites over time (three years); (4) gauge the magnitude and contribution of the major point and nonpoint toxic inputs into the drainage (that is, the wastewater treatment plant, urban runoff, coal washing and mining activities, septic tank seepages, and dumping of trash into the creek); and (5) determine the usefulness of sediment toxicity tests in assessing the environmental health of a stream.

Materials and Methods

Sediments and stream water were collected from eight localities in the Yellow Creek drainage, Bell County, Kentucky (see Fig. 1). The test sites were chosen with regard to their representation as possible sources of point and nonpoint pollution input into the drainage, as had been indicated by an U.S. Environmental Protection Agency report [*21*]. The sites were as follows: (1) a Yellow Creek control site located immediately upstream of the City of Middlesboro, at stream kilometre point (KP) 26.2 (kilometres upstream from mouth); (2) an in-town site established at KP 24.6, located above the old wastewater treatment plant (WWTP 1), but below a domestic sewage pumping station that had periodically overflowed raw sewage; (3) a site on Bennetts Fork, placed at 1.3 km from the confluence of the Fork with Yellow Creek, where mining and coal washing activities were located along the Fork's basin; (4) a site at KP 24.0, established below the observable mixing zone of WWTP 1 and 50-m upstream from the confluence of Bennetts Fork with Yellow Creek; (5) a site at KP 23.8, 50-m below the confluence of Bennetts Fork and Yellow Creek; (6) a site at KP 20.8; (7) a site at KP 16.0, 8.0-km downstream from WWTP 1, the town, and Bennetts Fork and a site that was subject to periodic refuse dumping; and (8) a site 100-m upstream from the confluence of the Yellow Creek with the Cumberland River (KP 0.1), the last 5 km of stream was impacted by septic tank seepage as well as general trash dumping.

Sediment samples were restricted to the upper 3 to 5 cm of bottom sediments and were collected at three locations across the stream per site. Sediment sampling containers were widemouth 3000-mL Pyrex bottles with Teflon-lined screw caps. Two of the samples were within 0.3 m from either bank at water depths under 0.5 m. The third sample of sediment was taken at midstream. Water samples were also taken from the midstream location. In addition, samples for chemical analysis were also collected. All water samples for chemical analysis were preserved with 10% nitric acid and held on ice; unpreserved sediment samples for chemical analysis and bioassays were iced for transport to the laboratory. The total metals in water and sediments were determined by atomic adsorption spectrometry (Perkin-Elmer Model 503) by procedures outlined by the U.S. Environmental Protection Agency (EPA) [*22*] and by Hergert [*23*], and these data are presented in greater detail in Ref *18*. In general, sediment samples were dried for 24 h at 60°C, sieved

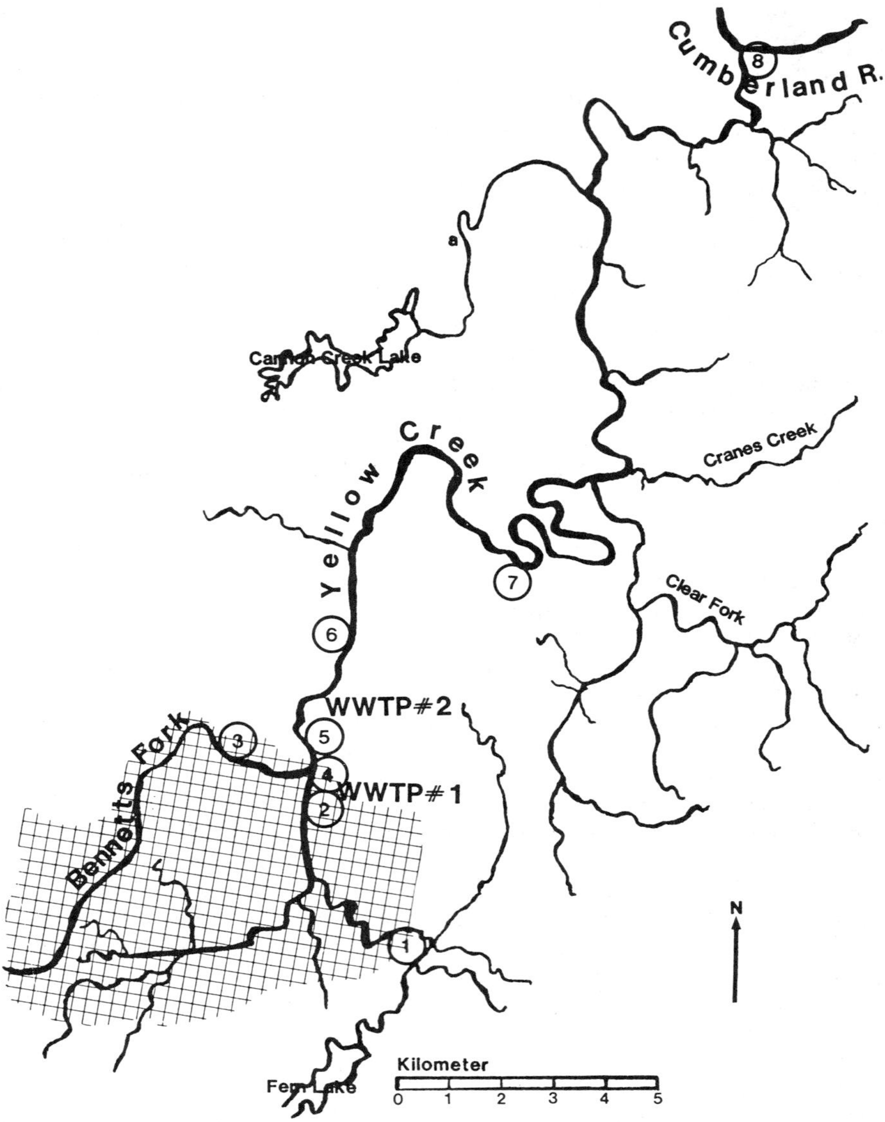

FIG. 1—*Sampling locations on the Yellow Creek system, labeled 1 through 8. Other designated areas include the city of Middlesboro (area with hatching), WWTP 1 (the discharge point of the old wastewater treatment plant), and WWTP 2 (the new plant which began operation in 1986).*

through a 0.5-mm sieve to ensure uniform particle size, weighed to the nearest 0.001 mg, digested in a 1:1 mixture of concentrated nitric and hydrochloric acids for 24 h at 80°C, filtered through a Whatman No. 541 filter paper, and analyzed by atomic absorption spectrometry graphite furnace Model HGA-2100. The water samples were analyzed with the same Perkin-Elmer Model 503 unit.

Static solid-phase sediment chronic tests were performed in replicate (three) per sample (three) per site (a total of nine test chambers per site) using deep Pyrex petri dishes which had a capacity of approximately 400 mL (Fig. 2). Sediment from each site was layered to a depth of 2 cm and covered with 350 mL of water collected from Site 1 (control). A stainless steel screen (mesh size 0.5 mm) was laid on the sediment to prevent the eggs from becoming buried. Aeration was provided by a capillary tube to within 10 mm of the screen. The control site was located on the edge of Cumberland Gap National Park (Yellow Creek KP 26.2). The site had previously been determined to support a diverse population of fish and invertebrates, including a federally listed endangered species of fish [*24*]. The water quality characteristics were monitored a minimum of four times during the experiments for every test chamber.

Stream water static-renewal chronic tests were performed in replicate (three) using deep Pyrex petri dishes (similar to those used in the sediment tests). Three replicate dishes were tested from each midstream site. In addition, a laboratory culture water was also tested as an additional control [*25*]. The test water was removed and replaced by fresh stream water every 24 h to preclude metabolic product buildup and to maintain toxicant concentrations. Water quality characteristics were monitored daily from the eight sites and laboratory control water, and the results are given in Table 1. All values represent a minimum of twelve measurements per site per year.

The egg sample size was 50 per replicate for the sediment and water tests (450 individuals per site for sediment and 150 for water tests). Eggs of the fathead minnow were obtained from the Environmental Protection Agency Fish Toxicology Laboratory at Newtown, Ohio. Exposure was initiated with eggs less than 6-h old and continued through Day 12. The majority of the eggs hatched on Day 5 at the test temperature of 25°C. The dissolved oxygen concentration was maintained at or near saturation by moderate aeration via the capillary tube [that is, in milligrams per litre, 8.35 ± 0.2 standard deviation (SD) for the year 1984; 8.40 ± 0.14 SD for 1985; and 8.39 ± 0.11 SD for 1987]. The fry were fed brine shrimp nauplii *ad libitum* twice daily. Dead or uneaten nauplii were removed from the test chambers daily. The cultures were inspected daily to determine the extent of embryonic development and to remove and record mortalities. The test responses

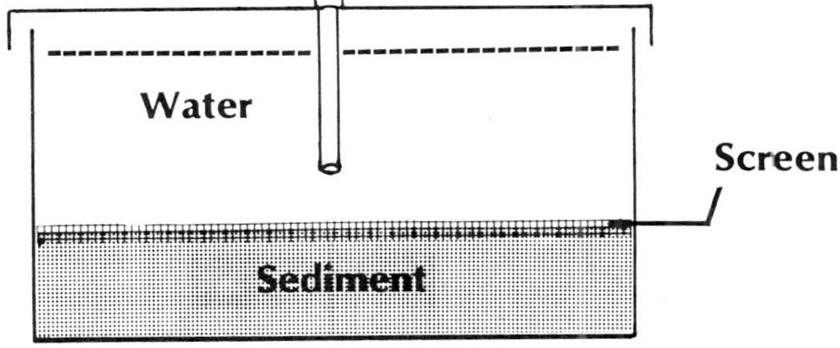

FIG. 2—*Sediment testing chamber.*

TABLE 1—*Water quality characteristics of the Yellow Creek collecting sites (± standard deviations).*[a]

Location[b]	Year	pH	Hardness, mg/L as $CaCo_3$	Alkalinity, mg/L as $CaCo_3$	Conductivity, μmho/cm
Site 1	1984	7.1 ± 0.2	24 ± 4.0	18 ± 2.0	40 ± 5.0
	1985	7.9 ± 2.0	20 ± 2.0	16 ± 1.0	64 ± 3.0
	1987	7.1 ± 0.5	12 ± 1.0	28 ± 0.0	70 ± 2.0
Site 2	1984	7.2 ± 2.0	48 ± 1.0	20 ± 1.0	64 ± 2.0
	1985	7.2 ± 1.4	44 ± 2.0	28 ± 2.0	80 ± 5.0
	1987	7.3 ± 0.8	52 ± 1.0	68 ± 1.0	162 ± 6.0
Site 3	1984	7.2 ± 0.8	36 ± 2.0	28 ± 2.0	48 ± 1.0
	1985	7.1 ± 0.6	32 ± 1.0	24 ± 2.0	48 ± 2.0
	1987	7.5 ± 0.2	232 ± 4.0	100 ± 4.0	450 ± 6.0
Site 4	1984	7.4 ± 2.0	490 ± 5.0	66 ± 2.0	2670 ± 25.0
	1985	7.4 ± 1.0	360 ± 2.0	64 ± 5.0	1270 ± 10.0
	1987	7.4 ± 0.5	64 ± 1.0	70 ± 2.0	250 ± 5.0
Site 5	1984	7.3 ± 2.0	340 ± 5.0	56 ± 2.0	1450 ± 10.0
	1985	7.2 ± 1.0	284 ± 3.0	52 ± 2.0	680 ± 5.0
	1987	7.4 ± 2.0	156 ± 1.0	86 ± 0.5	408 ± 5.0
Site 6	1984	7.2 ± 0.5	320 ± 4.0	52 ± 0.5	880 ± 2.0
	1985	7.0 ± 1.0	276 ± 2.0	48 ± 2.0	640 ± 5.0
	1987	7.7 ± 2.0	180 ± 1.0	152 ± 1.5	520 ± 4.0
Site 7	1984	7.0 ± 1.0	296 ± 2.0	44 ± 2.0	520 ± 2.0
	1985	6.9 ± 2.0	264 ± 1.0	40 ± 1.0	248 ± 3.0
	1987	7.7 ± 0.2	160 ± 5.0	88 ± 2.0	370 ± 8.0
Site 8	1984	7.3 ± 0.5	220 ± 2.0	56 ± 1.0	440 ± 5.0
	1985	7.6 ± 2.0	216 ± 4.0	52 ± 2.0	540 ± 5.0
	1987	7.8 ± 1.0	272 ± 4.0	84 ± 1.0	455 ± 10.0

[a] Measurements were made daily during the twelve testing days per year.
[b] Yellow Creek drainage sites were Site 1 at Yellow Creek KP 26.2, Site 2 at Yellow Creek KP 24.6, Site 3 on Bennett's Fork KP 1.3, Site 4 at Yellow Creek KP 24.0, Site 5 at Yellow Creek KP 23.8, Site 6 at Yellow Creek KP 20.8, Site 7 at Yellow Creek KP 16.0, and Site 8 at Yellow Creek KP 0.1.

included egg survival, percent larval survival, and percent gross anatomical deformities appearing in newly hatched individuals. Except that these experiments were longer, procedures were similar to those presented by Birge et al. [*4,26*] and by Horning and Weber [*27*].

Results and Discussion

Evaluation of the impacts of point (for example, wastewater treatment plant) and non-point (for example, urban runoff) pollution on a fourth-order stream was facilitated by direct toxicological monitoring of stream sediments and water as well as chemical analysis of the sediments and water for heavy metals. Eight sediment stations were chosen in the basin. Three sites were upstream from the old wastewater treatment plant (WWTP 1), which discharged into Yellow Creek at KP 24.1; these included a control site (Site 1), a site that received urban runoff and occasional raw domestic sewage from an old pumping station (Site 2), and a site located downstream from a coal washing facility (Site 3). The remaining five sites were established downstream of WWTP 1. The retired (1986) wastewater treatment plant had received approximately 10% of its waste load from an adjacent tannery. The wastes discharged from the tannery to the WWTP were determined in chronic toxicity tests to be highly toxic [a median lethal concentration (LC_{50}) of 0.3% and no-observable-effect concentration (NOEC) of 0.02% for fathead minnows] to fish and amphibians and were believed to be the major toxic component in the WWTP discharge [*17*]. Discharges to the creek from WWTP 1 were found to be chronically toxic to fathead minnows at NOECs of 0.05% chlorinated final effluent and 0.08% dechlorinated final effluent [*17*]. In the early fall of 1986, WWTP 1 ceased operation and a new expanded-capacity plant (WWTP 2) started operation; about the same time, the tannery cut its volume output by 33%. This combination of factors, a new WWTP and reduced tannery wastes, visibly improved the appearance of the stream. Chronic effects were determined in chlorinated effluents from WWTP 2 down to 64% effluent (NOEC), a 2000-fold improvement. No significant effects were found in 100% dechlorinated final effluent; 97% survival

TABLE 2—*Metal concentrations in water at various sites in the Yellow Creek system.*[a]

		Concentration, mg/L				
Site Number[b]	Year	Cadmium	Chromium	Copper	Iron	Zinc
1	1984	<0.001	<0.005	<0.001	<0.050	<0.001
	1985	<0.001	<0.005	<0.001	0.055	<0.001
	1987	<0.001	<0.005	<0.001	0.032	<0.001
2	1984	<0.001	<0.005	<0.001	0.194	0.052
	1985	<0.001	<0.005	<0.001	0.296	0.107
	1987	<0.001	<0.005	<0.001	0.213	0.123
3	1984	<0.001	<0.005	<0.001	<0.010	<0.001
	1985	<0.001	<0.005	<0.001	0.637	0.026
	1987	<0.001	<0.005	<0.001	0.082	0.028
4	1984	0.015	0.292	0.046	0.932	0.156
	1985	0.023	0.876	0.052	1.379	0.186
	1987	<0.001	0.102	0.010	0.102	0.026
5	1984	<0.001	0.250	0.032	0.495	0.075
	1985	0.011	0.524	0.026	1.196	0.101
	1987	<0.001	0.045	0.008	0.063	0.010
6	1984	<0.001	<0.005	<0.005	0.053	0.032
	1985	<0.001	0.049	<0.001	1.104	0.052
	1987	<0.001	0.023	0.011	0.113	0.122
7	1984	<0.001	<0.005	<0.005	0.053	0.032
	1985	<0.001	0.021	<0.001	0.605	0.038
	1987	<0.001	<0.005	<0.005	0.020	0.038
8	1984	<0.001	<0.005	<0.005	<0.010	0.019
	1985	<0.001	0.010	<0.001	0.503	0.041
	1987	<0.001	0.095	0.218	0.023	0.020
Laboratory culture water	1984	<0.001	<0.005	<0.001	<0.005	<0.001
	1985	<0.001	<0.005	<0.001	<0.005	<0.001
	1987	<0.001	<0.005	<0.001	<0.005	<0.001

[a] Five replicate water samples were analyzed daily, per year, per station. Lead, mercury, and nickel were also analyzed but remained at or near the detection limits of 0.001 for lead and nickel and 0.0001 for mercury.

[b] Yellow Creek drainage sites were Site 1 at Yellow Creek KP 26.2, Site 2 at Yellow Creek KP 24.6, Site 3 at Bennett's Fork KP 1.3, Site 4 at Yellow Creek KP 24.0, Site 5 at Yellow Creek KP 23.8, Site 6 at Yellow Creek KP 20.8, Site 7 at Yellow Creek KP 16.0, and Site 8 at Yellow Creek KP 0.1.

of fathead minnows was observed in the effluent in comparison with 96% survival in controls over the twelve-day tests.

Metal concentrations observed in the stream water taken from eight locations in the Yellow Creek basin are presented in Table 2. The levels of metals decreased downstream from WWTP 1 (KP 24.1). The water metal concentrations decreased in the system subsequent to the initiation of operation of WWTP 2.

Concentrations of eight metals in sediments from the eight locations in the Yellow Creek system are given in Table 3. Although the levels are considerably elevated over what would be expected in a nonpolluted stream, the concentrations of metals are down since the new wastewater treatment plant started operation. Metal levels decreased downstream from the WWTP 1 discharge point, but still remained elevated to the Cumberland River. Over the years, concentrations of copper, iron, and zinc were high below the coal washing facility. Cadmium and chromium showed the most pronounced increases of any metal measured below WWTP 1. These elevated levels of cadmium and chromium were persistent over time (that is, 1984 through 1987) and at the downstream sites. With the exception of higher levels of cadmium and nickel, sediments below the WWTP 2 outfall were comparable to the control station.

TABLE 3—*Metal concentrations in sediments at various sites in the Yellow Creek system.*[a]

Site Number[b]	Year	Concentrations, μg/g						
		Cadmium	Chromium	Copper	Iron	Lead	Nickel	Zinc
1	1984	<0.001	10.84	2.81	859.4	10.79	<0.01	8.56
	1985	<0.001	10.95	5.73	1 302.3	7.08	0.10	10.63
	1987	<0.001	9.83	2.73	847.9	6.20	<0.01	4.77
2	1984	0.013	10.32	11.14	1 149.3	14.86	<0.01	24.09
	1985	0.021	11.77	8.64	5 772.2	21.84	0.15	36.98
	1987	0.022	8.95	7.46	860.9	15.93	<0.01	39.07
3	1984	<0.001	9.56	14.00	6 000.9	13.53	<0.01	32.63
	1985	0.012	10.51	12.37	17 094.2	15.20	10.16	34.30
	1987	<0.001	6.42	26.99	5 732.8	16.08	<0.01	35.73
4	1984	0.709	21.01	6.56	5 367.3	19.19	<0.01	32.07
	1985	0.743	52.01	13.01	10 062.8	22.11	0.33	47.80
	1987	0.781	25.60	12.79	3 769.6	26.93	<0.01	35.40
5	1984	0.621	22.43	12.57	6 086.5	17.27	2.93	34.64
	1985	0.559	36.51	12.36	12 497.6	12.45	8.98	36.27
	1987	0.705	27.73	9.03	4 151.5	35.18	0.07	37.29
6	1984	0.317	12.53	10.53	5 046.7	12.92	5.81	29.19
	1985	0.288	23.31	13.97	11 546.8	14.76	6.90	39.21
	1987	0.275	5.95	3.64	3 691.5	13.25	0.09	28.46
7	1984	0.199	14.74	10.31	4 669.9	12.66	2.48	31.97
	1985	0.100	21.09	11.88	9 089.0	12.44	5.14	38.08
	1987	0.094	4.81	2.21	1 513.6	8.22	0.07	11.85
8	1984	0.123	20.79	12.83	5 612.9	14.57	2.54	31.66
	1985	0.118	19.43	12.67	7 892.9	16.52	8.87	36.27
	1987	0.064	2.05	1.71	971.9	6.67	0.03	7.12

[a] Mean metal concentration of nine samples per site. Mercury was analyzed but remained at or near detection limits of 0.001.

[b] Yellow Creek drainage sites were Site 1 at Yellow Creek KP 26.2, Site 2 at Yellow Creek KP 24.6, Site 3 at Bennett's Fork KP 1.3, Site 4 at Yellow Creek KP 24.0, Site 5 at Yellow Creek KP 23.8, Site 6 at Yellow Creek KP 20.8, Site 7 at Yellow Creek KP 16.0, and Site 8 at Yellow Creek KP 0.1.

The results of the three years of toxicity tests performed on the water taken from the eight locations along Yellow Creek and on laboratory control water are presented in Table 4. Control survivals from the laboratory and field site (Site 1) were high over the years (96 to 99% survival) and were considerably better than the minimum of 80% allowed by the EPA in their chronic testing manual [29]. Site 4, below the (old) WWTP 1, consistently demonstrated the strongest toxic responses and was the only site that had significantly lower survival than the control site for each testing period. Lethality decreased with dilution and distance downstream from WWTP 1. Unlike WWTP 1, WWTP 2, which discharged at KP 21.0, was not having a significant effect ($P < 0.5$) on Site 6 (KP 20.8). Nonpoint urban runoff (Site 2) from Middlesboro (population: 12 500) has shown a degrading effect over the years and in the future has the potential to be the primary toxic input into the system.

Direct toxicological monitoring of Yellow Creek sediments with embryo-larval stages of fathead minnows is presented in Table 5. Survival of animals tested on sediments from the control site (Site 1) were high for each of the years—98%, 96%, and 95% for 1984, 1985, and 1987, respectively. Control survival was also considerably better than the maximum allowable lethality for viable testing as established by EPA [27]. A relationship was apparent between the proximity of the sediment sample to WWTP 1 and its toxic response; the greater the distance downstream from the old plant, the less toxicity observed. In general, the toxic response was less after the new wastewater plant started operation. However, the toxic response from any given site was remarkably similar over the three years. For example, tests conducted on sediments collected at Site 4 produced lethalities of 82% in 1984, 81% in 1985, and 78% in 1987, and Site 5 had lethalities of 57%, 54%, and 56% in 1984, 1985, and 1987, respectively. The season of the year (that is, January, May, September) that samples were collected also did not effect the toxic responses.

Water-column toxicity tests conducted on the Yellow Creek drainage indicated periodic intervals of chronic in-stream toxic conditions over three years. Sediment solid-phase toxicity tests performed at the same sites indicated a persistent chronically toxic response over the years. Unless water-column toxicity testing has been performed during the toxic episodes, the long-term chronic effects on aquatic life may not be measurable. Such underestimations of long-term effects would be most prevalent with shorter-term toxicity tests (for

TABLE 4—*Comparison of yearly chronic fathead minnow* (Pimephales promelas) *stream water bioassays from stream sites in the Yellow Creek drainage system.*[a]

Site[b]	Mean Percent Survival ± Standard Deviation		
	23 Sept. 1984	11 May 1985	10 Jan. 1987
1 (Control site)	98 ± 0.6	97 ± 0.6	97 ± 1.2
2	94 ± 1.1	86 ± 2.0	82 ± 0.6
3	95 ± 0.6	97 ± 0.6	89 ± 1.5
4	79 ± 1.5	0	79 ± 1.5
5	83 ± 2.5	25 ± 3.5	85 ± 0.6
6	86 ± 1.0	62 ± 4.2	89 ± 1.0
7	92 ± 0.6	89 ± 0.6	94 ± 0.6
8	93 ± 0.6	90 ± 0.0	90 ± 2.6
Control laboratory water	99 ± 1.0	97 ± 0.0	96 ± 0.6

[a] Mean percent survival of three replicates per site compiled after twelve days exposure.

[b] Yellow Creek drainage sites were Site 1 at Yellow Creek KP 26.2, Site 2 at Yellow Creek KP 24.6, Site 3 at Bennett's Fork KP 1.3, Site 4 KP 24.0, Site 5 at Yellow Creek KP 23.8, Site 6 at Yellow Creek KP 20.8, Site 7 at Yellow Creek KP 16.0, and Site 8 at Yellow Creek KP 0.1.

TABLE 5—*Comparison of yearly chronic fathead minnow* (Pimephales promelas) *sediment bioassays from stream sites in the Yellow Creek drainage system.*[a]

Site[b]	Mean Percent Survival ± Standard Deviation		
	23 Sept. 1984	11 May 1985	10 Jan. 1987
1 (Control site)	98 ± 0.0	96 ± 0.6	95 ± 0.6
2	83 ± 1.1	73 ± 1.0	69 ± 1.2
3	90 ± 0.0	93 ± 2.5	87 ± 1.2
4	18 ± 4.0	19 ± 0.6	22 ± 1.0
5	43 ± 2.3	46 ± 1.0	44 ± 1.0
6	67 ± 1.2	72 ± 2.0	89 ± 1.2
7	89 ± 1.1	93 ± 0.6	92 ± 0.6
8	95 ± 1.1	88 ± 1.5	86 ± 1.0

[a] Mean percent survival of nine replicates per site compiled after twelve days exposure.
[b] Yellow Creek drainage sites were Site 1 at Yellow Creek KP 26.2, Site 2 at Yellow Creek KP 24.6, Site 3 at Bennett's Fork KP 1.3, Site 4 at Yellow Creek KP 24.0, Site 5 at Yellow Creek KP 23.8, Site 6 at Yellow Creek KP 20.8, Site 7 at Yellow Creek KP 16.0, and Site 8 at Yellow Creek KP 0.1.

example, 7 to 9 days), including the water-column tests in these experiments. These underestimations would remain a problem unless continuous testing was performed. Chronic tests generally are not used to measure releases of highly toxic substances (for example, they are too labor-intensive and costly). Such releases should produce acute toxic effects in the stream with resulting fish kills. A major use for sediment and water-column chronic toxicity testing is in assessing the impact of low concentrations of toxicants on stream organisms. Although sediment toxicity tests may not measure the severity of a particular toxic incident, they are shown to reflect the history of toxic incidents that have occurred. As the federal and state budgets for environmental protection are subject to continual constraints, and many areas are toxicologically monitored only once every three to five years, strong consideration should be given to utilization of solid-phase sediment toxicity testing to determine the long-term environmental health of our waters.

Conclusions

1. The chronic toxic response of stream water collected at eight stream locations decreased downstream of the point (for example, wastewater treatment plant) and nonpoint (for example, urban runoff) sources and was dependent on the timing of sample collection.

2. The chronic toxic response of sediments collected at the eight locations decreased downstream from the point and nonpoint sources of pollution, but the time of year (January, May, September) or the particular year (1984, 1985, 1987) did not affect the results.

3. In spite of extensive coal mining and washing activities, as well as other identifiable inputs into the Yellow Creek drainage, the predominant environmental effects were the result of a wastewater treatment plant discharge.

4. Solid-phase chronic sediment toxicity tests were shown to be useful for assessment of past episodes of toxic incidents and to be valuable in the determination of toxic conditions in areas that are infrequently monitored.

5. Toxicity tests conducted on the Yellow Creek basin demonstrate that sediments act as reservoirs of toxic contaminants that often are not apparent in water-column tests. Consequently, bottom sediments must be decontaminated before toxic conditions will be abated and the stream can be considered biologically healthy.

References

[*1*] Pionk, H. D. and Chesters, G., *Journal of Environmental Quality,* Vol. 2, No. 1, 1973, pp. 29–45.
[*2*] Potter, L., Kidd, D., and Standiford, D., *Environmental Science and Technology,* Vol. 9, 1975, pp. 41–46.
[*3*] Luoma, S. N. and Jenne, E. A., *Trace Substances in Environmental Health—X,* D. D. Hemphill, Ed., University of Missouri, Columbia, MO, 1976, pp. 343–351.
[*4*] Birge, W. J., Black, J. A., Westerman, A. G., Francis, P. C., and Hudson, J. E., "Embryopathic Effects of Waterborne and Sediment-Accumulated Cadmium, Mercury, and Zinc on Reproduction and Survival of Fish and Amphibian Populations in Kentucky," Research Report No. 100, U.S. Department of the Interior, Water Resources Research Institute, Lexington, KY, 1977.
[*5*] Patrick, W. H., Gambrell, R. P., and Khalid, R. A., *Journal of Environmental Science and Health,* Vol. A12, No. 9, 1977, pp. 475–492.
[*6*] Jungclaus, G. A., Lopez-Avila, V., and Hites, R. A., *Environmental Science and Technology,* Vol. 12, 1978, pp. 88–96.
[*7*] Nriagu, J. O. and Coker, R. D., *Environmental Science and Technology,* Vol. 14, 1980, pp. 443–446.
[*8*] Chapman, P. M., Dexter, R. N., Kocan, R. M., and Long, E. R., *Aquatic Toxicology and Hazard Assessment: Seventh Symposium, ASTM STP 854,* R. D. Cardwell, R. Purdy, and R. C. Bahner, Eds., American Society for Testing and Materials, Philadelphia, 1985, pp. 344–363.
[*9*] Matisoff, G., Fisher, J. B., and Matis, S., *Hydrobiologia,* Vol. 122, 1985, pp. 19–23.
[*10*] DiToro, D. M., Jeris, J. S., and Ciarcia, D., *Environmental Science and Technology,* Vol. 19, 1985, pp. 1169–1176.
[*11*] Young, T. C., DePinto, J. V., and Kipp, T. W., *Journal of Great Lakes Research,* Vol. 13, No. 3, 1987, pp. 353–366.
[*12*] "Water Quality Criteria 1972," NAS-NAE Committee on Water Quality Criteria, U.S. Environmental Protection Agency, Washington, DC, 1973.
[*13*] Handcock, H. M., "Heavy Metal Concentrations in Sediments and Waters of the Tennessee and Kentucky Rivers," Final Report No. 4-48R, Kentucky Department of Fish and Wildlife Resources, Frankfort, KY, 1976.
[*14*] Bondietti, E. A., Sweeton, F. H., Tamura, T., Perhac, R. M., Hulett, L. D., and Kneip, T. J., *Proceedings,* First Annual National Science Foundation Trace Contaminants Conference, Vol. 1, Fulkerson, W. D. Schultz, and R. I. Van Hook, Eds., 1984, pp. 211–224.
[*15*] "Yellow Creek Intensive Survey Bell County, Kentucky," Kentucky Division of Water Quality, Kentucky Department for Natural Resources and Environmental Protection, Frankfort, KY, 1978.
[*16*] Wilson, D. J., "Extensive Analysis for Heavy Metals in the Middlesboro, Kentucky Area," Report to the Yellow Creek Concerned Citizens, Inc., Middlesboro, KY, 1981.
[*17*] Birge, W. J. and Black, J. A., "*In Situ* Acute/Chronic Toxicological Monitoring of Industrial Effluents for NPDES Biomonitoring Program Using Fish and Amphibian Embryo-Larval Stages as Test Organisms," Office of Water Enforcement and Permits, U.S. Environmental Protection Agency, Washington, DC, 1981.
[*18*] Westerman, A. G., "Embryopathic Effects of Yellow Creek Stream Samples on Its Fish Populations: I—Sediment Effects," Report to the Yellow Creek Concerned Citizens, Inc., Middlesboro, KY, 1985.
[*19*] Westerman, A. G., "The Chronic Effect of Yellow Creek Sediments and Stream Waters, and Point Discharges into the Yellow Creek Drainage on the Fathead Minnow *(Pimephales promelas),*" Completion Report to Yellow Creek Concerned Citizens, Inc., Middlesboro, KY, 1987.
[*20*] Black, J. A., Roberts, R. F., Johnson, D. M., Minicucci, D. D., Mancy, K. H., and Allen, H. E., *Bioassay Techniques and Environmental Chemistry,* Ann Arbor Science Publishers, Ann Arbor, MI, 1973, pp. 259–275.
[*21*] "Middlesboro, Kentucky Yellow Creek Investigation, December 1982," U.S. Environmental Protection Agency, Region IV, Environmental Services Div., Athens, GA, 1983.
[*22*] "Research and Development," EPA 660/4-81-055, U.S. Environmental Protection Agency, Physical and Chemical Methods Branch, Cincinnati, OH, 1981.
[*23*] Hergert, G. W., *Soil Testing Methods,* Colorado Soil Testing Laboratory, Fort Collins, CO, 1971.
[*24*] Barbour, R. W., Davis, W. H., and Kuehne, R. A., "Vertebrate Fauna of Cumberland Gap National Historical Park," Final Report, CX-500-71-232, National Park Service, Washington, DC, 1 Nov. 1979.
[*25*] Peltier, W. and Weber, C., "Methods for Measuring the Acute Toxicity of Effluents to Aquatic Organisms," EPA-600/4-78-012, U.S. Environmental Protection Agency, Cincinnati, OH, 1978.

[26] Birge, W. J., Black, J. A., and Westerman, A. G., *Environmental Toxicology and Chemistry*, Vol. 4, 1985, pp. 807–821.

[27] Horning, W. B. and Weber, C. I., "Short-Term Methods for Estimating the Chronic Toxicity of Effluents and Receiving Waters to Freshwater Organisms," EPA/600/4-85/014, U.S. Environmental Protection Agency, Environmental Monitoring and Support Laboratory, Cincinnati, OH, 1985.

Dave Gruber,[1] *Jerry Diamond,*[1] *and Dennis Johnson*[2]

Performance and Validation of an On-Line Fish Ventilatory Early Warning System

REFERENCE: Gruber, D., Diamond, J., and Johnson, D., **"Performance and Validation of an On-Line Fish Ventilatory Early Warning System,"** *Aquatic Toxicology and Environmental Fate: Eleventh Volume, ASTM STP 1007,* G. W. Suter II and M. A. Lewis, Eds., American Society for Testing and Materials, Philadelphia, 1989, pp. 215–230.

ABSTRACT: A continuous, automated biological early warning system for rapidly detecting acutely toxic wastewater was developed and tested over a two-year period at an ammunition manufacturing plant. The computerized system simultaneously tracked and analyzed the ventilatory frequency of control and exposed bluegill *(Lepomis macrochirus),* generating either "warnings" or "alarms" depending on the toxicity of the exposure water. Standard acute flow-through toxicity tests (96 h) were concurrently performed on the wastewater to validate the results of the ventilatory system. Standard water quality and chemical monitoring were also performed during the study to interpret the results of the early warning system.

Several preliminary trials indicated that a data assessment interval of 10 min and a sampling rate of 60 Hz for each fish was generally sufficient for detecting acutely toxic conditions. Alarms (statistically significant changes in ventilatory rate over time) were usually generated within 30 min of exposure to effluent concentrations approximating the 96-h median lethal concentration (LC_{50}) value. Time to alarm was inversely related to the toxicity (percentage of the LC_{50}) of the effluent; longer times to alarm occurred with decreasing effluent toxicity ($R^2 = -0.88$). The system was capable of detecting 10% of the 96-h LC_{50} concentration within 24 h. Less toxic effluent concentrations were not reliably detected with the present system.

Water quality and chemical monitoring data generally contributed little information in this study. A general lack of analytical sensitivity may have contributed to this finding. It was thought that better analytical sensitivity would have yielded stronger correlations between the chemical composition of the effluent and toxicity effects. Regression analyses indicated that 35% of the variability in the toxicity tests and ventilatory data was explained by the available chemical data.

Two general problems were evident in this system. First, ventilatory rate alone was occasionally an insufficient parameter for detecting acute toxicity. In several runs, ventilatory amplitude and waveform patterns were more informative response variables than frequency. Second, due to the insufficient information generated at times by frequency data alone, false alarms or lack of alarms were evident, reducing the reliability of the system. Results of this study confirm that fish ventilatory systems can serve as useful early warning tools for monitoring wastewaters and that all facets of ventilatory behavior need to be monitored to ensure system reliability. This study also shows that validation of early warning systems, through traditional toxicity tests and chemical monitoring, is crucial to the development of sound approaches for rapidly detecting toxics in surface waters.

KEY WORDS: biological early warning systems, fish monitoring, ventilatory behavior, pollution monitoring, aquatic toxicology

[1] President and technical director, respectively, Biological Monitoring, Inc., Blacksburg, VA 24063.
[2] Contracting officer, U.S. Army Armament, Aberdeen Proving Grounds, MD 21010.

The field of biological monitoring has evolved to meet the societal need for a means of assessing changes in man's environment. The frequency, extent, and duration of unpleasant environmental perturbations have increased markedly during the past 20 years, threatening natural systems and human health. In an attempt to protect the environment and in turn human health, society has formulated various environmental standards. However, without proper surveillance [*1*] the status of compliance or environmental quality is unknown. Successful environmental management must utilize methodologies which ensure that standards are being met.

Historically, physiochemical sensors have been used as environmental management tools to assess the quality of the aquatic environment. Physiochemical measurements continue to provide valuable information on causes of toxicity to aquatic organisms and human health in many cases. However, when attempting to identify or monitor toxic effects as they occur, chemical sensors are often insufficient. In order to assess potential toxicity, it is apparent that biological evidence be gathered because toxicity has an adverse effect on the biological component of the system. Gathering biological data does not eliminate the necessity of physiochemical information. Physiochemical and biological data prove to be most valuable when used jointly, rather than in isolation.

In response to the limitations of chemical monitoring, biological monitoring protocols were developed to assess the toxic effects of pure compounds or wastewater discharges or both. The most common of these techniques include acute (short-term) and chronic (long-term) toxicity tests. A 48-h test period for invertebrates and fish is the minimum duration used in standard acute toxicity tests [*2*] to obtain a test endpoint (immobility or death). Chronic toxicity tests, in which the endpoints are more subtle, generally take seven days or more to conduct and require more skilled techniques than acute tests.

Although much useful information has been generated using standard toxicity test techniques, these methods have limitations. First, the information generated by these tests is often after the fact. If effluent toxicity problems are found (which may be hours after test initiation), severe damage may have already taken place in the receiving stream. In addition, fluctuations in water and wastewater quality over time may make representative sampling on a discontinuous basis impossible. Continuous sampling and testing with such methods is likewise impractical. As a result, standard toxicity tests are not applicable as tools for continuous, early warning or preventative monitoring of wastewater discharges.

In light of the limitations of traditional chemical and biological monitoring strategies as early warning tools for detecting toxic conditions, much interest has been generated in developing and testing continuous, on-line biological systems that are capable of rapidly detecting toxicants. Such systems are conceptually useful for continuously assessing the quality of raw drinking water supplies and wastewater discharges into lakes and streams. Due to the obvious benefits of such a system, if reliable, several different types of biological early warning systems have been developed to meet these needs.

A type of early warning system which has received wide attention by researchers and the industrial community is one based on fish ventilatory behavior. Fish ventilation has been successfully monitored as an indicator of environmental stress as early as 1907 [*3*]. The techniques utilized for such monitoring have incorporated visual observations and manometric, electromyographic, and noninvasive electrode methods. For biological early warning systems, most workers have utilized noncontact or noninvasive sensing electrodes (stainless steel or silver chloride) since this technique requires little skill to install and maintain in the system, and these electrodes do not restrict or stress the fish being monitored [*4*].

There is an extensive data base documenting significant changes in fish ventilatory behavior upon exposure to acutely toxic concentrations of various compounds and industrial effluents [*4–11*]. Common ventilatory responses to pollutants include changes in signal frequency, amplitude, and waveform.

Past research by personnel at Biological Monitoring, Inc. (BMI) has led to the development of the electronics necessary for interfacing fish ventilatory signals to a minicomputer system [*12,13*], thereby automating the collection and analysis of fish ventilatory data over time in several fish simultaneously. Similar systems have since been developed by van der Schalie [*11*], Morgan and Kuhn [*14*], and Evans et al. [*15*]. However, few of these studies have objectively examined and evaluated the reliability and sensitivity of fish ventilatory early warning systems by comparison with more standardized toxicity tests.

In this study, we tested the automated biological monitoring facility (BMF) at an ammunition and propellent manufacturing plant. This plant routinely treats waste products containing variable amounts of nitroglycerine derivatives, lead, and other heavy metals. The objective of this study was to calibrate, test, and evaluate the performance of the BMF with the aid of simultaneous toxicity tests and chemical monitoring. Since the effluent for both the ventilatory tests and acute toxicity tests came from the same source in a given run, we expected that the ventilatory response of the fish was related to the degree of effluent toxicity as measured by the 96-h median lethal concentration (LC_{50}) value in the corresponding toxicity tests. Specifically, we expected that the lower the LC_{50} value for a particular effluent sample (that is, a more toxic sample) the faster alarm conditions (a statistical departure in ventilatory behavior between control and exposed fish) should be generated. This paper presents the results of a two-year study using the automated BMF and discusses the advantages, present problems, and potentially useful applications of such a system.

Methods

A 10 by 40-m trailer housed the components of the monitoring system. Figure 1 depicts the 6-L monitoring/exposure chambers which housed individual 6 to 10-cm bluegill *(Lepomis macrochirus)* as the sensors. Noncontact, submerged, stainless steel wire electrodes received the microvolt bioelectric signals generated by ventilatory movements in each fish. These signals were amplified, filtered, and interfaced [*10*] to an on-line Digital Equipment Corporation PDP11/VO3 minicomputer system (Fig. 2). In this study, only the ventilatory rate of the fish was monitored due to the ease and reliability with which this parameter could be measured and analyzed. A modified peak-picking computer program computed the ventilatory rates of each fish at ≥ 1 min intervals. These values were stored on a floppy disk system and a hard copy of the values was printed. Figure 3 depicts the flow of information within the BMF.

Physiochemical values (conductivity, dissolved oxygen, pH, and temperature) of the undiluted wastewater were continuously monitored. These values were interfaced to the minicomputer system and were also stored and printed. Periodically, composite effluent samples were taken for additional chemical analyses. The chemicals analyzed included various derivatives of the propellant manufacturing process, that is, nitroglycerine, dinitroglycerine, dinitrotoluene, zinc, and lead, all of which are known to be acutely toxic to aquatic life at low concentrations (<5 ppm). After many such samples were taken, it was determined that most values were below the detectable limits of the plant's chemical capabilities (1 ppm). Therefore, fewer samples were subsequently taken.

Table 1 outlines the standard operating procedure followed. A test was initiated by first introducing the wastewater into the acute toxicity test dilutor system. This dilutor continuously fed a preset amount of diluent and effluent to each ventilatory and toxicity test chamber over a 96-h period via an electrically timed solenoid-valve system. There were

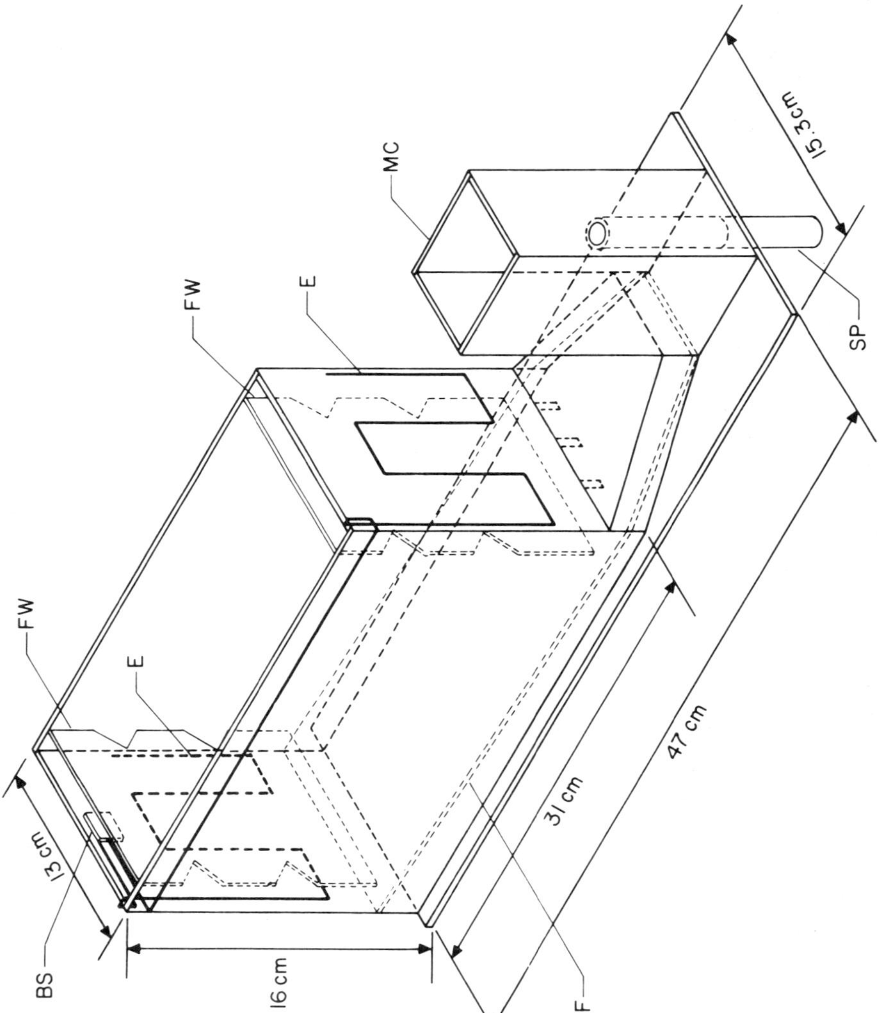

FIG. 1—*Exposure monitoring tank: BS = barrier strip; E = electrodes; FW = false walls; SF = sloping floor; SP = adjustable standing pipe; and MC = maintenance chamber.*

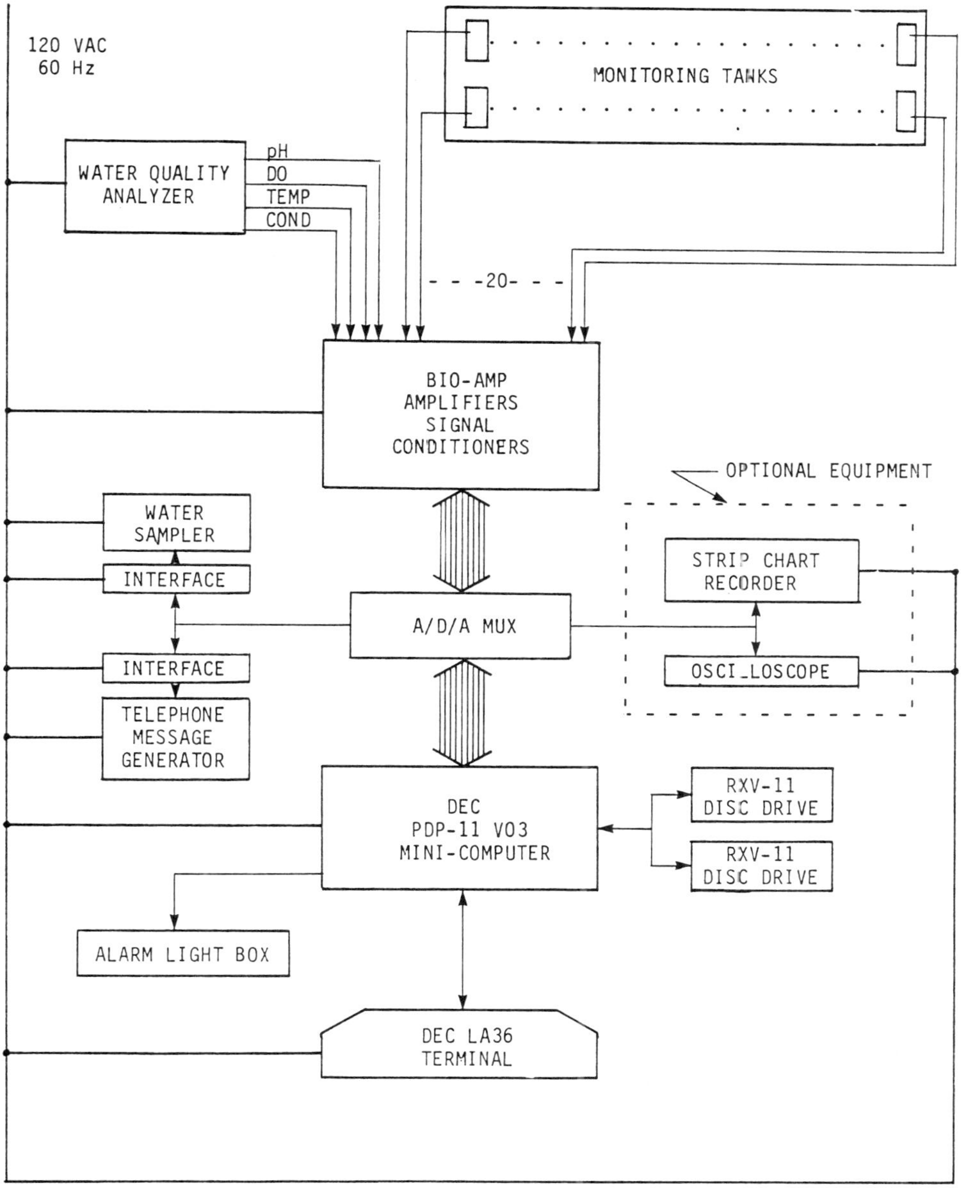

FIG. 2—*Schematic representation of computer interface with the ventilatory monitoring system.*

several water turnovers in the exposure chambers prior to introducing the fish in the acute toxicity test apparatus.

Five effluent concentrations, following a logarithmic dilution series (100, 56, 32, 18, and 10%), and a control were employed in each toxicity test. This range of effluent dilutions was usually adequate to determine a definitive 96-h LC_{50} value. The dilution water used

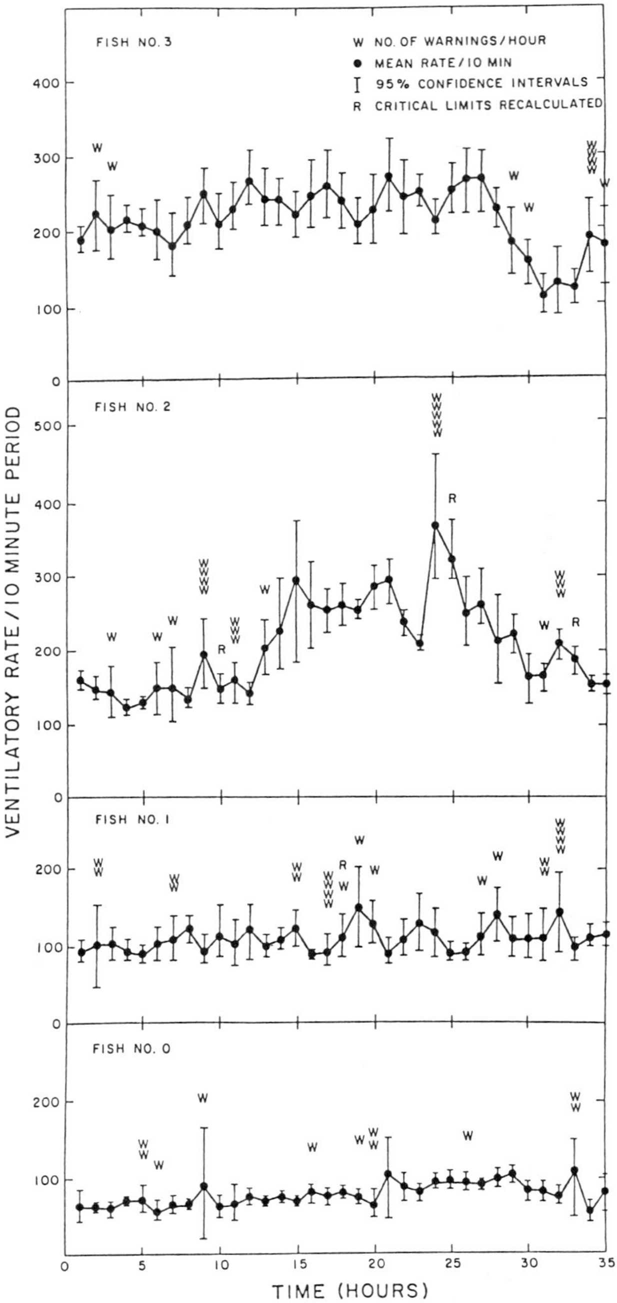

FIG. 3—*Ventilatory rate data for control (0 to 3) bluegill indicating warnings, means, and confidence intervals. Number of* w*'s over data print indicate the number of warnings generated during that time interval.*

TABLE 1—*Standard operating procedures for acute and ventilatory testing.*

Day/Time	Acute Toxicity Test	Ventilatory Test
Monday, 9 a.m.	drain dilutor	drain dilutor headbox
	spend standing water in feed pipe	introduce effluent to headbox
	introduce effluent	
	calibrate instruments	
	do water chemistries	check flow in monitor tanks
	take 100% water sample	start 24-h control, do not introduce effluent to exposure tanks
	introduce fish	start FISH program
Tuesday	check dilutor system count fish do water chemistries collect composite sample	introduce effluent 24-h after start of control period; record in log book
Wednesday	check dilutor system	
	count fish	
	do water chemistries	
	collect composite sample	
Thursday	check dilutor system	turn off effluent, allow sufficient time for subsequent 24-h control, record in log book
	count fish	
	do water chemistries	
	collect composite sample	
Friday	do water chemistries	end test, stop FISH program
	collect composite sample	
	end test	
	clean up: turn off effluent siphon and clean screens brush down introduce 100% diluent	clean up: clean effluent headbox and exposure tanks introduce diluent through effluent lines to exposure tanks

in all tests was filtered river water obtained upstream from the plant and subsequently pumped to the BMF. Two replicates per concentration were used in each toxicity test with ten bluegills (mean length = 8.2 cm) per replicate. Depending on the distribution of the data, several statistical techniques, including the arithmetic and log probit, Spearman-Karber, and moving average angle methods, were used to calculate an LC_{50} value in a given test.

In each experiment, twelve fish were also placed in separate ventilatory monitoring chambers several days prior to testing. One day of controlled (dilution water) exposure was monitored before introducing the wastewater. Wastewater was introduced to eight randomly assigned fish in the ventilatory monitoring chambers 24 h after initiation of the acute toxicity test. The remaining four fish (controls) continued to receive diluent only.

In each ventilatory experiment, the individual ventilatory rate of the four control (diluent-treated) bluegills and eight experimental (effluent-treated) bluegills were monitored simultaneously. Prior experimentation determined that this distribution of control and exposed fish offered a conservative yet sensitive response system to acutely toxic conditions at this facility. A time-series computer-based model [*9,10*] was used to determine if any of the twelve bluegill significantly altered its ventilatory rate over time. The algorithm provided user options for determining what was considered an abnormal ventilatory rate for a given fish (constituting a "warning") and how many simultaneous warnings were necessary to trigger an alarm condition.

Based on preliminary trials, the most sensitive and reliable algorithm utilized an assessment interval of 10 min with each sensor's ventilatory rate monitored every 16 ms within that interval. A departure of at least four standard deviations from the mean control value (based on the previous 1 h average of the four control fish) for each fish was the threshold level necessary to generate a warning, and at least four simultaneous warnings (from four different fish) were necessary to generate an alarm. In this model, a recalculation routine was automatically initiated once a warning was generated and prior to alarm determination in order to verify that warning conditions were present.

In this study, three levels of biologically based alarm conditions could be generated. Alarms A and B were considered high-priority alarms since at least four of the eight exposed fish were assessed as breathing abnormally. Alarm A was generated if four or more exposed fish were ventilating abnormally exclusively, whereas, under Alarm B conditions, one or more of the four control group fish were additionally assessed as abnormal. An Alarm C warning was considered a low-priority alarm since a total of at least four individuals, from both control and exposed groups, were assessed as breathing abnormally. When a Type A alarm was generated, the final effluent at this facility could automatically be diverted to a temporary holding pond for further treatment before discharging into the receiving stream.

The BMF was calibrated and tested using three different waste streams at the central wastewater facility of the ammunition plant: the influent to the equalizing lagoon at the facility (L-Series), the wastewater leaving the lagoon prior to entering the rotating biological contactors at the facility (R-Series), and the final discharge from the facility (F-Series). The present paper discusses the results of 31 concurrent ventilatory and toxicity tests performed on these waste streams over a two-year period.

Results and Discussion

Figure 3 depicts typical data collected from control fish in the ventilatory monitoring system. Figure 4 depicts the fish response to a 2-h exposure (between 19.3 and 21 h) of effluent. The effluent in this example was a 50% dilution of the simultaneous 96-h LC_{50} value of lagoon influent (L-Series). In this study, control ventilatory rate rarely exceeded 250 breaths/10 min and generally ranged between 50 to 150 breaths/10 min (Fig. 3). The ventilatory rate of exposed fish, based on the computer model, often decreased significantly and became more erratic during the period of effluent exposure. Following removal of the effluent, the ventilatory rate of exposed fish generally recovered rapidly (Fig. 4).

Subsequent strip-chart recordings taken under similar experimental conditions indicated that the fish ventilatory rate probably did not actually change during effluent exposure (Fig. 5). Rather, the amplitude of the ventilatory signal decreased significantly during effluent exposure such that the threshold voltage (amplitude) values used in the computer algorithm (± 1.0 V) were not reached. As a result, the computer model did not register as many fish breaths as actually occurred during the period of effluent exposure. Although the ventilatory rate may not have actually changed during these runs, a change was observed in the ventilatory response of exposed fish such that the computer algorithm correctly gen-

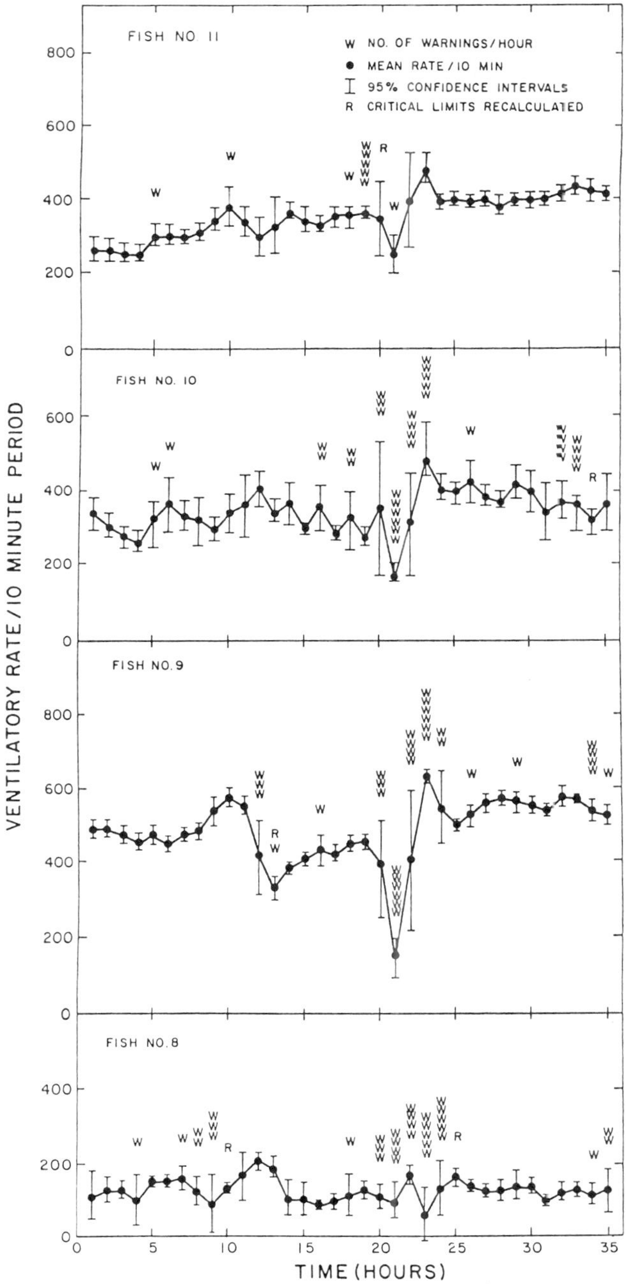

FIG. 4—*Ventilatory rate data for wastewater-exposed fish. Wastewater in this run was introduced at 19.3 h and removed at 21 h.*

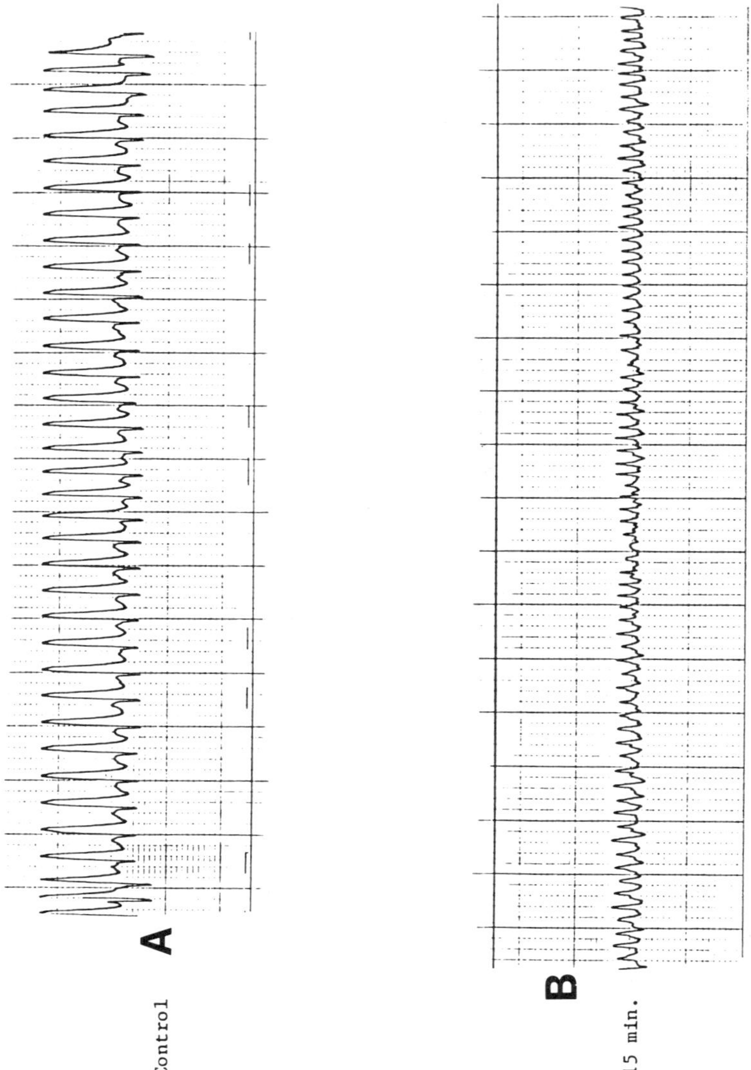

FIG. 5—*Strip-chart recordings of bluegill ventilatory response during control conditions just prior to effluent addition* (A) *and to a 15 min exposure of plant effluent* (B).

erated several warnings, resulting in Type A (high priority-hazard) alarms (Fig. 4). By contrast, few warnings and no alarms were generated by the control fish during these runs.

Table 2 summarizes the data from the ventilatory and acute toxicity tests. The average time required to generate an alarm is presented relative to the percentage of the LC_{50} value to which the fish were exposed. The results from these experiments indicate that the quickest time to an alarm was 10 min (the minimum assessment interval used in this study). The most rapidly generated alarms occurred when fish were exposed to the highest concentrations of wastewater (Table 2). These results indicate that the present early warning system could rapidly detect acutely toxic conditions ($\geq LC_{50}$ concentration). Sublethal concentrations of effluent (10 to 30% effluent) were detected over considerably longer time periods (several hours). Effluent samples below 10% of the 96-h LC_{50} concentration were not reliably detected in the present system. Regression analyses of the LC_{50} effluent concentration against the time of first response yielded the equation

$$LC_{50} = 0.226 \cdot (\text{EEC}) \cdot \ln (t) - 0.3716$$

TABLE 2—*Averaged time to alarm based upon percentage of LC_{50} exposure.*

Ventilatory Identification No.	Percentage of LC_{50} Exposed	Time to High-Priority Alarm, h	Averaged Time to High-Priority Alarm When an Alarm Occurred, h
129	10.0	12.33	19.61
123A	15.2	none	
115B	16.8	none	
120	19.1	26.00	
115	24.1	20.50	
121	27.6	9.83	9.50
119	29.2	9.17	
107	32.0	none	
111	41.7	none	
106	51.1	12.83	6.25
110	54.0	11.83	
113	55.8	none	
122	56.1	0.17	
114	68.8	none	
117	73.0	0.17	
124	76.3	0.33	0.46
123B	83.3	0.83	
109	85.5	0.33	
130	89.6	0.33	
118	98.8	6.33[a]	
126	100.0	42.83[a]	
112B	115.4	0.33	0.21
125	117.6	29.83[a]	
116	132.5	6.00[a]	
101–103	133.7	0.17	
104–105	139.7	0.17	
108	149.7	0.17	
111A	161.3	0.33	0.28
127	181.8	0.17	
128	181.8	34.83[a]	
112A	204.1	0.33	

[a] Not included in the averaged time to alarm. See the Evaluation section of the text.

where EEC is the effluent exposure concentration, and t is the time in hours. The regression coefficient, R^2, obtained using this equation and the data obtained in this study, was 0.88.

There was an occasional lack of system reliability as evidenced by two general problems. First, there were several ventilatory runs in which no alarm was generated, although the data from the respective toxicity tests indicated that the wastewater was acutely toxic (that is, ventilatory identification numbers 113 and 114, Table 2). Second, there were several runs in which the time to generate an alarm was surprisingly long, considering how acutely toxic the effluent was in the concurrent toxicity tests (that is, ventilatory identification numbers 126 and 128, Table 2). These problems are discussed subsequently in the evaluation section of this paper.

Given the historical importance of using chemical determinations as a water quality monitoring tool, an attempt was made to correlate the biological flow-through test data with the available chemical data. The mean concentrations of the various chemicals and the average LC_{50} values for each sample of undiluted effluent are presented in Table 3. A stepwise multivariate regression analysis on this data indicated that approximately 35% of the variance in the LC_{50} data was explained by the variability in pH and dinitroglycerine concentration in the effluent samples (Table 4). None of the other chemical parameters measured were significantly related to the observed LC_{50} data. Thus, less than half of the total variability in the LC_{50} data was explained by the available chemical data in this study.

The lack of a complete chemical data base was at least partly responsible for the low correlation. Several parameters were frequently reported as 0.0 mg/L, and the actual detection limit of the plant's chemical facilities was often at 1 ppm for most parameters. Therefore, potentially meaningful correlations between chemical parameters and toxicity in this study were masked by insufficient analytical capabilities. The lack of significant correlations between water quality parameters and 96-h LC_{50} values may also be due to the high variability observed in both correlates for any given effluent (coefficients of variation were generally $>100\%$). Some of this variability is probably attributable to real variations in the chemical composition of the effluent over time. Since the ammunition plant effluent could be chemically variable over time, and the experiment used a continuous flow system, an effluent obtained during the first 24 h of an experiment may not have been chemically similar to the effluent from the same source during the next 24 h.

The large number of samples with undetectable concentrations of potentially toxic substances, despite the acute toxicity observed in many of those samples, suggest that the chemical monitoring capabilities were insufficient. Given the chemical complexity of the effluent generated at this facility, adequate chemical monitoring on a continuous basis would appear to be an expensive and possibly impractical approach for determining water quality and potential toxicity of wastewater. Thus, for this facility, biological monitoring may be a more efficient means of monitoring water quality than chemical monitoring.

Evaluation of Ventilatory System

The computer algorithm used in this study appeared to provide a promising tool for monitoring water quality. Most warnings which occurred prior to effluent introduction were successfully dealt with by the recalculation subroutine in the present algorithm and did not contribute to false alarms. Also, consistent with the toxicity test data, the number of warnings issued in the ventilatory system increased dramatically following effluent introduction (2.4 times that with control water only).

The BMF was designed to rapidly generate high-priority alarms when the wastewater was acutely toxic. This was successfully achieved as verified by the data in Table 2. When the fish were exposed to wastewater concentrations approximating the 96-h continuous

TABLE 3—*Summary of the mean LC_{50} values and average values for the chemical parameters measured in the study.*[a]

Sample	LC_{50}	V1	V2	V3	V4	V5	V6	V7	V8	V9	V10	V11	V12	V13	V14	V15	V16
L101	37.4	0.950	0.00	0.00	4.100	0.00	0.035	0.00	0.00	0.00	0.000	7.20	100.00	210.0	11.40	8.50	96.0
R101	35.8	4.800	0.87	0.00	8.300	0.00	0.330	0.00	0.00	0.00	0.000	8.26	95.70	118.2	9.25	9.38	81.0
L102	58.7	0.530	0.66	0.27	4.960	0.00	0.010	0.13	0.00	0.00	0.000	9.12	97.10	94.0	9.30	11.20	85.0
R102	93.8	0.750	0.36	0.07	2.350	0.00	0.070	0.04	0.37	0.00	0.000	8.98	269.40	66.0	3.12	13.10	30.0
F101	33.4	1.020	3.23	0.01	3.430	0.00	0.000	0.00	0.23	0.00	0.000	7.90	277.60	99.6	5.48	13.60	53.0
F102	35.1	0.000	1.49	0.00	0.250	0.01	0.056	0.00	0.12	0.00	0.000	7.58	208.40	175.2	2.15	15.20	21.0
F103	27.8	8.810	1.42	0.00	0.020	0.00	0.000	0.00	0.00	0.00	0.000	7.80	216.00	160.0	6.00	20.00	66.0
R103	48.0	0.017	0.78	0.00	0.019	0.00	0.000	0.00	0.00	0.00	0.000	7.58	223.80	181.0	1.04	19.80	11.0
F104	18.6	0.400	1.21	0.00	0.490	0.00	0.000	0.00	0.19	0.01	0.000	7.60	234.00	94.4	2.32	20.10	26.0
R104	14.7	0.000	0.41	0.00	0.000	0.00	0.032	0.00	0.00	0.00	0.000	7.40	226.50	69.3	0.15	23.20	1.8
F105	26.9	0.000	0.00	0.00	0.000	0.00	0.000	0.00	0.00	0.00	0.000	7.33	222.75	71.3	0.00	1.49	22.6
F106	26.9	0.000	0.00	0.00	0.025	0.00	0.000	0.00	0.00	0.00	0.000	7.45	181.70	73.2	4.60	23.40	54.0
F107	72.7	0.000	0.00	0.00	0.000	0.00	0.000	0.00	0.00	0.00	0.000	7.13	140.00	75.7	3.24	24.10	39.0
R105	37.4	0.540	2.67	0.33	0.110	0.00	0.000	0.00	0.04	0.05	0.050	9.00	636.00	672.6	3.54	17.40	37.0
F108	56.5	0.000	0.54	0.00	0.010	0.00	0.000	0.00	0.00	0.00	0.063	7.27	383.30	233.0	0.80	15.90	8.1
L103	33.2	6.610	1.10	0.00	11.260	0.00	0.460	0.00	0.00	0.00	0.110	7.60	106.00	82.0	1.08	16.00	11.0
L104	27.4	1.490	1.27	0.17	1.580	0.00	0.000	0.00	0.87	0.00	0.000	7.00	151.30	102.0	8.00	14.00	78.0
R106	24.7	1.920	1.22	0.00	1.200	0.00	0.040	0.06	0.41	0.00	1.370	7.20	151.20	102.5	5.00	14.20	49.0

[a] Parameters V1 to V10 are in mg/L. V1 = nitroglycerine, V2 = dinitroglycerine, V3 = DEGON, V4 = dinitrotoluene, V5 = N−NDPA, V6 = diphenylamine, V7 = dibutylphthalate, V8 = diethylphthalate, V9 = ethylcentralite, V10 = lead, V11 = pH, V12 = alkalinity, V13 = hardness, V14 = dissolved oxygen, V15 = temperature, and V16 = percentage of oxygen saturation.

TABLE 4—*Summary of step-wise regression analysis of chemical and LC_{50} data presented in Table 3.*[a]

Step 1: variable entered is pH F value = 3.69 Probability > F = 0.073 R^2 = 0.187 $C(P)$ = −8.504
Step 2: variable entered is dinitroglycerine concentration, mg/L F value = 4.07 Probability > F = 0.039 R^2 = 0.352 $C(P)$ = −7.615

[a] No other variables met the 0.15 significance level for entry into the model.

flow LC_{50} value, alarms were generated in <30 min. As the exposure percentage increased, the time required to generate an alarm decreased. These data suggest that the automated ventilatory system used in these experiments could be a sensitive and useful early warning monitoring tool for potentially acutely toxic wastewaters. Furthermore, our results indicate that such a fish ventilatory system satisfies many of the criteria proposed by Cairns and van der Schalie [*6*] for useful biological early warning systems. Our system continuously and automatically monitored the toxicity of the wastewater and rapidly detected the presence of hazardous water quality conditions such that corrective action could be taken prior to the occurrence of environmental damage. Similarly, successful applications of fish ventilatory systems have been demonstrated in England [*15*], South Africa [*14*], and the Netherlands [*7*].

Results of this study indicate that while the BMF is useful as an environmental monitor, the present system is not entirely reliable. Several problems were evident. First, despite the fact that all the fish were held under identical conditions, and all were of the same age class, some fish appeared to breathe at a relatively constant rate when acclimated, while others breathed more erratically. In this study, this may not have been a serious problem since the model compared the ventilatory rate over time for each fish individually. Thus, the analysis of ventilatory behavior of one fish was independent of that from another fish. However, if the control ventilatory pattern for a given fish was highly erratic, then this would affect the ability of the model to detect differences over time for that fish, thus reducing the overall sensitivity of the system.

A second problem was related to the observation that a typical response to wastewater exposure in this study was a simultaneous decrease in ventilatory amplitude coupled with an increase in the variability of ventilatory frequency. This response presented a problem in view of the threshold voltage (amplitude) necessary in the model to filter out low-amplitude signal noise. It is thought that this ventilatory response may be related to the presence of heavy metal toxicants (particularly lead and zinc) in the effluent. A similar ventilatory response of bluegill to metals has been observed in laboratory experiments [*16–18*] and in other field studies [*11,19*]. These observations indicate that ventilatory frequency, under certain circumstances, may be a poor indicator of fish physiological stress and a poor predictor of environmental hazard. Other aspects of ventilatory behavior, not included in the present algorithm (gill purge rate and waveform amplitude changes), may need to be incorporated in order to make the ventilatory monitoring system more reliable and effective. Personnel at BMI are presently revising the algorithm to incorporate these changes into an on-line system.

A third problem evident in this study was that several warnings (individual statistical departure from a user-optioned set of variables) were generated by control fish and during preexposure conditions. An ideal biomonitoring system would be one which rapidly detects toxicity problems but not so sensitive as to generate false alarms [*6*]. Clearly, this is a conceptual problem requiring more research and development. Of the 70 ventilatory runs in this study, there were 4 runs (5.7%) in which false alarms were generated during control periods. One possibility is that these were not true false alarms and that the diluent was toxic at those times. The diluent in this system traveled through many miles of pipe before reaching the BMF. On at least one occasion, it was clear that the diluent was accidentally contaminated before it reached the BMF, killing fish in the holding tanks and in the ventilatory chambers. An earlier study, employing the same alarm parameters and control conditions, yielded a false alarm frequency of 0.4% based upon 1000 h of monitoring [*9*]. Thus, it is unclear whether alarms generated during control conditions in these experiments represented toxic conditions or whether they were, in fact, false alarms caused by improper calibration of the model. This is an issue which is being addressed by the authors under more controlled conditions.

A fourth type of problem identified in this study pertained to those runs in which no alarm was generated or the time to alarm was exceedingly long despite the demonstrated toxicity of the particular effluent (Table 2). These types of events occurred 11 times out of the 70 experimental runs or at a rate of 15.7%. One possible explanation for the discrepancy in response in these cases is that the wastewater toxicity varied over time in these experiments. In those experiments in which discrepancies of this type occurred, most of the deaths in the toxicity tests occurred within the first 24 h. Since the wastewater entered the ventilatory chambers 24 h later than in the toxicity test chambers, the wastewater may not have been acutely toxic in the ventilatory chambers, despite the fact that it was the "same" wastewater. Alternatively, the present model may not have been sensitive enough in those cases, perhaps due to some of the modeling problems raised above.

As already noted by Kingsbury and Rees [*20*], an automated biomonitoring facility, such as that used in this study, does not indicate the nature of the hazard, but rather that a hazard potentially exists. The advantage of biomonitoring systems is that they may serve as early warning devices which can be used to determine when and where intensive chemical monitoring is needed [*6*]. In this way, a biomonitoring system may help to reduce the cost of chemical monitoring and may aid in pinpointing the specific nature of toxicity problems. However, we caution that chemical data so derived may still not predict toxicity in certain chemically complex situations.

Results of this study indicate that an automated system based on monitoring fish ventilatory rate is useful for rapidly detecting acute toxicity in wastewaters. However, our results indicate that careful assessment of the entire ventilatory profile will be necessary in order to ensure maximum system reliability and sensitivity. Ventilatory frequency data alone may be misleading or unreliable under certain wastewater chemical conditions. While fish ventilatory systems have the potential for assessing subacute levels of toxicity [*18,21–23*], it remains to be demonstrated that such a system is reliable for detecting chronic toxicity in wastewater or drinking water. This is the present direction of our research.

Acknowledgments

We wish to thank Mr. Chris Thompson for his on-site technical assistance and Mr. William Sider for computer assistance. This study was performed under U.S. Army Contract No. DAAK11-81-C-0072.

References

[1] Hellawell, J. M., "Biological Surveillance of Rivers," Water Research Centre, Stevenage, United Kingdom, 1978, p. 322.

[2] Peltier, W. H. and Weber, C. I., *Methods for Measuring the Acute Toxicity of Effluents to Freshwater and Marine Organisms,* 3rd ed., EPA 600/4-85/013, U.S. Environmental Protection Agency, Washington, DC, 1985.

[3] Kuiper, T., *Pfugers Archiv. für Physiologie,* Vol. 177, 1907, p. 1.

[4] Gruber, D., Cairns, J., Jr., Dickson, K. L., and Hendricks, A. C., *Journal of Fisheries Biology,* Vol. 14, 1979, p. 429.

[5] Drummond, R. A. and Carlson, R. W., "Procedures for Measuring Cough (Gill Purge) Rates of Fish," EPA-600/3-77-133, U.S. Environmental Protection Agency, Washington, DC, 1977.

[6] Cairns, J., Jr., and van der Schalie, W. H., *Water Research,* Vol. 14, 1980, pp. 1179–1196.

[7] Sloof, W., *Bulletin of Environmental Contamination and Toxicology,* Vol. 23, 1979, p. 517.

[8] Rice, S. D., Thomas, R. E., and Short, J. W., "Effect of Petroleum Hydrocarbons on Breathing and Cough Rates and Hydrocarbon Uptake-Depuration in Pink Salmon Fry," *Physiological Responses of Marine Biota to Pollutants,* Academic Press, New York, 1977, p. 259.

[9] Gruber, D. and Cairns, J., Jr., *Water, Air, and Soil Pollution,* Vol. 15, 1981, p. 471.

[10] Gruber, D., Cairns, J., and Hendricks, A. C., *Journal of the Water Pollution Control Federation,* Vol. 53, 1981, p. 505.

[11] Cairns, J., Jr., and van der Schalie, W. H., "Early Warning Biological Monitoring," *Biological Monitoring in Water Pollution,* Cairns, J., Jr., and van der Schalie, Eds., Pergamon Press, New York, 1982.

[12] Cairns, J., Jr., and Gruber, D., *Bioscience,* Vol. 29, 1981, p. 665.

[13] Gruber, D., Cairns, J., Jr., Dickson, K. L., Hendricks, A. C., and Miller, W. R. III, *Journal of the Water Pollution Control Federation,* Vol. 52, 1980, p. 465.

[14] Morgan, W. S. G. and Kuhn, P. C., "Aspects of Utilizing Continuous Automatic Fish Biomonitoring Systems for Industrial Effluents Control," *Freshwater Biological Monitoring,* Pergamon Press, New York, 1984, pp. 65–73.

[15] Evans, G., Johnson, D., and Withell, C., "Development of the WRc Mark III Fish Monitor: Description of the System and its Response to Some Commonly Encountered Pollutants," Water Research Centre, Bucks, United Kingdom, 1986.

[16] Waller, W. T. and Cairns, J., Jr., *Water Research,* Vol. 6, 1972, p. 465.

[17] Sparks, R. E., Cairns, J., Jr., and Heath, A. G., *Water Research,* Vol. 6, 1972, p. 895.

[18] Diamond, J. M. and Gruber, D., "Rapid Detection of Chronic Toxicant Effects on Fish Ventilatory Behavior Using Heavy Metal and Chlorinated Hydrocarbon Substances," NIH/SBIR Phase I Final Report, National Institutes of Health, Washington, DC, 1986.

[19] van der Schalie, W. H., Dickson, K. L., Westlake, G. F., and Cairns, J., Jr., *Environmental Management,* Vol. 3, 1979, pp. 217–235.

[20] Kingsbury, R. W. and Rees, C. P., *Effluent and Water Treatment Journal,* Vol. 18, 1978, pp. 319–331.

[21] Cairns, M. A., Garton, R., and Tubb, R., *Transactions of the American Fisheries Society,* Vol. 111, 1982, pp. 70–77.

[22] Maki, A. W., *Aquatic Toxicology (Second Symposium), ASTM STP 667,* American Society for Testing and Materials, Philadelphia, 1979, p. 77.

[23] Sprague, J. B., Rowe, D. W., Westlake, G. F., Heming, T. A., and Brown, I. T., "Sublethal Effects of Treated Liquid Effluent from a Petroleum Refinery on Freshwater Organisms," *Fisheries Environment Canada,* 1978.

Jeffrey B. Frithsen,[1] *Diane Nacci,*[2] *Candace Oviatt,*[1] *Charles J. Strobel,*[2] *and Raymond Walsh*[2]

Using Single-Species and Whole Ecosystem Tests to Characterize the Toxicity of a Sewage Treatment Plant Effluent

REFERENCE: Frithsen, J. B., Nacci, D., Oviatt, C., Strobel, C. J., and Walsh, R., **Using Single-Species and Whole Ecosystem Tests to Characterize the Toxicity of a Sewage Treatment Plant Effluent,"** *Aquatic Toxicology and Environmental Fate: Eleventh Volume, ASTM STP 1007,* G. W. Suter II and M. A. Lewis, Eds., American Society for Testing and Materials, Philadelphia, 1989, pp. 231–250.

ABSTRACT: A four-month experiment was conducted to evaluate the toxicity of a sewage effluent using both single-species and whole ecosystem tests. Fresh effluent from the East Greenwich, Rhode Island, sewage treatment plant was added daily to six experimental ecosystems (mesocosms). Sewage effluent additions were made at 0.1, 1.0, and 10% of the daily seawater input (960 L/day) from Narragansett Bay. Reagent grade, inorganic nutrients were added to the control, 0.1% and 1.0% effluent treatments to normalize nutrient loadings in all treatments. The *Arbacia punctulata* sea urchin sperm cell test was conducted on effluent and mesocosm samples. In the mesocosms, measurements of responses at the population, community, and ecosystem levels of biological organization were made.

The single-species toxicity test indicated the mean EC_{50} of the sewage effluent was 1.1%. Toxicity decayed rapidly over time, and was unrelated to carbon, nutrient, residual chlorine, or metal concentrations. Toxicity in the mesocosms was variable due to short-term (4 to 5 h), incomplete mixing of the effluent. There was no evidence for a buildup of toxicity in the mesocosms. Mesocosm effluent additions decreased phytoplankton standing stock and produced an imbalance between total system production and respiration leading to hypoxia. At the 10% effluent loading, net system production was negative.

Results indicated both single-species and mesocosm approaches were useful to assess toxicity. The single-species test was best utilized to characterize the magnitude and persistence of toxicity, and the mesocosm experiments were best employed to identify sensitive communities and processes.

KEY WORDS: toxicity testing, microcosms, mesocosms, ecosystem-level effects, system production, system respiration, dissolved oxygen, *Arbacia punctulata,* aquatic toxicology

The assessment and management of hazardous waste disposal in the United States is the responsibility of the U.S. Environmental Protection Agency (EPA). This assessment includes efforts to predict the behavior of potential pollutants in order to define concentrations that will occur in the environment [*1*]. The use of large-scale, experimental ecosystems (mesocosms) [*2–7*] has been one approach used to study chemical behavior and

[1] Associate marine scientist and professor of oceanography, respectively, Marine Ecosystems Research Laboratory, Graduate School of Oceanography, University of Rhode Island, Narragansett, RI 02882-1197.

[2] Associate biologists, Science Applications International Corp., c/o U.S. Environmental Protection Agency, Environmental Research Laboratory, Narragansett, RI 02882.

fate, as have field studies of planned and accidental discharges of pollutants [*8,9*]. Extensive models have been developed describing chemical fate and behavior [*10,11*]. These approaches have been used to refine predictions made from smaller scale laboratory studies and theoretical investigations of pollutant behavior.

Formerly, EPA has relied upon a chemical-specific approach to regulating discharges of potentially hazardous wastes [*12*]. In this approach, biological effects were predicted from extrapolating the results from single-species toxicity tests to the real world environment using models and empirical data to predict exposure concentrations [*13–15*]. However, because toxicological data are not (and can never be) available for the myriad compounds discharged and because interactions between chemicals may ameliorate or magnify toxicity, the chemical-specific approach is being replaced by whole-effluent toxicity testing [*12*].

Single-species toxicity tests are by far the most common approaches used to measure the toxicity of whole effluents. Inspection of past ASTM volumes related to toxicology demonstrates wider use of and more experience with fresh water test species than with marine species, although a variety of tests using estuarine and marine test species have been developed. Microcosm tests, involving more complex systems with a greater level of biological organization, have not been conducted with whole effluents, but due to the past success of microcosm tests, studies using whole effluents can be expected. Single-species toxicity tests have been applied to the field to measure whole-effluent toxicity [*16*]. However, field studies of all but the most dramatic of pollution events [*17,18*] often lead to ambiguous results, confounded by a range of uncontrolled environmental variables.

There exists a strong need to define how well single-species toxicity tests applied to whole effluents can be used to predict responses at higher levels of biological organization [*19,20*]. Further, the single-species, multiple community mesocosm and the field measurement approaches to toxicity assessment have never been compared simultaneously for a discharge into a marine environment.

Such a comparison is one of the goals of a project conducted jointly by the Marine Ecosystems Research Laboratory (MERL) at the University of Rhode Island and the EPA Environmental Research Laboratory in Narragansett, Rhode Island (ERLN). The project is a part of the EPA Complex Effluent Toxicity Testing Program and involves laboratory, mesocosm, and field assessments of the toxicity of sewage effluent from the East Greenwich, Rhode Island, sewage treatment plant.

This paper will present results from a mesocosm experiment conducted at the Marine Ecosystems Research Laboratory using the East Greenwich sewage effluent. The MERL mesocosms are functional analogs of shallow, unstratified coastal systems (such as Narragansett Bay) and have become very useful tools with which to further the understanding of ecosystem structure and function, and the behavior, fate, and effects of pollutants (see Refs *6* and *21* through *26*, and references cited within).

The purpose of the mesocosm experiment was to compare toxicity as measured by the single-species tests to the detrimental effects measured in the MERL mesocosm ecosystems. This paper presents mesocosm results demonstrating detrimental and toxic responses and the results from one single-species toxicity test, the *Arbacia punctulata* sea urchin sperm cell test. We hope to demonstrate that the single-species test was most useful in characterizing the magnitude and persistence of toxicity, whereas the mesocosms were most useful in identifying sensitive ecosystem components and processes.

Methods

Single-Species Tests

A variety of single-species tests were used to measure toxicity (Microtox, and tests utilizing *Champia parvula, Laminaria saccharina, Mysidopsis bahia, Menidia beryllina,* and sea urchin sperm cells), but only results from the sea urchin *(Arbacia punctulata)* test will be reported here. This test was consistently the most sensitive in this investigation and was easily and reliably repeated under a broad range of test conditions.

The sea urchin sperm cell test was used (*a*) to compare the toxicity of 13 Rhode Island sewage treatment plant effluents, (*b*) to set a range for the effluent concentrations to be used in the experiment, (*c*) to characterize the decay of effluent toxicity, and (*d*) to assess toxicity directly in the mesocosm water columns. The *Arbacia punctulata* sperm cell toxicity test has been described in Refs *27* through *29*. Briefly, sea urchin sperm were directly exposed to effluent. Following a 1-h exposure, eggs were added and incubated for an additional 20 min. Eggs were then fixed by the addition of buffered formalin, and fertilization (defined as the presence of fertilization membranes) measured by microscopic observation. Results were expressed as the effluent concentration causing a 50% reduction in fertilization rates relative to Narragansett Bay seawater controls.

Description of the Experiment and MERL Mesocosms

The experiment was conducted from 15 July to 14 Nov. 1986 in eight mesocosms (Fig. 1 [*24,25,30–32*]) containing sediment and seawater from Narragansett Bay. Sediments were collected from the northwest corner of East Greenwich Bay (Fig. 2) using a 0.25 m^2 box corer and methods designed to preserve the vertical structure of the sediments (see Ref *33*). These sediments were chosen due to their proximity and expected similarity to sediments in Greenwich Cove, the receiving body of water for effluent from the East Greenwich, Rhode Island sewage treatment plant.

Two mesocosms were run as controls, receiving no addition of sewage effluent. Daily additions of fresh East Greenwich sewage effluent were made to six mesocosms at three loading levels (Table 1). Each tank received 960 L/day seawater (turnover time = volume/flow = 13.5 day). Turnover time was selected to ensure development of water column communities and not to simulate turnover at the East Greenwich Cove receiving water body.

The highest effluent loading level corresponded to an input rate of 10% of the daily seawater flow to the mesocosms and amounted to 96 L/day sewage effluent. Other effluent treatment levels corresponded to input rates of 1.0% (9.6 L/day) and 0.1% (0.96 L/day).

Mesocosm mixing was accomplished using a vertical plunger operating at 5 rpm in a 2 h on, 4 h off cycle. Seawater and sewage effluent additions were made four times daily. Seawater additions (8 L/min) commenced 15 min after the start of mixing and lasted for 30 min. Effluent additions (0.24, 2.4, and 24 L corresponding to the 0.1, 1.0, and 10% treatments) commenced 45 min later and were made in two equal pulses. Effluent additions ceased 14 min prior to the end of mixing. Effluent was refrigerated (4°C) during storage between additions to retard the decay of toxicity. In most cases, effluent was stored no longer than 24 h since delivery was made daily. In no case was storage longer than 48 h.

All treatments received approximately the same nutrient loading to separate nutrient from effluent toxicity effects. Reagent grade inorganic nutrients [ammonium chloride (NH_4Cl), potassium dihydrogen orthophosphate (KH_2PO_4), and sodium metasilicate ($Na_2SiO_3 \cdot 9H_2O$)] were added to the control tanks and the low and medium effluent treatments once per day to match nutrient loadings expected in the high-effluent treatment (Table 1). Reagent nutrient addition levels were held constant through the experiment. Effluent nutrient concentrations were determined before the start of the experiment. Nutrient concentrations were expressed as microgram-atoms ($\mu g \cdot at$), where 1 $\mu g \cdot at/L$ = 1 $\mu g/L$ divided by the atomic weight of nitrogen, phosphorus, or silicon depending upon the nutrient. Total nitrogen in these samples was 1117 $\mu g \cdot at$ nitrogen/L sewage effluent,

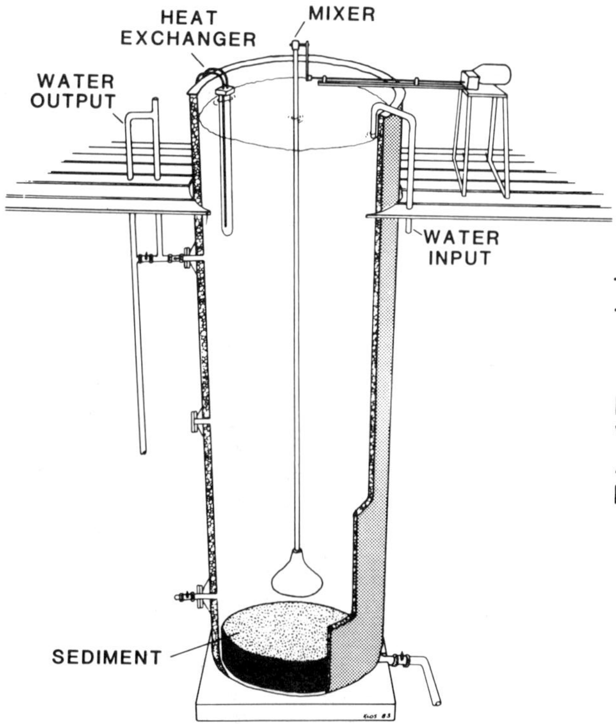

TANK DIAMETER	1.83m
TANK HEIGHT	5.49m
WATER SURFACE AREA	2.63m^2
DEPTH OF WATER	5.00m
VOLUME OF WATER	13.1m^3
AREA SEDIMENT	2.52m^2
DEPTH OF SEDIMENT	0.37m

FIG. 1—*A MERL mesocosm. Seawater was added at a rate of 960 L/day. Sewage effluent was added at the surface. Descriptions of the behavior of the mesocosms and their fidelity to natural coastal systems may be found in Refs* 24,25, *and* 30 *through* 32.

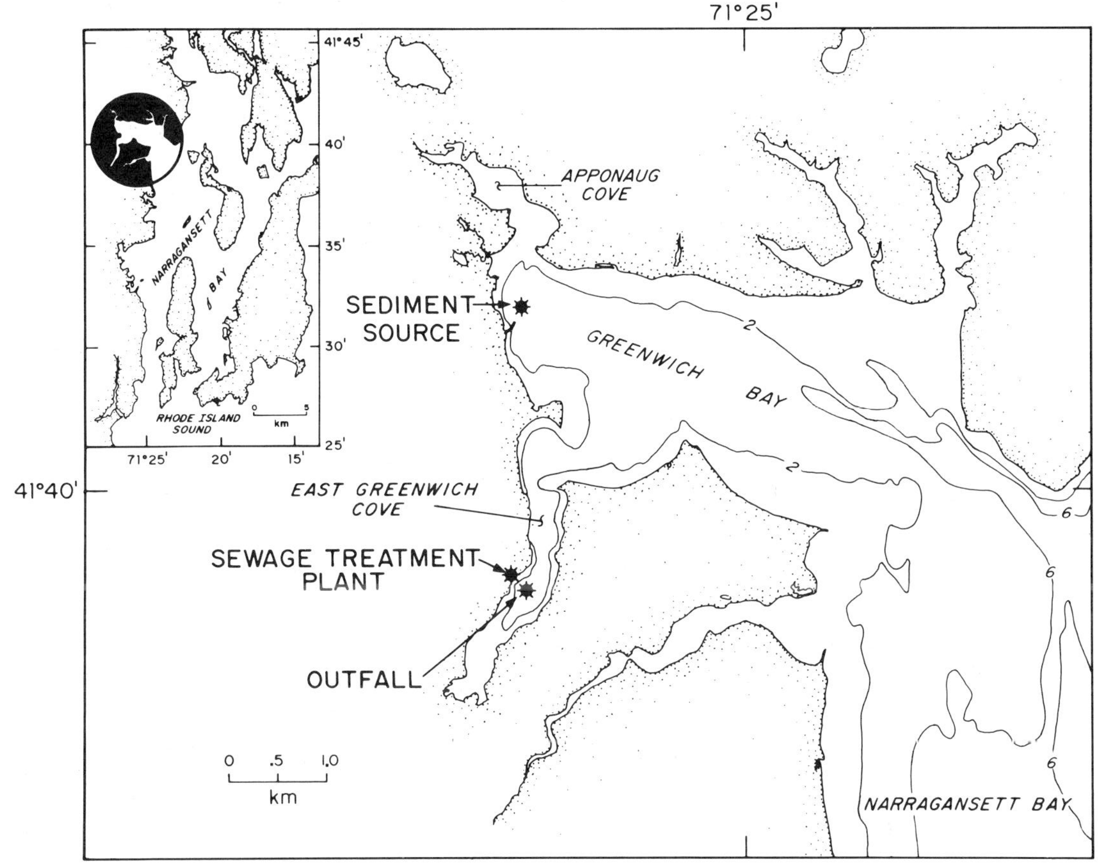

FIG. 2—*Greenwich Bay, Rhode Island. The East Greenwich sewage treatment plant discharges to East Greenwich Cove at a 3-m depth. Effluent was collected just prior to discharge into the cove. Sediments were collected south of Apponaug Cove at a water depth of 3 m. Depth contours are in meters.*

TABLE 1—*Experimental outline.*

	Nutrient Source, mmol/mesocosm/day					
	Nitrogen		Phosphorus		Silicate	
Treatment, % Effluent	SE[a]	RA[b]	SE	RA	SE	RA
10%, Tanks 2 and 4, 96 L effluent/day	77.9	0	32.	0	7.21	0
1.0%, Tanks 5 and 6, 9.6 L effluent/day	7.79	70.15	3.27	29.45	0.72	6.49
0.1%, Tanks 1 and 8, 0.96 L effluent/day	0.78	77.16	0.33	34.39	0.07	7.14
Controls, Tanks 0 and 9, 0 L effluent/day	0	77.94	0	32.72	0	7.21

[a] SE = sewage effluent.
[b] RA = reagent additions.

with 73% (812 μg·at nitrogen/L) present as dissolved, inorganic nitrogen (ammonia + nitrate + nitrite). Total phosphorus was 341 μg·at phosphorus/L, with almost 100% present as dissolved, inorganic phosphorus. The concentration of dissolved reactive silicate was 75 μg·at silicon/L. Nutrient additions to the mesocosms were made based upon the dissolved, inorganic nutrient concentrations since it was not known how biologically available other nutrient forms (particulate and dissolved organic) from sewage effluent would be.

Measurements were made to characterize daily batches of sewage effluent and to define ecosystem, community, and population level responses of the mesocosm ecosystems. Methods used are described in Refs *34* and *35*.

Effluent total suspended solids (TSS), particulate organic carbon (POC), and dissolved organic carbon (DOC) were sampled approximately daily. Daily samples were analyzed for TSS and POC. DOC samples from some weeks were pooled before analysis to reduce analytical time. Effluent nutrients were sampled five times per week during the first seven weeks of the experiment and once per week thereafter. All nutrient samples were analyzed using weekly pooled samples. Selected metals in the effluent were measured periodically by EPA ERLN using standard methods. Effluent residual chlorine was measured as total residual chlorine by amperometric back titration.

In the mesocosms, DCMU fluorescence [fluorescence after the addition of (3-(3,4-dichlorophenyl)-1-1-dimethylurea)] was taken daily as a measure of phytoplankton biomass. Size fractionated and dissolved fluorescence, chlorophyll *a* concentration, nutrients, and oxygen were measured weekly. Total system (total system = water column plus sediments) production and respiration were estimated from changes in oxygen concentration measured over a diel cycle once per week. Zooplankton biomass was measured once every two weeks. Zooplankton community composition and abundance were determined from separate samples taken at the same time. Sediment oxygen metabolism and inorganic nutrient fluxes across the sediment-seawater interface were measured approximately monthly. Sediment biota were collected once every three weeks. Preliminary results from the enumeration and identification of benthic macrofauna (organisms retained on a 300 μm sieve) are reported.

Statistical Analyses

Statistical analyses and data management were completed using SAS software [*36–38*]. Data for each month of the experiment were analyzed separately using a two-way analysis of variance (ANOVA). Factors were treatment with four levels (control, 0.1, 1.0, and 10% effluent) and replicate with two levels. Replicate mesocosms were nested within treatment. Student-Neuman-Kuels tests were used to separate treatment effects and statistical significance reported for $P < 0.05$ unless otherwise stated.

Results

Sewage Effluent Characterization

Sewage effluent from the East Greenwich sewage treatment plant was moderately toxic compared with effluents from twelve other municipal sewage treatment plants in Rhode Island (Table 2). This plant was built in 1928 and expanded to secondary treatment in 1956. The plant can efficiently treat 0.5 million gallons of sewage per day (MGD)(1514 m^3/day), but is currently discharging 0.7 MGD (2649 m^3/day). The plant is scheduled for enhanced secondary treatment within two years.

The composition of the East Greenwich effluent is given in Table 3. The toxicity of the sewage effluent was measured on 36 days during the experiment (Fig. 3). Toxicity [median effective concentration (EC_{50})] ranged from 0.3% effluent to nontoxic ($EC_{50} > 10\%$) and averaged 1.1% for all samples for which toxicity was detected. Toxicity was not significantly correlated with any measured parameters (Table 3). Additionally, toxicity could not

TABLE 2—*Relative toxicity of Rhode Island sewage treatment plant effluents.*[a]

Municipal Treatment Plant	Treatment Type[b]	Flow, million gallons/day[c]		EC_{50}, % effluent
		Design	Average[d]	
Bristol	3	3	2.1	0.15
New Shoreham	1	0.2	0.1	0.15
Blackledge Valley District	1	31	17.4	0.27
Providence	1	60	43.1	0.30
Warren	1	2	1.3	0.34
Quonset	3	2.3	0.5	0.36
Jamestown	1	0.7	0.4	0.42
East Greenwich	2	0.5	0.7	0.46
East Providence	1	10.4	5.3	0.62
Newport	3	5.8	8.1	0.99
Westerly	1	3.1	1.8	1.49
Narragansett	1	1.4	0.6	1.50
South Kingstown	1	4.1	2.7	1.63

[a] Based upon *Arbacia punctulata* sperm cell tests conducted on fresh samples.
[b] Treatment type:
1 = activated sludge and chlorination.
2 = trickling filter and chlorination.
3 = primary treatment and chlorination.
[c] One million gallons per day = 3 785 000 L/day = 3 785 m^3/day.
[d] Average for calendar year 1985 (R. Richardson, Rhode Island Department of Environmental Management, Providence, Rhode Island, personal communication).

TABLE 3—*Summary of East Greenwich sewage effluent characteristics 15 July–14 Nov. 1986.*

Parameter	Mean	Standard Deviation	Coefficient of Variation, %	Range
Sewage treatment plant flow, million gallons/day[a]	0.697	71	10	477 to 811
Total suspended solids, mg/L	68.9	34.9	51	9.0 to 207.6
Carbon, mg C/L				
Particulate organic	26.5	15.2	57	7.5 to 117.8
Dissolved organic	29.6	15.3	52	11.6 to 154.0
Nitrogen, μg·at N/L				
Total	1621	261	16	1200 to 2200
Total dissolved	1472	24	16	1100 to 2000
Total dissolved inorganic	1032	166	16	806 to 1339
Ammonia	915	195	21	613 to 1263
Nitrate + nitrite	117	78	67	12 to 290
Phosphorus, μg·at P/L				
Total	382	74	19	200 to 490
Total dissolved	329	60	18	180 to 430
Total dissolved inorganic	279	59	21	183 to 437
Silica, μg·at Si/L	205	32	15	77 to 233
Chlorine, residual, mg/L	2.09	2.08	100	−3.00 to 7.97
Metals, μg/kg				
Cadmium				
Particulate	0.44	0.38	86	0 to 2.01
Soluble	0.59	1.06	178	0 to 7.62
Chromium				
Particulate	25.65	14.58	57	0.41 to 67.16
Soluble	9.41	9.29	99	0 to 65.58
Copper				
Particulate	72.25	57.77	80	2 to 339.31
Soluble	101.90	38.15	37	0.66 to 189.74
Iron				
Particulate	4057.44	1795.56	44	466.21 to 10 571.89
Soluble	488.30	523.89	107	18.00 to 2523.69
Lead				
Particulate	54.70	56.74	104	3.21 to 446.52
Soluble	7.84	3.85	100	0 to 43.95
Nickel				
Particulate	8.41	5.79	69	0 to 35.47
Soluble	22.58	7.10	31	12.00 to 58.00
Zinc				
Particulate	115.88	108.80	94	12.05 to 945.68
Soluble	65.39	81.84	125	22.00 to 768.77

[a] One million gallons per day = 3 785 000 L/day = 3785 m^3/day.

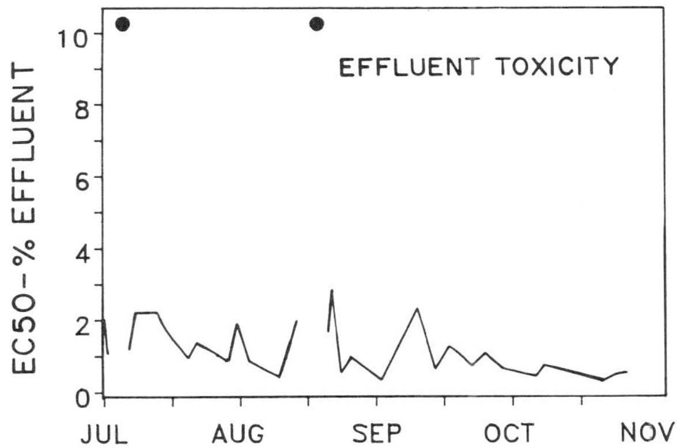

FIG. 3—*Toxicity of East Greenwich sewage treatment plant effluent. The EC_{50} from the sea urchin sperm cell test is expressed as percent effluent. The samples represented by the filled-in circles were considered nontoxic ($EC_{50} > 50\%$).*

be correlated with rainfall measured at T. F. Green Airport in Warwick, RI, approximately six miles north of the East Greenwich sewage treatment plant.

Toxicity decay was temperature dependent, but the functional relationship between decay rate and temperature could not be defined with the data available. Laboratory experiments indicated decay was negligible at 4°C, but a toxicity half-life of 1.8 h was calculated at 25°C. Toxicity decay was not related to residual chlorine concentrations.

Mesocosm Nutrient Additions, Temperature, and Salinity

The nitrogen concentration of the sewage effluent added to the mesocosms was greater than that of the effluent samples taken prior to the start of the experiment. Total nitrogen was 1621 μg·at nitrogen/L, [coefficient of variation (CV) = 16%], with 64% (1032 μg·at nitrogen/L) present as dissolved inorganic nitrogen. This was 27% higher than the concentrations of inorganic nitrogen selected for addition.

The total phosphorus concentration of the effluent added to the mesocosms was 382 μg·at phosphorus/L (CV = 21%), with 86% (329 μg·at phosphorus/L; CV = 19%) present as dissolved phosphorus. Dissolved inorganic phosphorus concentrations averaged 279 μg·at phosphorus/L (CV = 22%), 18% lower than concentrations of inorganic phosphorus selected for addition.

Silica concentrations in the effluent added to the mesocosms averaged 205 μg·at silicon/L (CV = 13%), almost three times higher than the initially determined concentrations. This large difference was traced to an error made in analyzing the initial samples. Freezing water can convert reactive silicate to a nonreactive, polymeric form [*39*]. Initial silica samples were run immediately after thawing, thus causing erroneously low values. Subsequent samples were allowed to thaw for approximately one week prior to analysis, thereby reversing the process. The bias introduced by freezing and thawing for one week is believed to be less than 5% [*39*].

Mesocosm water column temperatures during the experiment ranged from 7.8 to 22.2°C. The salinity of the inflowing seawater from Narragansett Bay ranged from 30.3 to 32.0 parts per thousand. Salinities in the control mesocosms were similar, but due to effluent additions, salinities in the 10% treatment ranged from 27.2 to 30.6 parts per thousand and averaged 28.5 parts per thousand. Single measurements made on the last day of the experiment indicated there was no significant salinity decrease compared with controls in the 0.1 and 1% treatments.

The temporal trend of salinity in the 10% treatment closely fit a predictive salinity model that assumed instantaneous mixing of seawater then sewage effluent, each followed by the removal of an equivalent volume of well-mixed water. The model included provisions for four daily additions of seawater and effluent. Measured salinities deviated from predicted salinities only slightly (ratio measured/predicted averaged 1.001, range 0.974 to 1.019, n = 17). These results indicated that mixing over the 6-h mixing cycle was very good. Vertical salinity profiles taken after sewage effluent additions and during the 4-h no-mixing cycle indicated stratification, and therefore, incomplete mixing occurred during shorter time periods. Mixing occurred primarily when the oscillating plunger (Fig. 1) was operating. Since this period lasted only 14 min after effluent addition, mixing was completed during the following mixing cycle, 4 h later.

Mesocosm Toxicity

Although the toxicity of whole water samples taken directly from the mesocosm systems was highly variable (Fig. 4), surprisingly good replication occurred between mesocosms within each treatment. Treatment replication this good has been rarely observed in past mesocosm experiments, wherein inorganic nutrient concentrations were usually the best replicated measured parameter.

Compared with Narragansett Bay seawater controls, significant toxicity was detected during most of the first 40 days of the experiment in the control mesocosms receiving no

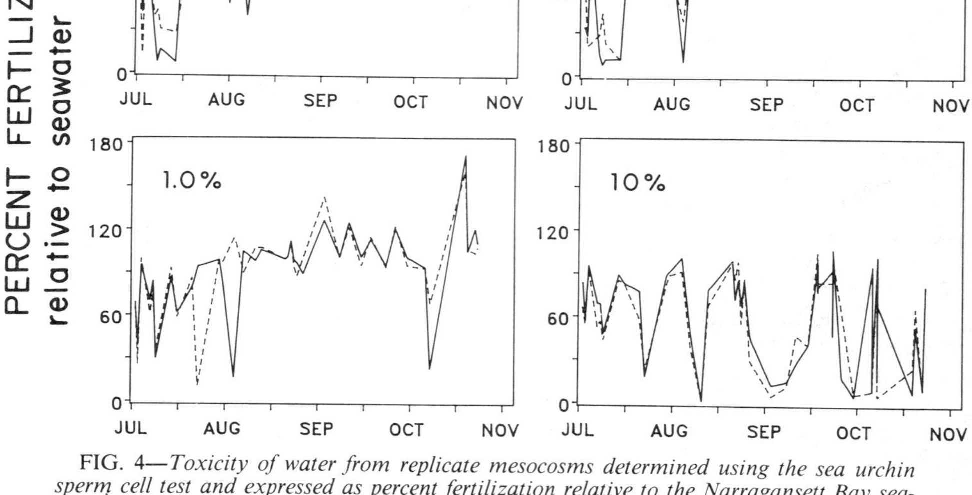

FIG. 4—*Toxicity of water from replicate mesocosms determined using the sea urchin sperm cell test and expressed as percent fertilization relative to the Narragansett Bay seawater controls.*

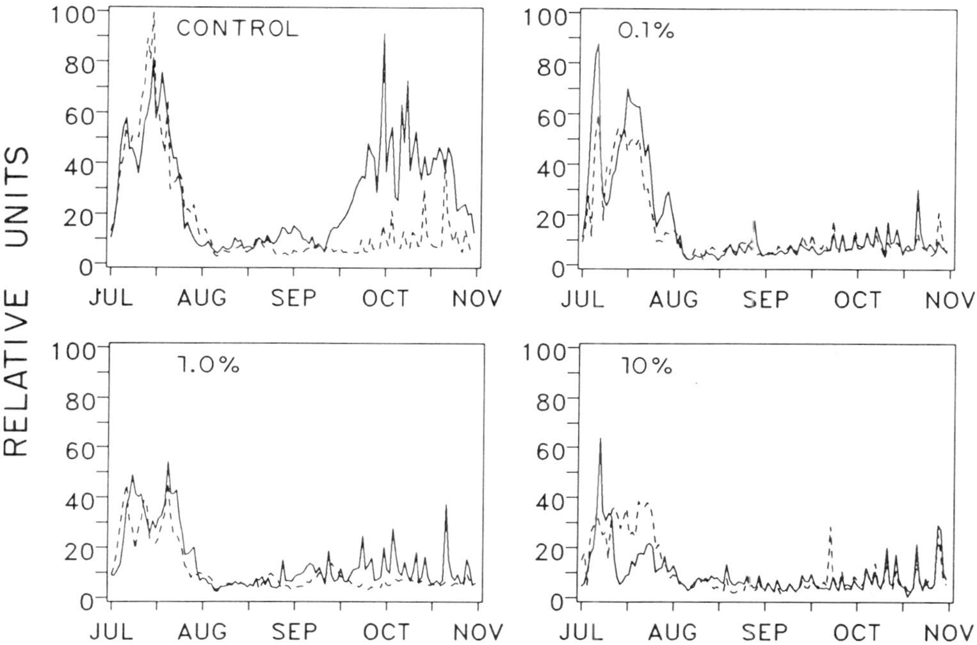

FIG. 5—*Total DCMU fluorescence in replicate mesocosms.*

sewage effluent additions (Fig. 4). Results were similar for the 0.1 and 1.0% sewage effluent treatment mesocosms, although the 1.0% treatment was less toxic than the control mesocosms during this period. Toxicity in the 10% sewage treatment was similar to that measured in the mesocosm controls during the first part of the experiment, but remained high (low percent fertilization) during the remainder of the experiment when toxicity decreased in the controls (Fig. 4). Vertical profiles of toxicity in the 10% effluent treatment indicated that toxicity was detected only in surface samples. Both the vertical profiles and the temporal trends of toxicity indicated that there was no residual component of toxicity that built-up in the mesocosms due to continued effluent additions. Toxicity measured in the mesocosms was not correlated with the EC_{50} of the sewage effluent.

Two experiments were conducted in the mesocosms during the study to investigate the persistence of toxicity. In one, toxicity persisted for >3 h and in the other, no toxicity was measured after 1 h. These differences were traced to inadequate mixing of the most recently added sewage effluent at the time samples were taken.

The persistence of effluent toxicity was also assessed following termination of the mesocosm experiment by adding 1000 L of fresh effluent, completely mixing the effluent, and measuring toxicity over time. At a temperature of 9°C, the half-life of the toxicity was 12.4 h, and no toxicity was detected after 99 h.

Mesocosm Responses

DCMU fluorescence was used as a daily measure of phytoplankton biomass (Fig. 5). In this and other studies [*40–42*], good correlations ($F = 667$, $P < 0.0001$, $R^2 = 0.685$) were made between DCMU fluorescence (Fig. 5) and chlorophyll *a* concentrations, a more traditional measure of phytoplankton biomass.

Daily measurements of DCMU fluorescence (Fig. 5) indicated phytoplankton biomass in the controls and all three sewage effluent treatments increased during the first month of the experiment. Increases were greatest in the control and 0.1% sewage effluent treatments ($P < 0.1$), but less in the 1.0 and 10% treatments. Fluorescence decreased in all mesocosms after the initial phytoplankton blooms. After the first month, treatment differences were not significant. Direct measurements of chlorophyll indicated the same trends in phytoplankton biomass as the fluorescence measurements. Weekly size-fractionated fluorescence measurements indicated there were no treatment differences in the proportion of total DCMU fluorescence less than 64 μm (overall mean 94%) or in the proportion of total DCMU fluorescence less than 10 μm (overall mean 83%).

Inorganic nutrient concentrations in the mesocosms generally showed treatment differences during the first half of the experiment, corroborating the phytoplankton results. Ammonia concentrations (Fig. 6), the dominant form of dissolved inorganic nitrogen, were greatest during the first month of the experiment in the 10% treatment (mean = 28.92 μg·at nitrogen/L) compared with the other treatments and the controls (mean = 12.92 μg·at nitrogen/L). Similar treatment differences continued into the second month of the experiment (mean for the 10% treatment = 51.33 μg·at nitrogen/L; mean for other mesocosms = 41.05 μg·at nitrogen/L).

Silica concentrations in the mesocosms showed trends similar to that of nitrogen, with highest concentrations being in the 10% treatment mesocosms (Fig. 7). Differences were significant only during the first month of the experiment.

Phosphate concentrations showed trends opposite to those of nitrogen and silica. During the first month, concentrations were highest in the controls and the 0.1% treatment (mean = 15.17 μg·at phosphorus/L). Phosphate concentrations in the 1% treatment were signif-

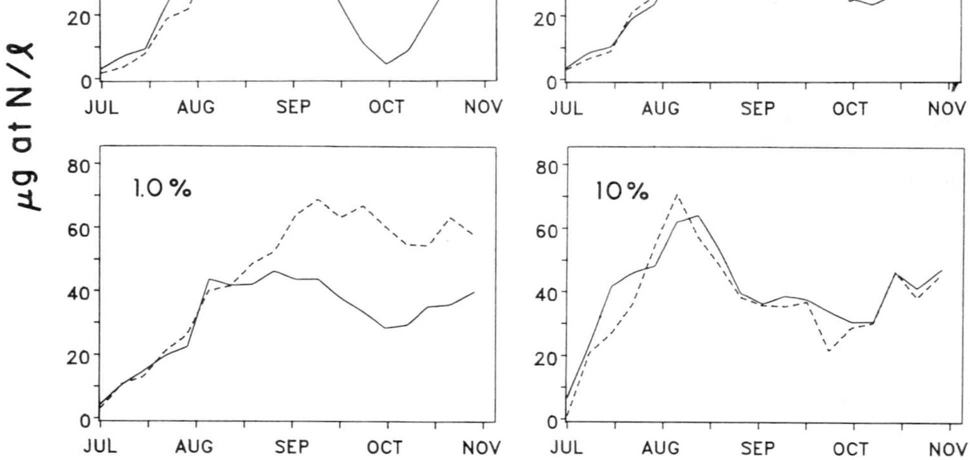

FIG. 6—*Ammonia concentrations in replicate mesocosms.*

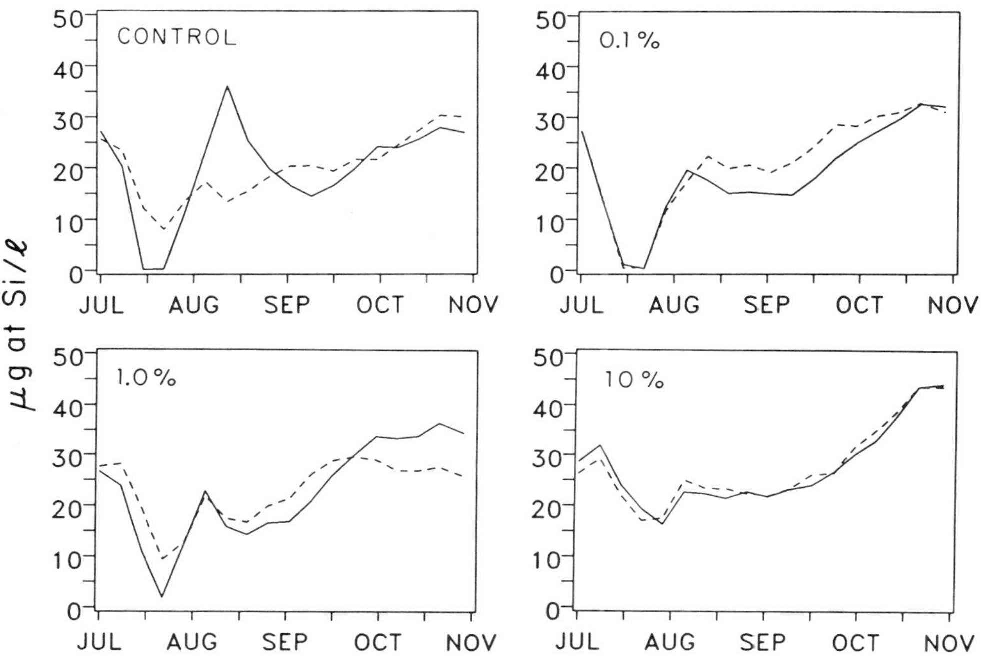

FIG. 7—*Silica concentrations in replicate mesocosms.*

icantly lower (mean = 14.34 μg·at phosphorus/L, and concentrations in the 10% treatment were still lower (mean = 12.23 μg·at phosphorus/L. Some treatment differences persisted into the second month when concentrations in the controls were higher than those in the 10% treatment, but no other differences could be discerned.

Dissolved oxygen concentrations demonstrated the most dramatic system response to sewage effluent additions. Oxygen concentrations, measured weekly over a diel cycle (dawn-dusk-dawn), were consistently lower throughout the experiment in the 10% treatment compared with other treatments (Fig. 8). Oxygen concentrations reached as low as 1.15 mg oxygen/L and were below 3 mg oxygen/L (an upper limit often used to define hypoxia) for 22% of the weekly measurements in one mesocosm and 39% in the other of the 10% effluent mesocosms. The control, 0.1%, and 1% treatments all had similar oxygen concentrations up until the last month when the controls had the highest oxygen concentrations (mean = 10.66 mg/L), the 0.1 and 1% treatments were significantly lower (mean = 8.80 mg/L), and the 10% treatment was still lower (4.76 mg/L).

Diel changes in water column oxygen concentrations were used as a measure of total system apparent daytime production and total system nighttime respiration. Total system metabolism is the net of all oxygen metabolism in the mesocosm ecosystem and includes (*a*) phytoplankton production and respiration, (*b*) zooplankton respiration, (*c*) the benthic respiration of sedimented organics, and (*d*) any respiration due to material on the mesocosm walls. Wall metabolism was not measured during this experiment but generally is 10% of the total system respiration [*43*].

There was only one significant treatment difference in total system production and respiration throughout the experiment. During the second month of the experiment, system respiration in the 10% treatment was significantly higher than the controls. No other treat-

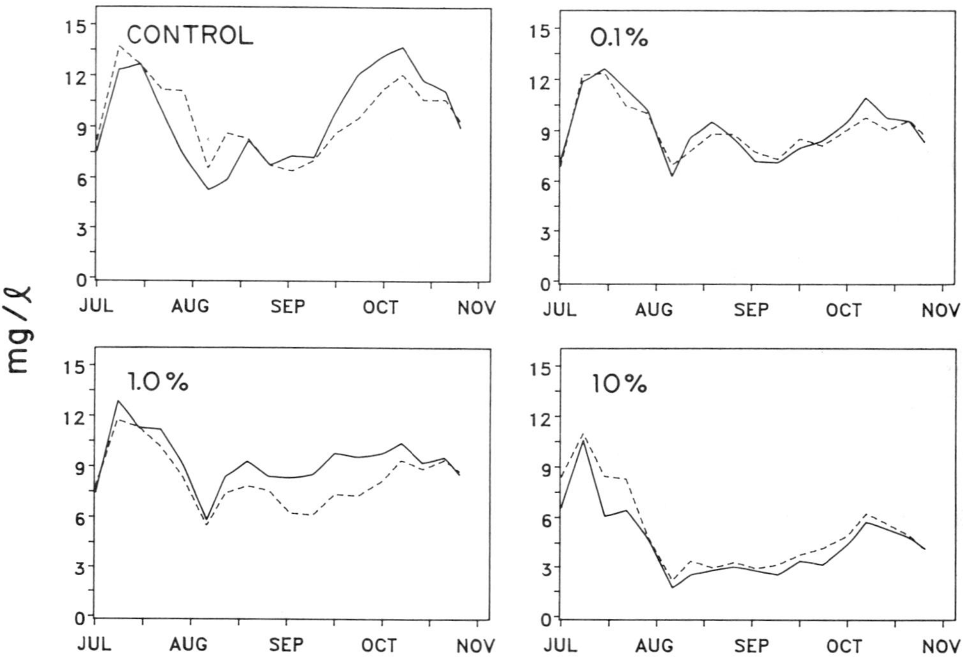

FIG. 8—*Mean oxygen concentrations in replicate mesocosms calculated as the mean dawn values plus the mean dusk values divided by two.*

ment differences could be distinguished. There was a tendency for production to be smallest and respiration greatest in the 10% treatment, but the trends were not statistically significant (Table 4).

As a further measure of the metabolic behavior of the mesocosms, the difference between production and respiration (P-R), a measure of net system production, was computed. Strong treatment differences could be detected by the last two months of the experiment. Net system production was lower during month three in the 10% treatment compared with the controls and the 1% treatment. During month four, net system production in the 10% treatment was lower than any other treatment. For the entire experiment, net system production was negative in only the 10% treatment (Table 4).

TABLE 4—*Total system production and respiration.*

Treatment, % effluent	Production[a]		Respiration		Difference	
	O_2, mg/m^2	C, g/m^2	O_2, mg/m^2	C, g/m^2	O_2, mg/m^2	C, g/m^2
Controls	477	179	413	155	64	24
0.1%	481	181	442	166	39	15
1.0%	501	188	439	165	62	23
10%	367	137	459	172	−92	−35

[a] Oxygen production converted to carbon production by assuming a photosynthetic quotient (PQ) of 1.

Other parameters measured in the mesocosm ecosystem showed few significant treatment effects. Treatment effects that were present did not indicate toxicity due to sewage effluent additions. Rather, stimulatory effects were indicated.

Zooplankton biomass was similar in all treatments, with treatment means for the entire experiment ranging from 71 to 81 mg dry weight/m^3. Enumeration of the zooplankton demonstrated few significant treatment effects for the numerically dominant species. The nauplii of *Acartia* sp., the dominant copepod species, were more abundant in the 10% effluent treatment compared with other treatments, but only during the third month. During the fourth month, meroplanktonic polychaete larvae were nearly ten times more abundant in the 10% effluent treatment (5635 individuals/m^3) when compared with other treatments (537 individuals/m^3). This reflected the activity of benthic organisms. The spionids *Polydora ligni* and *Streblospio benedicti* and the amphipod *Microdeutopus gryllotalpa* were more abundant in the 10% treatment compared with other treatments. No other treatment differences were observed for the numerically dominant macrofaunal species using the data available. There were no significant treatment differences for benthic oxygen metabolism and inorganic nutrient fluxes across the sediment-seawater interface.

Discussion

Sewage effluent additions caused prolonged and potentially detrimental effects in the mesocosm ecosystems. Effects were most dramatic in the 10% effluent treatment, but significant effects were also demonstrated at the 1% effluent treatment level. Effects could be related to direct toxic responses and to system responses to the organic carbon in the sewage effluent. The single-species tests and mesocosm measurements complimented each other, but contributed different types of information.

Toxicity Results

The sea urchin *(Arbacia punctulata)* sperm cell toxicity tests indicated that the East Greenwich sewage effluent was toxic with an EC_{50} averaging 1.1%. Toxicity was not detected in 2 of 36 samples. Daily additions to the mesocosms of up to 96 L of sewage effluent per day (10% treatment) gave more variable results.

Three major factors influenced sewage effluent toxicity in the mesocosms: incomplete mixing, toxicity persistence, and the initial phytoplankton bloom in all treatments. Short-term, incomplete mixing contributed to variations in the mesocosm toxicity results. During most of the experiment, toxicity samples were taken directly after sewage effluent additions at the surface of the mesocosms where, due to the lack of complete mixing, effluent concentrations would be expected to be highest. When samples were taken at different depths, only surface samples showed significant toxicity. We believe that if samples were taken when mixing was complete (directly before sewage effluent additions), effluent concentrations in the toxicity samples would have been too low to cause toxicity in the sperm cell tests.

Both laboratory and mesocosm experiments demonstrated that toxicity associated with the sewage effluent rapidly decayed with time. Toxicity in the mesocosms, as measured by the sperm cell test, did not increase over time. This indicated that there was no detectable fraction of the toxicity that persisted from one day to the next. We conclude that the rapid decay of effluent toxicity was most likely responsible for the absence of a buildup of toxicity in the mesocosms. However, the four-month experiment may not have been long enough to detect such a buildup, and toxic fractions may have accumulated in the sediments. A buildup of toxicity might have been observed if the same experiment had been conducted at lower temperatures because the rate of toxicity decay would have been slower.

The third factor influencing toxicity in the mesocosms was the initial phytoplankton bloom in all treatments during the first month of the experiment. Toxicity was detected during this period in the control mesocosms: that is, those receiving no additions of sewage effluent. Toxicity has been attributed to phytoplankton blooms in Narragansett Bay in the past [*44,45*]. Subsequent to this bloom, no toxicity was detected in the controls, and very little was detected in the 0.1 and 1.0% treatments. Toxicity in the 10% treatment continued to the end of the experiment, although toxicity was variable and attributed to the most recent addition of sewage effluent.

It is possible that the initial toxicity measured in all treatments at the start of the experiment was due to contaminants fluxing from the sediment to the water column. However, available mesocosm data for metals does not show evidence for this. Further, past experiments have demonstrated chemical fluxes from much more contaminated sediments were very small [*24,33*].

Mesocosm Responses

Evidence from the mesocosms suggests that sewage effluent additions limited some component of the phytoplankton. During the first month of the experiment, phytoplankton blooms occurred in the control and 0.1% sewage effluent treatments, but blooms were depressed in the 1.0 and 10% sewage effluent treatments.

Nutrient data corroborated the suppression of phytoplankton activity seen in the fluorescence and chlorophyll data. More nitrogen and silica were found in the 10% treatment than in other treatments and controls during this period. This can be linked to phytoplankton activity since phytoplankton primary production would have utilized and decreased nitrogen and silica concentrations. Nitrogen is thought to be the most limiting nutrient in marine systems [*46,47*], whereas silica is limiting only to those organisms, such as diatoms, that utilize silica in the formation of siliceous tests. Silica did not drop to near 0 μg·at silicon/L in the 10% treatment during the first month, as was seen in other treatments (Fig. 7). This suggests that it was diatoms that were particularly sensitive to the sewage effluent additions.

Since reagent nitrogen and silica additions were lower and phosphorus additions higher than nutrients from sewage effluent additions, attributing the higher nitrogen and silica concentrations in the 10% treatment to reduced phytoplankton activity may be criticized. We do not believe this to be a valid criticism for the following reasons.

Past experiments [*25,47*] have repeatedly shown that increased nutrient loading causes a greater phytoplankton response, not the suppressed response observed. Additionally, if total nutrient loadings (sewage effluent plus reagent) in each treatment were significantly different, nutrient concentrations would have approached different steady-state concentrations in each treatment. This was not the case, as there were very few significant treatment differences for total dissolved nitrogen, phosphorus, and silica after the first month of the experiment.

The mechanism whereby the early phytoplankton bloom was depressed in both the 10 and 1% treatments is not known. We believe that sewage effluent was directly toxic to some component of the phytoplankton, possibly diatoms. Previous studies have shown diatoms to be sensitive to sewage effluents [*48*]. Nutrient limitation is not a good explanation since there was an excess of all major nutrients. Light limitation was an unlikely factor since total suspended solids was similar in all treatments, averaging 25 mg/L. Predation by zoo-

TABLE 5—*Total carbon loading due to sewage effluent 15 July –14 Nov.*

Treatment, % effluent	Carbon, g/m^2
Controls	0
0.1%	2.5
1.0%	25
10%	249

plankton was probably not a controlling factor since the biomass of zooplankton was similar in all treatments, and no zooplankton group or species increased in abundance during the first month. Predation by microzooplankton is possible, but no measurements were made of these groups.

Carbon Loadings and System Metabolism

Sewage effluent additions constituted a significant carbon loading to the mesocosm ecosystem (Table 5). Approximately 249 g carbon/m^2 were added over 122 days to the 10% treatment mesocosms in the form of particulate and dissolved organic carbon. This is 93% of the mean annual primary productivity for Narragansett Bay [*49*]. Prior to the start of this experiment, we had thought this carbon would cause a considerable respiratory demand, increasing system respiration and lowering oxygen concentrations. Although oxygen concentrations were indeed lower in the 10% treatments (Fig. 8), total system respiration was only higher in this treatment during the second month of the experiment. We were not able to detect significant treatment differences in system respiration for any other time, nor were we able to detect differences in system production during the entire experiment. There was a trend for production to be smallest and respiration greatest in the 10% treatment. The cumulative effect of this was negative net system production (P-R) at the 10% sewage effluent treatment level.

A measure of the maximum amount of sewage effluent carbon metabolized during the experiment by the mesocosm system can be estimated by calculating the difference between net production in the controls and net production in the 10% treatment. This difference is 59 g carbon/m^2. The remainder of the sewage effluent carbon (249 g carbon/m^2 added − 59 g carbon/m^2 metabolized = 190 g carbon/m^2 remaining) can easily be accounted for by the loss due to water flow through the system. Assuming a completely mixed system, no deposition of sewage effluent carbon, and no biological utilization, nearly 90% of the added carbon could have been lost from the system simply due to flushing. This is obviously a maximum estimate since some particles were lost due to deposition and some carbon was respired. Nevertheless, since approximately 50% of the carbon was dissolved organic carbon, and 50% of the sewage effluent particulates were <5 μm in size, a significant portion of the sewage effluent carbon could have been lost simply due to flushing.

The net effect of the addition of sewage effluent carbon and the decreased phytoplankton activity in the mesocosms was to bring about an imbalance of system production and respiration. This imbalance reduced oxygen concentrations in the 10% treatment. This was brought about mainly by metabolism in the water column since benthic oxygen fluxes were similar in all treatments. Hypoxia resulted which, under the proper conditions, could have progressed to total system anoxia.

Conclusions

The purpose of this experiment was to compare toxicity as measured by single-species tests to those ecosystem level effects measured in mesocosm ecosystems. The single-species tests were useful in determining effluent concentrations used in the experiment and the magnitude and persistence of the effluent toxicity. Direct tests of effluent toxicity demonstrated toxicity at 1% effluent. However, few toxic effects were observed in this and the 0.1% mesocosm treatment levels. This lack of response was attributed to the rapid decay of effluent toxicity at summer temperatures. At the 10% effluent treatment level, where prior to this experiment massive toxic effects were expected, few toxic effects were observed directly. Greater toxicity might be expected during the winter months when the decay of toxicity would be slower.

The mesocosm measurements were useful in showing effluent effects in simulated coastal marine systems. Effluent effects were attributed partially to a toxic response of the phytoplankton and partially to the carbon associated with the effluent. Toxicity to the phytoplankton was inferred from fluorescence, chlorophyll, and nutrient data, and these data indicated diatom production might have been reduced in the 10% effluent treatment. Other mesocosm communities in the 10% treatment were simulated, suggesting a lack of toxicity. The combined effects on the phytoplankton and the organic carbon loading caused an imbalance in production and respiration that lead to hypoxia. Results suggest carbon loadings may be as much a problem to coastal ecosystems as direct toxic effects when toxicity decay is rapid.

Acknowledgments

This experiment was a cooperative effort between MERL and the EPA Environmental Research Laboratory in Narragansett, and there are many who played a part in its execution. Specifically, we would like to thank Steve Schimmel, Glen Thursby, and Eugene Jackim for helping with the *Arbacia* toxicity tests. Warren Boothman and Gerry Hoffman kindly provided the sewage effluent metal data. Viva Banzon, Lynn Beatty, Fran Brown, Betty Buckley, Nancy Craig, Peter Doering, Deborah Hayes, Aimee Keller, Steve Kelly, Eric Klos, Barbara Nowicki, Elizabeth Olson, Mike Perry, James Rentsch, Edwin Requintina, Karen Rudio, Barbara Sullivan, and Anna Thompson helped with the mesocosm experiment. Bob Richardson provided data for treatment plants within Rhode Island. Peter Doering kindly helped with the design of the statistical analyses.

This experiment was supported primarily by grants to the Marine Ecosystems Research Laboratory from the U.S. Environmental Protection Agency through the Narragansett Bay Project (NAR34.JCL) and the EPA Centers Program (CR812487-01). Additional funding supported D. Nacci, C. Strobel, and R. Walsh through EPA Contract 68-03-3226 to Science Applications International Corp.

References

[1] Bierman, V. J., Gentile, H. J., Paul, J. F., Miller, D. C., and Brungs, W. A., "Research Strategy for Ocean Disposal: Conceptual Framework and Case Study," *Environmental Hazard Assessment of Effluents,* H. G, Bergman, R. A. Kimerland, and A. W. Maki, Eds., Pergamon Press, NY, 1985.

[2] Gearing, P. J. and Gearing, J. N., *Marine Environmental Research,* Vol. 6, 1982, pp. 115–132.

[3] Gearing, P. J. and Gearing, J. N., *Marine Environmental Research,* Vol. 6, 1982, pp. 133–143.

[4] Hinga, K. R., Pilson, M. E. Q., Lee, R. F., Farrington, J. W., Tjessem, K., and Davis, A. C., *Environmental Science and Technology,* Vol. 14, 1980, pp. 1136–1143.

[5] Santschi, P. H., Adler, D., and Amdurer, M., "The Fate of Particles and Particle-Reactive Trace Metals in Coastal Waters: Radioisotope Studies in Microcosms," *Trace Metals in Sea Water,* C. S. Wong, E. Boyle, K. W. Bruland, J. D. Burton, and E. D. Goldberg, Eds., Plenum Press, NY, 1983, pp. 331–349.

[6] Santschi, P. H., *Environmental Technology Letters,* Vol. 6, 1985, pp. 335–350.

[7] Hunt, C. D., "Fate and Bioaccumulation of Soil-Associated Low-Level Natural Radioactivity Discharged into a Marine Ecosystem," Final report to the Environmental Protection Agency, Office of Radiation Programs, Washington, DC, 1986.

[8] Livingston, H. D., Bowen, V. T., and Kupferman, S. L., *Journal of Marine Research,* Vol. 40, 1982, pp. 253–272.

[9] Livingston, H. D., Bowen, V. T., and Kupferman, S. L., *Journal of Marine Research,* Vol. 40, 1982, pp. 1227–1258.

[10] Burns, L. A. and Cline, D. M., "Exposure Analysis Modeling System: Reference Manual for EXAMS II," EPA/600/3-85-038, U.S. Environmental Protection Agency, Athens, GA, 1985.

[11] Burns, L. A., Cline D. M., and Lassiter, R. R., "Exposure and Analysis Modeling System (EXAMS): User Manual and System Documentation," EPA-600/3-82-023, U.S. Environmental Protection Agency, Athens, GA, 1982.

[12] Wall, T. M. and Hanmer, R. W., *Journal of the Water Pollution Control Federation,* Vol. 59, 1987, pp. 7–12.

[13] Hansen, D. J., "Utility of Toxicity Tests to Measure Effects of Substances on Marine Organisms," *Concepts in Marine Pollution Measurements,* H. H. White, Ed., University of Maryland, Maryland Sea Grant, College Park, MD, 1984, pp. 33–56.

[14] Gentile, J. H. and Schimmel, S. C., "Strategies for Utilizing Laboratory Toxicological Information in Regulatory Decisions," *Concepts in Marine Pollution Measurements,* University of Maryland, Maryland Sea Grant Publication, College Park, MD, 1984, pp. 57–80.

[15] "Technical Support Document for Water Quality-Based Toxics Control," U.S. Environmental Protection Agency, Office of Washington, DC, 1985.

[16] Swartz, R. C., Cole, F. A., Schults, D. W., and DeBen, W. A., *Marine Ecology Progress Series,* Vol. 31, 1986, pp. 1–13.

[17] Sanders, H. L., Grassle, J. F., Hampson, G. R., Morse, L. S., Garner-Price, S., and Jones, C. C., *Journal of Marine Research,* Vol. 38, 1980, pp. 365–380.

[18] Mayer, G. F., *Ecological Stress and the New York Bight: Science and Management,* Estuarine Research Federation, Columbia, SC, 1982.

[19] Cairns, J., *Hydrobiologia,* Vol. 100, 1983, pp. 47–57.

[20] Cairns, J. and Buikema, A. L., "Verifying Predictions on Environmental Safety and Harm," *Concepts in Marine Pollution Measurements,* University of Maryland, Maryland Sea Grant, College Park, MD, 1984, pp. 81–111.

[21] Giesy, J. P., *Microcosms in Ecological Research,* DOE Symposium Series, CONF 781101, National Technical Information Service, Springfield, VA, 1980.

[22] Grice, G. D. and Reeve, M. R., *Marine Mesocosms: Biological and Chemical Research in Experimental Ecosystems,* Springer-Verlag, New York, 1982.

[23] Oviatt, C. A., Frithsen, J., Gearing, J., and Gearing, P., *Marine Ecology Progress Series,* Vol. 9, 1982, pp. 121–136.

[24] Oviatt, C. A., Pilson, M. E. Q., Nixon, S. W., Frithsen, J. B., Rudnick, D. T., Kelly, J. B., Grassle, J. F., and Grassle, J. P., *Marine Ecology Progress Series,* Vol. 16, 1984, pp. 203–217.

[25] Oviatt, C. A., Keller, A. A., Sampou, P. A., and Beatty, L. L., *Marine Ecology Progress Series,* Vol. 28, 1986, pp. 69–80.

[26] Frithsen, J. B., Elmgren, R., and Rudnick, D. T., *Marine Ecology Progress Series,* Vol. 23, 1985, pp. 1–14.

[27] Dinnel, P. A., Stober, Q. J., and DiJulio, D. H., *Marine Environmental Research,* Vol. 5, 1981, pp. 29–39.

[28] Dinnel, P., Stober, Q., Letourneau, M., Roberts, W., Felton, S., and Nakatani, R., "Methodology and Validation of a Sperm Cell Toxicity Test for Testing Toxic Substances in Marine Waters," Final Report, Grant R/TOX.FRI-UW-83, U.S. Environmental Protection Agency, Washington, DC, 1983.

[29] Nacci, D., Jackim, E., and Walsh, R., *Environmental Toxicology and Chemistry,* Vol. 5, 1986, pp. 521–525.

[30] Pilson, M. E. Q. and Nixon, S. W. in *Microcosms in Ecological Research,* J. P. Giesy, Ed., DOE Symposium Series, CONF 781101, National Technical Information Service, Springfield, VA., 1980, pp. 724–741.

[31] Pilson, M. E. Q., Oviatt, C. A., and Nixon, S. W. in *Microcosms in Ecological Research,* J. P.

Giesy, Ed., DOE Symposium Series, CONF 781101, National Technical Information Service, Springfield, VA, 1980, pp. 753–778.

[32] Pilson, M. E. Q., *Journal of Marine Research,* Vol. 43, 1985, pp. 849–873.

[33] Hunt, C. D. and Smith, D. L., *Canadian Journal of Fisheries and Aquatic Sciences,* Vol. 40, 1983, pp. 132–142.

[34] Lambert, C. E. and Oviatt, C. A., "Manual of Biological and Geochemical Techniques in Coastal Areas," MERL Series, Report No. 1, 2nd ed., University of Rhode Island, Kingston, RI, 1986.

[35] Frithsen, J. B., Oviatt, C. A., and Keller, A. A., "A Comparison of Ecosystem and Single-Species Tests of Complex Waste Toxicity: A Mesocosm Experiment Data Report," MERL Series, Report No. 7, University of Rhode Island, Kingston, RI, 1987.

[36] *SAS User's Guide: Basics, Version 5 Edition,* SAS Institute Inc., Cary, NC, 1985.

[37] *SAS User's Guide: Statistics, Version 5 Edition,* SAS Institute, Inc., Cary, NC, 1985.

[38] *SAS/GRAPH User's Guide: Version 5 Edition,* SAS Institute Inc., Cary, NC, 1985.

[39] Macdonald, R. W., McLaughlin, F. A., and Wong, C. S., *Limnology and Oceanography,* Vol. 31, 1986, pp. 1139–1142.

[40] Cullen, J. J., Zhu, M., and Pierson, D. C., *Limnology and Oceanography,* Vol. 13, 1986, pp. 1364–1373.

[41] Leftley, J. W., Bonin, D. J., and Maestrini, S. Y., *Oceanography and Marine Biology: An Annual Review,* Vol. 21, 1983, pp. 23–66.

[42] Keller, A. A., *Marine Biology,* Vol. 96, 1987, pp. 107–114.

[43] Oviatt, C. A., Rudnick, D. T., Keller, A. A., Sampou, P. A., and Almquist, G. T., *Marine Ecology Progress Series,* Vol. 28, 1986, pp. 57–67.

[44] Thomas, C. R. and Deason, E. E., *Marine Ecology,* Vol. 2, 1981, pp. 215–223.

[45] Verity, P. G. and Stoecker, D., *Marine Biology,* Vol. 72, 1982, pp. 79–87.

[46] Valiela, I., *Marine Ecological Processes,* Springer-Verlag, New York, 1984.

[47] Nixon, S. W., Pilson, M. E. Q., Oviatt, C. A., Donaghay, P., Sullivan, P., Seitzinger, S., Rudnick, D., and Frithsen, J., "Eutrophication of a Coastal Marine Ecosystem—An Experimental Study Using the MERL Microcosms," *Flows of Energy and Materials in Marine Ecosystems,* M. J. R. Fasham, Ed., Plenum, New York, 1984, pp. 105–135.

[48] Dunstan, W. M., *Environmental Science and Technology,* Vol. 9, 1975, pp. 635–638.

[49] Oviatt, C., Buckley, B., and Nixon, S., *Estuaries,* Vol. 4, 1981, pp. 167–175.

Harry C. Freeman,[1] *Gloria B. Sangalang,*[1] *and Laurie Sperry*[1]

Studies on Ameliorating the Effects of an Acidic River on the Atlantic Salmon *(Salmo salar)*

REFERENCE: Freeman, H. C., Sangalang, G. B., and Sperry, L., "**Studies on Ameliorating the Effects of an Acidic River on the Atlantic Salmon *(Salmo salar)*,**" *Aquatic Toxicology and Environmental Fate: Eleventh Volume, ASTM STP 1007,* G. W. Suter II and M. A. Lewis, Eds., American Society for Testing and Materials, Philadelphia, 1989, pp. 251–259.

ABSTRACT: The decline in the Atlantic salmon populations in the poorly buffered rivers in southwestern Nova Scotia, Canada, has been coincident with increased acidic precipitation making mitigation a necessity.

In our studies on mitigating the effects of acidic water, sexually maturing adult Atlantic salmon *(Salmo salar)* were held in the acidic Westfield River (pH 5.1 to 4.7) and the nearby less acidic Medway River (pH 5.7 to 5.2) as a control. Atlantic salmon fed a salt (sodium chloride)-enriched diet to compensate for blood salt losses that occur in acidic waters, had increased reproductive capacity. These fish had a larger percentage of spawners, a larger number of eggs per fish, increased fecundity (number of eggs per gram of fish), and a higher viability of fertilized eggs compared to fish that were fed a commercial trout diet. The female fish on the salt-enriched diet had higher peak levels of the female sex hormone, $17\alpha,20\beta$-dihydroxy-4-pregnen-3-one (17,20-diOHP), which was similar to levels in control fish.

When the pH of the acidic river was elevated using limestone, the salmon lost less weight and had lower egg mortalities compared to the fish in the untreated acidic river water. All fish in the limestone-treated water, however, still did not perform as well as the control fish, and they were highly stressed as indicated by the increased production of corticosteroids by the head kidneys (adrenal homologues) *in vitro.*

KEY WORDS: aquatic toxicology, aquatic biology, reproduction (biology), sublethal toxicity test, acid rain, steroidogenesis, Atlantic salmon

Atlantic salmon *(Salmo salar)* have disappeared from a number of rivers in southwestern Nova Scotia, and they are declining in other rivers as the water acidity approaches unacceptable levels [*1,2*]. There are now 13 rivers in southern Nova Scotia that no longer have salmon runs [*1*]. The decline in and ultimate loss of these runs has been coincident with increasing acidity, pointing out the need to ameliorate the effects of acidity in order to preserve native stocks of salmonids until abatement of acid precipitation occurs.

Our previous studies [*3–6*] have shown that Atlantic salmon held in cages in the acidic Westfield River (pH 4.7 to 5.3) gained less weight, produced fewer and smaller eggs, and had abnormal male sex and corticosteroid hormone metabolism, compared to similar fish held in the less acidic Medway River (pH 5.4 to 6.1).

In the present study, the effects of a diet enriched with sodium chloride on the reproductive physiology of the Atlantic salmon in the Westfield River was examined because fish in acidic waters are known to lose blood electrolytes, particularly sodium chloride, via

[1] Research scientist, research associate, and laboratory technician, respectively, Department of Fisheries and Oceans, Physical and Chemical Sciences Branch, Marine Chemistry Div., Halifax Fisheries Research Laboratory, P.O. Box 550, Halifax, Nova Scotia, Canada B3J 2S7.

the gills, and this salt loss is thought to be associated with reproductive failures [7,8]. The effects of elevating the pH of the acidic river water using crushed limestone was also evaluated. The Westfield River and Medway River waters have approximately the same water hardness (about 4-mg calcium carbonate per litre) and ionic strength.

Experimental Procedure

Atlantic salmon were purchased each year from a local fish farm (Ocean Farms Ltd., Tantallon, N.S.). These fish were progeny of River Phillip salmon that had been captured during their spawning migration and manually spawned. The eggs were hatched at the Mersey Hatchery, Department of Fisheries and Oceans, Milton, N.S. The alevins were raised to the smolt stage at which time they were transferred to the fish farm where they were acclimated to seawater and held until they were purchased in the following spring. For each year's investigation (1985 and 1986), 200 Atlantic salmon, weighing 600 to 1200 g, were selected, acclimated to freshwater in a nearby brook, and then transferred by tank truck to a floating cage (2.4 m by 3.7 m by 1.8 m deep, 25-mm mesh netting) in the Medway River. After a short holding period, the fish were anesthetized in tricaine methane sulfonate:freshwater (1:30 000; weight/volume) and tagged with Carlin tags attached through the dorsal muscle ventral to the anterior dorsal fin. Blood samples were obtained by caudal venipuncture into heparinized syringes. Blood plasma was separated by centrifugation and then analyzed by a steroid radioimmunoassay method [9] to determine the sex of the fish. The fish were also weighed and measured for length.

In 1985, 4 equivalent groups, each having 46 fish with the same number of males and females, were selected for experimentation. In mid-July, 3 of the groups were placed in cages in the Westfield River, and the remaining group was held in the Medway River. The fish in the Medway River were fed a commercial trout diet (6.2-mm pellet, Co-op diet) purchased from Co-op Atlantic Ltd., New Minas, N.S. The three groups in the Westfield River were treated as follows: Group I, not fed; Group II, fed the Co-op diet; Group III, fed a high-salt (sodium chloride) diet (DFO diet). The high-salt diet was formulated by a fish nutritionist, Dr. S. P. Lall of our laboratory. Except for the mineral premix, the vitamin premix, amino acid, herring oil, and crude protein (as herring meal and soybean meal) ingredients of the Co-op and DFO diets were the same. The mineral premix in the DFO diet had a higher sodium chloride (NaCl) content (2.3% compared to 0.3%) than the Co-op diet.

In the 1986 studies, salmon were divided into 3 groups of approximately 40 fish each, including an equal number of each sex. Again, one group was held in the Medway River as a control, while the other two groups were transported to the Westfield River in late August where one group was placed in a cage in a tank fed untreated river water, and the other was placed in a similar cage set in a tank where the incoming Westfield River water had been passed through 150 kg of limestone (marble chips) in a trough (30.5 cm wide by 30 m long) to elevate the pH. The water flow rate into both cages was maintained at approximately 23 L/min. Only a small amount of the limestone dissolved during several months of treatment. In the 1986 experiment, all fish were fed the Co-op diet daily.

The water pH, temperature, and height of each river were measured daily. At the beginning of the holding experiments and then at two to three week intervals until the last spawning in December (each year), the fish were weighed, measured for length, and sampled for blood [6,9]. Hormone levels in the blood plasma were determined by radioimmunoassay [9–11].

When sexually mature, the fish were spawned manually, and the eggs were fertilized with milt from the same group. The fertilized eggs were incubated at the Mersey Hatchery to determine their viability. The egg diameter (for measurement of egg volume) and the fecundity (number of eggs per unit weight of fish) of each ripe female salmon were also recorded.

At the end of spawning time, *in vitro* steroidogenesis studies on head kidneys (adrenal homologues), ovaries, and testes of six males and six females from each group were carried out using radioactive steroid precursors: tritium-labeled pregnenolone ([^{3}H]-pregnenolone) and carbon-14-radiolabeled progesterone ([^{14}C]-progesterone). The methodology is described elsewhere [*12*].

Results and Discussion

The pH of rain in Atlantic Canada ranges from 3.5 to 6.0 and is lowest (greatest acidity) in the area of our study in southwestern Nova Scotia [*13*]. The wet sulfate deposition in this area ranges from 20 to 40 kg/hectare/year [*13*]. In our studies, we observed that rain originating from the southwest caused a depression of river water pH while rain from the northeast caused a pH elevation. It is reasonable to assume that the high mortality of post-spawned salmon that was observed in earlier work [*5*] resulted from a decline in river water pH to below 4.7 due to heavy rains originating from the southwest.

The pH of the Westfield River and Medway River from July to December 1985 ranged from 5.14 to 4.83 and from 5.55 to 5.24, respectively (Fig. 1). In 1985 the pH of the Westfield River was lower than during the same period in 1984, but still higher than in 1981 when the pH at spawning time was 4.7 [*3–5*]. In general, the pH was low when the water level was high. The pH of the Medway River was also lower in 1985 than in 1984 and earlier years. Only once in 1984, after a heavy rain, was the pH of the Medway River as low as 5.4, but in 1985 the pH was below this value a number of times (Fig. 1). The water

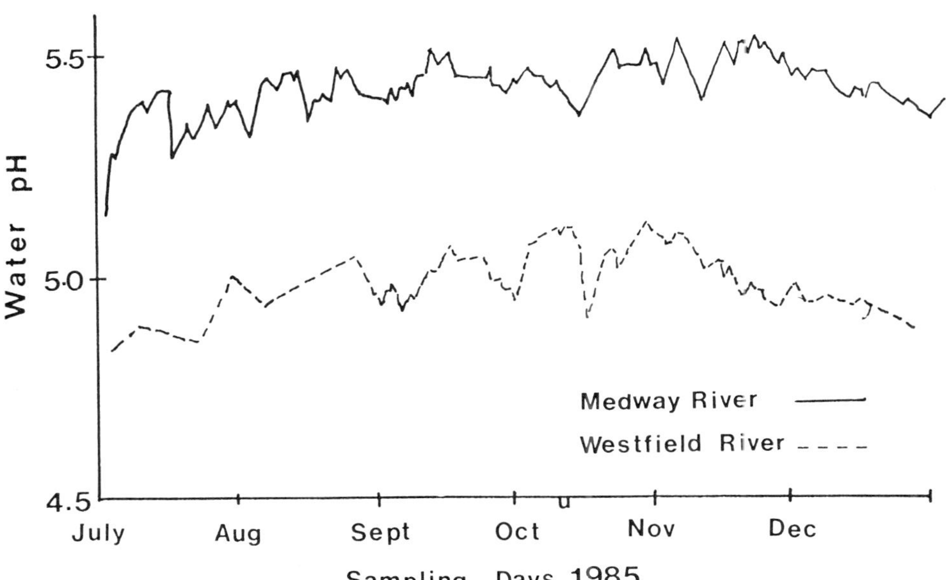

FIG. 1—*The pH of the Medway River and Westfield River during 1 July to 31 Dec. 1985.*

temperatures of the two rivers in 1985 were similar and ranged from a high of 25°C in July to a low of 0.5°C in December (Fig. 2).

In 1985, all groups, on average, showed weight losses from the time of transfer to the cages in August to before spawning in early November (Table 1). The unfed Westfield Group I fish had the greatest average weight loss (14.0%). All Westfield groups lost, on average, more body weight (8.6 to 14.0%) than the Medway group, which lost only 3.3%.

During the experiment, diseased and damaged fish were removed. In Westfield Groups I, II, and III, 10, 8, and 9 fish, respectively, were removed before spawning time. Only 3 fish were lost from the control group in the Medway River. The higher losses in the Westfield groups reflected the stressful conditions of the acidic water.

The fish in the acidic water on the Co-op diet (Westfield Group II) and the DFO diet (Westfield Group III) appeared to eat less and, like the unfed fish, were less active than the control fish. It is significant to note, however, that the fish on the DFO diet were more active than the other two Westfield groups. The Medway fish and Westfield Group III fish lost significantly ($P < 0.01$) less weight than the fish in the other two Westfield groups. This was the first year in our investigations, which started in 1982, that the salmon held in the Medway River have also lost weight.

In our 1984 study [*5*], the holding period was relatively short (10 Oct. to 10 Dec. 1984) and the Medway salmon experienced no weight change, while the fish in the acidic Westfield River lost 6.6% of their body weight. During our 1982 study [*6*], when the fish were held and fed daily under similar conditions, except from mid-July to late November, the Medway salmon gained 21.7% body weight compared to a gain of only 9.6% by the Westfield fish.

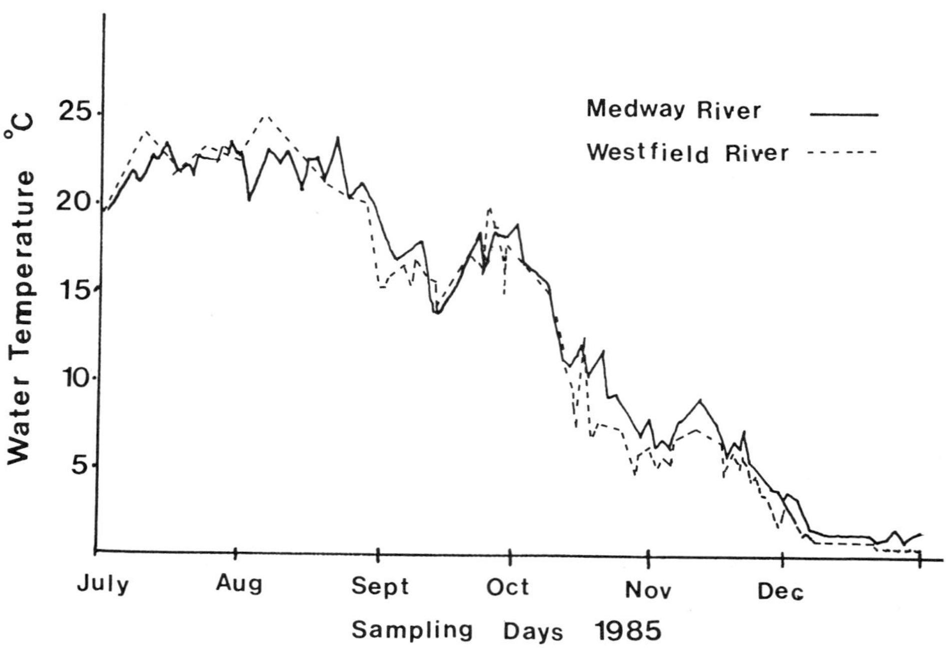

FIG. 2—*The water temperature of the Medway River and Westfield River during 1 July to 31 Dec. 1985.*

TABLE 1—*Mean weight loss [standard deviation (SD)] and weight range of female Atlantic salmon held in cages in the Medway River and Westfield River from mid-August to early November 1985.*

River and Group	Diet	Initial Weight, g, Mean (SD) and Range	Weight Loss, g, Mean (SD) and Range	Mean Weight Loss, %
Medway	Co-op[a]	775 (125) 575 to 1075	25 (45) −75 to +75	3.3
Westfield I	not fed[b]	700 (75) 550 to 750	100 (25) −170 to −65	14.0
Westfield II	Co-op	750 (100) 550 to 850	100 (50) −175 to 0	13.7
Westfield III	DFO[c]	850 (225) 575 to 1250	75 (75) −175 to −85	8.6

[a] Co-op diet contained 0.3% NaCl and 3% calcium.
[b] Unfed fish had access to natural foodstuffs in flowing water.
[c] High-salt diet contained 3% NaCl and 3% calcium.

The mean egg diameters of the Medway River salmon were much larger (t-test, $P <$ 0.001) than all Westfield group values (Table 2). It has been reported that egg size is not related to egg viability [*14*]. The significant difference in egg sizes between Medway and Westfield fish (Table 2) is worth investigating, however, as this difference may reflect other physiological conditions such as electrolyte balances and ionic compositions and, conceivably, fertilization or embryological development.

Fecundity was generally higher in the Medway fish than in the Westfield fish (Table 2). The fecundity (1.38 eggs per gram of body weight) of the Medway group (Co-op diet) was significantly ($P < 0.001$) higher than those of the Westfield Group I (unfed) and II (Co-op diet); 0.85 and 1.24 eggs per gram, respectively. However, the fecundity (1.34 eggs per gram) of the Westfield Group III (DFO diet) was not significantly different from that of the Medway fish, demonstrating the favorable effect of the high-salt diet on the fish in the acidic water. The favorable effect of the high-salt diet was also demonstrated when 78% of the female salmon on this diet became sexually mature compared to 85% of the fish in the Medway Group, and 54 and 60%, respectively, of the fish in Westfield Groups I (unfed) and II (Co-op diet).

The mean peak levels (nanograms per millilitre plasma ± SD) of the female sex hormone 17α,20β-dihydroxy-4-pregnen-3-one (17,20-diOHP) were 95.2 ± 22.4, 71.1 ± 21.0, 60.7 ± 21.0, and 53.9 ± 16.0 for the Medway group, Westfield Group III, Westfield Group

TABLE 2—*Egg production results for female salmon held in cages in the Medway River and Westfield River from mid-August to early November 1985.*

River and Group	Diet	Mature Females/ Surviving Females	Mean Weight of Surviving Females (SD)	Total Egg Production of Group	Mean Egg Diameter, mm, of Group (SD)	Mean Fecundity, eggs/g body weight (SD)
Medway	Co-op	17/20	700 (130)	16 093	5.32 (0.04)	1.38 (0.10)
Westfield I	unfed	7/13	625 (75)	3 778	4.37 (0.03)	0.85 (0.05)
Westfield II	Co-op	9/15	650 (100)	7 726	4.70 (0.04)	1.24 (0.08)
Westfield III	high salt	11/14	800 (200)	12 433	4.60 (0.05)	1.34 (0.08)

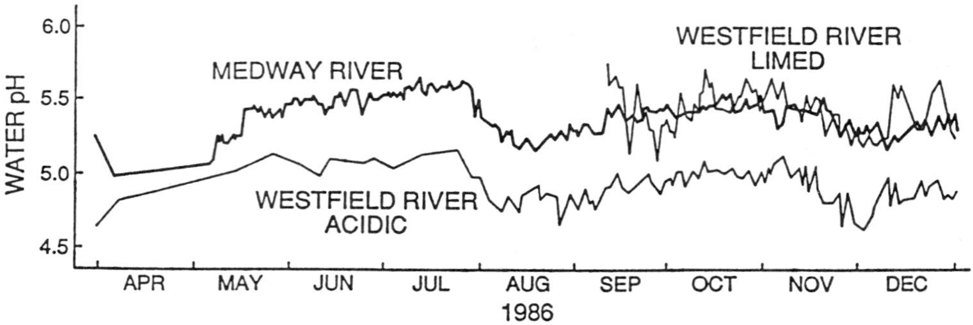

FIG. 3—*The water pH of the Medway River and the untreated and limestone-treated Westfield River during April to December 1986.*

II, and Westfield Group I, respectively, giving some indication of a relationship between fecundity in the females and their sex hormone levels, and reflecting the effect of the improved dietary and holding conditions. The hormone levels in the Westfield groups were significantly lower (F-test, $P < 0.001$) than those in the Medway group. Although 17,20-diOHP peaked in all four groups of fish at spawning, there was a two-week delay in 62% of the female salmon in the unfed Westfield Group I. The androgens, testosterone (T), and 11-ketotestosterone (11-KT) also peaked at, or just before, spawning time in all groups. Since peaking is coincident and imperative with the maturation process (that is, germinal vesicle breakdown or final ovulation) the two-week delay could have drastically affected the fertilization process in this group.

The egg mortalities, although high in all groups, was lowest for the Medway fish (79.1%) and only slightly higher (84.4%) in the Westfield Group III fish on the high-salt diet. The mortalities of the eggs in Westfield Groups I and II were 100 and 98.9%, respectively. These high mortalities also likely reflect a handling component between manual spawning and the transfer to incubators at the hatchery and greatly exceed the usual mortalities in salmon eggs at the facility.[2] The stress of handling the fish regularly, (every two to three weeks) was bound to have an adverse effect on them and may have also contributed to the low egg viability and body weight losses in all groups. It is obvious, however, since all groups of fish were in the same condition at the beginning of experimentation, that the control fish performed best followed closely by the fish in the acidic river that were fed the high-salt diet.

In all indices detailed above, the Westfield Group III fish on the high-salt diet performed nearly as well as the control fish in the Medway River. The addition of salt to the diet appears to have compensated to a large degree for the ion loss and other effects on the fish held in the acidic waters [*7,8*].

In the fall of 1986 the pH of the Westfield River was not as low as in 1983, 1984, and 1985. As in previous years, the pH of the Westfield River was lower at high-water levels. The water temperatures of the Westfield and Medway Rivers were the same or within 0.5°C of each other and ranged from 0.4°C in December to 22.6°C in August. Liming increased the water pH of the Westfield River to values observed in the Medway River (Fig. 3). Egg mortalities, until the end of January 1987, were 18.3% for the control Medway River fish

[2] Goff, T., Department of Fisheries and Oceans, Mersey Hatchery, Milton Queen's Co., Nova Scotia, Canada, personal communication.

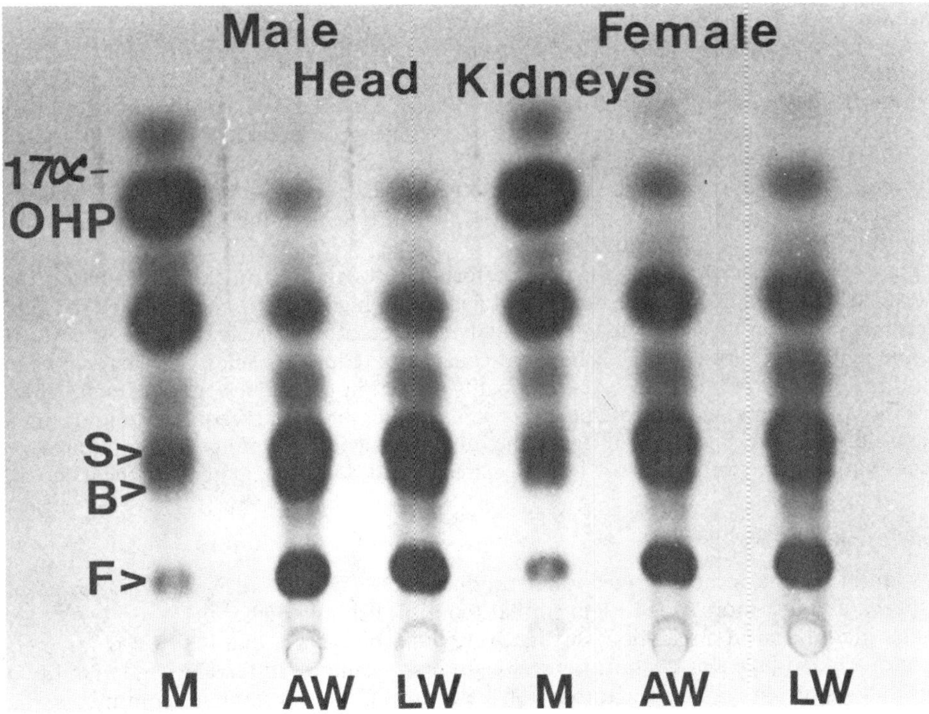

FIG. 4—*Autoradiogram of the thin-layer chromatogram of steroids biosynthesized in vitro from precursors* [^3H]*-pregnenolone plus* [^{14}C]*-progesterone by head kidneys (adrenal homologues) of male and female sexually mature Atlantic salmon held in the control Medway River* (M), *the acidic Westfield River* (AW), *and the limestone-treated Westfield River* (LW). F = *cortisol,* B = *11-deoxycorticosterone,* S = *11-deoxycortisol,* 17-OHP= *17 α hydroxy-4-pregnen-3-one.*

and 34.2 and 60.3% for the limestone-treated and untreated Westfield River fish, respectively.

The Medway River salmon gained 5.2% body weight from 6 Aug. 1986 to the first spawning on 21 Oct. 1986, while the salmon in the limestone-treated and untreated Westfield River water lost 4.6 and 11.7%, respectively. It is obvious that raising the pH has improved the performance of the Atlantic salmon in the Westfield River, although still not to the level observed in the Medway fish.

Both groups of fish in the Westfield River appeared highly stressed as shown by the autoradiograms of the steroid metabolic profiles in the head kidneys (Fig. 4). The *in vitro* steroidogenesis test developed in our laboratory [*12,15*] was able to detect this physiological difference between the Medway and Westfield fish. It is noted that the radioactive spots isopolar with cortisol (F in Fig. 4) and 11-deoxycortisol (S in Fig. 4) are very intense and prominent for both groups (limestone treated and untreated) in the Westfield River fish compared to those for the Medway control fish. The results indicate severe stress (that is, high production of cortisol).

The mortalities of the postspawned salmon maintained in the Westfield River during 1986 were not as high as in previous years. After spawning time, in December 1986, and until early April 1987 there was little rain but an abnormally high snowfall. During this period the river water level and flow rate were low, and the pH of the Westfield River was elevated (pH 4.85 to 5.00) compared to the three previous years. Even though the pH of the Westfield River was elevated relatively, 30% of the postspawned salmon held in the untreated Westfield River water died. The salmon ate poorly and lost weight. There were no mortalities in postspawned fish held over the same period in the limestone-treated Westfield River water where the pH was 5.3 to 5.8. These fish continued to eat well and gain weight. There were no weight losses and no mortalities in the Medway (control) fish over the winter holding period.

The severe stress condition and low performance of the fish in the limestone-treated Westfield River water compared to that of fish in the higher pH Medway River indicates limestone treatment was unable to completely compensate for the acidic river water environment. The limestone may only have the capacity to reduce the acidity of the river water, but it is not able to remove the other toxicants present in the acidic water or leached from the soils and sediments by acidic precipitation. Further research is required to understand this problem in order to better mitigate the impacts of acidic precipitation and to preserve stocks native to streams, possibly in refuge areas, until pollution abatement occurs.

Conclusions

Atlantic salmon in acidic river water produced fewer and smaller eggs, had lower egg viability, and lost more weight than similar (control) fish in a higher pH river.

Atlantic salmon in an acidic river that were fed a high-NaCl diet had test parameters, including blood levels of sex hormones, weight loss, fecundity of females, and viability of eggs and sperm, that approached those of the control fish fed a standard commercial diet but held in a river of higher pH.

Reproduction and survival of Atlantic salmon improved when the pH in the acidic river was elevated using limestone. The fish in the limestone-treated water did not, however, perform as well as the control fish in the higher pH river.

Both groups of Atlantic salmon in the acidic and limestone-treated water were under stress, as shown by the excessive production of cortisol in their head kidneys *in vitro*.

References

[1] Watts, W., *Water, Air, and Soil Pollution,* Vol. 31, 1986, pp. 775–789.

[2] Farmer, G. L., Goff, T. R., Ashfield, D., and Samant, H. S., "Some Effects on the Acidification of Atlantic Salmon Rivers in Nova Scotia," *Canadian Technical Report of Fisheries and Aquatic Sciences,* No. 972, Nov. 1980.

[3] Freeman, H. C., Sangalang, G. B., Burns, G., and McMenemy, M., *The Science of the Total Environment,* Vol. 32, 1983, pp. 87–91.

[4] Freeman, H. C., Sangalang, G. B., McMenemy, M., Burns, G., and Goff, T., "Studies on the Effect of a Low-pH River on Weight Gain, Sexual Maturation, Androgen Production, and Reproduction in the Atlantic Salmon *(Salmo salar),*" International Council for the Exploration of the Sea, Copenhagen, Denmark, 1983.

[5] Freeman, H. C. and Sangalang, G. B., "The Effects of an Acidic River on Weight Gain, Steroidogenesis, and Reproduction in the Atlantic Salmon *(Salmo salar),*" International Council for the Exploration of the Sea, Copenhagen, Denmark, 1985.

[6] Freeman, H. C. and Sangalang, G. B., *Aquatic Toxicology and Hazard Assessment,* (Eighth Symposium), *ASTM STP 891,* American Society for Testing and Materials, Philadelphia, 1985, pp. 339–349.

[7] Klaverkamp, J. F., Brown, S. B., Evans, R. E., Majewski, H. S., Metner, D., and Freeman, H. in *Proceedings,* Seventh Annual Meeting of the Society of Environmental Toxicology and Chemistry, Alexandria, VA, 4 Nov. 1986.

[8] Beamish, R. J., Lockhart, W. L., Van Loon, J. C., and Harvey, H. H., *Ambio,* Vol. 4, pp. 98–102.
[9] Sangalang, G. B., Freeman, H. C., and Flemming, R. B., *General and Comparative Endrocrinology,* Vol. 36, 1978, pp. 187–193.
[10] Sangalang, G. B. and Freeman, H. C., *General and Comparative Endocrinology,* Vol. 32, 1977, pp. 432–439.
[11] Sangalang, G. B. and Freeman, H. C., *General and Comparative Endocrinology,* Vol. 68, 1987, pp. 230–234.
[12] Freeman, H. C., Uthe, J. F., and Sangalang, G. B., *Rapports et Procès—Verbaux des Réunions Conseil International pour L'Exploration de la Mer,* Vol. 179, 1980, pp. 16–22.
[13] Eaton, P. B., Hildebrand, L. P., and D'Entremont, A. A., "Environmental Quality in the Atlantic Region," Catalog No. EN 40-336/1986E, Environment Canada Environmental Protection Service, Atlantic Region, Minister of Supply and Services, Dartmouth, Nova Scotia, Canada, 1986, pp. 30–31.
[14] Glebe, B. D., Appy, T. D., and Saunders, R. L., "Variation in Atlantic Salmon (*Salmo salar*) Reproduction Traits and Their Implications in Breeding Programs," International Council for the Exploration of the Sea, Copenhagen, Denmark, 1979.
[15] Lee, R., Davies, J. M., Freeman, H. C., Ivanovici, A., Moore, M. N., Stegman, J., and Uthe, J. F., *Rapports et Procès—Verbaux des Réunions Conseil International pour L'Exploration de la Mer,* Vol. 179, 1980, pp. 48–55.

Chronic Endpoints: Statistical Versus Biological Significance

Charles E. Stephan[1]

Topics on Expressing and Predicting Results of Life-Cycle Tests

REFERENCE: Stephan, C. E., "**Topics on Expressing and Predicting Results of Life-Cycle Tests,**" *Aquatic Toxicology and Environmental Fate: Eleventh Volume, ASTM STP 1007*, G. W. Suter II and M. A. Lewis, Eds., American Society for Testing and Materials, Philadelphia, 1989, pp. 263–272.

ABSTRACT: In April 1967, Mount and Stephan published results of two life-cycle toxicity tests with the fathead minnow and introduced the terms maximum acceptable toxicant concentration (MATC) and laboratory fish production index. Although the methodology for life-cycle tests has been improved substantially, and such tests can now be conducted with a variety of fishes and invertebrates, major issues exist concerning the ways in which results of these tests can be calculated and presented. The purpose of this paper is to present the rationale to answer three questions. First, is it better to calculate results by considering data for a smorgasbord of individual biological measurements or for one or a few integrated effects that are commercially, recreationally, or ecologically important? Second, is it better to calculate results using hypothesis testing or regression analysis? Third, what terminology is appropriate for expressing results of life-cycle chronic tests? The usefulness of various other tests for predicting results of life-cycle tests cannot be adequately assessed until some of these issues are resolved.

KEY WORDS: aquatic toxicology, chronic toxicity, ecological effects, hypothesis testing, maximum acceptable toxicant concentration, no-observed-effect concentration, regression analysis, toxicity tests

In 1962, Crandall and Goodnight published a paper titled "Effects of Sublethal Concentrations of Several Toxicants on Growth of the Common Guppy, *Lebistes reticulatus*" [*1*]. Most of the paper dealt with the effects of lead, zinc, and pentachlorophenol on survival and growth of guppies over a 90-day period, using weekly renewal of the test solutions. At the end of the 90 days, some of the fish were maintained for an unreported additional number of days to observe the effects of the test materials on reproduction. Although the control animals reproduced, the concentrations of the test materials were so high that they allowed little or no reproduction. Thus, even though their title mentions only "growth," Crandall and Goodnight did, in fact, publish in 1962 results of a study of the effects of three test materials on the survival, growth, and reproduction of a species of fish that is resident in North America.

Current interest in life-cycle toxicity tests began with the development of reliable flow-through diluters, starting with the serial diluter [*2*] and the proportional diluter [*3*]. In 1967, Mount and Stephan published results of two life-cycle tests with the fathead minnow [*4*]. These tests were begun with 25.4-mm (1-in.) fathead minnows and continued until the offspring were six-weeks old. Both tests were flow-through, and the concentrations of test material in the chambers were measured about every other week. No data were statistically

[1] Environmental scientist, U.S. Environmental Protection Agency, Duluth, MN 55804.

analyzed, so the reported results were based on judgments concerning acceptable and unacceptable effects on test organisms. The terms maximum acceptable toxicant concentration (MATC) and laboratory fish production index were introduced in this paper.

Several major advancements have occurred concerning life-cycle toxicity tests with aquatic animals in the 20 years since that publication by Mount and Stephan [*4*], including:

1. The methodology for life-cycle tests with the fathead minnow has been improved substantially: (*a*) Tests are now started with embryos; (*b*) results are calculated by statistical analysis of the raw data; (*c*) individual pairs can be isolated during reproduction [*5*]; and (*d*) stock solutions of slightly soluble organic materials are prepared using no surfactants and little or no organic solvent.
2. Life-cycle tests have been completed with a variety of other freshwater and saltwater fish and invertebrate species, including bluegills [*6*], snails [*7*], gammarids [*7*], brook trout [*8*], daphnids [*9*], flagfish [*10*], mysids [*11*], sheepshead minnows [*12*], bluntnose minnows [*13*], polychaetes [*14*], ceriodaphnids [*15*], caddis flies [*16*], and rainbow trout [*17*]. This is a fairly diverse group of species, and some of the species are recreationally important.
3. Several tests have been reported to be good predictors of the results of life-cycle tests, including the cough test [*18*], the early life-stage test [*19,20*], the Birge test [*21*], and the seven-day survival and growth test [*22*].

Although much progress has been made in the last 20 years in the measurement of chronic toxicity to aquatic animals, several issues exist concerning calculating, naming, and predicting results of life-cycle tests. The purpose of this paper is to discuss four ideas that can alleviate some of the concerns.

Intrinsic Rate of Population Growth

Several investigators have used the intrinsic rate of population growth in the calculation of results of life-cycle toxicity tests [*23–32*]. The symbol r is usually used for this quantity, which is also known as the intrinsic rate of natural increase, the instantaneous rate of population growth, the innate capacity for increase, the per capita rate of increase of a population, and the unrestricted rate of increase. The intrinsic rate of population growth is directly connected to the Euler and Lotka equations, to geometric (exponential, logarithmic) and logistic (Verhulst-Pearl, sigmoid) growth curves, and to life tables and Leslie matrices [*32–39*]. This interest in using r in the analysis of data from life-cycle tests is a step in the right direction for a number of reasons:

1. Use of r should provide a direct connection between life-cycle toxicity tests in aquatic toxicology and the branch of ecology concerned with population biology. Although there might be limitations on the usefulness of r in population biology, it does appear to be a useful concept [*35–39*]. Therefore, use of r in connection with life-cycle tests should increase the ability of laboratory tests to provide information that is ecologically relevant.
2. Use of r should provide an objective basis for determining the appropriate duration of life-cycle tests, at least as the duration relates to data concerning survival and reproduction. A consequence of the equation for calculating r is that what happens after the third brood has almost no effect on r [*40*], and this has been demonstrated experimentally with copepods [*27*], ceriodaphnids [*29*], and daphnids [*30,32*]. Because the first brood often affects r more than other broods [*40*], it might be acceptable to terminate life-cycle toxicity tests shortly after the first or second brood is produced by the organisms in the control treatment [*28*].

3. Use of *r* should help reverse the seemingly perpetual increase in the kinds of individual biological data that are analyzed, including such variables as percent hatch; percent deformed; percent survival to Day 30, 60, 90, and so forth; time to death; time to first brood; size of first, second, third and so forth, brood; number of broods, number of days between broods; and number of young per reproductive day. The proliferation of biological data that are suggested for analysis and the proliferation of statistical analyses that are applied to the data stem, in part, from the desire of some toxicologists to obtain the lowest possible number from a toxicity test. What aquatic toxicologists should be in search of are the data and analysis that produce the most relevant effects, that is, the effects that are most useful for assessing and managing risks [*41,42*]. Calculation of results using *r* will determine which data concerning survival and reproduction are used in the calculations.

4. Not only will use of *r* resolve questions about which data should be obtained concerning survival and reproduction, but its use will also help resolve some questions about how those data should be obtained and analyzed. Rather than looking for the experimental design and data analysis that give the lowest possible number for the most sensitive effect, aquatic toxicologists can look for experimental designs that give good estimates of *r* at the lowest cost. Calculation of *r* resolves questions about whether to separate or combine data on survival and reproduction because the calculation automatically combines the appropriate measurements concerning survival and reproduction into one variable. One possible advantage of this is that a concentration that causes a small reduction in survival and a small reduction in reproduction might cause a rather large reduction in *r*. Not only does *r* integrate data on survival and reproduction for an individual organism, but it also integrates it for all organisms in a treatment because *r* can be directly calculated on a per treatment basis [*32*]. Although *r* is usually thought of as a value for a group of females, the calculated value can be thought of as a mean of separate values for individual females, experimental units, or samples. Correct calculation of the individual values and their mean and variance might be difficult or impossible because the fundamental equation contains *r* as an exponent. If the mean and variance cannot be calculated correctly from the individual values, the variance can be calculated using the jackknife technique [*32*].

Aquatic toxicologists were initially interested in *r* because of the possibility of its providing a direct connection between toxicology and ecology, but it is obvious that there might be substantial fringe benefits to the use of *r*.

Although use of *r* is likely to provide several benefits, some work needs to be done before *r* can be used routinely, and its use will not answer all questions.

1. Calculation of *r* requires that the young be counted separately for each female and that the interval between counts be short. Many life-cycle tests are now designed so that young can be counted separately for each female, but some tests are not designed this way. The maximum acceptable interval between counts should be determined on a species-specific basis. For example, Meyer et al. [*32*] found that counting young *Daphnia pulex* three times a week provided as good an estimate of *r* as counting the young daily.

2. Use of *r* in the calculation of results does not require exposure and observation of the second generation. Exposure and observation of the second generation has been a routine part of life-cycle tests with fish, but not of tests with invertebrates. If life-cycle tests are begun with animals up to 24- or 48-hr old, exposure of the second generation would provide for exposure during the first days of the animals' lives and would also provide the opportunity for detecting effects on the young resulting from exposure of the parents. Both Gillespie and Baumann [*43*] and Landner et al. [*44*] found that offspring of fish were adversely affected by exposure of their parents to toxicants, but adverse effects have not

been observed during 96-hr exposures of second-generation mysids.[2] Additional studies should be performed to determine the usefulness of exposures of second-generation fish and invertebrates.

3. The intrinsic rate of population growth does not take into account growth of the test organisms, except in those cases in which growth is correlated with reproduction. At a minimum, the weights of animals that survive to the end of the test should be determined and included in the calculation of results. It is probably also desirable to weigh, if possible, any animals that die during the test to get a measurement of total biomass produced in each experimental unit.

4. It is hoped that use of *r* will provide an ecologically relevant way of combining survival and reproduction data. If so, the next step would be to find an ecologically relevant way of combining data concerning survival and reproduction with data concerning biomass, so that the results of life-cycle tests can be based on an analysis of only one biological variable. This is the approach Mount and Stephan [*4*] intended by the term laboratory fish production index. One way of implementing this idea has been used by McKim et al. [*45*], whereas Chapman [*46*] suggested an approach that might be more ecologically relevant.

Each of these four concepts has been used by at least one investigator. The next step is to develop an integrated application of these four concepts to life-cycle tests with at least one species.

Calculation of Results

Whether or not results of life-cycle tests are calculated using an ecologically relevant variable such as *r*, both hypothesis testing and regression analysis can be used to analyze the data. Mount and Stephan [*4*] simply drew conclusions without using either hypothesis testing or regression analysis. Since then, hypothesis testing has become the routine way of analyzing data, but Stephan and Rogers [*42*] discussed both computational and conceptual advantages of regression analysis. Data from a recent publication can be used to illustrate some of the major points made by Stephan and Rogers.

Hamelink et al. [*47*] reported the data presented in Table 1 concerning the effect of the herbicide Fluridone on the reproduction of *Daphnia magna.* The authors reported that reproduction at concentrations of 0.2 mg of Floridone or less per litre was not significantly less than in the controls. An alternative way of looking at the data is that 0.2 mg/L inhibited reproduction by 45%. Obviously the minimum detectable difference of the hypothesis test was >45%.

In the same publication the authors reported the data presented in Table 2 concerning the effect of the same herbicide on the growth of channel catfish. From these data the authors reported that both length and weight were significantly reduced by both 1.0 and 2.0 mg/L. On the other hand, the percent inhibitions in length and weight that actually occurred at these concentrations ranged from 6.5 to 19.6. For these sets of data, the minimum detectable differences of the hypothesis tests were <6.5 and 16%, respectively.

Stephan and Rogers [*42*] identified a number of publications in which various authors had reported that a small percent inhibition was considered unacceptable because it was statistically significant or had reported that a large percent inhibition was considered acceptable or had "no effect" because it was not statistically significant. The publication of

---

[2] Lussier, S., U.S. Environmental Protection Agency, Narragansett, RI, personal communication, 1987.

TABLE 1—*The effect of the herbicide Fluridone on the reproduction of* Daphnia magna.

Concentration, mg/L	Average Number of Young	Percent Inhibition from Control[a]
Control	375	0
0.06	430	−15
0.1	652[b]	−74
0.2	208	45
0.4	92[c]	75
0.8	18[c]	95
1.6	2[c]	99
3.4	0[c]	100

[a] Not reported in original publication, but calculated here as (375 − number of young)/375.
[b] Significantly greater than controls ($P \geq 0.05$).
[c] Significantly less than controls ($P \leq 0.05$).

TABLE 2—*The effect of a herbicide on the growth of channel catfish.*

Concentration, mg/L	Length, mm	Percent Inhibition in Length[a]	Weight, mg	Percent Inhibition in Weight[a]
Control	26.3	0.0	163	0.0
0.12	26.6	−1.1	176	−8.0
0.25	26.1	0.8	167	−2.5
0.5	26.5	−0.8	179	−9.8
1.0	24.6[b]	6.5	137[b]	16.0
2.0	24.2[b]	8.0	131[b]	19.6

[a] Not reported in original publication, but calculated here as (26.3 − length)/26.3 or as (163 − weight)/163.
[b] Significantly less than controls ($P \leq 0.01$).

Hamelink et al. [*47*] is somewhat unique because it provides recent examples of both problems in the same publication. The basic question is whether there is any reason to believe that a 6.5% inhibition in one variable is unacceptable, whereas a 45% inhibition in another is acceptable? If there is a reason, it is not presented by Hamelink et al. or in any of the publications referenced by Stephan and Rogers. As Stephan and Rogers [*42*] pointed out, results obtained using hypothesis testing depend not only on the amount of effect of the test material on the exposed organisms, but also on the number of replicates, the agreement between replicates, the hypothesis test used, and the value selected for α = the significance level.

A straightforward way of analyzing the data presented above concerning reproduction of *Daphnia magna* and the weight of channel catfish is to perform graphical semilog interpolation between two concentrations and estimate the concentration corresponding to a selected percent inhibition from the controls. Semilog interpolation is recommended because data from toxicity tests are usually analyzed in terms of the logarithm of the concentration or dose. When the data are monotonic in the region of interest and the bracketing concentrations are only a factor of two apart, the result will not depend very much on whether the interpolation is semilog or not. This procedure is not statistical and, therefore, does not raise such statistical issues as (*a*) what constitutes a valid experimental unit, (*b*) what is the distribution of the data within a treatment, and (*c*) what are the confidence limits. Such issues are still unresolved concerning the use of hypothesis tests, and there is no rationale for selection of α. In addition, after a hypothesis test is conducted, many people obtain a point estimate by calculating a geometric mean of the lower and upper limits. This is certainly not a statistical procedure, and it is not possible to obtain confidence limits on such a point estimate.

The graphical procedure described above does a better job [*42*] than hypothesis testing of summarizing the raw data and avoids some statistical issues that might not be adequately resolved in the near future. A procedure for performing regression analysis in a statistically sound manner would be preferable, and possible procedures have been described by Suter et al. [*48*] and DeGraeve et al. [*49*]. However, any statistical procedure will require at least two concentrations with percent inhibition between 0 and 100. Graphical procedures can be used to obtain approximate results when a high concentration or concentrations caused 100% inhibition and all lower concentrations caused 0% inhibition. Some data sets that can be analyzed using hypothesis tests cannot be analyzed using regression analysis, and vice versa, but both techniques should only be used on data sets that provide a clear indication of a concentration-effect curve.

The most persistent objection to the use of regression analysis for analyzing data from life-cycle tests is that its use requires the selection of a specific percent inhibition from the control for calculation of an endpoint. Any percent can be easily specified, and so the objection is really that the ecological importance of any specific percent inhibition is unknown. This objection is somewhat surprising because hypothesis tests are routinely used even though the ecological importance of the calculated endpoint is unknown. Discussion of the use of regression analysis has at least increased awareness of the issue of ecological importance.

Effect on 50% of the test organisms is routinely used to calculate endpoints for acute toxicity tests. For chronic tests, Yu and Hirayama [*31*] used 1 and 50%, Suter et al. [*48*] used 25%, DeGraeve et al. [*49*] and Spehar and Fiandt [*50*] used 50%, and Gentile et al. [*28*] used 100%, but none of these were suggested for routine use in the calculation of endpoints. Three of the easiest options are the following:

1. Continue to assume that results obtained using hypothesis tests have some ecological importance until enough field tests have been conducted to resolve the question of ecological importance.
2. Use regression analysis to estimate the concentration corresponding to a specified small percent inhibition, such as 10%. Ten percent inhibition can usually be determined by interpolation rather than extrapolation and is about as low as one can go without having to worry too much about noise in the tail of the concentration-effect curve. In addition, 10% inhibition in growth or reproduction might be a reasonable point of demarcation between acceptable and unacceptable effects.
3. Use regression analysis to estimate the concentration corresponding to 50% inhibition. On the average, this point can be determined better than any other using parametric or nonparametric statistical techniques or graphical techniques. Factors based on various rationalizations could be applied to the 50% endpoint in order to calculate concentrations to be used in risk assessments, water quality criteria, cost-benefit ratios, and so forth.

Those who feel that the status quo is unacceptable will oppose the first option, whereas some will oppose the second option because of the necessity for selecting the small percentage. The third option uses the point that is the easiest to calculate and has the fewest connotations to encourage investigators to begin using regression analysis while data and rationale are developed for the factors to be applied to the calculated endpoints.

Terminology for Results of Chronic Tests

Use of a regression-type approach to analyze results of life-cycle tests provides the opportunity of using a new approach to resolve the semantic problems concerning results of life-cycle tests. Mount and Stephan [*4*] introduced the term maximum acceptable toxicant concentration (MATC) for expressing the results of life-cycle tests. Various objections to the term have been raised, and the most widely used alternative is the pair of terms lowest observed effect concentration (LOEC) and no-observed-effect concentration (NOEC). Unfortunately, it is not unusual to find that the concentration designated as the NOEC, based on hypothesis testing, actually caused a 10 to 50% inhibition of the biological variable of interest in the toxicity test [*42*]. Thus in many cases, NOEC is a misnomer.

U. Cowgill,[3] T. Doane,[4] J. Hughes,[5] W. Landis,[6] M. Roberts,[7] and the author of this paper developed the term IC50 (IC_{50})[8] and recommended its use with the following definition to express the results of toxicity tests with microalgae:

> IC_{50}—a statistically or graphically estimated concentration that is expected to cause a 50% inhibition of one or more specified biological processes (such as growth or reproduction), for which the data are not dichotomous, under specified conditions.

This term, or one like it, can be applied just as well to life-cycle toxicity tests if results are calculated using graphical interpolation or regression analysis. It might appear that this could be called a median effective concentration (EC_{50}), but EC_{50}, median lethal concentration (LC_{50}), and median lethal dose (LD_{50}), have traditionally been defined in terms of dichotomous data, that is, data on an all-or-nothing effect such as alive or dead. Rather than adopt a new definition for EC_{50} or have two different definitions for EC_{50}, it seems more reasonable to adopt a new term and a new definition. Use of a new term and definition should also help make it clear that methods that are appropriate for calculating EC_{50}s, LC_{50}s, and 95% confidence limits from dichotomous data are not appropriate for calculating IC_{50}s, IC_{10}s, and their confidence limits from data that are not dichotomous.

Predictive Tests

Within a few years after the publication by Mount and Stephan [*4*], life-cycle tests had become an integral part of aquatic toxicology. Such tests with fish usually last at least nine months with small species and much longer with larger species. To save time and money, investigators began looking for quicker tests that would be good predictors of results of life-cycle tests. Partial life-cycle tests have long been accepted as useful predictors of life-cycle tests for fish, and most tests conducted with species such as brook trout have been partial life-cycle tests rather than life-cycle tests. Comparisons of results of life-cycle and partial life-cycle tests are available for only a few species with only a few test materials,

---

[3] Cowgill, U., Dow Chemical Co., Midland, MI, personal communication, 1986.

[4] Doane, T., U.S. Air Force, Brooks Air Force Base, TX, personal communication, 1986.

[5] Hughes, J., Malcolm Pirnie, Inc., White Plains, NY, personal communication, 1986.

[6] Landis, W., U.S. Army, Aberdeen Proving Ground, MD, personal communication, 1986.

[7] Roberts, M., Virginia Institute of Marine Science, Gloucester Point, VA, personal communication, 1986.

[8] Note that the author prefers the original term, with the number 50 on line (IC50); ASTM style, however, uses subscript type for the percentage numbers in such terms (IC_{50}), as is done throughout this volume.

but partial life-cycle tests are widely accepted because the exposure includes all portions of the life cycle that are considered important.

Decisions concerning the usefulness of other proposed predictive tests [*18–22*] have to be based on comparisons of results of the predictive test with results of life-cycle tests. For some proposed predictive tests, very few comparisons are available, and in some cases the comparisons are between results obtained with different species or in different waters or both. In addition, some of the comparisons are qualitative rather than quantitative.

On the basis of two publications [*19,20*], the early life-stage test has been accepted as such a good predictor that life-cycle tests with fish have become very rare. Although the early life-stage test might be useful in its own right, three recent publications [*44,48,51*] indicate that the earlier reports [*19,20*] might have been overly optimistic about the correlation between results of the early life-stage test and the life-cycle test. Further, Suter et al. [*48*] reported that comparisons based on regression analysis show that the early life-stage test is often less sensitive than the life-cycle test with the same species of fish.

Initial suggestions for useful predictive tests can be based on limited data, but a serious claim of the usefulness of a predictive test should take into account the following points:

1. Comparisons should be based on comparable data, preferably results of tests with the same species in the same water.
2. Generalizations to all chemicals should be based on data for chemicals with a variety of modes of action.
3. Extrapolation to such broad taxonomic groups as fishes should be based on data for a variety of species within that group.
4. Outliers should be examined to see if they provide information about species or chemicals for which the generalization is not valid.
5. Enough information should be given so that the comparison can be independently verified.
6. Comparisons should be quantitative, not qualitative. This does not mean, however, that they must be statistical. For example, if the predicted value is always a factor of 1.4 to 3.7 greater than the actual value, it is not necessary to be able to derive a statistically correct mean and variance to know that the predictor is useful. Similarly, if the predicted value ranges uniformly from 0.11 to 111 times the actual value, an elegant statistical analysis of the data will not make the predicted value look any more useful.
7. More life-cycle tests should be conducted with fish to widen the range of chemicals with which good comparisons can be made.

Predictive tests should not be accepted as useful unless a substantial amount of good data are available to justify the conclusion.

Some Recommendations

On the basis of the ideas discussed above, the following are recommended for consideration by aquatic toxicologists:

1. At least one life-cycle test with some species of fish should be conducted each year. This is the only way to expand the data base available for studying the usefulness of predictive tests. Each year there will be crucial chemicals (such as selenium and tributyltin in 1987) for which it would be desirable to have results of at least one life-cycle test with a species of fish.
2. Data from past and future tests should be analyzed to determine (*a*) whether the reproductive phase of life-cycle tests with fishes and invertebrates can be terminated after the

first, second, or third spawn or brood in the controls, and (*b*) how often young should be counted.

3. Data from past and future tests should be analyzed to determine whether the test methodology should include exposure of the second generation for a brief, species-specific period of time.

4. Each life-cycle test should be designed so that results can be analyzed in terms of *r*, in addition to any other analyses that are performed.

5. Until better procedures and endpoints are devised, one of the endpoints reported for each life-cycle test should be the IC_{50} calculated by graphical semilog interpolation of *r* versus concentration of test material.

6. A means should be devised for combining *r* and biomass into an ecologically relevant variable that can be used as a basis for calculation of results of life-cycle tests.

Implementation of these suggestions would ensure that the ideas that have been developed in the past few years receive appropriate consideration and would provide a basis for continued progress in the study of chronic toxicity to fishes and invertebrates.

References

[1] Crandall, C. A. and Goodnight, C. J., *Limnology and Oceanography,* Vol. 7, No. 2, 1962, pp. 233–239.

[2] Mount, D. I. and Warner, R. E., "A Serial-Dilution Apparatus for Continuous Delivery of Various Concentrations of Materials in Water," Public Health Service Publication No. 999-WP-23, U.S. Department of Health, Education, and Welfare, Cincinnati, OH, 1965.

[3] Mount, D. I. and Brungs, W. A., *Water Research,* Vol. 1, No. 1, 1967, pp. 21–29.

[4] Mount, D. I. and Stephan, C. E., *Transactions of the American Fisheries Society,* Vol. 96, No. 2, 1967, pp. 185–193.

[5] Benoit, D. A. and Holcombe, G. W., *Journal of Fish Biology,* Vol. 13, No. 6, 1978, pp. 701–708.

[6] Eaton, J. G., *Water Research,* Vol. 4, No. 10, 1970, pp. 673–684.

[7] Arthur, J. W. and Leonard, E. N., *Journal of the Fisheries Research Board of Canada,* Vol. 27, No. 7, 1970, pp. 1277–1283.

[8] McKim, J. M. and Benoit, D. A., *Journal of the Fisheries Research Board of Canada,* Vol. 28, No. 5, 1971, pp. 655–662.

[9] Beisinger, K. E. and Christensen, G. M., *Journal of the Fisheries Research Board of Canada,* Vol. 29, No. 12, 1972, pp. 1691–1700.

[10] Smith, W. E., *Journal of the Fisheries Research Board of Canada,* Vol. 30, No. 2, 1973, pp. 329–330.

[11] Nimmo, D. R., Bahner, L. H., Rigby, R. A., Sheppard, J. M., and Wilson, A. J., Jr. in *Aquatic Toxicology and Hazard Evaluation (First Symposium), ASTM STP 634,* American Society for Testing and Materials, Philadelphia, 1977, pp. 109–116.

[12] Hansen, D. J. and Parrish, P. R. in *Aquatic Toxicology and Hazard Evaluation (First Symposium), ASTM STP 634,* American Society for Testing and Materials, Philadelphia, 1977, pp. 117–126.

[13] Horning, W. B. and Neiheisel, T. W., *Archives of Environmental Contamination and Toxicology,* Vol. 8, No. 5, 1979, pp. 545–552.

[14] Oshida, P. S., Word, L. S., and Mearns, A. J., *Marine Environmental Research,* Vol. 5, No. 1, 1981, pp. 41–49.

[15] Mount, D. I. and Norberg, T. J., *Environmental Toxicology and Chemistry,* Vol. 3, No. 3, 1984, pp. 425–434.

[16] Nebeker, A. V., Savonen, C., Baker, R. J., and McCrady, J. K., *Environmental Toxicology and Chemistry,* Vol. 3, No. 4, 1984, pp. 645–649.

[17] Thurston, R. V., Russo, R. C., Luedtke, R. J., Smith, C. E., Meyn, E. L., Chakoumakos, C., Wang, K. C., and Brown, C. J. D., *Transactions of the American Fisheries Society,* Vol. 113, No. 1, 1984, pp. 56–73.

[18] Carlson, R. W. and Drummond, R. A., *Water Research* Vol. 12, No. 1, 1978, pp. 1–6.

[19] McKim, J. M., *Journal of the Fisheries Research Board of Canada,* Vol. 34, No. 8, 1977, pp. 1148–1154.

[20] Macek, K. J. and Sleight, B. H. III, in *Aquatic Toxicology and Hazard Evaluation (First Symposium), ASTM STP 634,* American Society for Testing and Materials, Philadelphia, 1977, pp. 137–146.
[21] Birge, W. J., Black, J. A., and Westerman, A. G., *Environmental Toxicology and Chemistry,* Vol. 4, No. 6, 1985, pp. 807–821.
[22] Norberg, T. J. and Mount, D. I., *Environmental Toxicology and Chemistry,* Vol. 4, No. 5, 1985, pp. 711–718.
[23] Marshall, J. S., *Ecology,* Vol. 43, 1962, pp. 598–607.
[24] Hummon, W. D., *American Midland Naturalist,* Vol. 92, No. 2, 1974, pp. 327–339.
[25] Winner, R. W. and Farrell, M. P., *Journal of the Fisheries Research Board of Canada,* Vol. 33, No. 8, 1976, pp. 1685–1691.
[26] Bertram, P. E. and Hart, B. A., *Environmental Pollution,* Vol. 19, No. 4, 1979, pp. 295–305.
[27] Daniels, R. E. and Allan, J. D., *Canadian Journal of Fisheries and Aquatic Sciences,* Vol. 38, No. 5, 1981, pp. 485–494.
[28] Gentile, J. H., Gentile, S. M., Hairston, N. G., Jr., and Sullivan, B. K., *Hydrobiologia,* Vol. 93, No. 1/2, 1982, pp. 179–187.
[29] McNaught, D. C. and Mount, D. I. in *Aquatic Toxicology and Hazard Assessment (Eighth Symposium), ASTM STP 891,* American Society for Testing and Materials, Philadelphia, 1985, pp. 375–381.
[30] van Leeuwen, C. J., Moberts, F., and Niebeek, G., *Aquatic Toxicology,* Vol. 7, No. 3, 1985, pp. 165–175.
[31] Yu, J. and Hirayama, K., *Bulletin of the Japanese Society of Scientific Fisheries,* Vol. 52, No. 9, 1986, pp. 1509–1513.
[32] Meyer, J. S., Ingersoll, C. G., and McDonald, L. L., *Environmental Toxicology and Chemistry,* Vol. 6, No. 2, 1987, pp. 115–126.
[33] Cheng, Z., Li, B., Zou, C., and Que, T., *Ecological Modelling,* Vol. 34, No. 3/4, 1986, pp. 191–196.
[34] Vandermeer, J., *Elementary Mathematical Ecology,* Wiley, New York, NY, 1981.
[35] Krebs, C. J., *Ecology,* Harper and Row, New York, NY, 1972, pp. 164–208.
[36] Caswell, H., *Limnology and Oceanography,* Vol. 17, No. 5, 1972, pp. 787–791.
[37] Edmondson, W. T., *Limnology and Oceanography,* Vol. 17, No. 5, 1972, pp. 792–795.
[38] Smith, R. L., *Ecology and Fish Biology,* 2nd ed., Harper and Row, New York, NY, 1974, pp. 316–328.
[39] Ginzburg, L. R. and Golenberg, E. M., *Lectures in Theoretical Population Biology,* Prentice-Hall, Englewood Cliffs, NJ, 1985.
[40] Cole, L. C., *Quarterly Review of Biology,* Vol. 29, No. 2, 1954, pp. 103–137.
[41] Mount, D. I., *Environmental Toxicology and Chemistry,* Vol. 6, No. 2 1987, p. 83.
[42] Stephan, C. E. and Rogers, J. W. in *Aquatic Toxicology and Hazard Assessment (Eighth Symposium), ASTM STP 891,* American Society for Testing and Materials, Philadelphia, 1985, pp. 328–338.
[43] Gillespie, R. B. and Baumann, P. C., *Transactions of the American Fisheries Society,* Vol. 115, No. 2, 1986, pp. 208–213.
[44] Landner, L., Neilson, A. H., Sorensen, L., Tarnholm, A., and Viktor, T., *Ecotoxicology and Environmental Safety,* Vol. 9, No. 3, 1985, pp. 282–293.
[45] McKim, J. M., Olson, G. F., Holcombe, G. W., and Hunt, E. P., *Journal of the Fisheries Research Board of Canada,* Vol. 33, No. 12, 1976, pp. 2726–2739.
[46] Chapman, D. W. in *Methods for Assessment of Fish Production in Fresh Waters,* 2nd ed., W. E. Ricker, Ed., IBP Handbook No. 3, Blackwell Scientific Publications, Oxford, United Kingdom, 1971, pp. 199–214.
[47] Hamelink, J. L., Buckler, D. R., Mayer, F. L., Palawski, D. U., and Sanders, H. O., *Environmental Toxicology and Chemistry,* Vol. 5, No. 1, 1986, pp. 87–94.
[48] Suter, G. W., Rosen, A. E., Linder, E., and Parkhurst, D. F., *Environmental Toxicology and Chemistry,* Vol. 6, No. 10, 1987, pp. 793–809.
[49] DeGraeve, G. M., Cooney, J. D., Pollock, T. L., Reichenbach, N. G., Dean, J. H., Marcus, M. D., and McIntyre, D. O., "Intra- and Interlaboratory Study to Determine the Reproducibility of the Seven-Day Fathead Minnow Larval Survival and Growth Test," Battelle, Columbus, OH, 1988.
[50] Spehar, R. L. and Fiandt, J. T., *Environmental Toxicology and Chemistry,* Vol. 5, 1986, pp. 917–931.
[51] Hansen, D. J. in *Concepts in Marine Pollution Measurements,* H. H. White, Ed., University of Maryland, College Park, MD, 1985, pp. 33–56.

Michael T. Barbour,[1] *Carol G. Graves,*[1] *and Wayne L. McCulloch*[1]

Evaluation of the Intrinsic Rate of Increase as an Endpoint for *Ceriodaphnia* Chronic Tests

REFERENCE: Barbour, M. T., Graves, C. G., and McCulloch, W. L., "**Evaluation of the Intrinsic Rate of Increase as an Endpoint for *Ceriodaphnia* Chronic Tests,**" *Aquatic Toxicology and Environmental Fate: Eleventh Volume, ASTM STP 1007,* G. W. Suter II and M. A. Lewis, Eds., American Society for Testing and Materials, Philadelphia, 1989, pp. 273–288.

ABSTRACT: Fecundity and survival data can be used to calculate a number of demographic measures which characterize populations including net replacement rate, generation length, life expectancy, and reproductive value. Another measure is r, alternately called the intrinsic or instantaneous rate of population increase. This statistic is a measure of the capacity of a population to increase in an unlimited environment and is an estimate of ecological growth potential. The equation of Lotka (1925) for calculating the intrinsic rate of increase is

$$1 = \sum_{x=0}^{n} l_x m_x e^{-rx}$$

Where n is the number of age classes, l_x is the probability of survival from birth to Age Class x, and m_x is the average number of births per individual in Age Class x.

The intrinsic rate of increase, r, has been calculated for >50 chronic toxicity tests on effluents from publicly owned treatment works (POTWs), metal refinishers, and acid-mine drainage. The hypothesis was that the intrinsic rate of increase is a sensitive endpoint for estimating toxicity to populations exposed to toxicants in laboratory experiments. Results of analyses performed on seven-day *Ceriodaphnia* tests indicate that r is actually less sensitive than the other chronic endpoints examined for toxicity tests with limited observation days which serve as age classes. These results suggest that for the type of toxicity test examined (short-term chronic), the calculation of r does not add to the interpretation of toxic response that can be made from evaluating fecundity and survival data as separate measures. Although r is a useful chronic endpoint, it is not cost effective to calculate when simpler estimates of toxicity are available.

KEY WORDS: intrinsic rate of increase, toxicity testing, chronic toxicity, lethality, complex effluents, fecundity, *Ceriodaphnia,* chronic endpoint, aquatic toxicology

Chronic toxicity tests are used to assess long-term responses to contaminants. Although lethality is used as an endpoint in chronic exposure tests, reproductive inhibition or growth retardation are generally considered more sensitive measurements, particularly for the estimation of sublethal response.

Reproductive inhibition is usually measured by examining the number of young produced during exposure. The toxicant effect on growth can also be quantified by measuring weight gain during the period of exposure. Other measures of sublethal response that are used less commonly include immobilization and respiratory stress. The intrinsic rate of

[1] Senior scientists, EA Engineering, Science, and Technology, Inc., Hunt Valley/Loveton Center, 15 Loveton Circle, Sparks, MD 21152.

natural increase is an ecological measure which expresses the growth potential of a population in an unlimited environment (optimum conditions where competition and predation are not influencing factors). The intrinsic rate of increase incorporates information of survival and reproductive success of a population and has been used in ecological theory since 1930 [*1*]. Its use in toxicity testing as a chronic endpoint is not unique [*2–5*]; however, to date, the intrinsic rate of increase has not received a great amount of attention in toxicity testing.

The intrinsic rate of increase is calculated from life table information according to the Euler equation

$$1 = \sum_{x=0}^{n} l_x m_x e^{-rx}$$

where

l_x = probability of population surviving to age x,
m_x = average fecundity at age x,
r = intrinsic (per capita) rate of increase, and
n = number of age classes in population.

We hypothesized that the intrinsic rate of increase, r, is a sensitive means of estimating toxicity to populations exposed to toxicants in laboratory experiments. Because r incorporates both survival, l_x, and reproductive success, m_x, r theoretically should be a more accurate estimate of toxicity than either measurement alone. Laboratory testing should be especially conducive to estimating r because environmental conditions of space, food, and species interactions are controlled, thereby providing a situation where the specific growth rate can be maximized and constant, and constrained only by a response of toxicity.

The calculation of r in toxicity testing is dependent on the use of a test organism capable of producing several broods. The relatively recent development of a short-term chronic test using the cladoceran *Ceriodaphnia* to evaluate reproductive potential [*6*] provided a means of examining the calculation of r in comparison to the standard endpoints of the number of young produced and survival. The advantage of the *Ceriodaphnia* reproductive potential test is that three broods are produced by individual females over a test period of seven days. Complex effluents from publicly-owned treatment works (POTWs), acid-mine drainage, and metal refinishing plants were used as toxicants to evaluate toxic response using the intrinsic rate of increase, fecundity, and survival as chronic endpoints.

Methods

Complex effluents from various sources were used as experimental test solutions. Effluents from POTWs, a metal-refinishing plant, and acid-mine drainage were used to evaluate differences and similarities in the chronic endpoints. Each toxicity test consisted of a control and a series of graded concentrations of the complex effluent. The seven-day reproductive potential test for *Ceriodaphnia* was used to calculate the intrinsic rate of increase, r, fecundity, and survival for this study. The procedures for conducting toxicity tests for *Ceriodaphnia* were as described by Mount and Norberg [*6*]. Daily observations were conducted for the tests involving effluents from one of the POTWs. All other tests had observations on Days 0, 3, 5, and 7.

Percent survival was calculated for the control and for each concentration of each test as an estimate of a lethal endpoint. Fisher's Exact Test [*7*] was used to determine which treatments caused effects that were statistically different from the controls at $P = 0.05$ for each test, based on mortality as the endpoint. The procedure utilizes a 2 by 2 contingency table to provide an estimate of equality of two survival proportions, assuming independence of responses from a binomial population.

Fecundity was calculated in two ways. The first method (young per total females) is the standard procedure of calculating young production and is based on the original number of females in each test concentration. If a female dies before producing young, she is still included in the calculation. Thus, this procedure incorporates both lethality and fecundity effects. The second method calculates young per female based only on young that are produced by females that survived the entire seven-day period. Young produced by females that die during the test are not used in this calculation. This method reflects toxicant effects on fecundity alone. Standard error was calculated for each mean, that is, for each concentration in each test, to give an estimate of the variation among the replicates of each test concentration. Statistical significance for both fecundity methods was calculated in the same way that statistical significance for r was calculated. Each concentration was compared with the corresponding control using a two-tailed t-test for the comparison of means from two independent samples of potentially unequal sizes.

The intrinsic rate of increase was calculated using the procedures of Meyer et al. [*8*]. Because r is part of the exponent in the summation term in the Euler equation, the equation cannot be solved explicitly for r. Therefore, the value of r is calculated through an iterative process. This "bootstrap" method recombines the original data to randomly select data for calculation of pseudovalues for an estimation of r. After 500 pseudovalues of r are determined in this way, the mean and standard error of the estimates of r are calculated. Conceptually, this random sampling of the original data serves as a hypothetical resampling of the entire population. In this way, an r and the corresponding standard error are calculated for the control and each concentration in a toxicity test.

Each concentration is compared in turn to the control using the following t-test for the comparison of means of two independent samples of unequal sizes.

$$t = \frac{r_c - r_i}{\sqrt{\left(\frac{(n_c - 1)n_c s_c^2 + (n_i - 1)n_i s_i^2}{n_c + n_i - 2}\right)\left(\frac{n_c + n_i}{n_c n_i}\right)}}$$

with $n_c + n_i - 2$ degrees of freedom, where

r_c = intrinsic rate of increase in the control group,
r_i = intrinsic rate of increase in the ith concentration,
n_c, n_i = number of replicates in the control group, n_c, and the ith concentration, n_i,
s_c, s_i = the standard errors of r in the control group, s_c, and ith concentration, s_i.

The r_i is considered significantly different from the r_c if the calculated t is greater than that of the tabled value of t for $n_c - n_i - 2$ degrees of freedom and for $P = 0.05$.

The no-observed-effect concentration (NOEC), defined as the highest test concentration which did not cause a statistically significant effect, was determined for each of the four chronic endpoints for 53 individual chronic *Ceriodaphnia* tests. For each test, the four endpoints were then ranked, based on their respective NOECs, with the most sensitive chronic endpoint (lowest NOEC) receiving a score of 1 and the least sensitive chronic endpoint receiving a score of 4. When more than one endpoint had the same NOEC, a mean

TABLE 1—*Demographic information for* Ceriodaphnia *used in toxicity tests performed on whole effluents.*

		Total Females			Surviving Females			
			Total Brood Size			Total Brood Size		
Concentration, percent effluent	Percent Survival	n	No. of Young	$\overline{X}$ (±SE)	n	No. of Young	$\overline{X}$ (±SE)	r (±SE)
FIRST POTW EFFLUENT—AUGUST 1983								
Control	100	9	101	11.2 (1.10)	9	101	11.2 (1.10)	0.59 (0.07)
1.0	100	10	131	13.1 (0.38)	10	131	13.1 (0.38)	0.68 (0.08)
3.0	100	10	127	12.7 (0.41)	10	127	12.7 (0.41)	0.81 (0.04)
10.0	20*	10	74	7.4 (1.17)*	2	13	6.5 (2.47)	0.71 (0.08)
30.0	0*	10	0	0*	0	0	0*	0.73 (0)
100.0	0*	10	0	0*	0	0	0*	0*
METAL-REFINISHER EFFLUENT—AUGUST 1983								
Control	77.8	9	72	8.0 (1.60)	7	72	10.3 (0.76)	0.62 (0.07)
1.0	70.0	10	76	7.6 (1.77)	7	76	10.9 (0.91)	0.56 (0.09)
3.0	90.0	10	105	10.5 (1.42)	9	105	11.7 (0.93)	0.69 (0.09)
10.0	88.9	9	85	9.4 (1.50)	8	85	10.6 (1.03)	0.68 (0.07)
30.0	33.3	9	29	3.2 (1.07)*	3	14	4.7 (1.33)*	0.35 (0.10)*
100.0	0*	10	0	0*	0	0	0*	0*
THIRD POTW EFFLUENT—MAY 1986								
Control	88.9	9	275	30.6 (3.43)	8	270	33.8 (1.41)	0.82 (0.03)
0.7	100	10	332	41.5 (0.95)	10	332	41.5 (0.95)	0.86 (0.01)
1.5	90	10	273	34.1 (2.72)	9	253	36.1 (2.30)	0.86 (0.01)
3.2	90	10	286	35.8 (1.83)	9	253	36.1 (2.03)	0.85 (0.01)
6.5	50	10	241	24.1 (1.90)	5	132	26.4 (1.03)*	0.80 (0.01)
13.5	0*	10	0	0*	0	0	0*	0*
40.5	0*	10	0	0*	0	0	0*	0*

				THIRD POTW EFFLUENT—OCTOBER 1986				
Control	90	10	339	33.9 (3.51)	9	334	37.1 (1.60)	0.81 (0.02)
0.2	100	10	350	35.0 (2.34)	10	350	35.0 (2.34)	0.77 (0.01)
0.8	100	10	303	30.3 (3.23)	10	303	30.3 (3.23)	0.77 (0.03)
1.6	50	10	222	22.2 (1.58)*	5	104	20.8 (2.01)*	0.74 (0.02)*
3.2	0*	10	48	4.8 (1.08)*	0	0	0*	0.46 (0.06)*
6.5	0*	10	0	0*	0	0	0*	0*
3.5	0*	10	0	0*	0	0	0*	0*
40.5	0*	10	0	0*	0	0	0*	0*
				SECOND POTW EFFLUENT—AUGUST 1983				
Control	90	10	114	11.4 (1.83)	9	114	12.7 (1.50)	0.77 (0.06)
1.0	100	10	104	10.4 (1.33)	10	104	10.4 (1.33)	0.80 (0.07)
3.0	100	10	137	13.7 (0.66)	10	137	13.7 (0.66)	0.89 (0.06)
10.0	90	10	127	12.7 (1.49)	9	127	14.1 (0.50)	0.87 (0.07)
30.0	80	10	62	6.2 (0.95)*	8	55	6.9 (1.06)*	0.75 (0.08)
100.0	80	10	30	3.0 (0.60)*	8	28	3.5 (0.60)*	0.23 (0.06)*
				ACID-MINE DRAINAGE—JULY 1984				
Control	80	10	212	21.2 (1.90)	8	190	23.8 (1.03)	1.61 (0.04)
0.032	100	10	202	20.2 (1.45)	10	202	20.2 (1.45)	1.47 (0.06)
0.056	100	10	223	22.3 (1.26)	10	223	22.3 (1.26)	1.54 (0.05)
0.10	100	10	236	23.6 (1.26)	10	236	23.6 (1.26)	1.63 (0.03)
0.18	100	10	199	19.9 (1.26)	10	199	19.9 (1.26)*	1.44 (0.05)*
0.32	90	10	124	12.4 (1.74)*	9	124	13.8 (1.20)*	0.93 (0.12)*
				ACID-MINE DRAINAGE—NOVEMBER 1984				
Control	100	10	121	12.1 (1.77)	10	121	12.1 (1.77)	0.94 (0.17)
5.0	70	10	75	7.5 (2.30)	7	73	10.4 (2.55)	0.51 (0.10)*
12.5	90	10	42	4.2 (1.34)*	9	36	4.0 (1.48)*	0.25 (0.06)*
20.0	90	10	13	1.3 (0.70)*	9	13	1.4 (0.77)*	0.09 (0.11)*
27.0	80	10	1	0*	8	1	0*	0*
34.0	78	9	0	0*	7	0	0*	0*

NOTE—An asterisk indicates that the treatment group is significantly different ($P \leq 0.05$) from the control.

of the scores was assigned to each endpoint. For instance, if three of the chronic endpoints tied as the most sensitive for a given test, the mean score was determined ($[1 + 2 + 3]/3 = 2$), and each chronic endpoint was assigned a score of 2, whereas the remaining chronic endpoint received a score of 4. The scores for the 53 individual tests were totaled and averaged.

Results

Toxicity tests that are representative of various effluents were selected for presentation in this paper. The demographic data for these seven tests discussed in this paper are summarized in Table 1. The test concentrations (percent effluent) ranged from a low concentration series of 0.032 to 0.32% for acid-mine drainage effluent to a high series of 1 to 100% for metal refinisher and POTW effluents. Calculated values for the intrinsic rate of increase, *r*, remained below 1.00 except for the acid-mine drainage effluent for July 1984 (Table 1). The value for *r* is dependent upon the number of females surviving to each age class and on the number of young produced by each female within each age class. In these studies, the relative importance of *r* is in the comparison of concentrations within each test series and not between test series.

The standard errors calculated for *r* were lower than those calculated for the fecundity endpoints. However, the coefficient of variation (SE/mean) were remarkably constant for all tests and for all measures of fecundity and *r*. For this reason and because the same *t*-test was used to determine statistical significance for the fecundity measures and *r*, the authors feel that the relative sensitivity of the endpoints is not due to differences in the magnitude or the variance of the endpoints.

Of the tests presented in this study, the results of four tests produced a similar response curve. In these tests, effects on *r* and fecundity were not significant ($P < 0.05$) until lethality occurred. The first POTW effluent produced a statistically significant ($P < 0.05$) response in young per surviving females at a concentration of 30% (Fig. 1). Survival and young per total females were significantly reduced at 10% effluent. At the 10% effluent concentration, the value for *r* was not significantly less than the control, although mortality was significant from the control (Fig. 1). This was a result of females producing young and then dying.

The values for *r* and both types of fecundity measures are statistically significant at a concentration of 30% for oganisms exposed to the effluent of a metal refinisher (Fig. 2). Lethality is significant at 100%. The third POTW effluent (May and October 1986) illustrated this same response (Figs. 3 and 4). In the May example, fecundity measured as young per surviving females was found to be more sensitive than young per total females because the control mortality of females lowered the average young per total female. The intrinsic rate of increase also accounted for the control mortality and reflected the same level of sensitivity as number of young per surviving female.

A sublethal response (reproductive inhibition) was obtained prior to lethality with *Ceriodaphnia* exposed to effluent from the second POTW (Fig. 5). Both estimates of fecundity were more sensitive than *r*. An effect level of 30% effluent was obtained with fecundity; *r* was not significantly affected until 100% effluent. No statistically significant effect on survival was noted. Exposure to an acid-mine drainage effluent (July 1984 test) also produced effects on fecundity prior to lethality, similar to that of the second POTW. In this test, the most sensitive endpoints were *r* and the number of young per surviving female (Fig. 6).

In the November 1984 acid-mine drainage effluent test (Fig. 7), the *r* endpoint was found to be the most sensitive, showing a significant effect at 5% effluent, while a significant effect at 12.5% was evident for both types of fecundity measures. Essentially, no adverse survival effects were noted. The lethality at 5% effluent was an aberrant factor and contributed to

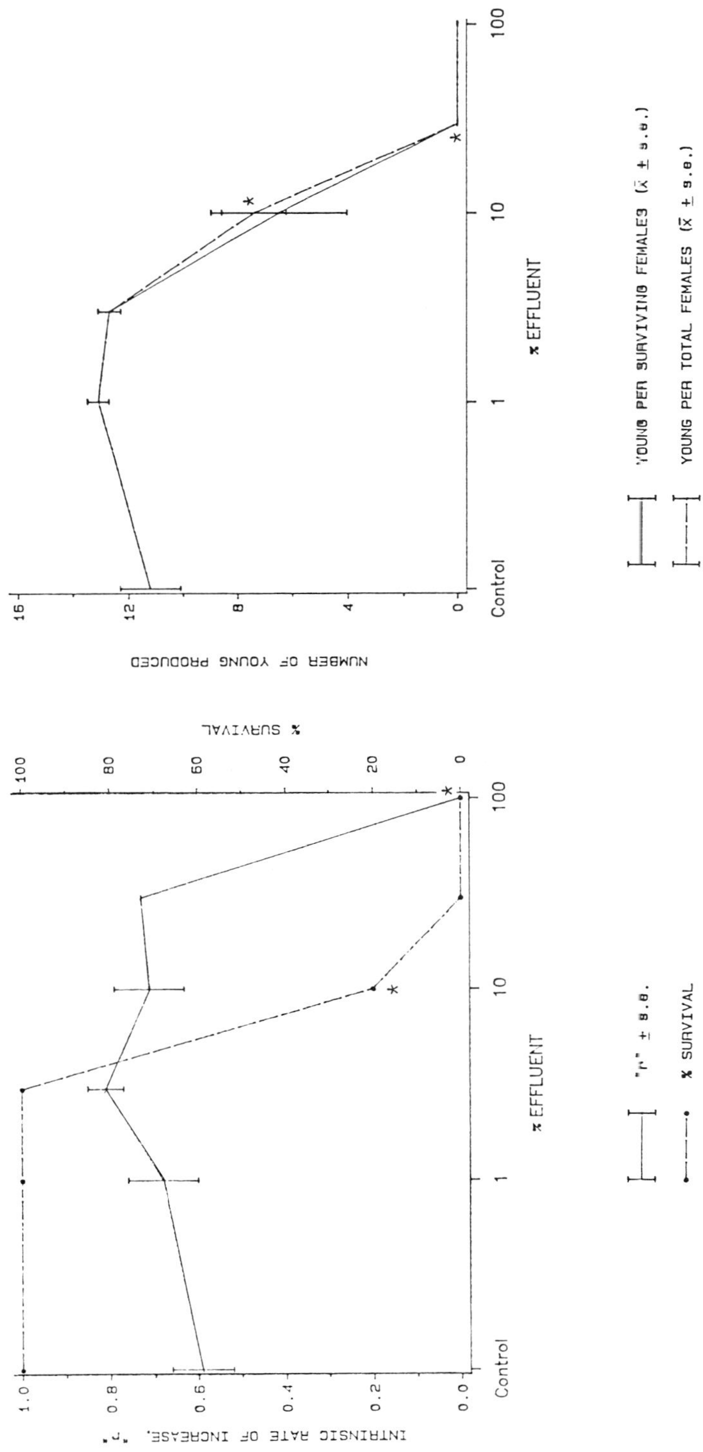

FIG. 1—*Toxicity test results from first POTW effluent—August 1983.*

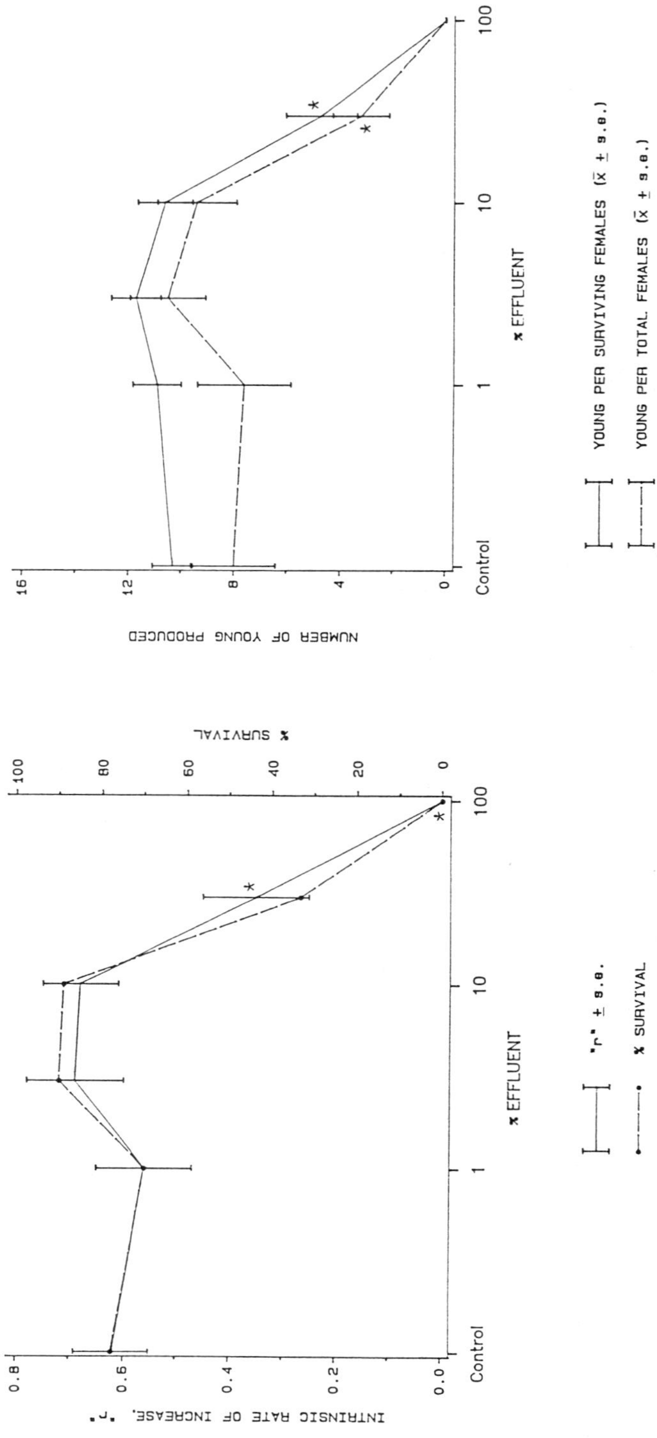

FIG. 2—*Toxicity test results from metal-refinisher effluent—August 1983.*

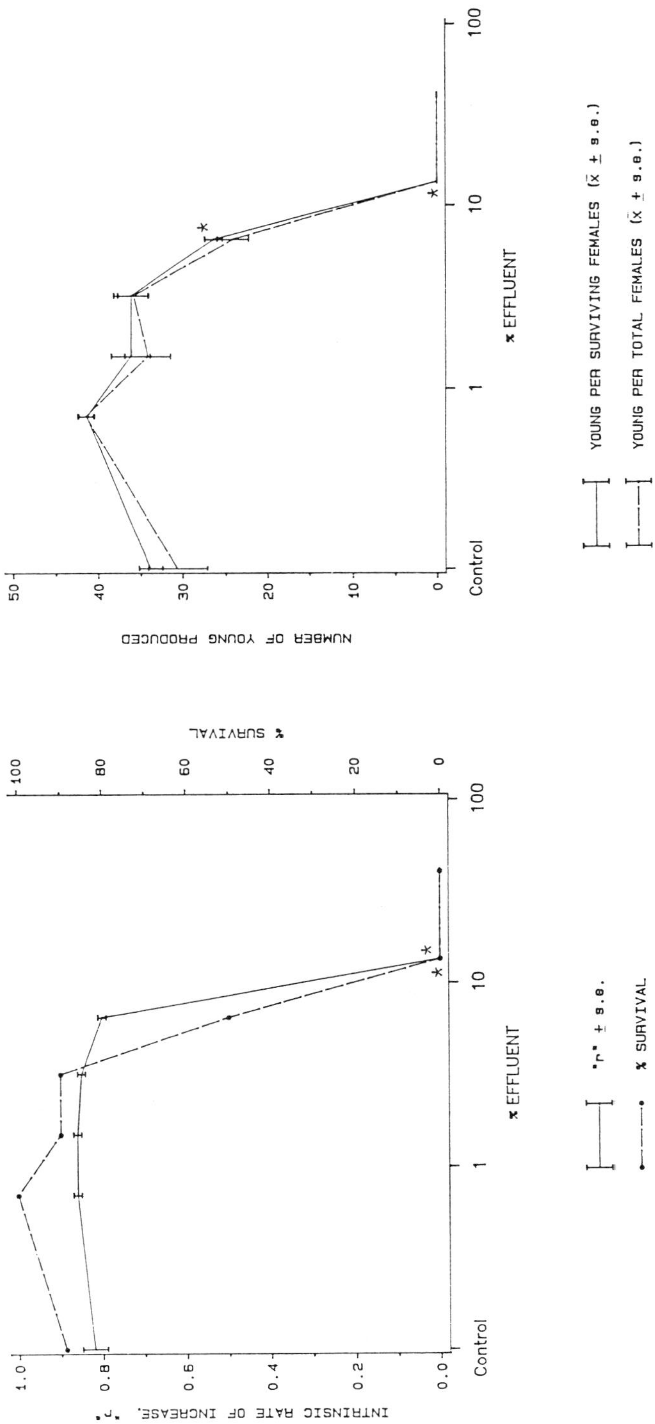

FIG. 3—*Toxicity test results from third POTW effluent—May 1986.*

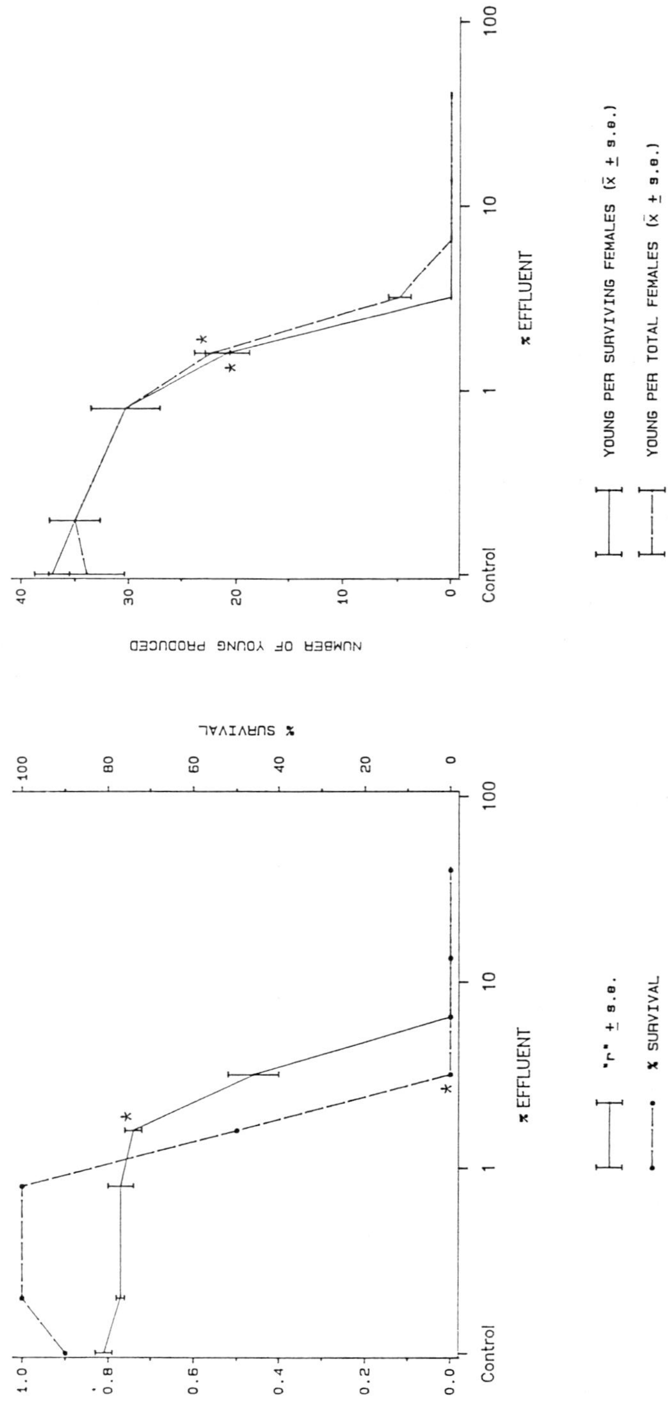

FIG. 4—*Toxicity test results from third POTW effluent—October 1986.*

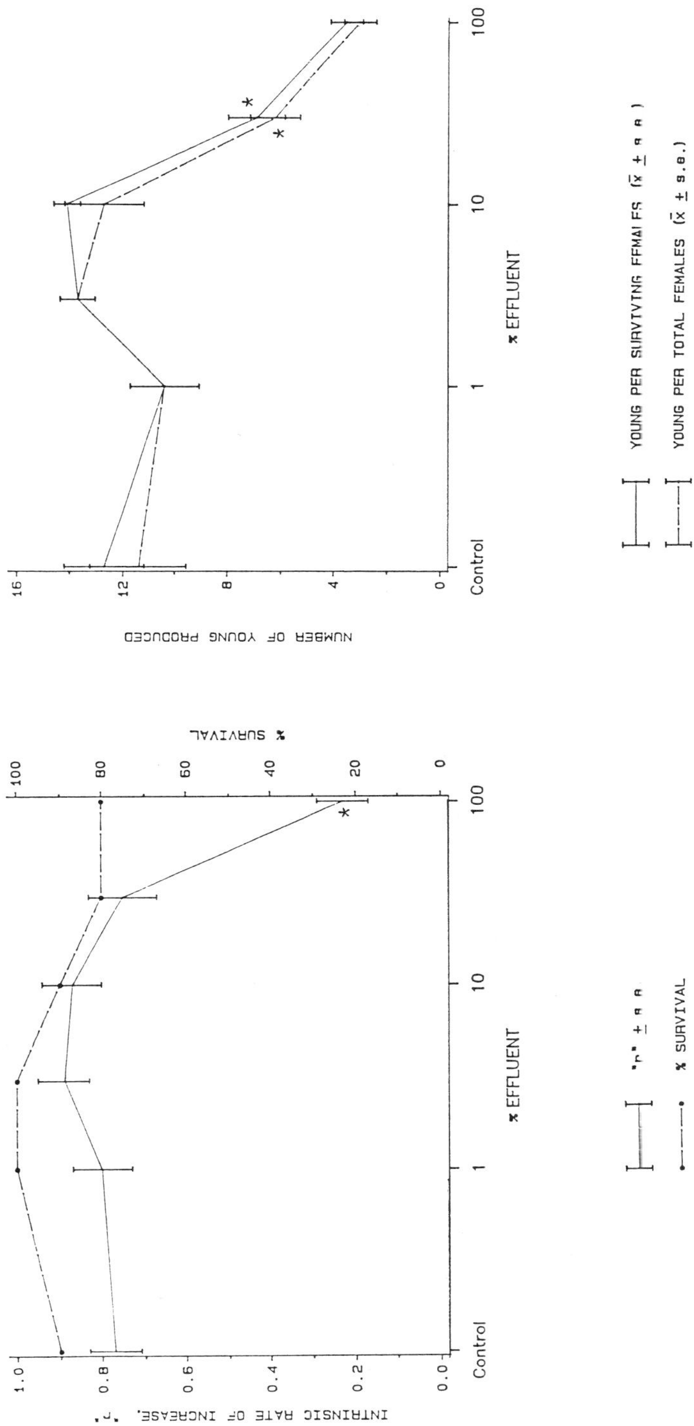

FIG. 5—*Toxicity test results from second POTW effluent—August 1983.*

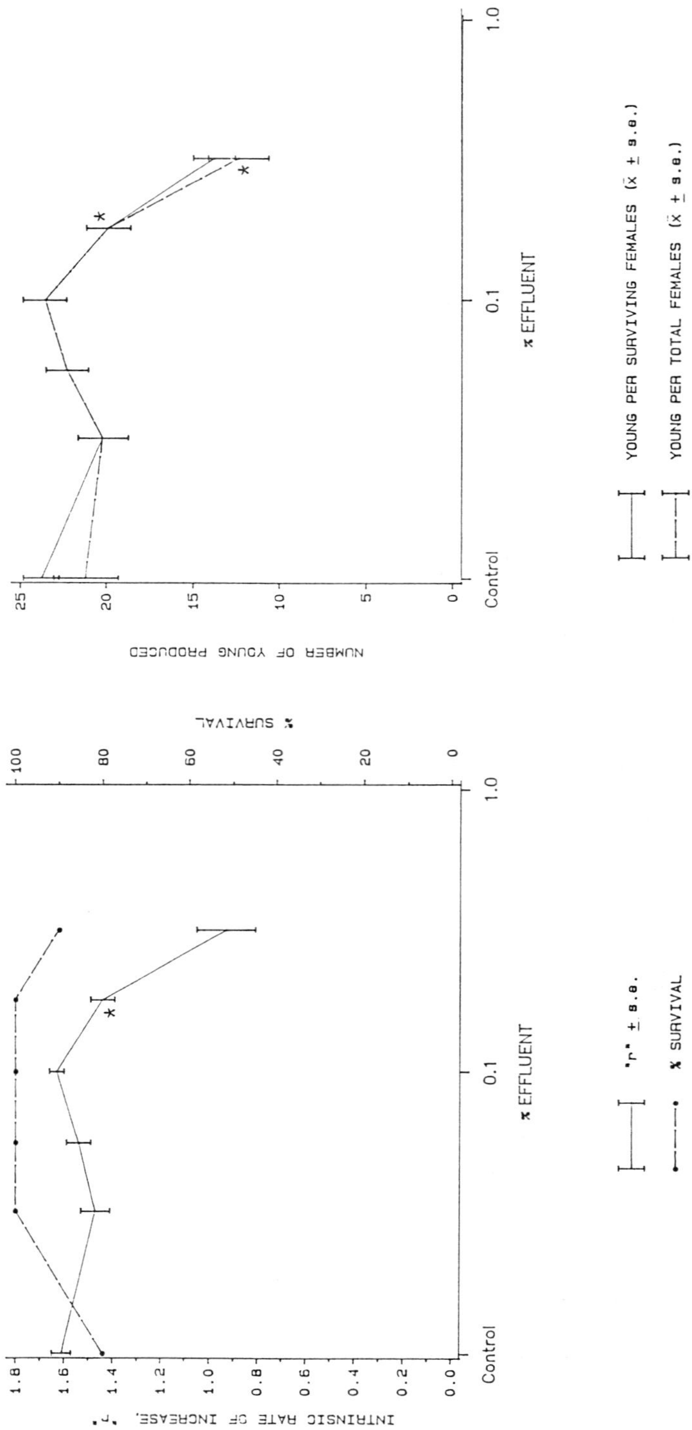

FIG. 6—*Toxicity test results from acid-mine drainage—July 1984.*

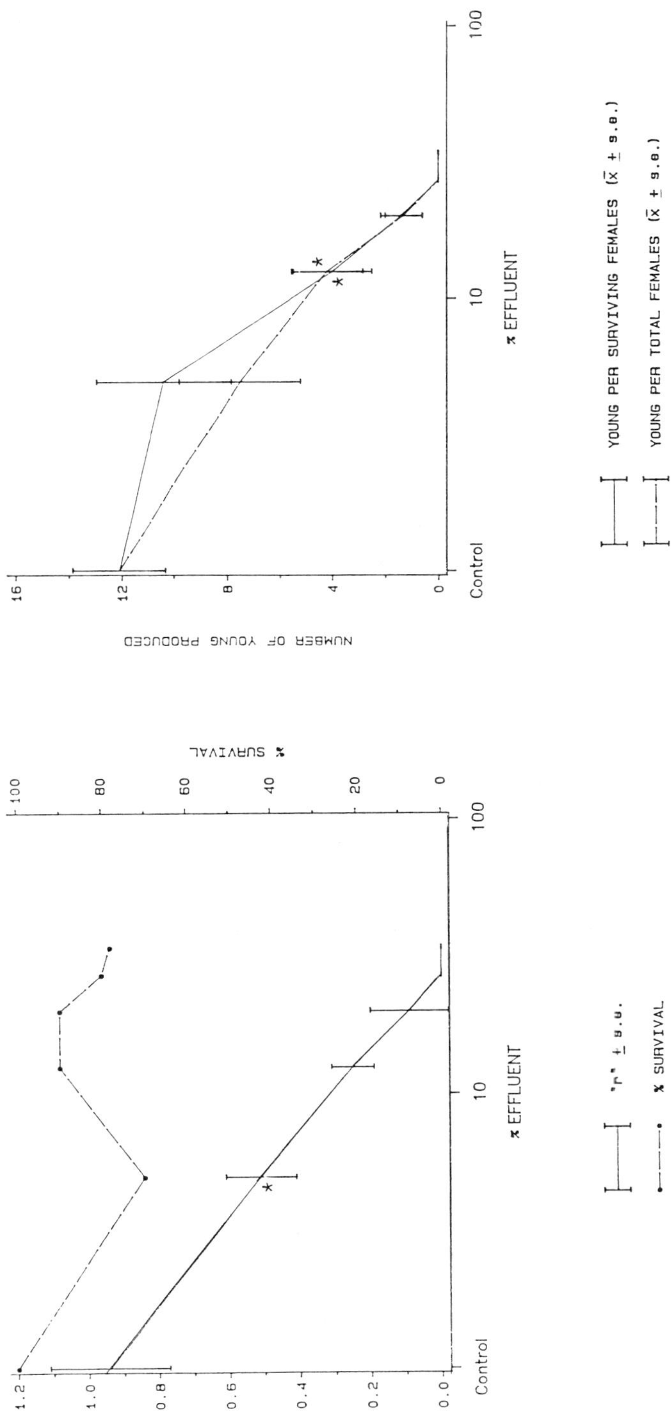

FIG. 7—*Toxicity test results from acid-mine drainage—November 1984.*

the sensitivity displayed by *r* for this test. Had survival been higher, as suggested by that obtained with higher concentrations of the effluent, the sensitivity of *r* may have been more similar to that of fecundity.

Results of 53 tests were evaluated for estimates of toxicity using *r*, mortality, and fecundity. These 53 tests were incorporated into a ranking scheme to compare relative sensitivities of the endpoints. In these tests, *r* was ranked as the least sensitive of the four endpoints (Table 2), having a mean score of 2.69 compared with 2.52 for mortality, 2.49 for fecundity per total females, and 2.30 for fecundity per surviving females. The intrinsic rate of increase was the most sensitive endpoint in only one test and was among the most sensitive in six other tests. The number of young produced (fecundity) per surviving females was calculated from these *Ceriodaphnia* tests to be the most sensitive endpoint (Table 2).

Discussion

Many investigators have used the intrinsic rate of increase, *r*, as a measurement of chronic effects from exposure to various chemicals [*2–5,9*]. These investigators found that a reduction in *r* was influenced by a combination of lowered survivorship, delayed onset of reproduction, and reduced fecundity. Daniels and Allan [*2*] found that *r* appeared to be an ecologically realistic measure of sublethal stress and required an equivalent or shorter time to conduct than conventional long-term tests. McNaught and Mount [*10*] also suggested that the use of the intrinsic rate of increase as a chronic endpoint might be used to abbreviate the period of observation necessary for toxicity testing.

The intent of this study was to compare various chronic endpoints for relative sensitivities in exposure experiments using complex effluents. In most of these tests, the intrinsic rate of increase was not more sensitive than the number of young per surviving female or young per total females. However, *r* was more sensitive than the other points in one test (Fig. 7). Because *r* integrates survival and fecundity, it was expected that *r* would be a more sensitive endpoint which would enable a closer approximation to chronic effects than either fecundity or survival alone. Based on the chronic data we analyzed, the calculation of *r* did not add to the estimation of chronic toxicity over current techniques.

TABLE 2—*Summary of a comparison of relative sensitivities of chronic endpoints.*[a,b]

		Chronic Endpoints		
	r	Fecundity per Total Females	Fecundity per Surviving Females	Mortality
Mean score[c]	2.69	2.49	2.30	2.52
Overall rank[d]	4	2	1	3
Single most sensitive	1	0	0	4
Most sensitive with one other	4	3	8	2
Most sensitive with two others	2	9	8	6
Least sensitive alone	8	0	0	5

[a] Number of tests = 53.
[b] Entire database available in table form (as in Table 1) from authors upon request.
[c] Average of scores attained for each endpoint from all 53 tests.
[d] Ranked in comparison among the four endpoints.

One possible explanation for the relative insensitivity of r as a measure of toxicity may be the manner in which components of the Euler equation are calculated by the method of Meyer et al. [*8*]. The calculation of m_x does not take into account mortality in the last age class, that is, the number alive in the last age class is not used in the calculation. This decreases the value of m_x in the equation. Thus, in tests where the onset of reproduction is delayed, or where fecundity production is maximum at the end of the test, r may be artificially low. A more pronounced effect on the value of r is the way in which the midpoints of the age classes are calculated. In most of the toxicity tests analyzed for this paper, observations were made on Days 0, 3, 5, and 7. Because there is more than one time unit (that is, more than one day) between observations, midpoints of age classes need to be used. In the computer program used for these calculations, age class midpoints are not calculated in the same way for all age classes. In these computations, the first age class midpoint is smaller than it should be in relation to later age class midpoints. Because age class enters the calculation as a factor in a negative exponent in each summation term, the larger the exponent, the smaller the term in the summation. This means that the term for the earlier age class is larger and, therefore, dominates the calculation of r. If results are similar on Day 3 for the control and for a concentration and then diverge on Day 5, this calculation of r will not be sensitive to this later difference in response to toxicity.

As discussed above, the calculation of r gives more weight to the early broods of young and less weight to the later broods. From a long-term population dynamics perspective, this is an important feature of the equation. In *Ceriodaphnia* tests for both control and exposed organisms, the early broods are generally consistent in their timing (Day 3 to 4) and size (3 to 5 young). Conversely, the later broods are considerably larger (6 to 16 young), tend to vary more than the early broods between exposure concentrations, and have a greater contribution to the total young produced. The small early broods that are more comparable across concentrations are emphasized by the calculation, while the larger, later broods are deemphasized. For this reason, we feel that the use of r as a chronic endpoint requires frequent observations (age classes) and may not be an accurate predictor of chronic effects in shortened toxicity tests. The authors are reevaluating the methodology used to calculate r to see if refinements will make r a more sensitive endpoint.

Acknowledgments

We are indebted to Joe Meyer for sending us a copy of his bootstrap program for calculating r and for giving us helpful suggestions as we progressed with this study. This study was supported by EA Engineering, Science, and Technology, Inc. We are especially appreciative of the graphics done by Annie Deckret at Imagic, Inc., and for the extensive revisions to our manuscript made by Ann Carr.

References

[*1*] Fisher, R. A., *The Genetical Theory of Natural Selection,* Oxford University Press, Oxford, England, 1930.

[*2*] Daniels, R. E. and Allan, J. D., "Life Table Evaluation of Chronic Exposure to a Pesticide," *Canadian Journal of Fisheries and Aquatic Sciences,* Vol. 38, 1981, pp. 485–494.

[*3*] Allan, J. D. and Daniels, R. E. "Life Table Evaluation of Chronic Exposure of *Eurytemora affinis* (Copepoda) to Kepone," *Marine Biology,* Vol. 66, 1982, pp. 179–184.

[*4*] Walton, W. E., Compton, S. M., Allan, J. D., and Daniels, R. E., "The Effect of Acid Stress on Survivorship and Reproduction of *Daphnia pulex* (Crustacea: Cladocera)," *Canadian Journal of Zoology,* Vol. 60, 1982, pp. 573–579.

[*5*] Winner, R. W. and Farrell, M. P., "Acute and Chronic Toxicity of Copper to Form Species of *Daphnia,*" *Journal of the Fisheries Research Board of Canada,* Vol. 33, 1976, pp. 1685–1691.

[6] Mount, D. I. and Norberg, T. J., "A Seven-Day Life-Cycle Cladoceran Toxicity Test," *Environmental Toxicology and Chemistry,* Vol. 3, 1984, pp. 425–434.

[7] Horning, W. B., II and Weber, C. I., "Short-Term Methods for Estimating the Chronic Toxicity of Effluents and Receiving Waters to Freshwater Organisms," U.S. Environmental Protection Agency, Environmental Monitoring and Support Laboratory-Cincinnati, EPA/600/4-85-014, December 1985, p. 102.

[8] Meyer, J. S., Ingersoll, C. G., McDonald, L. L., and Boyce, M. S., "Estimating Uncertainty in Population Growth Rates: Jackknife vs. Bootstrap Techniques," *Ecology,* Vol. 67, No. 5, 1986, pp. 1156–1166.

[9] Hummon, W. D., "Effects of DDT on Longevity and Reproductive Rate in *Lepedodermella squammata* (Gastrotricha, Chaetonotida)," *American Midland Naturalist,* Vol. 92, No. 2, 1974, pp. 327–339.

[10] McNaught, D. C. and Mount, D. I., "Appropriate Durations and Measures for *Ceriodaphnia* Toxicity Tests," *Aquatic Toxicology and Hazard Assessment (Eighth Symposium), ASTM STP 891,* R. C. Bahner and D. J. Hansen, Eds., American Society for Testing and Materials, Philadelphia, 1985, pp. 75–381.

Lawrence W. Barnthouse,[1] Glenn W. Suter II,[1] and Aaron E. Rosen[1]

Inferring Population-Level Significance from Individual-Level Effects: An Extrapolation from Fisheries Science to Ecotoxicology

REFERENCE: Barnthouse, L. W., Suter, G. W. II, and Rosen, A. E., **"Inferring Population-Level Significance from Individual-Level Effects: An Extrapolation from Fisheries Science to Ecotoxicology,"** *Aquatic Toxicology and Environmental Fate: Eleventh Volume, ASTM STP 1007,* G. W. Suter II and M. A. Lewis, Eds., American Society for Testing and Materials, Philadelphia, 1989, pp. 289–300.

ABSTRACT: This paper discusses the concepts and operational definitions of significance used in fisheries management and environmental impact assessment, and demonstrates their applicability to contaminant hazard assessment. Methods of linking chronic test data to models commonly used in fisheries assessments are demonstrated, using the Chesapeake Bay striped bass population as a representative species of regulatory interest. These models can (1) integrate concentration-response functions for the separate components of life cycle chronic tests into a single measure of population-level effects, and (2) translate experimental variability into uncertainty concerning consequences of contaminant exposure. We compared population-level risk functions derived from five life-cycle toxicity data sets to maximum acceptable toxicant concentrations (MATCs) derived from the same five data sets. The comparison showed the MATC to be an inadequate measure of significant population-level effects because (1) several life stages frequently contribute to the effect associated with a given exposure, and (2) test responses such as fecundity, which are both highly sensitive and highly variable, are systematically undervalued in the calculation of the MATC. We discuss several approaches for enriching ecotoxicology through the use of population models and other methods outside the current domain of the field.

KEY WORDS: risk assessment, hazard assessment, population models, chronic toxicity, aquatic toxicology

What kinds of experimental tests, environmental data, and decision criteria are required to determine whether a toxic chemical poses a significant risk to aquatic biota? This question was first systematically addressed by ecotoxicologists at an ASTM-sponsored workshop ten years ago [*1*]. The workshop developed a conceptual scheme for hazard assessment based on the premises that (1) regulatory decisions should be based on comparisons between estimated exposure concentrations and concentrations shown to cause toxicological effects, and (2) toxicological threshold concentrations can be estimated using a sequentially graded series of tests of increasing completeness and realism. Most existing hazard assessment schemes for pesticides and toxic chemicals are based on the outline developed at the ASTM workshop.

This hazard assessment system depends on the assumption that the standard test endpoints represent, with some error, thresholds for significant effects. Therefore, it is sensitive

[1] Research staff members, Environmental Sciences Division, Oak Ridge National Laboratory, Oak Ridge, TN 37831.

to the formulation of test endpoints. The conventional formulation of test endpoints in terms of thresholds for statistically significant differences from controls has increasingly been challenged [*2,3*]. Suter et al. [*4*] pointed out inconsistencies in the assumption that conventional chronic test endpoints are protective of all life stages of fish in chronic exposures. Even more fundamentally, the wisdom of basing toxicity testing and hazard assessment schemes strictly on the detection of effects, with no consideration of importance or acceptability, has been challenged [*5*].

Controversies over the proper objectives of assessment and the data and methods needed to address those objectives are not unique to ecotoxicology. Because fish are the primary focus of most aquatic toxicology and hazard assessment, the similar controversies among scientists involved in fisheries management and environmental impact assessment are especially relevant. The problems of (1) determining optimal fishing effort, and (2) identifying "significant" impacts of power plants on fish populations are analogous to the problem of determining the significance of effects of toxic chemicals on fish populations. In this paper we show how the concepts and operational definitions developed to address the impacts of fishermen and power plants can be adapted and used to determine the significance (biological, societal, or both) of impacts caused by toxic chemicals.

Lessons from Fisheries Management and Environmental Impact Assessment

Determination of Optimal Fishing Effort

Thirty years ago most fisheries scientists believed that the objective of fisheries management was (or should be) to obtain the greatest possible yield of fish consistent with long-term maintenance of the populations. Most of the classic works in fish population biology (for example, Refs *6* and *7*) are devoted to developing the analytical tools needed to estimate the maximum sustained yield of fish populations. However, for reasons discussed by Larkin [*8*], this simplistic utilitarian view has now been abandoned.

Some of the difficulties with the sustainable yield approach to fisheries management reflect the needs of modern fisheries managers to consider interactions between exploited species, the effects of pollution and habitat modification, and the rehabilitation of depressed populations [*8,9,10*]. In addition, fisheries scientists have been forced to acknowledge the inevitable influence of economic [*11*] and even moral concerns on management decisions. With recognition of the complexity of fisheries management has come pessimism that coherent, value-free management objectives can be formulated.

Regardless of how they are determined, management objectives must be operationally defined in terms of the controllable aspects of the fishery. These may include the catch, the fishing effort, and the minimum size or age caught. One of the principal roles of fisheries scientists involved in the management process is to forecast the effects of the chosen controls (or of alternative controls being considered) on the population characteristics of ultimate management interest such as lifetime yield per recruit, exploitation rate, and spawning stock size. Performing this task almost inevitably requires the use of a mathematical model of the fished population. This is necessary because (1) the effects of fishing on a fished population cannot be experimentally measured, and (2) only a few of the properties of fish populations that influence their responses to exploitation are directly observable. The inadequacies of the commonly used fish population models and their management applications have been widely documented [*10,12,13*]. Nonetheless, there appear to be few alternatives to models for relating the available control measures to their effects on fish populations.

Defining the Significance of Environmental Impacts

The objectives of environmental impact assessment are (1) to identify adverse impacts of existing or proposed human activities on the environment, and (2) to communicate the significance of those impacts to decision makers and to the general public. Because of the critical importance of the second objective, much effort has been devoted to defining the term "significant."

The most straightforward definition is in terms of statistical significance, for example, between unimpacted and impacted stations or areas, between before and after samples from the same stations, or some combination of both. Despite their apparent objectivity and rigor, purely statistical definitions of significance are universally recognized as at best incomplete and at worst misleading. Zar [*14*] showed, in a manner similar to Skalski's [*2*] critique of hypothesis testing in ecotoxicology, that the probability of finding a given magnitude of difference between "impacted" and "control" data sets to be statistically significant is inversely related to the quantity and quality of the data. For typical fisheries data sets, where sample size is limited and variability is high, changes of 50% or more may be undetectable using conventional statistical criteria [*15,16*]. Conversely, given extremely precise data, very small and biologically inconsequential changes might be found statistically significant.

Most scientists who have discussed the problem of significance [*17–22*] have addressed the relationship between statistical, biological, and societal significance. Some [*19,20*] have argued that statistical significance is, if not a sufficient definition, at least a prerequisite for biological and societal significance. Others [*18,23*] have argued that statistical tests are usually of questionable value because meaningful replication is impossible. Some [*19*] have considered societal significance to be a secondary issue, to be determined by decision makers and the public after scientists have identified the statistically and biologically significant impacts. Others [*21,24*] have argued that scientists must assist in the determination of societal significance.

The one point on which all of the above authors appear to agree is that impacts worthy of being called biologically or societally significant involve adverse changes in populations, communities, and ecosystems. If changes in the relative abundance of benthic taxa between perturbed and control stations are noted, it is necessary to relate the observed changes to the function of the benthic community. If fish are predicted to be killed at a power plant, it is necessary to relate the predicted mortality to the abundance and persistence of the populations affected. As is the case in fisheries management, mathematical models often provide the only objective link between the phenomena that can be directly observed and the phenomena of interest in impact assessment. The history of mathematical modeling in environmental impact assessment is rather short and success has been uneven [*25*]. However, given their necessity, it can be expected that both the use of models and the quality of the models used in impact assessment will continue to grow.

Using Fish Population Models to Interpret Toxicity Test Results

The above discussion suggests that, because contaminant hazard/risk assessment is an integral part of the regulatory process, Stephan [*5*] is correct to emphasize the need to consider societal values in determining the significance of toxicity test data. Because societal values relate to the abundance and persistence of populations, communities, and ecosystems, it is necessary for significance to be determined at these higher levels of organization. Therefore, it is necessary not simply to extrapolate test results from the laboratory to the field, but to the abundance and persistence of populations in the field. As in fisheries management and impact assessment, models of the affected populations must be used.

Barnthouse et al. [*26*] described a general method for linking toxicity test data to fish population models. Although many types of fish population models can be coupled to these data, we used a simple model of proven utility in fisheries management and environmental impact assessment. The population-level variable employed here is the reproductive potential of a female recruit (a one-year-old female fish), defined as the expected contribution of that female to future generations, taking into account her annual probability of survival at different ages; her expected fecundity at different ages, provided that she survives; the probability that a spawned egg will hatch; and the probability that a newly hatched fish will survive to age one. The reproductive potential index could be defined in terms of any age (it would be numerically identical for any age prior to sexual maturity). Definition in terms of one-year-olds is convenient because this is the youngest age that can be studied using conventional sampling gear.

The ability of a fish population to sustain harvesting, to survive contaminant-induced mortality, and to persist in a variable environment is directly related to the reproductive potential of female fish. The reproductive potential of a one-year-old female recruit, P, is

$$P = S_0 \sum_{i=1}^{n} S_i E_i M_i \tag{1}$$

where

S_0 = probability of survival of eggs from spawning to age one year,
S_i = probability of survival of female fish from age one to age i,
E_i = average fecundity per mature female at age i,
M_i = fraction of age i females that are sexually mature, and
n = number of age classes in the population.

Toxic contaminants may reduce both the survival and fecundity of fish, and may affect all ages. The reproductive potential of a female recruit exposed to a toxic contaminant throughout her life cycle, P_s, is given by

$$P_s = S_0(1 - m_0) \sum_{i=1}^{n} S_i(1 - m_r)^{i-1} M_i f_i E_i \tag{2}$$

where

m_0 = probability of contaminant-induced mortality during the first year of life,
m_r = probability of contaminant-induced mortality for one-year-old and older fish, assumed equal for all age classes, and
f_i = fractional reduction in fecundity due to contaminant exposure, assumed equal for all reproducing age classes.

The age-specific measures of contaminant-induced mortality used in Eq 2 are directly comparable to measures used in power plant impact assessment and fisheries management. The parameter m_0 is equivalent to a fractional reduction in year-class abundance due to contaminant-induced mortality. Impacts of power plant cooling systems on fish populations are commonly expressed in terms of fractional reductions in year class abundance [*26,27*]. The logarithm of $(1 - m_r)$ is an instantaneous rate (per year) of contaminant-induced mortality equivalent to the instantaneous rate of fishing mortality used in fisheries models [*6,7*].

Barnthouse et al. [*26*] developed a method for estimating life-stage-specific mortality rates by fitting the logistic concentration-response (logit) model to chronic concentration-response data sets. When the logit model is fitted to data, estimates of the parameters α and β and their variances and covariances define a function and associated uncertainty band expressing the percent mortality as a function of exposure concentration. When test data for several life stages within the first year of life (for example, eggs, larvae, and juveniles) are available, these can be combined according to

$$(1 - m_0) = \prod_{i=1}^{n}(1 - m_i) \tag{3}$$

where m_i is the probability of mortality for life stage i.

Analogous concentration-response functions and uncertainty bands for effects of contaminants on fecundity can be obtained by fitting a quadratic model to data on the number of eggs produced by fish exposed to different contaminant concentrations

$$E(X) = aX^2 + bX + c \tag{4}$$

where

$E(X)$ = number of eggs produced by a female fish exposed to concentration X, and
a,b,c = fitted coefficients.

The variances and covariances obtained from this fit define a trivariate normal distribution. This distribution can be sampled iteratively to obtain random values of a, b, and c. The mean value of the intercept coefficient, c^*, is the expected egg production at zero contaminant exposure. For concentration X and any randomly chosen set of coefficients (a', b', c'), the estimated fractional reduction in egg production is given by

$$f_i = \frac{E(X)}{c^*} = \frac{a'X^2 + b'X + c'}{c^*} \tag{5}$$

Concentration-response functions and uncertainty bands for the effects of chronic contaminant exposures on female reproductive potential can readily be obtained from Eq 2 by iterative sampling from the distributions of fitted parameters: estimates of α and β from fits of the logit model to mortality data, and estimates of a, b, and c from fits of the quadratic model to fecundity data. The results can be normalized and expressed as fractional reductions in reproductive potential, R_s

$$R_s = \frac{P - P_s}{P} \tag{6}$$

This form facilitates interpretation and eliminates the need for estimates of the natural young-of-the-year survival, S_0.

Application to a Representative Fish Population

The Chesapeake Bay striped bass population provides a good example for demonstrating the value of a population-level approach to defining the significance of toxicity test data.

It has until recently supported intensive commercial and recreational fisheries. It has been declining in abundance since 1970 for reasons that are not completely understood, but may involve both overfishing and environmental pollution [*28,29*]. Given these circumstances, a regulatory scientist performing a hazard assessment for a potential contaminant release into Chesapeake Bay or its tributaries would be expected to pay especially close attention to potential effects on this population. A life table for the Chesapeake Bay striped bass population containing all the information necessary to apply our methods is available [*29*].

Few toxicity tests are performed using striped bass as a test species, and it is unfeasible to perform life cycle tests on this large fish. Therefore, we used a number of the commonly used test species as surrogates. The chemicals and test species were trifluralin/sheepshead minnow [*30*], methylmercury/brook trout [*31*], captan/fathead minnow [*32*], chloramine/fathead minnow [*33*], and malathion/sheepshead minnow [*30*]. For the purposes of this paper, we assumed that striped bass would respond identically to the tested species. Although methods exist that could be used to account for the differences in sensitivity between striped bass and the above test species [*34*], we did not use them here because they would only have added variance to the outcome without changing the relationship between the endpoints.

All of the above studies report toxicity test data in which the following criteria are satisfied: (1) all life stages of the fish were exposed to contaminants, (2) the fecundity of test fish was reported, (3) mean control mortality for all life stages was 30% or less, (4) the range of test concentrations for some or all tested life stages spanned the LC_{25} (concentration lethal for 25% of organisms), and (5) life stages for which the fitted value of β was not significantly greater than zero showed no response over the test range.

To examine the relative importance of different testing strategies and of assumptions made in matching the test results to the striped bass life history parameters, we performed four types of extrapolations for each data set. The Type 1 extrapolation simulated an early life stage test, that is, used only data for egg and larval stages. The Type 2 extrapolation used data for egg, larval, and parental mortality, but assumed that two-year old and older fish are unaffected (mortality data are unavailable for fish after their first spawning). The Type 3 extrapolation assumed that all ages are subject to contaminant-induced mortality at the same rate observed in the parental generation of test fish, but that the fecundity of the surviving adults is unaffected. The Type 4 extrapolation assumed, in addition, that the fecundity of adults is reduced by the amount observed in the tests.

Concentration-response functions and uncertainty bands (based on 1000 iterations for each concentration) for three of the five chemicals and for all four extrapolation types are plotted in Figs. 1 through 3. Upper and lower 90% confidence bounds on population-level effective concentrations producing 5% and 10% effects (EC_5 and EC_{10}) for all five chemicals and for all four extrapolation types are listed in Table 1. The maximum acceptable toxicant concentrations (MATC) reported by the authors of the three studies are also plotted on the figures and listed in the table.

Figures 1 to 3 illustrate the diversity of population-level concentration-response functions and uncertainty bands that are produced by typical chronic test data. For trifluralin, juvenile and adult mortality (extrapolation Types 2 and 3) is very low in comparison to early life stage mortality (Type 1), so that Figs. 1*a* to 1*c* are nearly identical. The MATC corresponds to a $<$10% reduction in female reproductive potential for these extrapolation types. When fecundity data are included (Type 4, Fig. 1*d*), however, the confidence band for population effects becomes very broad at low concentrations, and the MATC corresponds to a $>$10% reduction in female reproductive potential. For methylmercury (Fig. 2), the juvenile and adult mortality observed in the life cycle test was substantial and rather variable; therefore, extrapolation Types 2 and 3 (Figs. 2*b* and 2*c*) estimate substantial pop-

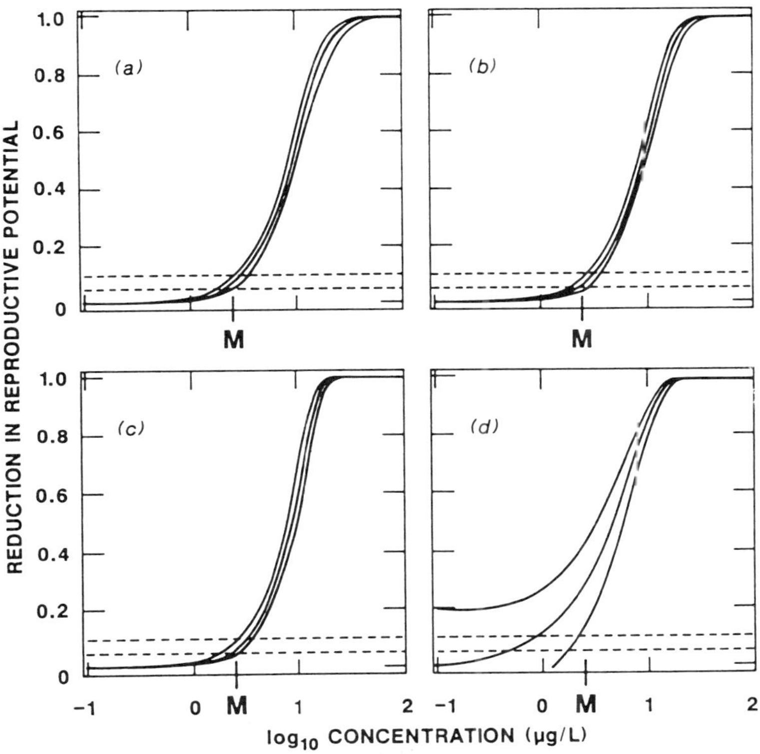

FIG. 1—*Estimated reduction in reproductive potential of a female striped bass as a function of exposure to trifluralin, based on data of Parrish et al.* [30] *on chronic toxicity to sheepshead minnow* (Cyprinodon variegatus). *The three curves on each graph are the 10th percentile, median, and 95th percentile values for the projected effects at each exposure concentration. Figure 1*a *is based on data for embryolarval mortality only. Figure 1*b *includes embryolarval and juvenile/adult mortality and assumes that adults that survive their first spawning are unaffected thereafter. Figure 1*c *includes the same data as Fig. 1*b *but assumes that mortality is repeated annually throughout the entire life cycle. Figure 1*d *is based on the same mortality data and assumptions used in Fig. 1*c *but, in addition, includes data on reductions in fecundity. The MATC estimated from the original data set* [8] *is denoted as* **M** *on the concentration axis.*

ulation-level effects at much lower concentrations than if only early life stage data are included (Fig. 2*a*). The fecundity data for methylmercury (Fig. 2*d*) are so variable as to be virtually meaningless. For all extrapolation types, the MATC corresponds to a 30% or greater reduction in female reproductive potential. The data set for captan (Fig. 3) is the most precise of the data sets examined. As with trifluralin, the effects on juveniles and adults are small in comparison to the effects on early life stages, so that Figs. 3*a* to 3*c* nearly coincide. The effects on fecundity are, however, substantial, and when these are included (Fig. 3*d*) the concentration-response function is shifted to the left by nearly an order of magnitude. Again, the MATC corresponds to surprisingly high reductions in reproductive potential for all extrapolation types.

Comparison of the EC_5, EC_{10}, and MATC values in Table 1 confirms that in the majority of cases examined, the MATCs correspond to >10% reductions in female reproductive

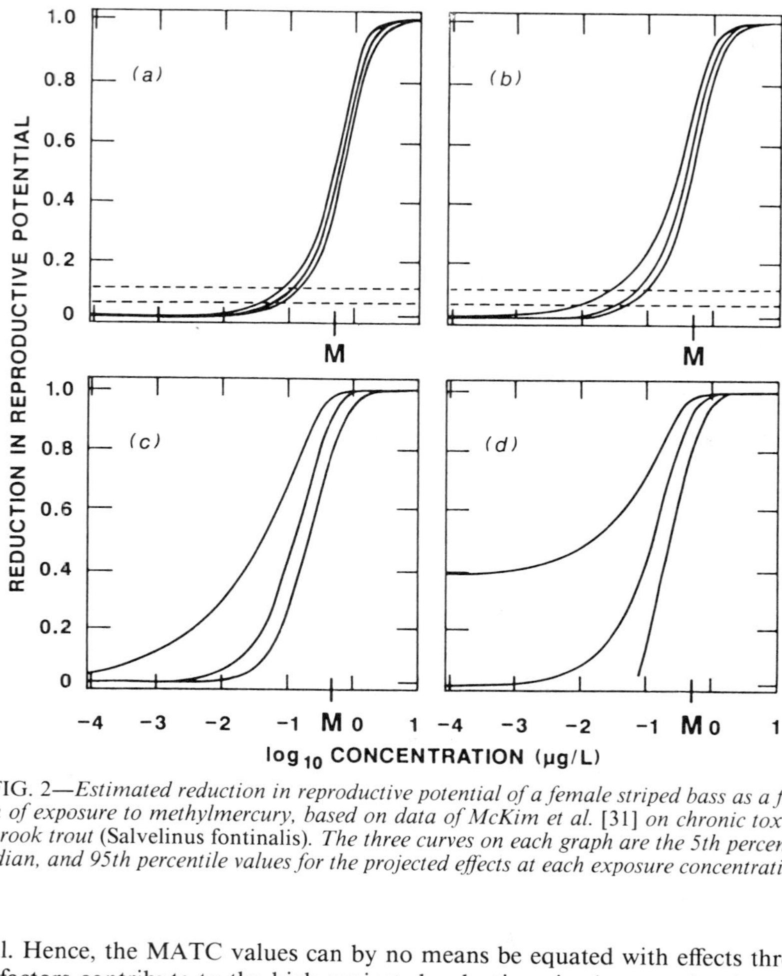

FIG. 2—*Estimated reduction in reproductive potential of a female striped bass as a function of exposure to methylmercury, based on data of McKim et al.* [31] *on chronic toxicity to brook trout* (Salvelinus fontinalis). *The three curves on each graph are the 5th percentile, median, and 95th percentile values for the projected effects at each exposure concentration.*

potential. Hence, the MATC values can by no means be equated with effects thresholds. Several factors contribute to the high projected reductions in the reproductive potential associated with the MATC values. When the test data indicate significant effects on juveniles and adults (for example, methylmercury in Fig. 2*c*), the MATC necessarily underestimates the effects on reproductive potential because it cannot account for repeated annual mortality to adults. When the observed concentration-response relationship is extremely steep (for example, captan in Fig. 3), the MATC can correspond to a high reduction in reproductive potential simply because of its intermediate position between the lowest concentration producing a statistically significant effect and the highest concentration producing no effect. Perhaps the most important contributor to the poor performance of MATC as an effects threshold is its undervaluing of effects on fecundity. In all of the data sets examined, inclusion of fecundity greatly increased both the width of the uncertainty band and the expected reduction in reproductive potential at MATC. This result is consistent with the observation of Suter et al. [*4*] that fecundity has generally been undervalued in conventional hazard assessment because of high experimental variability.

Mortality to juvenile and adult fish (extrapolation Types 2 and 3) appears to be somewhat less important on average than either fecundity or early life stage mortality. Upper

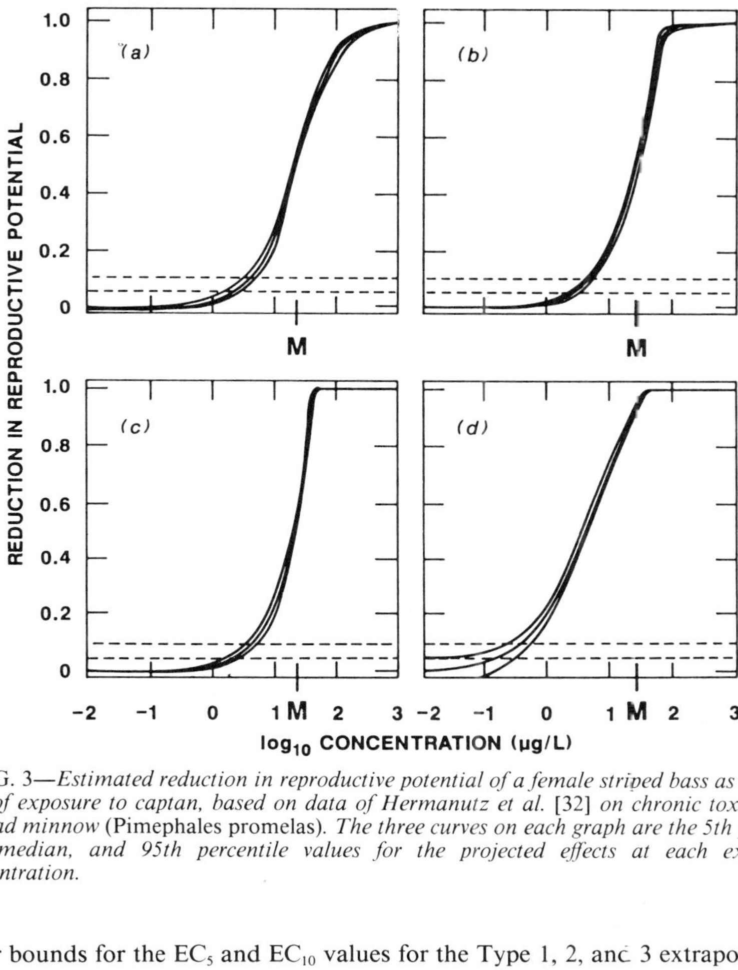

FIG. 3—*Estimated reduction in reproductive potential of a female striped bass as a function of exposure to captan, based on data of Hermanutz et al.* [32] *on chronic toxicity to fathead minnow* (Pimephales promelas). *The three curves on each graph are the 5th percentile, median, and 95th percentile values for the projected effects at each exposure concentration.*

and lower bounds for the EC_5 and EC_{10} values for the Type 1, 2, and 3 extrapolations are nearly identical for trifluralin and captan, and are markedly (that is, an order of magnitude or more) different only for malathion and methylmercury. Except for methylmercury (Fig. 2), differing assumptions concerning the effect of contaminants on two-year-old and older fish (extrapolation Types 2 and 3) have essentially no effect on the concentration-response functions and confidence intervals.

Discussion

The approach demonstrated in this paper has several distinct advantages over conventional hazard assessment schemes. First, results of tests performed on different life stages and measures of different effects are integrated into a single measure of effects. Second, experimental error, rather than being confounded with contaminant effects, is explicitly expressed as a confidence interval. Third, the particular level chosen as indicating regulatory significance can reflect both biological and societal concerns.

Integration of effects on different life stages is important because our results suggest that there is no single "most sensitive" life stage that can provide a reliable surrogate for life

TABLE 1—*Comparison of the population-level EC_5 and EC_{10} values to the MATC values derived from the same data sets.*

	Extrapolation Type[a]			
Benchmark, μg/L	1	2	3	4
Trifluralin [*30*]				
EC_5	1.7/2.6[b]	1.7/2.6	1.7/2.6	0/1.7
EC_{10}	2.6/3.6	2.6/3.6	2.6/3.7	0/2.1
MATC	2.5	2.5	2.5	2.5
Methylmercury [*31*]				
EC_5	0.042/0.081	0.007/0.052	0.0002/0.034	0/0.079
EC_{10}	0.089/0.15	0.026/0.095	0.0009/0.055	0/0.089
MATC	0.5	0.5	0.5	0.5
Captan [*32*]				
EC_5	2.2/3.3	2.2/3.3	2.2/3.3	0.025/0.89
EC_{10}	4.0/5.5	4.0/5.4	4.0/5.4	0.21/0.52
MATC	25.5	25.5	25.5	25.5
Malathion [*30*]				
EC_5	. . .[c]	7.4/12	4.3/8.0	0/7.4
EC_{10}	. . .	10/13	5.4/9.3	0/8.0
MATC	6	6	6	6
Chloramine [*33*]				
EC_5	21/55	16/43	8.0/40	0/2.3
EC_{10}	32/64	23/52	13/43	0/3.6
MATC	24	24	24	24

[a] Type 1 = effects on early life stages only. Type 2 = Type 1 + juvenile or parental mortality applied once. Type 3 = Type 1 + juvenile/parental mortality applied annually throughout life cycle. Type 4 = Type 3 + reduction in fecundity.

[b] Upper and lower 90% confidence bounds.

[c] No mortality to early life stages observed over test range.

cycle tests in chronic toxicity studies. It might be objected that our sample size is too small for meaningful extrapolation; however, Suter et al. [*4*] reached the same conclusion after examining a much larger data set. Our finding that fecundity is both the most sensitive and the most variable test response (also supported by Suter et al. [*4*]) demonstrates the necessity of developing risk assessment models that account for the statistical power associated with different test types. Calculation of the EC_5 or the EC_{10} values rather than the MATC for the individual test components would eliminate the confounding of experimental error and contaminant toxicity. However, a population model would still be necessary to integrate the responses of different life stages.

We believe it is feasible to use the reproductive potential model described in this paper to develop operational measures of biological and societal significance analogous to those that have evolved in fisheries management and environmental impact assessment. To accomplish this task, reproductive potential must be related to properties of greater management relevance such as population size, sustainable yield, or risk of catastrophic decline. It may well be possible to estimate approximate levels of sustainable reduction in reproductive potential from analyses of levels of fishing mortality sustained by exploited populations. However, for toxic contaminants such estimates may be unnecessary. From the steep slopes of the curves in Figs. 1 to 3, it can be seen that concentrations that cause near 100% reductions in reproductive potential often differ by no more than an order of magnitude from those that cause 10% or smaller reductions. This result is not an artifact of small sample size. Barnthouse et al. [*26*] found the same result when analyzing 77 fitted concentration-response functions. The observed order-of-magnitude spread between inconsequential and catastrophic reductions is well within the range of uncertainty due to variability in interspecies sensitivity [*26,34*]. As long as testing programs are restricted to a few standard test species of uncertain sensitivity relative to the species of regulatory interest, it would appear that a high degree of conservatism is appropriate. A 10% or lower threshold for determining regulatory significance (far smaller than the 50% or greater reductions commonly sustained by exploited populations) would appear to accord more with both societal preferences and with the reality of biological responses.

In this paper we have described only one of many possible modeling approaches that could enrich ecotoxicology. We expect that other approaches will find applications in this field, once it is accepted that (1) considerations of risk and societal acceptability must be integrated into the hazard assessment paradigm, and that (2) mathematical models are a necessary link between toxicity test data and the biological phenomena that are of societal interest. Stephan [*5*] proposed the following goal for applied aquatic toxicology:

> A major goal of applied aquatic toxicology is to be able to make a useful prediction concerning whether or not a specific addition of a toxic agent to a particular aquatic ecosystem will cause any unacceptable effect on that ecosystem.

Achieving Stephan's goal will require contributions from resource management, limnology, and other disciplines that are outside the current boundaries of ecotoxicology. Our optimistic message is that these contributions can greatly increase our ability to understand and regulate the effects of toxic chemicals on aquatic biota.

Acknowledgments

We gratefully acknowledge the comments of S. W. Christensen and J. E. Breck. Research was sponsored by the U.S. Environmental Protection Agency, under Interagency Agreement No. DW89931959-01-0 with the U.S. Department of Energy, under Contract No. DE-AC05-840R21400 with Martin Marietta Energy Systems, Inc. Although the research described in this article has been funded wholly or in part by the U.S. Environmental Protection Agency (EPA), it has not been subjected to EPA peer review and, therefore, does not necessarily reflect the views of EPA and no official endorsement should be inferred. Publication No. 3194, Environmental Sciences Div., Oak Ridge National Laboratory.

References

[*1*] Cairns, J., Dickson, K. L., and Maki, A. W., Eds., *Estimating the Hazard of Chemical Substances to Aquatic Life, ASTM STP 657,* American Society for Testing and Materials, Philadelphia, 1978.

[*2*] Skalski, J. in *Aquatic Toxicology and Hazard Assessment (Fourth Conference), ASTM STP 737,* D. R. Branson and K. L. Dickson, Eds., American Society for Testing and Materials, Philadelphia, 1981, pp. 377–387.

[*3*] Gelber, R. D., Lavin, P. T., Mehta, C. R., and Schoenfeld, D. A., in *Fundamentals of Aquatic Toxicology,* G. M. Rand and S. R. Petrocelli, Eds., Hemisphere Publishing, Co., Washington, DC, 1985, pp. 110–123.

[*4*] Suter, G. W., Linder, E., Rosen, A., and Parkhurst, D. F., *Environmental Toxicology and Chemistry,* Vol. 6, 1987, pp. 793–910.

[*5*] Stephan, C. E., in *Aquatic Toxicology and Environmental Fate: Ninth Volume, ASTM STP 921,* T. M. Poston and R. Purdy, Eds., American Society for Testing and Materials, Philadelphia, 1986, pp. 3–10.

[6] Ricker, W. E., *Journal of the Fisheries Research Board of Canada,* Vol. 11, 1954, pp. 559–623.
[7] Beverton, R. J. H. and Holt, S. J., "On the Dynamics of Exploited Fish Populations," *Fisheries Investigations (Series 2),* Vol. 19, U.K. Ministry of Agriculture and Fisheries, 1957.
[8] Larkin, P. A., *Transactions of the American Fisheries Society,* Vol. 106, 1977, pp. 1–11.
[9] Regier, H. A., in *Fish Population Dynamics,* J. A. Gulland, Ed., Wiley, London, 1977, pp. 134–155.
[10] Rothschild, B. J., *Dynamics of Marine Fish Populations,* Harvard University Press, Cambridge, MA, 1986.
[11] Clark, C. W., *Mathematical Bioeconomics: the Optimal Management of Renewable Resources,* Wiley, New York, 1976.
[12] Walters, C. J. and Ludwig, D., *Canadian Journal of Fisheries and Aquatic Science,* Vol. 38, 1981, pp. 704–710.
[13] Fletcher, R. I. and Deriso, R. B., *American Fisheries Society Monographs,* Vol. 4, 1988, pp. 232–244.
[14] Zar, J. H., in "The Biological Significance of Environmental Impacts," Technical Report NR-CONF-002, U.S. Nuclear Regulatory Commission, Washington, DC, 1977, pp. 285–294.
[15] Van Winkle, W., Vaughan, D. S., Barnthouse, L. W., and Kirk, B. L., *Canadian Journal of Fisheries and Aquatic Science,* Vol. 38, 1981, pp. 627–632.
[16] Vaughan, D. S. and Van Winkle, W., *Canadian Journal of Fisheries and Aquatic Science,* Vol. 39, 1982, pp. 782–785.
[17] Sharma, R. K., in "The Biological Significance of Environmental Impacts," Technical Report NR-CONF-002, U.S. Nuclear Regulatory Commission, Washington, DC, 1977, pp. 3–10.
[18] Christensen, S. W., Van Winkle, W., and Mattice, J. S., in "The Biological Significance of Environmental Impacts," Technical Report NR-CONF-002, U.S. Nuclear Regulatory Commission, Washington, DC, 1977, pp. 191–220.
[19] Buffington, J. D., in "The Biological Significance of Environmental Impacts," Technical Report NR-CONF-002, U.S. Nuclear Regulatory Commission, Washington, DC, 1977, pp. 319–327.
[20] Voightlander, C. W., in *Issues Associated with Impact Assessment: Proceedings of the Fifth National Workshop on Entrainment and Impingement,* L. D. Jensen, Ed., EA Communications, Sparks, MD, 1981, pp. 3–12.
[21] Beanlands, G. E. and Duinker, P. N., "An Ecological Framework for Environmental Impact Assessment in Canada," Institute for Resource and Environmental Studies, Dalhousie University, Halifax, 1983.
[22] Westman, W. E., *Ecology Impact Assessment and Environmental Planning,* Wiley, New York, 1985.
[23] Eberhardt, L. L., "The Biological Significance of Environmental Impacts," Technical Report NR-CONF-002, U.S. Nuclear Regulatory Commission, Washington, DC, 1977, pp. 307–316.
[24] Van Winkle, W., Christensen, S. W., and Mattice, J. S., *International Journal of Environmental Studies,* Vol. 9, 1976, pp. 247–254.
[25] Barnthouse, L. W., O'Neill, R. V., Bartell, S. M., and Suter, G. W. II, in *Aquatic Toxicology and Environmental Fate: Ninth Volume, ASTM STP 921,* T. M. Poston and R. Purdy, Eds., American Society for Testing and Materials, Philadelphia, 1986, pp. 82–96.
[26] Barnthouse, L. W., Suter, G. W., II, Rosen, A. E., and Beauchamp, J. J., *Environmental Toxicology and Chemistry,* Vol. 6, 1987, pp. 811–824.
[27] Vaughan, D. S., *American Fisheries Society Monographs,* Vol. 4, 1988, pp. 121–123.
[28] Mehrle, P. M., Haines, T. A., Hamilton, S., Ludke, J. L., Mayer, F. L., and Ribick, M. A., *Transactions of the American Fisheries Society,* Vol. 111, 1982, pp. 231–241.
[29] Goodyear, C. P., *Transactions of the American Fisheries Society,* Vol. 114, 1985, pp. 107–113.
[30] Parrish, P. R., Dyar, E. E., Enos, J. M., and Wilson, W. G., "Chronic Toxicity of Chlordane, Trifluralin, and Pentachlorophenol to Sheepshead Minnows *(Cyprinodon varietgatus),*" Technical Report EPA-600/3-78-010, U.S. Environmental Protection Agency, Gulf Breeze, FL, 1978.
[31] McKim, J. M., Olson, G. F., Holcomb, G. W., and Hunt, E. P., *Journal of the Fisheries Research Board of Canada,* Vol. 33, 1976, pp. 2726–2739.
[32] Hermanutz, R. O., Mueller, L. H., and Kempert, K. D., *Journal of the Fisheries Research Board of Canada,* Vol. 30, 1973, pp. 1811–1817.
[33] Arthur, J. W. and Eaton, J. G., *Journal of the Fisheries Research Board of Canada,* Vol. 28, 1971, pp. 1841–1845.
[34] Suter, G. W., II, Vaughan, D. S., and Gardner, R. H., *Environmental Toxicology and Chemistry,* Vol. 2, 1983, pp. 369–378.

Statistical Procedures

Jeffrey F. Bromaghin[1] *and Richard M. Engeman*[1]

Head-to-Head Comparison of SAS and ASTM-Proposed Probit Computer Programs

REFERENCE: Bromaghin, J. F. and Engeman, R. M., "**Head-to-Head Comparison of SAS and ASTM-Proposed Probit Computer Programs,**" *Aquatic Toxicology and Environmental Fate: Eleventh Volume, ASTM STP 1007,* G. W. Suter II and M. A. Lewis, Eds., American Society for Testing and Materials, Philadelphia, 1989, pp. 303–307.

ABSTRACT: Estimates of the median lethal dose (LD_{50}) [or the median lethal concentration (LC_{50})] are used in toxicological and hazard assessment studies to indicate the dosage of a substance that produces a 50% response (death) rate in a population. The calculation of LD_{50} estimates is frequently accomplished through probit analysis. Probit analysis is based on certain underlying assumptions about the tolerance of the population in question to the substance being tested. The response rate is assumed to follow a cumulative normal distribution with respect to the dosage or the logarithms of the dosage.

Recently we reviewed ASTM's proposed standard practice for using probit analysis (Draft No. 2). A BASIC computer program for performing probit analysis was included with the proposed standard. We evaluated this program using the PROBIT procedure in the well-known SAS Institute package as a standard of comparison. The SAS procedure offers a default option that assumes the response rate of the population follows the cumulative normal distribution. In addition, two logarithmic options are also offered (natural and base 10), whereby the response rate is assumed cumulative normal relative to the logarithm of the dose. The BASIC and SAS programs give very similar results when the appropriate logarithmic option is used with the SAS procedure. Thus, the ASTM program would be a subset of the SAS procedure in that it is comparable to one of the SAS options, employing one set of possible assumptions. We believe that the assumptions one would make about the distributional response of a population are not necessarily standard, and that careful examination of the data relative to analytical assumptions prior to analysis is recommended.

KEY WORDS: probit computer programs, aquatic toxicology

Estimates of the median lethal dose (LD_{50}) [or median lethal concentration (LC_{50})] are used or required in toxicological and hazard assessment studies to indicate the dosage of a substance that produces a 50% response (death) rate in a population. Various analytical methods are available for calculating LD_{50} estimates [*1*]. Probit analysis is one of the most frequently used of the parametric methods and is based on the assumption that the response rate of the population in question to the substance being tested follows a cumulative normal distribution with respect to dosage or the logarithm of the dosage.

Recently, we reviewed ASTM's proposed standard practice for using probit analysis (Draft No. 2). A BASIC computer program for performing probit was included with the proposed standard. This program was installed on a Radio Shack Model 16 microcomputer. Because estimates from the ASTM program calculated LD_{50} values that were different than those calculated in our routine data analyses, we proceeded to evaluate this pro-

[1] Statisticians, U.S. Department of Agriculture/Animal and Plant Health Inspection Service, Animal Damage Control Program, Denver Wildlife Research Center, Denver, CO.

gram using the PROBIT procedure in the well-known SAS package [2] as a standard of comparison.

The SAS procedure provides three options for analyses corresponding to two assumptions about the distributional relationship between the response rate and dosage level. The default option assumes that the response rate follows the cumulative normal distribution with respect to the dosage. The other two options, which must be specified to be implemented, assume that the response rate is cumulative normal with respect to a logarithm of dosage. One option uses the base 10 logarithm of the dosage while the other uses the natural (base *e*) logarithm.

Our initial attempts at using the ASTM program to analyze data previously analyzed using SAS indicated that for calculating an LD_{50}, the ASTM program probably was equivalent to one of the SAS options. To determine this, exact data generated from known distributions with known LD_{50}s were used for comparing the ASTM-proposed program to the PROBIT procedure in SAS.

Test Data

Two artificial data sets were generated based on specified normal distributions. These data sets provide a controlled basis of comparison between the two programs because the properties (including the LD_{50}) of the data are known.

One data set was constructed using a normal distribution with a mean (LD_{50}) of 4.0 and a standard deviation of 1.25. Eleven dosage levels, ranging from 1.5 to 6.5 in increments of 0.5, were selected for inclusion in the analyses.

The second data set was based on a normal distribution with a mean (LD_{50}) equal to 8 and a standard deviation equal to 10. Five dosage levels, ranging from 2 to 32 by multiples of 2, were considered for the analyses.

Specification of distributions and dosage levels of interest allows exact population response (death) rates for each dosage to be calculated. Calculation of these response rates for theoretical populations requires the use of the cumulative normal distribution function (see, for example, Ref *3*). The standard normal cumulative distribution function is well tabulated in many statistics texts and is denoted by $\Phi(z)$, the probability that a standard normal random variable, Z, is less than or equal to z.

Response rates for populations with normal tolerance distributions with respect to dosage are calculated using the following equation

$$\text{response rate} = \Phi\left[\frac{\text{dosage} - \text{mean}}{\text{standard deviation}}\right] \tag{1}$$

For example, in our first data set (mean = 4.0, standard deviation = 1.25) we calculate the response rate for the 5.5 dosage by

$$\Phi\left[\frac{5.5 - 4.0}{1.25}\right] = \Phi(1.2) = 0.88$$

To calculate the population response rates when distributed cumulative normal with respect to the logarithm (base 10 or natural) of the dosage, Eq 1 is employed using logarithms of dosage, mean, and standard deviation. For the purposes of our analyses we assumed that 100 subjects were tested at each dosage. The data that we produced for comparing the two probit programs are given in Table 1.

TABLE 1—*Test data generated for comparing two probit programs (the number responding is based on 100 subjects tested at each dosage).*

Data Set	Dosage Level	Number Responding: Normal with Respect to Dosage	Number Responding: Normal with Respect to $\log_{10}$ (Dosage)	Number Responding: Normal with Respect to ln (Dosage)
1	1.5	2	0	0
	2.0	5	0	0
	2.5	12	2	2
	3.0	21	10	10
	3.5	34	27	27
	4.0	50	50	50
	4.5	66	70	70
	5.0	79	84	84
	5.5	88	92	92
	6.0	95	97	97
	6.5	98	99	99
2	2	38	32	32
	4	42	41	41
	8	50	50	50
	16	66	59	59
	32	88	68	68

Program Runs

The variables input into both of the programs are dosage, sample size at each dose, and number responding at each dose. For each run on each data set, the dosage values given in Table 1 were input to the program. We always assumed a sample size of 100 for each dosage level in both data sets. Three runs were made on each data set using the ASTM program, one run for each category of response rates (corresponding to the two underlying assumptions about the tolerance distributions). The same three runs were made on each data set for each of the three options in the SAS PROBIT procedure (thus, three times as many runs were made using SAS). The output information from each run that was used for comparative purposes included the point estimate of the LD_{50}, the 95% fiducial limits, and the chi-square value for testing goodness-of-fit.

As an additional test of the ASTM program, we input a new set of dosage levels with the response rates that are normal with respect to dosage. The new dosage levels were calculated as 10 to the power of the original dosages (10^{dosage}). Base 10 logarithms of the LD_{50} estimate and fiducial limits were calculated for comparison with the results produced by the analysis of the original dosage levels and the same response rates when the default in the SAS PROBIT procedure was used.

Results

Analysis of the sets of response rates that are normally distributed with respect to the base 10 and natural logarithms of dosage yielded identical results, because the response rate data in Table 1 input into the programs were identical. Also, the results from using the base 10 and natural logarithm options in the SAS PROBIT procedure always gave identical results for each data set analyzed using these options. For this reason, we will only examine the results from using the default and base 10 logarithm options in SAS. Similarly, we will not present separate results for data analyses where the response rates were calculated for the response rates that are cumulative normal with respect to the natural logarithm of dosage. The results from the analyses of the data in Table 1 are presented in Table 2.

The ASTM program and the option in SAS that assumes the response rate to be normally distributed with respect to the logarithm of dosage produce nearly identical results. When the sets of response rates that satisfied this distributional assumption were submitted to these routines, both produced the correct point estimate of the LD_{50}, small chi-square goodness-of-fit values, and fiducial limits that contained the true LD_{50} (see Table 2). Results produced by these routines were always identical to at least two decimal places. When the data normally distributed with respect to the base 10 logarithm of dosage were submitted, the default option in SAS overestimated the LD_{50}. Correspondingly, larger goodness-of-fit values were also produced, and the 95% fiducial limits just missed containing the true LD_{50}.

When the sets of response rates normally distributed cumulatively with respect to dosage were analyzed, the default option in the SAS PROBIT procedure was the only routine to produce correct results. The ASTM program and the base 10 logarithm option in SAS underestimated the LD_{50} in this case. Again, the wrong underlying assumption produced larger chi-square goodness-of-fit values. The 95% fiducial limits barely contained the true LD_{50}.

In all cases the choice of routine incorporating the appropriate distributional assumption produced the correct LD_{50} estimates. If the appropriate program (or option) was not used, the results ranged from being biased a small amount in the first data set to a large amount of relative bias in the second data set.

The data using 10 to the power of the dosage and the set of response rates cumulatively normally distributed with respect to dosage were analyzed with the ASTM program (Table 3). Transforming the estimate of the LD_{50} and the fiducial limits in the 10 to the power of dose scale back to the original scale produced estimates nearly identical to those obtained using the default option in the SAS PROBIT procedure. The chi-square goodness-of-fit statistics produced in these two runs were also nearly identical.

Discussion

The ASTM-proposed probit program produces results very similar to those given by the option in the SAS PROBIT procedure that assumes the response rate to be normally distributed with respect to the logarithm of the dosage. Both of these routines appear to follow the standard logarithmic transformation as found in Ref *1*. Use of this transformation appears to be the most common approach to probit analyses. SAS is one of the most widely used and respected program packages available, and the use of a default option in the PROBIT procedure, where no transform is calculated, indicates a widespread application of the assumption that response rate is frequently considered cumulative normal with respect to dosage rather than the logarithm of dosage. A standard program incorporating only one of the possible distributional assumptions can be considered too restrictive. This limitation is emphasized by the poor results obtained in the present study when the distributional assumption is violated.

The results in Table 3 indicate that the ASTM program can be "tricked" into producing the same estimates as the default option in the SAS PROBIT procedure. If the response rates are suspected to be normally distributed with respect to dosage, then raising 10 to the power of the dosage and inputting these values as the dosages into the ASTM program will produce results that can be back transformed to the values that the SAS default option

TABLE 2—*Results from analyses of the data in Table 1.*

Data Set	True LD_{50}	Normal with Respect to Dose or Log_{10} Dose	Program	Estimated LD_{50}	95% Fiducial Limits	Chi-Square
1	4.0	dose	SAS default	4.000	3.88, 4.12	0.1772
			SAS log 10	3.800	3.59, 4.02	20.5966
			ASTM	3.800	3.58, 4.02	20.5946
		log_{10} dose	SAS default	4.112	4.01, 4.21	5.9737
			SAS log 10	3.999	3.90, 4.10	0.3706
			ASTM	3.999	3.90, 4.10	0.3706
2	8.0	dose	SAS default	8.021	5.35, 10.36	0.0277
			SAS log 10	5.365	1.46, 10.91	8.8362
			ASTM	5.365	1.46, 10.85	8.8364
		log_{10} dose	SAS default	12.283	8.04, 16.57	4.1715
			SAS log 10	8.000	5.61, 11.42	0.0039
			ASTM	8.000	5.61, 11.42	0.0039

TABLE 3—*Results from using 10 to the power of dosage and response rates distributed normally with respect to dosage.*

Data Set	True LD_{50}	Value Estimated	ASTM Estimate 10^{dosage} Scale	ASTM Estimate Transformed to Original Scale	SAS Default Estimate, (Table 2)
1	4	LD_{50}	10^4	4.000	4.000
		fiducial limits	7628.5, 13 108.7	3.88, 4.12	3.88, 4.12
		chi-square	0.1771	...	0.1772
2	8	LD_{50}	1.0494×10^8	8.021	8
		fiducial limits	223 450, 2.308×10^{10}	5.35, 10.36	5.35, 10.36
		chi-square	0.0277	...	0.0277

would have produced. The options available in the SAS PROBIT procedure can be duplicated in the ASTM program through prior data manipulation.

Although the SAS PROBIT procedure gives more information on output (such as LD values from LD_{01} to LD_{99} and plots), we feel that the primary limitation of the ASTM program is the lack of options corresponding to different possible distributional assumptions. The rigid use of a particular routine without considering the underlying assumptions can produce poor results. Our analyses indicate that these methods, at least in these large sample cases (100 subjects per dose), may not be particularly robust to departures from the underlying assumptions. We recommend using analytical methods based on examination of the data to determine the most appropriate underlying assumptions. When using the ASTM program this would require a run for the original dosages and one using transformed dosages. Examination of the goodness-of-fit results helps to provide a means of comparing the underlying assumptions prior to commitment to a final analytical technique.

References

[1] Finney, D. J., *Statistical Method in Biological Assay,* Griffin, London, 1971.
[2] SAS Institute Inc., *SAS User's Guide: Statistics,* Cary, NC, 1985.
[3] Mood, A. M., Graybill, F. A., and Boes, D. C., *Introduction to the Theory of Statistics,* McGraw-Hill, New York, 1974.

Morris H. Roberts, Jr.[1]

Comparison of Several Computer Programs for Probit Analysis of Dose-Related Mortality Data

REFERENCE: Roberts, M. H., **"Comparison of Several Computer Programs for Probit Analysis of Dose-Related Mortality Data,"** *Aquatic Toxicology and Environmental Fate: Eleventh Volume, ASTM STP 1007,* G. W. Suter II and M. A. Lewis, Eds., American Society for Testing and Materials, Philadelphia, 1989, pp. 308–320.

ABSTRACT: Five readily available computer programs for probit analysis were evaluated for input/output options and reliability of output: (1) DULUTH-TOX from Charles Stephan of the Environmental Protection Agency's Environmental Research Laboratory, Duluth; (2) ASTM-PROBIT from the ASTM Guide for Probit Analysis, Draft 2; (3) UG-PROBIT, which was written by statisticians at the University of Guelph, Canada; and (4) SPSSx; and (5) SAS. SPSSx and SAS are probit procedures available in standard statistical packages.

Except for UG-PROBIT, the programs yielded essentially identical median lethal concentration (LC_{50}) values for the 20 data sets tested. DULUTH-TOX and ASTM-PROBIT include an objective evaluation of the validity of input data and calculated results, but they lack graphical output. SAS-PROBIT provides graphical output superior to that from SPSSx-PROBIT. SPSSx-PROBIT yields inappropriate fiducial limits in some cases.

UG-PROBIT, SPSSx-PROBIT, and SAS-PROBIT purport to consider control or "natural" responses. Only SPSSx-PROBIT and SAS-PROBIT actually adjusted observed treatment responses for the control response.

SPSSx-PROBIT and SAS-PROBIT are currently available only for mini and mainframe computers. All other programs can be implemented on microcomputers as well. DULUTH-TOX and ASTM-PROBIT, in all implementations, have a major advantage in that objective tests are included to determine the validity of input data and to guide the interpretation of output. The commercial statistical programs have the advantages of graphical output and a method for handling control mortality.

KEY WORDS: probit analysis, LC_{50}, fiducial limits, computer programs, evaluation, aquatic toxicology

There are a variety of methods to estimate the lethal concentration for 50% of a test population (LC_{50}) based on different statistical models [*1,2*]. Each method has specific assumptions which may make a particular model more or less appropriate for specific data sets. The probit method has become widely accepted as applicable to many data sets in aquatic toxicology, although one must be alert to any failure to meet the test's assumptions. Formerly the probit method was often applied using a graphic approximation technique [*3*]; now mathematical solutions have been facilitated by the increased availability of computer programs which can quickly perform the tedious computations.

There are a considerable number of programs available to facilitate probit computations on both micro- and mini-computers; virtually everyone can gain access to at least one program. The specific program used will differ based on the computer or computers avail-

[1] Professor of marine science, Virginia Institute of Marine Science, School of Marine Science, College of William and Mary, Gloucester Point, VA 23062.

able and the user's degree of familiarity with software and its use. There is no certainty that various programs which implement probit analysis yield the same or a similar result for a given data set. Further, programs differ in the type of user interface for data input and option selection, the amount of error trapping, the objective evaluation of data adherence to the test's assumptions, and the nature and extent of output.

The objective of this study was to compare the output of five readily available probit programs for micro- and mini-computers when used to analyze several data sets of varying quality.

Procedures

Programs Compared

The five programs tested differ in (*a*) data input and output format, (*b*) implementation of algorithms, and (*c*) the format and substance of the output including the graphical presentation. Three of the programs, written in BASIC, were evaluated on microcomputers [an AT&T PC 6300 operating under MS DOS 2.11 and a Kaypro personal computer operating under MS DOS 3.2] using the GWBASIC interpreter. Two of these programs (DULUTH-TOX and ASTM-PROBIT) have been used previously in versions which operate under MBASIC on CP/M based machines, and DULUTH-TOX has been used on a PRIME 9955 computer with a BASIC interpreter, although neither of these implementations were used for the present study. The third program written in BASIC (UG-PROBIT) runs only under PC-BASIC or GWBASIC. The remaining programs tested were procedures from two major commercially available statistical packages (SPSSx and SAS) operating on a PRIME 9955 minicomputer. Neither commercial program can presently be implemented on microcomputers.

DULUTH-TOX, originally written in BASIC by Charles Stephan of the Duluth U.S. Environmental Protection Agency's Environmental Research Laboratory (EPA-ERL), automatically estimates an LC_{50} and fiducial limits by several methods: (1) nonlinear interpolation using values bracketing the 50% response [*1*], (2) the binomial test, (3) the moving average method [*4*], and (4) the probit method using the maximum likelihood protocol [*1,5*]. Only the results of the probit analysis will be considered in this report. The original implementation required data entry by cards, and data were output to the screen only. The version used in this study allows keyboard entry with editing and printer output. DULUTH-TOX includes a variety of objective error checks and outputs error messages to alert the user to problems or inadequacies of the data output. No provision is made for graphical output or adjustment for control mortality in the calculation of the LC_{50}.

ASTM-PROBIT, under development by ASTM Committee E-47[2] and also written in BASIC, implements only probit analysis using the maximum likelihood method. Like DULUTH-TOX, ASTM-PROBIT was originally written to use cards for data entry, but was modified for this study to allow keyboard entry. This program is very similar to the probit portion of DULUTH-TOX, although the algorithms used are slightly different in a few places. ASTM-PROBIT is being modified by Charles Stephan for a future ASTM release.

UG-PROBIT was originally developed as a teaching tool by the Department of Mathematics and Statistics at the University of Guelph, Canada [*2*]. It provides for the keyboard

---

[2] Proposed "New Standard Practice for Using Probit Analysis," Draft No. 2. ASTM Committee E-47, October 1984.

entry of data and the selection of either maximum likelihood or least squares methods of the probit regression. For this study, only results from the maximum likelihood method are presented. The final output includes the LC_{50}, 95% fiducial limits, and the slope of the probit line with 95% fiducial limits. No error checking or warning messages are included in the output. During keyboard entry of the data, control responses can be entered, but these data are not actually used in the calculations.

SPSSx-PROBIT is the probit procedure of the commercially available SPSSx statistical program [*6*]. Data are entered from a file called by a batch program which must be written by the user. LC_{50}s are estimated by the maximum likelihood method. Data output includes (1) a scatter plot of probit values against log dose (but no fitted line); (2) parameter estimates including the slope and intercept, each with standard errors; (3) the Pearson goodness-of-fit chi-square statistic; (4) a statement about heterogeneity of variance; and (5) a list of the probability of response from 1 to 99%, corresponding detransformed dose, and 95% fiducial limits. One can read the LC_{50} and fiducial limits from the final listing. This program also includes as an option a "natural response" parameter which either accepts a fixed control mortality value or estimates an optimum background (control) response from data for control treatments and low doses, adjusting the response probability results accordingly. SPSSx-PROBIT is the only program of those evaluated which can estimate relative potency.

SAS-PROBIT is the probit procedure in the commercially available SAS statistical package [*7*]. Data are entered from a data file called from a user-written batch program (as in the SPSSx package) or directly through the keyboard. As each iteration is calculated under the maximum likelihood ratio method, the intercept, slope, mean log dose (log LC_{50}), and standard deviation are printed for up to 100 iterations. A covariance matrix is then printed for intercept versus slope and log LC_{50} versus standard deviation followed by the chi-square goodness-of-fit estimate. Two plots are provided, a probit versus log(dose) plot with both data points and the probit line plotted, followed by a probability versus log(dose) plot with data points and fitted line. The final output is a table of log(dose) and dose for response probabilities between 1 and 99%, along with 95% fiducial limits for both transformed and detransformed doses.

Data Sets Tested

Control Response Equals Zero—Twenty data sets for which control mortalities are assumed to be zero were analyzed with each program. These data sets, found in the proposed ASTM document for probit analysis,[2] include one example taken directly from Finney [*5*] plus 19 other sets designed to evaluate the robustness of the program. These include sets with (*a*) replicate treatments (Data Sets 9, 17, and 20 in Table 1), (*b*) variable population sizes among treatments (Data Sets 1, 13, and 17), (*c*) limited numbers of doses producing results meeting the assumption of the test (Data Sets 2, 7, 11, and 12), and so forth.

Control Response Greater than Zero—An additional seven data sets with control responses greater than zero were analyzed (Table 2). Two data sets were taken from Finney (1971) [*5*]; the remainder were selected to evaluate the effect of control mortality responses from 10 to 50% and variable initial numbers. While not an exhaustive set, it is believed to encompass many situations one is likely to encounter.

---

[2] Proposed "New Standard Practice for Using Probit Analysis," Draft No. 2, ASTM Committee E-47, October 1984.

TABLE 1—*Twenty data sets with control mortality equal to zero used for program evaluation.*[a]

Data Set	Dose	Animals, No.	
		Exposed	Dead
1	10.2	50	44
	7.7	49	42
	5.1	46	24
	3.8	48	16
	2.6	50	6
2	50	20	18
	40	20	0
	30	20	1
	20	20	0
	10	20	0
3	16	20	20
	8	20	20
	4	20	9
	2	20	9
	1	20	5
4	16	20	15
	8	20	11
	4	20	11
	2	20	0
	1	20	0
5	4.39	100	57
	3.00	100	50
	1.61	100	43
6	100	20	13
	60	20	11
	36	20	9
	22	20	7
7	32	20	20
	16	20	20
	8	20	16
	4	20	0
	2	20	2
	1	20	0
8	10	100	61
	6	100	39
9	100	24	24
	100	24	20
	60	24	20
	60	24	16
	36	24	10
10	100	48	44
	60	48	36
	36	48	16
	22	48	4
11	100	20	20
	60	20	6
	36	20	5
	22	20	0
12	2000	20	20
	1500	20	20
	1000	20	20
	100	20	20
	60	20	6
	36	20	5
	22	20	0
13	100	20	18
	80	2	1
	25	4	1
	22	20	4
14	100	20	18
	80	20	10
	25	20	5
	22	20	4
15	32	20	20
	16	20	12
	8	20	10
	4	20	8
	2	20	6
	1	20	0
16	32	20	20
	16.05	20	12
	8	20	10
	4	20	8
	2	20	6
	1	20	0
17	100	20	20
	100	20	20
	60	20	11
	60	20	9
	36	24	0
	36	24	0
18	8	20	20
	4	20	9
	2	20	9
	1	20	0
19	32	20	0
	16	20	1
	8	20	7
	4	20	11
	2	20	17
	1	20	0
20	9	30	19
	9	30	9
	6	30	18
	6	30	10

[a] Proposed "New Standard Practice for Using Probit Analysis," Draft No. 2, ASTM Committee E-47, October.

TABLE 2—*Seven data sets with control mortality greater than zero used for evaluations.*

Data Set	Dose	Animals, No. Exposed	Animals, No. Dead
1	148	142	142
	100	127	126
	48	128	115
	12	126	58
	0	129	21
2	62	125	125
	46	117	115
	31	127	114
	14.8	51	40
	3.7	132	37
	0	129	21
3	100	20	20
	56	20	16
	32	20	9
	18	20	5
	10	20	3
	0	20	3
4	100	20	19
	56	20	15
	32	20	8
	18	20	4
	10	20	2
	0	20	2
5	100	20	20
	56	20	16
	32	20	9
	18	20	5
	10	20	0
	0	20	3
6	100	20	20
	56	20	17
	32	20	15
	18	20	13
	10	20	12
	0	20	10
7	100	18	18
	56	22	17
	32	17	10
	18	21	6
	10	23	4
	0	20	3

Results and Discussion

Evaluation of Data Sets with Zero Control Responses

As the first indicator of agreement in results generated by the various programs, the slope, LC_{50}, and fiducial limits produced by the five programs were compared (Table 3). In 19 of 20 cases, DULUTH-TOX and ASTM-PROBIT produced identical LC_{50} and fiducial limits. The sole exception to total agreement was in response to Data Set 20; DULUTH-

TOX calculated an extreme LC_{50} with fiducial limits, whereas ASTM-PROBIT more correctly indicated "slope = 0." SAS-PROBIT was next best, with agreement in 19 of 20 cases for the LC_{50} (the exception is Data Set 20 for which it produced an LC_{50} of 0).

There was less agreement in fiducial limits than in the LC_{50}. At the degree of arithmetic precision shown in Table 3, DULUTH-TOX and ASTM-PROBIT often yielded intervals of "0" and "+infinity" for which SAS-PROBIT gave no interval estimates; this is tantamount to the same thing. In many cases, additional differences between the outputs of all five programs are apparent when the fiducial limits are compared to three or more decimal points. These differences in fiducial limits arise, in some cases, simply from the numeric precision or rounding by the program and BASIC interpreter. More importantly, the programs differ in the algorithms for calculation of fiducial limits (and possibly for implementation of the maximum likelihood method). To determine fiducial limits for a data set with a poor fit as indicated by the chi-square value, DULUTH-TOX and ASTM-PROBIT apply the heterogeneity factor (H) when $P(\chi^2) < 0.05$, SPSSx-PROBIT when $P(\chi^2) < 0.15$ (user adjustable), and SAS-PROBIT when $P(\chi^2) < 0.10$ (user adjustable). To calculate the results presented in Table 3, the default criterion for the χ^2 value was used for each program; therefore, in some cases, the fiducial limits differ because H was not applied by all the programs.

In addition, SPSSx-PROBIT and SAS-PROBIT would appear to have a restrictive criterion which prevents reporting fiducial limits when limits can, in fact, be calculated as in Data Sets 3, 4, 8, and 14; the nature of the restriction is not immediately obvious. Data Sets 3 and 4 have one or more pairs of adjacent treatments with equal responses. Data Set 8 includes only two treatments, so one could argue that there is insufficient information to calculate fiducial limits. The problem with Data Set 14 is not obvious, especially since fiducial limits were calculated for Data Set 1, which is the same except for the small population size at intermediate doses. DULUTH-TOX and ASTM-PROBIT are less restrictive in accordance with the recommendations of Finney [5].

The LC_{50} derived by SPSSx-PROBIT agreed with DULUTH-TOX in 15 of 20 cases, but the fiducial limits agreed in only 4 cases. No LC_{50} was produced for Data Set 8 (gave the message "not enough cases; skipped") or for Data Set 20 (program crashed). For Data Set 17, an intuitively acceptable LC_{50} was given, although assumptions of the probit analysis were seriously violated. In many cases, the fiducial limits reported were equal to the LC_{50} when DULUTH-TOX indicated values of 0 and +infinity.

UG-PROBIT was least in accord with DULUTH-TOX, agreeing precisely in LC_{50} only rarely. Lower and upper fiducial limits were produced which closely bracket the LC_{50}, even when 0 and +infinity would be more appropriate.

ASTM-PROBIT gave the same slope as DULUTH-TOX in all cases, SAS-PROBIT in all but one case, and SPSSx-PROBIT in all but four cases. For Data Set 8, SPSSx-PROBIT failed to converge in 100 iterations. UG-PROBIT showed the least agreement, with absolutely equivalent slopes in only three cases, and varying degrees of deviation in all other cases.

The most reliable programs were DULUTH-TOX, ASTM-PROBIT, and SAS-PROBIT. Each program was robust and able to handle a variety of aberrant data sets to yield an estimate of the LC_{50} or provided an appropriate message to explain the failure. Data Set 20, for which there is no valid LC_{50}, was handled appropriately only by ASTM-PROBIT, which recognized the slope to be too close to zero and did not attempt to produce an LC_{50}; DULUTH-TOX calculated both slope and an unrealistic LC_{50} estimate, while SAS-PROBIT correctly reported a slope = 0, but then also reported an LC_{50} = 0.

SAS-PROBIT has the great advantage over all other programs in that it provides a graphical output from which one can evaluate the quality of the fitted line visually, both as a

TABLE 3—*Comparison of slope, LC_{50}, and fiducial intervals*

Data Set	DULUTH-TOX				ASTM-PROBIT			
	Slope	LC_{50}	LFL	UFL	Slope	LC_{50}	LFL	UFL
1	4.21	4.85	4.36	5.35	4.21	4.85	4.36	5.35
2	16.70	44.85	0.00	+infinity	16.70	44.85	0.00	+infinity
3	2.44	2.40	0.00	30.94	2.44	2.40	0.00	30.85
4	2.44	6.66	0.51	+infinity	2.44	6.66	0.52	+infinity
5	0.79	2.77	6×10^{-39}	3×10^{30}	0.79	2.77	0.00	2×10^{30}
6	1.17	46.68	2.51	880.17	1.17	46.68	2.51	880.22
7	4.65	5.86	4×10^{-8}	1×10^{9}	4.65	5.86	4×10^{-9}	1×10^{10}
8	2.51	7.75	6.29	9.53	2.52	7.75	6.29	9.53
9	4.35	44.73	35.28	56.40	4.35	44.73	35.28	56.40
10	4.35	44.73	39.78	50.22	4.35	44.73	39.78	50.22
11	5.59	57.58	0.00	+infinity	5.59	57.58	0.00	+infinity
12	5.59	57.59	49.16	68.14	5.59	57.58	49.16	68.14
13	3.06	41.91	29.12	60.35	3.06	41.91	29.12	60.35
14	2.37	51.11	37.51	72.09	2.37	51.11	37.51	72.09
15	1.83	6.56	4.68	9.33	1.83	6.56	4.68	9.33
16	1.83	6.56	2.61	18.68	1.83	6.56	2.61	18.68
17	...[d]	...[d]	...[d]	...[d]	...[d]	...[d]	...[d]	...[d]
18	3.60	3.02	0.00	+infinity	3.60	3.02	0.00	+infinity
19	0.60	7.89	0.00	+infinity	0.60	7.88	0.00	+infinity
20	-2×10^{-6}	6×10^{-39}	0.00	+infinity	0.00	...	...	...

[a] Lower fiducial limits.
[b] Upper fiducial limits.
[c] Not enough cases.
[d] Less than two different concentrations with the number of dead strictly between 0 and 100%.
[e] Program crashed.

TABLE 4—*Summary of results for seven data sets*

Data Set	SPSSx-PROBIT												
	Fixed Control Model[a]						Control-Response Specified, Optimized						
	MCM,%[a]	Slope	SES[d]	LC_{50}	LFL_m[e]	UFL_m[f]	OCM[b]	SE_{ocm}[g]	Slope	SES	LC_{50}	LFL_m	UFL_m
1	16.3	2.77	0.18	16.62	13.80	19.68	17.5	3.3	2.79	0.21	16.78	13.36	20.37
2	16.3	2.77	0.18	9.54	7.81	11.39	17.5	3.3	2.79	0.21	9.66	7.46	11.96
	RP[h] = 1.742						RP = 1.738						
1	16.3	2.82	0.27	16.77	13.60	19.93	17.0	3.4	2.83	0.28	16.82	13.29	20.50
2	16.3	2.73	0.32	9.44	7.51	11.38	17.1	3.9	2.75	0.28	9.56	7.08	11.99
3	15.0	4.84	1.19	37.45	27.67	46.33	17.8	5.8	4.89	1.36	37.57	24.44	47.53
4	10.0	3.91	0.87	39.79	29.95	50.09	9.9	5.0	3.91	0.96	39.78	27.24	50.95
5	15.0	4.99	1.25	37.77	27.98	46.58	−0.01	...[i]	...	...	...	...	
6	50.0	2.97	1.03	29.71	11.45	44.87	54.7	9.8	3.00	1.32	29.87	0.57	51.65
7	15.0	3.82	0.87	33.43	24.01	42.69	15.9	6.5	3.82	1.01	33.45	20.08	44.14

[a] In the fixed control models, the measured (observed) control mortality (MCM) was used in the model.
[b] In the control-response specified, optimized model, the MCM was used as a specified seed value for the optimization; only the optimized control mortality is reported in the table (OCM).
[c] For the simply optimized model available with SAS-PROBIT only, no observed control mortality was specified as a seed value.
[d] SES = standard error of the slope.
[e] LFL_m = lower fiducial limit for LC_{50}.
[f] UFL_m = upper fiducial limit for LC_{50}.
[g] SE_{ocm} = standard error of OCM.
[h] RP = relative potency.
[i] Program crashed after producing the OCM reported.

(LFL[a], UFL[b]) for data sets 1 to 20 estimated with all 5 probit programs.

UG-PROBIT				SPSSx-PROBIT				SAS-PROBIT			
Slope	LC_{50}	LFL	UFL	Slope	LC_{50}	LFL	UFL	Slope	LC_{50}	LFL	UFL
4.15	4.87	4.40	5.39	4.21	4.85	4.36	5.35	4.21	4.85	4.36	5.35
13.17	43.74	40.42	47.33	15.09	44.90	44.90	44.90	16.70	44.85	...	...
2.05	2.56	1.83	3.59	2.44	2.40	2.40	2.40	2.44	2.40	...	...
2.05	6.24	4.45	8.75	2.44	6.66	6.66	6.66	2.44	6.66	...	...
0.79	2.77	1.83	4.18	0.79	2.77	6×10^{-22}	4×10^{16}	0.79	2.77	0.00	3×10^{17}
1.17	46.68	26.94	80.89	1.17	46.68	2.51	878.86	1.17	46.68	2.49	880.76
3.80	5.33	4.21	6.74	4.65	5.86	5.86	5.86	4.65	5.86	1.56	56.17
2.52	7.75	6.59	9.10	...[c]	...[c]	...[c]	...[c]	2.52	7.75	...	...
3.73	44.94	39.41	51.26	4.35	44.73	44.73	44.73	4.35	44.73	39.77	50.21
4.31	44.84	40.00	50.25	4.35	44.72	39.78	50.22	4.35	44.73	39.77	50.21
4.53	59.29	49.14	71.53	5.59	57.58	57.58	57.58	5.59	57.58	...	...
3.84	61.96	50.41	76.16	5.13	58.20	58.20	58.20	5.59	57.58	41.65	89.02
2.99	42.59	30.43	59.62	3.06	41.91	29.12	60.35	3.06	41.91	29.14	60.42
2.97	53.95	39.17	74.30	2.37	51.11	51.11	51.11	2.37	51.11	...	...
1.49	6.79	4.54	10.16	1.83	6.56	6.56	6.56	1.83	6.56	2.58	19.49
1.49	6.80	4.55	10.17	1.83	6.56	6.56	6.56	1.83	6.56	2.58	19.57
13.30	60.00	56.18	64.08	17.90	60.00	...	...	...[d]	...[d]	...[d]	...[d]
2.84	3.06	2.31	4.05	3.60	3.02	3.02	3.02	3.60	3.02	...	...
...	...	...	...	0.60	7.89	7.89	7.89	0.60	7.89	...	...
0.02	1.8×10^{4}	0.00	1.5×10^{46}	0.00	...[e]	...[e]	...[e]	0.00	0.00	...	...

with observed control mortality greater than zero.

SAS-PROBIT															
	Fixed Control Model				Control-Response Specified, Optimized Model					Optimized Model[c]					
MCM, %	Slope	LC_{50}	LFL_m	UFL_m	OCM	Slope	LC_{50}	LFL_m	UFL_m	OCM	Slope	LC_{50}	LFL_m	UFL_m
...	...	...	...	...	...	...	...	...	...	...	...	...	...	...
...	...	...	...	...	...	...	...	...	...	...	...	...	...	...
16.3	2.82	16.77	13.59	19.93	16.50	2.82	16.82	13.22	20.40	16.50	2.82	16.82	13.22	20.40
16.3	2.73	9.44	7.49	11.39	16.58	2.74	9.51	7.02	11.93	16.58	2.74	9.51	7.02	11.93
15.0	4.84	37.46	26.33	46.44	16.41	5.03	38.22	21.16	48.25	16.41	5.03	38.22	21.16	48.25
10.0	3.91	39.79	29.89	50.12	9.95	3.91	39.76	26.78	51.03	9.95	3.91	39.76	26.78	51.03
15.0	4.99	37.77	27.89	46.66	8.91	4.47	35.07	26.01	43.91	8.91	4.47	35.07	26.01	43.91
50.0	2.97	29.72	4.96	45.02	55.92	3.88	36.50	...	57.97	55.92	3.88	36.50	...	57.97
15.0	3.82	33.43	23.8	42.72	15.61	3.87	33.75	19.65	44.47	15.60	3.87	33.75	19.65	44.47

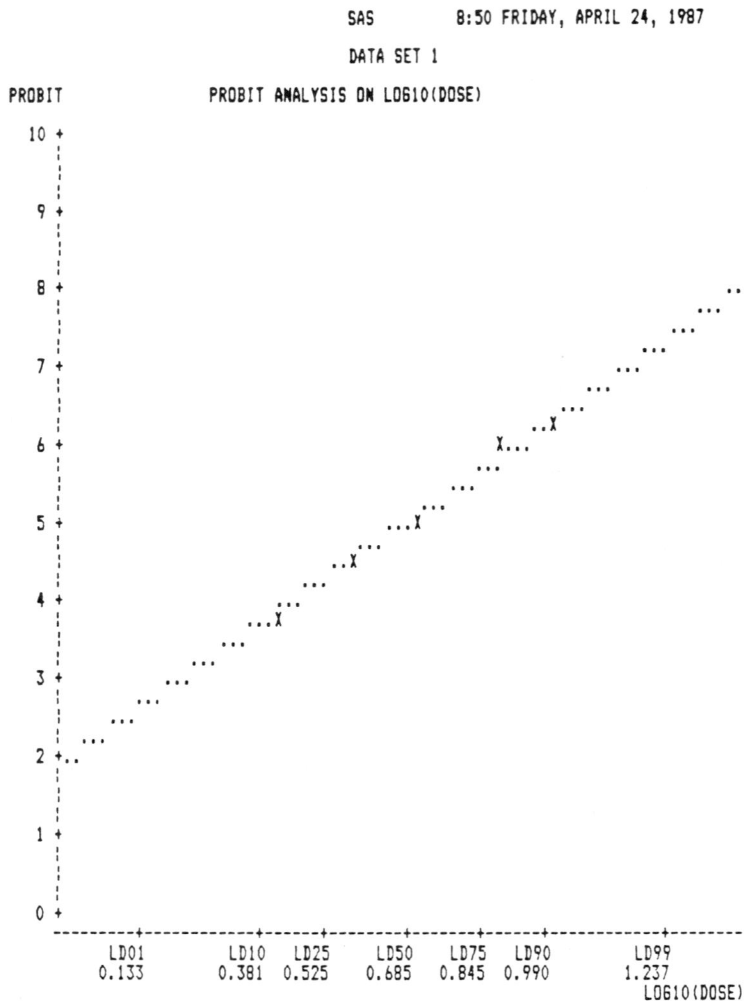

FIG. 1—*Graphic output from SAS Probit Procedure.* (a) *Plot of probit values against log10(dose).*

probit plot and a log-probability plot (Fig. 1*a* and *b*). SPSSx-PROBIT provides a scatter plot of the data, but no fitted line (Fig. 2).

SPSSx-PROBIT showed significant lack of agreement in a number of cases, but generally, the estimates were reasonable and not recklessly different from those of the other programs; these results would be acceptable. One quirk of this program is the duplication of the LC_{50} as the upper and lower fiducial limits in those cases for which there are no fiducial limits; that is, the limits are 0 and +infinity. No warning is given in the data output or the manual about this anomaly.

UG-PROBIT is deemed unacceptable since it consistently estimated the slope, LC_{50}, and fiducial limits incorrectly even for "good" data sets. In addition, it is somewhat awkward to use, despite reasonably good data entry and editing routines, requiring the operator to

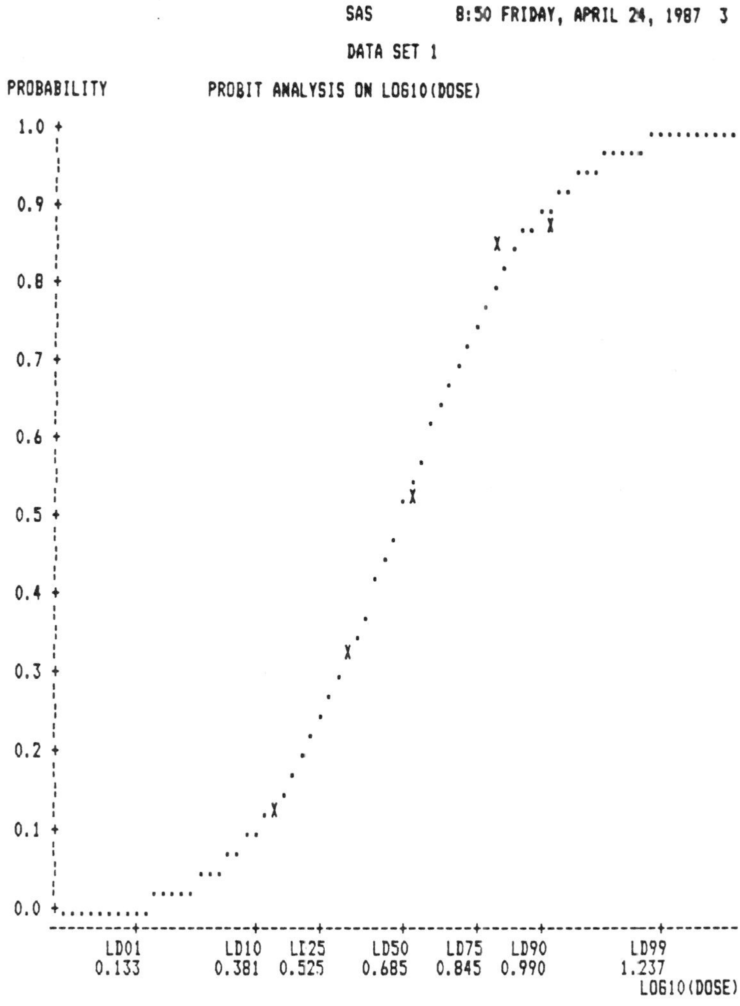

FIG. 1—*Continued:* (b) *Plot of equivalent percentages against log10(dose).*

decide after each iteration whether another iteration is appropriate rather than providing an objective routine. While this may be desirable as a teaching tool, it is not desirable when analyzing large numbers of data sets.

Evaluation of Data Sets with Nonzero Control Responses

UG-PROBIT, SPSSx-PROBIT, and SAS-PROBIT all allow for inclusion of nonzero control responses in the data set. UG-PROBIT does not adjust dosed treatment responses for the threshold mortality and, therefore, will not be discussed further. SPSSx-PROBIT and SAS-PROBIT may be instructed either to perform all calculations with a user-defined control mortality or to optimize the threshold mortality based on responses for the lowest

```
PROBIT ANALYSIS SPSS-X Release 2.1+ --- VIMS   Prime 9955        PAGE #

* * * * * * * * * * * * *  P R O B I T   A N A L Y S I S  * * * * * * * * * * * *
*
```

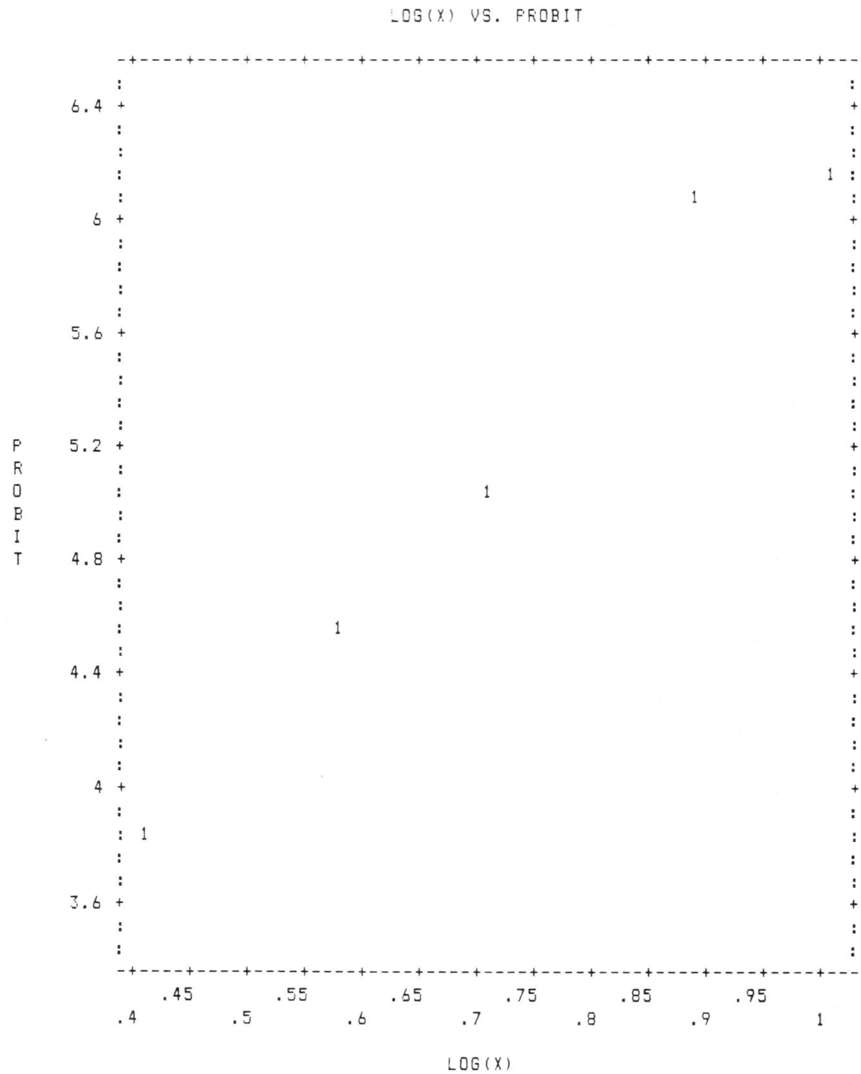

FIG. 2—*Graphic output from SPSSx probit procedure.*

doses and control treatment data included in the data files. SAS-PROBIT also provides for LC_{50} calculations with a control response specified by the user prior to optimization.

When analyzed in such a way that the relative potency could be estimated by SPSSx-PROBIT, the LC_{50} values calculated using the fixed control mortality did not agree precisely with the values derived under SAS-PROBIT for Data Sets 1 and 2; the relative potency estimate, however, is reasonably close to that given by Finney [*5*]. Under this model, SPSSx-PROBIT assumes equal slopes for the two data sets, whereas SAS-PROBIT, which does not include relative potency estimation as an option, estimated different slopes (Table 4). Agreement in LC_{50} between the outputs for the two programs was closer for the control-response specified, optimized model, but the optimal control mortality estimates differed dramatically. When Data Sets 1 and 2 were reanalyzed with SPSSx-PROBIT treating the data sets as totally independent (as in the SAS-PROBIT model), the LC_{50} and slope results using both models were nearly identical to those for SAS-PROBIT.

For Data Sets 3 through 7, the programs agreed in both slope and LC_{50} estimates within rounding error when calculated using the fixed control model (Table 4). The slope and LC_{50} from SPSSx-PROBIT were somewhat higher than those for SAS-PROBIT when applying the optimized model for Data Sets 3 and 4. For Data Set 5, in which one dosed treatment exhibited no mortality, SPSSx-PROBIT was unable to calculate a reasonable optimum control mortality and slope, and crashed during the attempt to calculate the LC_{50}. For Data Set 6 with 50% control mortality, SPSSx-PROBIT grossly underestimated the optimum control mortality, although the LC_{50} was comparable to that obtained using the fixed control model. SAS-PROBIT handled all data sets in an entirely reasonable manner, although

TABLE 5—*Evaluation of each program for selected characteristics.*

	Program				
	DULUTH-TOX	ASTM-PROBIT	UG-PROBIT	SPSSx	SAS
Type of data input	cards/screen	cards/screen	screen	cards/file	cards/file
Ease of data input	awkward/easy	awkward/easy	easy	easy/depends on editor	easy/depends on editor
Allows data edit	awkward/easy	awkward/easy	easy	depends on editor	depends on editor
Method of operation	keyboard interactive	keyboard interactive	keyboard interactive	batch program	batch program
Type of data output	screen/printer	screen/printer	screen	file	file
Ability to handle "poor" data	excellent	excellent	poor	fair	excellent
Ability to handle control mortality greater than zero	none	none	none	fair	excellent
Ability to calculate relative potency	no	no	no	yes	no
Ability to plot data points	no	no	no	yes	yes
Ability to plot probit line	no	no	no	no	yes
Ability to plot survivorship curve	no	no	no	no	yes
Documentation	limited within program	ASTM draft practice	none	manual	manual
Support	none	none	none	phone	phone

the LC_{50} for Data Set 6 with the optimized models was somewhat higher than that obtained with the fixed control model.

SAS-PROBIT was also used to analyze the data using the optimization model with the observed control mortality given explicitly as an initial value. In every case the LC_{50}, slope, optimum control mortality, and fiducial intervals were identical to the simple optimization model without an explicit initial control mortality. Although the results during the initial iterations were different in some cases, generally the number of iterations to reach an optimum was not affected. There seems to be no benefit to providing an initial value.

Conclusions

These five programs have been evaluated both for characteristics related to the user interface (Table 5) and for agreement of the calculated output with published results (in three cases) and among the programs (Tables 3 and 4).

Most of these programs yield useful and interchangeable LC_{50} estimates; only UG-PROBIT yielded consistently deviant results. DULUTH-TOX and ASTM-PROBIT have a major advantage in that objective tests are included to insure that the input data meets the assumptions of the test and to guide the interpretation of output. The commercial statistical programs have the advantages of graphical output and a method for handling control mortality.

DULUTH-TOX, ASTM-PROBIT, and UG-PROBIT can be implemented on microcomputers, whereas SPSSx-PROBIT and SAS-PROBIT are presently only available for use on mini and mainframe computers. DULUTH-TOX and ASTM-PROBIT can be used on a mainframe with various BASIC interpreters; some minor modifications may be necessary depending on the computer and interpreter used. UG-PROBIT would require considerable modification in order for it to be implemented on a mini or mainframe computer.

Acknowledgments

Mr. G. Piehler encouraged me to undertake this project; without his support, this work would not have been attempted. The presentation of the data benefits greatly from the suggestions of R. Lukins (Virginia Institute of Marine Science), D. Weiner (Statistical Consultants, Inc., Lexington, KY), as well as two anonymous reviewers. Any errors of data or interpretation remain, however, the responsibility of the author. I also thank C. E. Stephan who several years ago made available to me the source code for the program herein referred to as DULUTH-TOX and with whom I have discussed data analysis numerous times over the past several years.

This paper is Contribution No. 1480 from the School of Marine Science, College of William and Mary, Gloucester Point, VA.

References

[1] Stephan, C. E. in *Aquatic Toxicology and Hazard Evaluation (First Symposium), ASTM STP 634,* F. L. Mayer and J. L. Hamelink, Eds., American Society for Testing and Materials, Philadelphia, 1977, pp. 65–84.

[2] Hubert, J. J., *Bioassay,* 2nd ed., Kendall/Hunt Publishing Co., Dubuque, IA, 1984.

[3] Litchfield, J. T., Jr. and Wilcoxon, F., *Journal of Pharmacology and Experimental Therapeutics,* Vol. 96, No. 2, June 1949, pp. 99–113.

[4] Thompson, W. R., *Bacteriological Reviews,* Vol. 11, No. 2, June 1947, pp. 115–145.

[5] Finney, D. J., *Probit Analysis,* third ed., Cambridge University Press, London, 1971.

[6] *SPSSx User's Guide, Edition 2,* SPSS Inc., Chicago, IL, 1986, pp. 602–619.

[7] DeLong, D. M. and Goodnight, J. H. in *SAS User's Guide: Statistics, Version 5 Edition,* SAS Institute, Inc., Cary, NC, 1985, chapter 29, pp. 639–645.

David Groggel,[1] Robert Schaefer,[1] and John Skillings[1]

Effects of Withdrawals on Tests Involving Censored Data in Toxicology Experiments

REFERENCE: Groggel, D., Schaefer, R., and Skillings, J., **"Effects of Withdrawals on Tests Involving Censored Data in Toxicology Experiments,"** *Aquatic Toxicology and Environmental Fate: Eleventh Volume, ASTM STP 1007,* G. W. Suter II and M. A. Lewis, Eds., American Society for Testing and Materials, Philadelphia, 1989, pp. 321–338.

ABSTRACT: Censoring can occur with some response variables that are frequently of interest in toxicology studies (for example, time to maturation or lifetime). This censoring can be in the form of fixed right censoring or a withdrawal, and its presence can have a major impact on the analysis of the data. In this paper we consider the effect of censoring on statistical procedures that are commonly used in toxicology experiments. The tests examined in this study are the Gehan and log rank tests which are available on the major statistical packages (SAS, SPSSX, and BMDP). The application of these tests for $k \geq 2$ samples and for multiple comparisons is considered. The effect of withdrawals on the performance of these tests is found to be quite different than the effect of fixed right censored observations. Our study results indicate that these tests can lead to incorrect or misleading inferences when there is a high percentage of withdrawals. The tests may have very poor sensitivity to treatment differences, and the actual significance levels may be quite different from what the user specifies.

KEY WORDS: censoring, withdrawals, Gehan test, log rank test, multiple comparisons, analysis of variance, toxicity studies, aquatic toxicology

In toxicology experiments it is often the case that final measurements cannot be taken on some subjects that are being studied. This can occur because the study was terminated before the measurement could be completely determined or because the subject was removed from the study before the measurement could be completely determined. For example, consider a study where the goal is to determine the effect of several levels of copper concentrations on the lifetime of an aquatic species. Suppose that the study must be ended in no later than 60 days. Any organism that lives longer than 60 days will not have their actual lifetime recorded. Only a lower bound on the lifetime can be observed. Consider the same copper study again, but suppose now that the goal of the study is to compare the mean time to maturation for organisms in the various copper concentration groups. If an organism dies before reaching maturity, then a measurement on the time to maturation cannot be made. Such examples are illustrations of problems in which censoring can occur.

A data observation is censored if the actual measurement cannot be made on the subject of interest. The goal of this paper is to consider the effect of censored observations on nonparametric statistical procedures that are commonly used in analyzing data from toxicology experiments. We will consider the effects of censoring on the comparison of two

[1] Associate professors and professor, respectively, Department of Mathematics and Statistics, Miami University, Oxford, OH 45056.

groups (for example, treatment versus control and multiple comparisons) and on the comparison of several groups (the one-way analysis of variance).

The most common types of censoring that occur in toxicology experiments are fixed right censoring and withdrawal censoring. Fixed right censoring occurs in studies that have a set duration time. An observation on a subject is called a fixed right censored observation if the subject is still alive at the end of the study and the measurement of interest has not yet been made for the subject. The number recorded for the subject is the time at the end of the study. For example, in the 60-day copper experiment described previously, if the measurement of interest is lifetime, then organisms that are still alive at 60 days will be right censored. The observation will be recorded as 60 days. Notice that we have assumed that all subjects start the study at the same time. We will continue to assume this throughout our discussion.

Another type of censoring that is common in toxicology experiments is withdrawal censoring. A subject that leaves the experiment for some reason before a measurement can be made is called a withdrawal. The data observation that is made is the time of withdrawal, namely, the time that the subject left the study. This type of censoring occurs mainly when the measurement of interest is something other than the lifetime. For example, in the experiment studying the effects of copper levels in water on an organism, the measurement of interest may be the time to maturation. Any organism that dies before this measurement can be completely determined is a withdrawal. Even if the lifetime is the measurement of interest, any organism that exits the study for some other reason (that is, dies because of equipment failure or is lost from the study) is a withdrawal.

Toxicology experiments can involve both fixed right censoring and withdrawals. Also, the degree of censoring may vary from group to group. For example, if one is comparing the effects of several copper concentrations in a lifetime study with a fixed duration time, then the organisms in the heavy copper concentrations may all die before the study is finished. On the other hand, most of the organisms in the control group (no copper) may still be alive at the end of the study. In this case, the heavy copper concentration group will have no fixed right censored observations, while the control group will have several.

Some previous studies (for example, Refs *1* to *4*) have determined that censored observations can greatly affect the performance of the commonly used nonparametric statistical procedures. These studies have concentrated on the two-sample problem and described little about the effects on other common analyses. Our goal is to consider the effect of censoring on these other analyses that are common to toxicology studies.

Common Tests

In this section we discuss some of the statistical tests and procedures that are commonly used in toxicology studies. We also consider the availability of computer programs on the three main statistical computer packages (SAS, SPSSX, and BMDP) that perform these tests. (We consider the mainframe version of these packages only since the PC (personal computer) versions typically do not include the "survival" programs.) Each package has a survival analysis program; however, all three use different tests and have different provisions for pair-wise comparisons of the populations.

In the analysis of survival data, one of the primary questions of interest is whether or not two or more populations have the same survival distribution. The survival distribution is the probability of survival beyond a certain time. Typically, the hypothesis of interest is whether the probability of survival at any time is the same for all of the populations; therefore, the null hypothesis states that the survival distributions are identical. The tests we investigate are primarily location tests, meaning that they will detect differences in the mean or center of the distribution, but may not detect differences in the variances or shapes of the distributions.

In this paper we will distinguish between the two-sample case, in which we only wish to compare two populations, and the k-sample case, in which $k > 2$ populations are to be compared. In both cases the hypothesis of interest is whether the survival distributions of the populations are the same. In the k-sample case, if the analysis indicates that the populations differ, multiple comparisons of the populations are often performed to determine which of the populations are different. In toxicology experiments we are primarily interested in comparing each of the $k - 1$ treatments to the control population. The $k - 1$ treatments, however, are not usually compared among themselves. Hence, the multiple comparison procedures we concern ourselves with are the treatment versus control variety.

When the data are complete (not censored) an analysis of variance (ANOVA) could be used to test the hypothesis. However, when the data are subjected to censoring, ANOVA techniques are no longer appropriate and other tests must be used. Two popular and widely used tests for comparing two or more survival distributions are the Gehan/Breslow and log rank tests. Both of these tests are nonparametric tests since they make no assumption (other than that the distributions are continuous) about the shape or form of the survival distribution in the populations. Both tests will detect differences in the mean or center of the survival distribution for two or more groups, but may not detect differences in the scale or shape or both.

The Gehan [*5*] test is a two-sample test, whereas the Breslow [*6*] test is the k-sample or one-way ANOVA extension of the Gehan test. (These tests are also referred to as the Wilcoxon test.) In the Gehan test, two survival distributions are compared as follows. Take each observation in one of the samples and count how many observations are *known* to be less than the true lifetime corresponding to this observation and subtracting the number of observations that are *known* to exceed the true lifetime. (Recall that with censored data we do not actually know the true lifetime but rather a lower limit. Hence, any observation greater than a censored lifetime *may or may not exceed* the true lifetime.) The Gehan test statistic is then based on the sum of all the counts corresponding to the observations in this sample. If the survival distributions for the two groups are equal, then we would expect the survival times to be similar for the two groups; hence, this sum would be close to zero. If the survival distributions are not equal, then the sum of the counts would tend to be positive or negative (depending on whether the sample we use is the one with larger or smaller survival times) indicating that the two survival distributions are not the same. As mentioned earlier, the Breslow test is an extension of the Gehan test to more than two samples.

The Gehan/Breslow test is a nonparametric test. No assumption is made about the form of the survival distributions in any of the populations; however, the test does assume that, when the null hypothesis is true, the censoring distribution is equal in all of the populations. Hence, as we will see later, if the percent censoring or pattern of censoring is different when the null hypothesis is true, the Gehan/Breslow test results may be misleading. Finally, we note that there are several different versions of the Gehan/Breslow test. Their differences lie in the different variance forms of the test statistic that can be derived. All of these tests, when appropriately standardized and transformed, have asymptotic chi-square distributions with degrees of freedom (df) equal to the number of populations minus one. That is, for large sample sizes the test can be approximated using a chi-square distribution with df $= k - 1$. In later sections we show that it is the number of uncensored observations in each sample that must be sufficiently large and not simply the sample size for these approximations to be valid.

The log rank test also assigns scores to each lifetime, but these scores are more complicated computationally than the Gehan/Breslow scores. The computational form of these scores can be found in Miller [*7*]. The log rank test is applicable in both the two- or k-sample cases. As with the Gehan/Breslow test, the log rank test is nonparametric. The test does assume, however, that the censoring distribution is the same for the populations when the null hypothesis is true.

The log rank test can also be computed in several different ways. One way involves obtaining the sum of the scores for each of the groups or populations and the variances and covariances of these group scores. The log rank test is then the quadratic form of the vector of these scores. The second, simpler, method of computing the log rank statistic involves computing observed (O) and expected (E) scores for each group and forming the "usual" chi-squared statistic by summing over the groups the observed group score minus the expected group score squared and divided by the expected group score ($\Sigma_i (O_i - E_i)^2/E_i$). It can be shown that this second form of the log rank test is conservative (its value is less than or equal to) when compared to the log rank test described above. For large sample sizes, the log rank test statistic will have an approximate chi-square distribution with degrees of freedom given by $k - 1$ regardless of which log rank computation is used. Computational formulas for both the "quadratic form" and the "conservative" versions of the log rank test are given in Miller [*7*].

Recently there have been other nonparametric tests proposed for comparing two or more survival distributions. For example, in Refs *3* and *4*, a test due to Peto and Prentice (see Ref *3* for a description of the test) was investigated and found to have merit as an alternative test procedure.

Although not quite as common as the nonparametric tests given above, parametric tests are also available. In a parametric test, the distributional form for the survival distribution must be specified. For example, if the data have a known distributional form, then a likelihood ratio test can be used. (The details of the likelihood ratio test can be found in most graduate level theory texts; see Ref *8*, for example.) Such tests work very well if the survival times have the stated distributional form, and they can work very poorly if the survival times have a different distributional form. These parametric tests are usually carried out using a large sample approximation based on the chi-square distribution.

We now turn our attention to which tests are available on the three major statistical packages. In SAS [*9*], three tests are available: the Gehan/Breslow (using the permutation variance, see Miller [*7*]), the log rank, and the likelihood ratio tests. The program that performs the analysis is PROC SURVTEST. The appropriate version of the test (two sample or k sample) is automatically selected, and one can select any or all three test statistics. The Gehan/Breslow test performed is as described above. The version of the log rank test given by SAS is the conservative version. The likelihood ratio test assumes that the survival times are exponentially distributed. The output provides the sample size, number of uncensored observations, and total score for each group. Also given is the asymptotic test result (based on the chi-square approximation), the degrees of freedom, and the P-value. SURVTEST does not offer any multiple comparison procedure. However, by forming data sets comprised of pairs of groups, the two-sample statistics could be obtained and used to compare the groups on a pair-wise basis.

In SPSSX [*10*], the survival distribution analysis program is SURVIVAL. Only the Gehan/Breslow test is available. The output gives the sample size, number of censored and uncensored observations, the percent censored, and the mean score for each group. The asymptotic chi-square test statistic, degrees of freedom, and P-value are also given. Of the three packages, only SPSSX's SURVIVAL will give the pair-wise comparisons. If the pair-wise option is specified, the two-sample test results are given for each pair of groups or treatments. The format of the pair-wise output is identical to that given above.

In the BMDP [*11*] program BMDP1L, two tests can be requested: the Generalized Wilcoxon test, which is a quadratic form version of the Gehan/Breslow test (see Ref *7*), and the Generalized Savage test, which is the quadratic form version of the log rank test. The output includes the sample size, number of censored and uncensored observations, and the percent censored in each group. The asymptotic chi-square test statistic, the degrees of freedom, and the *P*-value are given. Also, the graphed patterns of censored and uncensored observations can be obtained for each group. Like SAS, pair-wise comparisons cannot be requested, but rather must be obtained by forming data sets comprised of pairs of groups and using BMDP1L on each data set separately.

Results for Comparison of More Than Two Samples

When there are more than two populations of interest, one usually performs first an overall test for the equality of the $k > 2$ survival distributions. In this section we investigate the performance of the four k-sample test procedures that are available on SAS, SPSSX, and BMDP; these are the conservative log rank, the quadratic form log rank, the quadratic form Gehan, and the permutation Gehan. Recall that SAS provides the conservative log rank and permutation Gehan tests, SPSSX performs the permutation Gehan test, and BMDP performs the quadratic forms of both the log rank and Gehan tests. These four tests all test the null hypothesis that the k populations have the same survival distribution.

Using simulation techniques, we investigated the effect sample size, different censoring patterns in the populations, and different types and combinations of censoring in the populations had on these tests. The number of populations investigated was fixed at four throughout the simulation. We investigated the Weibull distribution in this simulation study since previous results indicated that the distribution of the lifetimes did not have as large an impact on the tests as the other factors investigated. The Weibull is a unimodal, skewed right distribution for which the mean and standard deviation both depend on the shape *and* scale parameters. We kept the shape parameter fixed at four throughout the study and varied the scale parameter, θ, to simulate different lifetime distributions. Thus, the Weibull distribution used in the simulations study has density function

$$f(x) = (4x^3/\theta) \exp(-x^4/\theta) \text{ for } x > 0$$

(Note, for the time to maturation the same Weibull distribution was used with the scale parameter denoted by δ.)

The sample sizes considered included: (10, 10, 10, 10), (20, 20, 20, 20), and (20, 10, 10, 10), where the first value is the number in the control group followed by the number in the 3 treatment groups. Previous studies have shown that for samples >25, most of the survival tests available perform reasonably well, so we investigated the smaller, more common sample sizes. We considered two different types of censoring: fixed right and withdrawal. Our simulations included cases with fixed right censoring only, withdrawal censoring only, and both fixed right and withdrawal censoring. An observation was censored by fixed right censoring if the generated lifetime was larger than the duration of the study. An observation was censored by withdrawal censoring if the withdrawal time generated was less than the response value generated. The Weibull distribution with shape parameter equal to four was also used as the withdrawal distribution.

The percentage censoring in each sample was also varied and was determined by changing the parameters of the lifetime or withdrawal distributions or both. For example, in the fixed right censoring only case, let the scale parameters of the Weibull lifetime distributions of the four populations be denoted by (θ_1, θ_2, θ_3, θ_4). The expected percentage censored for any population is the probability that a lifetime exceeds the endpoint of the study. Since the θ's can be different for the different populations, the expected percentage censored in each population can differ. The expected percentage censored was controlled by changing the parameters of the lifetime or withdrawal distribution or both. Note that when withdrawals are considered, the expected percentage of withdrawals can be altered by changing the parameters of the lifetime *or* withdrawal distribution. For example, consider the case where the response of interest is the time to maturation. However, if the organism dies before maturing, the observation is censored. Thus, the distribution of the lifetime of the organism serves as the censoring distribution. Consequently, if the mean time to maturation *or* the lifetime of an organism is affected by the toxin, the censoring distribution will change.

In the simulation we considered several degrees of censoring, ranging from 0% censoring to 97%. We considered cases where the censoring distributions were the same and also cases where they were not the same in the four populations. In general, when changing the parameters, we tried to simulate what happens in real toxicology experiments. For example, we considered cases where the mean lifetimes or survival times changed linearly, indicating a dose-response type of relationship. In others we considered a threshold type of model in which the mean lifetime or survival time would change value after a certain threshold dose. Finally, we considered the case where the mean time to maturation would remain unchanged until it jumped severely as a result of a high level of toxin.

We also point out that by changing the censoring distribution in the different populations, we have violated the assumption of equal censoring distributions in all populations. Our goal was to determine how the four tests react to this violation of assumption that sometimes occurs in toxicology data.

We now present selected results of the simulation study for the fixed right censoring only and withdrawal censoring only cases. The results when both fixed right and withdrawal censoring are present were very similar to those with withdrawal censoring only and, therefore, are not presented here.

Table 1 provides the design parameters for the fixed right censoring only case. Presented in the table are the scale parameters used for the four distributions of lifetimes (the first group is always the control), the expected percentage of fixed right censored observations

TABLE 1—*Expected percent censoring and comparison of population mean lifetimes for the various Weibull parameters investigated (fixed right only censoring).*

Lifetime Distribution Scale Parameters/ (θ_1, θ_2, θ_3, θ_4)	Level of Censoring	Expected Percent Censoring[a]	Number of Standard Deviations Below Control Mean Lifetime[b]
Null case/	light	(0, 0, 0, 0)	(−, 0, 0, 0)
(4, 4, 4, 4)	moderate	(50, 50, 50, 50)	
	heavy	(90, 90, 90, 90)	
Dose-response	light	(0, 0, 0, 0)	(−, 0.25, 0.57, 1.04)
alternative/	moderate	(50, 40, 25, 6)	
(4, 3, 2, 1)	heavy	(90, 87, 81, 66)	
Threshold alternative/	light	(0, 0, 0, 0)	(−, 0, 1.04, 1.04)
(4, 4, 1, 1)	moderate	(50, 50, 6, 6)	
	heavy	(90, 90, 66, 66)	

[a] Using fixed right censoring with endpoints of plus infinity, 1.2904, and 0.8057, respectively.
[b] These values were calculated using the standard deviation of the control.

TABLE 2—*Expected percent censoring and comparison of population mean time to maturation for the various Weibull parameters investigated (withdrawal only censoring).*

Time to Maturation Distribution Scale Parameters/ (δ_1, δ_2, δ_3, δ_4)	Level of Censoring	Expected Percent Censoring[a]	Number of Standard Deviations Above Control Mean Time to Maturation[b]
Null Case/ (0.5, 0.5, 0.5, 0.5)	light	(7.7, 7.7, 7.7, 7.7)	(−, 0, 0, 0)
	moderate	(7.7, 7.7, 50.0, 50.0)	
	heavy	(7.7, 83.3, 83.3, 83.3)	
Dose-response alternative/ (0.5, 1.0, 1.5, 2.0)	light	(7.7, 14.3, 20.0, 25.0)	(−, 0.67, 1.13, 1.48)
	moderate	(7.7, 25.0, 50.0, 80.0)	
	heavy	(5.7, 90.8, 93.6, 95.1)	
Threshold alternative/ (0.5, 0.5, 2.0, 2.0)	light	(7.7, 7.7, 25.0, 25.0)	(−, 0, 1.48, 1.48)
	moderate	(7.7, 7.7, 80.0, 80.0)	
	heavy	(7.7, 7.7, 95.2, 95.2)	
Jump alternative/ (0.5, 0.5, 0.5, 4.0)	light	(7.7, 7.7, 7.7, 40.0)	(−, 0, 0, 2.43)
	moderate	(7.7, 7.7, 50.0, 88.9)	

[a] The expected percent censoring is determined by the parameters of the lifetime (or withdrawal) and time to maturation distributions.

[b] These values were calculated using the standard deviation of the control.

using various endpoints for the study, and the magnitude of the difference in the mean lifetimes (in terms of standard deviations) between each population and the control population. In order to see the effects of various amounts of censoring, the endpoint of the study was changed to yield what will be referred to as light, moderate, and heavy amounts of censoring. Table 2 provides the same type of information for the withdrawal censoring only case. Recall that in this case the variable of interest is time to maturation with the lifetime distribution acting as the censoring distribution. The jump alternative (described above) is not considered until the next section.

Figure 1 gives the results of the simulation study when equal sample sizes of 10 were used. Each bar represents the proportion of times (out of 5000 runs) that the null hypothesis of equality of the population distributions is rejected using an α value of 0.05.

We first examine how the four tests control the error level when the null hypothesis is true. These entries are the observed probabilities of rejecting the null hypothesis when it should not be rejected and the target value is 0.05. Problems arise when the observed error values are liberal (>0.05) or conservative (<0.05). Liberal tests are undesirable because they lead to incorrect rejection of the null hypothesis too frequently. Conservative tests protect well against incorrectly rejecting the null hypothesis, but at the expense of a loss of power (lower probability of correctly rejecting the null hypothesis).

We see from Fig. 1 that for both fixed right and withdrawal censoring the quadratic log rank is very liberal for light and moderate amounts of censoring. For the same cases, the conservative log rank and quadratic Gehan are slightly liberal, while the permutation Gehan gives observed error values near the target value. For heavy amounts of fixed right censoring, all of the tests become very conservative. For heavy amounts of withdrawal censoring, there is also a drop in the observed error levels, but only the permutation Gehan drops far below the 0.05 level.

Though not presented in Fig. 1, similar relationships among the four tests existed for sample sizes of (20, 20, 20, 20) and (20, 10, 10, 10). In all of these cases the observed error levels were generally highest for the quadratic log rank, and lowest for the permutation

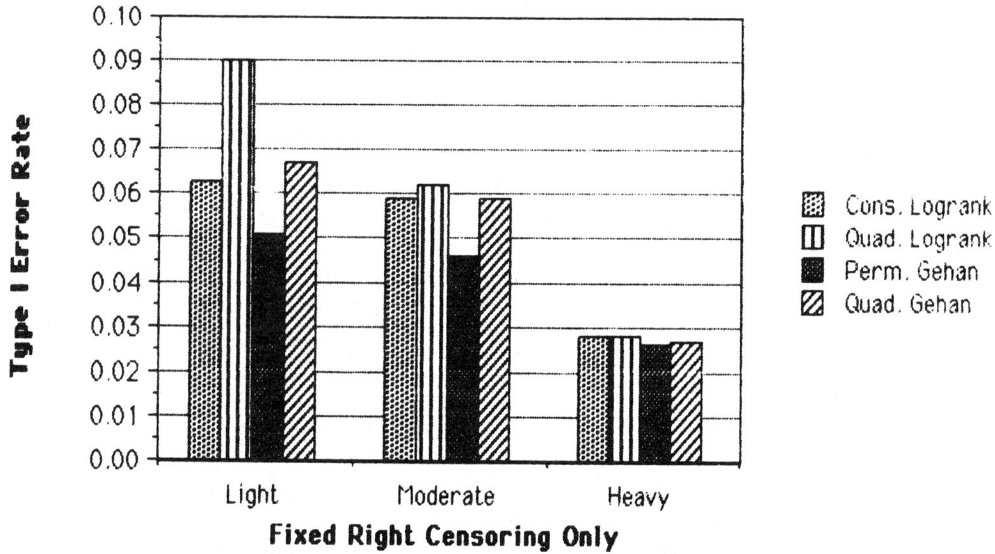

FIG. 1—*Proportion of rejections of the null hypothesis (out of 5000 runs) of equality of population distributions using* $\alpha = 0.05$*, four populations, and sample sizes of 10 for each sample:* (a) *null hypothesis true, fixed right censoring only.*

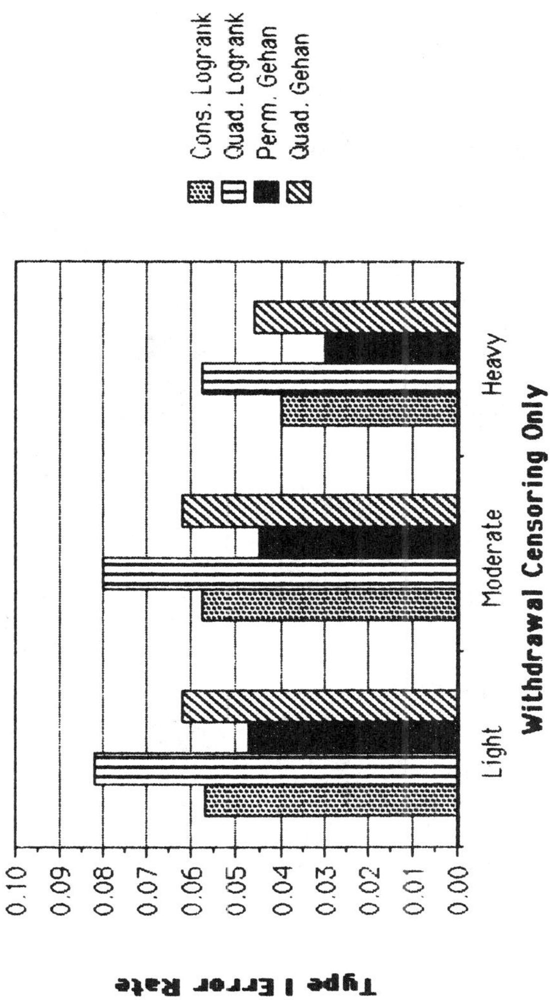

FIG. 1—*Continued:* (b) *null hypothesis true, withdrawal censoring only.*

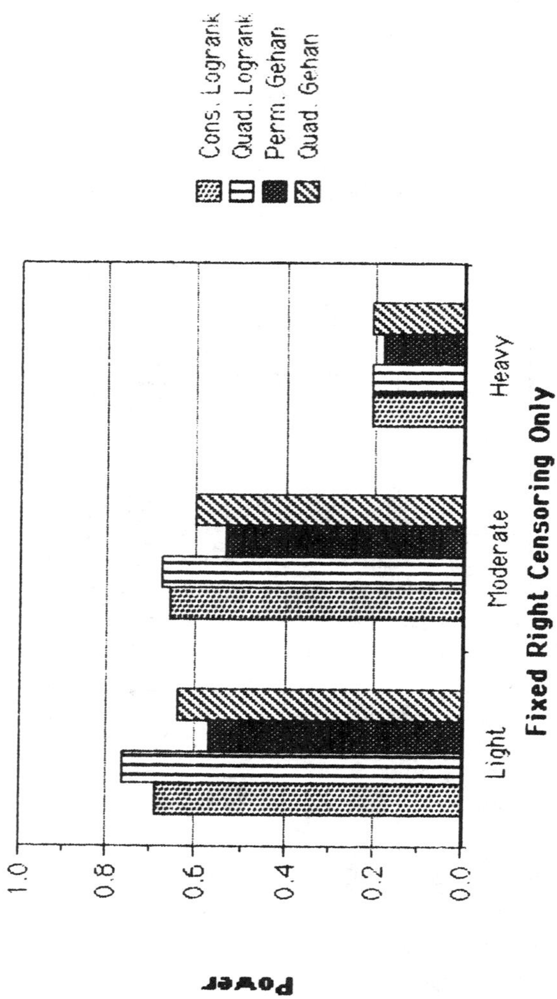

FIG. 1—*Continued: (c) dose-response alternative, fixed right censoring only.*

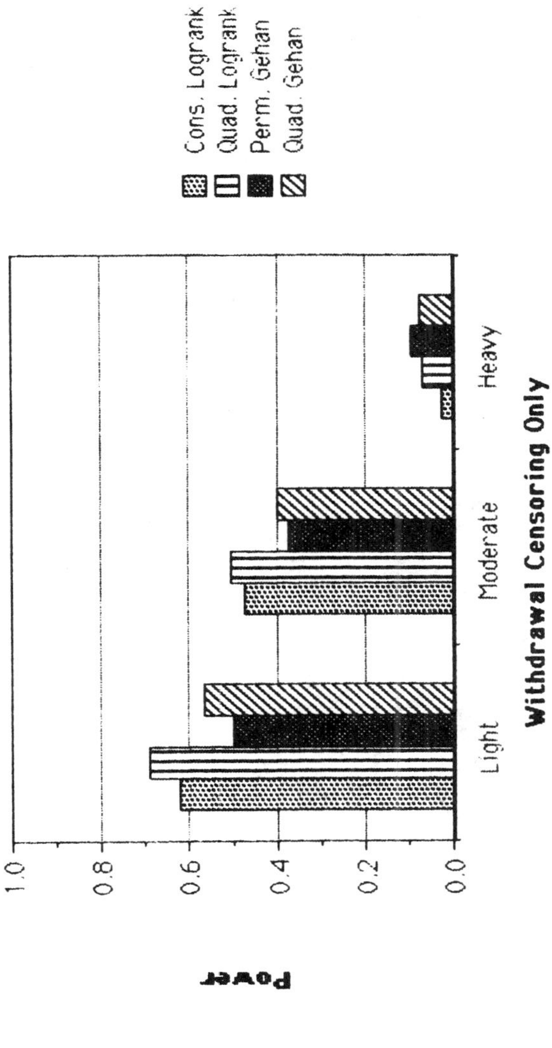

FIG. 1—*Continued:* (d) *dose-response alternative, withdrawal censoring only.*

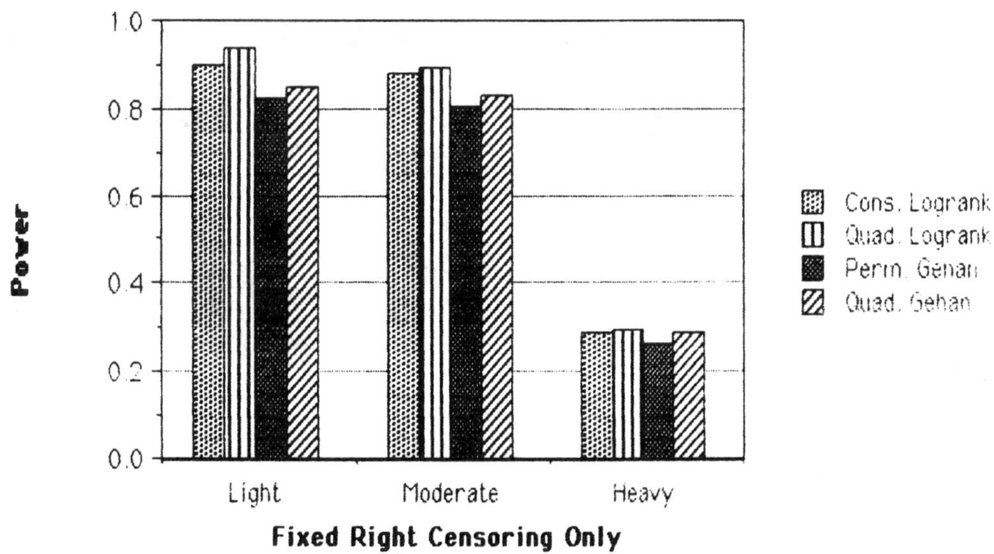

FIG. 1—*Continued:* (e) *threshold alternative, fixed right censoring only.*

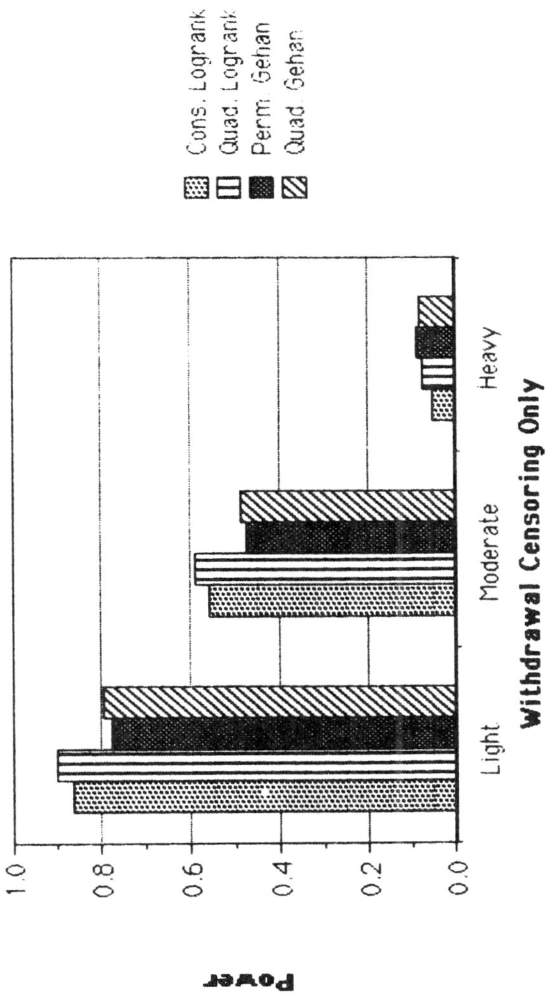

FIG. 1—*Continued:* (f) *threshold alternative, withdrawal censoring only.*

Gehan, with the values for the other two tests falling in between. For the larger sample sizes of (20, 20, 20, 20), the observed error levels for all the tests were consistently closer (less affected by censoring amounts) to the target value of 0.05 than they were for the smaller sample sizes of (10, 10, 10, 10). For the sample sizes of (20, 10, 10, 10), all of the tests produced observed error levels similar to those seen in Fig. 1 except for the permutation Gehan. In this situation of unequal sample sizes, the permutation Gehan produced error levels that were consistently conservative, sometimes severely so. Again the performance of the other tests was more consistent for the various censoring amounts, possibly due to the added observations in the control group.

Figure 1 also gives the observed powers of the four tests for the two alternatives. The first thing we see is that the power decreased as the amount of censoring increased. This is true for all four tests and for both fixed right and withdrawal censoring. It is also quite obvious that increasing amounts of withdrawal censoring decreases the power more so than does increasing amounts of fixed right censoring. This is evidenced by the fact that for fixed right censoring the powers for heavy censoring are about 25 to 35% as big as the powers for light censoring, while for withdrawal censoring the powers for heavy censoring are only about 4 to 12% as big as those for light censoring. We see that the powers for heavy withdrawal censoring are not much larger than the desired error value of 0.05.

We also note that the powers for the threshold alternatives are consistently higher than the corresponding powers for the dose-response alternatives. It appears that these tests are better able to detect a sharp, sudden change between population distributions than a more gradual change.

Since the powers are closely tied to error levels, it is not surprising that the quadratic log rank produces the highest powers and the permutation Gehan the lowest, since this was the ordering of the observed error levels. We also see that the conservative log rank consistently gives higher power than the quadratic Gehan, and it is not too far behind the quadratic log rank.

Similar patterns and relationships were also seen for the powers of the tests when other sample sizes were used. Larger samples, however, produced larger powers and powers that did not decrease as much, proportionally, as the amount of censoring increased. Thus, a partial solution to the loss of power due to heavy censoring would be to take larger samples.

Results for the simulations that included both fixed right and withdrawal censoring are not discussed here since they were very similar to those for the withdrawal censoring only cases. The addition of fixed right censoring to withdrawal censoring increases the expected percentage of censoring so the error levels and powers go down even further.

We have seen from Fig. 1 that heavy amounts of censoring of any type can affect both the error level and power of tests for comparing k population distributions. If the censoring is heavy enough, the tests become quite conservative in error levels. If the null hypothesis is not true, the presence of heavy censoring can mask the differences between the populations to such an extent that the powers are very small. We have also seen that the effect of heavy censoring on the power is more pronounced for smaller sample sizes and for withdrawal censoring and that threshold alternatives are more easily identified than dose-response alternatives. Finally, we have seen that taking into account both error level and power, the conservative log rank test performed by SAS is probably as good, if not better, in most cases than the other tests; therefore, if an overall choice is to be made, the conservative log rank would probably be the best.

Results of Multiple Comparisons

If the overall test for equality rejects the null hypothesis, then it is of interest to determine which populations are significantly different. Our present objective is to discuss multiple comparison techniques that can be used for this purpose. We will restrict our attention to cases where the first population is a control group, and the other populations are treatment groups; our interest will be in comparing the control group to each of the treatment groups.

Two error levels that are commonly considered in multiple comparison procedures are the per comparison error level and the type one (or overall) error level. The per comparison error level is the probability that two particular equal populations are incorrectly declared different by the procedure. The type one error level is the probability that at least one pair of equal populations (among all those considered) are incorrectly declared different by the procedure.

Two procedures for making multiple comparisons of censored data are applicable using SAS, SPSSX, or BMDP. An overall test is commonly performed first before considering differences between just two populations. As a consequence, we will only compare the treatments versus the control if the overall test rejects. If the overall test does not reject, then we conclude that there are no significant differences in the populations. In this way, the overall test serves as a screen for the multiple comparison procedure. It is customary to refer to multiple comparisons screened in this way as *protected* procedures.

The first procedure we discuss, the protected LSD (least significant difference) procedure, is similar in philosophy to the protected LSD used for ANOVA problems (see Ref *12* for details). This procedure is very simple to use. We perform the overall test at level α. If the overall test fails to reject, then all populations are declared not significantly different. If the overall test rejects, then each treatment is compared with the control using the same test procedure at level α as was used for the overall test but using the data from only these two groups. This technique will theoretically maintain the per comparison error levels at α or less; however, there is no guarantee of control for the type one error level.

The second procedure, the protected Bonferroni procedure, will theoretically control the type one error level. The name Bonferroni is used since the theoretical justification for the type one error level uses the Bonferroni inequality (see Ref *12*). As with the LSD procedure, we only perform the treatment versus control comparisons if the overall test performed at the α level rejects. Each treatment is compared with the control, using the same test procedure as was used for the overall test and using only the data from these two groups; however, these pair-wise tests are carried out using a level α/t, where t is the total number of treatments to be compared to the control. The purpose for this unusual error level for the test is to insure a type one error level of α or less. This procedure also controls the per comparison error level as well; in fact, it is usually much less than α.

To determine how well these two multiple comparison procedures work for censored data, we performed a computer simulation using the same cases and the same specifications as presented in the previous section, as well as also including the jump alternative. We used only the conservative log rank test for the overall test and for the pair-wise comparisons. Calculated in the simulation were the per comparison and type one error levels, and the powers for appropriate comparisons.

A few selected results from the simulation study are contained in Table 3. Only sample sizes of ten are included since the results were found to be similar for other sample sizes. The overall test is performed at the 0.05 level as a screen for all cases. Only the $\alpha = 0.05$ values for the LSD are included. For the Bonferroni, the $\alpha = 0.10$ (Bon-10%) and the $\alpha = 0.05$ (Bon-5%) values are included. The 0.10 level was included for the Bonferroni since some users believe that a protected Bonferroni with $\alpha = 0.10$ may produce a type one error level nearer to 0.05 than would be obtained using $\alpha = 0.05$. The results in Table 3 are for withdrawal censoring only. The alternatives and censoring values considered here are the same as those given in Table 2.

After inspecting the results in Table 3, several interesting things can be observed. We

TABLE 3—*Results of multiple comparison simulation using the conservative log rank test and sample sizes of ten (withdrawal only censoring).*

True State/ Censoring Case (See Table 2)	Treatments Compared to Control[a]	Shift in Standard Deviations	LSD[b]	BON-10%[c]	BON-5%[d]
Per comparison error[e]					
1. null/light	2	0	0.020	0.017	0.011
2. null/moderate	2	0	0.023	0.017	0.011
	3	0	0.017	0.014	0.010
3. null/heavy	2	0	0.008	0.006	0.004
4. threshold/light	2	0	0.046	0.026	0.009
5. threshold/moderate	2	0	0.043	0.029	0.015
6. threshold/heavy	2	0	0.027	0.021	0.012
7. jump/light	2	0	0.050	0.031	0.015
8. jump/moderate	2	0	0.042	0.029	0.015
	3	0	0.038	0.026	0.011
Type one error[f]					
9. null/light	2, 3, 4	0, 0, 0	0.043	0.038	0.027
10. null/moderate	2, 3, 4	0, 0, 0	0.045	0.039	0.027
11. null/heavy	2, 3, 4	0, 0, 0	0.023	0.017	0.011
12. jump/light	2, 3	0, 0	0.091	0.058	0.032
13. jump/moderate	2, 3	0, 0	0.076	0.052	0.025
Power[e]					
14. dose-response/light	2	0.67	0.220	0.173	0.111
	3	1.13	0.442	0.386	0.287
	4	1.48	0.557	0.522	0.433
15. dose-response/ moderate	2	0.67	0.193	0.152	0.097
	3	1.13	0.358	0.319	0.246
	4	1.48	0.317	0.279	0.201
16. dose-response/ heavy	2	0.67	0.010	0.008	0.003
	3	1.13	0.009	0.007	0.003
	4	1.48	0.010	0.006	0.004
17. threshold/light	4	1.48	0.649	0.581	0.462
18. threshold/moderate	4	1.48	0.330	0.280	0.200
19. threshold/heavy	4	1.48	0.009	0.006	0.002
20. jump/light	4	2.43	0.910	0.888	0.826
21. jump/moderate	4	2.43	0.454	0.424	0.368

[a] The treatment groups are labeled as populations 2 to 4.
[b] LSD represents the protected LSD at a 0.05 level.
[c] BON-10% represents the protected Bonferroni at a 0.10 level.
[d] BON-5% represents the protected Bonferroni at a 0.05 level.
[e] Tabled values for per comparison error and power are the proportion of times the treatment group was declared different from the control group.
[f] For type one error the tabled values are the proportion of times at least one treatment was declared different from the control group.

first discuss the error levels. For all of the cases the per comparison error levels are <0.05, even for the BON-10% procedure. The type one error levels are all <0.05 for the BON-5% procedure which is as expected. The type one error levels are near 0.05 for the BON-10% procedure. For the LSD procedure, the type one error levels exceed 0.05 in some cases. For example, in Cases 12 and 13 the type one error levels are well above 0.05. This shows that the LSD controls the per comparison errors but may not control the type one error.

As with the overall test, the presence of censoring decreases the power of all three procedures. This is most evident in Cases 14 through 16 and 17 through 19. In fact, the powers in Cases 16 and 19, for a large shift alternative, are <0.05. This shows that heavy censoring can completely mask a large treatment effect. Reversals in the natural order of powers is also possible due to heavier censoring in one of the samples. For example, in Case 15 the comparison of the control to Treatment 3 has a somewhat higher power than the power for the comparison with Treatment 4.

The power seems to be better for threshold type alternatives than for dose-response alternatives. For example, the power in the dose-response Case 14 for the comparison with Treatment 4 is less than the power in the threshold Case 17 for the same comparison. This is due partly to the fact that the overall test has higher power for the threshold alternative than for the dose-response alternative (see Fig. 1).

As expected, the LSD has higher power than the Bonferroni procedure, but this is clearly due to the higher type one error level. If control of the type one error level is crucial, then the Bonferroni procedure is superior to the LSD. The BON-10% gives type one error levels near 0.05, while the BON-5% insures that the type one error level is controlled at the 0.05 or less level theoretically.

In summary, the Bonferroni procedure seems to be an easy-to-implement procedure for comparing populations. The type one error levels are controlled, but the power is substantially less than that observed for the LSD procedure. Using a 0.10 level for the Bonferroni may be appropriate in some cases. It should also be kept in mind when analyzing studies with withdrawals that the presence of heavy withdrawals causes a substantial loss in power. Treatment effects can be masked by the presence of censoring, and, in some cases, reversals in the power may occur because of the censoring patterns.

We have only considered treatment versus control type comparisons in this section mainly because this is a common goal in a toxicology experiment. It should be clear, however, that similar results will hold if one compares the treatments among themselves. In fact, the power will become worse in some cases because there is often more censoring in the treatments than in the control.

Summary

In this paper we have examined the effects of fixed right and withdrawal censoring on nonparametric statistical tests used for comparing k populations and for making multiple comparisons in toxicology experiments. The following conclusions can be drawn from this study.

1. Heavy censoring of any type can effect the performance of the k-sample procedures by causing incorrect observed error levels (too high or too low) and greatly decreased power.
2. Heavy withdrawal censoring can mask differences or produce reversals in multiple comparison procedures.
3. Withdrawal censoring is more harmful to statistical performance than is fixed right censoring.
4. Threshold alternatives are more readily detected than are dose-response alternatives in both the k-sample and multiple comparison procedures.
5. Taking larger samples can help offset the adverse effects of heavy censoring.
6. The conservative log rank statistic on SAS seems to be the best choice among the four possible tests.

References

[*1*] Gehan, E. A. and Thomas, D. G., "The Performance of Some Two-Sample Tests in Small Samples With and Without Censoring," *Biometrika,* Vol. 56, 1969, pp. 127–132.

[*2*] Lee, E. T., Desu, M. M., and Gehan, E. A., "A Monte Carlo Study of the Power of Some Two-Sample Tests," *Biometrika,* Vol. 62, 1975, pp. 425–432.

[*3*] Latta, R. B., "A Monte Carlo Study of Some Two-Sample Rank Tests With Censored Data," *Journal of the American Statistical Association,* Vol. 76, 1981, pp. 713–719.

[*4*] Groggel, D. J., Schaefer, R. L., and Skillings, J. H., "The Effect of Withdrawal Censoring on Two-Sample Tests," *Journal of Statistical Computation and Simulation,* Vol. 28, 1987, pp. 275–302.

[*5*] Gehan, E. A., "A Generalized Wilcoxon Test for Comparing Arbitrarily Singly-Censored Samples," *Biometrika,* Vol. 52, 1965, pp. 203–223.

[*6*] Breslow, N. E., "A Generalized Kruskal-Wallis Test for Comparing *K* Samples Subject to Unequal Censorship," *Biometrika,* Vol. 57, 1970, pp. 579–594.

[*7*] Miller, R. G., *Survival Analysis,* John Wiley and Sons, New York, 1981.

[*8*] Hogg, R. V. and Craig, A. T., *Introduction to Mathematical Statistics,* fourth ed., Macmillan, New York, 1978.

[*9*] *SUGI Supplemental Library User's Guide, 1983 Edition,* SAS Institute Inc., Cary, NC.

[*10*] *SPSSX User's Guide,* McGraw-Hill, New York, 1983.

[*11*] *BMDP Statistical Software,* University of California Press, Berkeley, CA, 1983.

[*12*] Miller, R. G., *Simultaneous Statistical Inference,* 2nd ed., Springer-Verlag, New York, 1980.

Charles A. Staples[1] *and Jeanne L. Sebaugh*[1]

Sensitivity and Uncertainty Analyses for Use with Chemical Fate and Transport Models

REFERENCE: Staples, C. A. and Sebaugh, J. L., "**Sensitivity and Uncertainty Analyses for Use with Chemical Fate and Transport Models,**" *Aquatic Toxicology and Environmental Fate: Eleventh Volume, ASTM STP 1007,* G. W. Suter II and M. A. Lewis, Eds., American Society for Testing and Materials, Philadelphia, 1989, pp. 339–349.

ABSTRACT: A method for evaluating sensitivity and uncertainty analyses is presented here for use with chemical environmental exposure models. Sensitivity analysis allows one to assess the relationship between a set of potentially important model input variables and the model output. Uncertainty analysis allows one to assess the set of output values obtained and estimate various statistical properties such as the mean, median, standard deviation, and range. Absence of the ability to quantify parameter value uncertainty and model output uncertainty limits the validity of predictive model use. Use of the methods presented in this paper can increase the confidence in predictive environmental chemical fate model use.

The present investigation employed the Latin hypercube sampling method developed by Iman and Shortencarier at Sandia National Laboratories, Albuquerque, NM. It requires that a set of input variables be identified, and the probability distribution of each be specified. The range of each input variable is divided into intervals having equal probabilities, and one value is randomly selected from each interval. Sets of model input variables are then generated which cover the range of each variable. Some model inputs may then be calculated as functions of these key input variables. The computer model is run for each set of input variables and the model output is obtained.

Regression techniques can then be employed to examine the differential sensitivity of the model to the various input variables and functions thereof, for example, cross products of input variables. Since the model input values are chosen based on probabilities, the set of model outputs can also be interpreted on a probabilistic basis.

The techniques discussed in this paper respond to the growing need by chemical exposure model users for better defined sensitivity analysis procedures which can aid in explaining and interpreting results. An example problem data set is presented which highlights the methodological techniques and their utility.

KEY WORDS: modeling, sensitivity, uncertainty, Latin hypercube sampling, aquatic toxicology

Environmental and human health risk analyses and hazard assessments for chemicals require consideration of two critical components: exposure concentrations and effects concentrations [*1*]. Effects analyses require determination of appropriate affected species or targets; chemical dose response relationships; acute effects, chronic effects, mutagenicity, teratogenicity, or carcinogenicity; routes of exposure (inhalation, drinking, eating, and so forth); and metabolic effects. Effects analyses can then yield "safe" chemical exposure concentrations, where safe is equivalent to no-effect levels, acceptable daily intakes, or acceptable risk doses. Exposure analysis is used to determine the chemical concentrations that organisms are exposed to and requires careful determination and consideration of routes, spatial and temporal factors, fate processes, mixtures, and abiotic environmental factors.

[1] Research specialist and statistical consultant, respectively, Monsanto Co., St. Louis, MO 63167.

The current uses of solute transport and fate models seldom sufficiently quantify the appropriate uncertainty associated with model use.

Chemical exposure can be either directly measured or estimated by mathematical modeling. The variability of measured chemical exposure can be quite large because of spatial and temporal variability, sampling variability, and analytical variability. The "true" measured concentration cannot be determined practically but only described statistically to an acceptable level of uncertainty. Modeling of the fate and transport of a chemical in environmental media relies on the physical and chemical properties and characteristics of the media, which also have a measurable degree of uncertainty. Modeling of exposure is frequently the method of choice when addressing questions prior to environmental exposure [for example, the Toxic Substances Control Act (TSCA) new chemical registration] [*2*]. Typical questions involving future exposure patterns of existing chemical concentrations in an environment also rely on mathematical modeling. Determination of future exposures from known existing exposures is, for instance, important for surface water quality or National Pollutant Discharge Elimination System (NPDES) questions [*3*], Resource Conservation and Recovery Act (RCRA) closures and cleanup [*4*], and Comprehensive Environmental Response, Compensation, and Liability Act (CERCLA) remediations [*5*].

An important part of chemical exposure modeling is thorough verification of model results [*6*]. The verification process consists of two key parts: (1) correct identification and determination of input variables and model assumptions and the effects of each variable on model output, and (2) comparison of output with measured data. The first part can be more difficult than the second. Any site or environmental medium may have a wide range of possible values for a particular variable. For instance, groundwater microbial decay is dependent on the redox potential, which is an indicator of oxygen presence, the population of microorganisms, and many other factors [*7*]. Redox potential and microbial populations may vary by one to four orders of magnitude across a small site. Effective modeling of chemical decay, therefore, requires quantification of the range and frequency distribution of the variables across the site. Use of appropriate sensitivity analyses that use ranges and frequency distributions for model input values allows the best explanation and, hence, verification of exposure modeling output.

This paper describes a sensitivity analysis method based on sampling procedures. There are many examples of this general approach [*8–11*]. McKay et al. [*12*] compared three methods of generating the required sample of model inputs: random sampling (or Monte Carlo), stratified sampling, and Latin hypercube sampling. Their results showed that if a model's output is a monotonic function of its input, better results with fewer runs of the model could be obtained with the Latin hypercube sampling method.

Therefore, a computer program developed by Iman and Shortencarier at Sandia National Laboratories, Albuquerque, NM, [*13*] to perform the Latin hypercube sampling was used to generate the model inputs for performing a sensitivity analysis on a groundwater solute transport model.

The method requires identification of a set of input variables, each with a specified range of values and probability distribution. Based on these probability distributions, several representative sets of input values may be generated. The computer-based model may be run for each set of input variables and the model outputs obtained. Regression techniques can then be employed to examine the differential sensitivity of the model to the variables.

Groundwater Model

Model Description

The groundwater solute transport model employed in these studies is called Random-Walk [*14,15*]. Random-Walk is composed of two parts, a flow model and a solute transport model. The flow model provides the velocity or convective vectors of an aquifer system. A finite difference scheme is used to generate a head or water level distribution. Aquifer characteristics (porosity, conductivity, and geometry) are combined with the head distribution to yield a velocity field and, hence, flow. The flow distribution is then used to move a set of "particles" of chemical that have units of mass. The solute transport may be driven by both the convective flow and dispersion. Dispersion is handled as a random process within the porous subsurface medium.

The specific governing equation for non-steady-state two-dimensional solute transport may be expressed as follows

$$\frac{d}{dx}\left(\frac{D}{R_f}\right)\left(\frac{dC}{dx}\right) + \frac{d}{dy}\left(\frac{D}{R_f}\right)\left(\frac{dC}{dy}\right) - \frac{d}{dx}\left(\frac{V}{R_f}\right)(C) = \frac{dC}{dt} + Q \tag{1}$$

where

x, y = longitudinal and lateral directions,
D = dispersion coefficient, L^2/T,
R_f = retardation factor, without units,
C = chemical concentration, M/T,
V = average linear velocity, L/T, and
Q = source and sink term.

The hydrodynamic dispersion coefficient, D, is a function of dispersivity, velocity, and molecular diffusion (Eq 2), where d = dispersivity, L,

$$D = dV + D^* \tag{2}$$

and D^* = molecular diffusion, L^2/T.

Retardation factors are a function of soil/water chemical partition coefficients, K_p, porosity, P, and subsurface material bulk density, B, (Eq 3)

$$R_f = 1 + \left(\frac{B}{P}\right) K_p \tag{3}$$

Flow velocity alone (disregarding solute transport) can be represented by an interpretation of Darcy's Law (Eq 4)

$$V = K\frac{dh}{P} \tag{4}$$

where K = hydraulic conductivity and dh = head gradient. In two-dimensional form, the flow may be represented as follows

$$dx\left(K_x b\,\frac{dh}{dx}\right) + dy\left(K_y b\,\frac{dh}{dy}\right) = \frac{dh}{dt} + Q \tag{5}$$

where b = aquifer thickness, L, and K_b = transmissivity, L^2/T, a property of the aquifer (in x or y directions).

Model Output

A typical model output is shown in Table 1. The output contains a number equal to the particle load or mass per node. Plume shape and definition are controlled by the distribution of the particles in the nodes as the particles are moved with each time step. The output in Table 1 is 1 of the 20 outputs based on the specified input data sets discussed in a later section of this paper.

Independent Variables

The seven independent variables which exert the greatest influence on model results are hydraulic conductivity, K, retardation factor, R_f, bulk density of the subsurface material, B, porosity, P, head gradients, dh, longitudinal dispersivity, D_l, and transverse dispersivity, D_t. The possible range and probability distribution for each of these variables were estimated from several sources [*16,17*] and are shown in Table 2.

Hydraulic conductivity, K, tends to range lognormally from 2 to 8 m/day for many unconfined aquifer systems commonly dealt with, although K can range over several orders of magnitude over all geological formation types. Retardation factors range from 1 to 100 in most applications and are distributed lognormally. Bulk densities typically range from about 1.5 to 2.0 g/cm^3, distributed uniformly. Porosities may typically range from 0.2 to 0.5 or 20 to 50%, distributed uniformly. Hydraulic gradients (local or regional) may range from 0.005 to 0.050 m/m and be distributed normally in many aquifer systems. Longitudinal and lateral dispersivities may typically range lognormally from 3 to 300 m and from 0.3 to 30 m, respectively.

Dependent Variables

Four types of dependent variables or solute modeling output types can be used to explain results. The assumption is made here of a continuing chemical source input to the aquifer system. The peak or maximum chemical concentration is represented by Random-Walk to be the maximum number of particles in an individual finite difference cell (cell = area with a node in the center). The peak concentration allows estimation of the highest possible chemical concentration encountered. A second output result type or dependent variable is

TABLE 1—*A typical model output showing chemical being transported in a grid.*

Particle mass = 0.100 00 E + 00
Number of particles in system = 951
Time = 3650 days

0[a]	0	0	0
0	0	0	0
0	0	0	0
0	0	0	0
0	0	0	0
0	0	0	0
0	1	1	0
0	12	7	0
0	47	42	0
0	144	127	0
0	296	274	0

[a] Numbers represent relative concentrations.

TABLE 2—*Model input variables, ranges, and distributions.*

Variable	Range	Distribution
Hydraulic conductivity	2 to 8 m/day	lognormal
Retardation factor	1.0 to 100	lognormal
Bulk density	1.5 to 2.0 g/cm^3	uniform
Porosity	0.2 to 0.5	uniform
Head gradient	0.005 to 0.050 m/m	normal
Longitudinal dispersivity	3 to 300 m	lognormal
Lateral dispersivity	0.3 to 30 m	lognormal

that of the average chemical concentration within the plume defined by >1 particle per node or area around a node. Average chemical concentrations are used to determine areas within which potential health effects could occur.

The third and fourth variables examined explained the solute or chemical plume shape. The leading edge of the plume is defined by the distance farthest downgradient of the nonzero nodes or cells. The leading edge defines how far the plume has traveled and is needed to determine when or if a plume has reached some critical distance. Total plume spread allows definition of chemical concentrations in two dimensions. The two-dimensional view of solute concentrations allows better interpretation of consequences of predicted chemical levels.

For the example in Table 1, the maximum number of particles per cell is 296, the average number of particles per nonzero cell is 95, the leading edge of the plume is 5 nodes forward, and the number of nonzero cells is 10.

Sensitivity Analyses

As mentioned previously, a set of potentially important model input variables, along with the range of values and probability distribution for each, was first selected. The computer program which generated the set of model input values [*13*] assumes the lower range to be the 0.001 quantile [that is, P (value $\leq$ lower range) $=$ 0.001] and the upper range to be the 0.999 quantile [that is, P (value $\geq$ upper range) $=$ 0.001]. The choice of probability distributions includes the normal, lognormal, uniform, log uniform, triangular, beta, and user specified. The ranges and probability distributions for the seven model input variables considered here are shown in Table 2. The following sections describe the generation of actual model inputs, the resultant model outputs, and the analysis of the relationship between the model inputs and outputs.

Model Inputs

Presentations by McKay et al., Iman et al. [*18–19*], and Helton et al. [*12,19–21*] have described and developed both the Latin hypercube sampling (LHS) and subsequent regression analyses used here. Among other reasons, the LHS method for generating model input values was used because it forces the sampling of the full range of each variable and, assuming that model output is a monotonic function of model input, fewer model runs are required to generate enough data for fitting a response surface analysis than with other methods. The computer program used [*18*] also generates sets of model input variable values with either specified or zero correlations between pairs of variables. The only correlations used in this analysis involved lateral and longitudinal dispersivities.

It is generally recommended that the number of model runs be approximately 1.5 times the number of model input variables [*21*]. However, the required 10 to 11 runs for 7 input variables would not have supplied much information about the distribution of output variables. Therefore, the authors decided to generate 20 sets of model input and output variables. Similar reasoning was used by Helton et al. [*20*] when generating a sample size of 50 for 20 model input variables.

To generate the 20 sets of model input variables, the LHS program divides the specified range of each model input variable into 20 subranges of equal probability. These subranges only correspond to subranges of equal size in the case of the uniform distribution. A value is randomly chosen within each of the subranges. The chosen values for each of the variables are randomly combined into sets in such a way as to maintain the specified correlations among the variables.

Descriptive statistics on the LHS-generated values for the 20 sets of model inputs are shown in Table 3. As expected, the medians for the lognormally distributed variables are less than the means. Also, there is some variability in the extent to which upper and lower tail values were generated which approached the specified upper and lower ranges. This is not unexpected as the size of the intervals from which a model input value is randomly selected is much larger in the upper and lower tails of a normal distribution and, particularly, the upper tail of a lognormal distribution.

Model Outputs

The four model outputs to be used in the analysis were described previously as (*a*) the maximum number of particles per cell, (*b*) the average number of particles per cells containing particles, (*c*) the number of cells containing particles, and (*d*) the number of cells over which the plume has advanced. Table 4 shows the mean, median, minimum, maximum, and standard deviation for these 4 variables over the 20 runs. By inspecting frequency distributions for each set of 20 values, it could be seen that (*a*) the values for the maximum number of particles per cell were approximately normally distributed, (*b*) the values for the average number of particles per cell were lognormally distributed, and (*c*) the values for the number of cells containing particles and the number of cells over which the plume had advanced both tended to be bimodally distributed at low and high values, with a greater density at the lower values.

The maximum number of particles per cell and the average number of particles in nonzero cells are significantly positively correlated (Spearman $r = 0.94$, $P < 0.001$) as are the number of cells containing particles and the number of cells over which the plume has advanced (Spearman $r = 0.94$, $P < 0.0001$). The first two variables are significantly negatively correlated with the last two variables.

TABLE 3—*Statistics for the 20 sets of LHS-generated model variables.*

Variable	Mean	Median	min	max
Hydraulic conductivity, m/day	4.10	4.02	2.72	6.24
Retardation factor	12.83	9.83	2.43	38.20
Bulk density, g/cm^3	1.75	1.76	1.52	2.00
Porosity	0.35	0.34	0.22	0.50
Head gradient, m/m	0.028	0.028	0.015	0.041
Longitudinal dispersivity, m	38.90	30.55	3.23	149.30
Transverse dispersivity, m	4.54	2.99	0.85	25.46

TABLE 4—*Statistics for the 20 sets of model-generated output variables.*

Variable	Mean	Median	min	max	Standard Deviation
Maximum number of particles per cell	221.85	222.0	72	422	109.60
Average number of particles per cell	67.60	55.0	23	155	41.73
Number of cells containing particles	20.15	17.5	6	39	11.37
Number of cells plume has advanced	6.35	6.0	3	10	2.52

Relationship Between Model Inputs and Outputs

Correlation and regression techniques were used to study the relationships between model input and output variables. The analyses presented here were performed on the ranks of the raw data since it was found to be consistent with and often more revealing than the analyses on the raw data. The rank transform has been found useful in other similar studies [*20*] because it tends to linearize relationships which are monotonic, as is the case with this Random-Walk model. It also reduces the effect of extreme values that

TABLE 5—*Simple correlation coefficients between ranks of model output variables (*N = *20).*

	Model Output Variables			
	Particles/Cell		Number of Cells	
Model Input Variables	max	Average	Particles	Plume Advance
Hydraulic conductivity				
r	−0.27	−0.17	0.18	0.17
P	0.2511	0.4772	0.4535	0.4797
Retardation factor				
r	0.81[a]	0.80[a]	−0.80[a]	−0.80[a]
P	0.0001	0.0001	0.0001	0.0001
Bulk density				
r	0.01	0.03	−0.02	0.05
P	0.9599	0.9072	0.9447	0.8207
Porosity				
r	0.38	0.27	−0.29	−0.32
P	0.0951	0.2545	0.2175	0.1688
Head gradient				
r	−0.28	−0.18	0.18	0.25
P	0.2350	0.4305	0.4574	0.2938
Longitudinal dispersivity				
r	−0.22	−0.35	0.33	0.39
P	0.3490	0.1258	0.1617	0.0930
Transverse dispersivity				
r	−0.09	−0.31	0.31	0.04
P	0.6958	0.1831	0.1876	0.8507

[a] Correlation coefficient significantly different from zero.

occur with lognormally distributed variables. Although analyses with quadratic and cross-product terms were also considered, these derived variables were not found to be significant when the rank-transformed values were used.

Table 5 shows the simple correlation coefficients between the rank-transformed model input and model output variables. The retardation factor was the only model input variable which by itself, ignoring the values of the other six model input variables, was significantly related to the model output variables. Larger values of the retardation factor are related to a larger maximum and average number of particles per cell and a smaller number of cells having particles and a less advanced plume. These outputs are to be expected, as retardation of chemicals in an aquifer implies a "holding" back of a plume by definition.

Simple correlation demonstrates the effect of a parameter on model performance when no control over other parameters exists. Partial rank correlation coefficients between the model input and output variables are shown in Table 6. These partial correlation coefficients indicate the relationship between the model output variables and each model input variable after having taken into account the influence of the other six model input variables. Again, the strongest relationships with the model output variables were found for the retardation factor input variable. The only model input variable not having a significant relationship with the model output variables was bulk density. Bulk density probably has the least impact on the input/output relationship due to its reasonably small uniformly distributed range.

Multiple regression analyses were performed using the maximum R^2 improvement technique in SAS [*21*]. Multiple regression analysis was performed in order to study the rela-

TABLE 6—*Partial correlation coefficients between ranks of model input and model output variables (N = 20).*

	Model Output Variables			
	Particles/Cell		Number of Cells	
Model Input Variables	max	Average	Particles	Plume Advance
Hydraulic conductivity				
r	−0.73[a]	−0.49	0.49	0.43
P	0.0028	0.0749	0.0757	0.1217
Retardation factor				
r	0.97[a]	0.96[a]	−0.95[a]	−0.96[a]
P	0.0001	0.0001	0.0001	0.0001
Bulk density				
r	0.09	0.05	0.00	0.34
P	0.7708	0.8522	0.9948	0.2287
Porosity				
r	0.85[a]	0.67[a]	−0.67[a]	0.79[a]
P	0.0001	0.0089	0.0083	0.0008
Head gradient				
r	−0.81[a]	−0.60[a]	0.60[a]	0.75[a]
P	0.0004	0.0117	0.0240	0.0018
Longitudinal dispersivity				
r	−0.71[a]	−0.85[a]	0.81[a]	0.86[a]
P	0.0046	0.0001	0.004	0.0001
Transverse dispersivity				
r	−0.27	−0.80[a]	0.77[a]	0.12
P	0.3527	0.0006	0.0013	0.6728

[a] Correlation coefficient significantly different from zero.

TABLE 7—*Regression analyses for the four-model output variables using rank-transformed data (N = 20).*

Output Variable	Input Variable	SRC[a]	Rank,[b] R^2
Maximum number of particles per cell	retardation factor	0.796	66.2
	porosity	0.337	77.9
	head gradient	−0.281	86.5
	hydraulic conductivity	−0.215	91.8
	longitudinal dispersivity	−0.200	95.6
Average number of particles per cell	retardation factor	0.768	63.6
	longitudinal dispersivity	−0.383	75.0
	transverse dispersivity	−0.299	85.0
	porosity	0.210	89.5
	head gradient	−0.198	93.4
Number of cells containing particles	retardation factor	−0.771	64.2
	longitudinal dispersivity	0.354	73.8
	transverse dispersivity	−0.290	83.3
	porosity	−0.231	88.7
	head gradient	0.188	92.2
Number of cells for plume advancement	retardation factor	−0.787	64.4
	longitudinal dispersivity	0.376	78.1
	porosity	−0.290	86.7
	head gradient	0.257	93.3

[a] Standardized regression coefficients for rank regression.
[b] R^2 value for rank regression.

tionship between significant combinations of the model input variables and each model output variable, as well as to provide further information on the relative importance of the model input variables. For inclusion in the regression equation, each variable was required to be significant at the 0.05 level. The regression was performed on the rank-transformed data.

As a reflection of their importance in explaining the variability in each model output variable, Table 7 shows the order in which the model input variables entered the regression equation and the corresponding increase in the R^2 value. Also shown are the standardized regression coefficients from the final equation. In cases, like the present one, in which the independent variables are not intercorrelated, the magnitude of the standardized regression coefficients reflects the relative importance of each independent variable. These results agree with the order in which the input variables entered the regression equation. The sign of the standardized regression coefficients indicates whether the model output variable increases (positive coefficient) or decreases (negative coefficient) as the corresponding input variable increases.

In all cases, the retardation factor was the most important variable. The variables longitudinal dispersivity, porosity, and head gradient entered all four regression equations. Bulk density was not useful in explaining the variability in any of the model output variables, and hydraulic conductivity entered only one model. It must be remembered, however, that these results are site specific and dependent on both the range of values and the probability distribution specified for each of the seven model input variables.

Discussion

The examination of data inputs and outputs has shown the importance of retardation factors. However, many important groundwater pollutants are common laboratory solvents such as benzene. Most common solvents have reasonably high water solubilities and, hence, low soil sorption constants, and hence, low retardation factors. Negligible retardation factors would greatly enhance the relative importance of dispersivities, conductivity, porosity, and head gradients.

Use of the LHS model and the various correlations discussed here would have to be redone for chemicals with negligible retardation factors. Positive results would then emphasize variables governing the flow of water (conductivity and head gradient) rather than solute retardation and distance traveled (retardation factor, dispersivity, and porosity). It should be realized from Eqs 1 through 5 that most of the important variables readily come into play and are affected differently by both solute and aquifer properties.

Conclusions

A sensitivity and uncertainty analysis is presented here that allows a better understanding and explanation of chemical fate and transport modeling results. A Latin hypercube sampling method uses input variable statistical properties to choose input values for a series of model runs. Model run output variables (dependent variables) are then examined for significant influence by or correlation with input values. Significance is determined using rank correlation and regression methods.

The value to the chemical fate and transport modeler (whether in air, surface water, or groundwater media) is that the quality of any model output can be explained. A ready explanation of model output variables is necessary for the model verification process.

Chemical fate and transport models play a vital role in many chemical exposure investigations. Use of sensitivity and uncertainty analyses, such as that explained in this paper, will aid chemical exposure modeling utility and increase our confidence in modeling as a useful tool for exposure evaluations.

References

[*1*] *Analyzing the Hazard Evaluation Process,* Dickson, K. L., Maki, A. W., and Cairns, J., Jr., Eds., American Fisheries Society, Washington, DC, 1979.

[*2*] Toxic Substances Control Act (TSCA), PL94.469 1976, United States Government, 1976.

[*3*] Clean Water Act (CWA), PL92-500 1972, United States Government, 1972.

[*4*] Resource Conservation and Recovery Act (RCRA), PL94-580 1976, United States Government, 1976.

[*5*] Comprehensive Environmental Response, Compensation and Liability Act (CERCLA), PL96-510 1980, United States Government, 1980.

[*6*] *Modeling the Fate of Chemicals in the Aquatic Environment,* K. L. Dickson, A. W. Maki, and J. Cairns, Jr., Eds., Ann Arbor Science, Ann Arbor, MI, 1982.

[*7*] Bitton, G. and Gerba, C. P., *Ground Water Pollution Microbiology,* Wiley, New York, 1984.

[*8*] Gardner, R. H., Huff, D. D., O'Neill, R. V., Mankin, J. B., Carney, J., and Jones, J., "Application of Error Analysis to a Marsh Hydrology Model," *Water Resources Research,* Vol. 16, 1980, pp. 659–664.

[*9*] Iman, R. L. and Conover, W. J., "Small Sample Sensitivity Analysis Techniques for Computer Models, with an Application to Risk Analysis," *Communication in Statistics: Theory and Methods,* Vol. 9, 1980, pp. 1749–1842.

[*10*] O'Neill, R. V., Gardner, R. H., and Mankin, J. B., "Analysis of Parameter Error in a Nonlinear Model," *Ecological Modeling,* Vol. 8, 1980, pp. 297–311.

[*11*] Summers, J. K. and McKellon, H. N., Jr., "A Sensitivity Analysis of an Ecosystem Model of Estuarine Carbon Flow," *Ecological Modeling,* Vol. 13, 1981, pp. 283–301.

[*12*] McKay, M. D., Conover, W. J., and Beckman, R. J., "A Comparison of Three Methods for Selecting Values of Input Variables in the Analysis of Output from a Computer Code," *Technometrics,* Vol. 21, No. 2, 1979, pp. 239–245.

[*13*] Iman, R. L. and Shortencarier, M. J., "A Fortran 77 Program and User's Guide for the Generation of Latin Hypercube and Random Samples for Use with Computer Models," NUREG/CR-3624, Sand 83-2365, Sandia National Laboratories, Albuquerque, NM, 1984.

[*14*] Prickett, T. A. and Lonnquist, C. G., "Computer Techniques for Ground Water Resource Evaluation," Illinois State Water Survey Bulletin, No. 55, 1981.
[*15*] Prickett, T. A., Naymik, T. G., and Lonnquist, C. G., "A Random-Walk Solute Transport Model for Selected Ground Water Quality Evaluations," Illinois State Water Survey Bulletin, No. 65, 1981.
[*16*] Freeze, R. A. and Cherry, J. A., *Ground Water,* Prentice-Hall, Inc., Englewood Cliffs, NJ, 1979.
[*17*] "Development of Land Disposal Banning Decisions Under Uncertainty—Draft," Environmental Research Laboratory, U.S. Environmental Protection Agency, Athens, GA, 1985.
[*18*] Iman, R. L., Helton, J. C., and Campbell, J. E., "An Approach to Sensitivity Analysis of Computer Models, Part I: Introduction, Input Variable Selection and Preliminary Variable Assessment," *Journal of Quality Technology,* Vol. 13, No. 3, 1981, pp. 174–183.
[*19*] Iman, R. L., Helton, J. C., and Campbell, J. E., "An Approach to Sensitivity Analysis of Computer Models, Part II: Ranking of Input Variables, Response Surface Validation, Distribution Effect and Technique Synopsis," *Journal of Quality Technology,* Vol. 13, No. 4, 1981, pp. 232–240.
[*20*] Helton, J. C., Iman, R. L., and Brown, J. B., "Sensitivity Analyses of the Asymptotic Behavior of a Model for the Environmental Movement of Radionuclides," *Ecological Modeling,* in press.
[*21*] "Statistical Analysis System," SAS Institute Inc., Cary, NC, 1986.

Microcosms

Wayne G. Landis,[1] Nancy A. Chester,[1] Mark V. Haley,[1] Dennis W. Johnson,[1] William T. Muse, Jr.,[1] and Renee M. Tauber[1]

Utility of the Standardized Aquatic Microcosm as a Standard Method for Ecotoxicological Evaluation

REFERENCE: Landis, W. G., Chester, N. A., Haley, M. V., Johnson, D. W., Muse, W. T., Jr., and Tauber, R. M. **"Utility of the Standardized Aquatic Microcosm as a Standard Method for Ecotoxicological Evaluation,"** *Aquatic Toxicology and Environmental Fate: Eleventh Volume, ASTM STP 1007,* G. W. Suter II and M. A. Lewis, Eds., American Society for Testing and Materials, Philadelphia, 1989, pp. 353–367.

ABSTRACT: The standardized aquatic microcosm (SAM) as developed by Taub is a 64-day multispecies assay designed to model a generic freshwater ecosystem. An extensive protocol is available that covers every aspect of the experiment. In addition, a series of computer programs is available at the University of Washington, Seattle, Washington for data analysis. The SAM has recently undergone a round-robin test series, in which good repeatability was demonstrated among the four laboratories that took part. These factors have led to our use of the SAM as an ecosystem-level assay for assessment of the effects and fate of disparate materials in aquatic systems.

To date, three materials have been assayed in our laboratory. Copper sulfate was assayed at two series of concentrations to act as an in-house control and as part of the round-robin SAM testing. A brass dust and graphite dust have also been assayed using the SAM protocol. In setting test concentrations for the SAM, *Daphnia magna* 48-h and algal 96-h assays were performed using ASTM methodologies.

The SAM has proven to be a useful assay. Currently, the disadvantages are the amount of labor required in the preparation of the test media, culturing of the organisms, and data analysis; the breakage of test vessels was also a problem. The authors are planning to automate much of the data handling to reduce repeated manual manipulation and the concomitant errors. Breakage of the test vessels has been reduced considerably by modifications of the sterilization procedure. The advantages of the SAM are the ability of the system to handle a variety of test compounds and the ability to simultaneously look at both effects and fate. Results varied among the compounds tested from very strong acute toxic effects to classical signs of eutrophication. Interactions among the biota were observable. The fate of the toxicant was also altered by the complex community, in contrast to simpler abiotic fate studies.

KEY WORDS: standardized aquatic microcosm, microcosm, trophic interactions, ecotoxicology, community structure, aquatic toxicology, *Daphnia magna,* algae

Over the last several years, a variety of single-species assays have been developed—using vertebrates, invertebrates, algae, and microorganisms—that are useful in determining the effects of xenobiotics on individual species. Although these assays have proven to be very useful in protecting a variety of species, no information is available on the impacts of toxicants on many aspects of the dynamics and metabolism of an ecosystem. The goal of a multispecies assay is to model the characteristics of naturally occurring ecosystems that

[1] Environmental Toxicology Branch, Toxicology Division, Chemical Research, Development, and Engineering Center, Aberdeen Proving Ground, MD, 21010-5423.

can be reproduced on a smaller scale. In the search for a multispecies assay, the authors of this paper evaluated several of the microcosms in the literature by the criteria listed below.

In a properly designed system, interspecific interactions both within and between trophic levels can be analyzed. Parameters of ecosystem metabolism such as nutrient uptake, production, and respiration can be important clues to the long-term impact of a material and should be included in the protocol. In addition, a multispecies assay should be able to provide data on the transformation and fate of the material in natural ecosystems.

Other, perhaps more practical, criteria were also incorporated in our evaluation process and are similar to the criteria of Hammons [*1*]. Generality was an important criterion since a generic freshwater assay was desired. The assay had to be well documented as to its behavior under the impact of at least one well-understood toxicant. Rejection standards, or at least some first approximations of rejection standards, had to exist so that a poor assay could be identified. Replicability within an experiment had to have been demonstrated. A standardized protocol had to exist that was sufficiently detailed for the preparation of a standard operating procedure (SOP). Lastly, but equally important, a multispecies assay had to be highly repeatable regardless of the time of year, laboratory, or geographical location when used to screen materials for long-term ecosystem level effects.

Using the above criteria as a guide, we selected the standardized aquatic microcosm (SAM) as developed by Taub and colleagues at the University of Washington as an ecosystem-level assay. The SAM protocol has undergone an extensive and lengthy period of research and development [*2–6*]. During the last three years we have participated in the U.S. Food and Drug Administration (FDA) supported round-robin evaluation of the method using copper sulfate as the toxicant and have evaluated two very different materials for long-term ecosystem effects. This report is a summary of our experience in using the SAM as a routine ecosystem-level ecotoxicological assay.

The SAM was found to be a reliable and repeatable assay in the round-robin testing [*7*]. A deviation from the protocol was made in the sterilization procedure to reduce breakage of the test vessels during the course of the experiment. Several minor changes were made in the counting of organisms that had no effect on the repeatability of the assays, as determined by the round-robin testing. The required numbers of amphipods were also difficult to obtain at the beginning of the experiment.

We examined two materials in addition to the copper sulfate ($CuSO_4$), brass dust and graphite dust. Brass dust produced signs of severe environmental and population stress. Graphite dust was not acutely toxic, but the dynamics of the system were significantly altered. Isolated parameters, such as species diversity, were poor indicators of environmental effects. Suggestions for further improvement of the protocol include a change in the sterilization procedure and the transfer of the data handling and analysis software to available microcomputer systems. In our experience, the SAM has an excellent potential of becoming a useful and standard generic ecosystem level assay.

Methods and Materials

Training

Three members of the ecological toxicology group received training on the SAM protocol in the laboratory of Dr. Taub at the University of Washington, in Seattle, Washington, during a three-day workshop. Subsequently, the training of new employees was conducted in house. Dr. Taub's laboratory and especially Dr. Andrew Kindig were unrelenting in their support of this research program.

SAM Protocol

The 64-day SAM protocol has been described previously [*8*]. The microcosms were prepared by the introduction of ten algal, four invertebrate, and one bacterial species into 3 L of sterile defined medium. The test containers were 4-L glass jars. An autoclaved sediment consisting of 200 g of silica sand and 0.5 g of ground chitin originally was autoclaved in the jar, but experience demonstrated that the resultant culture jars were very fragile, making the loss of a replicate very likely. Two different processes were tried, and both were effective in reducing the fragility of the vessels. First, the autoclaved sediment consisting of silica sand and ground chitin was added after the separate sterilizations of medium and vessel. We found that separate sterilization improved the durability of the glass jars and did not add to the contamination of the microcosms. Another modification of the process was to follow the SOP but to immerse the culture vessel in a water bath to a point above the sand and chitin level during sterilization.

The numbers of organisms, dissolved oxygen (DO), and pH were determined twice weekly. Nutrients (nitrate, nitrite, ammonia, and phosphate) were sampled and measured twice weekly for the first four weeks, then only once weekly thereafter. The room temperature was 20 ± 2°C. Illumination was 79.2 μeinsteins m^{-2} s^{-1}, with a range of 78.6 to 80.4 and a 16-h day/8-h night cycle.

A stock suspension of the graphite dust or brass dust was prepared for distribution to the sample vessels. The material was weighed on a Cahn-28 electrobalance and dispensed into disposable polycarbonate tubes, and diluent was added to make a 1-mg/mL suspension. The sealed tube was placed in an ultrasonic water bath and manipulated until all the particles were suspended as uniformly as possible. The exposure concentrations were set so that a no-effect and an effect level could be expected, estimated from *Daphnia magna* 48-h and algal 96-h growth assay results. The concentrations of brass were 0.01, 0.5, and 1.0 mg/L. Graphite was added to make concentrations of 0.01, 1.00, and 10.0 mg/L.

Chemistry

The sampling and analytical methods have previously been reported [*9–11*]. Samples for dissolved metal analysis were collected weekly from each SAM vessel. The sample was passed through a Swinex-25 mm filter holder and filtered with a prewashed 0.45-μm filter. Sample analysis for soluble copper was conducted on an atomic absorption spectrophotometer set at a wavelength of 324.7 nm and slit width of 0.7 nm, using an air acetylene plume and a lamp of copper (hollow cathode lamp) 18 mV. The analytical sensitivity was 0.03 mg/L copper.

Data Analysis

All data were recorded onto standard computer entry forms and checked for accuracy. The forms were then sent to the University of Washington for input and analysis. Parameters calculated included the concentrations of each of the species, DO, DO gain and loss, nutrient concentrations, net photosynthesis/respiration ratio (P/R) (calculated as the ratio of gain in oxygen during the light period to loss in oxygen during the dark period), pH, algal species diversity using the Shannon-Weaver index, daphnid fecundity, algal biovolume, and biovolume of available algae. The statistical significance of each of these parameters in comparison with the controls was also computed for each sampling day.

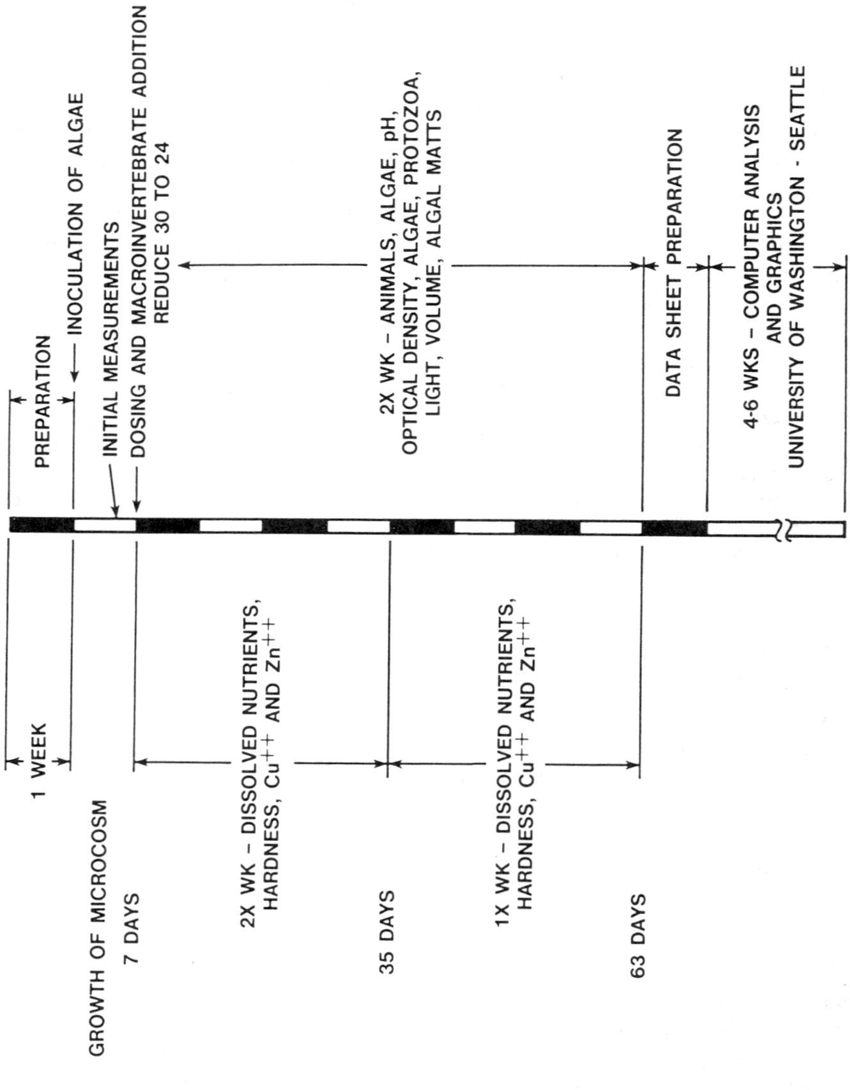

FIG. 1—*Time line for the standardized aquatic microcosm (SAM).*

Time and Manpower

Approximately 30 days before the start-up date, the culture of the test organisms was begun to ensure that enough were available at the start of the SAM. As illustrated in Fig. 1, a week of preparation was allocated for the preparation of the test vessels, media, equipment, and facilities before the initiation of the SAM. On sampling days, half a day was required from three to four people experienced in the sampling and counting techniques. Even during the SAM, considerable effort had to be made in the preparation of cultures for reinoculation.

After keypunching and initial analysis by the University of Washington, the output was sent to our laboratory and spot checked for accuracy. Eventually, the extensive computer printout was delivered with graphs of selected parameters included. The process of analysis took one to two months, portions of which included contracting procedures and data corrections.

Results

In this section we show a sample of the types of data and some of the interactions actually available, using the SAM to illustrate the toxic and eutrophic responses of the system. Also included are brief descriptions of the learning curve and an evaluation of the modification of the sterilization procedure. Detailed reports and evaluation of the fate and effects of the brass and graphite have been previously published [*9*].

Interspecific Interactions

The interaction between the daphnids and algae is typical of a predator-prey situation altered by the sensitivity of the predator to the xenobiotic (Fig. 2). The *Daphnia* were more sensitive to the brass than the algae. Consequently, algal peak biomass increased with an increase in the dosage of the brass. Graphite does not exhibit acute toxicity to either *Daphnia* or algae [*10*]. In the graphite SAM, the algal biomass remained remarkably consistent, except for Days 56 and 60, during which the medium treatment group (1.0 mg/L) showed an immense peak, which appeared to some extent in each of the replicates at that concentration. The bloom was due to a sharp increase in *Nitzschia kutsigiana* and to a lesser extent *Ankistrodesmus falcatus.*

Species Diversity

Algal species diversity (Shannon-Weaver) demonstrated the variability of this measure of community structure (Fig. 3). In the brass SAM, the diversity in all of the treatment groups was similar for the first 17 days of the experiment. The medium treatment group (0.5 mg/L) deviated markedly downward from all other treatment groups until midway through the experiment. At the end of the experiment, the diversity of the 0.5-mg/L group had recovered to the control values. In contrast, the group with highest toxicant concentration (1.0 mg/L) tracked the control values until after Day 40 of the experiment. At Day 64, algal diversity in the highest treatment group was far below that of the other groups. The picture of diversity in the graphite SAM was also varied. The highest concentration of graphite did consistently exhibit the lowest species diversity throughout the experiment. In general, the other treatment groups tracked the controls well, especially the medium treatment group.

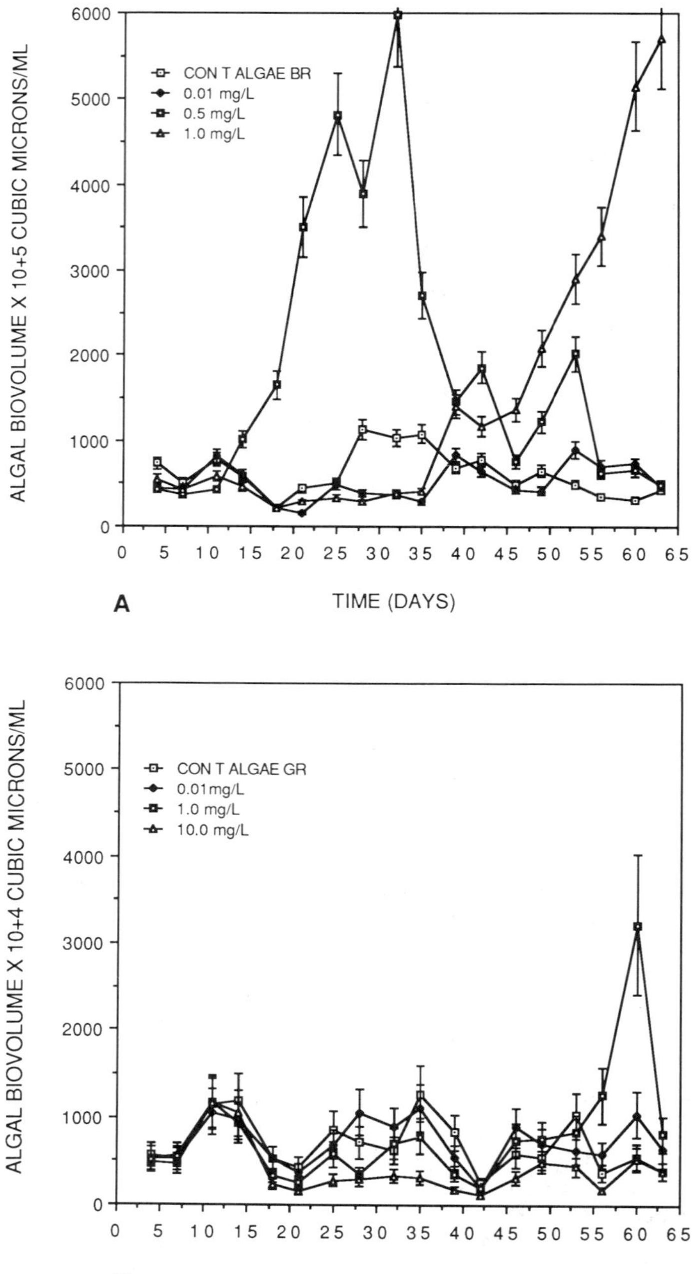

FIG. 2—*Comparison of algal growth patterns in the SAM:* (a) *The algal biomass in the brass SAM demonstrates a proportional dose-response relationship due to the sensitivity of the daphnid grazers.* (b) *In the graphite SAM, all of the treatment groups track the controls except the medium treatment group (1.0 mg/L). This group saw a bloom in* Nitzschia *on Days 56 and 60.*

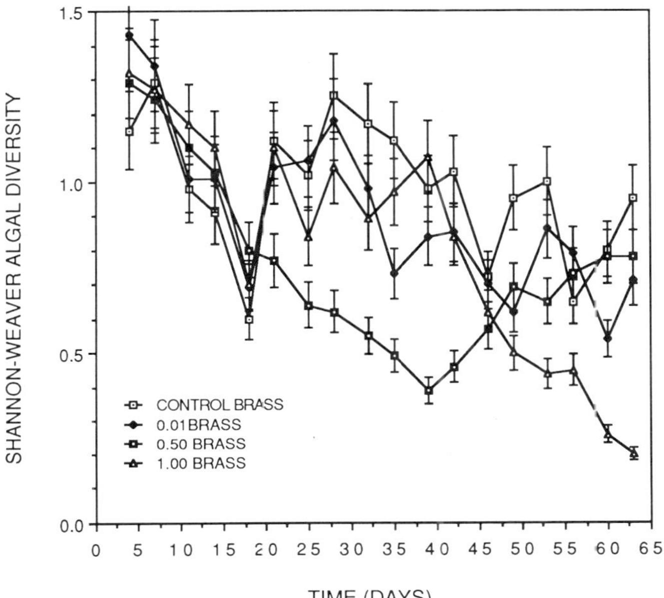

A

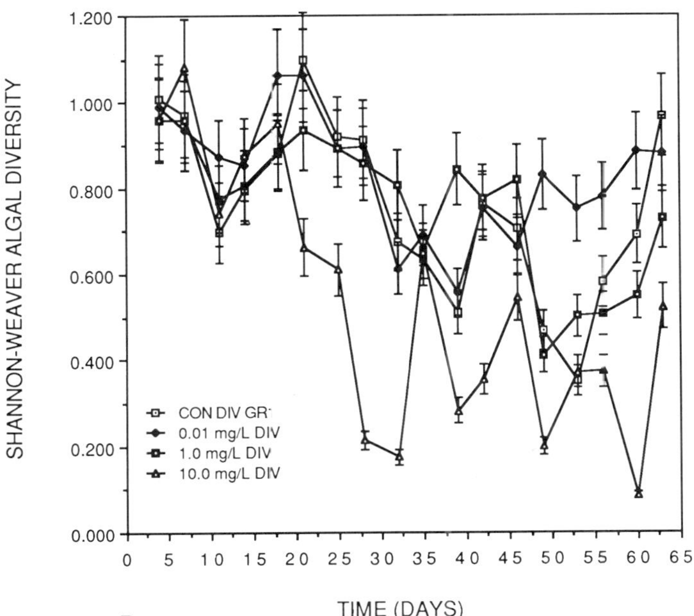

B

FIG. 3—*Species diversity:* (a) *In the brass SAM, the 0.5-mg/L concentration deviates sharply from the control value. However, by the end of the assay, diversity in this group has become statistically indistinguishable from the control. In contrast, the highest concentration of brass tracks the control until near the end of the assay when diversity plummets.* (b) *The graphite SAM illustrates a decrease in species diversity at the highest concentration.*

Nutrients

Nutrients demonstrated a dose-response relationship in both experiments. For example, phosphate deviated significantly from the control values in both the brass and graphite SAMs (Fig. 4). In the brass SAM, phosphate in the 1.0-mg/L replicates followed the declining trend of the other treatment groups but remained significantly higher until near the end of the experiment, when an apparent equilibrium was reached. The graphite SAM demonstrated a markedly different pattern. The highest concentration tracked the control and the two lower treatment groups until three weeks into the experiment. At that time, the phosphate steadily increased in the high treatment group.

Photosynthesis/Respiration Ratio

The photosynthesis/respiration ratio fluctuated and often ranged to below one in the highest concentrations for both the brass and the graphite SAMs (Fig. 5). In the brass SAM, all but the highest concentration tracked the control replicates well. In the highest concentration, the P/R ratio dropped below one three times before again returning to the equilibrium ratio. In the graphite SAM, the P/R ratio stayed below one for the first 46 days of the experiment. Large deviations above one were also exhibited by the other treatment groups, although the large deviation on Day 33 in the 0.01-mg/L group may have been spurious.

Fate of the Toxicant

The fate of the toxicant was observed to follow a very different pattern in the SAM than in a simpler abiotic system (Fig. 6). In the brass SAM, we followed the evolution of the copper from the brass dust. After an initial increase, the concentration of the copper declined, reaching an equilibrium for all concentrations by the end of the experiment. In the medium prepared by Muse [*11*] at a hardness and pH similar to those of the SAM, the amount of copper steadily increased. In this prepared abiotic medium, an equilibrium was rapidly reached with the kinetics of the disassociation approximating a first-order equation.

SAM Operations

As experience was gained, the manpower requirement of the SAM rapidly decreased to about half of that of the first SAM conducted for the round robin. Approximately a 0.65-man year is currently budgeted for each SAM.

Modification of the sterilization procedure led to a rapid decrease in the breakage of the test vessels. In the first SAM for the round robin, more than ten vessels broke during the course of the experiment, with complete loss of two of the replicates. In the graphite SAM, only two of the vessels cracked, and these were easily replaced.

Discussion

The SAM appeared to be capable of distinguishing both toxic effects and the onset of biological and chemical parameters associated with eutrophication. The brass SAM demonstrated the differential toxicity of the toxicant to the daphnid population and to the algae. The proportional increase in biomass of the algae in relation to the concentration of the toxicant was a direct outcome of the differential toxicity. Resuspension of the brass during sampling also served to make the toxicant repeatedly available to the filter-feeding organisms, but in temperate lakes, the fall and spring turnovers mix bottom material with

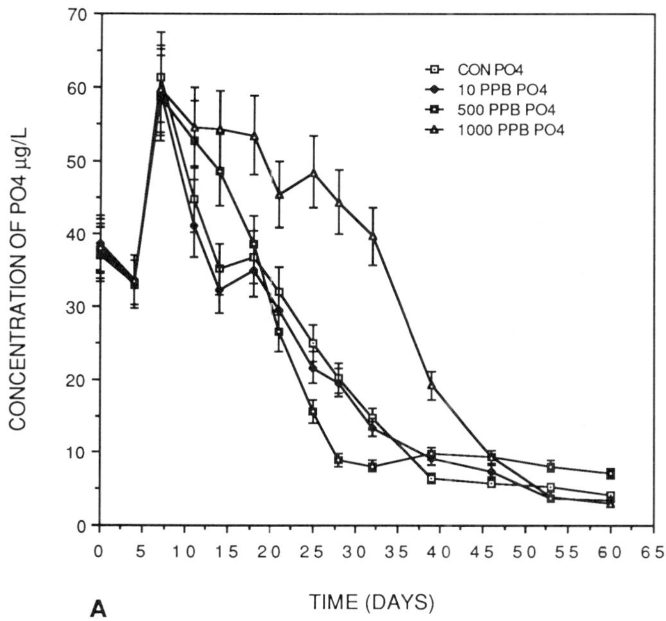

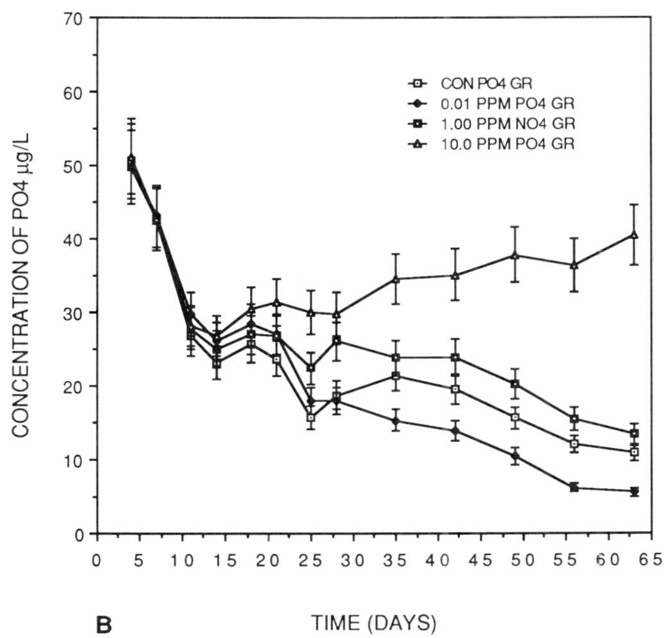

FIG. 4—*Phosphate metabolism:* (a) *In the brass SAM, phosphate concentration in all treatment groups is indistinguishable from the control at the end of the 64 days.* (b) *In sharp contrast, the treatment groups diverge dramatically in the amount of phosphate in the graphite SAM.*

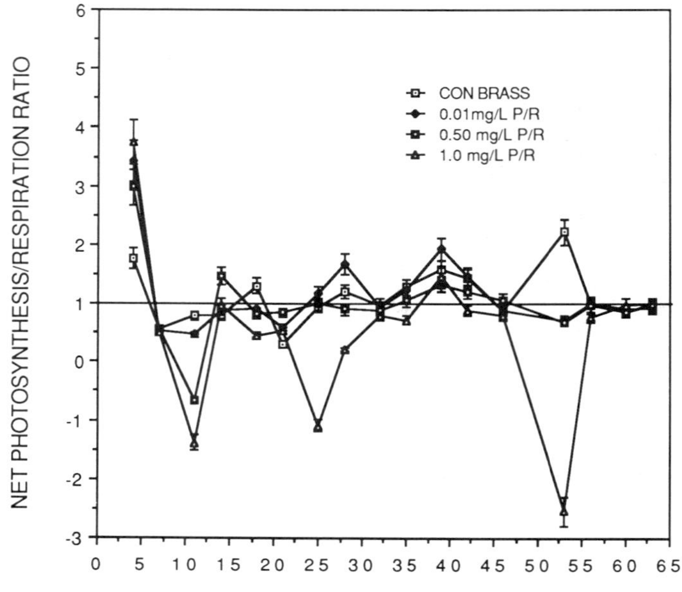

A

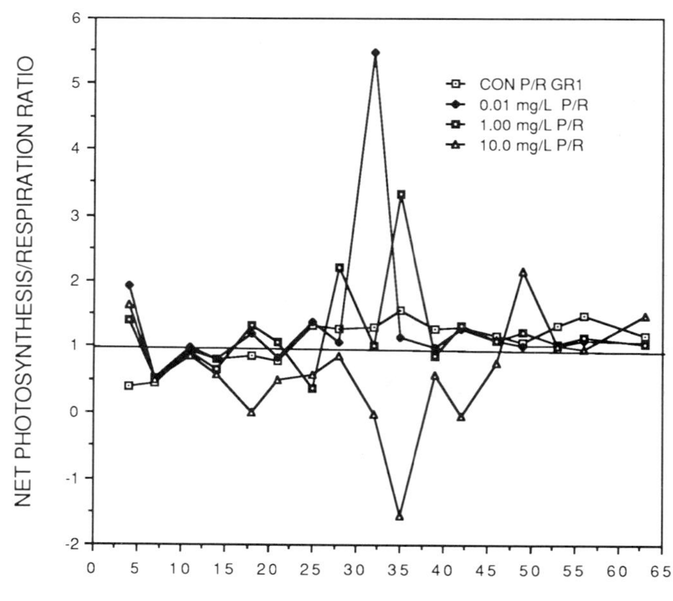

B

FIG. 5—*Photosynthesis/respiration ratio (P/R):* (a) *The P/R ratio in the highest concentration of the brass SAM appears unstable, often deviating to below one. The negative P/R ratio indicates that DO actually decreased during the light period.* (b) *In the graphite SAM, the P/R ratio of the highest concentration stayed below one for the first 46 days of the experiment.*

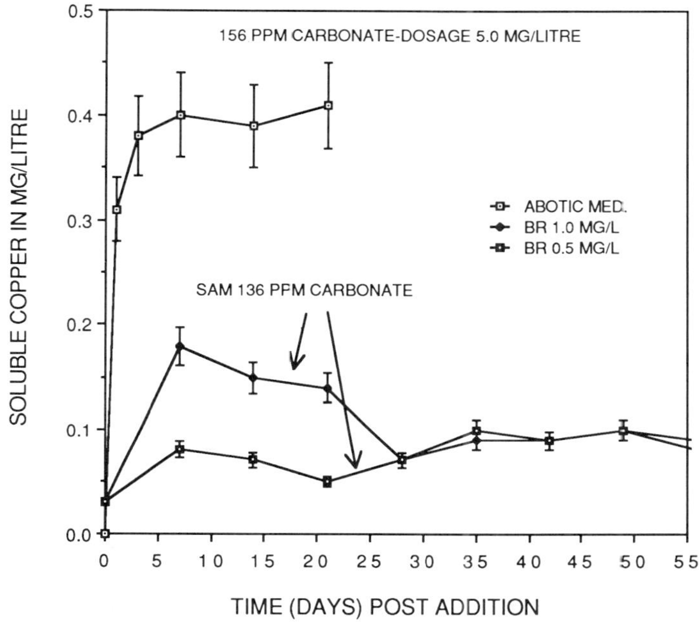

FIG. 6—*Fate of copper in the brass SAM and an abiotic system: In the abiotic experiment, the copper concentration increases until an equilibrium is reached. The kinetics can be approximated using a first-order equation. Soluble copper in the brass SAM increases initially, but then decreases throughout the course of the experiment until an equilibrium in concentration is reached.*

the water column. Nutrient cycling in the brass SAM also demonstrated stress in a dose-response manner. The graphite SAM illustrated the effects of a material that has the potential of causing eutrophication. Species diversity in the highest concentration was lower than that in the controls. Ammonia was elevated. The negative P/R ratio indicated that a great deal of respiration was occurring, more than the rate of photosynthesis during the lighted periods.

The components of the P/R ratio have generally been demonstrated to be a more sensitive indicator of toxic effects. This is probably due to the covariance of the oxygen gain and oxygen loss during the addition of a toxicant.[2] Toxicants that directly affect photosynthesis or are very selective in their effect on target organisms would have less of a covariance than broad-spectrum toxicants. The P/R ratio must be considered a summary of these effects, a concise way of looking at the overall result but lacking in the resolution and sensitivity available from its components.

In evaluating toxic effects, criteria such as species diversity, biomass, P/R ratio, and nutrient cycling should not be used individually to identify an impact. Traditional theory suggests that stressed ecosystems are forced to earlier successional stages and, therefore, species diversity drops. In the brass microcosm, only the highest concentration demonstrated a decrease in algal diversity, and then only after the halfway point of the experi-

[2] Taub, F. B., School of Fisheries, University of Washington, Seattle, WA, personal communication, August 1985.

ment. Although the 0.5-mg/L concentration demonstrated effects in algal and daphnid density attributable to the brass, no significant decrease in the algal diversity was apparent. Two explanations come to mind. First, the SAM protocol is examining the system during its early development and maturation. Kindig et al. [*12*] has previously shown that immature microcosms, with an excess of nutrients, are more sensitive to insults than mature microcosms in equilibrium. However, the old, mature control microcosms demonstrated poorer replicability and stability, discounting the hypothesis that older systems are inherently more stable. The use of developing microcosms is justified, moreover, in that many aquatic systems undergo seasonal development with a succession of biota depending on temperature, light, and colonization. Also, the idea that a toxic effect automatically manifests itself in a decrease in diversity may be an erroneous assumption. Landis [*13*], using derivations of the resource competition models of Tilman [*14*], has demonstrated in the two-species case that, in certain circumstances, a toxicant can increase species diversity. The brass SAM illustrates the insensitivity of diversity, as measured by the Shannon-Weaver information index, as an indicator of toxicant stress. On the other hand, the graphite SAM did show a decrease in diverstiy at the highest concentration of the test material. In contrast to the brass SAM, the graphite SAM exhibited many of the responses characteristic of eutrophication and other processes; a decrease in diversity, a photosynthesis/respiration ratio of less than one, and a severe drop in DO during the night.

The fate of the toxicant could be followed, and the results mimicked those of a natural ecosystem, as opposed to a laboratory abiotic experiment. In laboratory experiments using waters of varying hardness, the brass disassociated into copper and zinc. The concentrations of these materials increased over time. In the case of the brass SAM, the concentration of copper decreased after an initial period following the initial disassociation. Copper concentration reached an equilibrium in a fashion very similar to the copper in the $CuSO_4$ SAMS run as part of the interlaboratory round robin. Certainly this equilibrium of concentration does not imply the cessation of disassociation of the copper and zinc from the brass. Complexation of metals due to biologically derived chelates and binding to substrates has been extensively documented [*15–17*]. Copper can also be eliminated as a result of the formation of insoluble precipitates in the presence of carbonate [*18*]. In the fate of the copper at least, the SAM mimicked larger ecosystems.

The manpower requirements of the SAM are high, but the data output perhaps can be used to replace chronic assays, thus producing an overall savings. In the case of the daphnids, data on reproduction, age class, and productivity can be derived from the SAM data. Similar kinds of data can also be derived for the other invertebrates and protozoa. Several species can be assayed simultaneously. Also, as the toxicant degrades or is transformed, the effects can be recorded. A drawback is that the food supply is constantly varying in quantity and quality. In addition, the toxicant concentration is altered as degradation and transformation takes place. However, the data gathered from a SAM experiment should make it possible to estimate chronic effects on a particular species.

Among the drawbacks to the current SAM system are (*a*) the necessity of hand-recording and checking the large amount of data, (*b*) the data are processed remotely, and (*c*) there is dependence on the University of Washington software. It is our recommendation that comparable software for data analysis and graphics be prepared for the currently popular microcomputer systems. As data could be input directly to the computer during the course of the experiment, data analysis and interpretation could begin on completion of the experimental protocol.

As demonstrated, in many ways the SAM mimics natural ecosystem responses without being a deliberate model of any particular type of freshwater aquatic ecosystem. A potential that needs to be further explored is that the SAM can be used for many kinds of basic research in the impacts of xenobiotics on the dynamics of ecosystem behavior. A tremendous advantage of the SAM is that the vast majority of the members of the community are known beforehand, and colonizers are conspicuous by their rarity. The repeatable nature of the initial inoculum lends itself to repeatability of the community, a prime criterion for basic research as well as for routine testing. In addition, basic research on the impact of xenobiotics using the SAM can have a direct application to the risk assessment process. A standardized, stable, and reproducible community has the potential of serving the ecotoxicologist interested in community dynamics and metabolism in the same way that *Escherichia coli* serves the bacteriologist interested in molecular genetics.

Conclusions

1. The SAM is a potentially excellent tool for the evaluation of the impact of a xenobiotic on an ecosystem.
2. Chemical fate can be followed as the assay is being conducted and processes similar to natural ecosystems appear to occur.
3. Toxicity and the effects of nutrient input can both be observed using the SAM protocol.
4. Modifications of the sterilization procedure have led to an increase in the durability of the test vessels.
5. Direct input to a computer data base is desirable.
6. The SAM has excellent potential to serve as a tool for investigation of the impacts of xenobiotics on ecosystem interactions and metabolism.

References

[*1*] Hammons, A. S., *Methods for Ecological Toxicology,* Ann Arbor Science, Ann Arbor, MI, 1981, pp. 5–6.

[*2*] Taub, F. B., "Synthetic Microcosms as Biological Models of Algal Communities," *Algae as Ecological Indicators,* L. E. Shubert, Ed., Academic Press, New York, Chapter 13, pp. 363–394.

[*3*] Taub, F. B., "Demonstration of Pollution Effects in Aquatic Microcosms," *International Journal of Environmental Studies,* Vol. 10, 1976, pp. 23–33.

[*4*] Taub, F. B. and Crow, M. E., "Loss of a Critical Species in a Model (Laboratory) Ecosystem," *Verh. Internat. Verein. Limnol,* 1978, pp. 1270–1276.

[*5*] Crow, M. E. and Taub, F. B., "Designing a Microcosm Bioassay to Detect Ecosystem Level Effects," *International Journal of Environmental Studies,* 1979, pp. 141–147.

[*6*] Taub, F. B., Crow, M. E., and Hartmann, H. J., "Responses of Aquatic Microcosms to Acute Mortality," *Microcosms in Ecological Research,* J. P. Giesy, Jr., Technical Information Center, U.S. Department of Energy, Washington, DC, 1980, pp. 513–535.

[*7*] Taub, F. B., Kindig, A. C., Conquest, L. L., and Meador, J. P., "Results of Interlaboratory Testing of the Standardized Aquatic Microcosm Protocol," this publication, pp. 368–394.

[*8*] Taub, F. B. and Read, P. L., "Standardized Aquatic Microcosm Protocol," Final Report, Contract No. 223-80-2352, Vol. II, U.S. Food and Drug Administration, Washington, DC, 1983.

[*9*] Haley, M. V., Johnson, D. W., Muse, W. T., Jr., and Landis, W. G., "The Aquatic Toxicity and Fate of Brass Dust," *Aquatic Toxicology and Hazard Assessment: Tenth Volume, ASTM 971,* W. Adams, G. Chapman, and W. G. Landis, Eds., American Society for Testing and Materials, Philadelphia, 1987, p. 468.

[*10*] Johnson, D. W. and Landis, W. G., "The Acute Aquatic Toxicity of Selected Mineral Particles," CRDEC-TR-870000, Chemical Research, Development, and Engineering Center, Aberdeen Proving Grounds, MD, in press.

[*11*] Muse, W. T., Jr., "The Chemical Fate of Brass Dust in Waters of Varying Hardness," CRDEC-TR, Chemical Research, Development, and Engineering Center, Aberdeen Proving Grounds, MD, in press.

[*12*] Kindig, A. C., Conquest, L. L., and Taub, F. B., "Differential Sensitivity of New Versus Mature Synthetic Microcosms to Streptomycin Sulfate Treatment," *Aquatic Toxicology and Hazard*

Assessment: Sixth Symposium, ASTM 802, W. E. Bishop, R. D. Cardwell, and B. B. Heidolph, Eds., American Society for Testing and Materials, Philadelphia, 1983, pp. 192–203.

[*13*] Landis, W. G., "Resource Competition Modeling of the Impacts of Xenobiotics on Biological Communities," *Aquatic Toxicology and Environmental Fate: Ninth Volume, ASTM 921,* T. M. Poston and R. Purdy, Eds., American Society for Testing and Materials, Philadelphia, 1986, pp. 55–72.

[*14*] Tilman, D., *Resource Competition and Community Structure,* Princeton University Press, Princeton, NJ, 1982, pp. 11–138.

[*15*] Riley, J. P. and Chester, R., "Primary and Secondary Production in Marine Environments," *Introduction to Marine Chemistry,* 5th ed., Academic Press, New York, 1971, pp. 246–247.

[*16*] Fitzgerald, G. P. and Faust, S. L., "Factors Affecting the Algicidal and Algistatic Properties of Copper," *Applied Microbiology,* Vol. 11,1963, pp. 345–351.

[*17*] Van Den Berg, C., Wong, P., and Chau, Y., "Measurement of Complexing Materials Excreted from Algae and Their Ability to Ameliorate Copper Toxicity," *Journal of the Fisheries Research Board of Canada,* Vol. 36, 1979, pp. 901–905.

[*18*] Rose, K. A., "Evaluation of the Nutrient-Phytoplankton-Zooplankton Models and Simulation of the Ecological Effects of Toxicants Using Laboratory Microcosm Ecosystems," M.S. Thesis, University of Washington, Seattle, WA, 1986.

DISCUSSION

J. R. Pratt[1] *(written discussion)*—Is the sensitivity of the standard aquatic microcosm (SAM) similar to that of acute or chronic species tests?

W. G. Landis, N. A. Chester, M. V. Haley, D. W. Johnson, W. T. Muse, Jr., and R. M. Tauber (authors' closure)—In our experience, the sensitivity of the SAM has varied depending upon the effect. In the experiments using $CuSO_4$ and brass dust, the SAM was less sensitive on a pure dosage-by-dosage effect basis. An effect on the daphnid population was seen at the 0.5-mg/L brass concentration. Although the daphnid population did not become extinct, the rate of daphnid reproduction was certainly slowed to the point that an algal bloom occurred. In the latter stages of the brass SAM, the daphnids recovered even though the aqueous copper concentration remained relatively constant. Yet the 48-h EC_{50} for brass had previously been determined to be 0.022 mg/L, or about 19 times less. However, the actual exposure of the organisms to the toxic materials in the SAM is much less than the initial dose delivered to the microcosm. Many components of the microcosm are able to chelate and sequester the copper. Also, the pH fluctuations in the microcosm cause the actual species of ion to be changed, thereby altering the toxicity. In the case of the brass microcosm, the 0.5 mg/L of brass had a maximum concentration of only 0.09 mg/L copper. Compared with this concentration is an expected concentration of 0.3 mg/L copper if the brass was completely disassociated, and the resultant material completely solubilized.

On the other hand, the SAM is sensitive to effects that cannot be described using acute and chronic toxicity tests. Graphite in our acute assays demonstrated little if any toxicity. However, the SAM assays responded, demonstrating effects that the acute assays are not equipped to show. Alterations in species diversity, oxygen metabolism, and nutrient cycling occurred at concentrations that did not demonstrate any toxic effect in the acute assays.

J. R. Pratt (written discussion)—How could the SAM be used as a predictive tool to estimate "safe" concentrations of a particular toxicant?

W. G. Landis, N. A. Chester, M. V. Haley, D. W. Johnson, W. T. Muse, Jr., and R. M. Tauber (authors' closure)—First, I would like to know the definition of a "safe" concentration. Starting from the premise that a "safe" concentration is a concentration that does not significantly alter the functional characteristics of an ecosystem, the standardized aquatic microcosm has several uses. The fate of a material and its realized concentration can be

[1] Pennsylvania State University, University Park, PA.

studied using the microcosm. As in the brass and $CuSO_4$ experiments, the concentration delivered to the system may not be the dose realized by the organisms. The population dynamics of invertebrates can often be seen to follow a dose-response pattern that may be extrapolated. The patterns of ecosystem response can also be examined. In addition, effects such as decreases in diversity and productivity and alterations in metabolism can be examined using these systems. As far as setting safe levels, all of this information could be incorporated in the judgement call that eventually decides upon an acceptable concentration.

Frieda B. Taub,[1] Andrew C. Kindig,[2] Loveday L. Conquest,[3] and James P. Meador[1]

Results of Interlaboratory Testing of the Standardized Aquatic Microcosm Protocol

REFERENCE: Taub, F. B., Kindig, A. C., Conquest, L. L., and Meador, J. P., **"Results of Interlaboratory Testing of the Standardized Aquatic Microcosm Protocol,"** *Aquatic Toxicology and Environmental Fate: Eleventh Volume, ASTM STP 1007,* G. W. Suter II and M. A. Lewis, Eds., American Society for Testing and Materials, Philadelphia, 1989, pp. 368–394.

ABSTRACT: The "standardized aquatic microcosm" (SAM) protocol was performed by four laboratories, including the authors' laboratory, and the results of seven experiments are presented. Each laboratory tested the toxicant copper at three concentrations (500, 1000, and 2000 ppb) plus a control.

Nitrate depletion and an early algal bloom which was terminated by an increase in grazers (mainly *Daphnia*) were consistently observed in the control microcosm. In comparison with controls, the copper-treated microcosms were associated with reductions in *Daphnia* and in sensitive algal populations. In all toxicant treatments, copper eliminated the *Daphnia* and inhibited primary production; the duration of inhibition was prolonged at the higher copper concentrations. At the lowest concentration, recovery was characterized by an algal bloom of copper-resistant green algae, which was followed by a *Daphnia* bloom. Recovery occurred in most of the 1000-ppb copper-treated microcosms but was usually absent at the 2000-ppb concentration until late in the experiment.

All experiments exhibited the same sequence of events, but the timing of these events varied among experiments. Experiments provided similar statistical differences between control and treatment microcosms within the same experiment and gave the same rank order of the day-weighted-by-variable" (DWV) statistic, defined in the text. Comparison of the variable means on a day-by-day analysis of variance did show the experiments to be statistically different because of differences in the timing of events.

KEY WORDS: microcosms, copper, interlaboratory testing, standardized aquatic microcosm, toxicity, *Daphnia,* algae, toxicity testing, aquatic toxicology

"The ability of any test method, whether biological or chemical, to provide reproducible data within and among laboratories must be known before the method is adopted for regulatory purposes" [*1*]; this principle has been stated as the justification for an interlaboratory test of *Daphnia* acute toxicities, but is as valid for more complex ecosystem level tests. Data used to allow or forbid the manufacture and use of a chemical must be subject to verification at a later time, and possibly at a different laboratory. Uncertainties concern-

[1] Professor and predoctoral research associate, respectively, School of Fisheries, University of Washington, Seattle, WA 98195.

[2] Senior scientist, Beak Consultants, Bellevue, WA 98005.

[3] Associate professor, Center for Quantitative Science, School of Fisheries, University of Washington, Seattle, WA 98195.

ing the reproducibility of ecosystem responses have limited the use of multispecies tests in the regulatory process.[4]

The literature contains many studies on the interlaboratory or intralaboratory variability of single-species toxicity tests, for example, on *Daphnia* or a marine amphipod [*1,2*]. The replicability of aquatic multispecies test systems has been reviewed [*3*]; in general, the replicability of microcosm results is adequate to demonstrate statistically valid differences among controls and treatment groups. A review of several different types of microcosm bioassays can be found in Ref *4,* specifically the standardized aquatic microcosm [*5*], shallow-water microcosm [*6*], mixed flask culture microcosm [*7*], and generic aquatic ecosystem [*8*]. A flow-through microcosm system treated with copper has been studied by Hedtke [*9*]. Information on replication is provided in each of these studies; all but two [*6,9*] provide information on repeatability of results within the same laboratory. However, we do not know of any other studies that address the question of interlaboratory comparisons of a microcosm protocol.

Microcosm data share many statistical attributes of pond and lake studies which complicate statistical comparison. Statistical analyses of repeated measurements of related variables present special problems. First, the responses are statistically related; that is, they are not statistically independent. Nitrate is depleted as algae grow, *Daphnia* graze down the algae, and community metabolism is related to all processes. Second, even for a single response from a single microcosm, the value on Day $t + 1$ will be correlated with the value on Day t. These kinds of data usually call for the application of time series analysis, the set of statistical methods designed to address repeated measures data with a correlation structure. But the use of time series analysis requires at least 30 sampling times, preferably more. The time traces for all the variables measured from a standard aquatic microcosm (SAM) have at most 19 sampling days; chemical analyses have fewer. Requiring more samples would make the protocol more labor-intensive.

The goals of this study were to determine (*a*) if the protocol could be performed in other laboratories, (*b*) the degree to which results were reproducible, (*c*) the causes of differences, and (*d*) the appropriateness of the provisional quality control criteria from the draft SAM protocol [*10*]. Earlier results from this study [*5,11,12*] and reviews of the method development have been published [*13–15*]. More detailed results of this study and a revised protocol are available [*16*]. This paper differs from our earlier publications in that it (*a*) provides interlaboratory analyses on *all* treatment groups (0, 500, 1000, and 2000 ppb copper, not just controls and 500 ppb), (*b*) replaces Experiment UW No. 1 with UW No. 1R (ME64 is replaced with ME89), and (*c*) includes the ranked analysis of variance (ANOVA) test of the day-weighted-by-variable (DWV) statistic.

Methods

The interlaboratory study consisted of experiments performed at the University of Washington and three participating laboratories, including University of Minnesota—Duluth with the U.S. Environmental Protection Agency (EPA), Aberdeen Proving

---

[4] After submission of this paper, the U.S. Environmental Protection Agency (EPA) proposed new test guidelines for the Toxic Substances Control Act that included laboratory microcosms. Within the section on the generic freshwater microcosm test, aspects of the standardized aquatic microcosm, naturally derived mixed flask culture microcosm, and naturally derived pond microcosm are included. (EPA, 1987 Toxic Substance Control Act Test Guidelines, proposed rule, *Federal Register,* Vol. 52, No. 187, pp. 36334–36371.)

TABLE 1—*Interlaboratory testing experiments.*

Laboratory	First Experiment	Second Experiment
Accepted Experiments (Included in DWV Statistical Comparison)		
University of Washington	UW No. 1R	UW No. 2R
Date initiated	2 May 1986	24 Jan. 1986
Experiment No.[a]	ME 89	ME 82
Copper deviation[b]	+10%	+5%
EPA—Duluth	DU No. 1	DU No. 2
Date initiated	2 Dec. 1983	6 April 1984
Experiment No.	ME 65	ME 67
Copper deviation	−3%	+6%
Aberdeen—Army	AB No. 1	AB No. 2
Date initiated	3 May 1985	24 Jan. 1986
Experiment No.	ME 74[c]	ME 87[d]
Copper deviation	3	−10%
MBL	MB No. 1	...
Date initiated	15 Nov. 1985	...
Experiment No.	ME 79	...
Copper deviation	−1%	...
Rejected Experiments (Documented)		
University of Washington	UW No. 1 (original)	UW No. 2 (original)
Date initiated	2 Dec. 1983	12 Oct. 1984
Experiment No.	64	69
Copper deviation	−13	+18%
University of Washington	...	...
Date initiated	1 Feb. 1985	...
Experiment No.	72	...
Copper deviation	+25%	...

[a] Experiment numbers indicate the model ecosystem number under which the original data are stored in the host laboratory.
[b] Copper deviation indicates the difference between the nominal concentration of copper and the measured total copper on Day 7.
[c] For this experiment only, the treatments consisted of control, 127, 255, and 509 μg copper per litre; all other experiments consisted of control, 500, 1000, and 2000 μg copper per litre.
[d] Nutrient analyses are incomplete for this experiment at this time.

Grounds, and Marine Bioassay Laboratory (MBL) (Table 1). The experiments were performed between 1983 and 1986 and were initiated at different times of the year. Cultures of the organisms were initially supplied by the University of Washington, but the organisms used were reared at the laboratory performing the test. All solutions, including copper sulfate ($CuSO_4$), were prepared by each laboratory. All attempted to follow the protocol [*10*], and any differences were reported. The data were sent to the University of Washington where they were statistically analyzed; drafts of the data and statistical interpretations were returned to the participating laboratory for assurance that their data were accurately represented. A variety of statistical techniques were used to compare the experiments; these are described in the results section.

Within each experiment, the purpose of the SAM was to detect ecological differences between the control and treatment means that can be attributed to the test chemical; the statistic used is the interval of nonsignificant difference (IND) [*5*]. The IND is based on a Dunnett-like procedure with a pooled variance estimate from a one-way ANOVA for each variable on each sampling day. The experimental work involved preparing more than 24 three-litre microcosms with chemically defined medium [the limiting nutrient was nitrate (NO_3) at 500 μg·at] and sediment, and inoculating the microcosms on Day 0 with 10^3 cells each of ten algal species and on Day 4 with known numbers of five animal species (*Daphnia,* ostracods, amphipods, rotifers, and protozoa). On Day 7, 24 microcosms are selected, randomly assigned to four treatment groups, and treated with the appropriate chemicals (in this case, control and 500, 1000, and 2000 ppb copper as $CuSO_4$). Thereafter, organism abundances, three-point oxygen concentrations (dawn, afternoon, and dawn), and pH were measured twice a week. Nutrients were measured twice weekly for the first month and once a week until Day 63. Reinoculation was made once each week, after counts had been taken; a small number of microorganisms was added, and *Daphnia,* amphipods, and ostracods were added only if necessary to provide three animal species per microcosm.

The protocol [*10*] included provisional criteria for acceptable tests, based on the concept that control microcosms must exhibit "normal" behavior and have low enough coefficients of variation (CVs) to allow demonstration of statistical differences within a biologically feasible range. The actual values were determined from previous SAM microcosm data. The provisional criteria applied to the controls only and included Day 28 and the total experiment performance. By Day 28, the following criteria applied:

1. The nitrate should be reduced from the initial concentration of approximately 500 μg·at per litre to less than 50, and the CV should not exceed 0.5 more than 50% of the observations (CV should not be calculated if the nitrate mean is less than 2 μg·at per litre).
2. Algal biovolume should have achieved a mean peak volume of more than 2000×10^4 μm^3 per mL sometime during the 0 to 28 day period, and CV should not exceed 0.5 more than 50% of the observations.
3. Oxygen gain should have achieved a mean peak value of more than 4 ppm sometime during the Day 0 to 28 period; the mean CV should not exceed 0.5 more than 50% of the time (CV should not be calculated if the oxygen gain is <1 ppm).
4. *Daphnia* should have achieved an abundance of >100 individuals per 100 mL, and the CV should not exceed 0.5 more than 50% of the time.
5. The pH should be between 6 and 10.3, and the CV should not exceed 0.05 more than 50% of the time.
6. In addition, during the remainder of the experiment, the provisional quality control criteria require that the algal biovolume exceed 100×10^4 μm^3 per mL, and the CV not exceed 0.5 more than 50% of the time.
7. The oxygen gain should be positive throughout the experiment, and the CV should not exceed 0.5 more than 50% of the time.
8. *Daphnia* should remain >25 individuals, and the CV should not exceed 0.5 more than 50% of the time.
9. The pH should remain between 6 and 9, and the CV should not exceed 0.05 more than 50% of the time.

Problems

It was not always possible for the laboratories to meet all of the requirements of the protocol. Aberdeen omitted amphipods and did not measure *in vivo* fluorometry; in their second experiment they also omitted *Ulothrix,* one of the filamentous green algae. Different methods of nutrient analyses were used: in DU No. 1, nitrate + nitrite was measured, and in AB No. 2, the nutrients have not been analyzed yet. Stock cultures of protozoa and rotifers had different animal densities, and different amounts of medium were added during inoculation and reinoculation. In some experiments, this was a major source of phosphate. Variations in the amounts of copper differed slightly; with the exception of Aberdeen No. 1, all of the accepted experiments were −10 to +6% of the nominal 500, 1000, and 2000 ppb copper. Aberdeen No. 1 used the weight of $CuSO_4$ (instead of the weight of copper) to calculate concentrations, and the corresponding copper concentrations were control, 127, 254, and 509 ppb copper. Only the control and the 509 ppb treatments were subject to comparison.

Three experiments performed at the University of Washington were rejected because they failed to meet the provisional quality control criteria since they had poor *Daphnia* population development; also, the copper concentrations were somewhat off (Table 1). Results on these rejected experiments will be discussed but not included in the statistical treatment here; more complete documentation is available [*16*]. The three rejected experiments (ME 64, 69, and 72) used *Daphnia* that had been cultured in T82MV, the microcosm medium, rather than in autoclaved Lake Washington water, as had been done in our earlier studies. More recent results suggest that the T82MV is deficient in selenium and possibly other trace metals. A new chemically defined medium (T85MVK) that includes several trace metals from Keatings medium [*17*] was used to rear the *Daphnia* used in the replacement experiment, UW No. 1 (ME 89). Waters used to rear *Daphnia* in the other experiments included UW No. 2R (Lake Washington water), DU No. 1 and DU No. 2 (Lake Superior water), AB No. 1 and AB No. 2 (treated well water), and MB No. 1 (aged microcosm medium). The relationships between the *Daphnia* rearing medium and the results will be discussed in the results section, as will problems encountered in the other laboratories.

Results

Responses to Copper Within an Experiment (UW No. 1R)

In the controls, nitrate was depleted from Days 7 to 20, concurrent with increasing algal biovolume; the developing algal bloom was reduced by grazing of the increasing *Daphnia* population (Figs. 1*a,* 1*b,* and 1*c*). The *Daphnia* overgrazed the algal food supply and decreased during the experiment. The oxygen gained during the lighted period (an index of net primary production), and the pH reflected the abundance and activity of the primary producers and grazers (Figs. 1*d* and 1*e*).

The effects of the 500 ppb copper were that it delayed nitrate depletion, temporarily eliminated the *Daphnia,* and allowed the development of an algal bloom concurrent with the absence of *Daphnia.* Damage to the algal population was shown by reduced pH, oxygen gain, and the change of algal species to those relatively insensitive to copper, especially *Ankistrodesmus* and *Scenedesmus* (and at Duluth, *Oocystis*). Sensitive species were eliminated (*Lyngbya* and *Anabaena*). After the bloom of copper-resistant algae, *Daphnia* populations became reestablished from reinoculated animals and grazed down the algal bloom.

At 1000 ppb copper, the effects were similar but more severe; nitrate depletion was further delayed, the algal bloom took longer to develop, and *Daphnia* populations became established later and remained sparse. At 2000 ppb copper, nitrate had not been depleted by the end of the experiment, algal populations were very low, and *Daphnia* never reestablished populations.

The degree of replication can be judged by the width of the interval of nonsignificant difference (IND) shown in Figs. 1*a* through 1*e*. Slight differences in timing expanded the IND in some cases. This graphical representation allows the control and treatment means to be displayed together with the distance that would be required for a statistical difference [*5,12*]. More detailed data analyses for this experiment are available [*16*].

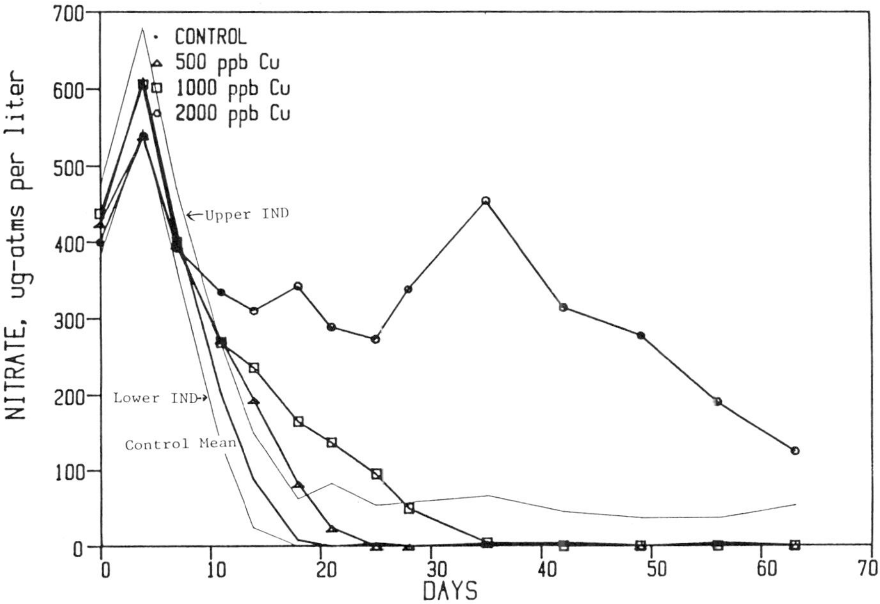

FIG. 1—*Microcosm responses to copper within a single experiment (UW No. 1R). The control means (heavy black line without symbols) are bracketed by the interval of nonsignificance (IND) shown in thinner lines. The other symbols indicate the following concentrations of copper: triangles, 500 ppb; squares, 1000 ppb; and circles, 2000 ppb.* (a) *nitrate.*

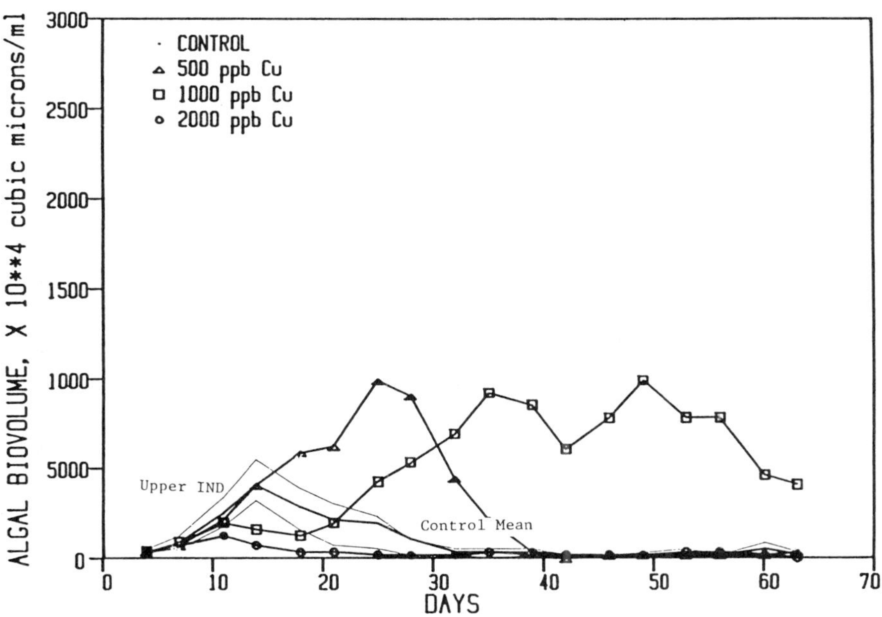

FIG. 1—*Continued:* (b) *algal biovolume.*

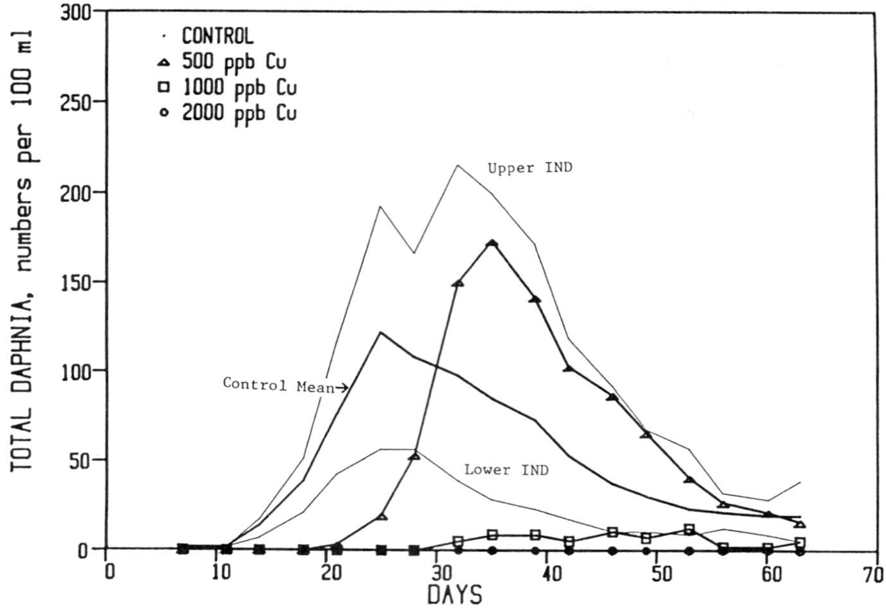

FIG. 1—*Continued:* (c) *total* Daphnia.

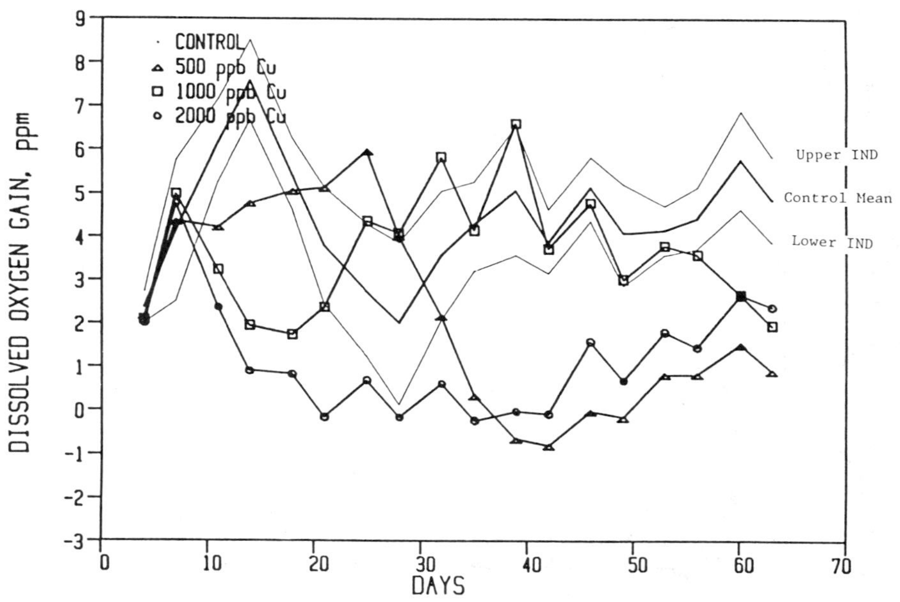

FIG. 1—*Continued:* (d) *dissolved oxygen gain.*

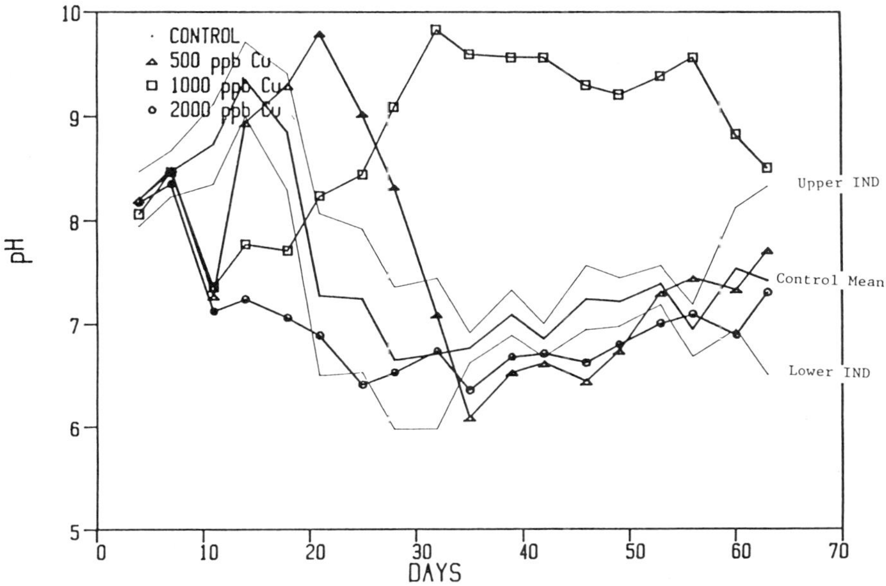

FIG. 1—*Continued:* (e) *pH.*

Comparison Among Experiments Day by Day

The experiment described above (Figs. 1*a* through 1*e*) is presented with six other experiments. The means and IND values for the other six experiments have been previously presented [*12*]. In the controls of all of the experiments, the events were the same; nitrate was depleted during its conversion into algal cells, and the algal cells were consequently converted into grazers (Figs. 2*a* through 2*l*).

Nitrate—Nitrate concentrations remained high during the first week, during which the algal density was low and the cells were probably growing on their initially high internal stores (Fig. 2*a*). Although the nominal nitrate concentration was 500 μg·at per litre, some differences occurred as a result of experimental techniques and evaporation during autoclaving. Occasional increases in nitrate concentration during the first week were probably an artifact of dilution of these highly concentrated samples. After Day 7, nitrate was depleted rapidly in all of the experiments, and was generally low by Day 28. The DU No. 1 nitrate value represents the sum of nitrite and nitrate; a small nitrite peak occurred in other experiments as the nitrate was being depleted. Recycled ammonia is the algal nitrogen source after the nitrate depletion.

In spite of the similarities of the curves representing nitrate in the controls, these experiments are often statistically different from each other in a two-way ANOVA on a given sampling day. This represents the sensitivity of the day-by-day analyses of the data; a similar problem was found in the comparison of Marine Ecosystem Research Laboratory (MERL) mesocosm replicates [*18*].

In all of the experiments, the effect of the copper was to delay the nitrate depletion; the greater the copper concentration, the longer the delay (Figs. 2*b* through 2*d*). At 2000 ppb copper, nitrate was proceeding toward depletion by the end of the experiment, and probably would have occurred if the experiments had been longer. Figure 2*d* demonstrates the

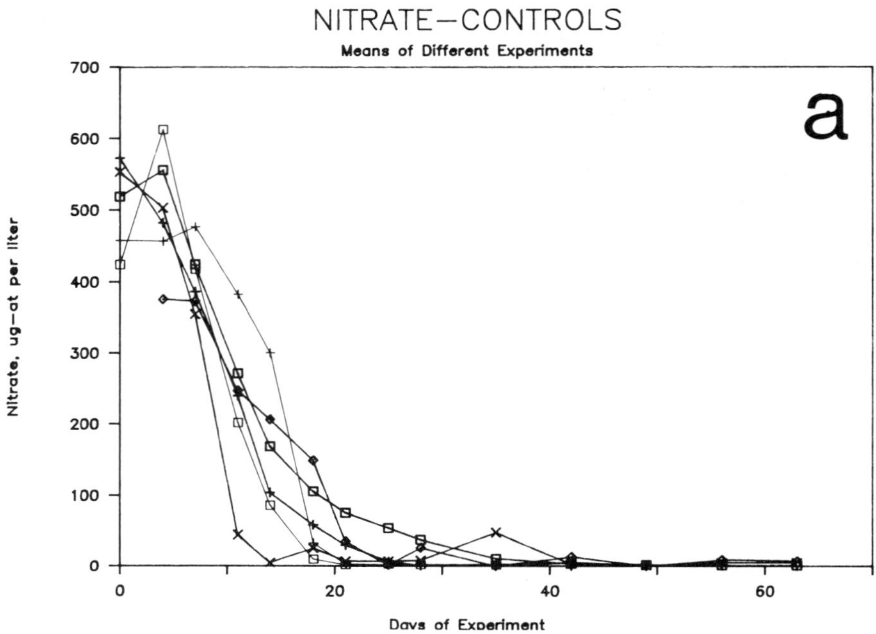

FIG. 2—*Means of seven SAM experiments, performed in four laboratories: thin line (square), UW No. 1R; light plus mark, UW No. 2R; bold line (open square), DU No. 1; bold plus mark, DU No. 2; open diamond, AB No. 1; open triangle, AB No. 2; cross mark, MB No. 1.* (a) *Nitrate, controls.*

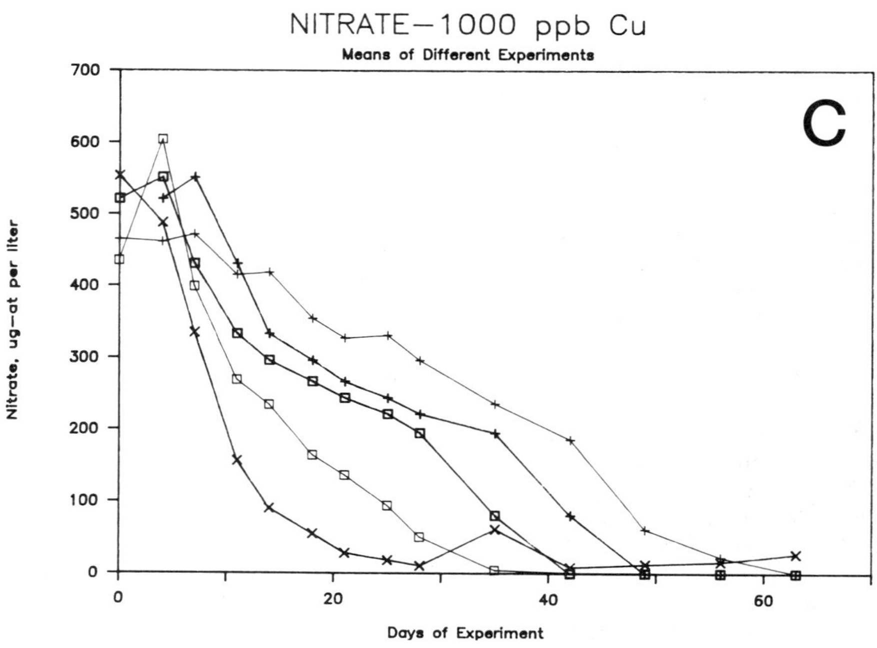

FIG. 2—*Continued:* (c) *nitrate, 1000 ppb copper.*

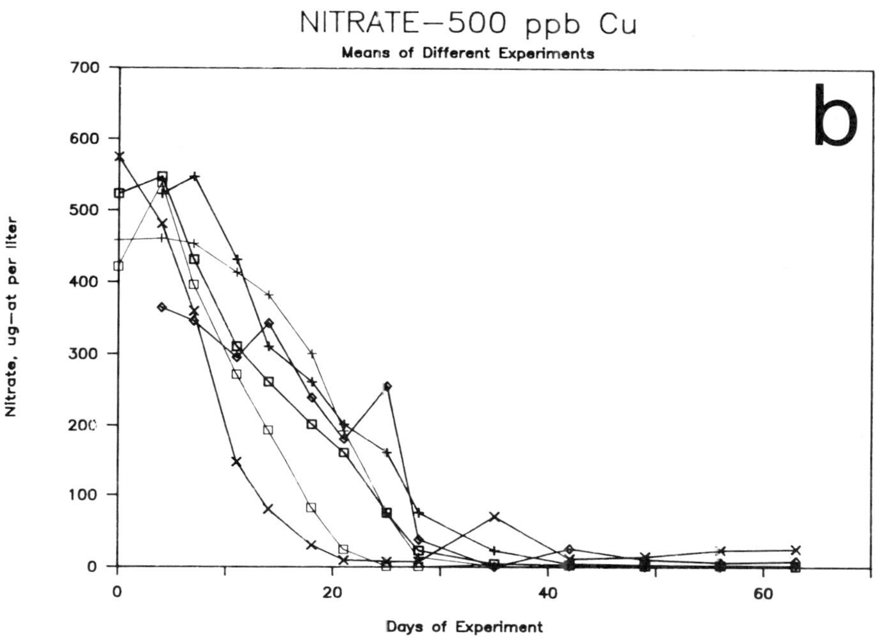

FIG. 2—*Continued:* (b) *nitrate, 500 ppb copper.*

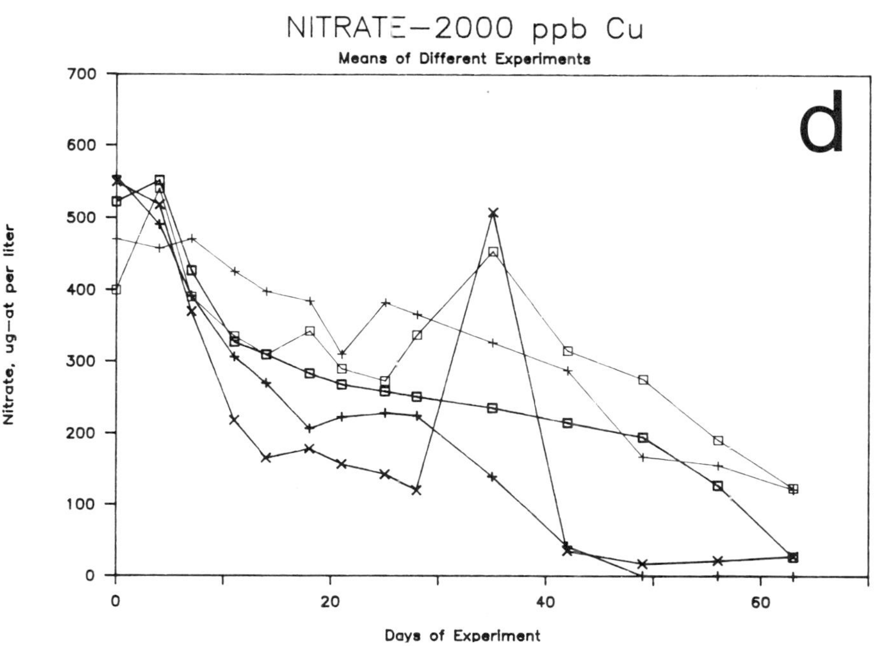

FIG. 2—*Continued:* (d) *nitrate, 2000 ppb copper.*

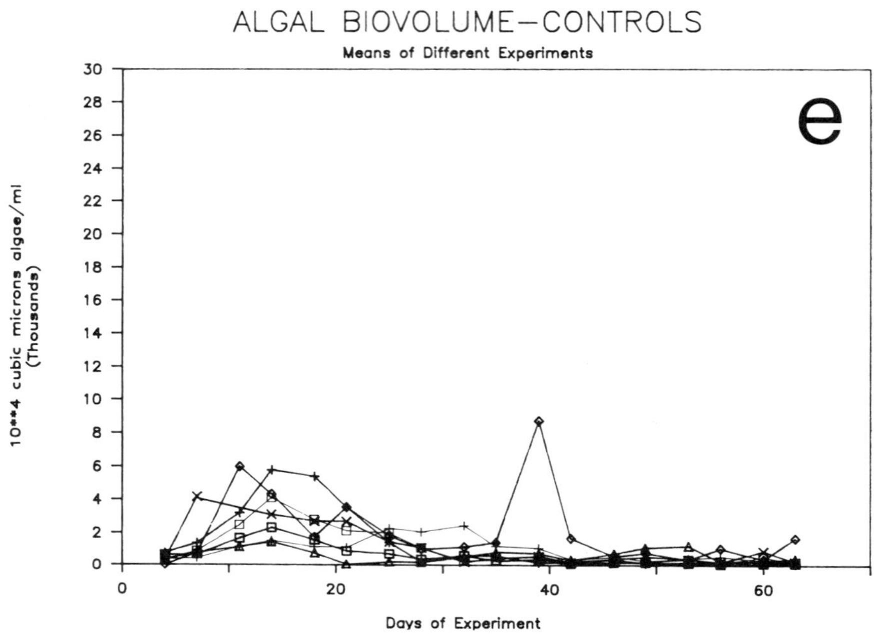

FIG. 2—*Continued:* (e) *algal biovolume, controls.*

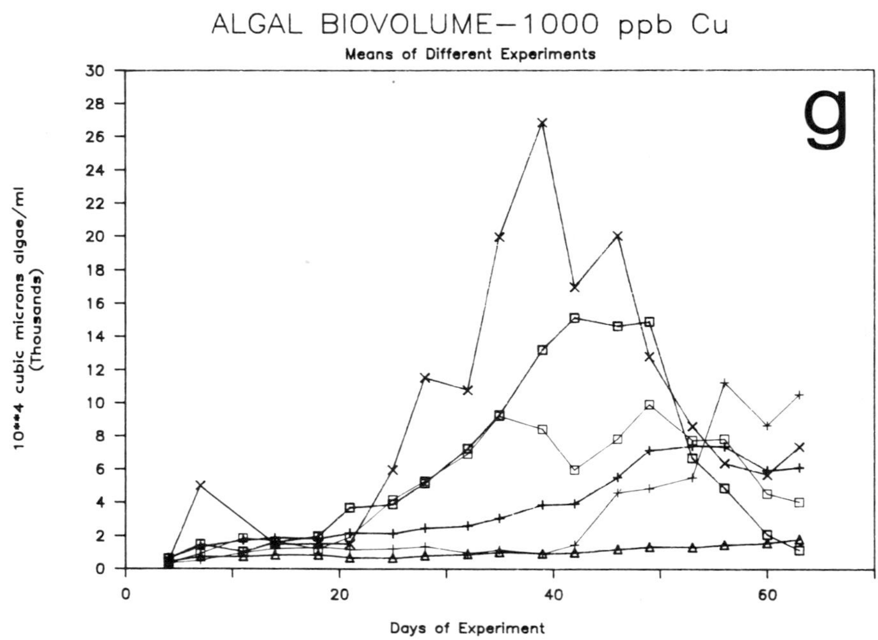

FIG. 2—*Continued:* (g) *algal biovolume, 1000 ppb copper.*

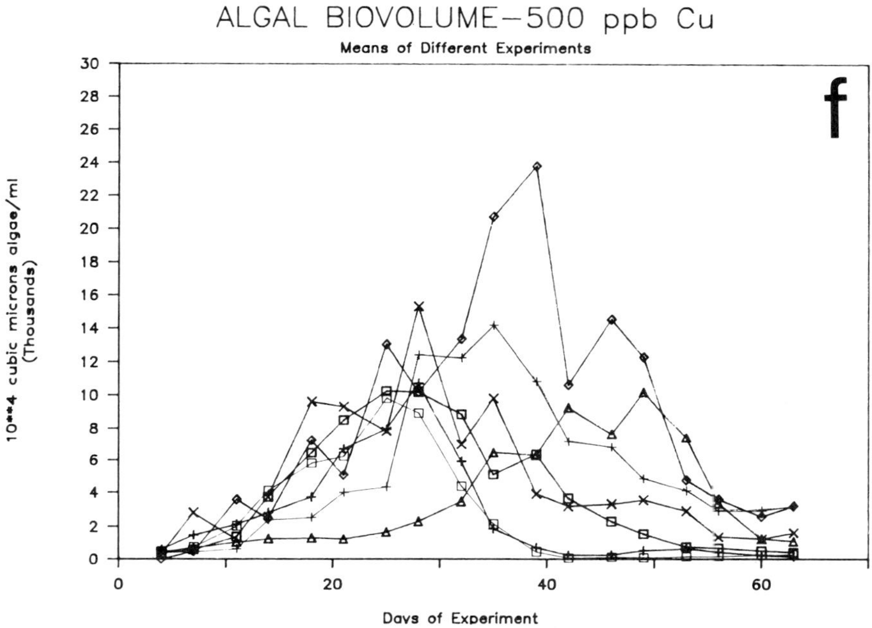

FIG. 2—*Continued:* (f) *algal biovolume, 500 ppb copper.*

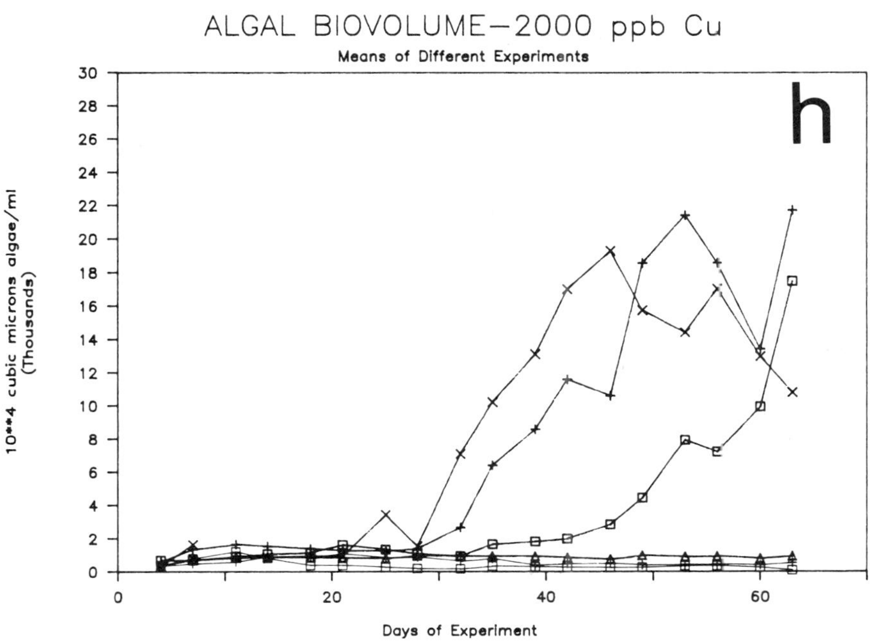

FIG. 2—*Continued:* (h) *algal biovolume, 2000 ppb copper.*

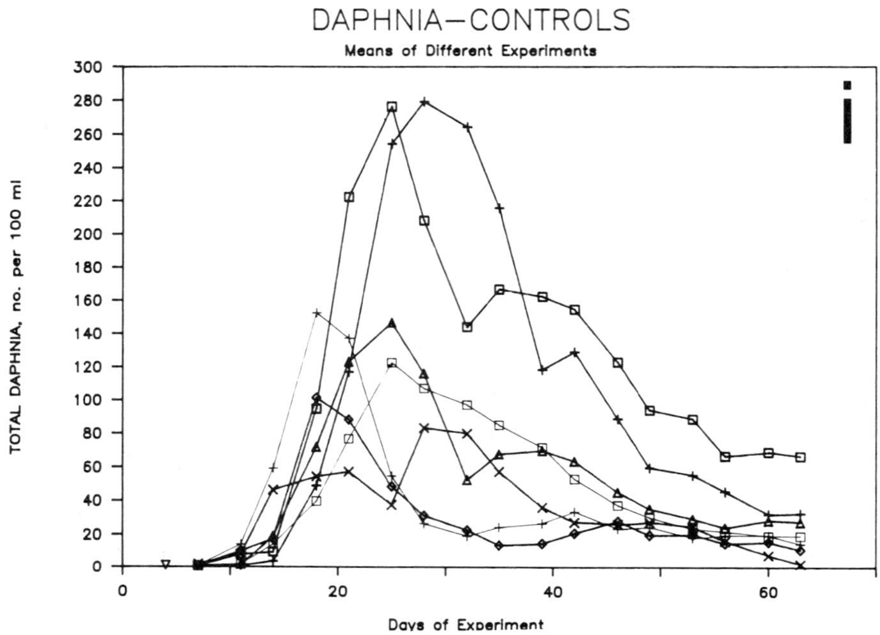

FIG. 2—*Continued:* (i) Daphnia, *controls.*

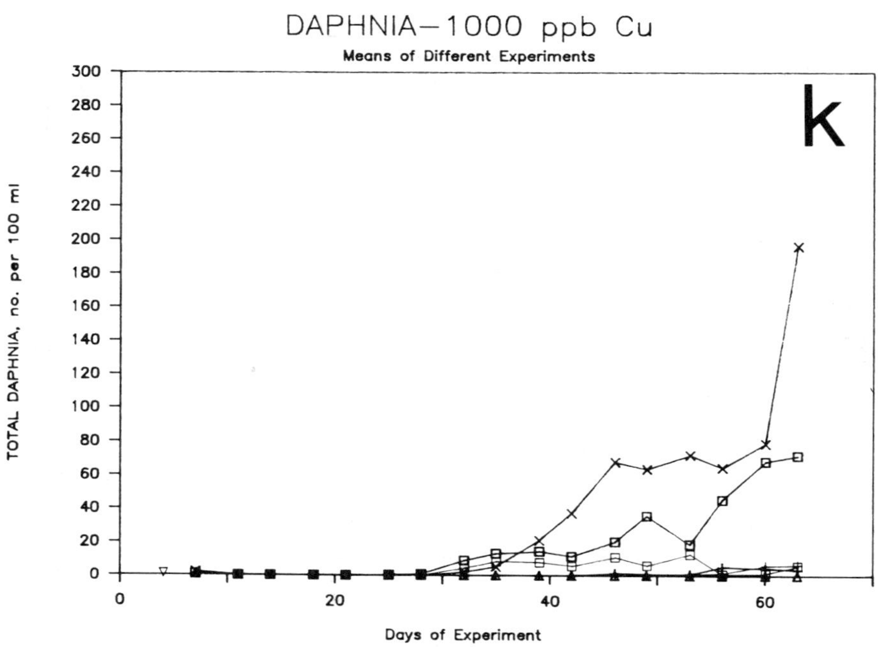

FIG. 2—*Continued:* (k) Daphnia, *1000 ppb copper.*

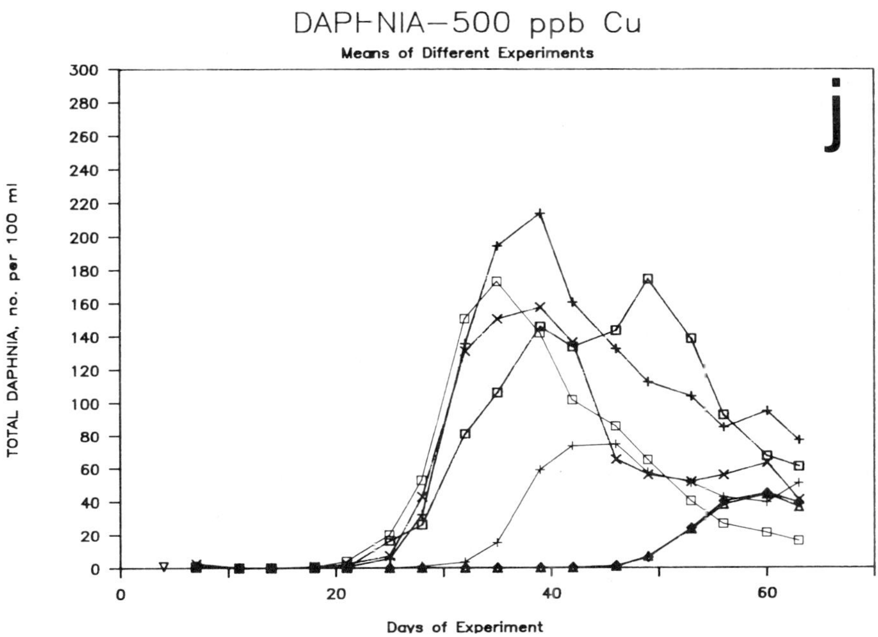

FIG. 2—*Continued:* (j) Daphnia, *500 ppb copper.*

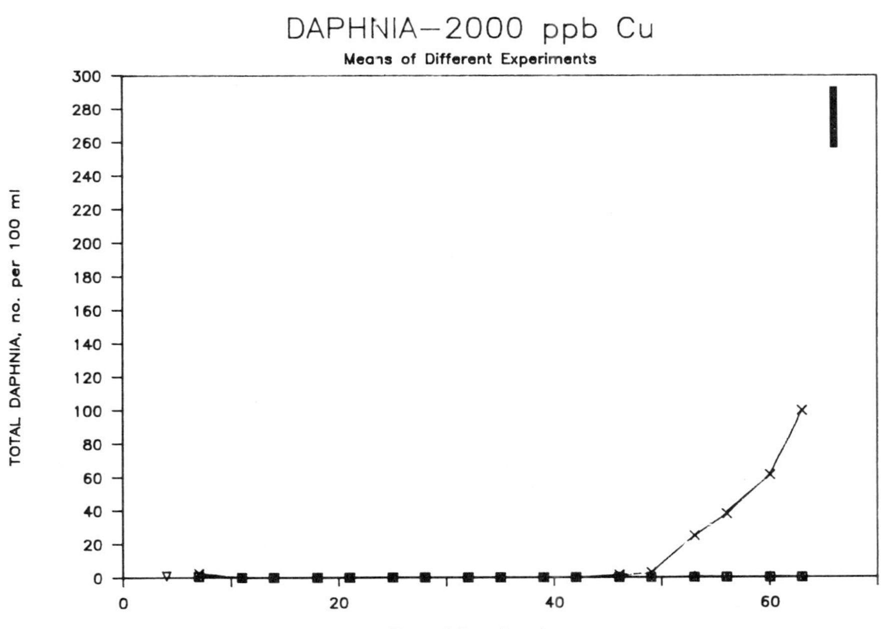

FIG. 2—*Continued:* (l) Daphnia, *2000 ppb copper.*

sensitivity to a single, very large value: in MB No. 1, on Day 35, one microcosm (No. 21) had a nitrate value an order of magnitude greater than that of the previous sampling day; on the following sampling day, the nitrate was only very slightly greater than in other microcosms, and the nitrite and phosphate concentrations were similar to those of other replicates. It cannot be determined if this was a miscalculation or a temporary contamination event.

Algal Biovolume—The algal biovolume increased to a modest density (in comparison to the 500-ppb copper treatments) in all of the controls, and the bloom was reduced by the development of the *Daphnia* populations (Figs. 2*e* through 2*l*). In Fig. 2*e* (Experiment AB No. 1), an unusually dense clump of *Ulothrix* in the count of microcosm No. 14 is responsible for the single high value late in the experiment. The algal biovolume represents the sum of the algal species, and the successional sequences were similar in the various experiments, with *Ankistrodesmus* usually being the dominant and *Scenedesmus* a slightly less abundant co-dominant genus. Blue-greens and diatoms were a modest part of the algal community.

The effect of copper on the algal biovolume (Figs. 2*e* through 2*h*) is a balance between the damage to the algal cells and the reduced grazing because the copper was more toxic to the *Daphnia* than to some species of algae. At 500 ppb, the net effect is an increase in the abundance of copper-resistant algal cells. At 1000 ppb, there is a long lag before algal density begins to increase (until Day 20), but then it increases rapidly in some experiments and slowly in others. At 2000 ppb, the lag occurs at least until Day 28, and the algae either increase rapidly thereafter (three experiments) or fail to grow (three experiments). (Only six experiments tested concentrations of 1000 and 2000 ppb; the AB No. 1 experiment did not use these concentrations.) In the experiments in which dense algal blooms eventually occurred in the 2000 ppb copper treatment, *Ankistrodesmus, Selanastrum,* and *Nitzschia* were at densities of 10^5 or 10^6 cells/mL. In Duluth's experiments, *Oocystis,* a contaminant alga, was the major part of the bloom, sometimes reaching densities of 10^7 cells/mL.

Daphnia—All control *Daphnia* populations increased to a peak, during which they depleted the algal food supply and thereafter declined (Fig. 2*i*). *Daphnia* that had been inoculated from artificial medium (T85MVK in UW No. 1R and aged T82MV in MB No. 1) either increased more slowly (UW No. 1R) or did not become as dense (MB No. 1) as in the other experiments in which natural waters had been used. The *Daphnia* population in the Duluth experiments was especially dense.

In all experiments, the copper killed all of the *Daphnia* on the day of the treatment (Day 7). At 500 ppb, the *Daphnia* populations became reestablished from reinoculated specimens after Day 20 in all experiments. At 1000 ppb copper, the toxicity persisted longer, at least until Day 30. In two experiments, some replicates developed very dense *Daphnia* populations late in the experiment, while other replicates had no *Daphnia* survive. In all cases, algae had bloomed before *Daphnia* survival occurred. At 2000 ppb, the microcosms remained toxic to *Daphnia* for an even longer time, *Daphnia* populations developed in only a very few replicates of one experiment. More recent work suggests that given a long enough time (200 days), detoxification does occur.

In spite of the general similarities among the controls of the seven experiments, there were enough differences in timing and magnitude that statistical evaluations, on a day-by-day basis, show these experiments to be different from each other. The differences were not entirely statistical variation within each variable; rather, experiments in which *Daphnia* populations developed very rapidly tended to display less dense algal blooms (for example, UW No. 2R, AB No. 1). Similar differences in timing could be produced by varying the initial algal density [*19*].

Results in Rejected Experiments

The three experiments that were rejected as having failed to meet the *Daphnia* population quality control criterion had results much like those of the other experiments in nitrate behavior. Algal biovolumes tended to be greater in the controls, blue-green algae were especially abundant, and *Daphnia* densities were lower. The responses to copper were very similar to those in the accepted experiments. In the rejected experiment that used lower copper concentrations (−13% of the nominal, ME 64), recovery from copper toxicity occurred earlier than in the other experiments. In the rejected experiment that used greater copper concentrations (+25% of nominal, ME 72), the recovery was delayed longer. In those experiments that failed to maintain *Daphnia* populations during the latter part of the experiment in the controls, it was difficult to show a statistically significant difference between the control and a treatment microcosm in which the *Daphnia* had been eliminated by toxicity.

Comparison of the Day-Weighted-by-Variable Statistics

Although the comparison of means of the different experiments is interesting, the differences demonstrated between the controls and treatment microcosms within a single experiment are more important. The critical question should be, "Did the various experiments yield the same conclusions?" Given that the different experiments demonstrated day-to-day differences, the comparison needs to be made on a different basis. One approach is to compare the pattern of statistical differences (whether a treatment was significantly higher or lower than the controls for each sampling day) between the different experiments. All experiments have shown the delay of nitrate depletion, the delay in algal biovolume bloom, and the delay of *Daphnia* recovery with increased copper concentration. This yields a common pattern of differences, but is not amenable to statistical analyses and requires 15 pages of tables to display the data. Since the effect of the copper involved a delay in events, a new statistic was developed (described below) to express the timing of events. Other authors have used accumulative techniques [*9*] or hazard rankings [*7*] to handle the problem of different phases within or between experiments.

For ease of analysis, it was desirable to obtain a single value as a summary statistic of a time trace for a measured variable. Yet simple averages of nutrient measurements or organism abundance over a time trace will tend to mask a great deal of information and may not distinguish, for example, between early blooms and late blooms.

The concept of the day-weighted-by-variable (DWV) has been developed to allow comparison of results from different experiments. Its principal use is to measure the amount of "lateral shift" or "phase lag" (for example, the amount by which *Daphnia* bloom is delayed, or the amount by which the nitrate depletion is delayed) between otherwise similar time traces. For this reason, its present use is limited to copper or other chemicals that cause delayed development. By not depending upon day-by-day comparisons of time traces, we avoid the time lag problems caused when two microcosms exhibit nearly identical behavior, but for a slight time shift. The DWV for a given microcosm for a given response variable is computed as follows

$$\underset{\text{(in days)}}{\text{DWV}} = \frac{\Sigma(\text{day} \times \text{value})}{\Sigma(\text{value})}$$

where value equals the organism abundance, nutrient concentration, and so forth, and where the summation is performed over all sampling days of a variable. The DWV is equivalent to computing a mean for a bona fide probability distribution. In that sense, it characterizes the "center of gravity" in terms of days for each time trace. The DWV is a familiar concept used to describe shifts in the mean of otherwise similarly shaped probability distributions; we use it here to describe the time displacements of events in different treatment groups. By weighting each sampling day with its associated value, the DWV can reflect differences between an early bloom (low DWV) and a period of inhibition followed by a later bloom (high DWV). This is quite different from simply taking an average count over the experiment.

For a given response variable, the DWV then produces a single value for each microcosm, the experimental unit. By having the simplicity of a summary value to represent each microcosm's time trace, one can then do both treatment group comparisons and interlaboratory comparisons using a standard two-way ANOVA approach, with the treatment group as one factor and the experiment as the other factor.

There are occasions in which no organism recovery occurs. In this case, organism recovery clearly has been delayed well beyond Day 63, the end of the experiment. The DWV here is best characterized as simply having a value of >63. A value like >63 is not acceptable in a parametric ANOVA but may still be used in a nonparametric type of analysis based upon ranks, as will be shown below. DWV results for algal biovolume and *Daphnia* for the seven experiments are shown in Figs. 3*a* through 3*c*. In each experiment, the effect of the increasing concentrations of copper has been to delay the development of the algal and *Daphnia* populations. The greater the copper concentration, the greater the delay, sometimes beyond the end of the experiment. Friedman's nonparametric randomized block analysis by ranks [*20*] on the five experiments with complete rankings demonstrates that the events happened later as the copper concentration increased (Tables 2, 3, and 4). The rank orderings display this phenomenon clearly, and the Kendall coefficients of concordance (a measure of agreement in rankings among experiments [*21*]) were values of 1.0 and 0.97. The only case requiring the assignment of tied ranks occurred for Aberdeen-Army No. 2, where the two higher concentrations delayed organism recovery beyond the experiment duration.

The duration of the delay was significantly different in different experiments, as shown by a two-way ANOVA of treatment and experiment having a significant interaction term (Table 5). Differences were less obvious at the control and 500 ppb copper concentration for which algal and *Daphnia* recovery occurred in all seven experiments. At 1000 ppb copper, algal recovery occurred in five of the six experiments that tested this concentration; at 2000 ppb copper, algal recovery occurred in three of the possible six experiments. *Daphnia* recovered only after algal recovery at 500 ppb copper and eventually established populations in all of the seven experiments. At 1000 ppb copper, recovery occurred in five of the six experiments that tested that concentration; at 2000 ppb copper, *Daphnia* recovered in only two of the possible experiments. The difference in the duration of the delay does not change the general conclusion (evidenced by the Friedman rank test) that the copper delayed both the algal and *Daphnia* population development in all seven experiments.

The DWV has also been used to detect shifts in trends for nitrate, which is depleted by algal incorporation. As the copper concentration increases, and the nitrate depletion is further delayed, the DWV increases (Fig. 3). The delay in nitrate depletion at the higher concentrations is confirmed by Friedman's nonparametric randomized block test and the Kendall coefficient of concordance of 0.97 (Tables 2, 3, and 4). The actual amount of the delay with increased concentration does vary enough between experiments to result in a significant treatment-by-experiment interaction term when subjected to a parametric two-way ANOVA (Table 5). By the use of both parametric and nonparametric analyses, we have

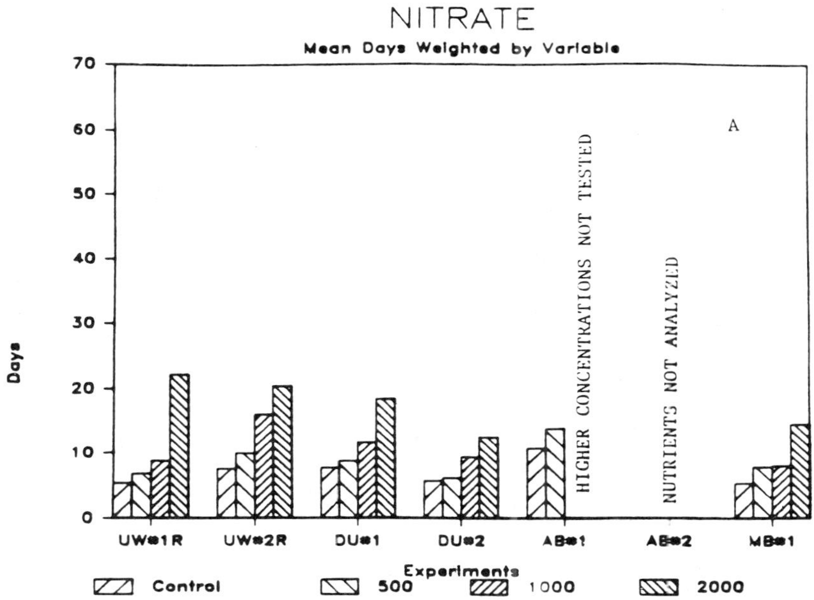

FIG. 3—*Comparison of results in seven SAM experiments. The day-weighted-by-variable (DWV) statistic is a measure of the central distribution of the nitrate concentration; a higher number indicates a delay.* (a) *Nitrate.*

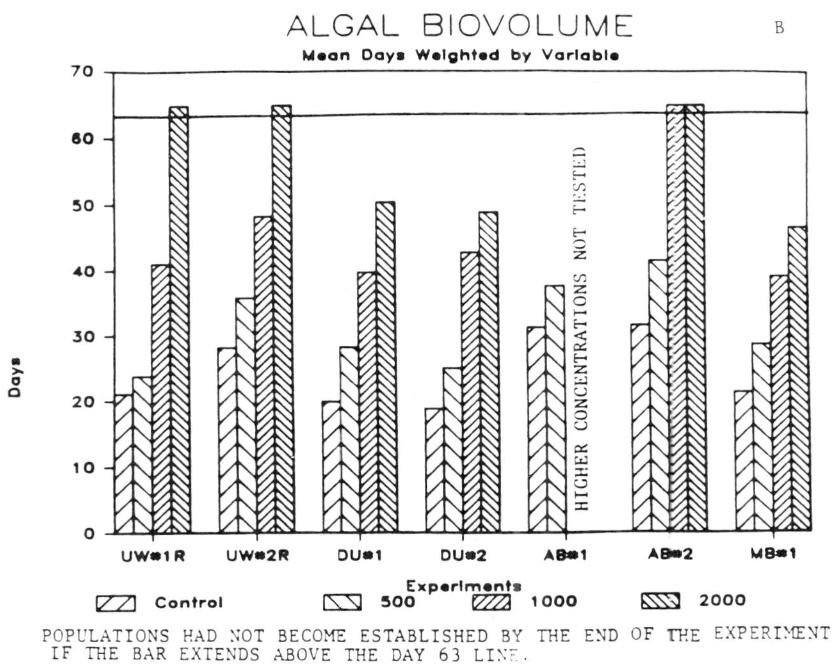

FIG. 3—*Continued:* (b) *algal biovolume.*

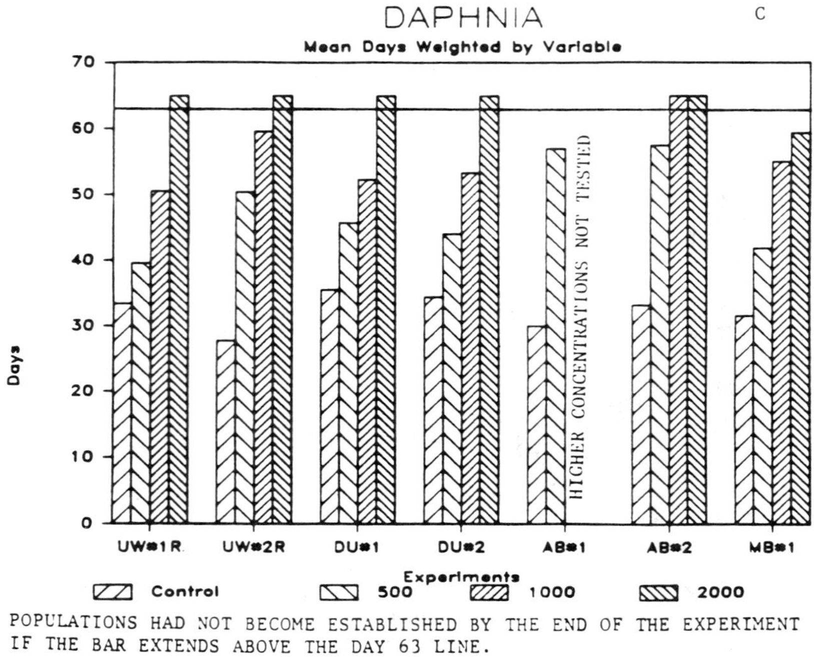

FIG. 3—*Continued:* (c) Daphnia

TABLE 2—*Ranked test of interlaboratory experiments using the mean days-weighted-by-variable.*[a]

Nitrate	Control	500 μg/L Copper	1000 μg/L Copper	2000 μg/L Copper
UW No. 1R	1	2	3	4
UW No. 2R	1	2	3	4
DU No. 1	1	2	3	4
DU No. 2	1	2	3	4
AB No. 1[b]	...	...	...	...
AB No. 2[b]	...	...	...	...
MB No. 1	1	2	3	4
Sum of ranks	5	10	15	20

[a] For the five experiments with complete rankings the value of Friedman's nonparametric test statistic = 15.00. Compared to chi-square (degrees of freedom = 3) $P < 0.005$. Kendall's coefficient of concordance = 1.0.

[b] Excluded from analysis.

been able to state statistically that all the experiments agreed in the nature and the direction of the effect, but that the exact timing of such effects differed.

Evaluation and Modification of Quality Control Criteria

Comparing the provisional quality control criteria listed in the 1983 protocol with the interlaboratory experiments, we recommend several modifications. The pH criteria for means were never exceeded and are so broad as to be meaningless. The pH criterion for

TABLE 3—*Interlaboratory comparisons (ranked mean days-by-weighted-variable).*[a]

Algal Biovolume	Control	500 μg/L Copper	1000 μg/L Copper	2000 μg/L Copper
UW No. 1R	1	2	3	4
UW No. 2R	1	2	3	4
DU No. 1	1	2	3	4
DU No. 2	1	2	3	4
AB No. 1[b]	...	...	...	...
AB No. 2	1	2	3.5	3.5
MB No. 1	1	2	3	4
Sum of ranks	6	12	18.5	23.5

[a] For the six experiments with complete rankings the value of Friedman's nonparametric test statistic = 17.45. Compared to chi-square (degree of freedom = 3) $P < 0.001$. Kendall's coefficient of concordance = 0.97.

[b] Excluded from analysis.

TABLE 4—*Interlaboratory comparisons (ranked mean days-by-weighted-variable).*[a]

Daphnia	Control	500 μg/L Copper	1000 μg/L Copper	2000 μg/L Copper
UW No. 1R	1	2	3	4
UW No. 2R	1	2	3	4
DU No. 1	1	2	3	4
DU No. 2	1	2	3	4
AB No. 1[b]	...	...	...	...
AB No. 2	1	2	3.5	3.5
MB No. 1	1	2	3	4
Sum of ranks	6	12	18.5	23.5

[a] For the six experiments with complete rankings the value of Friedman's nonparametric test statistic = 17.45. Kendall's coefficient of concordance = 0.97.

[b] Excluded from analysis.

the CV is so narrow that entirely satisfactory experiments occasionally exceeded this criterion by one or two sampling days. Inappropriate pH would probably be indicated by biological variables. We therefore suggest that pH be removed from the quality control criteria.

The algal biovolume mean criterion of 2000 × 10^4 μm^3 cells per millilitre was not met by only one experiment, AB No. 2; however, the *Daphnia* population development was entirely adequate, suggesting that adequate primary production had occurred but was being grazed. A few experiments failed to meet the CV criteria, but this had no major effect on the ability of the IND to show a significant effect. We suggest that the algal biovolume criterion be modified or dropped. If modified, the CV should not be calculated if the mean is below 1000 units.

Mean *Daphnia* populations occasionally failed to meet the 0 to 28 day criterion of 100 animals per 100 mL; we suggest this be reduced to 80. The 28 to 63 day criterion of 25 animals per 100 mL was often not met; we suggest this be reduced to 15.

The most important criterion should be the ability to detect a significant difference between the control and a treatment; this will be compromised if the mean is so low and the IND so wide that total extinction would not be statistically different from the control results, or if the IND is so wide that it would be unlikely that its value could be exceeded.

TABLE 5—*Two-way analysis of variance of interlaboratory experiments using the mean days-weighted-by-variable statistic.*[a]

Source[b]	Sum of Squares	Degrees of Freedom	Mean Square	F	Tail Probability
Nitrate					
Mean	901.8	1	901.8	...	...
EXP	264.3	4	66.1	...	...
TRT	29.0	2	14.5	...	...
ET	86.8	8	10.8	15.9	.0000
Error	51.2	5	0.7	...	...
Algal Biovolume					
Mean	15 935.3	1	15 935.3	...	...
EXP	1 011.2	4	252.8	...	...
TRT	934.3	2	467.2	...	...
ET	172.8	8	21.6	2.4	0.0222
Error	669.9	75	8.9	...	...
Daphnia					
Mean	33 102.5	1	33 102.5	...	...
EXP	153.8	4	38.5	...	...
TRT	1 674.3	2	837.2	...	...
ET	582.5	8	72.8	4.5	0.0002
Error	1 064.5	66	16.1	...	...

[a] AB No. 1 was omitted from analysis since 1000 and 2000 μg/L copper concentrations were not done. AB No. 2 was omitted from analysis due to no organism recovery in the 1000 and 2000 μg/L copper groups. The 2000 μg/L copper groups were omitted from analysis since two experiments showed no organism recovery for algal biovolume; four showed no organism recovery for *Daphnia*.

[b] Abbreviations: EXP = experiment; TRT = treatment; ET = experiment plus treatment interaction.

The suggested quality control criteria are the following:

For Days 0 to 28:

1. Mean nitrate must have been reduced from the initial concentration of 500 μM to less than 50 μM by Day 28.
2. The mean oxygen gain must reach at least 4 mg/L sometime during this period.
3. The mean *Daphnia* populations must have reached at least 80 animals per 100 mL sometime during this period.
4. The algal biovolume should exceed 2000×10^4 μm^3 of cells per millilitre sometime during this period. If this criterion is not met, the experiment may be accepted if the *Daphnia* population has exceeded 100 animals per 100 mL. (Low algal abundance may occur in a satisfactory experiment if the grazers eat the algal cells as fast as they grow.)

If the behavior of the controls has not conformed to the criteria of this section by Day 28, it is recommended that the microcosm test be discontinued, as the experiment is probably too flawed to be successful.

The following criteria refer to the means of the control treatment group for each sampling day:

For Days 0 to 63:

1. The control mean oxygen gain must be positive throughout the experiment; the lower bound of the interval of nonsignificance must be greater than zero for more than 50% of the sampling periods after the test chemical addition.

2. The control mean *Daphnia* populations must have greater than 15 animals per 100 mL, and the lower bound of the interval of nonsignificance must be greater than zero for more than 50% of the sampling periods after the test chemical addition.

If the experiment fails to meet these minimum means, it is likely that the medium is inadequate or toxic, or that the organisms are impaired. Unless this level of activity is present, it is unlikely that negative effects can be expressed. If the experiment fails to meet these criteria but demonstrates statistical differences due to treatments, and if these differences appear to be biologically reasonable and meaningful, the results may be accepted as a special case.

Problems Encountered by the Participating Laboratories

One laboratory prematurely terminated an experiment because the control *Daphnia* populations were developing poorly; this was attributed to a pesticide contamination in their distilled water supply. Another laboratory reported that it was difficult to get all of the algal and animal cultures at high enough densities at the same time to initiate an experiment. Two laboratories expressed concern about the excessive time required to perform protozoan and rotifer counts and were advised on easier, faster methods that are satisfactory because those measurements are not intended to be very precise. The amount of time required for algal counts was reported to be a problem in a laboratory that did not have a ×40 objective with an adequate depth of field; they counted the large algal cells with one microscope and the small cells with another. Excessive additions of phosphorus occurred in some experiments because the protozoan and rotifer stocks were not dense, and large volumes of medium (including phosphate buffer) were added to obtain the initial concentrations of organisms; however, this did not compromise the results. The directions for the interlaboratory test should have made clear the correct amount of $CuSO_4$ that should have been added to obtain the test concentrations of copper; this procedure would have avoided the error of using the weight of $CuSO_4$ instead of copper. All of the laboratories found the techniques to be labor-intensive, although less than they feared on first reading the protocol. Shannon et al. [7] compared the SAM protocol with the mixed flask culture protocol and found that the SAM takes more time, but provides more information.

The data handling/statistical package which does the within-experiment statistics and replicate graphics is on the CDC CYBER mainframe computer at our University and is not easily transported to another mainframe. We are in the process of developing a microcomputer spread-sheet handling package that will allow laboratories to handle their own data.

Discussion

Even with the similarities of responses in the different experiments, it is interesting to examine the causes of differences in the time of response. Differences between experiments can be caused by (*a*) differences in initial or operating conditions, and (*b*) copper recovery phenomena. Differences in initial or operating conditions, for example, the exact numbers and physiological conditions of the organisms usually vary to some degree, and temperature and light conditions were undoubtedly somewhat different. More nutrient may have been introduced in the Duluth experiments from volumes of the media introduced with rotifers and protozoa. These experimental conditions could probably be reduced somewhat, although some differences would always occur. Greater standardization is possible, but requires more effort.

Differences in the initial physiological condition of the *Daphnia* obviously made a difference. We strongly recommend that *Daphnia* be reared in natural waters or in chemically defined media that have been shown to produce multiple generations of healthy animals [*17*]. *Daphnia* that are compromised do not develop populations in a synchronous fashion, and the populations do not persist throughout the experiments. This compromises their ability to show effects of a chemical that is toxic to *Daphnia*.

Differences in the timing of recovery from copper toxicity depend on a large number of factors. Other studies in our laboratory suggest that the numbers of algal cells present at the time of the copper addition (Day 7), along with the associated dissolved organic carbon, may be a factor. Also, the detoxification seems to be associated with the growth of copper-resistant algae; the survival of the *Daphnia* tends to occur about 10 days after the algal bloom. The algal bloom causes the pH to be very high, which by itself would precipitate much of the copper. Given the large number of factors that influence copper toxicity, it is not unreasonable that it is highly variable in timing in different experiments and even in different replicates within an experiment.

Recovery from copper toxicity is not an artifact of the microcosms but has been seen in many natural communities. Generally, the sensitive algal species are reduced or eliminated; zooplankton are eliminated, and a bloom of copper-resistant algal species occurs during the absence of grazers [*22,23*]. The microcosm responses have been compared with responses of several natural systems [*24*].

There are shortcomings in analyzing each variable separately, thus giving the appearance of simultaneous single-species tests, but current statistical techniques do not recognize many of the realities of ecosystem data, as stated in the introduction to this paper.

We tentatively hypothesize that variables that are controlled by a single process will be less variable than those controlled by the balance of processes. For example, nitrate has high reproducibility and is effectively controlled by algal uptake (nitrification does not appear to be occurring); nitrate can only decrease. Algal biovolume tends to be more variable and is the net effect of growth, toxicity, and grazing. It is Taub's impression that the variables that we have trouble reproducing experimentally are the same as those that we have trouble fitting in the mathematical model [*25*] that is under development.

Conclusions

The feasibility of performing the protocol in other laboratories was demonstrated by the ability of the three participating laboratories, having attended a 3 to 4 day training session in our laboratory and received a copy of the protocol and the stock cultures, to set up and collect data on the microcosms. At Duluth-EPA, they had the advantage of having one of Taub's former students (M. Harrass), but he was assigned mostly to other duties.

The seven experiments showed similar results for the algal biovolume, *Daphnia,* and nitrate responses in that the behavior of the four treatment groups was always the same in terms of the rankings of the centers of gravity (DWVs) of the time traces. The differences among the experiments occurred in terms of the duration of the delay. The seven experiments have also shown the same behavior with respect to the pattern of "sign differences" (whether the treatment was statistically greater or less than the control on a given day) for each experiment.

Due to differences in magnitude among controls and difference in the duration of delays, performing a sequence of two-way ANOVAs for each variable on each sampling day (with "treatment" as one factor and "experiment" as another) made all the experiments appear statistically different. Even experiments that appeared quite similar graphically were statistically different when the events were offset by a single period between samplings (3 to 4 days). We have resolved this issue by using the center of gravity of the time trace (DWV) in the between-experiment statistical analyses. In addition, use of nonparametric tests on ranked DWVs has allowed us to use data like >63 days, where no organism recovery occurred by the end of the experiment.

The unworkable two-way ANOVAs on a day-by-day basis between experiments are not to be confused with the one-way day-by-day ANOVAs within an experiment, which form the basis of the IND plots in order to tell when treatments are statistically different from controls. The one-way ANOVA approach within an experiment has been successfully used by other authors also.

Evaluation of the provisional quality control criteria suggested that they could be simplified and reduced. However, it is important that experiments meet the criteria or demonstrate that a significant difference could have been demonstrated within reasonable biological values. In the absence of satisfactory controls, the ability to display effects is compromised.

We hope that these studies will aid in the acceptance of multispecies toxicological tests for regulatory purposes. These results do provide evidence that experimental results could be verified in other laboratories since all the experiments yielded similar results. The use of ecological information has been impeded by the perception that the results would be too complex, not subject to objective or statistical interpretation, and nonreproducible. We look forward to the interlaboratory testing of other microcosm protocols so that we can compare the degrees of reproducibility of different methods.

Acknowledgments

Special thanks are due the staffs of the participating laboratories: Drs. David Yount, Lyle Shannon, Michael Harrass (during these studies), and Mr. Chuck Walbridge at Duluth—EPA; Dr. Wayne Landis, Mark Haley, and Dennis Johnson at U.S. Army—Aberdeen; and Dr. Allan Thum, Skip Newton, Sherrie Hamer, and John Hardin at Marine Bioassay Laboratory. We wish to thank Dr. Thomas Sibley for supervision of the copper chemistry studies and Marcus Duke for manuscript preparation.

The project has been supported by the U.S. Food and Drug Administration under Contract No. 223-83-7000. We also acknowledge the help of the project officers, Dr. Buzz Hoffmann and John Matheson, who have contributed to the general development of the project and helped to stimulate progress.

This paper is Contribution No. 741 of the School of Fisheries, University of Washington, Washington.

References

[1] Grothe, D. R., and Kimerle, R. A., *Environmental Toxicity and Chemistry,* Vol. 4, 1985, pp. 189–192.

[2] Mearns, A. J., Swartz, R. C., Cummins, J. M., Dinnel, P. A., Plesha, P., and Chapman, P. M., *Marine Environmental Research,* Vol. 19, 1986, pp. 13–37.

[3] Giesy, J. P. and Allred, P. M. in *Replicability of Aquatic Multispecies Test Systems,* J. Cairns, Jr., Ed., Pergamon Press, New York, 1985.

[4] *Community Toxicity Testing, ASTM STP 920,* Cairns, Jr., J., Ed., American Society for Testing and Materials, Philadelphia, 1986.

[5] Taub, F. B., Kindig, A. C., and Conquest, L. L., "Preliminary Results of Interlaboratory Testing of a Standardized Aquatic Microcosm," *Community Toxicity Testing, ASTM STP 920,* J. Cairns, Jr., Ed., American Society for Testing and Materials, Philadelphia, 1986, pp. 93–120.

[6] Giddings, J. M., "A Microcosm Procedure for Determining Safe Levels of Chemical Exposure in Shallow-Water Communities," *Community Toxicity Testing, ASTM STP 920,* J. Cairns, Jr., Ed., American Society for Testing and Materials, Philadelphia, 1986, pp. 121–134.

[7] Shannon, L. J., Harass, M. C., Yount, J. D., and Walbridge, C. T., "A Comparison of Mixed

Flask Culture and Standardized Laboratory Model Ecosystems for Toxicity Testing," *Community Toxicity Testing, ASTM STP 920,* J. Cairns, Jr., Ed., American Society for Testing and Materials, Philadelphia, 1986, pp. 135–157.
[8] Sheehan, P. J., Axler, R. P., and Newhook, R. C., "Evaluation of Simple Generic Aquatic Ecosystem Tests to Screen the Ecological Impacts of Pesticides," *Community Toxicity Testing, ASTM STP 920,* J. Cairns, Jr., Ed., American Society for Testing and Materials, Philadelphia, 1986, pp. 158–179.
[9] Hedtke, S. F., *Aquatic Toxicology,* Vol. 5, 1984, pp. 227–244.
[10] Taub, F. B. and Read, P. L., "Standardized Aquatic Microcosm Protocol," Vol. II, Final Report, FDA Contract No. 223-80-2352, University of Washington, Seattle, WA, 1983.
[11] Taub, F. B., "Toward Interlaboratory (Round-Robin) Testing of a Standardized Aquatic Microcosm," *Multispecies Toxicity Testing,* J. Cairns, Jr., Ed., Pergamon Press, UK, chapter 13, pp. 165–186.
[12] Taub, F. B., Kindig, A. C., and Conquest, L. L., "Interlaboratory Testing of a Standardized Aquatic Microcosm," *Aquatic Toxicology and Hazard Assessment: Tenth Volume,* American Society for Testing and Materials, Philadelphia, 1988, pp. 384–405.
[13] Taub, F. B., "Measurement of Pollution in Synthesized Aquatic Microcosms," *Concepts in Marine Pollution Measurements,* H. H. White, Ed., University of Maryland, College Park, MD, 1984, pp. 159–192.
[14] Taub, F. B., "Synthetic Microcosms as Biological Models of Algal Communities," *Algae as Ecological Indicators,* L. E. Shubert, Ed., Academic Press, New York, 1984, pp. 363–394.
[15] Taub, F. B., "Standardized Aquatic Microcosm—Development and Testing," *Aquatic Ecotoxicology,* A. Boudor and F. Ribeyre, Eds., CRC Press, Inc., Boca Raton, FL, in press.
[16] Taub, F. B., Kindig, A. C., Conquest, L. C., and Meador, J. P., Final Report, FDA Contract 223-83-7000, U.S. Food and Drug Administration, Washington, DC, 1986.
[17] Keating, K. I., "A System of Defined (Sensu Stricto) Media for Daphnid (Cladocera) Culture," *Water Research,* Vol. 19, No. 1, 1985, pp. 73–78.
[18] Pilson, M. E. Q., and Wilson, S. W., "Marine Microcosms in Ecological Research," *Microcosms in Ecological Research,* J. Giesy, Ed., National Technical Information Service, Springfield, VA, 1980.
[19] Taub, F. B., Harrass, M. C., Hartmann, H. J., Kindig, A. C., and Read, P. L. "Effects of Initial Algal Density on Community Development in Aquatic Microcosms," *Verh. Internat. Verein. Limnol.,* Vol. 21, 1981, pp. 197–204.
[20] Zar, J. H., *Biostatistical Analysis,* Prentice-Hall, New Jersey, 1984.
[21] Conover, W. J., *Practical Nonparametric Statistics (Second Edition),* Wiley, New York, 1980.
[22] McKnight, D., "Chemical and Biological Processes Controlling the Response of a Freshwater Ecysystem to Copper Stress: A Field Study of the $CuSO_4$ Treatment of Mill Pond Reservior, Burlington, Massachusetts," *Limnology and Oceanography,* Vol. 26, 1981, pp. 518–531.
[23] Thomas, W. H., Holm-Hansen, O., Siebert, D.L.R., Azam, F., Hodson, R., and Takahashi, M., "Effects of Copper on Phytoplankton Standing Crop and Productivity: Controlled Ecosystem Pollution Experiment," *Bulletin of Marine Science,* Vol. 27, 1977, pp. 34–43.
[24] Harrass, M. C. and Taub, F. B., "Comparison of Laboratory Microcosms and Field Responses to Copper," *Validation and Predictability of Laboratory Methods for the Assessment of Fate and Effects of Environmental Contaminants in Aquatic Ecosystems, ASTM STP 865,* American Society for Testing and Materials, Philadelphia, 1985, pp. 57–74.
[25] Swartzman, G. L. and Rose, K. A. "Simulating the Biological Effects of Toxicants in Aquatic Microcosm Systems," *Ecological Modeling,* Vol. 22, 1984, pp. 123–134.

DISCUSSION

J. R. Pratt[1] *(written discussion)*—(1) Is the sensitivity of the standard aquatic microcosm (SAM) similar to that of acute or chronic single-species tests? Why were such high concentrations of copper used in the interlaboratory testing?

(2) What would be the advantages and disadvantages of using regression procedures to estimate effective concentrations (for example, EC_{20}, EC_{50}) for a given variable at a particular point in time?

[1] Hancock Biological Station, Murray State University, Murray, KY.

F. B. Taub, A. C. Kindig, L. L. Conquest, and J. P. Meador (authors' closure)—(1) Microcosms may be more or less sensitive than single-species tests. In the case of copper, the microcosm is probably less sensitive than a single-species test because sorption by other organisms probably reduces exposure to dissolved cupric ion. In the case of a biologically active transformation product or the bioconcentration of a toxic chemical in food, the microcosm is likely to be more sensitive than single-species tests.

Such high concentrations of copper were used in the interlaboratory testing because (*a*) the goal was to compare the results from different laboratories and we didn't want to be near a threshold, and (*b*) the concentrations used were similar to those used to treat eutrophic ponds and lakes for control of algal blooms.

(2) There is much that can be said about a regression approach versus a hypothesis testing approach in the design and analysis of experiments like these. With a hypothesis testing approach, one is looking for statistically significant differences between means. With a regression approach, one is looking for information on trends. The objectives are slightly different, and so the experimental designs would be different. Since the regression approach is trend oriented, it would not depend upon finding significant differences among response means for distinct concentration values. It *would,* however, depend upon the existence of a good fitting trend any time a regression procedure was performed (that is, a high value of the correlation coefficient). The hypothesis testing approach is designed to uncover significant differences between means (and hence requires more intensive replication for each concentration), but it does not require a good fitting trend with a high correlation coefficient.

There are many items for consideration if one wanted to estimate, say, an EC_{50} for a given variable at a particular point in time.

(*a*) This is a day-by-day procedure and any phase lag phenomenon would tend to throw the procedure off. That is, if one microcosm reaches some critical event on Day 28 and another one does the same thing a few days later, a comparison at Day 28 would show increased variability and complicate any regression fitting procedure.

(*b*) To compensate for any complications introduced by phase lag phenomena, why not do a regression procedure at every given sampling day? This gets rather complex because then one has to worry about the particular shape of the regression and how good the regression fit is for *each sampling day.* In order to get the most information about any trend (which could be linear, quadratic, or have a strange nonlinear shape that would be hard to fit), textbooks often recommend about 25 distinct data points—this would suggest that the experimenter assign each microcosm to a different copper concentration, but then there would be absolutely no replication, a dangerous situation (no information on within-group variance). So, one might think of having about twelve distinct concentrations with two replicates each. But, in order to obtain any information about goodness of fit, one would have to assume homogeneity of variance across the twelve distinct concentrations, and there would not be enough information to test this assumption with only two replicates per treatment group. One could try eight distinct concentrations with three replicates each or six distinct concentrations with four replicates each. Increasing the number of replicates in each group yields more information with which to test homogeneity of variance assumptions, but one still faces the problem of what to do if the correlation coefficient for the best fitting function (and the nature of that function has to be guessed by eye) is not very good. One simply cannot put much faith in estimated regression parameters from an ill-fitting function. Depending upon the sampling day, the shape of the regression function may change for inverse linear or negative exponential to quadratic, cubic , or something more complex. And for any polynomial fits, there is still the issue of trying to give biological meaning to the polynomial coefficients.

(*c*) Unless the fits were very good and consistent across sampling days, there would be a lot of uncertainty associated with, say, an estimated EC_{50}. Estimations of EC_{50}'s usually assume that a *linear* relationship is present between the response variable and the predictor variable (concentrations) or that some transformation can produce a linear relationship. If the relationship were a quadratic one, then the whole meaning of an EC_{50} for this type of functional form would be called into question. The data exist to test these relationships, but we have not seen a regression procedure used on a multitrophic level study with multiple sampling times.

The authors plan to address this and other issues in more detail at upcoming ASTM meetings.

QSARs and Other Data Extrapolations

Kurt Enslein,[1] Thomas M. Tuzzeo,[1] Benjamin W. Blake,[1] Jeffrey B. Hart,[1] and Wayne G. Landis[2]

Prediction of *Daphnia magna* EC_{50} Values from Rat Oral LD_{50} and Structural Parameters

REFERENCE: Enslein, K., Tuzzeo, T. M., Blake, B. W., Hart, J. B., and Landis, W. G., **"Prediction of *Daphnia Magna* EC_{50} Values from Rat Oral LD_{50} and Structural Parameters,"** *Aquatic Toxicology and Environmental Fate: Eleventh Volume, ASTM STP 1007,* G. W. Suter II and M. A. Lewis, Eds., American Society for Testing and Materials, Philadelphia, 1989, pp. 397–409.

ABSTRACT: We describe a structure-activity model of *Daphnia magna* median effective concentration (EC_{50}) based on rat oral median lethal concentration (LD_{50}) and structural parameters. Based on 182 compounds for which both biological end points were available, a regression equation explaining 84% of the variation in *D. magna* EC_{50} was developed. The independent parameters used in the equation consist of substructural keys and molecular connectivity indexes (MCI). Alternate models based on octanol/water partition coefficients did not perform as well as those based on MCIs and keys. An equation for twelve cholinesterase-inhibiting compounds explained 80% of the EC_{50} variation in those compounds, based on one MCI, the molecular weight, and rat oral LD_{50}.

KEY WORDS: toxicology, *Daphnia magna,* structure-activity relationships, regression, rat oral median lethal dose (LD_{50}), median effective concentration (EC_{50}), prediction, aquatic toxicology

The goal of this research was to examine the potential for predicting the toxicity of chemicals from data of an evolutionarily distant organism and structure. The major factor that drove this development was the need to predict environmental toxicity from a toxicological data base consisting mostly of mammalian information.

Estimation of environmental toxicity from a limited data base is a practical problem of considerable dimensions. Several attempts have been made to derive structure-activity (SAR) models for fish and *Daphnia magna* using the log of octanol/water partition coefficient (log *P*) for compounds whose mode of action is narcosis [*1–9*]. Birge and Cassidy [*10*] have also examined the relationship between bioconcentration potential and toxicity. The number of compounds for which mammalian toxicity data exist outweighs by far that of environmental toxicity. Over 8000 compounds are incorporated into our rat oral median lethal dose (LD_{50}) data base, yet less than 300 compounds are available for which a reliable *D. magna* median effective concentration (EC_{50}) can be found. To make matters worse, the intersection of the two data sets results in less than 200 compounds.

An algorithm for the prediction of environmental toxicity may be represented by an SAR equation incorporating rat oral LD_{50} as a parameter. The inclusion of a biological end point

[1] President, research associate, manager of QSAR department, and programmer/analyst, respectively, Health Designs, Inc., Rochester, NY 14604.

[2] Environmental scientist, Chemical Research, Development, and Engineering Center, Aberdeen Proving Grounds, MD.

may both increase the accuracy of the SAR estimate and expand its utility to more classes of compounds.

In summary, it has proven feasible to construct interspecies SAR models that provide accurate estimations. With the current data base, the predictions including the biological end point have proven to be as accurate as SAR models based on five to six times the number of compounds.

Methods

Data Base

The *D. magna* EC_{50} data were acquired from the following sources:

G. A. LeBlanc	*Bulletin of Environmental Contamination and Toxicology,* Vol. 24, 1980, pp. 684–691 [*11*].
F. L. Mayer, Jr.,	Acute Toxicity Data for Freshwater Animals from the Columbia National Fisheries Laboratory, 1965 to 1983 [*12*].
ACQUIRE	Chemical Information System, Aquatic Information Retrieval Database, Baltimore, MD, 1984 [*13*].
B. G. Dowden and H. J. Bennett	*Journal of the Water Pollution Control Federation,* Vol. 37, 1965, pp. 1308–1316 [*14*].
U.S. Army	Chemical Research, Development, and Engineering Center (CRDEC), Aberdeen Proving Grounds, MD.
W. W. Johnson and M. T. Findley	*Handbook of Acute Toxicity of Chemicals to Fish and Aquatic Invertebrates,* 1980 [*15*].

The rat oral LD_{50} data were acquired from the following sources:

R. D. Lewis, Sr., *Registry of Toxic Effects of Chemical Substances,* (after review) [*16*].
Chemical Information System, Baltimore, MD, 1984.
U.S. Army, CRDEC, Aberdeen Proving Grounds, MD.

While we had originally intended to utilize only 48-h EC_{50}s, due to the scarcity of data, compounds with only 24-h values were also utilized. These compounds were coded with an indicator variable in the SAR equation to partially adjust for the difference between 24-h and 48-h readings. We also used only compounds that were not permanently charged, and we used only organics, due to the scarcity of data on organometallics and metals.

In view of the fact that cholinesterase (ChE)-inhibiting compounds operate by totally different mechanisms than the other compounds in the data base, the 14 compounds in this class were used for the development of a separate SAR equation.

In Fig. 1 we show the steps involved in obtaining the matched EC_{50}/LD_{50} data sets from which the major models were calculated.

Starting with about 8000 rat oral LD_{50}s and 233 *D. magna* EC_{50}s, the compounds for which EC_{50}s exist were matched with the available LD_{50} values, and the data removed from assays which had more than 5% impurities, mixtures, and like problems. After this process, we were left with 182 chemicals. Of these, log *P* could not be obtained with sufficient accuracy for 15 compounds, thus resulting in a data set of 167 chemicals for models using log *P* or its simulation with MCIs and keys. (See below for definition of these parameters).

In Table 1, we show the classes of compounds included in the *D. magna* data base. Note the considerable variety of chemical classes.

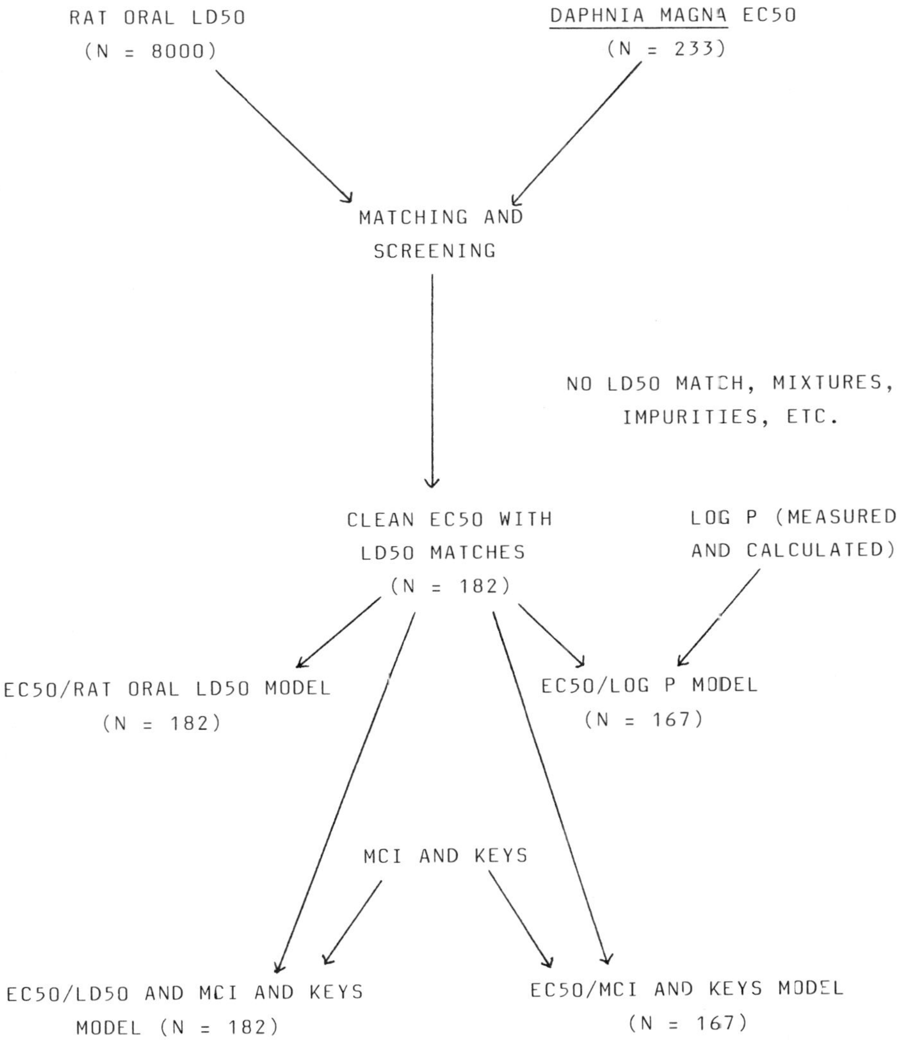

FIG. 1—*Procedure for constructing the four models.*

Parameters

Log *P*—Log *P*s were obtained from the collection of values in the Medchem [*17*] data base. For those values not listed, they were calculated with the CLOGP3 computer program from the same source. For 15 compounds, the confidence in the calculated values was insufficient for their inclusion in the SAR model. These compounds were therefore not used in the log *P* models.

Molecular Weight—This parameter was calculated from the molecular formula.

Molecular Connectivity Indexes (MCI)—These indexes, due to Kier and Hall [*18*], were calculated up to order 7, including sums and differences between simple and valence values.

TABLE 1—*Structures in the* D. magna EC_{50} *data base.*

Compounds	Number of Compounds
Aromatic compounds	
Nitrobenzenes	16
Phenols	11
Alkylbenzenes	9
Chlorobenzenes	15
Arylnitrogen compounds	6
Nonaromatic compounds (ring and nonring)	
Hydrocarbon	
Alkenes and alkanes	7
Carbonyl	
Aldehydes	4
Carboxylic acid	10
Esters	12
Ketones	7
Amide, carbamate, and urea compounds	13
Halide	
Haloalkanes	17
Chloroalkenes	4
Nitrogen	
Amines	12
Cyanides	4
Oxygen	
Ethers	10
Alcohols	16
Sulfur	
Thioethers, thiols, and thioureas	6
Phosphorus	
Oxides of phosphorus	8

Substructural Keys—Health Design's (HDI) MOLSTAC(c) substructural key system was used to generate these parameters. MOLSTAC(c) consists of five different classes of substructural descriptors:

1. Identification of the longest continuous chain of atoms (excluding hydrogen) in the molecule.
2. Identification of carbon chain fragments.
3. Identification of ring systems, including combinations such as the rings forming the bay region in certain carcinogens.
4. Identification of chemically or biologically or both functional substructural fragments.
5. Identification of electron-donating and electron-withdrawing substructural keys.

Statistical Methodology

We use multiple regression equations in this research [*19*]. The regression equations take the following form:

$$\begin{aligned} \textit{Daphnia}\ \mathrm{EC}_{50} = \text{constant} &+ (\text{parameter } 1 \times \text{coefficient } 1) \\ &+ (\text{parameter } 2 \times \text{coefficient } 2) + \ldots \\ &+ (\text{parameter } n \times \text{coefficient } n) \end{aligned}$$

Model Development—Model development can be portrayed as shown below:

Assembly of DATA BASE
↓
Selection of POTENTIAL PARAMETERS
↓
SIMPLE STATISTICS
↓
STEPWISE REGRESSION ANALYSIS
(up and down)
↓
Identification of INFLUENTIAL OBSERVATIONS
and their removal
↓
Diagnostics for the identification and
removal of POORLY BEHAVED PARAMETERS
↓
FINAL REGRESSION ANALYSIS
↓
VALIDATION

The data base is assembled as indicated above. One then selects a set of parameters which are presumed to be related to *D. magna* EC_{50}. Simple statistics on these parameters are then obtained to detect problematical distributions so that these can be corrected through transformations. This is followed by regression analysis [*19*]. Any parameters that have the slightest chance of contributing to the explanation of the endpoint are at first given an opportunity to enter the equation. After the list of potentially useful candidates has been exhausted, parameters are removed one at a time, starting with those which contributed least to the explanation of the *D. magna* EC_{50} and continuing in this fashion until the variable with the lowest "power" still included has an *F*-to-remove (variance ratio) greater than 1.7 [*20*].

In the selection of candidate parameters, one needs to exercise care that the number does not exceed approximately half the number of chemicals, in order to avoid spurious correlation problems. It is also important to not allow variables to enter the equation which have a correlation of greater than 0.9 with a variable already in the equation, to avoid collinearity difficulties. When such pairs of variables exist, the one with the lower partial correlation with the endpoint is deleted.

From the regression analysis one obtains a putative equation which is examined for robustness from the standpoint of both influential objects (chemicals) and poorly behaved parameters. A variety of techniques is used for this purpose, including ridge regression [*21*], Cook's distance, partial correlations, and principal component analysis [*22*]. After the objects or parameters or both that are in question have been removed, another regression analysis is performed, this time using only the parameters which survived the robustness tests. It is common to find that several more parameters are eliminated, thus resulting in a final equation.

While it would be desirable to validate the regression equation by evaluating it with newly tested compounds, too few compounds that fit within the chemical space of the compounds from which the model was developed have become available since that development. Because of the fact that the data base consists of a variety of chemical classes, split-sample validation is also impractical. Consequently, we have used the cross-validation method. In that method, one chemical is removed from the data set, the equation is recalculated, and the omitted compound estimated with the equation. This process is repeated for all compounds, one at a time, and the statistics calculated on the estimates [*23*].

The estimation procedure for untested compounds is sketched below:

STRUCTURE to be evaluated

↓

Look up and/or compute PARAMETERS

↓ ↓ ← Rat LD_{50}

Calculate REGRESSION SCORE

↓

Calculate EC_{50} ESTIMATES

↓

EVALUATE in terms of similar compounds and statistical RELIABILITY

Rather than reporting the estimated EC_{50} values without comment, the results are examined in terms of the parameters that contributed to the estimate and similar compounds for which results were found in the literature. This results in an indication of the reliability of the estimates.

Rather than modeling *D. magna* EC_{50} directly, following standard practice in the SAR field, we have transformed the EC_{50} (and LD_{50}) values to molar form as follows:

$$\log \frac{1}{C} = \log \left(\frac{\text{molecular weight} \times 1000}{EC_{50}} \right)$$

where C = molar concentration.

Results

Five separate SAR relationships were developed:

1. A simple correlation between *D. magna* EC_{50} and rat oral LD_{50}.
2. *Daphnia* EC_{50} predicted from log P and log P^2.
3. *Daphnia* EC_{50} predicted from MCIs and substructural keys.
4. *Daphnia* EC_{50} predicted from MCIs, substructural keys, and rat oral LD_{50}.
6. *Daphnia* EC_{50} for ChE-inhibiting compounds.

Table 2 summarizes the coefficient of determination, R^2, and F-values obtained for relationships one through four. Note the increasing value of R^2 as one proceeds from the simpler to the more complex models, as well as the attendant reduction in standard error.

TABLE 2—*Comparison statistics for the four EC_{50} models.*

	R^2	Standard Error of Estimate	N	F	P
Log $1/C$ rat oral LD_{50}	0.452	1.116	182	...	$\ll 0.001$
Log P and log P^2	0.523	1.000	167	92.1	$\ll 0.001$
MCI and keys	0.796	0.709	167	26.1	$\ll 0.001$
Log $1/C$ rat oral LD_{50}, MCI, and keys	0.835	0.658	182	30.2	$\ll 0.001$

During the equation development process, five compounds were removed for a variety of different reasons from the full model as follows:

CAS No.	Compound
107-02-8	Acrolein: severe statistical outlier
35367-38-5	Diflubenzuron: low EC_{50}/LD_{50} ratio
680-31-9	Hexamethylphosphoric acid triamide: severe statistical outlier
75-35-4	1,1-Dichloroethylene: statistically unduly influential
95-94-3	1,2,4,5-Tetrachlorobenzene: after this compound was noted to be a severe statistical outlier, it was learned that the experiment in which it was used was faulty

Two compounds were removed from the ChE-inhibiting compounds model:

2642-71-9 Azinphos-ethyl
56-38-2 Parathion

Both compounds were detected to be severe outliers. Both compounds need to be metabolized before they are absorbed; thus they should really not be included in the model in their unmetabolized state.

Daphnia *EC₅₀ Versus Rat Oral LD₅₀*

The R^2 of 0.452 indicates that 45% of the variation in *D. magna* EC_{50} can be explained by the variation in rat oral LD_{50}. The simple statistics for the two parameters are shown below.

	Log 1/*C* *D. magna*	Log 1/*C* rat
N	182	182
Mean	3.767	2.174
Standard deviation	1.450	0.830
Minimum	0.613	0.477
Maximum	8.492	5.104

It is evident that the values for both species span a large range. Examination of the cross-tabulation between the two endpoints does not indicate large outliers that could be responsible for the surprisingly large correlation.

Daphnia *EC₅₀ from Log* P

Table 3 shows the results for this relationship. Log P is the parameter most often used for SAR studies of aquatic species. In addition to log P, we also used log P^2 as well as the indicator variable for the 24-h observations. Note that the contribution of both log P^2 and the 24-h variable are relatively minor. An examination of the residuals did not indicate that any other log P terms would be useful.

TABLE 3—Daphnia EC_{50} *equations for model based on log* P *and log* P^2 *(*N = *167).*

	Coefficients				
	Intercept	Log P	Log P^2	24-h	R^2
Log P	2.5294	0.5768	...	...	0.483
Log P + 24-h	2.8731	0.5084	...	−0.737 5	0.529
Log P + log P^2 + 24-h	2.8937	0.4447	0.0155	−0.745 55	0.523

Daphnia EC_{50} *from MCIs and Substructural Keys*

Table 4 shows the best equation for this relationship. This equation was developed in order to permit a comparison of the fit obtained with log P and the MCIs and keys. Note that the 15 compounds which could not be used in the log P equation have also been omitted for this model in order to provide a more meaningful comparison.

It is evident that a better fit to *D. magna* log $1/C$ can be achieved with these independent parameters than with log P. This is really not surprising as the substructural keys and MCIs have a greater opportunity to fit the particular data base, whereas log P is necessarily constrained by the components which contribute to it.

Daphnia *Log 1/C from MCIs, Keys, and Rat Oral 1/C*

The regression equation is shown in Table 5. This model explains nearly 84% of the variance in *D. magna* log $1/C$ on the basis of 26 independent parameters, 5 of which are continuous, including rat oral log $1/C$. The cross-validation results are as follows:

Mean square error from equation	0.4334
Mean square error from cross validation	0.4909
Pooled standard error	0.0464
t	1.24
P	>0.20

TABLE 4—Daphnia EC_{50} *equations for model with MCIs and substructural keys (*N = *167).*

Primary amine bound to an aromatic ring atom	1.0167
Primary amine bound to aliphatic or alicyclic carbon	1.0343
Aliphatic alcohol	−0.5294
Oxygen-substituted aryl ester	−0.7801
Benzene	1.0320
Secondary or tertiary diphatic alcohol	−1.0058
1,1-Dichloro (non-beta phenyl)	0.8091
1,1-Divinyl chloride (non-beta phenyl)	1.0021
Secondary or tertiary amine bound to electron-releasing groups only	1.3375
One or more electron-releasing groups and four or more electron-withdrawing groups on a single benzine ring	0.7820
Three carbon fragments between two functional groups (electron withdrawing, electron releasing, or combination)	0.9442
NH substituted with one electron releasing and one electron withdrawing group	1.3467
Ethane or ethylene between two electron-releasing groups	−0.1819
Valence path MCI, order 2	0.3515
Valence path-cluster, order 4	0.1198
Sum simple and valence chain MCI, order 6	0.3621
Intercept	2.2578

TABLE 5—Daphnia EC_{50} *equations for model with MCIs, keys, and rat oral* LD_{50} *(N = 182).*

Primary amine bound to an aromatic ring	0.6751
1,1,1-Trichloro fragment (non-beta phenyl)	0.5482
Aliphatic alcohol	−0.3197
Aliphatic ether	−0.5129
Primary or secondary aliphatic amine	0.8676
Aliphatic three-branched nitrogen (non-amide)	−0.7273
Aliphatic amide	−0.8091
Oxygen-substituted aryl ester	−0.6702
Aliphatic ketone	0.5717
Benzene	0.7267
Two electron-releasing groups on a single benzene (para)	0.5122
Single benzene ring without releasing groups and at least four withdrawing groups in the 1,2,3,5, positions	−0.5751
1,1-Dichloro (non-beta phenyl)	0.8726
1,1-Divinyl chloride (non-beta phenyl)	1.2474
Aryl nitro	−0.2905
One methyl, one ethanol, or two propanol or tertiary butanol fragments	−0.6745
Ketone fragment	−1.0588
One or two electron-withdrawing groups bound to alpha carbon of an ethane fragment	0.6179
NH substituted with one electron-releasing and one electron-withdrawing group	0.6349
N substituted with two electron-releasing and one electron-withdrawing group or groups	0.4726
24-h EC_{50} value	−0.5988
Log valence path MCI, order 2	0.8097
Log valence path MCI, order 4	2.6081
Valence chain MCI, order 3	0.6093
Simple—valence cluster MCI, order 5	−0.7027
Log $1/C$ LD_{50}	0.4350
Intercept	1.8025

Other cross-validation parameters are shown below:

	Actual	Cross-Validation Estimate
Mean	3.7669	3.7627
Median	3.6955	3.8078
Standard deviation	1.4995	1.3683
Quantiles		
100	9.3552	8 2862
95	6.3207	6 2536
90	5.7779	5 0235
10	1.9232	2.1054
5	1.1766	1.4568
1	0.6634	0.5655

With the exception of a less than ideal fit for the largest log $1/C$, it is evident that the equation displays very little bias and that the fit is not spurious.

ChE-Inhibiting Compound Model

Based on twelve such compounds, a model was developed with rat oral log $1/C$, molecular weight, and the valence path MCI, order 2 (VALPAT2) as independent variables. The statistics are shown in Table 6. The R^2 between the two species is 0.407. Adding molecular

TABLE 6—Daphnia EC_{50} *equations for cholinesterase compounds model* (N = *12).*

$1/C\ LD_{50}$	Molecular Weight	VALPAT2	R^2	Standard Error	Cp[a]	F	P
X			0.407	0.666	15.7	...	...
	X		0.384	0.678	16.6	...	...
X	X		0.726	0.477	5.0	...	...
X		X	0.495	0.648	14.2	...	...
X	X	X	0.800	0.432	4.0	10.66	<0.01

	Mean	Standard Deviation	Minimum	Maximum
Log $1/C\ EC_{50}$	8.338	0.824	7.266	9.837
Log $1/C\ LD_{50}$	3.740	0.788	2.906	5.256

[a] Cp = Mallow's Cp [*22*].

weight (MW) raises it to 0.726, and further adding the MCI raises it to 0.800. The resulting equation is

$$\log \frac{1}{C_{EC50}} = \left(0.738 \log \frac{1}{C_{LD50}}\right) + 6.399\text{MW} - 0.147\text{VALPAT2} - 9.29$$

Discussion

These results demonstrate that it is possible to construct SAR models incorporating a biological endpoint and structural parameters to estimate toxicity in an evolutionarily distant species. Metabolism, absorption and excretion, and interaction of the toxiphores with their active sites are all incorporated into an experimentally determined EC_{50} or LD_{50}. The structural parameters and intercept in part describe the differences in metabolism and scaling factors between the two species. These components of the equation account for the fact that the models describe a large fraction of the total variation, although the number of compounds incorporated in the data base is relatively low.

The equation describes with considerable accuracy the relationship of chemical structure and toxicity in n-dimensional space, in much the same way a topographer describes shapes. It is not necessary for the topographer to understand how the shape was formed from underlying physical laws; he can still provide an accurate description. Since relatively few of the molecular mechanisms of toxicity are understood, statistical relationships are used to describe the chemical structure-toxicity topography.

Another way to compare the fit from the four different SAR models is to estimate the fraction of compounds predicted within certain factors. Table 7 shows this comparison, which is more graphically displayed in Fig. 2. It is evident that the model using log P does not perform much better than the simple relationship with rat oral LD_{50}. It is also clear that the use of MCIs and substructural keys produces a considerable improvement over the use of log P, so much so in fact that the addition of rat oral LD_{50} to the MCI-keys model only accounts for an incremental improvement in the prediction of *D. magna* EC_{50}. These comparisons also serve to illustrate the fact that for a heterogeneous data base, at least, substantially better models can be produced with MCIs and keys than log P and its square. This is really not surprising since log P is calculated from the contribution of fragments, and the fragments can be fitted more precisely to the compounds and the endpoint

TABLE 7—*Comparison of performance of the four EC_{50} models.*

Percentage of Compounds Predicted Within Factor of	Rat Oral LD_{50}, %	Log P and P^2, %	MCI and Keys, %	LD_{50}, MCI, and Keys, %
1.2	6.0	9.0	10.2	12.6
1.5	12.6	16.8	19.8	27.5
2	19.2	25.7	34.7	39.0
5	48.4	53.3	70.7	71.4
10	68.1	70.1	85.6	89.6
20	80.0	80.2	94.6	98.4
50	90.0	92.8	99.4	99.5
100	94.5	97.6	100.0	100.0
100	100.0	100.0	100.0	100.0

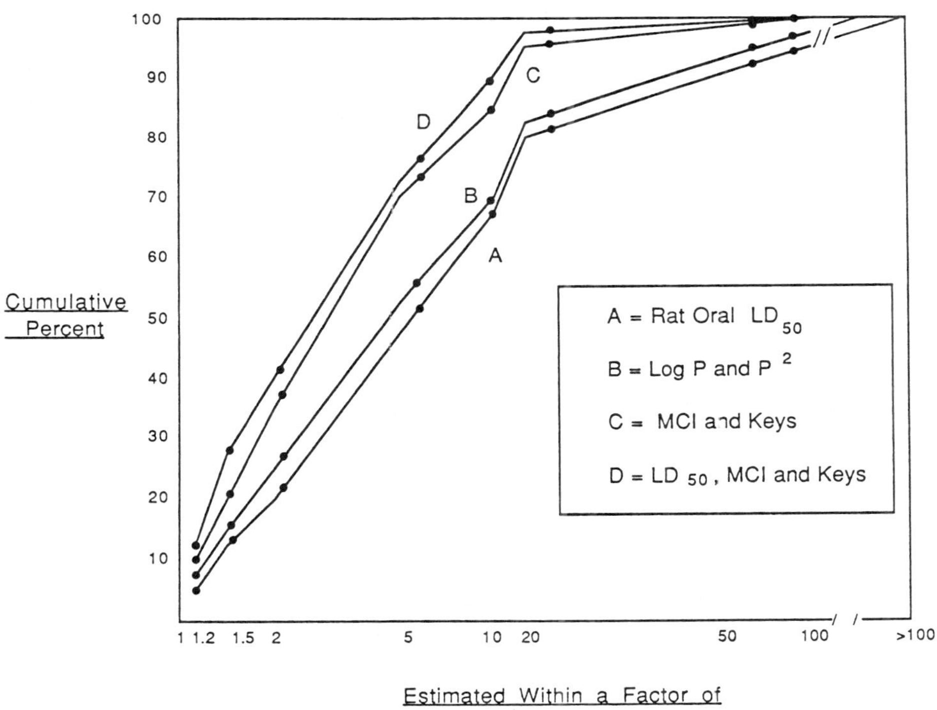

FIG. 2—*Comparison among four EC_{50} models.*

if one is not constrained by having to estimate log P. This is another way of saying that one might as well use the fragments that contribute to log P rather than log P itself.

The full equation is being incorporated into the TOPKAT program, our software system for the estimation of toxic end points based on structure. The TOPKAT program is able to provide toxicity estimates of untested compounds from a variety of SAR equations, and has been implemented on several personal microcomputers.

A major drawback of the model is the fact that there is such a low intersection of the sets of compounds assayed in the rat and *D. magna.* This fact means that we have described toxicity in a relatively small part of chemical space. Expansion of the data base is currently under way with a steady increase in determination of *D. magna* EC_{50}s for compounds already in the rat oral LD_{50} data base.

The applications of any SAR model are limited to those chemicals which fit within the "space" of the compounds used for the design of the model. In other words, one must be sure not to extrapolate (except in a minor way) outside the model, but rather interpolate within it. The following were the ranges for some parameters of the compounds used in the model:

	Minimum	Maximum
Molecular weight	41	546
Log *P*	−2.19	6.91

Five compounds were removed from the design set. Two were severe statistical outliers, one was unduly influential, one had an unusual EC_{50}/LD_{50} ratio, and one was found to have been tested incorrectly. The removal of these compounds from the design set had only a minor effect on the performance of the equation.

As can be seen from Table 1, the variety of chemicals in the data base is limited, mostly due to the lack of compounds for which both rat oral LD_{50} and *D. magna* EC_{50} values are available. It stands to reason that many groups represented in the rat oral LD_{50} collection are not represented in the model. Such groups include the following: epoxides and iodides; aromatic azos, azoxy, and hydrazo compounds; benzimidazoles; dioxins; fluorenes and fluorenones; hydrazides; lactones; nitro-alkanes, -imidazoles, -furans, -amides, -amines, and -ureas; organo-lead and -mercury compounds; phenothiazines; purine derivatives; saturated alkyl halides, acyl halides, unsaturated alkyl bromides, and fluorides; xanthenes and thioxanthenes; amino acids and derivatives; amine-*N*-oxides; alicyclic and aromatic sulfur rings; alicyclic and aromatic sulfur-nitrogen rings; alicyclic and aromatic oxygen-nitrogen rings; organic sulfhydryls, sulfides, and disulfides; quinones and anthraquinones; alpha-beta diketones; alpha hydroxy carbonyls; and anhydrides.

The user should, of course, consider the validity of an EC_{50} estimate in terms of known parameters about the compound. For example, it would not be sensible to attribute a high degree of confidence to an EC_{50} estimate which is greater than its maximum water solubility. An estimate should also be evaluated in terms of the EC_{50} values for similar compounds in the data base. In other words, it is important to relate the obtained estimates to related information for the particular chemical.

Several applications of these models come readily to mind in the environmental arena. Many of the compounds for which rodent toxicity information is available can be evaluated for environmental toxicity within the limitations (stated above) of the model. Compounds for which a high level of toxicity is estimated can be further examined. Although the model is not meant to replace human judgment, it can certainly aid in the evaluation of environmental risk.

References

[1] Veith, G. D., DeFoe, D., and Knuth, M., "Structure-Activity Relationships for Screening Organic Chemicals for Potential Ecotoxicity Effects," *Drug Metabolism Reviews,* Vol. 15, 1984, pp. 1295–1303.

[*2*] Veith, G. D., Call, D. J., and Brooke, L. T., "Structure Toxicity Relationships for the Fathead Minnow, *Pimephales promelas:* Narcotic Industrial Chemicals," *Canadian Journal of Fisheries and Aquatic Sciences,* Vol. 40, 1983, pp. 743–748.
[*3*] Veith, G. D., Call, D. J., and Brooke, L. T., "Estimating the Acute Toxicity of Narcotic Industrial Chemicals to Fathead Minnows," *Aquatic Toxicology and Hazard Assessment: Sixth Symposium, ASTM STP 802,* W. E. Bishop, R. D. Cardwell, and B. B. Heidolph, Eds., American Society for Testing and Materials, Philadelphia, 1983, pp. 90–97.
[*4*] Hermens, J., Koenemann, H., Leeuwanggh, P., and Musch, A., "Quantitative Structure-Activity Relationships in Aquatic Toxicity—Studies of Chemicals and Complex Mixtures of Chemicals," *Environmental Toxicology and Chemicals,* Vol. 4, 1985, pp. 273–279.
[*5*] Janardan, S. K. and Olson, C. S., "Quantitative Comparisons of Acute Toxicity of Organic Chemicals to Rat and Fishes," *Ecotoxicology and Environmental Safety,* 1984.
[*6*] Bobra, A. M., Shiu, W. Y., and Mackay, D., "A Predictive Correlation for the Acute Toxicity of Hydrocarbons and Chlorinated Hydrocarbons to the Water Flea, *Daphnia magna,*"*Chemosphere,* Vol. 12, 1983, pp. 1121–1129.
[*7*] Bobra, A., Shiu, W. Y., and Mackay, D., "Quantitative Structure-Activity Relationships for the Acute Toxicity of Chlorobenzenes to *Daphnia magna,*" *Environmental Toxicology and Chemistry,* Vol. 4, 1985, pp. 297–305.
[*8*] Koenemann, H., "Quantitative Structure-Activity Relationships and Fish Toxicity Studies, Part I: Relationships for 50 Industrial Pollutants," *Toxicology,* Vol. 19, 1981, pp. 209–221.
[*9*] Govers, H., Ruepert, C., and Aiking, H., "Quantitative Structure-Activity Relationships for Polycyclic Aromatic Hydrocarbons: Correlation between Molecular Connectivity, Physical-Chemical Properties, Bioconcentration and Toxicity in *Daphnia pulex,*" *Chemosphere,* Vol. 13, 1984, pp. 227–236.
[*10*] Birge, W. J. and Cassidy, R. A., "Structure-Activity Relationships in Aquatic Toxicology," *Fundamentals of Applied Toxicology,* Vol. 3, 1983, pp. 359–368.
[*11*] LeBlanc, G. A., "Acute Toxicity of Priority Pollutants to Water Flea *(Daphnia magna),*" *Bulletin of Environmental Contamination and Toxicology,* Vol. 24, 1980, pp. 684–691.
[*12*] Mayer, F. L., Jr., and Ellersieck, M. R., *Manual of Acute Toxicity: Interpretation and Data Base for 410 Chemicals and 66 Species of Freshwater Animals,* Resource Publication No. 160, Fish and Wildlife Service, U.S. Department of the Interior, Washington, DC, 1986.
[*13*] Chemical Information System, Aquatic Information Retrieval Database (AQUIRE), Baltimore, MD, 1984.
[*14*] Dowden, B. G. and Bennett, H. J., "Toxicity of Selected Chemicals to Certain Animals," *Journal of the Water Pollution Control Federation,* Vol. 37, 1965, pp. 1308–1316.
[*15*] Johnson, W. W. and Finley, M. T., *Handbook of Acute Toxicity of Chemicals to Fish and Aquatic Invertebrates,* Resource Publication 137, Fish and Wildlife Service, U.S. Department of the Interior, Washington, DC, 1980.
[*16*] Lewis, R. D., Sr., *Registry of Toxic Effects of Chemical Substances,* (DHEW (NIOSH) Publication No. 79-100, U.S. Department of Health, Education, and Welfare, Cincinnati, OH, 1981.
[*17*] Leo, A. and Weininger, D., *User Reference Manual,* CLOGP3, Version 3.2, Medicinal Chemistry Project, Pomona College, Claremont, CA, 1984.
[*18*] Kier, L. B. and Hall, L. H., *Molecular Connectivity in Structure-Activity Analysis,* Wiley, New York, 1986.
[*19*] *Statistical Methods for Digital Computers,* Enslein, K., Ralston, A., and Wilf, H. S., Eds., Wiley, New York, 1977.
[*20*] Costanza, M. C. and Afifi, A. A., "Comparison of Stopping Rules in Forward Stepwise Discriminant Analysis," *Journal of the American Statistical Association,* Vol. 74, 1979, pp. 777–785.
[*21*] Marquardt, D. W. and Snee, R. D., "Ridge Regression in Practice," *American Statistician,* Vol. 29, 1975, pp. 3–20.
[*22*] Belsley, D. A., Kuh, E., and Welsch, R. E., *Regression Diagnostics,* Wiley, New York, 1980.
[*23*] Gray, H. L. and Schucany, W. R., *The Generalized Jackknife Statistic,* Marcel Dekker, New York, 1972.

T. Wayne Schultz,[1] *Mariely Cajina-Quezada,*[1] *Muwon Chang,*[1] *David T. Lin,*[1] *and Rajit Jain*[1]

Structure-Toxicity Relationships of Para-Position Alkyl- and Halogen-Substituted Monoaromatic Compounds

REFERENCE: Schultz, T. W., Cajina-Quezada, M., Chang, M., Lin, D. T., and Jain, R., **"Structure-Toxicity Relationships of Para-Position Alkyl- and Halogen-Substituted Monoaromatic Compounds,"** *Aquatic Toxicology and Environmental Fate: Eleventh Volume, ASTM STP 1007,* G. W. Suter II and M. A. Lewis, Eds., American Society for Testing and Materials, Philadelphia, 1989, pp. 410–423.

ABSTRACT: The relative toxicity of 46 para-position alkyl- and halogen-substituted pyridines, cyanobenzenes, nitrobenzenes, phenols, benzyl alcohols, and benzaldehydes was determined by measuring the population growth impairment (log BR) of the ciliate *Tetrahymena pyriformis*. The 1-octanol/water partition coefficient (log K_{ow})-dependent quantitative structure-activity relationship (QSAR) for the pyridines, cyanobenzenes, and nitrobenzenes is log BR = 0.8222 (log K_{ow}) − 1.7889; n = 19, r^2 = 0.937, and s = 0.155. A similar QSAR for the phenols, benzyl alcohols, and benzaldehydes is log BR = 0.6360 (log K_{ow}) − 1.2012; n = 27, r^2 = 0.726, and s = 0.284 (where n = the number of data points; r^2 = the coefficient of determination; and s = root mean square.) The predictability of the latter QSAR is improved by the addition of the orthogonal field electronic parameter F. The resulting QSAR is log BR = 0.7891 (log K_{ow}) + 0.6793 (F) − 1.7457; n = 27, r^2 = 0.925, and s = 0.163.

KEY WORDS: quantitative structure-activity relationships, toxicity, monoaromatic chemicals, 1-octanol/water partition coefficient, electronic parameters, *Tetrahymena pyriformis,* narcosis, polar narcosis, aquatic toxicology

For the past several years, the principles of quantitative structure-activity relationships (QSAR) have been used to model correlations between aquatic toxicity measured in different test systems and descriptors of molecular structure or physiochemical properties or both. Most of the studies describe linear models of the toxic response in terms of the 1-octanol/water partition coefficient (log K_{ow}). It has been widely assumed that a different QSAR exists for each mode of toxic action. Most industrial organic chemicals which are nonionic and nonreactive have narcosis as the mechanism of toxic action [*1*]. This reversible retardation of cytoplasmic activity or membrane perturbation was modeled in the *Tetrahymena* system for a series of aromatic compounds including several alkyl-substituted pyridines and nitrobenzenes [*2*]. Hereafter, this mode of action will be referred to as Narcosis I. Recently Veith et al. [*3*] have developed a log K_{ow}-dependent linear model for esters which differs from baseline toxicity or Narcosis I. This is for the polar narcosis or the Narcosis II mode of action. Veith and co-workers suggest that the predictive ability of this polar narcosis model can be improved by the addition of an orthogonal electronic

[1] Associate professor, research assistant, and participants (high school student programs), respectively, College of Veterinary Medicine, The University of Tennessee, Knoxville, TN 37901-1071.

parameter [*3*]. At about the same time, Schultz and Riggin [*4*] noted that the subdivision of alkyl- and halogen-substituted phenols, based on substituent field or polar electronic effects, shows that electron-releasing alkyl derivatives are slightly less toxic than electron-withdrawing halogen derivates at similar log K_{ow} values. This theme was expanded in a more recent study by Schultz et al. [*5*] of a heterogeneous set of 34 para-substituted phenols. This latter study revealed that the addition of an electronic term to a log K_{ow}-dependent model improved its predictability. These investigations suggest that although the hydrophobic parameter, log K_{ow}, is consistently a highly significant descriptor of toxicity, there are classes of compounds, such as the esters and phenols, where an electronic parameter can significantly aid in explaining toxicity.

The present paper further explores such relationships by studies of 48 compounds, including alkyl and halogen derivatives of pyridine, cyanobenzene, nitrobenzene, phenol, benzyl alcohol, and benzaldehyde. Relative toxicity was determined using the *Tetrahymena* growth assay [*6*]. Structure-toxicity modeling was undertaken using least-squares regression analysis with log K_{ow} and three electronic parameters or a discrete variable as independent variables.

Materials and Methods

Bioassay

Tetrahymena pyriformis strain GL-C under static conditions was the test system of choice. This chronic bioassay, which uses population density of axenic cultures as its endpoint, has been described in detail by Schultz [*6*]. Each monoaromatic was tested in duplicate for a minimum of three replicates following range-finding experiments. Each replicate was, at a minimum, a five-step graded concentration series using freshly prepared stock solutions. *Tetrahymena* cultures without monoaromatics served as controls. Cell population levels were estimated spectrophotometrically at 48 h as absorbance at 540 nm. Only replicates with control absorbances equivalent to late log-growth-phase, 0.6 to 0.9 optical density, were used in the analysis. The 50% inhibitory growth concentration (IGC_{50}) and 95% fiducial limits were determined for each monoaromatic using probit analysis of the Statistical Analysis System (SAS) software in conjunction with an IBM 3081 computer. In these analyses, Y was the absorbance normalized to percent control, and X was the concentration of tested compound in g/m^3.

Test Monoaromatics

The monoaromatics selected for testing were a series of 48 four-position alkyl or halogen derivatives of either pyridine, cyanobenzene, nitrobenzene, phenol, benzyl alcohol, or benzaldehyde. They were purchased from commercial sources (Aldrich Chemical Co., Milwaukee, WI, or Pfaltz and Bauer Inc., Waterbury, CT) and not repurified. Individual stock solutions were prepared in dimethylsulfoxide (DMSO) at several concentrations, including 5, 10, 25, or 50 kg/m^3. In all cases, the volume of stock solution added to the culture medium was limited so that DMSO did not exceed a final concentration of 0.75%, a level shown to have no effect on *Tetrahymena* growth [*7*].

Structure-Activity Analysis

For predictive correlation determination, the log BR (biological response) was generated for each monoaromatic tested by taking the log of the inverse of the IGC_{50} expressed in moles per cubic metre (mol/m^3). Log K_{ow} values were computer calculated by the fragment method, by the CLOG P-3.2 program, or retrieved as measured values from the select list for comparison [*8,9*]. The Hammett sigma constants for the para position, σ_p, were secured from Hansch and Leo [*10*]. The polar or field parameter, F [*11*], and the resonance parameter, R [*11*], were retrieved as the redefined values of Swain and co-workers [*12*]. Mathematical models were generated using the general linear model of multiple regression analysis of SAS with log BR as the dependent variable and log K_{ow} alone or in combination with one of the other parameters as the independent descriptors. Model adequacy was measured as the coefficient of determination, r^2.

Chemical Persistence

Since the toxicity tests were static in design, there existed a potential for abiotic loss of toxicants. Therefore, chemical persistence studies were undertaken with the aid of high-performance liquid chromatography (HPLC). The concentration of each test compound was approximately equal to that of its IGC_{50} value. The analyzed solutions consisted of the appropriate aliquot of stock added to 50 μm^3 of sterile distilled water in a foam stoppered 250 μm^3 Erlenmeyer flask. For analysis, a Waters Model 840 HPLC with a C-18 reverse phase column in conjunction with a Waters Model 600 Multisolvent Delivery System and a Model 712 Waters Intelligent Sample Processor was used. Test chemicals were eluted using a degassed 65/45 mixture of methanol and 0.5 mol ammonium acetate buffer adjusted to a pH of 6.7. The solvent flow rate was set at 1.0 μm^3/min and the absorbance detector set at 254 nm and recorded directly onto a Digital Professional 350 microprocessor. At t = 0 h a 10 to 20 nm^3 aliquot was injected into the HPLC and eluted for 10 min. Peaks were integrated using Waters Expert Software. The sample-containing flasks were then placed under the same environmental conditions as those for the bioassay. At t = 48 h, the test solutions were injected onto the HPLC and analyzed using the same method. Percent loss was measured as the difference between the t = 0 and t = 48 h concentration. Each test chemical was analyzed for abiotic loss twice or, in the case of those showing a loss greater than 10%, three times.

Results

Table 1 is a summary of toxicity data for the 48 compounds assayed. Two compounds, 4-phenylnitrobenzene and 4-iodonitrobenzene, did not elicit the measured response of 50% growth inhibition at saturation. With the exception of 4-fluoronitrobenzene, the data for each compound fit the probit model extremely well with $P < \chi^2$ being greater than 0.9000 based on >30 data points.

The log BR and molecular descriptor values are presented in Table 2. The results of least-squares regression analysis of log BR versus log K_{ow} for the pyridines, cyanobenzenes, nitrobenzenes, phenols, benzyl alcohols, and benzaldehydes are presented in Eq 1 through Eq 6, respectively (Table 3). An examination of Table 3 reveals that Eqs 1 and 2 are not statistically different. Similarly, Eqs 4, 5, and 6 are not statistically different from each other. Combining the data for the pyridines and cyanobenzenes, subsequent reanalysis results in Eq 7 (Table 3). Combining the phenol, benzyl alcohol, and benzaldehyde data and recalculating the regression analysis results in Eq 8 (Table 3). A plot of these data is shown in Fig. 1. Subdivision of the data of Eq 8 into electron-releasing substituents and electron-withdrawing substituents followed by regression analysis results in Eq 9 and 10, respectively, (Table 3). While the nitrobenzene data, Eq 3, does not appear to be similar to log K_{ow}-dependent regression for any of the other monoaromatic groups, a combined plot of the pyridine, cyanobenzene, and nitrobenzene data suggests colinearity (Fig. 2).

TABLE 1—*Toxicity to* Tetrahymena *of Alkyl and Halogen Monoaromatics.*

Compound	CAS No.	Probit Regression Equation	$P > \chi^2$	n	IGC_{50} (range), g/m^3	IGC_{50}, mol/m^3
Pyridine	110-86-1	$Y = 7.2330 - 0.0013X$	0.9635	32	1678.58 (1289.02 to 2781.77)	21.221
4-Methyl	108-89-4	$Y = 7.9078 - 0.0040X$	0.9917	30	731.04 (583.62 to 959.59)	7.850
4-Ethyl	536-72-4	$Y = 12.9632 - 0.0375X$	0.9999	30	212.26 (194.13 to 256.98)	1.981
4-*(tert)*butyl	3978-81-2	$Y = 10.1272 - 0.0554X$	0.9993	30	92.56 (79.33 to 105.49)	0.685
4-Phenyl	939-23-1	$Y = 7.8748 - 0.0853X$	0.9977	21	33.69 (17.70 to 43.07)	0.217
4-Chloro	19524-06-2	$Y = 7.2989 - 0.0028X$	0.9996	36	826.03 (567.15 to 1013.69)	7.275
4-Bromo	7379-35-3	$Y = 6.8385 - 0.0057X$	1.0000	40	322.62 (214.11 to 536.52)	2.042
Cyanobenzene	100-47-0	$Y = 7.5706 - 0.0075X$	0.9996	32	324.55 (248.88 to 423.10)	3.322
4-Methyl	104-85-8	$Y = 7.7509 - 0.0185X$	0.9993	42	149.07 (113.82 to 182.00)	1.272
4-Phenyl	2920-38-9	$Y = 7.4534 - 0.2391X$	0.9999	43	10.26 (8.05 to 12.54)	0.057
4-Fluoro	1194-02-1	$Y = 7.3426 - 0.0107X$	0.9998	62	218.16 (181.18 to 266.96)	1.801
4-Chloro	623-03-0	$Y = 7.1564 - 0.0157X$	1.0000	39	137.47 (95.28 to 168.72)	0.999
4-Bromo	623-00-7	$Y = 6.5950 - 0.0162X$	0.9999	33	98.72 (59.32 to 136.88)	0.542
Nitrobenzene	98-95-3	$Y = 6.8833 - 0.0120X$	0.9640	28	156.33 (100.33 to 219.35)	1.160
4-Methyl	99-99-0	$Y = 7.7840 - 0.0342X$	0.9685	28	81.36 (62.77 to 107.14)	0.593
4-Ethyl	100-12-9	$Y = 7.4130 - 0.0482X$	0.9597	50	50.07 (40.09 to 61.63)	0.331
4-Phenyl	92-93-3	not active at saturation	...	...	...	...

TABLE 1—*(Continued).*

Compound	CAS No.	Probit Regression Equation	$P > \chi^2$	n	IGC_{50} (range), g/m^3	IGC_{50}, mol/m^3
4-Fluoro	350-46-9	$Y = 6.2779 - 0.0073X$	0.7742	48	174.39 (103.74 to 236.06)	1.236
4-Chloro	100-00-5	$Y = 6.2000 - 0.0161X$	0.9574	44	74.44 (38.33 to 104.57)	0.472
4-Bromo	586-78-7	$Y = 6.1565 - 0.0133X$	0.9996	32	86.81 (51.57-not calculated due to solubility)	0.430
4-Iodo	636-98-6	not active at saturation	...	...	...	...
Phenol	108-95-2	$Y = 7.0118 - 0.0079X$	0.9982	56	253.90 (204.99 to 330.17)	2.698
4-Methyl	106-44-5	$Y = 7.8487 - 0.0169X$	0.9974	40	168.25 (140.84 to 224.53)	1.556
4-Ethyl	123-07-9	$Y = 8.8206 - 0.0502X$	0.9928	34	76.06 (63.49 to 88.17)	0.623
4-Isopropyl	99-89-8	$Y = 9.4377 - 0.0969X$	0.9887	35	45.81 (39.88 to 52.82)	0.336
4-*(tert)* butyl	98-54-4	$Y = 7.7422 - 0.1493X$	0.9999	35	18.37 (14.08 to 22.20)	0.122
4-Phenyl	92-69-3	$Y = 8.6606 - 0.5190X$	1.0000	36	7.05 (5.70 to 8.26)	0.042
4-Fluoro	371-41-5	$Y = 7.0653 - 0.0192X$	0.9842	34	107.81 (42.67 to 133.75)	0.962
4-Chloro	106-48-9	$Y = 7.2339 - 0.0634X$	0.9776	32	36.68 (26.23 to 47.05)	0.285
4-Bromo	106-41-2	$Y = 7.5923 - 0.0718X$	0.9860	40	36.10 (28.52 to 43.84)	0.209
4-Iodo	540-38-5	$Y = 7.0683 - 0.0673X$	0.8896	45	30.76 (21.64 to 37.61)	0.140
Benzyl alcohol	100-51-6	$Y = 7.3594 - 0.0028X$	0.9998	46	853.47 (686.42 to 1057.02)	7.892

4-Methyl	589-18-4	$Y = 7.5024 - 0.0066X$	0.9997	49	379.80 (305.38 to 456.64)	3.109
4-Ethyl	768-52-9	$Y = 6.9619 - 0.0173X$	0.9996	53	113.50 (85.61 to 145.25)	0.833
4-Isopropyl	536-60-7	$Y = 7.6659 - 0.0263X$	0.9933	53	101.32 (83.91 to 119.00)	0.674
4-*(tert)* butyl	877-65-6	$Y = 7.7679 - 0.0474X$	1.0000	56	58.44 (47.97 to 67.61)	0.356
4-Phenyl	3597-91-9	$Y = 7.4227 - 0.1094X$	1.0000	36	22.15 (16.24 to 29.77)	0.120
4-Fluoro	459-56-3	$Y = 7.2771 - 0.0050X$	0.9999	53	451.10 (328.91 to 530.65)	3.576
4-Chloro	873-76-7	$Y = 7.1636 - 0.0270X$	0.9947	40	80.24 (57.94 to 99.66)	0.563
4-Bromo	873-75-6	$Y = 7.0095 - 0.0359X$	0.9702	56	55.94 (38.95 to 72.65)	0.299
Benzaldehyde	100-52-7	$Y = 7.8327 - 0.0177X$	0.9915	38	159.70 (119.45 to 190.12)	1.505
4-Methyl	104-87-0	$Y = 7.6090 - 0.0194X$	0.9342	32	134.51 (95.82 to 164.65)	1.120
4-Ethyl	4748-78-1	$Y = 7.7837 - 0.0397X$	1.0000	35	79.19 (53.58 to 84.51)	0.523
4-Isopropyl	122-03-2	$Y = 7.2254 - 0.0685X$	0.9711	52	32.47 (24.59 to 39.23)	0.219
4-Phenyl	3218-36-8	$Y = 7.1241 - 0.1461X$	0.9983	51	13.86 (10.76 to 17.79)	0.076
4-Fluoro	459-57-4	$Y = 8.2108 - 0.0224X$	0.9926	38	143.55 (113.03 to 168.18)	1.157
4-Chloro	104-88-1	$Y = 7.3940 - 0.0428X$	0.9993	36	55.92 (39.88 to 68.54)	0.398
4-Bromo	1122-91-4	$Y = 6.8962 - 0.0396X$	0.9997	25	47.94 (27.86 to 67.45)	0.259

TABLE 2—*Toxicity and molecular descriptors of alkyl and halogen monoaromatics.*

Compound	log BR	log K_{ow}[a]	F[b]	R[b]	σ_p[c]	Group[d]
Pyridine	−1.327	0.65	...	...	...	...
4-Methyl	−0.895	1.22	...	...	...	...
4-Ethyl	−0.297	1.84	...	...	...	...
4-*(tert)*-butyl	0.164	2.64	...	...	...	...
4-Phenyl	0.664	2.59	...	...	...	...
4-Chloro	−0.862	1.28	...	...	...	...
4-Bromo	−0.310	1.52	...	...	...	...
Cyanobenzene	−0.521	1.56	...	...	...	...
4-Methyl	−0.106	2.12	...	...	...	...
4-Phenyl	1.242	3.52	...	...	...	...
4-Fluoro	−0.256	1.70	...	...	...	...
4-Chloro	0.000	2.27	...	...	...	...
4-Bromo	0.313	2.42	...	...	...	...
Nitrobenzene	−0.064	1.885	...	...	...	...
4-Methyl	0.227	2.534	...	...	...	...
4-Ethyl	0.480	3.063	...	...	...	...
4-Phenyl	...	3.773	...	...	...	...
4-Fluoro	−0.092	2.028	...	...	...	...
4-Chloro	0.326	2.598	...	...	...	...
4-Bromo	0.367	2.748	...	...	...	...
4-Iodo	...	3.008	...	...	...	...
Phenol	−0.431	1.49	0.00	0.00	0.00	R
4-Methyl	−0.192	1.94	−0.01	−0.41	−0.17	R
4-Ethyl	0.206	2.58	−0.02	−0.44	−0.15	R
4-Isopropyl	0.473	3.052	−0.06[e]	−0.36[e]	−0.15	R
4-*(tert)*-butyl	0.913	3.31	−0.11	−0.29	−0.20	R
4-Phenyl	1.383	3.20	0.25	−0.36	−0.01	R
4-Fluoro	0.017	1.77	0.74	−0.60	0.06	W
4-Chloro	0.545	2.39	0.72	−0.24	0.23	W
4-Bromo	0.681	2.59	0.72	−0.18	0.23	W
4-Iodo	0.854	2.91	0.65	−0.12	0.18	W
Benzyl alcohol	−0.897	1.10	0.00	0.00	0.00	R
4-Methyl	−0.493	1.66	−0.01	−0.41	−0.17	R
4-Ethyl	0.079	2.12	−0.02	−0.44	−0.15	R
4-Isopropyl	0.171	2.63	−0.06[e]	−0.36[e]	−0.15	R
4-*(tert)*-butyl	0.449	3.08	−0.11	−0.29	−0.20	R
4-Phenyl	0.420	2.99	0.25	−0.36	−0.01	R
4-Fluoro	−0.553	1.24	0.74	−0.60	0.06	W
4-Chloro	0.250	1.81	0.72	−0.24	0.23	W
4-Bromo	0.524	1.96	0.72	−0.18	0.23	W
Benzaldehyde	−0.178	1.45	0.00	0.00	0.00	R
4-Methyl	−0.049	2.01	−0.01	−0.41	−0.17	R
4-Ethyl	0.281	2.47	−0.02	−0.44	−0.15	R
4-Isopropyl	0.659	2.98	−0.06[e]	−0.36[e]	−0.15	R
4-Phenyl	1.119	3.41	0.25	−0.36	−0.01	R
4-Fluoro	−0.063	1.59	0.74	−0.60	0.06	W
4-Chloro	0.400	2.16	0.72	−0.24	0.23	W
4-Bromo	0.587	2.31	0.72	−0.18	0.23	W

[a] From CLOG P 3.2 computer program.
[b] From Swain et al. [*12*].
[c] From Hansch and Leo [*10*].
[d] R equals electron-releasing, and W equals electron-withdrawing.
[e] Estimated value.

TABLE 3—*Log* K_{ow}*-dependent QSAR.*

Equation	Data Set	Intercept	Slope	n	r^2	Root Mean Square Error	$Pr > f$
1	Pyridine (Pyr)	−1.9023	0.8904	7	0.920	0.211	0.0006
2	Cyanobenzene (CN)	−1.8724	0.8761	6	0.977	0.104	0.0002
3	Nitrobenzene (NO_2)	−1.0813	0.5204	6	0.963	0.051	0.0005
4	Phenol (Phen)	−1.1249	0.6198	10	0.705	0.318	0.0024
5	Benzyl alcohol (MeOH)	−1.3797	0.6926	9	0.707	0.339	0.0045
6	Benzaldehyde (CHO)	−1.5478	0.7909	8	0.829	0.256	0.0017
7	PyrCn	−1.8921	0.8846	13	0.952	0.156	0.0001
8	PhenMeOHCHO	−1.3522	0.6993	27	0.755	0.289	0.0001
9	PhenMeOHCHO electron-releasing	−1.8497	0.8391	17	0.916	0.196	0.0001
10	PhenMeOHCHO electron-withdrawing	−1.0213	0.6635	10	0.826	0.186	0.0003
11	$PyrCNNO_2$	−1.7889	0.8222	19	0.937	0.155	0.0001

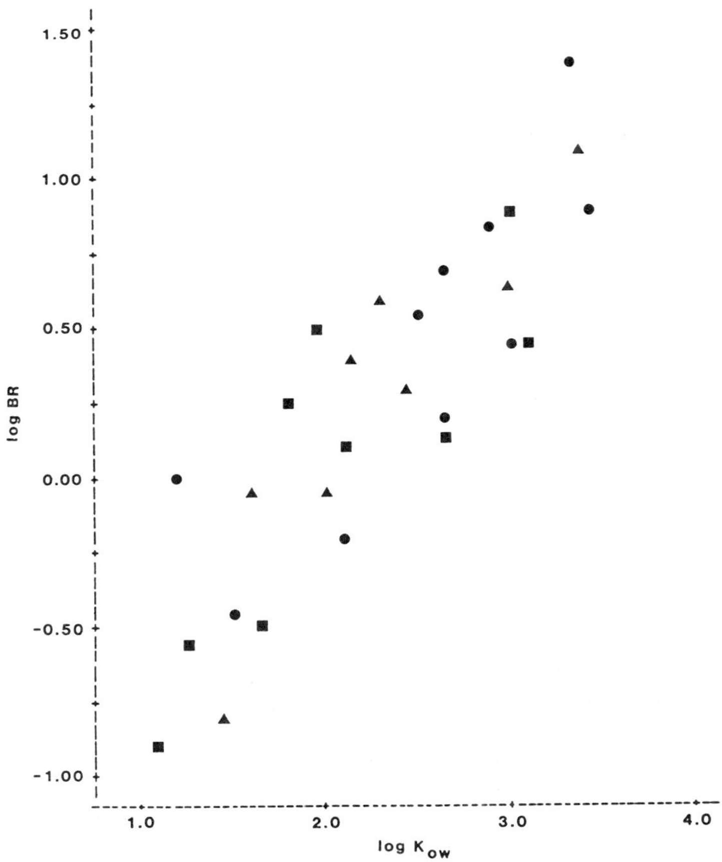

FIG. 1—*Scatter plot of log BR versus log* K_{ow} *for phenols (circles), benzyl alcohols (squares), and benzaldehydes (triangles).*

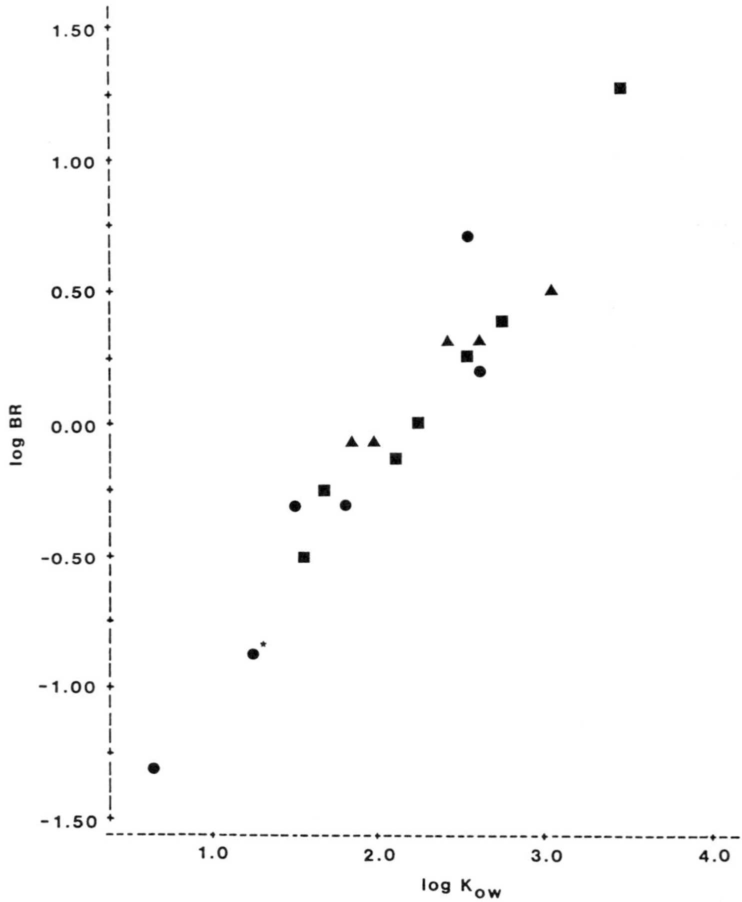

FIG. 2—*Scatter plot of log BR versus log* K_{ow} *for pyridines (circles), cyanobenzenes (squares), and nitrobenzenes (triangles).*

Regression analysis of these combined data results in Eq 11 (Table 3). An analysis of residual values based on Eq 11 shows them not to be distributed significantly different from normal ($W = 0.959$; $P < W = 0.545$). A comparison of observed toxicity with the 95% confidence intervals reveals no outliers.

In an effort to explore the possibility that the addition of an orthogonal electronic parameter would improve the predictability of Eq 8, the three electronic substituent constants were evaluated as second independent variables by multiple regression analysis. The regression of log BR versus log K_{ow} and F, σ_p, or R for the combined phenol, benzyl alcohol, and benzaldehyde results in Eqs 12, 13, and 14, respectively (Table 4). Based on the r^2 values, F is the best electronic descriptor. An analysis of residual values based on Eq 12 shows them not to be distributed significantly different from normal ($W = 0.957$; $P < W = 0.363$). A comparison of observed toxicity with the 95% confidence limits of Eq 12 reveals no outliers.

TABLE 4—*Log* K_{ow} *and electronic parameter-dependent QSAR for PhenMeOHCHO.*

Equation	Intercept	K_{ow}	Second Term	n	r^2	Root Mean Square Error	$Pr > f$ for K_{ow}	$Pr > f$ for Second Term
12	−1.7457	0.7891	0.6793 F	27	0.925	0.163	0.0001	0.0001
13	−1.5187	0.7736	1.3844 σ_p	27	0.895	0.193	0.0001	0.0001
14	−1.3936	0.6986	0.1390 R	27	0.756	0.294	0.0001	0.6865
15	−2.1941	0.7952	0.4528 G	27	0.892	0.196	0.0001	0.0001

With the exception of phenyl, where $F = 0.25$, the F values in Table 2 fall into two distinct clusters, the alkyls with values of about 0.00 and the halogens with values of about 0.70. These clusters mimic the electronic groups (Table 2). Following the assigning of $G = 1$ to the electron-releasing compounds and $G = 2$ to the electron-withdrawing compounds, the data were analyzed using the discrete variable group, G, along with log K_{ow} as molecular descriptors. This regression analysis resulted in Eq 15 (Table 4). A comparison of observed toxicity with the 95% confidence limits of Eq 15 reveals 4-phenylphenol to be an outlier.

Data from the chemical persistence studies are summarized in Table 5. While the phenols, benzyl alcohols, and pyridines show minimal abiotic loss, the cyanobenzenes, nitrobenzenes, and benzaldehydes reveal significant abiotic loss, especially for the lower molecular weight derivatives.

Discussion

In the present study, the toxicity of all the pyridine, cyanobenzene, and nitrobenzene derivatives can be modeled by a single log K_{ow}-dependent QSAR. However, the toxicity of the phenol, benzyl alcohol, and benzaldehyde derivatives is not modeled by the same QSAR nor any other simple log K_{ow}-dependent QSAR. A second descriptor is required along with log K_{ow} to accurately model the toxic response of these chemicals.

The substituents used in this study, that is hydrocarbons and halogens, are all non-H-polar; they neither accept or donate hydrogens [*10*]. However, the tested substituents do exhibit varied electronic effects, for example, Hammett sigma constants. Positive sigma values indicate electron withdrawal by the substituent, while negative values of sigma represent electron donating. Alkyl and phenyl substituents are therefore electron releasing, while halogen substituents are electron withdrawing. Sigma constants are also position dependent, with the value for a given substituent being different for the meta- and para-position [*10*]. Ortho-position electronic effects are difficult to quantitate because of overlap and steric hindrance [*13*]. However, recent efforts in our laboratory suggest that at least in the case of phenols, both σ_p and F are potential estimators of ortho-electronic effects [*14*].

Problems have arisen in using σ values for para-substituted compounds because such electronic effects include both a resonance and field component. Swain and Lupton [*11*] have attempted to segregate these effects. The quantitation of these effects were improved by Swain and others [*12*]. Since all the derivatives tested here were para-oriented, such a separation of effects is feasible. As noted by Carey and Sundberg [*15*], all the substituents examined in this study express electron-releasing resonance effects. While the alkyl groups have electron-releasing field effects, the halogens and phenyl groups have electron-withdrawing field effects. This probably accounts for the improved predictability of toxicity

TABLE 5—*Abiotic loss of selected monoaromatics.*

Compound	Percent Loss ± S_x[a]
Pyridine	11.5 ± 5.4
4-Methyl	15.7 ± 1.9
4-Ethyl	12.8 ± 4.1
4-*(tert)*butyl	31.8 ± 2.8
4-Phenyl	0
4-Chloro	14.9 ± 0.9
4-Bromo	9.0 ± 1.1
Cyanobenzene	26.6 ± 1.7
4-Methyl	36.6 ± 3.9
4-Phenyl	0
4-Fluoro	29.1 ± 7.7
4-Chloro	29.2 ± 5.5
4-Bromo	16.0 ± 6.7
Nitrobenzene	NT[b]
4-Methyl	29.2 ± 5.2
4-Ethyl	36.6 ± 2.0
4-Phenyl	NT
4-Fluoro	29.7 ± 1.4
4-Chloro	25.7 ± 5.9
4-Bromo	15.4 ± 6.7
4-Iodo	NT
Phenol	1.0 ± 0.9
4-Methyl	1.9 ± 0.6
4-Ethyl	2.4 ± 0.7
4-Isopropyl	2.6 ± 0.3
4-*(tert)*butyl	0
4-Phenyl	0
4-Fluoro	1.5 ± 0.9
4-Chloro	0
4-Bromo	1.5 ± 0.6
4-Iodo	0.1 ± 0.1
Benzyl alcohol	2.5 ± 3.5
4-Methyl	0.4 ± 0.6
4-Ethyl	0
4-Isopropyl	1.3 ± 1.8
4-*(tert)*butyl	10.0 ± 1.6
4-Phenyl	0
4-Fluoro	3.0 ± 0.9
4-Chloro	0.1 ± 0.1
4-Bromo	8.0 ± 1.1
Benzaldehyde	33.1 ± 3.6
4-Methyl	24.2 ± 2.5
4-Ethyl	33.2 ± 2.7
4-Isopropyl	45.6 ± 4.3
4-Phenyl	0
4-Fluoro	27.0 ± 1.7
4-Chloro	28.3 ± 2.0
4-Bromo	20.4 ± 1.2

[a] Standard deviation.
[b] NT = not tested.

observed when F is used as an additional independent variable for the phenols, benzyl alcohols, and benzaldehydes (see Eq 12). The strong colinearity between F and σ_p (r^2 = 0.958) accounts for σ_p being almost as good a molecular descriptor as F (see Table 4). Recent investigations in this laboratory [*5*] with a series of 34 diverse, four-position derivatives of phenol has lent further support to this idea of a need for a second descriptor, an electronic parameter, to model the toxicity of certain types of aromatic compounds.

The efforts to combine Eqs 9 and 10 into a single model, Eq 15, by using the discrete variable G as a selector did not improve the predictability. Therefore, Eq 12 is considered the better model.

Biological effects resulting from the exposure of an animal to chemicals can be grouped into three major cause-effect relationships. The most common effects are nonspecific actions on biological membranes, resulting in behavioral alterations including hyperexcitability and narcosis. Such nonspecific actions, if allowed to persist, will eventually result in death. However, removal of the organism from the chemical-containing environment prior to death results in rapid recovery of normal function. Such reversible toxicity depends upon the physical properties of the toxicant rather than their chemical properties. This physical toxicity as described in several papers [*16–20*] occurs when the concentration in the biophase reaches a finite level. This level is relatively constant for all species when compared on a per unit lipid basis. It is this idea of a constant biophase concentration for similar organisms and modes of action that has lead to the recent successful use of QSARs in aquatic toxicology to predict the relative toxicity of a wide variety of chemicals, based solely on an index of their hydrophobicity (log K_{ow}). Such QSARs, of course, have as their roots the classic Meyer-Overton observations (for a review see Ref *21*).

The ability to predict biological activity such as relative toxicity, based on partitioning, that is log K_{ow}, provides a quick and inexpensive means of prescreening organic toxicants, assuming the compound in question has the same mode of toxic action as the model. Studies over the past several years, both with the acute, lethal, flow-through Fathead minnow assay system [*22*] and the chronic, sublethal, static *Tetrahymena* assay system [*6*] suggest that reversible, physical toxicity may not be a single mechanism of action defined by a single group of symptoms or signs and thus able to be predicted by a single log K_{ow}-dependent relationship [*1–3,23*]. Instead, reversible physical toxicity appears to be best modeled by a series of log K_{ow}-dependent linear equations. These equations are nonparallel and tend to converge at a log K_{ow} value limited by aquatic solubility. To date, at least three modes of toxic action have been identified and modeled by QSARs in both systems: (1) Narcosis I [*1*] or general membrane perturbation [*2*], (2) polar narcosis or Narcosis II [*3,23*], and (3) uncoupling of oxidative phosphorylation [*23*].

The Narcosis I QSAR predicts the biological activity of nonionic, nonreactive chemicals which are not metabolized or at least metabolized very slowly. This activity is directly related to hydrophobicity and is linearly related to log K_{ow} so long as equilibrium has been attained. If the experimental protocol (for example, length of exposure, size of test organism, and so forth) is such that high log K_{ow} compounds do not attain equilibrium, the relationship takes on a curvilinear form as observed by Veith et al. [*1*]. Equation 11 is the Narcosis I QSAR in the *Tetrahymena* assay.

The Narcosis II QSAR predicts the activity of more polar compounds [*16*]. As suggested by Veith et al. [*3*] and demonstrated by Purdy [*24*] and Schultz et al. [*5*], the addition of an orthogonal electronic term enhances the predictability of log K_{ow}-dependent QSAR for Narcosis II. Similarly, the addition of the ionization term pK_a to the log K_{ow}-dependent Narcosis II QSAR improves the predictability [*25–27*]. Equation 12 is a Narcosis II QSAR in the *Tetrahymena* assay.

The uncoupling QSAR [*23*] is the least explored and newest of the three. It models the activity of, among others, dinitro-substituted and polyhalogenated polar compounds, for example, 2,4-dinitrophenol and pentachlorophenol.

McKim and co-workers have shown that a series of specific behavioral, biochemical, hematological, and physiological response sets or a fish acute toxic syndrome (FATS) can all be used with stepwise discriminant function multivariant analysis to discriminate between modes of toxic action [*28*]. They have defined four modes of action including anesthetics (Narcosis I), uncoupling of oxidative phosphorylation, acetylcholine inhibition, and respiratory irritation [*28,29*]. In similar fashion, two other mechanisms of toxic action, respiratory blocking and convulsants, have been defined [*30*]. Veith and Broderius [*31*] in their recent study of polar narcotics have shown in joint toxicity studies that binary mixtures of chemicals with the same modes of action are concentration additive, while binary mixtures of chemicals with dissimilar modes of action are nonadditive. These studies support the premise that a separate QSAR is required to predict activity for each mode of toxic action. In contrast to this, the studies of Saarikoski and Viluksela [*25*], Schultz and Riggin [*4*], and Lipnick [*21*] show that, at least for phenols and the mechanisms of Narcosis II and uncoupling, a single QSAR can be used to model toxicity. This observation is due to the fact that the log K_{ow}-dependent QSAR for polar narcotics and uncouplers intersect at a log K_{ow} value of about 5 [*23*]. Therefore, the toxicity of chemicals such as tetrachlorophenols and pentachlorophenol is modeled well by either QSAR.

The results reported here must be tempered with the fact that the cyanobenzenes, nitrobenzenes, and benzaldehydes show significant abiotic loss with time. The impact of such loss, however, is difficult to evaluate and may vary from one chemical class to another or even from one derivative to another. The chemical persistence test protocol by design maximizes abiotic loss. These experiments are conducted in sterile water rather than more hydrophobic cell-containing medium. The latter retards volatilization, the primary mechanism of abiotic loss of these chemicals. Based on our current knowledge of mechanisms of toxic action, the lack of persistance of cyanobenzenes and nitrobenzenes in the present studies may be of little or no consequence. However, in light of the recent work on FATS [*28–30*], the abiotic loss of the benzaldehydes cannot be ignored. Therefore, the inclusion of the benzaldehydes in the Narcosis II QSAR is at present questionable. The presence of the reactive carbonyl group may cause the benzaldehydes to elicit their toxic response via the membrane irritation mode of action [*29*].

Acknowledgment

This work was supported in part by the U.S. Environmental Protection Agency (EPA) through Grants R-810791-01-0 and R-813190-01-0. (The analyses and conclusions herein, however, are those of the authors and do not necessarily reflect views of the EPA, hence, no EPA endorsement should be inferred.)

M. Chang was supported in part by the Participant National Institutes of Health (NIH) Minority High School Student Research Apprentice Program NIH-RR03326-03, and D. T. Lin was supported in part by the NIH Minority High School Student Research Apprentice Program NIH-RR03326-05.

References

[*1*] Veith, G. D., Call, D. J., and Brooke, L. T., *Canadian Journal of Fish Aquatic Science,* Vol. 40, 1983, pp. 743–784.

[*2*] Schultz, T. W. and Moulton, B. A. in *QSAR in Environmental Toxicology,* K. L. E. Kaiser, Ed., D. Reidel Publishing Co., Dordrecht, Holland, 1983, pp. 337–352.

[3] Veith, G. D., DeFoe, D., and Knuth, M., *Drug Metabolism Review,* Vol. 15, 1985, pp. 1295–1303.
[4] Schultz, T. W. and Riggin, G. W., *Toxicological Letters,* Vol. 25, pp. 47–54.
[5] Schultz, T. W., Riggin, G. W., and Wesley, S. J. in *QSAR in Environmental Toxicology—II,* K. L. E. Kaiser, Ed., D. Reidel Publishing Co., Dordrecht, Holland, 1987, pp. 335–345.
[6] Schultz, T. W. in *Aquatic Toxicology,* J. O. Nriagu, Ed., Wiley, New York, 1983, pp. 579–612.
[7] Schultz, T. W. and Cajina-Quezada, M., *Archives of Environmental Contamination and Toxicology,* Vol. 11, 1982, pp. 353–361.
[8] Leo, A. and Weininger, D., "CLOGP Version 3.2: Estimation of the *n*-Octanol/Water Partition Coefficient for Organics in the TSCA Industrial Inventory," Pomona Medicinal Chemistry Project, Pomona College, Claremont, CA, 1984.
[9] Leo, A., *Environmental Health Perspective,* Vol. 61, 1985, pp. 275–285.
[10] Hansch, C. and Leo, A., *Substituent Constants for Correlation Analysis in Chemistry and Biology,* Wiley, New York, p. 339.
[11] Swain, C. G. and Lupton, E. C., Jr., *Journal of American Chemical Society,* Vol. 90, 1968, pp. 4328–4337.
[12] Swain, C. G., Unger, S. H., Rosenquist, N. R., and Swain, M. S., *Journal of American Chemical Society,* Vol. 105, 1983, pp. 492–502.
[13] Fujita, T. and Nishioka, T. in *Progress in Physics of Organic Chemistry,* Wiley Interscience, New York, 1976, pp. 49–119.
[14] Schultz, T. W. and Cajina-Quezada, M., *Toxicological Letters,* Vol. 37, 1987, pp. 121–130.
[15] Carey, F. A. and Sundberg, R. J., *Advanced Organic Chemistry Part A: Structure and Mechanisms,* Plenum Press, New York, 1977, p. 583.
[16] Ferguson, J., *Proceedings of the Royal Society of London, (B),* Vol. 127, 1939, pp. 387–404.
[17] Ferguson, J. and Piri, H., *Annals in Applied Biology,* Vol. 35, 1948, pp. 532–550.
[18] McGowan, J., *Journal of Applied Chemistry,* Vol. 1, 1951, pp. 120–126.
[19] McGowan, J., *Journal of Applied Chemistry,* Vol. 2, 1952, pp. 323–328.
[20] McGowan, J., *Journal of Applied Chemistry,* Vol. 2, 1952, pp. 651–658.
[21] Lipnick, R. L., *Trends in Pharmacological Sciences,* Vol. 7, 1986, pp. 161–164.
[22] U.S. Environmental Protection Agency (USEPA) Committee on Methods for Toxicity Tests with Aquatic Organisms, EPA 660/3-75-009, Duluth, Minnesota, 1975.
[23] Schultz, T. W., Holcombe, G. W., and Phipps, G. L., *Ecotoxicology and Environmental Safety,* Vol. 12, 1986, pp. 146–153.
[24] Purdy, R. in *QSAR in Environmental Toxicology—II,* K. L. E. Kaiser, Ed., D. Reidel Publishing Co., Dordrecht, Holland, 1987.
[25] Saarikoski, J. and Viluksela, M., *Ecotoxicology Environmental Safety,* Vol. 4, 1982, pp. 501–512.
[26] Lipnick, R. L., Bickings, C. K., Johnson, D. E., and Eastmond, D. A., "Comparison of QSAR Predictions with Fish Toxicity Screening Data for 110 Phenols," *Aquatic Toxicology and Hazard Assessment (Eighth Symposium), STP 891,* American Society for Testing and Materials, Philadelphia, 1986.
[27] Schultz, T. W., *Bulletin of Environmental Contamination Toxicology,* Vol. 38, 1987, pp. 994–999.
[28] McKim, J. M., Schmeider, P. K., Carlson, R. W., Hunt, E. P., and Niemi, G. J., *Environmental Toxicology and Chemistry,* Vol. 6, 1987, pp. 295–312.
[29] McKim, J. M., Schmeider, P. K., Niemi, G. J., Carlson, R. W., and Henry, T. R., *Environmental Toxicology and Chemistry,* Vol. 6, 1987, pp. 313–329.
[30] McKim, J. M., Bradbury, S. P., and Niemi, G. J., *Environmental Health Perspective,* Vol. 1987, pp. 171–186.
[31] Veith, G. D. and Broderius, S. in *QSAR in Environmental Toxicology—II,* K. L. E. Kaiser, Ed., D. Reidel Publishing Co., Dordrecht, Holland, 1987, pp. 385–391.

Klaus L. E. Kaiser[1] *and Kathleen M. Gough*[1]

Predictability of Unusually High Acute Toxicity to *Photobacterium phosphoreum* of 1,4-Di-Substituted Benzene Derivatives

REFERENCE: Kaiser, K. L. E. and Gough, K. M., **"Predictability of Unusually High Acute Toxicity to *Photobacterium phosphoreum* of 1,4-Di-Substituted Benzene Derivatives,"** *Aquatic Toxicology and Environmental Fate: Eleventh Volume, ASTM STP 1007,* G. W. Suter II and M. A. Lewis, Eds., American Society for Testing and Materials, Philadelphia, 1989, pp. 424–441.

ABSTRACT: The authors report here on the relationships of the acute toxicities to *Photobacterium phosphoreum* of 1,4-di-substituted benzenes with their octanol/water partition coefficients and with the toxicities of the corresponding mono-substituted benzene derivatives and also identify the types of derivatives for which such toxicities can or cannot safely be predicted. "Outliers" to the previously established quantitative structure-activity relationships (QSARs) are identified, and reasons for their aberrant behavior are discussed in terms of specific substituent groups and their electronic effects on the basic molecule and on the other substituents present, as measured by the substituent-induced chemical shifts of carbon-13 (^{13}C) nuclear magnetic resonance data of the substituent-bearing carbon atoms of the benzene ring. With the combination of these parameters, the toxicity of highly toxic compounds can be predicted within an error of approximately 0.35 logarithmic toxicity units over more than four orders of magnitude. Nitrobenzene derivatives require the use of a parameter for the charge buildup on the oxygen atoms of the nitro group.

KEY WORDS: *Photobacterium,* Microtox, nitrobenzenes, anilines, phenols, benzonitriles, fluorobenzenes, chlorobenzenes, bromobenzenes, iodobenzenes, trifluoromethylbenzenes, *N,N*-dimethylanilines, toxicity, chemical shifts, molar refractivity, octanol/water partition coefficient, structure-activity relationships, quantitative structure-activity relationships, carbon-13 nuclear magnetic resonance, oxygen-17 nuclear magnetic resonance, aquatic toxicology

Quantitative structure-activity relationships (QSARs) of physicochemical characteristics with biological effects of organic chemicals are useful for the prediction of such effects for compounds and conditions that are not readily accessible to measurement. In particular, QSARs are evolving as an important tool for the screening of industrial chemicals for potentially hazardous or undesirable environmental properties and effects [*1*]. Although many compounds are amenable to such quantitative predictions, it is not unusual to find outliers whose effects, when measured, far exceed the values estimated from QSAR analysis. (Typically, an outlier is defined as any compound for which the experimentally measured acute toxicity differs from that calculated by the QSAR by more than two standard errors; in the present case, we estimate that the practical limits to our predictions should be on the order of 0.3 to 0.4 log units; thus a difference of greater than about 0.7 log units

[1] Project chief and research scientist, respectively, Lakes Research Branch, National Water Research Institute, Burlington, Ontario L7R 4A6, Canada.

is regarded as an outlier.) This "aberrant" behavior is usually explained—*a posteriori*—in terms of specific compound-active site interactions which are different from the normal interactions for which such QSAR models are valid. Until now, no mechanism has evolved to predict—*a priori*—which types of compounds are likely to exhibit effects outside these normal structure-activity relationships.

In recent publications [*2,3*] we have reported the acute toxicity of a large number of mono- and di-substituted benzene derivatives to *Photobacterium phosphoreum* (the Microtox test). Satisfactory QSARs were derived, but in both cases it was necessary to break down the original general group into subgroups, according to chemical similarities among the substituents, in order to achieve a reasonable degree of accuracy in predicting toxicity. For the 133 1,4-di-substituted benzene derivatives [*3*], a general dependence of acute toxicity on the octanol/water partition coefficient, log *P*, was found, but the standard error (SE) of the estimate was 0.75, which is too large to be of practical use. Consequently, this set was divided into subsets (1-R-C_6H_4-4-X) for which R was a constant (R = NH_2, NO_2, OH, Cl, and $N(CH_3)_2$), but a satisfactory correlation between the acute toxicities of the di-substituted and the mono-substituted derivatives (C_6H_5-X) was found only for the series with R = Cl. After the elimination of 27 outliers from the general group, a highly significant correlation (SE = 0.41) was found between the acute toxicities of the mono- and di-substituted derivatives. Inclusion of the octanol/water partition coefficient reduced the SE to 0.34, which is a very acceptable value and is probably near the practical limit achievable for this type of data.

The problem of explaining and predicting the acute toxicities of the outliers remains, although the initial indications are that they occur (*a*) for compound-specific toxicant-receptor interactions [*4–8*] and (*b*) for R,X pairs that can interact strongly (that is, those with a large value of sigma$_p$, the Hammett sigma [*9*], or the field (*F*) and resonance (*R*) parameters of Swain and Lupton [*10*]. However, there is no overall consistency between these latter parameters and the occurrence of unexpectedly high acute toxicity. Therefore, additional parameters have been sought for this purpose.

The chemical shifts in the nuclear magnetic resonance (NMR) spectra of magnetic nuclei are influenced by molecular substituents, since the latter can alter the electron density in the vicinity of the magnetic nuclei and thus affect the amount of shielding experienced by these nuclei [*11*]. It is well known that it is possible to correlate the substituent-induced chemical shift (SCS) in the carbon-13 NMR (^{13}C-NMR) spectrum of substituted benzene rings with sigma$_I$ and sigma$_R$, the inductive and resonance parts of the Hammett sigma [*12*]. It is also well known that large nonadditive effects are observed for 1,4-di-substituted derivatives with strongly interacting groups [*13*]. The SCS values are, therefore, a direct experimental probe of the electronic interaction between substituents in 1,4-di-substituted benzene derivatives, and may be of use in the correlation and prediction of the acute toxicities of these compounds. We have now performed such an investigation for both mono- and 1,4-di-substituted benzene derivatives, and the results are presented here.

Materials and Methods

General

Most of the toxicity data used in these correlations has been published elsewhere [*2,3*], with the exception of some derivatives of the general formula 1-R-C_6H_4-4-X, where R = CN, CF_3, F, Br, and I. The test chemicals were purchased in the best grade available from Aldrich Chemical Co., Inc., or Fluka Chemical Corp. and were used without further purification, except where noted. The acute toxicities to *Photobacterium phosphoreum* were determined with the Microtox toxicity analyzer, following the procedure described previously. In some cases, up to 5% methanol was used to increase substrate solubility, since this concentration does not affect the toxicity determination [*14*].

Experimental Values

All toxicity values reported here are the negative logarithms to base ten ("p") values of the millimolar concentrations at which a 50% light reduction (gamma = 1) was observed on 30-min exposure. Each value is the mean of at least three independent determinations, usually performed with different bacterial suspensions to reduce any systematic variations or biases. The standard deviations of such triplicate analyses were normally in the order of 0.05 (logarithmic) toxicity units, and as high as 0.10 units for a few compounds of either high toxicity or sensitivity to light, oxygen, or water or for those of very low solubility, where extrapolation to gamma = 1 was necessary from lower gamma values.

Parameters

Octanol/water partition coefficients, log *P*, were taken from the compilation by Hansch and Leo [*15*]. Where no experimental values were reported, log *P* was calculated from pi or fragment values and related compounds. If neither was available, log *P* was estimated on the basis of the partition coefficients of similar compounds or groups [*15*].

SCS values for C-1 and C-4 of mono- and di-substituted benzene derivatives were obtained from the literature for as many of the molecules studied in Refs *2* and *3* as possible. The values reported here, d1 and d4, give the ^{13}C shifts at C-1 and C-4, respectively, relative to the ^{13}C shift of benzene. In the regression analyses, these parameters were included independently, as well as their difference, d1 − d4, and the absolute value of their difference, |d1 − d4|. For some molecules, SCS values were reported by several authors, and most values agree to within 1%. Larger discrepancies were typically due to solvent or concentration effects. In these cases, the SCS measured in the most inert solvent, at the most dilute concentration, was used in the regression analysis.

The substituents' molar refractivity contributions, dMR_X, were also considered as parameters [*15*]. This parameter is commonly used in QSAR analysis as a measure of the molar volume of the substituent and, to a lesser degree, its hydrophobicity.

A detailed study of the relationship between the substituent-induced chemical shift in the oxygen-17 (^{17}O) NMR spectra of 1,4-di-substituted nitrobenzenes and the change in the electron population at that oxygen atom, as determined from *ab initio* molecular orbital calculations at the STO-3G level, has recently been published [*16*]. The absolute value of the latter values, |dQox|, were also used as parameters for the nitrobenzene subset.

Table 1 lists the additional di-substituted chemicals investigated for which SCS data were available: their acute toxicity values, $pTm_{R,X}$, and those of the corresponding mono-substituted compounds, $pTm_{H,X}$ [*2*], and their log *P* values are also listed. Table 2 [Refs *13,17–34*] gives the substituent-induced chemical shift data at C-1 (d1) and C-4 (d4), the dMR values, and the |dQox| values, where available. There are 124, in total, 1,4-di-substituted benzene derivatives and 38 mono-substituted derivatives. For the combined set of 124 di-substituted derivatives, there are 30 duplicate entries; that is, the compounds appear in two subsets. All of the 1-R-C_6H_4-4-H compounds also appear in the mono-substituted series. For the combined halogen subset of 60 derivatives, there are 8 duplicate entries.

TABLE 1—*Measured toxicity values,* $pTm_{R,X}$*, of 1,4-di-substituted benzene derivatives of the general formula 1-R-*C_6H_4*-4-X to* Photobacterium phosphoreum*; their Chemical Abstract Service (CAS) numbers: the octanol/water partition coefficients, log* P*; and the toxicity values,* $pTm_{H,X}$*, of the corresponding mono-substituted benzene derivatives of the general formula* C_6H_5*-X* [2].

X	CAS No.	$pTm_{R,X}$ [a]	$pTm_{H,X}$	log *P*
R = CF_3				
Br	402-43-7	1.55	1.22	3.87
$COCH_3$	728-86-9	1.40	0.89	2.46
CN	455-18-5	1.33	0.95	2.22
F	402-44-8	0.30	−0.28	3.15
R = CN				
C_6H_5	2920-38-9	2.99	1.91	3.38 [b]
CH_3	104-85-8	1.41	0.60	2.12
OCH_3	874-90-8	1.35	0.76	1.54 [b]
CF_3	455-18-5	1.33	0.66	2.22
Br	623-00-7	1.32	1.22	2.42
CHO	105-07-7	1.02	1.34	0.91
$COCH_3$	1443-80-7	0.45	0.89	1.01
F	1194-02-1	0.30	−0.28	1.70 [b]
CN	623-26-7	0.26	0.95	0.99 [b]
R = I				
OCH_3	696-62-8	2.27	0.86	3.06 [b]
COOH	619-58-9	0.45	0.86	3.23 [b]
R = Br				
Br	106-37-6	1.96	1.22	3.85
OCH_3	104-92-7	1.86	0.76	2.97
$COCH_3$	99-90-1	1.52	0.89	2.44
F	460-00-4	0.78	−0.28	3.13
R = F				
NCS	1544-68-9	2.28	1.81	3.42
COC_6H_5	345-83-5	1.57	1.31	3.32
Br	460-00-4	0.78	1.22	3.13
OCH_3	459-60-9	0.76	0.76	2.25
CH_3	352-32-9	0.46	0.60	2.88
$COCH_3$	403-32-9	0.42	0.89	1.87
CHO	459-57-4	0.21	1.34	1.62
F	540-36-3	0.04	−0.22	2.41
$CONH_2$	640-17-7	−0.31	0.31	0.91

[a] The pTm values are the negative logarithm of the effective toxicant concentrations (in millimoles per litre) which produce a 50% light reduction in 30 min.
[b] Indicates derived log *P* values; for details, see Refs *2* and *3*.

Computations

All toxicity value computations from the concentration/light reduction plots were done with the COMPUTOX program on an HP 86 computer. The statistical calculations were performed on the same computer with prerecorded linear and multiple regression analysis programs. Although there are duplicate entries, duplicate enumeration occurs only for the log *P* values. The significance of the regressions was ascertained from the *F*-values calculated with each regression. This test incorporates the number of independent variables as well as the size of the data set.

TABLE 2—*Substituent-induced chemical shifts on C-1 (d1) and C-4 (d4) of mono- and di-substituted benzene derivatives of the general formula 1-R-C_6H_4-4-X, with substituent molecular refractivity contribution (dMR_X) and, for nitro-substituted compounds, |dQox| (see text).*

X	d1	d4	dMR_X	|dQox|	Reference
R = NH_2					
CN	22.12	−28.5	6.33		*13,17*
CF_3	21.04	−8.3	5.02		*13,18*
NO_2	24.12	10.82	7.36		*13*
$COCH_3$	22.77	−6.3	11.18		*13*
Cl	16.55	−5.19	6.03		*13*
C_6H_5	19.69	1.14	25.36		*19*
CH_3	15.39	−0.58	6.65		*13*
COC_6H_5	23.09	−1.21	30.33		*20*
OCH_3	11.56	24.38	7.87		*13*
NH_2	10.24	10.24	5.42		*13*
F	14.05	27.97	0.92		*13*
H	17.91	−9.8	1.03		*13*
R = OH					
OH	20.93	20.93	2.85		*21*
I	25.63	−45.75	13.94		*21*
CF_3	30.32	−5.57	5.02		*21*
CN	31.58	−24.98	6.33		*21*
C_6H_5	29.73	4.03	25.36		*19*
CH_3	21.47	2.29	5.65		*17,21,22*
$COCH_3$	32.29	2.02	11.18		*21*
COC_6H_5	32.69	0.99	30.33		*20*
Cl	25.63	−2.94	6.03		*21*
NO_2	33.43	13.24	7.36		*21*
CHO	32.89	2.02	6.88		*21*
F	24.17	28.18	0.92		*21*
H	27.50	−7.75	1.03		*21*
$NHCOCH_3$	24.87	2.76	14.93		*23*
R = NO_2					
NO_2	22.72	22.72	7.36	108.5	*13*
NH_2	10.82	24.12	5.42	81.7 (average)	*13*
CN	21.70	−10.02	6.33	73.3	*13,17*
I	18.99	−24.18	13.94		*24*
COC_6H_5	24.59	14.59	30.33		*20*
$N(CH_3)_2$	8.85	25.70	15.55		*13*
CHO	22.82	12.08	6.88	34.3	*13*
$OCOCH_3$	16.94	26.99	12.47		*29*
OH	13.24	33.43	2.85	50.7	*21*
$COOCH_3$	12.08	22.82	12.87		*13*
CF_3	21.65	7.60	5.02	43.5	*13,18*
Br	18.74	1.65	8.88		*13,18,25*
CH_3	17.81	17.58	5.65	15	*13,18,22*
COOH	21.59	7.79	6.96		*28*
OCH_3	13.19	36.26	7.87		*13*
Cl	18.13	13.00	6.03		*13*
$COCH_3$	22.04	13.06	11.88		*13*
H	19.95	6.18	1.03	0	*13*
$CONH_2$	20.72	11.65	9.81		*29*
F	16.11	37.89	0.92	9.5	*13,18*
R = Cl					
NCS	4.60	1.68	17.24		*30*
COC_6H_5	10.49	1.89	30.33		*18,20*
I	5.87	−37.15	13.94		*25*

TABLE 2—*(Continued)*.

X	d1	d4	dMR_X	|dQox|	Reference
$CH{=}CH_2$	5.22	7.38	10.99		*31*
OCH_3	−2.81	29.85	7.87		*13,18*
CN	11.17	−17.55	6.33		*13,17,18*
Br	4.86	−7.28	8.88		*13,18*
Cl	4.22	4.22	6.03		*13,18*
NH_2	−5.19	16.55	5.42		*13*
COOH	9.59	0.89	6.96		*28*
$COCH_3$	11.07	6.99	11.18		*13*
CH_3	2.72	7.87	5.65		*13*
OH	−2.94	25.63	2.85		*21*
CHO	12.67	9.95	6.88		*13*
CF_3	9.81	0.77	5.02		*13,18*
H	5.93	−1.90	1.03		*13*
NO_2	13.00	18.13	7.36		*13*
$CONH_2$	7.71	4.66	9.81		*24,29*
$NHCOCH_3$	−1.72	10.03	14.93		*23*
F	0.78	32.91	0.92		*13,18*
R = $N(CH_3)_2$					
CN	24.07	−31.07	6.33		*13,17*
NO_2	25.70	8.85	7.36		*13*
H	22.28	−11.69	1.03		*13,18*
R = CF_3					
OH	−5.57	24.68	2.85		*21*
NH_2	−8.30	21.04	5.42		*13,18*
Br	1.41	−1.91	8.88		*13,18*
$COCH_3$	6.02	11.31	11.88		*13,18*
CN	6.22	−12.27	6.33		*13,18*
NO_2	7.60	21.65	7.36		*13,18*
Cl	0.77	9.81	6.03		*13,18*
H	2.31	3.67	1.03		*13,18*
F	−1.86	36.30	0.92		*13,18*
R = CN					
C_6H_5	−17.38	17.01	25.36		*19*
$N(CH_3)_2$	−31.07	24.07	15.55		*13*
NH_2	−28.50	22.13	5.42		*13*
OH	−24.98	31.58	2.85		*21*
NO_2	−10.02	21.70	7.36		*13*
Cl	−17.55	11.17	6.03		*13,18,32*
CH_3	−19.10	15.30	5.65		*13,18,22*
OCH_3	−24.44	34.45	7.87		*13,18*
CF_3	−19.65	6.22	5.02		*13,18*
Br	−17.09	−0.36	8.88		*13,18*
CHO	−10.67	1.39	6.88		*13*
H	−15.96	4.35	1.03		*13,18*
$COCH_3$	−11.91	11.69	11.88		*17,13*
F	−19.77	36.65	0.92		*13,18*
CN	−11.65	−11.65	6.33		*13*
R = I					
OH	−45.25	26.09	2.85		*21*
OCH_3	−45.71	30.99	6.96		*24*
Cl	−37.15	5.87	6.03		*25*
H	−33.61	−1.01	1.03		*26,27*
NO_2	−24.01	18.99	7.36		*25*
COOH	2.0	−27.41	7.87		*28*

TABLE 2—*(Continued).*

X	d1	d4	dMR_X	|dQox|	Reference
R = Br					
Br	−7.28	−7.28	8.88		*13,18*
OCH_3	−15.54	30.32	7.87		*13,25*
CF_3	−1.91	1.41	5.02		*13,18*
$COCH_3$	−0.04	7.43	11.18		*13,25*
Cl	−7.28	4.86	6.03		*13,18*
CN	−0.36	−17.09	6.33		*13*
H	−5.87	−1.51	1.03		*13,18*
NO_2	1.65	18.74	7.36		*13,18*
F	−11.83	33.46	0.92		*13,18*
R = F					
NCS	32.75	−0.98	17.24		*30*
COC_6H_5	37.19	5.49	30.33		*20*
Br	33.46	−11.83	8.88		*13,18*
OCH_3	28.88	27.37	7.87		*13,18*
OH	28.68	24.67	2.85		*21*
CH_3	31.71	4.96	5.65		*13*
$COCH_3$	37.33	5.15	11.18		*13*
CF_3	36.30	−1.86	5.02		*13*
CN	36.65	−19.77	6.33		*13,18*
CHO	38.17	4.65	6.88		*13*
NH_2	27.97	14.05	5.42		*13*
NO_2	37.89	16.11	7.36		*13,18*
F	29.44	29.44	0.92		*13,18*
Cl	32.91	0.78	6.03		*13,18*
H	34.79	−4.51	1.03		*13,18*
$CONH_2$	35.55	2.37	9.81		*29*
R = H					
C_6H_4-*p*-CN	0.39	10.70	30.66		*20*
C_6H_5	−0.31	9.99	25.36		*24,27*
C_6H_4-*p*-CH_3	−1.31	12.69	29.98		*20*
C_6H_4-*p*-OH	−1.91	12.79	27.18		*20*
NCS	−1.05	2.91	17.24		*30*
I	−1.01	−33.61	13.94		*33*
OC_6H_5	−5.11	29.19	27.68		*27*
$COOCH_3$	1.99	4.49	12.87		*24*
C_6H_4-*p*-NH_2	−2.41	13.19	29.75		*21*
CHO	5.49	8.59	6.88		*13,32*
COC_6H_5	3.59	9.39	30.33		*20*
$CH{=}CH_2$	−0.81	9.09	10.99		*34*
Br	−1.50	−5.87	8.88		*13*
$OCOCH_3$	−2.31	22.99	12.47		*27*
COCl	6.19	4.59	10.44		*27,32*
CH_2CH_3	−2.81	15.89	10.30		*24,31*
Cl	−1.91	6.19	6.03		*13,32*
CN	4.49	−15.81	6.33		*13,17*
$N(CH_3)_2$	−11.81	22.59	15.55		*13,27*
$NHCH_3$	−11.61	21.89	10.33		*24*
$COCH_3$	4.19	8.69	11.18		*13,27,32*
COOH	5.39	2.09	6.96		*24,28,32*
CH_2NH_2	−1.81	15.09	9.09		*24*
NCO	−2.81	6.09	8.82		*32*
OCH_3	−7.71	31.39	7.87		*13,24,32*
CF_3	3.19	−9.01	5.02		*13,32*

TABLE 2—*(Continued).*

X	d1	d4	dMR_X	|dQox|	Reference
$\dot{C}SNH_2$	2.76	11.11	18.28		*29*
OH	−7.31	26.89	2.85		*21,32*
CH_3	−3.01	9.29	5.65		*13,17*
NO_2	5.79	19.99	7.36		*13,32*
$COCF_3$	6.69	−5.61	11.17		*32*
$CONH_2$	2.81	5.87	9.81		*29*
$NHNH_2$	−9.41	22.99	8.44		*24*
CH_2OH	−1.31	12.29	7.19		*27*
NH_2	−9.81	17.99	5.42		*32*
H	0.00	0.00	1.03		*13*
F	−4.51	34.79	0.92		*13,32*
$NHCOCH_3$	−5.61	11.09	14.93		*27*

Results

In previous studies on these benzene derivatives, the following conclusions were researched:

1. In general, the toxicity of the mono-substituted benzenes is a function of log *P*, while some outliers can be explained by the inclusion of other variables such as dMR_X; that is, they can be described by the general equation

$$pTm_{H,X} = a + b dMR_X + c \log P \tag{1}$$

2. After removal of the extreme outliers, the toxicity of the di-substituted benzene derivatives is a function of $pTm_{H,X}$ and log *P*; that is, it can be described by the general equation

$$pTm_{R,X} = a + b pTm_{H,X} + c \log P \tag{2}$$

Since the data set to be investigated here is smaller than those in Refs *2* and *3*, and since some new data are included, the dependence of $pTm_{R,X}$ on these parameters was determined initially (*a*) to set a minimum standard for the quality of the correlations here, (*b*) to identify any major differences with previous results, and (*c*) to identify the outliers. These results appear in Table 3, which also includes the correlation coefficient, r^2, the SE of the estimates, and the statistical *F*-value for each data set or subset

Additional descriptors, including the SCS values, were then tested in an attempt to obtain a better fit. In the case of the nitrobenzene derivatives, the substituent-induced change in the electron population at the oxygen atom of some nitrobenzene derivatives, as determined by *ab initio* molecular orbital calculations, was also included as a parameter where available. The relationships between $pTm_{R,X}$ of the nitrobenzenes and $pTm_{H,X}$ or |dQox| are shown in Figs. 1 and 2. It has already been observed [*3*] that the chlorobenzene derivatives correlate very well with $pTm_{H,X}$ alone. A larger set consisting of all compounds for which either R or X is a halogen was compiled from the data in Tables 1 and 2, and a regression analysis was performed. Because of a similarity between these results and those of the CF_3 subset, the latter data were included in the halogen subset, and the correlation

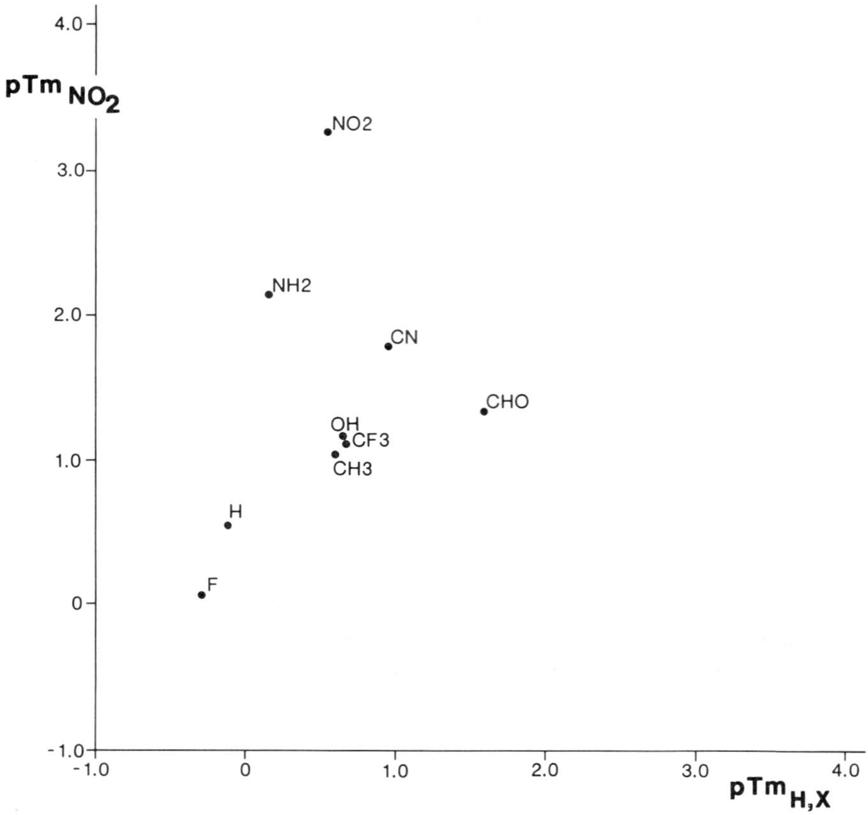

FIG. 1—*Toxicities* ($pTm_{R,X}$) *of nine* p-*nitro-substituted benzene derivatives versus the toxicities of the corresponding mono-substituted benzenes* ($pTm_{H,X}$).

was repeated. The best results from all of these correlations appear in Table 4, along with the statistical data, as in Table 3.

Discussion

The primary goal of the current investigation is to predict accurately the acute toxicities of compounds which have been found to be outliers in previous studies. However, as noted above, we must first establish a base correlation which will define the outliers.

Of the original 38 mono-substituted benzenes, 2 were not included in the fit. The first is X = *para*-C_6H_4-CN, which is really a 1,4-di-substituted benzene and is fit in the CN subset. The toxicity of the second (X = $NHCOCH_3$) is quite low and has been discussed previously. For the remaining 36 mono-substituted benzenes, the toxicity can be described by log *P* and dMR_X (dMR of substituent X) with a reasonable SE of 0.38 and only 2 outliers (Eq 1).

In the earlier work, the toxicities of the 1,4-di-substituted benzenes were found to correlate generally with $pTm_{H,X}$ and log *P*, Eq 2. From the results in Table 3, it can be seen that the data set studied here retains this relationship. There is less dependence on log *P*

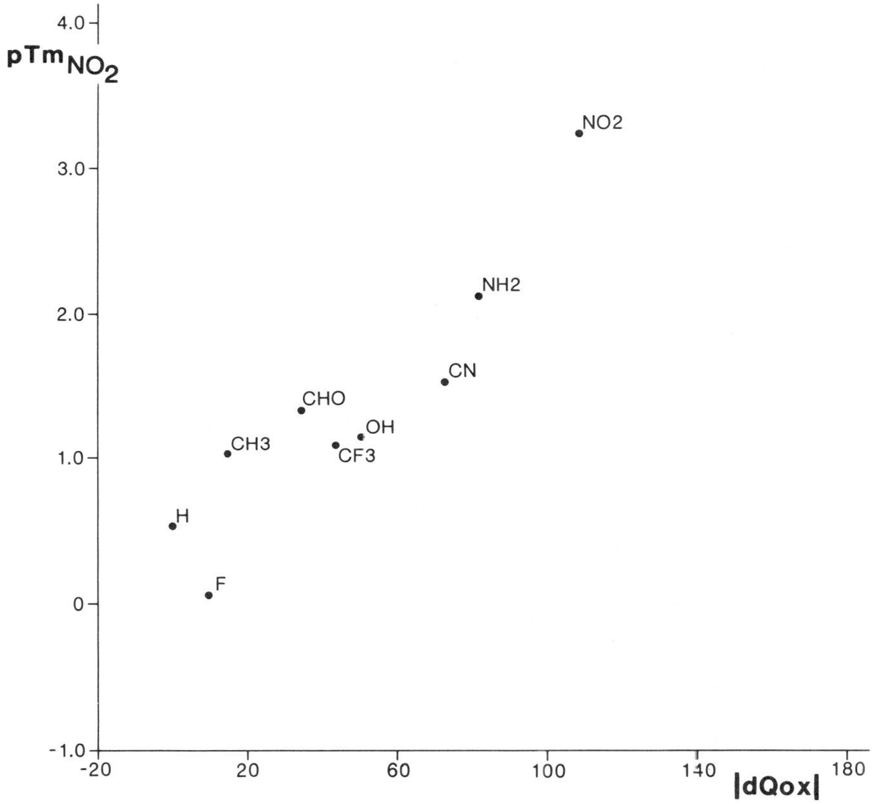

FIG. 2—*Toxicities* ($pTm_{R,X}$) *of nine* p-*nitro-substituted benzene derivatives versus the absolute values of the charge difference on the nitro group oxygen atoms* (|*dQox*|).

than was found for the original set of 133 points, or for that set minus outliers (106 points). It should be noted, however, that of the original 133 points, 40 have a log *P* value of greater than 2.5, and 18 of these are absent from the present set. The same is true of the new data for the CF_3, CN, and F subsets: SCS data are not available for many substituents with log $P > 2.5$. The reduced dependence of $pTm_{R,X}$ on log *P* is probably due to the elimination of a disproportionate number of data points with large log *P* values. Apart from this change, the results are similar for the two data sets. The most significant parameter is still $pTm_{H,X}$, and while the overall correlation is quite significant, the SE is 0.74, which is not acceptable for practical purposes.

We have proceeded intuitively with the supposition that unexpectedly high toxicities are associated with substituents that are capable of interacting strongly with each other through the benzene ring. The correlation between $pTm_{R,X}$ and the other variables in Table 2 was determined for each subset. In every case but that of the NO_2 subset, there is a significant dependence on d1, the substituent-induced chemical shift at C-1 of the benzene ring. There is also some dependence on $pTm_{H,X}$ and dMR_X in most cases. Because the parameter $pTm_{H,X}$ is itself dependent on dMR_X, it may be concluded that the latter is incorrectly weighted by the coefficient of $pTm_{H,X}$ and must be altered. At the beginning of each regres-

TABLE 3—*Summary of multiple linear regression analyses of* $pTm_{R,X}$ *according to Eq 1 for mono-substituted benzenes and Eq 2 for di-substituted benzenes.*

R	Equation	Regression Coefficients							
		a	*b* ($pTm_{H,X}$)	*c* (log *P*)	*n*	r^2	SE	F[a]	Outliers[b]
H	1	0.029	0.038[c]	0.193	36	0.60	0.38	24.4	2
I	2*a*	3.081	−0.0002	−0.0432	6	0.16	0.99	0.3	2
Br	2*b*	0.559	0.525	1.711	9	0.59	0.25	5.8	0
Cl	2*c*	0.310	0.715	0.164	20	0.74	0.33	24.4	0
F	2*d*	−0.747	0.693	0.309	16	0.63	0.44	11.2	2
CF_3	2*e*	2.850	0.704	−0.660	9	0.47	0.69	2.7	2
NH_2	2*f*	0.924	0.820	−0.186	12	0.27	0.77	1.7	3
OH	2*g*	1.251	1.072	−0.280	14	0.34	0.95	2.8	4
CN	2*h*	0.935	0.361	0.181	15	0.10	0.92	0.7	7
NO_2	2*i*	1.259	0.412	−0.191	20	0.09	0.68	0.9	4
Halogen	2*j*	−0.348	0.463	0.422	51	0.40	0.61	16.2	10
Halogen + CF_3	2*k*	−0.035	0.491	0.316	60	0.30	0.66	12.2	16
All compounds	2*l*	0.790	0.628	0.0203	124	0.19	0.74	14.0	34
All, less duplicates	2*m*	0.510	0.689	0.102	94	0.32	0.68	21.4	22

[a] Significance of results: Eqs 2*a*, *e*, *f*, *h*, and *i*, $P > 0.25$; Eq 2*g*, $P = 0.10$; Eq 2*b*, $P = 0.05$; Eqs 1, 2*c*, *d*, *j*, *k*, *l*, *m*, $P < 0.005$.
[b] Outliers are defined as those with a residual greater than $|0.7|$.
[c] This is the coefficient of dMR_X in Eq 1.

sion calculation there was frequently some dependence on log P, but this was typically eliminated upon the inclusion of $pTm_{H,X}$, which clearly acts as a "blanket" parameter by including all the factors which contribute to the toxicity of the mono-substituted species. The acute toxicity of the compounds in each subset, except for R = NO_2 and OH, follow the general equation

$$pTm_{R,X} = a + b\text{d}1 + c\,pTm_{H,X} + d\,\text{dMR}_X \tag{3}$$

From the data in Table 2, it is clear that the SCS values for C-1, when R = Cl, are quite small, and there is little variation regardless of the nature of the X substituent. While the magnitude of the SCS at C-1 is quite different for the other halogens, there is again little variation within the Br and F subsets. Moreover, the general degree of acute toxicity decreases in the order I > Br,Cl > F, as does the shift relative to benzene, d1. Because the regression equations for the halogens and for CF_3 were found to be fairly similar (Eqs 3*a* through 3*e*, see Table 4), these subsets were combined into a single group of 60 data points. The best fit again follows the format of Eq 3, but the dMR value of the group R, dMR_R, is now included

$$pTm_{R,X} = a + b\text{d}1 + c\,pTm_{H,X} + d\,\text{dMR}_X + e\,\text{dMR}_R \tag{4}$$

It might have been anticipated that the molar refractivity contributions would be similar for both R and X; however, the dMR_X term is related to the $pTm_{H,X}$ term, as noted above, and this affects the apparent dMR_X contribution. The SE is acceptably small (0.35), and the actual error in calculating $pTm_{R,X}$ is less than 0.6 logarithmic units for all but 5 of the 60 points. The five are (F,NCS), (F,Cl), (CF_3,OH), (CF_3,NH_2), and (CF_3,NO_2). Only (F,NCS) and (CF_3,OH) are significantly more toxic than calculated. The OH, NH_2, and NO_2 subsets are found to correlate in a manner different from that of the halogens, *vide infra*; in fact, all three of these (CF_3,X) points are fit within the CF_3 subset, so the extension of the halogen set to include CF_3 may not be acceptable. The remaining outlier is (F,NCS). The phenylisothiocyanate (H,NCS) is quite toxic by itself, and the isothiocyanate group may be one of the "strongly interacting" species, such as NO_2, but, at present, there are insufficient data on this group to enable us to draw any firm conclusions.

As stated above, the other subsets correlate in a different fashion, but excellent correlations may be obtained in almost every case. The NH_2 subset is fit very well by d1 and $pTm_{H,X}$, with a small adjustment of dMR_X, Eq 3*f* (Table 4). There are no outliers in this subset; the highest observed residual is for X = CN, residual = 0.52.

The NO_2 subset does not correlate well with any of the original variables, and the toxicity of the very toxic outliers, with X = NO_2, NH_2, and CN, cannot be predicted from them. Literature data are available for |dQox| at the oxygen atom for 9 of the 20 data points here (Table 2), including the outliers. It was found that the toxicity of all nine compounds is highly correlated with this parameter, and $pTm_{R,X}$ can be predicted with no outliers

$$pTm_{R,X} = a + b|\text{dQox}| \tag{5}$$

We can only conclude that this change renders the molecule more capable of reacting with receptors in the bacteria; however, it would be most helpful to have more data to test this conclusion. The regression analysis showed that there could be a small dependence on dMR_X, but the inclusion of this parameter was not justified, given the already small value of the SE and the limited size of the physical data set. Consequently, this equation is not given.

TABLE 4—*Summary of results of multiple linear regression analyses for subsets or combined sets of data.*

		Regression Coefficients									
R	Equation	*a*	*b* (d1)	*c* ($pTm_{H,X}$)	*d* (dMR_X)	*e* (dMR_R)	*n*	r^2	SE	F[a]	Outliers[b]
I	3*a*	0.512	−0.0445	...	...		6	0.92	0.27	43.1	0
Br	3*b*	0.890	−0.0254	0.626	...		9	0.74	0.22	8.4	0
Cl	3*c*	0.759	−0.0302	0.953	...		20	0.78	0.30	29.8	0
F	3*d*	2.540	−0.0883	0.692	0.0478		16	0.81	0.32	17.3	1[c]
CF_3	3*e*	2.537	−0.0940	1.040	−0.584[d]		9	0.88	0.32	12.3	0
NH_2	3*f*	−0.768	0.1037	1.357	−0.085		12	0.78	0.45	9.6	0
OH	3*g*	1.172	...	1.342	−0.0676		14	0.52	0.80	6.0	4
CN	3*h*	−0.641	−0.083	...	0.0747		15	0.64	0.58	10.8	3
CN	3*i*	−0.969	−0.105	...	0.0637		13	0.79	0.45	18.6	1[e]
Halogen	4*a*	0.737	−0.265	0.864	...		51	0.79	0.36	90.6	3
	4*b*	1.299	−0.0442	0.729	0.0288	−0.0931	51	0.85	0.35	66.0	0
Halogen + CF_3	4*c*	0.799	−0.0275	0.824	...		60	0.73	0.41	76.9	6
	4*d*	1.444	−0.0481	0.715	0.0287	−0.112	60	0.80	0.35	56.4	2

NO_2	5	0.272	0.0240 (|dQox|)		9	0.88	0.35	49.2	0
OH	6	0.811	−0.0248 (d4)	0.339 (log P)	12	0.60	0.50	6.7	2[f]
All, no duplicates	7*a*	0.412 ($pTm_{R,H}$)	0.437 ($pTm_{H,X}$)	0.746	94	0.37	0.66	26.5	18
All non-interacting	7*b*[g]	0.233	0.650 ($pTm_{R,H}$)	0.706 ($pTm_{H,X}$)	76	0.74	0.31	102.2	1

[a] Significance of results: $P < 0.005$ for all equations except Eq 3*b*, for which $P = 0.02$, and Eq 3*g*, for which $P = 0.01$.
[b] Outliers are defined as those with a residual greater than $|0.7|$.
[c] This outlier (F,Cl), with a residual of −0.70, is fit in Eq 3*c*, with a residual of −0.49.
[d] The parameter in this term is log P, not dMR_X.
[e] This outlier (CN,OCH_3) has a residual of −0.74; it may possibly be fit in an OCH_3 subset, but there are insufficient data at this time.
[f] The outlier (H,OH), with a residual of −1.08, is fit in Eq 1, with a residual of 0.23; the outlier (OH,CF_3), with residual 1.03, is fit in Eq 3*e*, with a residual of 0.34.
[g] For the purpose of calculating the regression coefficients here, the data were regrouped to give approximately equal numbers of substituents R and X; this procedure does not affect the statistical results of the regression analysis, but affords more similar regression coefficients.

The OH subset still has outliers which cannot be resolved with the parameters here. The correlation given by Eq 3*g* (Table 4) has an unacceptably high SE. In particular, the hydroquinone (OH,OH) has an unexpectedly high toxicity (pTm = 3.46) [*3*], which does not appear to be related to any of the parameters here. Similarly, the toxicity of the (OH,$NHCOCH_3$) derivative is unexpectedly low. It is necessary to remove these two from the data set, after which a fairly good correlation

$$pTm_{R,X} = a + bd4 + c \log P \tag{6}$$

predicts the toxicity of all but the (OH,CF_3) and (OH,H) compounds; the latter is overestimated. The toxicity of the (OH,CF_3) compound is fit adequately in the CF_3 subset, and that of phenol (OH,H) in the mono-substituted set, Eq 1.

The CN subset is fairly well fit with d1 and dMR_X, Eq 3*h* (Table 4), but on examination of the residuals, it is apparent that the value for the *p*-benzonitrile (CN,NO_2) point is not fit. This is one of the compounds which must be fit with |dQox| at the NO_2 oxygen and, therefore, should be deleted from the CN set. Similarly, the (CN,F) point is too low, but is fit in the F subset, and is therefore also deleted. The remaining points are fit with no outliers and an acceptable SE of 0.45, Eq 3*i* (Table 4).

The entire set of observations of 1,4-di-substituted compounds (R = NH_2, NO_2, OH, CF_3, $N(CH_3)_2$, CN, F, Cl, Br, and I) contains 30 compounds which are present in two different subsets, for example, (Cl, F) and (F, Cl). After deletion of these duplicates, the remaining set of nonredundant observations is fit in one equation, (n = 94), in which the dependence on electronic parameters is diminished relative to the dependence on *both* $pTm_{R,H}$ and $pTm_{H,X}$ (Eq 7*a*, Table 4). In fact, the latter two variables explain close to 40% of the variation, and no significant improvement is obtained from the inclusion of any of the other variables. However, this general equation

$$pTm_{R,X} = a + b\,pTm_{R,H} + c\,pTm_{H,X} \tag{7}$$

can only be viewed as a first approximation of values as it does not distinguish between interacting and noninteracting substituents. Consequently, Eq 7*a* has a total of 18 outliers. However, these outliers are correctly fit in one of the specific subset equations, Eqs 3*a* through 4*d* (Table 4), except for the two compounds (OH,OH) and (OH,$NHCOCH_3$). The removal of these outliers results in Eq 7*b* (Table 4), which is representative of all compounds (n = 76; SE = 0.31; r^2 = 0.74) for which no strong interaction of the substituents occurs. This equation can therefore be regarded as a baseline toxicity equation which shows the toxicity of any noninteracting substituent bearing 1,4-di-substituted benzene (R,X) to be an additive function of the toxicities of the respective mono-substituted benzenes, namely, R-C_6H_5 and X-C_6H_5.

The important results obtained here are that (1) Eq 4*d* and 7*b* are highly significant and cover a broad range of substituents, and (2) virtually all of the outliers of the previous correlations have been included in one of the obtained correlations and can therefore be predicted with a high degree of confidence.

Conclusions

The present work demonstrates the predictability of highly toxic outliers from the previously observed relationships of the toxicities of 1,4-di-substituted benzenes ($pTm_{R,X}$) with those of the corresponding mono-substituted derivatives ($pTm_{H,X}$ and $pTm_{R,H}$). The following rules can be made for the calculation of these toxicity values of mono- and di-substituted benzenes. If more than one of the specified groups is present, both equations should be tested, and the higher predicted toxicity should be assumed to be correct.

Mono-Substituted Benzenes—Eq 1

The results found here are similar to those of the earlier study [2]. Of the 38 mono-substituted benzenes, 36 could be fit by Eq 1 (Table 3) with an SE of 0.38 and an r^2 value of 0.60, which is highly significant. The two derivatives which were omitted are X = *para*-C_6H_4-CN and $NHCOCH_3$, the reasons for which are given.

Di-Substituted Benzenes—General

*1-R-C_6H_4-4-X—Eq 7*b—For compounds of the general formula 1-R-C_6H_4-4-X, acute toxicity is a function of $pTm_{H,X}$ and $pTm_{R,H}$ (the acute toxicities of the mono-substituted benzenes, C_6H_5-X and R-C_6H_5). Substituents that have strongly electron-withdrawing or donating effects on the benzene ring, particularly when in conjugation with a substituent of similar strong field, resonance effects, or both, require a more specific correlation, as indicated below.

Di-Substituted Benzenes—Specific

*X = Halogen or CF_3—Eq 4*d*; Eqs 3*a *to 3*e—Because of their chemical similarity, and only a minor dependence on electronic interactions, such as $|d1 - d4|$, the toxicity of these compounds may be determined from Eq 4*d*. However, d1 is still an important parameter, and predictions may not work well in cases where X is strongly interacting. In this event, the equation appropriate for X = halogen should be used (Eqs 3*a* through 3*d*) or the separate equation for the CF_3 group (Eq 3*e*).

R = NO_2—Eq 5—There is a good indication that the toxicity of the compound is related to the charge buildup at the oxygen atom ($|dQox|$), as all nine points for which these data are available could be fit, with no outliers, with this single parameter. More data for this correlation would be most useful to confirm this conclusion.

*R = NH_2—Eq 3*f*: R = CN—Eq 3*i—For both of these sets, the parameters are the same as for the halogen subset, but the coefficients vary in size and sign, reflecting the different degrees to which the toxicity is affected by these parameters. When either of these substituents is present, the appropriate subset equation should be used to predict the toxicity.

R = OH—Eq 6—There is still some uncertainty in the prediction of the toxicity of members of this subset, since the two remaining outliers with higher (OH,OH) and lower than predicted toxicity (OH,$NHCOCH_3$) could not be fit with the available parameters. The pH of the system and the possibility of ion or zwitterion formation remain to be investigated. Further studies of this group are warranted.

R = NCS or $NHCOCH_3$—The former appears to be more toxic than expected, the latter less toxic. The NCS group may possibly be another special case like NO_2 or OH, or it may act by a specific mechanism [*4–8*], but insufficient data are available to confirm this at this time.

With the above guidelines, it is possible to predict the acute toxicity of all of the compounds studied, with virtually no outliers, and an SE of the estimate of 0.35 to 0.45 logarithmic units, depending on the substituent R. The present results confirm the supposition that the previous outliers occurred because of strong electronic interactions between the substituents. Their higher than expected toxicity is chiefly a function of an electronic parameter, d1 or |d1 − d4|, and the "base" toxicity, $pTm_{H,X}$, of the corresponding monosubstituted benzene derivative. With these physicochemical parameters, the toxicities of such compounds with strongly interacting substituents, for example, 1,4-dinitrobenzene, can also be predicted from these equations.

These results raise some interesting questions. In particular, it appears that, at least within the mentioned subsets, there is no need to exclude any compounds on account of specific compound-receptor mechanisms. This finding should be investigated further with respect to the effects of the same compounds on other organisms and of the effects of more complicated compounds, such as multiple-substituted benzene derivatives, on the same bacteria.

Acknowledgment

We thank Virginia S. Palabrica for the determination of some of the toxicity values reported here and Dr. T. Wildman, McMaster University, Hamilton, for making available computing facilities.

References

[1] Kaiser, K. L. E., Ed., *QSAR in Environmental Toxicology,* D. Reidel Publishing Co., Dordrecht, Holland, 1984.
[2] Kaiser, K. L. E., Palabrica, V. S., and Ribo, J. M. in *QSAR in Environmental Toxicology—II,* Kaiser, K. L. E., Ed., D. Reidel Publishing Co., Dordrecht, Holland, 1987, pp. 153–168.
[3] Kaiser, K. L. E. in *QSAR in Environmental Toxicology—II,* K. L. E. Kaiser, Ed., D. Reidel Publishing Co., Dordrecht, Holland, 1987, pp. 169–188.
[4] Lipnick, R. L., Watson, K. R., and Strausz, A. K., *Xenobiotica,* 1987.
[5] Roberts, D. W., in *QSAR in Environmental Toxicology—II,* K. L. E. Kaiser, Ed., D. Reidel Publishing Co., Dordrecht, Holland, 1987, pp. 295–308.
[6] Schultz, T. W., Riggin, G. W., and Wesley, S. K. in *QSAR in Environmental Toxicology—II,* K. L. E. Kaiser, Ed., D. Reidel Publishing Co., Dordrecht, Holland, 1987, pp. 333–345.
[7] Veith, G. D. and Broderius, S. J. in *QSAR in Environmental Toxicology—II,* K. L. E. Kaiser, Ed., D. Reidel Publishing Co., Dordrecht, Holland, 1987, pp. 385–391.
[8] Newsome, L. D., Johnson, D. E., Cannon, D. J., and Lipnick, R. L. in *QSAR in Environmental Toxicology—II,* K. L. E. Kaiser, Ed., D. Reidel Publishing Co., Dordrecht, Holland, 1987, pp. 231–250.
[9] Hammett, L. P., *Physical Organic Chemistry,* McGraw-Hill, New York, 1940.
[10] Swain, C. G. and Lupton, E. C., *Journal of the American Chemical Society,* Vol. 90, 1968, pp. 4328–4337.
[11] Stothers, J. B., *Carbon-13 NMR Spectroscopy,* Academic Press, New York, 1972.
[12] Hehre, W. J., Taft, R. W., and Topsom, R. D., *Progress in Physical Organic Chemistry,* Vol. 12, 1976, pp. 159–187.
[13] Bromilow, J., Brownlee, R. T. C., Craik, D. J., Sadek, M., and Taft, R. W., *Journal of Organic Chemistry,* Vol. 45, 1980, pp. 2429–2438.
[14] Ribo, J. M. and Kaiser, K. L. E., *Chemosphere,* Vol. 12, 1983, pp. 1421–1442.
[15] Hansch, C. and Leo, A. J., *Substituent Constants for Correlation Analysis in Chemistry and Biology,* Wiley, New York, 1979.
[16] Craik, D. J., Levy, G. C., and Brownlee, R. T. C., *Journal of Organic Chemistry,* Vol. 48, 1983, pp. 1601–1606.
[17] Inamoto, N., Masuda, S., Tokumaru, K., Tori, K., Ysohida, M., and Yoshimura, Y., *Tetrahedron Letters,* No. 41, 1976, pp. 3707–3710.
[18] Hugel, H. M., Kelly, D. P., Spear, R. J., Bromilow, J., Brownlee, R. T. C., and Craik, D. J., *Australian Journal of Chemistry,* Vol. 32, 1979, pp. 1511–1519.

[*19*] Shulman, E. M., Christensen, K. A., Grant, D. M., and Walling, C., *Journal of Organic Chemistry,* Vol. 39, 1974, pp. 2686–2690.
[*20*] Shapiro, M. J., *Tetrahedron,* Vol. 33, 1977, pp. 1091–1094.
[*21*] Guillaume, F., Seguin, J. P., Nadjo, L., Uzan, R., Membrey, F., and Doucet, J. P., *Journal of the Chemical Society, Perkin Transactions II,* 1984, pp. 1139–1144.
[*22*] Olah, G. A. and Forsyth, D. A., *Journal of the American Chemical Society,* Vol. 97, 1975, pp. 3137–3141.
[*23*] O'Connor, C. J., MacLennan, D. J., Calvert, D. J., Lomak, T. D., Porter, A. J., and Rodgers, D. A., *Australian Journal of Chemistry,* Vol. 37, 1984, pp. 497–510.
[*24*] Johnson, L. F. and Janowski, W. C., *Carbon-13 NMR Spectra,* Wiley-Interscience, New York, 1972.
[*25*] Lynch, B. M., *Canadian Journal of Chemistry,* Vol. 55, 1977, pp. 541–546.
[*26*] Smith, W. B. and Deavenport, D. L., *Journal of Magnetic Resonance,* Vol. 7, 1972, pp. 364–369.
[*27*] Maciel, G. E. and Natterstad, J. J., *The Journal of Chemical Physics,* Vol. 42, 1965, pp. 2427–2435.
[*28*] Kosugi, Y. and Furuya, Y., *Tetrahedron,* Vol. 33, 1980, pp. 2741–2744.
[*29*] Brownlee, R. T. C. and Sadek, M., *Australian Journal of Chemistry,* Vol. 34, 1981, pp. 1593–1602.
[*30*] Jones, R. G. and Allen, G., *Organic Magnetic Resonance,* Vol. 19, 1982, pp. 196–203.
[*31*] Hamer, G. K., Peat, I. R., and Reynolds, W. F., *Canadian Journal of Chemistry,* Vol. 51, 1973, pp. 897–914.
[*32*] Nelson, G. L., Levy, G. C., and Cargioli, J. D., *Journal of the American Chemical Society,* Vol. 94, 1972, pp. 3089–3094.
[*33*] Coulson, D. R., *Journal of the American Chemical Society,* Vol. 98, 1976, pp. 3111–3119.
[*34*] Reynolds, W. F., Gomes, A., Maron, A., MacIntyre, D. W., Maunder, R. G., Tanin, A., and Wong, H. E., *Canadian Journal of Chemistry,* Vol. 61, 1983, pp. 2367–2375.

Gary W. Holcombe,[1] *Gary L. Phipps,*[1] *and Gilman D. Veith*[1]

Use of Aquatic Lethality Tests to Estimate Safe Toxicant Concentrations for Initial Ecological Risk Assessments

REFERENCE: Holcombe, G. W., Phipps, G. L., and Veith, G. D., **"Use of Aquatic Lethality Tests to Estimate Safe Toxicant Concentrations for Initial Ecological Risk Assessments,"** *Aquatic Toxicology and Environmental Fate: Eleventh Volume, ASTM STP 1007,* G. W. Suter II and M. A. Lewis, Eds., American Society for Testing and Materials, Philadelphia, 1989, pp. 442–458.

ABSTRACT: This article presents an approach which allows the body of comparative toxicity data to be used in initial ecological risk assessments to extrapolate from an acute test with an indicator species to an estimate of the no-effect concentration in the environment. The fathead minnow *(Pimephales promelas)* acute value was selected as the reference value since this ecotoxicity endpoint has the largest data base for comparative toxicity comparisons. Comparative toxicity endpoints for fish and invertebrates were collected from various sources. When data for all ecotoxicity endpoints are plotted for all chemicals, this plot can be analyzed statistically using regression analysis to calculate an equation defining the upper 95 percentile prediction limit. The upper 95% prediction limit uses the reference test (fathead 96-h acute value) to calculate a concentration that would be safe for 95% of the species and chemicals, assuming that enough comparative toxicological data (especially chronic endpoints) are included in the data set.

KEY WORDS: risk assessment, comparative toxicity, fish, invertebrates, fathead minnow, aquatic toxicology

Decisions concerning the evaluation of chemical safety carry with them considerable uncertainty. In the probabilistic language of risk assessment, the uncertainty of parameter estimation unavoidably becomes associated with the cost of generating data for the risk assessment. Hence, the management of chemical safety programs encompassing many chemicals seeks to prevent too much data generation on safe chemicals as well as too little on chemicals that may represent true chemical hazards. Decisions to regulate or require testing, or both, often involve initial risk assessments to determine if more comprehensive risk assessments are justified. This paper addresses the use of acute toxicity data in the extrapolation of initial hazard assessments to potential adverse effects from long-term exposure of aquatic populations in the environment.

The literature is replete with the search for acceptable, sensitive test organisms that can serve as "surrogate" or "alternative" species [*1,2*]. Several methods of analyzing and summarizing comparative toxicity data have been detailed by previous authors. Suter et al. [*3*] previously described a new method of risk assessment that consisted of extrapolation between taxa and between toxicity test types by regression analysis. In addition, several

[1] Research aquatic biologists and director, respectively, Environmental Research Laboratory—Duluth, U.S. Environmental Protection Agency, Duluth, MN 55804.

investigators [4,5] have used regression analysis to make taxonomic extrapolations of acute and chronic data sets for various fish species, including fathead minnows. This system [4] allows extrapolation to species that have rarely or never been tested by assuming that they are represented by tested species that are members of some common higher taxonomic level. These investigators [3–5] concluded that using the prediction interval on regressions of various toxicity data sets can be used in risk analysis to predict a chronic endpoint when a measured value is not available. Kenaga and Moolenaar [6] compared the relative degree of static acute toxicity of thousands of chemicals of heterogenous structure to four species of fish, one species of *Daphnia,* five species of vascular aquatic plants, and one species of algae. From their review of toxicity testing done by Dow Chemical Co., they concluded that animals are generally more sensitive indicators of acute toxicity than plants, and that water quality limitations based on fish and daphnids should be sufficiently restrictive to protect algae and plants. LeBlanc [7] also analyzed acute interspecies relationships in aquatic organisms. He made acute sensitivity comparisons for several groups of chemicals consisting of nonpesticide organics, pesticides, and metals. These comparisons included warm-water versus cold-water fish; freshwater fish, invertebrates, and algae versus salt-water fish, invertebrates, and algae; and species differences that occur between fish species from the same family, between fish to different families, and between fish and daphnids.

However, systematic comparative toxicity studies have shown that relative species susceptibility is not constant but varies with the mode of action of the chemical being considered [8]. Faced with large variations in sensitivity, a battery of tests using representative organisms from various taxa are recommended [1]. The guidelines for deriving numerical water quality criteria in the United States [9] require acute toxicity tests on freshwater animals in at least eight different families, chronic toxicity tests on three animals including a fish and an invertebrate, one test of an alga or vascular plant, and a bioconcentration test. The result of these test requirements is that, after ten years of testing, water quality criteria have been determined for fewer than 30 of the more than 1000 chemicals commonly found in natural waters. Reducing the uncertainty of individual chemical evaluations by increasing data requirements for each chemical is not only costly, but also increases the uncertainty of overall environmental protection by preventing even preliminary evaluation of hundreds of other potential chemical hazards.

The criticisms of selections of toxicological endpoints and organisms for batteries of tests often center on the sensitivity and relevance of the tests. Our criticism of this approach is that the amount of data needed in a chemical evaluation is not equal for all chemicals, but varies with the specificity of the mechanisms leading to adverse effects. Thurston et al. [8] have shown that little new information concerning toxic concentrations is gained by increasing the number of test organisms from one to ten for nonspecific narcotic chemicals. More important, the requirement of fixed batteries of tests precludes the judicious use of relationships between the distribution of susceptibilities of test organisms from one chemical to the next. Such relationships are based on data of comparative physiology, biochemistry, and toxicology and must be integrated into cost-effective protocols for test requirements. In addition, the estimate of no-effect concentrations from single-species tests must be extrapolated to concentrations that are safe for entire aquatic ecosystems.

This paper presents an approach which allows the body of comparative toxicology data to be used to extrapolate from an acute value with an indicator species to an estimate of no-effect concentration in the environment. The approach is intended for use in initial regulatory decisions, in the absense of extensive data, and is not meant to replace thorough testing on priority hazardous chemicals. To accomplish this, we have assumed that all acute and chronic tests on all aquatic species have equal weight as ecotoxicological endpoints. This assumption is based on the difficulty in distinguishing acute and chronic tests for many species, and the data showing acute tests on some organisms can be more sensitive to a chemical than chronic tests with another species. If safe concentrations are to protect mixed populations in aquatic ecosystems, each test on each species should be considered a valid measurement of potential ecotoxicity. Although any species could be used as a reference organism in this approach, we have used the fathead minnow *(Pimephales promelas)* acute value. This ecotoxicity endpoint has the largest data base to use as a reference, and it is also the endpoint being estimated from structure-toxicity relationships.

Methods

Data Compilation

A literature search was conducted to obtain the currently available freshwater comparative toxicity data dealing with organic compounds. For requirements were placed on comparative toxicity references for inclusion in this report. The first requirement was that toxicity tests for a specific chemical had to be conducted with two or more aquatic species, and one of the tested species had to be the fathead minnow. In most cases, all data for a particular toxicant were from studies conducted by the same investigator or by a group of investigators within the same laboratory or university. However, separate tests on individual species, conducted by different investigators at a specific laboratory or university, were also considered acceptable for inclusion provided that (*a*) one species tested was the fathead minnow, (*b*) the dilution water was similar between the tests, and (*c*) the prescribed methodology was used. Comparative toxicity data contained in an interlaboratory comparison report [*10*] were treated as separate endpoints for endosulfan since the tests were conducted at different laboratories that had dilution water that varied in water quality parameters. A second requirement was that toxicity values for all fish species, including the fathead minnow, were collected under flow-through conditions. Static toxicity values for several invertebrate species (mainly cladoceran and midge species) were included since static testing is currently recommended for these species. The third requirement for the included toxicity data was that water quality conditions be similar between the species tested. The fourth and final requirement was that the tests accepted for inclusion be conducted using acceptable methodology. Most toxicity data included in this report were derived using methods prescribed by standard test methods [*11–15*].[2]

This report contains comparative toxicity data extracted from 39 references for 64 organic chemicals. Twenty-three of these 64 represented chemicals are classified as pesticides. The common name of all the chemicals, the number and types of endpoints plotted for each, and the references that include each chemical are listed in Table 1 [*16–52*].[3]

Toxicity data for 45 separate aquatic species are included in this report. Of the 45 freshwater species represented, 18 are fish, 19 are invertebrates, 7 are amphibians, and 1 is an alga. The common and specific names of the represented species and the number of times each species was plotted in the comparative toxicity data set are given in Table 2.

All data presented for fathead minnows are acute median lethal concentration (LC_{50}) values, and for most chemicals, these are 96-h LC_{50} values. However, fathead minnow data for acrolein, endosulfan (in one instance), heptachlor, lindane, trifluralin, and 1,3,5-trinitrobenzene (Table 3) are incipient LC_{50} values. Acute endpoints for all other fish species

[2] ASTM Practice for Conducting Acute Toxicity Tests with Fishes, Macroinvertebrates, and Amphibians (E 729-80).

[3] U.S. Environmental Protection Agency, unpublished laboratory data, Environmental Research Laboratory—Duluth, MN.

and many invertebrates are 96-h LC_{50} values. Acute endpoints for cladoceran and midge species are 48-h LC_{50} values since these are the recommended test durations for these species. Data for amphibian species and fathead minnows from tests conducted by Black et al. [*16*] are LC_{50} values based on exposures initiated at fertilization and maintained through four days posthatch. Several long-term endpoints available for other aquatic species were also plotted against fathead minnow acute values for 20 represented chemicals (Table 1). These long-term endpoints ranged from LC_{50} values at prolonged time periods to maximum acceptable toxicant concentrations (MATCs) based on embryolarval, partial life-cycle, or full life-cycle exposures for both fish and invertebrates.

Data Analysis

Toxicity endpoints extracted from the various references (expressed as milligrams per litre were first divided by the molecular weight of each individual chemical to make the toxicity equivalent on a millimoles-per-litre basis. These toxicity endpoints were then divided by 1000 to convert them to moles per litre, and these were converted to logarithms for plotting and statistical comparisons. However, these toxicity endpoints were negative numbers at this point. All endpoints were converted to positive numbers by expressing the endpoints as the inverse of the log of the endpoint concentration in moles per litre. The converted acute toxicity values for fathead minnows and converted acute and long-term endpoints for all other aquatic species tested in conjunction with that particular fathead minnow test for specific chemicals are presented in Table 3. Converted toxicity endpoints become larger as the toxicity and molecular weight of the chemical being tested increases. Some "greater than" toxicity values were used in this report since these endpoints are useful in species comparisons and in setting water quality criteria for chemicals. In comparative toxicity testing, greater than toxicity values are usually reported in cases where a particular species is less sensitive than other species being tested and where less than 50% of the test organisms in a specific test are killed in the highest exposure concentration. Species with greater than toxicity values were always less sensitive than fathead minnows, and these greater than values appear (Table 3) as "less than" values for the converted toxicity endpoints.

Results and Discussion

The format for relating the toxicity measured in the reference test, such as the fathead minnow acute test, to other ecotoxicity endpoints is a two-dimensional plot with equally scaled axes for effect concentration. On this plot, the line of exact correlation is a 45° line through the origin. When the reference value is plotted on the x-axis and the corresponding value on the y-axis, values less sensitive than the reference value fall below the line and those more sensitive fall above the line. Figure 1 presents the data for 309 acute toxicity tests on all species and all chemicals for which reference test (fathead minnow) data are also available.

The range of toxicity values in Fig. 1 is almost nine orders of magnitude, yet the LC_{50}s for the majority of chemicals and species are less than an order of magnitude of that for the fathead minnow. As has been observed previously [*8*], there appears to be greater deviations for the more toxic chemicals. Since greater toxicity is generally associated with greater specificity of toxicity mechanisms, it is reasonable that physiology and biochemistry differences among fish and invertebrates would be important for toxicants with more specific mechanisms. At the same time, the data [*8*] show that no one species is consistently more susceptible to a broad range of chemical structures. If a straight line was fitted

TABLE 1—*The chemical names, Chemical Abstracts Service (CAS) registry numbers, molecular weights, the number and type of endpoints represented, and the number of references that include data for each chemical for the various comparative toxicity tests.*

Molecular Weight	Chemical Names	CAS No.	Number of Acute Endpoints Plotted Besides Fatheads	Number of Chronic Endpoints Plotted Besides Fatheads	Number of References That Include This Chemical	Reference Number
154.2	Acenaphthene	83 329	4	0	1	*17*
56.1	Acrolein	107 028	1	1	1	*18*
500.0	Alcohol ether sulphate	. . .	5	0	1	*19*
400.0	Alcohol ethoxylate	. . .	5	0	1	*19*
215.7	Atrazine	1 912 249	5	5	1	*20*
106.1	Benzaldehyde	100 527	6	0	1	*21*
108.1	*p*-Benzoquinone	106 514	1	0	1	*22*
302.0	S-Bioallethrin	584 792	1	0	1	*23*
259.1	γ-Bromo-2′,5′-dimethoxyacetophenone	1 204 213	6	0	1	*21*
300.6	Captan	133 062	2	0	1	*24*
153.8	Carbon tetrachloride	56 235	7	0	1	*16*
110.1	Catechol	120 809	1	0	1	*22*
409.8	Chlordane	57 749	4	5	1	*25*
80.5	2-Chloroethanol	107 073	18	0	3	*8,26,21*
126.6	*o*-Chlorotoluene	95 498	1	0	1	*27*
108.1	*o*-Cresol	95 487	2	0	1	*22*
108.1	*m*-Cresol	108 394	2	0	1	*22*
108.1	*p*-Cresol	106 445	2	0	1	*22*
304.4	Diazinon	333 415	3	1	1	*28*
228.3	Dibutylfumarate	105 759	4	0	1	*29*
147.0	1,3-Dichlorobenzene	541 731	2	1	1	*30*
147.0	1,4-Dichlorobenzene	106 467	1	0	1	*30*
237.0	1,3-Dichloro-4,6-dinitrobenzene	3 698 837	6	0	1	*21*
99.0	1,2-Dichloroethane	107 062	1	1	1	*30*
96.9	Dichloroethylene	75 354	1	0	1	*31*
73.1	Dimethylformamide	68 122	4	0	1	*32*
274.4	Disulfoton	298 044	1	0	1	*33*
350.6	Dursban	2 921 882	7	0	2	*33,21*

407.0	Endosulfan	115 297	15	1	3	*10,18,34*
380.9	Endrin	72 208	9	0	1	*8*
134.2	2-(2-ethoxy-ethoxy)-ethanol	111 900	9	0	1	*8*
317.3	Guthion	86 500	5	0	1	*29*
373.4	Heptachlor	76 448	1	1	1	*18*
260.8	Hexachlorobutadiene	87 683	3	0	2	*30*
236.7	Hexachloroethane	67 721	21	0	4	*30,26,21,8*
32.1	Hydrazine hydrate	10 217 524	1	0	1	*35*
110.1	Hydroquinone	123 319	2	0	1	*22*
370.5	Kelthane	115 322	1	1	2	*33,36*
290.8	Lindane	58 899	5	5	1	*37*
32.0	Methanol	67 561	2	0	1	*32*
345.7	Methoxychlor	72 435	1	0	1	*38*
84.9	Methylene chloride	75 092	3	0	1	*16*
118.2	2-Methyl-2,4-pentanediol	107 415	9	0	1	*8*
74.1	2-Methyl-1-propanol	78 831	9	0	1	*8*
128.2	Naphthalene	91 203	2	0	2	*39,40*
391.3	Permethrin	52 645 531	7	1	4	*41,42,33,8*
202.3	Pentachloroethane	76 017	1	0	1	*30*
266.3	Pentachlorophenol	87 865	28	3	6	*43,44,45,26,21,8*
100.1	2,4-Pentanedione	123 546	16	0	3	*8,26,21*
94.1	Phenol	108 952	11	0	3	*16,22,46*
419.9	Pydrin	66 827 381	1	0	1	*33*
110.1	Resorcinol	108 463	2	0	1	*22*
201.2	Sevin	63 252	6	0	1	*21*
257.1	Sodium nitrilotriacetic acid	5 064 313	2	1	2	*47,48*
258.3	Systox	8 065 483	5	0	1	*29*
167.9	1,1,2,2-Tetrachloroethane	79 345	1	1	1	*30*
165.8	Tetrachloroethylene	127 184	2	1	1	*30*
413.8	Toxaphene	8 001 352	3	1	2	*49,50*
302.0	d-Transallethrin	584 792	8	0	1	*51*
181.5	1,2,4-Trichlorobenzene	120 821	3	1	2	*30*, footnote 3
133.4	1,1,2-Trichloroethane	79 005	1	1	1	*30*
149.4	2,2,2-Trichloroethanol	115 208	9	0	1	*8*
335.2	Trifluralin	1 582 098	1	1	1	*18*
213.1	1,3,5-Trinitrobenzene	99 354	3	2	1	*52*

TABLE 2—*The common and scientific names of the represented species, the number of endpoints plotted with each species, and the number of references that include data for each species.*

Common Name	Scientific Name	Number of Acute Endpoints Plotted with This Species	Number of Chronic Endpoints Plotted with This Species	Number of References That Included This Species
Fathead minnow	*Pimephales promelas*	all	...	all
Rainbow trout	*Salmo gairdneri*	61	1	18
Brook trout	*Salvelinus fontinalis*	8	7	6
	Salmo trutta	1	0	1
Lake trout	*Salvelinus namaycush*	1	0	1
Coho salmon	*Oncorhynchus kisutch*	1	0	1
Bluegill	*Lepomis macrochirus*	36	6	14
Channel catfish	*Ictalurus punctatus*	25	0	6
Goldfish	*Carassius auratus*	27	3	6
Flagfish	*Jordanella floridae*	1	0	1
Northern pike	*Esox lucius*	1	0	1
White sucker	*Catostomus commersoni*	1	0	1
Smallmouth bass	*Micropterus dolomieui*	1	0	1
Yellow perch	*Perca flavescens*	2	0	2
Convict cichlid	*Cichlasoma nigrofasciatum*	1	0	1
Zebra danio	*Brachydanio rerio*	1	0	1
Guppy	*Lebistes reticulatus*	1	0	1
Mosquito fish	*Gambusia affinis*	10	0	1
Cladoceran	*Daphnia magna*	38	14	12
	Daphnia pulicaria	6	0	2
	Daphnia pulex	4	0	2
	Ceriodaphnia reticulata	3	0	1
	Simocephalus vetulus	3	0	2

Amphipod	*Hyalella azteca*	1	2	2
	Crangonxy pseudogracilis	1	0	1
Midge	*Chironomus tentans*	2	2	2
	Chironomus No. 51	1	1	1
	Tanytarsus dissimilis	10	0	2
Isopod	*Asellus racovitzai*	1	0	1
Scud	*Gammarus fasciatus*	2	2	2
	Gammarus pseudolimnaeus	1	0	1
Caddis fly	*Philarctus quaeris*	1	0	1
Mayfly	*Callibaetis skokianus*	1	0	1
Stone fly	*Pteronarcys dorsata*	0	1	2[a]
Snail	*Physa gyrina*	1	0	1
	Aplexa hypnorum	12	0	3
Crayfish	*Orconectes immunis*	17	0	2
Tadpole	*Rana catesbiana*	10	0	1
European common frog	*Rana temporaria*	3	0	1
Leopard frog	*Rana pipiens*	2	0	1
Pickerel frog	*Rana palustris*	2	0	1
Northwestern salamander	*Ambystoma gracile*	3	0	1
Fowler's Toad	*Bufo fowleri*	2	0	1
African clawed frog	*Xenopus laevis*	2	0	1
Algae	*Selenastrum capricornutum*	1	0	1

[a] Different papers from Environmental Research Laboratory—Duluth.

TABLE 3—*Converted comparative toxicity data expressed as 1/log of the end point in moles per litre for the various chemicals and species included in this report.*

Chemical	Fathead X	Acute Values for y for Fish and Amphibians	Acute Values for y for Invertebrate Species and Algae	Long-Term End Points with Fish	Long-Term End Points with Invertebrates
Acenaphthene	4.984	4.953 5.362 5.425	<4.8 <4.878		
Acrolein	5.825[a]		5.9 5.993		6.372
Alcohol ether sulphate	5.301	5.301 5.420 5.319 5.222 5.699			
Alcohol ethoxylate	5.648	5.456 5.602 5.301 5.155 5.699			
Atrazine	4.158[b]	4.495 5.477 4.578	4.5 4.508 4.535	6.012 6.431	6.130 6.055 6.380
Benzaldehyde	3.932	3.977 3.886 4.294 4.996	<3.8 <3.827 <3.827		
p-Benzoquinone	6.381	5.937			
s-Bioallethrin	6.756	7.316			
α-Bromo-2′,5′-dimethoxyacetophenone	6.689	6.689 6.215 6.770 6.413	<6.057 <6.057		
Captan	6.665	6.621 6.947			
Carbon tetrachloride	4.585[c]	5.122[c] 4.972[c] 4.892[c] 4.890[c] 4.812[c]			
	4.735[c]	3.836[c]			
Catechol	4.498	4.092			
Chlordane	7.046	6.842 6.940	7.159 <6.625[d]	8.398 9.017[e]	8.575 7.719 7.404
2-Chloroethanol	3.309		2.338 2.068 2.429		
	3.309	3.354 3.625 3.484 3.573	<2.998 <2.998		
	3.318	3.170 3.357 3.716 3.609 3.486	2.580 2.834 2.546 3.636		
o-Chlorotoluene	4.227	4.741			
o-Cresol	3.774	4.110	<3.061		
m-Cresol	3.286	4.084	<3.036		
p-Cresol	3.577	4.136	3.678		
Diazinon	4.591	5.821 5.597 5.279		8.743[c]	
Dibutylfumarate	5.520	5.677 5.409 5.414 5.566			
1,3-Dichlorobenzene	4.276	4.960	4.296		5.167
1,4-Dichlorobenzene	4.548	5.118			
1,3-Dichloro-4,6-dinitrobenzene	6.795	6.960 6.703 6.928 6.506	<6.274 <6.274		
1,2-Dichloroethane	2.924		2.568		3.825
1,1-Dichloroethylene	2.953		3.922		

Dimethylformamide	0.835	0.864	1.013				0.702	0.308							
Disulfoton	4.836	4.958													
Dursban	6.237	7.642													
	5.811	7.591	<5.639	5.639	7.545		7.767	<5.639							
Endosulfan	8.419	8.752					6.212								
	9.041	9.248					5.966								
	8.735	9.147					5.862								
	8.701	9.078					6.283								
	8.400	9.041					6.122								
	8.665	9.212					6.066								
	8.675[a]						6.389								
Endrin	8.775	9.062	9.259	8.742	8.958	8.603	6.382	8.662	<6.631	8.183			7.976		
2-(2-Ethoxy-ethoxy)-ethanol	1.056	1.001	0.797	0.978	1.349	0.810	1.525	0.854	0.587	0.808					
Guthion	6.692	7.542	5.484	4.994	7.532		<4.934								
Heptachlor	7.726[a]						6.680						7.324		
Hexachlorobutadiene	6.416	5.911													
	6.408	5.903	5.911												
Hexachloroethane	5.284	5.387	<5.052	5.192	5.387		<5.052	<5.052							
	5.277	5.302	5.442	5.234	5.058	5.222	5.241	5.284	4.943	4.926					
	5.190	5.450					4.912								
	5.284						4.943	<4.374	4.856	4.611					
Hydrazine hydrate	3.730						5.251								
Hydroquinone	6.398	6.055					5.832								
Kelthane	5.861	6.247											7.171		
Lindane	6.624[a]	6.745[a]	6.817				5.778	6.148	6.873		7.434	7.381	7.945	7.302	7.678
Methanol	0.056	0.202	0.315												
Methoxychlor	7.664	7.238													
Methylene chloride	3.397[c]	3.810[c]	3.700[c]	3.678[c]											
2-Methyl-2,4-pentanediol	1.134	1.097	0.965	1.108	0.979	0.993	1.224	1.438	0.678	1.001					
2-Methyl-1-propanol	1.691	1.746	1.666	1.615	1.706	1.562	1.825	1.540	1.893	1.390					
Naphthalene	4.210	4.904													
	4.418	4.756													
Permethrin	7.399	7.747											10.050		
	7.786	7.855	7.877	7.792	8.153	<6.235	6.532								

TABLE 3—(*Continued*).

Chemical	Fathead X	Acute Values for y for Fish and Amphibians	Acute Values for y for Invertebrate Species and Algae	Long-Term End Points with Fish	Long-Term End Points with Invertebrates
Pentachloroethane	4.443		4.441		
Pentachlorophenol	6.026	6.235 6.693 6.282	6.249		
	6.026		6.270 6.034 6.211 6.089		
	6.001	6.365 6.120 5.974 6.305 6.004	6.264 4.024 <3.163 6.109		
	6.124	6.124	5.599 5.562 5.346 5.195 5.147 5.064		
	6.103	6.083			
	5.971	6.285 5.994 5.944		6.353[g] 6.149[g] 6.093[g]	
2,4-Pentanedione	2.810	3.146 2.971 3.078 3.129	<2.664 2.810 3.465		
	2.848	3.086 3.198 2.750 2.975 2.918	3.323 2.802 2.293 2.940		
Phenol	3.144	4.024	<2.936		
	3.309	4.024	<2.936		5.558
	3.468[c]	6.372[c] 5.797[c] 5.542[c] 5.394[c] 4.584[c]			3.857
	4.088[c]	3.979[c]			
Pydrin	7.891	8.301			7.902
Resorcinol	3.042	<3.042	<3.042		5.558
Sevin	4.604	5.369 4.081 4.210 4.460	4.846 <3.872		
Sodium nitrilotriacetic acid	3.306	3.419		3.300	
	3.353		3.419		
Systox	4.211	6.260 <3.991 4.643 6.669	<3.991		
1,1,2,2-Tetrachloroethane	3.918		3.432		
Tetrachloroethylene	4.089	4.521	3.962		

Toxaphene	7.759	7.399									9.846	
	7.839	7.804	8.010[h]									
d-Trans allethrin	6.799	8.065	7.493	7.276	7.961	7.387						
	7.128	7.594	7.484									
1,2,4-Trichlorobenzene	4.801	5.083										
	4.818	5.077					4.939					
1,1,2-Trichloroethane	3.213						2.856					
2,2,2-Trichloroethanol	2.936	2.766	2.871	2.718	2.785	2.679	3.004	2.630	2.854	2.807		4.228
Trifluralin	6.465[a]						6.240					
1,3,5-Trinitrobenzene	5.666[a]	5.695[a]					4.716	5.302[i]			6.249	5.345

[a] Incipient LC_{50} values.
[b] Renewal test.
[c] Embryo plus 96-h exposures.
[d] 168-h LC_{50}.
[e] Some effects observed at the lowest concentration tested.
[f] Based on an approximation of the no-effect concentration.
[g] 336-h LC_{50} values.
[h] Means of LC_{50} values at 20 and 25°C.
[i] Algae values.

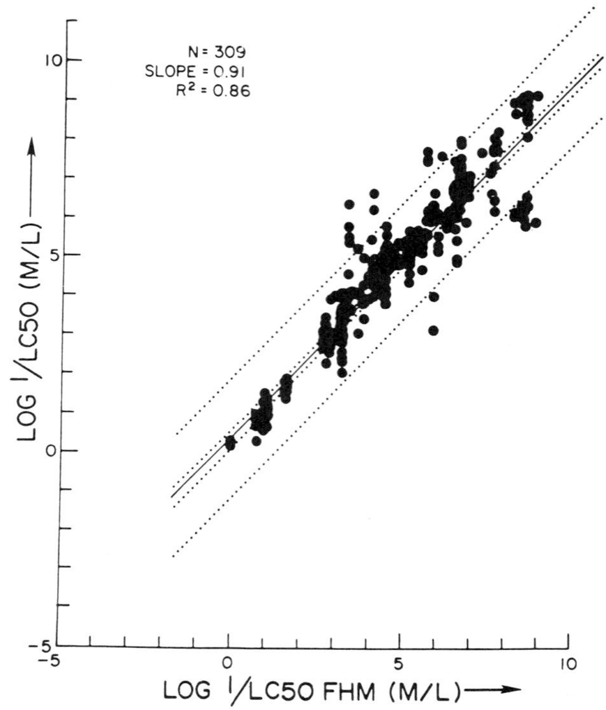

FIG. 1—*Variation of 309 acute toxicity endpoints for all species with the acute toxicity for fathead minnow. The dotted lines nearest to the regression lines represent the 95% confidence limits. The outermost dotted lines above and below the regression lines represent the upper and lower 95% prediction limits, respectively.*

through the data presented in Fig. 1 by least squares, the slope of 0.91 and intercept of 0.42 indicate that the bias in selecting the fathead minnow as a reference organism is small. Other investigators [5] that have used the fathead minnow as a reference organism in predictive models have also concluded that this species appears to be a reasonably representative freshwater fish. They [5] found that fathead minnows appeared to be somewhat less sensitive on the average than other tested fish species, but the difference was small relative to the variance among species.

If a chemical being evaluated has been tested with an invertebrate and a fish and the data are substantially different, it is difficult to determine whether one organism should be considered unusually sensitive or the other unusually resistant to the chemical. In either case, significant errors in judging the potential hazards of the chemical would occur. Figure 1 is useful in demonstrating that the fathead minnow reference test is not an invalid test for estimating the mean acute toxicity to a variety of aquatic organisms, thereby offering a perspective on relative sensitivity. However, the purpose of this effort is not to suggest the use of the fathead minnow as a surrogate species for other aquatic organisms. Rather, the intent is to develop a statistical expression in which the reference test value can be used as an independent variable to estimate a concentration of toxicant that would be low enough to protect a specified percentage of test organisms. With respect to Fig. 1, this would be equivalent to calculating the upper 95, 98, or 99 percentile prediction limit, which is the line computed from the reference test value as an independent variable but highly dependent on the variance of toxicology data for all species and all chemicals. The upper 95%

prediction limit uses the reference test value to calculate a concentration that would be safe for 95% of the species and chemicals, assuming enough comparative toxicological data are included in the data set.

Obviously, using this approach to evaluate ecotoxicological hazards of chemicals must involve many chronic endpoints as well as many dissimilar aquatic organisms. If all of the data in Table 3 are included in a single composite of ecotoxicity endpoints versus the fathead minnow acute value, the 344 data points create Fig. 2. This plot is still heavily biased by the acute toxicity data; however, for many species the distinction between acute and chronic toxicity endpoints is obscure. Figure 2 shows the confidence interval for a regression of the *y* on the *x* axis as well as the upper and lower 95% prediction limit. The upper prediction limit (UPL) is given by the equation

$$95\%\ \text{UPL} = 0.44 - 0.93 \log \text{LC}_{50} + 1.97 \sqrt{0.72\left[1.00 - \frac{(\log \text{LC}_{50} + 4.96)^2}{1450}\right]}$$

This equation for the 95% upper prediction limit was derived from the comparative toxicity data set presented in Fig. 2 using a statistical method described by Sokal and Rohlf [*53*]. In this equation, the 95% upper prediction limit is the concentration that would be expected to protect 95% of the test combinations involving species, test type, and chemicals. The independent variable is the LC_{50} test for fathead minnows, which is a reference

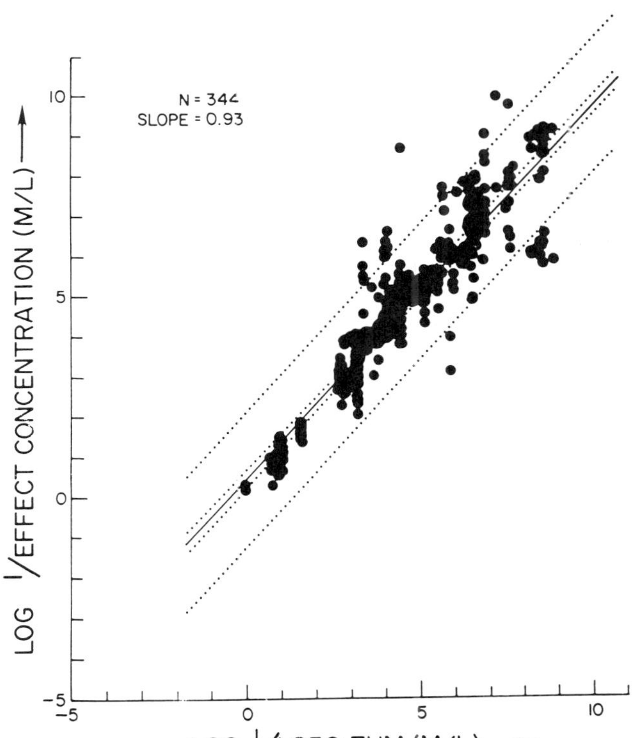

FIG. 2—*Variation of acute and chronic toxicity endpoints for all species with acute toxicity for fathead minnow. The dotted lines nearest to the regression line represent the 95% confidence limits. The outermost dotted lines above and below the regression line represent the upper and lower 95% prediction limits, respectively.*

test for the U.S. Environmental Protection Agency (EPA). This equation integrates all literature ecotoxicity data on species and chemicals into each new estimate by the variance and sum of squares terms in the equation. If the LC_{50} for the fathead minnow is known, the equation simply estimates the concentration below which 95% of ecotoxicity test data with other species are likely to fall.

This approach is obviously heavily biased by acute toxicity data. The regression approach is only appropriate if the acute and chronic data are plotted as paired data points, which is precluded by the lack of chronic test data on important vertebrates. Nonetheless, the approach provides a framework for a structured research program aimed at using the immense body of ecotoxicity data to make initial risk assessments. The estimates from the above equation will be accurate when sufficient chronic tests are included in the data set. The estimates will be ecologically relevant when the data base contains a representative sampling throughout the taxonomic spectra. Finally, the basic principles of toxicology will be incorporated when the major toxicity mechanisms are represented by the chemicals selected for testing. As it stands, the current body of data provides estimates of safe concentrations which are within a factor of three of those estimated by the "Guidelines for Deriving National Water Quality Criteria." We have embarked on a chronic toxicology program to improve these estimates as quickly as possible.

References

[1] Blanck, H., "Species Dependent Variation Among Aquatic Organisms in Their Sensitivity to Chemicals," *Ecological Bulletins,* 1984, pp. 107–119.

[2] Goss, L. B. and Sabourin, T. D., "Utilization of Alternative Species for Toxicity Testing: An Overview," *Journal of Applied Toxicology,* Vol. 5, 1985, pp. 193–219.

[3] Suter, G. W., II, Vaughan, D. S., and Gardner, R. H., "Risk Assessment by Analysis of Extrapolation Error: A Demonstration for Effects of Pollutants on Fish," *Environmental Toxicology and Chemistry,* Vol. 2, 1983, pp. 369–378.

[4] *User's Manual for Ecological Risk Assessment,* Barnthouse, L. W. and Suter, G. W., II, Eds., ORNL-6251, Oak Ridge National Laboratory, Oak Ridge, TN, 1986, pp. 1–215.

[5] Suter, G. W. II and Rosen, A. E., "Endpoints for Responses of Fish to Chronic Toxic Exposures," *Environmental Toxicology and Chemistry,* Vol. 6, 1987, pp. 793–809.

[6] Kenaga, E. E. and Moolenaar, R. J., "Fish and Daphnia Toxicity as Surrogates for Aquatic Vascular Plants and Algae," *Health and Environmental Sciences,* Vol. 13, 1979, pp. 1479–1480.

[7] LeBlanc, G. A., "Interspecies Relationships in Acute Toxicity of Chemicals to Aquatic Biota," *Environmental Toxicology and Chemistry,* Vol. 3, 1984, pp. 47–60.

[8] Thurston, R. V., Gilfoil, T. A., Meyn, E. L., Zajdel, R. K., Aoki, T. I., and Veith, G. D., "Comparative Toxicity of Ten Organic Chemicals to Ten Common Aquatic Species," *Water Research,* Vol. 19, 1985, pp. 1145–1155.

[9] Stephan, C. E., Mount, D. I., Hansen, D. J., Gentile, J. H., Chapman, G. A., and Brungs, W. A., "Guidelines for Deriving Numerical National Water Quality Criteria for the Protection of Aquatic Organisms and Their Uses," NTIS-PB85-227049, National Technical Information Service, Springfield, VA, 1985.

[10] Lemke, A. E., "Interlaboratory Comparison—Acute Testing Set," *Ecological Research Series,* EPA-600/3-81-005, U.S. Environmental Protection Agency, Duluth, MN, 1981.

[11] Sprague, J. B., "Measurement of Pollutant Toxicity to Fish, I—Bioassay Methods for Acute Toxicity," *Water Research,* Vol. 3, 1969, pp. 793–821.

[12] "The Committee on Methods for Toxicity Tests with Aquatic Organisms—Methods for Acute Toxicity Tests with Fish, Macroinvertebrates, and Amphibians," *Ecological Research Series,* EPA 660/3-75-009, U.S. Environmental Protection Agency, Duluth, MN, 1975.

[13] *Standard Methods for the Examination of Water and Wastewater,* 13th ed., American Public Health Association, Washington, DC, 1971.

[14] *Standard Methods for the Examination of Water and Wastewater,* 14th ed., American Public Health Association, Washington, DC, 1975.

[15] *Standard Methods for the Examination of Water and Wastewater,* 15th ed., American Public Health Association, Washington, DC, 1980.

[*16*] Black, J. A., Birge, W. J., McDonnell, W. E., Westerman, A. G., Ramey, B. A., and Bruser, D. M., "The Aquatic Toxicity of Organic Compounds to Embryo-Larval Stages of Fish and Amphibians," Project Number B-071-KY, Research Report No. 133, University of Kentucky, Water Resources Research Institute, Lexington, 1982.

[*17*] Holcombe, G. W., Phipps, G. L., and Fiandt, J. T., "Toxicity of Selected Priority Pollutants to Various Aquatic Organisms," *Ecotoxicity and Environmental Safety,* Vol. 7, 1983, pp. 400–409.

[*18*] Macek, K. J., Lindberg, M. A., Sauter, S., Buxton, K. S., and Costa, P. A., "Toxicity of Four Pesticides to Water Fleas and Fathead Minnows," *Ecological Research Series,* EPA-600/3-76-099, U.S. Environmental Protection Agency, Duluth, MN, 1976.

[*19*] Newsome, C. S., "Susceptibility of Various Fish Species at Different Stages of Development to Aquatic Pollutants," *Environment and Quality of Life: Principles for the Interpretation of the Results of Testing Procedures in Ecotoxicology,* EUR. 7549, Commission of the European Communities, Luxembourg, Belgium, 1982, pp. 284–295.

[*20*] Macek, K. J., Buxton, K. S., Sauter, S., Gnilka, S., and Dean, J. W., "Chronic Toxicity of Atrazine to Selected Aquatic Invertebrates and Fishes," *Ecological Research Series,* EPA-600/3-76-047, U.S. Environmental Protection Agency, Duluth, MN, 1976.

[*21*] Phipps, G. L. and Holcombe, G. W., "A Method for Aquatic Multiple Species Toxicant Testing: Acute Toxicity of Ten Chemicals to Five Vertebrates and Two Invertebrates," *Environmental Pollution (Series 1),* Vol. 38, 1985, pp. 141–157.

[*22*] DeGraeve, G. M., Geiger, D. L., Meyer, J. S., and Bergman, H. L., "Acute and Embryo-Larval Toxicity of Phenolic Compounds to Aquatic Biota," *Archives of Environmental Contamination and Toxicology,* Vol. 9, 1980, pp. 557–568.

[*23*] Mauck, W. L., Olson, L. E., and Marking, L. L., "Toxicity of Natural Pyrethrins and Five Pyrethroids to Fish," *Archives of Environmental Contamination and Toxicology,* Vol. 4, 1976, pp. 18–29.

[*24*] Hermanutz, R. O., Mueller, L. H., and Kempfert, K. D., "Captan Toxicity to Fathead Minnows *(Pimephales promelas),* Bluegills *(Lepomis macrochirus),* and Brook Trout *(Salvelinus fontinalis),*" *Journal of the Fisheries Research Board of Canada,* Vol. 30, 1973, pp. 1811–1817.

[*25*] Cardwell, R. D., Foreman, D. G., Payne, T. R., and Wilbur, D. J., "Acute and Chronic Toxicity of Chlordane to Fish and Invertebrates," *Ecological Research Series,* EPA-600/3-77-019, U.S. Environmental Protection Agency, Duluth, MN, 1977.

[*26*] Mount, D. I. and Norberg, T. J., "A Seven-Day Life Cycle Cladoceran Toxicity Test," *Environmental Toxicology and Chemistry,* Vol. 3, 1984, pp. 425–434.

[*27*] Casseri, N. A., Ying, W., and Sojka, S. A., "Use of a Rapid Bioassay for Assessment of Industrial Wastewater Treatment Effectiveness," *Proceedings,* 38th Industrial Waste Conference. Purdue University, West LaFayette, IN, 1983, pp. 867–878.

[*28*] Allison, D. T. and Hermanutz, R. O., "Toxicity of Diazinon to Brook Trout and Fathead Minnows," *Ecological Research Series,* EPA-600/3-77-060, U.S. Environmental Protection Agency, Duluth, MN, 1977.

[*29*] Holcombe, G. W., Phipps, G. L., Sulaiman, A. H., and Hoffman, A. D., "Simultaneous Multiple Species Testing: Acute Toxicity of 13 Chemicals to 12 Diverse Freshwater Amphibian, Fish, and Invertebrate Families," *Archives of Environmental Contamination and Toxicology,* Vol. 16, 1987, pp. 697–710.

[*30*] Ahmad, N., Benoit, D., Brooke, L., Call, D., Carlson, A., DeFoe, D., Huot, J., Moriarity, A., Richter, J., Shubat, P., Veith, G., and Walbridge, C., "Aquatic Toxicity Tests to Characterize the Hazard of Volatile Organic Chemicals in Water: A Toxicity Data Summary—Parts I and II," *Ecological Research Series,* EPA-600/3-84-009, U.S. Environmental Protection Agency, Duluth, MN, 1984.

[*31*] Dill, D. C., Alexander, H. C., McCarty, W. M., and Bartlett, E. A., "Toxicity of 1,1-Dichloroethylene (Vinylidene Chloride) to Aquatic Organisms," *Ecological Research Series,* EPA 600/3-80-057, U.S. Environmental Protection Agency, Duluth, MN, 1980.

[*32*] Call, D. J., Brooke, L. T., Ahmad, N., and Richter, J., "Toxicity and Metabolism Studies with EPA Priority Pollutants and Related Chemicals in Freshwater Organisms," *Ecological Research Series,* EPA-600/3-83-095, U.S. Environmental Protection Agency, Duluth, MN, 1983.

[*33*] Holcombe, G. W., Phipps, G. L., and Tanner, D. K., "The Acute Toxicity of Kelthane, Dursban, Disulfoton, Pydrin, and Permethrin to Fathead Minnows *(Pimephales promelas)* and Rainbow Trout *(Salmo gairdneri),*" *Environmental Pollution (Series A),* Vol. 29, 1982, pp. 167–178.

[*34*] Nebeker, A. V., McCrady, J. K., Mshar, R., and McAuliffe, C. K., "Relative Sensitivity of *Daphnia magna,* Rainbow Trout, and Fathead Minnows to Endosulfan," *Environmental Toxicology and Chemistry,* Vol. 2, 1983, pp. 69–72.

[35] Velte, J. S., "Acute Toxicity of Hydrazine Hydrate to the Fathead Minnow *(Pimephales promelas)* and Daphnia *(Daphnia pulex)*," *Bulletin of Environmental Contamination and Toxicology,* Vol. 33, 1984, pp. 598–604.

[36] Spehar, R. L., Tanner, D. K., and Gibson, J. H., "Effects of Kelthane and Pydrin on Early Life Stages of Fathead Minnows *(Pimephales promelas)* and Amphipods *(Hyalella azteca)*," *Aquatic Toxicology and Hazard Assessment (Fifth Conference), ASTM STP 766,* J. G. Pearson, R. B. Foster, and W. E. Bishop, Eds. American Society for Testing and Materials, Philadelphia, 1982, pp. 234–244.

[37] Macek, K. J., Buxton, K. S., Derr, S. K., Dean, J. W., and Sauter, S., "Chronic Toxicity of Lindane to Selected Aquatic Invertebrates and Fishes," *Ecological Research Series,* EPA-600/3-76-046, U.S. Environmental Protection Agency, Duluth, MN, 1976.

[38] Merna, J. W., Bender, M. E., and Novy, J. R., "The Effects of Methoxychlor on Fishes: I—Acute Toxicity and Breakdown Studies," *Transactions of the American Fisheries Society,* Vol. 2, 1972, pp. 298–301.

[39] DeGraeve, G. M., Elder, R. G., Woods, D. C., and Bergman, H. L., "Effects of Naphthalene and Benzene on Fathead Minnows and Rainbow Trout," *Archives of Environmental Contamination and Toxicology,* Vol. 11, 1982, pp. 487–490.

[40] U.S. Environmental Protection Agency, "Effects of Aqueous Effluents from In-Situ Fossil Fuel Processing Technologies on Aquatic Systems," Contract No. EY-77-C-04-3913, Annual Progress Report, Rocky Mountain Institute of Energy and Environment, University of Wyoming, Laramie, WY, 1977.

[41] Anderson, R. L., "Toxicity of Fenvalerate and Permethrin to Several Nontarget Aquatic Invertebrates," *Environmental Entomology,* Vol. 11, 1982, pp. 1251–1257.

[42] Spehar, R. L., Tanner, D. K., and Nordling, B. R., "Toxicity of the Synthetic Pyrethroids, Permethrin and AC 222, 705 and Their Accumulation in Early Life Stages of Fathead Minnows and Snails," *Aquatic Toxicology,* Vol. 13, 1983, pp. 171–182.

[43] Adelman, I. R. and Smith, L. L., Jr., "Standard Test Fish Development, Part I: Fathead Minnows *(Pimephales promelas)* and Goldfish *(Carassius auratus)* as Standard Fish in Bioassays and Their Reaction to Potential Reference Toxicants," *Ecological Research Series,* EPA-600/3-76-061a, U.S. Environmental Protection Agency, Duluth, MN, 1976.

[44] Cardwell, R. D., Foreman, D. G., Payne, T. R., and Wilbur, D. J., "Acute Toxicity of Selected Toxicants to Six Species of Fish," *Ecological Research Series,* EPA-600/3-76-008, U.S. Environmental Protection Agency, Duluth, MN, 1976.

[45] Hedtke, S. F. and Arthur, J. W., "Evaluation of a Site-Specific Water Quality Criterion for Pentachlorophenol Using Outdoor Experimental Streams," *Aquatic Toxicology and Hazard Assessment (Seventh Symposium), ASTM STP 854,* R. D. Cardwell, R. Purdy, and R. C. Bahner, Eds., American Society for Testing and Materials, Philadelphia, 1985, pp. 551–564.

[46] DeGraeve, G. M., Overcast, R. L., and Bergman, H. L., "Toxicity of Underground Coal Gasification Condenser Water and Selected Constituents to Aquatic Biota," *Archives of Environmental Contamination and Toxicology,* Vol. 9, 1980, pp. 543–555.

[47] Arthur, J. W., Lemke, A. E., Mattson, V. R., and Halligan, B. J., "Toxicity of Sodium Nitrilotriacetate (NTA) to the Fathead Minnow and an Amphipod in Soft Water," *Water Research,* Vol. 8, 1974, pp. 187–193.

[48] Macek, K. J. and Sturm, R. N., "Survival and Gill Condition of Bluegill *(Lepomis Macrochirus)* and Fathead Minnows *(Pimephales promelas)* Exposed to Sodium Nitrilotriacetate (NTA) for 28 Days," *Journal of the Fisheries Research Board of Canada.* Vol. 30, 1973, pp. 323–325.

[49] Mayer, F. L., Jr., Mehrle, P. M., Jr., and Dwyer, W. P., "Toxaphene: Chronic Toxicity to Fathead Minnows and Channel Catfish," *Ecological Research Series,* EPA-600/3-77-069, U.S. Environmental Protection Agency, Duluth, MN, 1977.

[50] Johnson, W. W. and Julin, A. M., "Acute Toxicity of Toxaphene to Fathead Minnows, Channel Catfish, and Bluegills," *Ecological Research Series,* EPA-600/3-80-005, U.S. Environmental Protection Agency, Duluth, MN, 1980.

[51] Johnson, W. W. and Finley, M. T., "Handbook of Acute Toxicity of Chemicals to Fish and Aquatic Invertebrates," Resource Publication 137, U.S. Department of the Interior, Fish and Wildlife Service, Washington, DC, 1980.

[52] van der Schalie, W. H., "The Acute and Chronic Toxicity of 3,5-Dinitroaniline, 1,3-Dinitrobenzene, and 1,3,5-Trinitrobenzene to Freshwater Aquatic Organisms," Technical Report 8305, U.S. Army Medical Bioengineering Research and Development Laboratory, Fort Detrick, Frederick, MD, 1983.

[53] Sokal, R. R. and Rohlf, F. J., *Biometry,* 2nd ed., W. H. Freeman and Co., San Francisco, CA, 1981.

Gerald J. Niemi[1] *and Gilman D. Veith*[2]

An Approach for Development of Structure-Biodegradation Relationships of Organic Chemicals

REFERENCE: Niemi, G. J. and Veith, G. D., **"An Approach for Development of Structure-Biodegradation Relationships of Organic Chemicals,"** *Aquatic Toxicology and Environmental Fate: Eleventh Volume, ASTM STP 1007,* G. W. Suter II and M. A. Lewis, Eds., American Society for Testing and Materials, Philadelphia, 1989, pp. 459–467.

ABSTRACT: Success in the development of predictive methods for relating biodegradation rates in water to chemical structure has been elusive. We present a strategy for evaluation of biodegradability of chemicals that involves a combination of structure-biodegradation relationships (SBR) and laboratory testing arranged in a four-tier heirarchy. The first tier involves a search of a data base on biodegradation information. Each subsequent tier consists of a query of potential SBR for the chemical or an experimental laboratory test of the chemical using a tier-specific procedure. The tiered hierarchy proceeds from a data base search, the simplest and cheapest level, to a river die-away test, the most complex and expensive level.

The main purpose of the first three tiers in this strategy is to discriminate readily biodegradable from persistent chemicals. The fourth tier involves the measurement of a biodegradation rate constant under experimental conditions. If a chemical is determined to be degradable, then metabolites of the chemical are also evaluated to identify possible recalcitrant by-products. The approach (*a*) satisfies the critical need to develop a systematic reference data base on biodegradation endpoints for existing and new chemicals, (*b*) incorporates the metabolites of biodegradation into the process, and (*c*) provides an economical means to evaluate chemical biodegradation.

KEY WORDS: biodegradation, structure-activity relationships, chemical evaluation, biochemical oxygen demand, hierarchy, expert system, river die-away, metabolites, aquatic toxicology

Predictions of the potential for organic chemicals to biodegrade in aqueous environments have been elusive, primarily because of the large number of factors potentially involved in the biodegradation process [*1–4*]. Yet, understanding biodegradation remains a major concern because the degree of hazard posed by a chemical is related to its concentration and often to its residence time in the environment. Chemicals that do not biodegrade have a greater chance of reaching higher concentrations and, concomitantly, provide longer exposure times to organisms.

Our perspective on biodegradation is based on the fundamental assumption that the biodegradation potential of a molecule is related to its chemical structure. Niemi et al. [*5*]

[1] Associate director, Natural Resources Research Institute, University of Minnesota, Duluth, MN 55811.

[2] Director, Environmental Research Laboratory—Duluth, U.S. Environmental Protection Agency, Duluth, MN 55804.

described a structure-biodegradation relationship (SBR) approach that identified features of chemicals that were associated with their degradability or persistence. However, further progress in development of SBR will be hampered by lack of a systematic approach to gathering additional information for new and diverse chemical structures [*6*]. This would be especially useful for identifying or substantiating structural features associated with the biodegradability of chemicals [*3,5*] Here we describe an approach that could be used to further the development of predictive relationships between chemical structure and biodegradation. In addition, the approach fulfills the need to develop a systematic data base on endpoints of biodegradation.

Our approach for assessing the relative degradability of chemicals is cast in a hierarchical, tiered framework. The tiered framework is structured to proceed from a relatively simple, qualitative, and inexpensive query of biodegradation potential at the first tier to a relatively complex, quantitative, and expensive evaluation at the fourth tier. The evaluation of a chemical proceeds down the tiered framework until a reasonable determination can be made about the biodegradability established by the user, who would consider such factors as complexity of the molecule, information available for analogues, and quality of comparative data.

Here we illustrate a preliminary application of this approach to the second level of the tier. Currently, the approach is conceptual because most aspects of the protocol have not been fully developed, tested, or implemented. For instance, there are many aspects of the biodegradation process that have not been adequately examined, such as the acclimation process of inocula, water quality, nutrient conditions in the aqueous environment, and chemical concentration. Examination of the importance of these factors and how to standardize them properly for developing a useful biodegradation test must be a parallel effort, along with the refinement in this generalized approach.

Definition of Biodegradation

Chemicals are transformed by a variety of means including microbial degradation, photolysis, and hydrolysis [*3,7*]. Here, we are concerned with biological processes that lead to changes in the molecular structure of a parent compound. These processes primarily include transformations induced by microorganisms as the primary mechanism for the degradation of most organic compounds [*1,8–10*]. Within each level of the framework we are interested in primary degradation. This has been defined as any biologically induced structural transformation in the parent compound that changes its molecular integrity [*3*]. The whole framework, however, is concerned with ultimate degradation of the compound, which is defined as biologically mediated conversion of an organic compound to inorganic compounds and products associated with normal metabolic processes [*3*].

Chemical Evaluation

There are numerous philosophical approaches to assessments of biodegradation in water and, consequently, different methods for assessing biodegradation of chemicals [*6*]. These include the shake-flask, activated sludge, river die-away, biochemical oxygen demand (BOD), and carbon dioxide (CO_2) evolution [*3*] methods. For cost-effective examination of biodegradation, we propose that the selection of techniques to be used is a function of the desired objectives of the test, the degree of resolution needed for the test, and monies available.

Fortunately, these factors are interrelated, so that if the objectives require a detailed assessment of the biodegradation process with accurate predictions, the cost for the evaluation will also be high. If only a qualitative estimate of the process is required, then the cost may be relatively low. Thus, information on biodegradation follows a gradient from general to detailed objectives, qualitative to accurate resolution, and low to high monetary costs, depending on the user's need.

Who are the users? We foresee two primary users of chemical evaluation regarding biodegradation. One is the regulatory agencies who are responsible for evaluating chemicals in order to protect the integrity of ecological systems and human health. A primary interest of this user is the need to know how persistent a chemical might be if released into the environment. Another user is the chemical producer who invests millions of dollars in synthesis, testing, and marketing of chemicals. This user has the need to know whether a chemical is degradable under a variety of conditions including conventional wastewater treatment. This does not imply that degradation is the only endpoint of interest, because it represents only one of many parameters that need to be evaluated. Furthermore, the desired activity of a chemical might be related to its biodegradation pattern. For example, a desirable property of an insecticide is to kill the intended pest quickly but not be so persistent as to cause undue harm to nontarget organisms. In this case, the chemical must be persistent enough to achieve the desired effect, yet degradable enough to be eliminated from the environment. Because biodegradation potential is an important benchmark, knowing that the chemical is readily degradable or persistent can help in making economically important decisions in product development of chemicals.

It is imperative to expand briefly on the different necessities of these two users. In a regulatory setting, there is a need to make an independent, unbiased evaluation of a chemical that has been selected by industry from a variety of potentially desirable chemicals. The number of chemicals from all industries that must be evaluated is relatively large and the structures diverse. In contrast, industry is faced with identifying chemicals that are environmentally safe, yet effective enough to achieve a specific purpose. The identification of these chemicals can be pursued from a variety of potential lead structures, such as the hundreds of benzene derivatives, each of which result in hundreds of potential structural variations. Each structural analog will result not only in different chemical behavior for the intended purpose, but also in different environmental behavior. Synthesis of all the potentially useful chemicals is impossible, but predictions of the biodegradation potential of possible structures based on SBR would be useful for objectively ranking lists of synthetically possible chemicals according to potential risks. Therefore, this approach can be useful both to a regulatory agency for chemical evaluation and to industry in the selection of appropriate chemicals with desirable industrial and environmental qualities.

Overview of the Approach

In Fig. 1 we illustrate our conceptual approach to the problem of biodegradation. In its most simplistic form, the approach consists of four tiers of information in which each succeeding level represents greater effort, more time for assessment, and, subsequently, more detailed knowledge of the chemical. Briefly, the approach is organized as follows. At the first level, a data base is searched for existing information on biodegradation using a variety of substructural features. The second level uses a test which potentially simulates the treatability of the chemical in a modern sewage treatment facility; and the third level uses a test that potentially simulates longer term degradation (for example, up to one month) of a chemical in aqueous environments. Finally, the fourth level considers long-term degradation (for example, 2 to 4 months) in aqueous environments. All levels except the first consist of (1) an expert system based on SBR for data derived from previous experiments from that level, and (2) an experimental protocol using the biodegradation test for that level.

The first level in this approach is a review of test data already available for the chemical

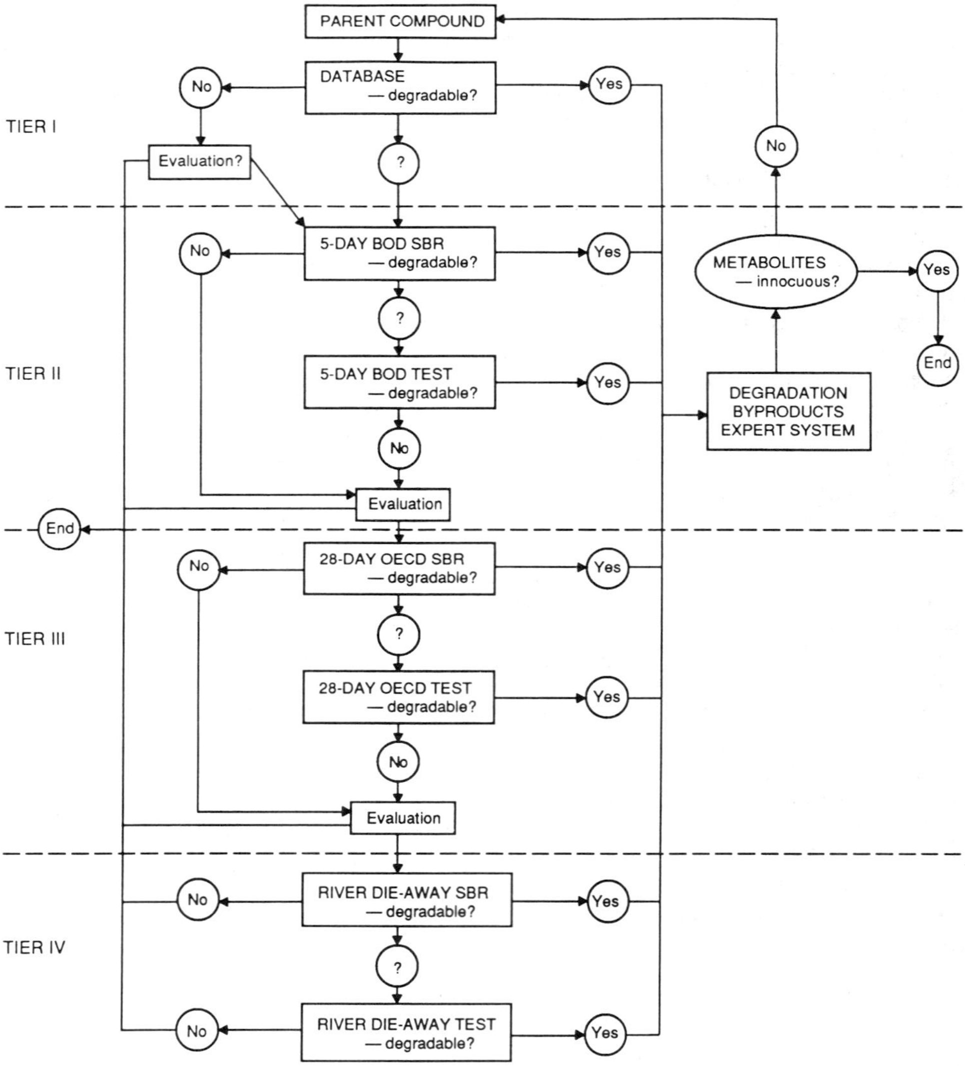

FIG. 1—*Conceptual flow diagram of the tiered framework for assessing the biodegradation potential of organic compounds.*

or similar chemicals. We envision that at this level a weight-of-evidence approach like that suggested by Howard et al. [*6*] could be used. In succeeding levels, SBR are first applied to the chemical of interest and a decision is made on whether information is adequate, inadequate, or inconclusive to determine biodegradability for the chemical. If the chemical is determined to be biodegradable, then the chemical is evaluated with respect to possible metabolites of the chemical, based on known biochemical pathways. For all levels of the approach, the metabolites of the chemical are also subjected to the biodegradation model as well as the parent compound. In this fashion, potentially recalcitrant metabolites are also identified (see the section below for a more detailed discussion of the metabolite identification step).

If SBR information for the chemical is judged to be inadequate or inconclusive to make a decision, then the chemical must be subjected to a laboratory test. If SBR predicts that the chemical is not biodegradable, then the chemical must be reevaluated with respect to further testing. For example, it would be monetarily inefficient to subject a chemical to further testing in Tier II if a valid 5-day BOD test had already been performed on the chemical. However, if only a river die-away test existed for the chemical, then further testing in the 5-day BOD test may be desirable, depending on the proposed use for the chemical. An evaluation phase should always precede entry of the chemical into the next tier of evaluation. The second tier consists of a conventional acclimated 5-day BOD test, a test to determine whether the chemical is potentially treatable under conditions similar to a modern sewage treatment facility. This does not suggest that the 5-day BOD test produces the same result as a chemical would in a sewage treatment facility; however, it is likely that a chemical that has a high 5-day BOD value will also degrade in a modern sewage treatment facility. This relationship should be tested, but some evidence suggests that the degradation rate is faster in a modern sewage treatment facility than in the 5-day BOD test [*11*]. In the last two levels, a similar process occurs. If the information from the SBR system is inadequate or inconclusive to make a decision, then the chemical must be subjected to a test. If SBR predicts a chemical to be nonbiodegradable, then an evaluation must be made on whether further testing is warranted.

Structure-Biodegradation Relationships

Development of SBR at the current time is limited, but several papers have expounded on the potential for development of such relationships [*5,6,9,12,13*]. The reasons for the lack in development of these relationships have primarily been the lack of a systematic reference data base on biodegradation [*3,5,6,9*] and the complexity of the biodegradation process itself [*1,3,9*]. Several potential data bases have been developed [*14,15*], but the data bases tend to be biased toward relatively small compounds (for example, those with a molecular weight of less than 500) [*5,12*]. The approach outlined here would promote the development of a systematic reference data base in which SBR can be expanded. For instance, a natural consequence of the approach suggested here is that novel, diverse structures will be considered in the evaluation. By virtue of their novelty, SBR will not be available and, hence, the structures will need to be tested. The results of those tests will then be incorporated into the SBR data base. SBR is then a dynamic data base, growing in knowledge with each new chemical and test. In addition, information obtained for one level of the tier through actual testing can be used to revise and refine SBR at other levels of the framework.

5-Day BOD

The 5-day BOD test is a relatively standardized procedure for testing whether a chemical is readily degradable [*16*]. The test has been used for making industrial decisions on the treatability of chemicals, and a comparatively large data base has been developed on endpoints of 5-day BOD [*5*]. Although there are several tests that could be used at this stage, Means and Anderson [*16*], in their comparison of five different tests, including the 5-day BOD, showed that the test method made little difference in determining whether a chemical was readily degradable. Similarly, Howard and Banerjee [*9*] stated that different screening tests appear to give fairly reproducible qualitative results for determining chemicals that are either very resistant or very biodegradable. Here the BOD test is only suggested for use in determining chemicals that are readily degradable. Furthermore, the BOD test is simple and relatively inexpensive to perform.

For the 5-day BOD level of screening, we have developed an SBR method that could be used to classify chemicals as readily degradable or persistent [5]. In the classification protocol, chemicals which exhibit a BOD greater than 15% of the theoretical demand in a conventional acclimated 5-day BOD test are considered readily degradable. The protocol is designed to prevent false-negative classifications, that is, to avoid classifying persistent chemicals as readily treatable. False-positive chemicals that biodegrade in the 5-day BOD test are used to revise SBR, and persistent chemicals are designated for the third tier of testing. If SBRs for the chemical are inadequate or inconclusive, then the chemical must be subjected to an actual test.

At the current time, the SBRs identified above are one of the few published studies that consider a wide variety of heterogeneous chemicals. Yet, the breadth of chemicals included in the SBR are relatively limited. For example, there are few chemicals with the following configurations that have been considered: (*a*) sulfonated compounds, (*b*) polyhalogenated compounds, or (*c*) multicyclic compounds. Therefore, data bases need to be expanded to increase the breadth of structural types to be considered.

To examine the potential for use of this system at this level, we sought to test the system on about 3000 structurally diverse industrial chemicals. We chose the chemicals which appear on both the Toxic Substances Control Act (TSCA) inventory of industrial chemicals (20 000) and a commercial chemical supplier (10 000). The chemicals in common totaled 3420 and served as our evaluation data base of structures. Each chemical was analyzed with an expert system that uses SBR on the basis of a set of heuristic rules on the biodegradability of chemicals [5]. With these data, 1292 of the 3420 (37.7%) chemicals would have been identified as potentially persistent chemicals, while no prediction would have been made for 246 (7%) of the chemicals because of inadequate or inconclusive data on their structures. Most importantly, about 1881 or 55% of the chemicals would have been predicted to be readily degradable in a BOD test. The SBRs are not yet ready for use in screening because they need to be tested with an independent data set and, as mentioned previously, the breadth of chemicals included is still relatively restricted. However, this exercise indicated that such a system might be efficient for evaluating chemicals since it only takes seconds to evaluate the chemical through the use of SBR. Actual laboratory testing would require a minimum of a week of effort.

In this evaluation of 3420 organic chemicals, we concluded that SBR for screening chemicals using chemical structure will not be improved until the data base is expanded to include reference data on a wider variety of chemical structures. In contrast, the evaluation process could prove especially useful for screening a large number of chemical structures. Furthermore, even though as many as 55% of commercially available chemicals may be readily degradable, as suggested by the 5-day BOD model, the potential metabolites of these chemicals must also be evaluated.

28-Day Organization for Economic Cooperation and Development (OECD) Methodology

The 5-day BOD test has the capability for qualitatively determining readily degradable chemicals, and the last test proposed in this approach, the die-away test, can assess the persistence of chemicals after release into surface waters. However, in the first and second tiers it is likely that too many chemicals would be passed to lower tiers, which require more detailed and expensive tests. Because the die-away test is the most expensive test of those proposed here, an intermediately priced test is needed to delineate further the biodegradation potential of chemicals, particularly those with longer acclimation times. Therefore, we propose a 28-day unacclimated test using the modified Ministry of International Trade and Industry (MITI) method suggested by OECD as an intermediate test [*17*]. The major purpose of including this test is to gain insight on moderately degradable chemicals that may appear nondegradable in short-term tests but are degradable in aqueous environments after being discharged from a sewage treatment facility on a continuous basis. As with the previous level, one of the major reasons for selecting this method was that more than 1000 chemical structures have been tested using this technique.[3] Although the raw data are not yet available, MITI and the Kurume Chemical Inspection and Testing Institute (CITI) laboratory have permitted us to examine the data base. Because this MITI data required substantial time and resources, it will probably remain the largest systematic data base on biodegradation when it is published. This test is similar to the 5-day BOD test because it uses oxygen consumption as an indirect measure of biodegradation [*3*] but more expensive because it is of longer duration.

Progress has already been made in the development of SBR with the 28-day BOD test. For example, Kitano[3] summarized a series of relationships on chemical structure as related to biodegradation, while Takatsuki [*18*] expanded on some of these relationships using a quantum chemical approach. What is needed next is a coordination of efforts and synthesis of potential SBR that have been developed using this technique.

The process for using this third level is the same as for the second level. A chemical is first subject to SBR using relationships identified from previous 28-day tests. If the SBRs are inadequate or inconclusive, then the chemical is subject to the 28-day test.

Die-Away Test

The fourth level we propose is the die-away test, which is also the most expensive test to perform of the three suggested here. The first two tests make use of first-order kinetics in which the rate is simply the product of the rate constant and the substrate concentration [*3*]. In contrast, the die-away test makes use of the Monod kinetics rate equation in which a second-order rate expression is calculated as a function of both population and substrate concentration [*3,19*]. The use of this kinetic equation requires that biodegradation be directly measurable in terms of microbial growth and the disappearance of the chemical over time [*3*].

A variety of chemicals have been tested with the die-away test, and some SBRs have been developed using this technique. For instance, Scow [*3*] summarized data on about 40 different chemicals that were presumably all based on the river die-away test for second-order rate constants. At the current time, it is difficult to develop SBRs from these data because the units of measurement are different, and the testing protocols were relatively unstandardized. SBRs, however, have been presented for several chemical structures (for example, see Refs *20* through *22*), of which most are based on selected homologous series of chemicals. Hence, SBRs are currently limited to a restricted set of chemicals, and the breadth of extrapolation potential to a larger group of chemicals is relatively small. A coordinated effort for standardizing the method and subsequently expanding the chemicals in SBR data for river die-away data are needed to improve our understanding of chemicals with relatively long-term biodegradation tests.

For the fourth and final tier of testing, we have proposed the same process of evaluation as in the previous two levels: SBR evaluation and then subject the chemical to a test if SBR indicates that the biodegradation potential of the chemical is inadequate or inconclu-

[3] Kitano, M., Chemical Inspection and Testing Institute, personal communication, June 1985.

sive. If the chemical is predicted to be nondegradable, then the end of this tiered approach is obtained. The development of the SBR for this level is in a preliminary stage and has limited usefulness at the present time. However, as the data base grows, and if methods are standardized, then data suitable for SBR could be developed.

Degradation Metabolites and Pathways

Many researchers have commented on the necessities of including the entire biodegradation process or ultimate degradation of a chemical into the evaluation process (for example, see Refs *23* and *24*). We have illustrated this need in Fig. 1 by the review of metabolic pathways and metabolites formed from each degradable chemical. If a chemical is predicted to be degradable at any level of the protocol, the likely metabolites are evaluated for potentially hazardous or persistent intermediates. At present, the prediction of metabolites must be done by experts. However, there are several commercially available computer programs that identify metabolites for many structures. Moreover, we have developed an expert system to simulate metabolism and which can be fully integrated with the biodegradation model proposed herein.[4] After the potential metabolites are predicted, each is evaluated as to whether it is innocuous. Here innocuous is minimally defined as the absence of toxicophores or structural fragments of unusual persistence. If metabolites are innocuous, then ultimate biodegradation is presumed and the evaluation process ends. If a metabolite is not innocuous, the metabolite is reevaluated as a new parent compound. From our knowledge in biochemistry, much is known on the degradation pathways for these organic compounds (for example, see Scow [*3*] for a review).

Conclusions

The greatest potential for progress in using the approach outlined here will probably be in the first three levels. There is already progress in the development of biodegradation data bases [*6,13,14*] and SBR [*3,5,10,12*]. A critical need is in development of a coordinated effort to assemble and analyze the available information. Because of limited syntheses of available data, few comprehensive analyses have been completed for biodegradation data bases in which to develop SBR. The data currently available are limited to qualitative judgments on relative biodegradability or on more limited estimations for homologous series of chemicals [*6*]. The data are especially inadequate with regard to the breadth of chemicals that have been evaluated and very limited in applications to developing SBR estimates of rate constants from structure.

To improve development of SBR, we have selected a set of 300 chemical structures from the TSCA inventory of industrial chemicals. These chemicals were selected because they are probably not degradable in a 5-day BOD test as evaluated by the 5-day BOD SBR and were randomly selected from those chemicals that are commercially available. Progress on expanding SBR is inhibited until a systematic effort to assemble data on structures with little information or on structures with currently nebulous or conflicting endpoints are actually tested. If these chemicals were tested systematically, this would enable further SBR to be developed. Furthermore, as the systematic effort for gathering biodegradation information proceeds, a parallel effort is necessary to examine the effects of how inocula, acclimation, microorganism concentration, chemical concentration, and different environmental exposures affect the process of biodegradation.

---

[4] Bradbury, S., unpublished report, Environmental Research Laboratory—Duluth, U.S. Environmental Protection Agency, Duluth, MN, November 1896.

Acknowledgments

Development of the ideas and framework for this paper were funded in part by a cooperative agreement extended to G. J. Niemi (CR 810824) and to R. R. Regal (CR 811981) by the U.S. Environmental Protection Agency. Although the research described has been funded wholly or in part by the U.S. Environmental Protection Agency, it has not been subjected to EPA review and therefore does not necessarily reflect the view of the EPA, and no official endorsement should be inferred.

References

[*1*] Alexander, M., *Science,* Vol. 211, 1981, pp. 132–139.
[*2*] Painter, H. A., *Proceedings of the Royal Society of London, Series B, Biological Sciences,* Vol. 185, 1974, pp. 149–158.
[*3*] Scow, K. M. in *Handbook of Chemical Property Estimation Methods,* W. J. Lyman, W. F. Reehl, and D. H. Rosenblatt, Eds., McGraw-Hill, New York, 1982, chapter 9, pp. 9-1–9-85.
[*4*] Howard, P. H., Saxena, J., and Sikka, H., *Environmental Science and Technology,* Vol. 12, 1978, pp. 398–407.
[*5*] Niemi, G. J., Veith, G. D., Regal, R. R., and Vaishnav, D. D., *Environmental Toxicology and Chemistry,* Vol. 6, 1987, pp. 515–527.
[*6*] Howard, P. H., Hueber, A. E., and Boethling, R. S., *Environmental Toxicology and Chemistry,* Vol. 6, 1987, pp. 1–10.
[*7*] Mill, T., *Environmental Toxicology and Chemistry,* Vol. 1, 1982, pp. 135–141.
[*8*] Howard, P. H. and Banerjee, S., *Environmental Toxicology and Chemistry,* Vol. 3, 1984, pp. 551–562.
[*9*] Banerjee, S., Howard, P. H., Rosenberg, A. M., Dombrowski, A. D., Sikka, H., and Tullis, D. L., *Environmental Science and Technology,* Vol. 18, 1984, pp. 416–422.
[*10*] Enslein, K., Tomb, M. E., and Lander, T. R. in *QSAR in Environmental Toxicology.,* K. L. E. Kaiser, Ed., D. Reidel Publishing, New York, 1984, pp. 89–109.
[*11*] Larson, R. J., *Residue Reviews,* Vol. 95, 1983, pp. 159–171.
[*12*] Boethling, B., *Environmental Toxicology and Chemistry,* Vol. 5, 1986, pp. 797–806.
[*13*] Howard, P. H., Sage, G. W., Lamacchia, A., and Colb, A., *Journal of Chemical Information and Computer Sciences,* Vol. 22, No. 1, 1982, pp. 38–44.
[*14*] Vaishnav, D. D., Element 16, Vol. 1, Biochemical Oxygen Demand Data Base, Technical Report, EPA/CR809234, University of Wisconsin, Superior, WS, 1984.
[*15*] *Standard Methods for the Examination of Water and Wastewater,* 14th ed., American Public Health Association, Washington, DC, 1975.
[*16*] Means, J. L. and Anderson, S. J., *Water Air and Soil Pollution,* Vol. 16, 1981, pp. 301–315.
[*17*] "OECD Guidelines for Testing Chemicals," Organization for Economic Cooperation and Development, OECD Publications Office, Paris, France, Vol. 301, No. C, 1981.
[*18*] Takatsuki, M., "Relationship Between the Biodegradability of Chemicals and Parameters Calculated from LCAO Method," Preprint Extended Abstract, Division of Environmental Chemistry, American Chemical Society, Denver, CO, April 1987.
[*19*] Baughman, G. L., Paris, D. F., and Steen, W. C., "Quantitative Expression of Biotransformation Rate," *Biotransformation and Fate of Chemicals in the Aquatic Environment,* A. Maki, K. Dickson, and J. Cairns, Jr., Eds., American Society for Microbiology, Washington, DC, 1980, pp. 105–111.
[*20*] Wolfe, N. L., Paris, D. F., Steen, W. C., and Baughman, G. L., *Environmental Science and Technology,* Vol. 14, 1980, pp. 1143–1144.
[*21*] Paris, D. F., Wolfe, N. L., and Steen, W. C., *Applied Environmental Microbiology,* Vol. 44, 1982, pp. 153–158.
[*22*] Paris, D. F., Wolfe, N. L., Steen, W. C., and Baughman, G. L., *Applied Environmental Microbiology,* Vol. 45, 1983, pp. 1153–1155.
[*23*] Dagley, S. in *Proceedings,* Workshop on Microbial Degradation of Pollutants in Marine Environments, April 1979, EPA-600/9-79-012, U.S. Environmental Protection Agency, Washington, DC, pp. 534–542.
[*24*] Chapman, P. in *Proceedings,* Workshop on Biodegradation Kinetics, Navarre Beach, FL, October 1983, U.S. Environmental Protection Agency, Washington, DC, pp. 118–120.

Robert L. Lipnick[1]

A Quantitative Structure-Activity Relationship Study of Overton's Data on the Narcosis and Toxicity of Organic Compounds to the Tadpole, *Rana temporaria*

REFERENCE: Lipnick, R. L., **"A Quantitative Structure-Activity Relationship Study of Overton's Data on the Narcosis and Toxicity of Organic Compounds to the Tadpole, *Rana temporaria*,"** *Aquatic Toxicology and Environmental Fate: Eleventh Volume, ASTM STP 1007,* G. W. Suter II and M. A. Lewis, Eds., American Society for Testing and Materials, Philadelphia, 1989, pp. 468–489.

ABSTRACT: A quantitative structure-activity relationship (QSAR) study has been performed on the toxicity of organic compounds to the tadpole, *Rana temporaria.* The biological test data were taken from Charles Ernest Overton's classic monograph *Studien über die Narkose,* which has been widely cited by researchers studying the correlation of biological activity with partition coefficient and the mechanism of anesthesia. A baseline narcosis QSAR model, $\log (1/C) = 0.909 \log P + 0.727$, was derived using data on 18 alcohols, ketones, and hydrocarbons, where C is the lowest narcosis-inducing concentration in moles per litre, and P is the *n*-octanol/water partition coefficient. The corresponding lowest tadpole test concentration producing narcosis or other toxic effects for the remaining 100 compounds was compared with that predicted by the above QSAR equation. In general, very good agreement was obtained between the predicted and measured values for those nonelectrolytes producing solely a narcosis physiological response. In contrast, acetamide, succinimide, urea, and ethylene glycol produced lethality without narcosis at considerably lower concentrations than predicted. These compounds, which have low lipoid solubility and undergo little or no membrane transport, produce osmotic effects. The observed and predicted narcotic concentrations for the two hypnotic agents, sulfonal and trional, are in general agreement, based upon log P values calculated from the water solubility and melting point. The lack of observed effect for cetyl alcohol, anthracene, anthraquinone, parachloralose, and terpin hydrate, even in tadpole tests of long duration, can be attributed to a water solubility limitation, or cutoff, that is due to a high log P, a high melting point, or a combination of these factors. Although coumarin is 17 times more toxic than predicted, it produces reversible narcosis effects only. This behavior may reflect a reversible Michael-type addition by nucleophilic moieties present in molecules comprising the membrane lipoid site of action. Alkaloids and strongly basic amines are all more toxic than predicted.

KEY WORDS: quantitative structure-activity relationships (QSARs), toxicity, tadpole, narcosis, anesthesia, Overton, structure-activity relationships, aquatic toxicology

Correlations between the partition coefficient and the toxicity of organic chemicals acting by a narcosis mechanism to aquatic organisms can be traced to the studies of Charles Ernest Overton [*1,2*] at the University of Zurich and studies published independently by

[1] Leader, Structure-Activity Group, Environmental Effects Branch, Health and Environmental Review Division, Office of Toxic Substances, U.S. Environmental Protection Agency, Washington, DC 20460.

Hans Horst Meyer [3] and his collaborator Fritz Baum [4] at the University of Marburg. Most of Overton's studies, which were performed using tadpoles of the species *Rana temporaria,* were reported in his classic monograph *Studien über die Narkose* [2], which was the subject of a recent review [5]. Hansch and co-workers [6,7] used data on 53 of the compounds tested by Overton to derive a quantitative structure-activity relationship (QSAR) model (Eq 1)

$$\log\left(\frac{1}{C}\right) = 0.901 \log P + 0.909 \quad (n = 57;\ r = 0.962;\ s = 0.312) \tag{1}$$

where C is the lowest concentration in moles per litre found to produce narcosis in tadpoles, P is the n-octanol/water partition coefficient, n is the number, r is the correlation coefficient, and s is the standard error.

Hansch and co-workers found that, in contrast, the parameters molecular weight, polarizability, molar attraction constant, parachor, and adjusted parachor, gave much poorer correlations (Table 1). Hansch and co-workers concluded that the Overton data set was the best they had examined for comparing the utility of these properties, due to the widely varying structures and the degree of biological activity of Overton's compounds.

The present study was initiated to assemble Overton's entire set of data on nongaseous compounds, including those that produce other than narcotic effects, and to compare the lowest observed narcosis or lethal concentration with that predicted by a narcosis QSAR model, from a much smaller subset of alcohols and ketones known to be well associated with a narcosis mechanism (Table 1).

Procedures

Data on the lowest concentration producing reversible narcosis or other toxic responses in tadpoles of the species *R. temporaria* were obtained from Overton's 1901 monograph *Studien über die Narkose* [2]. Overton reported that these concentrations produced narcosis within as little as several minutes for simple monohydric alcohols and other compounds, but much longer times were required for others. Log P values, where P is the n-octanol/water partition coefficient, were calculated from the chemical structure using the CLOGP3.3 computer program [8] and compared with measured log P values [8,9] when data were available. The n-octanol/water partition coefficient values were also compared with the olive oil/water partition coefficient values of Overton [2] and Macy [10].

TABLE 1—*Correlation of Overton's tadpole narcosis data with various parameters.*[a]

Parameter	r^2	s
Log P	0.913	0.343
Molecular weight	0.567	0.765
Polarizability	0.683	0.654
Molar attraction	0.758	0.571
Parachor	0.556	0.775
Adjusted parachor	0.861	0.434

[a] Data were fitted [6] to linear models of the type log $(1/C) = aX + b$, where $n = 53$ for all correlations.

A good correlation exists between log P and water solubility (Eq 2) [*11*]

$$\log P = 6.5 - 0.89 \log S - 0.015\text{MP} \qquad (2)$$
$$(n = 27;\ r = 0.96)$$

where S is the water solubility in micromoles per litre, and MP is the melting point in degrees centigrade (nominal value of 25°C for liquid solutes). Equation 3 was obtained by algebraic transformation of Eq 2

$$S = \text{MW}10^{(4.30 - 1.12 \log P - 0.017\text{MP})} \qquad (3)$$

where MW is the molecular weight, and the solubility is in milligrams per litre. Equation 3 was used to compare water solubility with Overton's test concentrations as well as with Overton's measured water solubilities. For carbon disulfide and trimethylthiourea, CLOGP3.3 did not provide an estimated log P value due to missing fragment values, and no measured log P was available. For these compounds, as well as for several compounds for which the CLOGP3.3 value appeared anomalous, a log P value was calculated from water solubility (Eq 2) and from melting point data obtained from the literature.

Results and Discussion

A QSAR model (Eq 4) was derived from data on alcohols, ketones, and aromatic hydrocarbons (Table 2) [*12–14*], which are chemical classes known to be associated with a narcosis mechanism [*2*].

$$\log\left(\frac{1}{C}\right) = 0.909 \log P + 0.727 \qquad (4)$$
$$(n = 18;\ r = 0.986;\ s = 0.227)$$

C is the lowest concentration in moles per litre reported by Overton to produce narcosis or anesthesia in tadpoles *(R. temporaria)*. The slope and intercept of this QSAR are similar to those values found previously for fish toxicity QSARs [*15–17*]. Since the slope is close to unity, it is consistent with the hypothesis that (*a*) *n*-octanol provides a reasonable model for the physicochemical properties of the lipoid site of action, (*b*) steady-state thermodynamic partitioning occurs between the aqueous toxicant donor phase and the lipoid site, and (*c*) narcosis is observed when a common molar concentration of toxicant is achieved within the tadpole at the lipoid site of action. The observed narcotic concentrations, C_{obs}, were then compared with those predicted, C_{pred}, according to Eq 5 [*17*]

$$T_e = \frac{C_{pred}}{C_{obs}} \qquad (5)$$

where values of the excess toxicity, T_e, close to unity are considered to be consistent with a narcosis mechanism. In general, all nonelectrolytes are expected to have T_e values ≥ 1. Values of $T_e > 1$ provide evidence of a more specific molecular mechanism of action.

The T_e values for these alcohols, ketones, and hydrocarbons range from 0.4 to 1.6 and provide a baseline by which to compare the predicted and observed effect concentrations for the remaining chemicals studied by Overton. The predicted water solubilities (Eq 3) compare favorably with those for the five alcohols, three ketones, and four hydrocarbons measured by Overton, and this consistency provides further confidence in the utility of the

QSAR calculations themselves. Overton found that cetyl alcohol (Compound 19)[2] (Table 3) produced no narcosis, even if tadpoles were exposed for 5 to 8 days to a supersaturated solution prepared by vigorously shaking the melted alcohol with warm water. Under these conditions, tadpoles became somewhat sluggish after a few hours, but this condition decreased with time. The reversibility of this partial narcosis may be considered to reflect a decreasing aqueous concentration of cetyl alcohol as the degree of supersaturation decreases with time. This result is consistent with the prediction that the narcotic concentration (0.0138 mg/L) exceeds the water concentration (0.0063 mg/L) by only a small factor. Phenanthrene, with a log *P* value of 4.49, produced narcosis at 2 mg/L, which is just within Overton's reported water solubility of 3.3 mg/L. In contrast, he found that anthracene (Compound 20) (Table 3), an isomer of phenanthrene, has no narcotic effect. Anthracene has the same calculated log *P* value but has a predicted water solubility of only 0.0063 mg/L, over two orders of magnitude lower than the predicted narcosis concentration. This large difference in water solubility between the two isomers is due to the much greater melting point of anthracene (218°C), reflecting a much higher degree of symmetry and corresponding stability in the crystal lattice for anthracene.

Much longer test durations were required for compounds producing narcosis at low concentrations. Overton attributed this behavior, not to the lack of permeability of epithelia of the gills and ganglia cells, but to the slow rate of uptake through the gills to the site of action. He concluded that even if complete equilibrium could be achieved between the tadpole blood plasma concentration and that in the aqueous circumambient phase, the blood can transport only a small amount of toxicant to the ganglia cells in each circulation of the blood through the body. Overton proposed that other hydrocarbons including retene ($C_{18}H_{18}$), fluoranthrene ($C_{16}H_{10}$), and pyrene ($C_{16}H_{10}$) may also produce narcosis, given a sufficiently long experimental test duration and a sufficiently large volume of test solution so that no appreciable change in concentration occurs during the experiment. Pyrene and fluoranthrene each have a calculated log *P* value of 4.95 and a predicted minimum narcosis concentration of 1.2 mg/L. The predicted water solubility for fluoranthrene (0.149 mg/L), with a melting point of 111°C, is only slightly less than the predicted narcosis concentration. This margin is greater for pyrene (melting point 156°C), where the predicted water solubility is only 0.026 mg/L. Retene, with a calculated log *P* of 6.72 and melting point of 99°C, has a predicted narcosis concentration (0.035 mg/L) that exceeds the predicted water solubility (0.0029 mg/L) by only one order of magnitude.

Acetone and methyl ethyl ketone do not produce sustained narcosis without side effects, as in the case of the alcohols. For acetone, complete narcosis requires a concentration of 1.5% by weight (16 700 mg/L), which quickly leads to death through cessation of the heart. At 0.8%, only incomplete narcosis was produced, but weakening of the circulation was observed in 3 h, and death occurred in 7 h. At a 0.4% concentration, the tadpoles continued movement for 14 h, but all were dead after 19 h. The cardiotoxic side effects reported by Overton probably reflect the transformation of acetone in the tadpole bloodstream to one or more toxic metabolites. Rats administered acetone at high doses metabolize only a portion of the toxicant. A metabolic study using labeled acetone indicated that it may be split into formate and acetate [*18*], which may be responsible for the side effects noted by Overton.

Equation 4, which predicts the minimum concentration for narcosis, was applied to data on 100 additional compounds compiled from Overton's monograph. The majority have T_e values of <2 and produce narcosis for varying periods of time. Overton regarded as

---

[2] The compound numbers indicate the numbers used to refer to the chemical compounds in the tables.

TABLE 2—*Tadpole narcosis data for alcohols, ketones, and hydrocarbons used to derive baseline narcosis QSAR Eq 4.*

Compound No.	Chemical Name	Chemical Abstracts Registry No.	Molecular Formula	Molecular Weight	Log P^a	Narcosis Concentration, mg/L		T_e^d	Solubility (Overton), mg/L		Predicted Water Solubility, mg/L[f]	Melting Point, °C[g]	Log K^h
						Observed[b]	Predicted[c]		Water	Oil[e]			
	Saturated Monohydric Alcohols												
1	Methyl alcohol	67-56-1	CH_4O	32.04	−0.764 (−0.77)	18 200	29 700	1.64	∞	<20 000	2×10^6	liquid	(−1.96)
2	Ethyl alcohol	64-17-5	C_2H_6O	46.07	−0.235 (−0.31)	13 300	14 100	1.061	...	...	645 000	liquid	−1.48 (−1.34)
3	*n*-Propyl alcohol	71-23-8	C_3H_8O	60.09	0.294 (0.25)	6 670	6 090	0.914	...	...	214 000	liquid	−0.90 (−0.85)
4	*iso*-Propyl alcohol	67-63-0	C_3H_8O	60.09	0.074	7 690	9 660	1.26	...	...	378 000	liquid	...
5	*n*-Butyl alcohol	71-36-3	$C_4H_{10}O$	74.12	0.823 (0.88)	2 850	2 480	0.871	83 000	∞	67 200	liquid	(−0.28)
6	*iso*-Butyl alcohol	78-83-1	$C_4H_{10}O$	74.12	0.693 (0.76)	3 330	3 260	0.978	100 000	...	94 000	liquid	0.78 (−0.36)
7	*tert*-Butyl alcohol	75-65-0	$C_4H_{10}O$	74.12	0.473 (0.35)	10 000	5 170	0.517	∞	∞	166 000	liquid	(−0.64)
8	*iso*-Amyl alcohol	123-51-3	$C_5H_{12}O$	88.15	1.222 (1.42)	2 000	1 280	0.641	20 000	∞	28 500	liquid	(0.26)
9	*tert*-Amyl alcohol	75-85-4	$C_5H_{12}O$	88.15	1.002 (0.89)	5 000	2 030	0.406	125 000	∞	50 300	liquid	(−0.21)
10	Capryl alcohol (octyl alcohol)	111-87-5	$C_8H_{18}O$	130.22	2.939 (2.97)	50.0	52.1	1.04	500	∞	495	liquid	...

					Saturated Monoketones								
11	Acetone	67-64-1	C_3H_6O	58.08	−0.240 (−0.24)	16 700	18 000	1.08	∞	...	824 000	liquid	(−0.83)
12	Methyl ethyl ketone	78-93-3	C_4H_8O	72.11	0.261 (0.29)	6 670	7 830	1.18	200 000	∞	280 000	liquid	...
13	Diethyl ketone	96-22-0	$C_5H_{10}O$	86.13	0.790	2 500	3 090	1.24	50 000	∞	85 000	liquid	...
14	Methyl propyl ketone	107-87-9	$C_5H_{10}O$	86.13	0.790 (0.91)	1 670	3 090	1.85	40 000	∞	85 000	liquid	...
					Aromatic Hydrocarbons								
15	Benzene	71-43-2	C_6H_6	78.11	2.142 (2.15)	166	166	1.00	1 000	∞	2 300	liquid	(2.21)
16	Xylene	1330-20-7	C_8H_{10}	106.16	3.440 (3.15)	40.0	14.9	0.372	125	∞	110	liquid	...
17	Napthalene	91-20-3	$C_{10}H_8$	128.16	3.316 (3.30)	8.33	23.3	2.80	33 to 50	∞	22	80.22	...
18	Phenanthrene	85-01-8	$C_{14}H_{10}$	178.22	4.490 (4.46)	2.00	2.78	1.39	3.3	140 000	0.71	97.5 to 98.5	...

[a] Calculated using the CLOGP3.3 computer program [*8*]; values in parentheses are measured and were obtained from the STARLIST file supported in Medchem Software Release 3.33 [*8*] or from Ref *9*.

[b] Observed concentration producing complete narcosis in tadpoles, *R. temporaria,* from Overton [*2*].

[c] Predicted narcosis concentration based upon Eq 4.

[d] Excess toxicity (predicted narcosis concentration)/(observed narcosis concentration) (Eq 5).

[e] Olive oil.

[f] Water solubility prediction based upon CLOGP3.3 calculated log P values and literature melting points using Eq 3.

[g] Melting point data were taken from Refs *12* through *14*.

[h] Logarithm to the base 10 of the measured olive oil/water partition coefficient from Overton [*2*]. Values in parentheses are from Macy [*10*].

TABLE 3—*Data on chemicals producing neither narcosis nor toxicity in saturated aqueous solution.*

Compound No.	Chemical Name	Chemical Abstracts Registry No.	Molecular Formula	Molecular Weight	Log P^a	Predicted Narcosis Concentration, mg/L[b]	Solubility (Overton), mg/L		Predicted Water Solubility, mg/L[d]	Melting Point, °C[e]	Log K^f
							Water	Oil[c]			
19	Cetyl alcohol	36653-82-4	$C_{16}H_{34}O$	242.43	7.171	0.0138	...	...	0.0063	49.3	...
20	Anthracene	120-12-7	$C_{14}H_{10}$	178.22	4.490	2.77	<1	...	0.0068	218	...
21	Parachloralose	...	$C_8H_{11}Cl_3O_6$	309.54	−0.558	27 000	...	...	660	227 to 230	...
22	Terpin hydrate	2451-01-6	$C_{10}H_{20}O_2 \cdot H_2O$	190.29	1.026	4 170	10 000 (4 000)	1430	2870	116 to 117	−0.85
23	Anthraquinone	84-65-1	$C_{14}H_8O_2$	208.20	2.720	132	...	...	0.051	286	...

[a] Calculated using the CLOGP3.3 computer program [*8*]; values in parentheses are measured and were obtained from the STARLIST file supported in Medchem Software Release 3.33 [*8*] or from Ref *9*.
[b] Predicted narcosis concentration based upon Eq 4.
[c] Olive oil.
[d] Water solubility prediction based upon CLOGP3.3 calculated log *P* values and literature melting points using Eq 3.
[e] Melting point data were taken from Refs *12* through *14*.
[f] Logarithm to the base 10 of the measured olive oil/water partition coefficient from Overton [*2*]. Values in parentheses are from Macy [*10*].

good narcotic agents only those substances which produced undisturbed anesthesia for an extended period of time and whose effects were completely reversible on transfer to pure water. Some of the chemicals he tested produced only death without narcosis.

Among the chemicals Overton tested are a number of anesthetic and hypnotic agents (Table 4). The T_e values for diethyl ether, chloroform, paraldehyde, chloral hydrate, chloral formamide, sulfonal, and trional are all close to unity. Chloral hydrate is almost seven times more toxic than predicted, based upon the calculated log P. The measured log P value, which may reflect a considerable portion of the aldehyde form in the octanol phase, gives a more consistent prediction (T_e = 1.1). Sulfonal and trional yield consistent predictions using log P values derived from water solubility data. The CLOGP3.3 values for sulfonal and trional give anomalous results, suggesting that measured log P values for these hypnotic agents should be obtained. The anomalously high value (T_e = 7.4) for acetophenone appears suspect. The measured log P value for this hypnotic, upon which the CLOGP3.3 calculated value is probably based, appears to be too low. The predicted narcosis concentration for chloralose is consistent, based upon the reported log P value of 1.6 [*19*]. By contrast, a second literature log P of 0.8 [*20*] and calculated log P values yield T_e values much greater than unity and inconsistent with a simple narcosis mechanism. Further work is in progress to resolve these apparent discrepancies.[3]

Overton found that parachloralose, an isomer of chloralose formed during the acid-catalyzed synthetic condensation of glucose and chloral, exhibits no activity at saturation to tadpoles, even though it differs only in the configuration of the acetal carbon atom [*21*]. Parachloralose, unlike chloralose, also produces little or no response when administered to mammals [*22*]. Hanriot and Richet [*23*] attributed this lack of activity to its very low solubility. They reported also that a small tench fish showed no effects when immersed for two weeks in water containing dissolved and undissolved parachloralose which, in total amount, was ten times that required for chloralose to produce death in 24 h. The lack of response by fish is consistent with Overton's findings for tadpoles. The water solubility for parachloralose (Table 3) is predicted to be close to the minimum narcotic concentration of chloralose of 1000 mg/L, consistent with water solubility limited toxicity, placing it in the same category as other compounds in Table 3.

No CLOGP3.3 value could be calculated for carbon disulfide (Table 5). A log P value calculated from Overton's water solubility and Eq 2 gives a predicted narcosis concentration (61 mg/L) in excellent agreement with Overton's reported value (40 mg/L). Pentane was predicted to produce narcosis at 12.5 mg/L, but actually produced this effect at a much higher concentration (200 mg/L, T_e = 0.06). This result can be accounted for based upon the expected rapid rate of volatilization of this C_5 hydrocarbon from water. In fact, Overton observed that methane and ethane do not produce narcosis at atmospheric pressure, and he predicted that at 10°C, wherein the water solubility would be greater, 18 to 30 atm of pressure would be required to produce narcosis by methane.

Overton reported results on the tadpole narcosis for twelve simple carboxylic acid esters (Table 7). The predicted narcosis concentrations and those observed for each of these esters agree well (T_e = 0.7 to 1.8). Overton found, however, that the duration of undisturbed narcosis for these compounds varied in a systematic manner dependent upon the nature of the parent acid and chain length of the parent alcohol. He found a regularity between these times and the corresponding rates of chemical saponification of these esters. Thus, in contemporary terms, the esters undergo *in vivo* esterase cleavage with delayed side effects yielding greater toxicity from metabolism to the corresponding carboxylic acid and

[3] Lipnick, R. L., "A QSAR Study of Overton's Data," *Quantitative Structure-Activity Relationships in Drug Design*, J. L. Fauchere, Ed., Alan Riliss Inc., in press.

TABLE 4—*Comparison of observed and predicted narcosis concentrations for anesthetics and hypnotics.*

Compound No.	Chemical Name	Chemical Abstracts Registry No.	Molecular Formula	Molecular Weight	Log P^a	Narcosis Concentration, mg/L: Observed[b]	Narcosis Concentration, mg/L: Predicted[c]	T_e^d	Solubility (Overton), mg/L: Water	Solubility (Overton), mg/L: Oil[e]	Predicted Water Solubility, mg/L[f]	Melting Point, °C[g]	Log K^h
24	Diethyl ether	60-29-7	$C_4H_{10}O$	74.12	0.870 (0.89)	2500	2 250	0.901	66 000	∞	60 000	liquid	0.65 (0.58)
25	Chloroform	67-66-3	$CHCl_3$	119.39	1.952 (1.95)	169	377	2.23	7 200	...	5 830	liquid	1.5 (1.9)
26	Chloralose	15879-93-3	$C_8H_{11}Cl_3O_6$	309.54	−0.558[i] (1.6)[j] (0.8)[m] [−0.163][o] [−0.004][q]	1000	187 000 (2 040)[k] (10 900)[n] [81 700][p] [58 600][r]	187 (2.04)[k] (10.9)[n] (81.7)[p] (58.6)[r]	6 700 (4 440)[l]	625	17 300 66 520	187	−1.03
27	Paraldehyde	123-63-7	$C_6H_{12}O_3$	132.16	0.998	3330	3 070	0.92	100 000	∞	76 000	liquid	0.48 (0.28)
28	Chloral hydrate	302-17-0	$C_2H_3Cl_3O_2$	165.42	0.715 (1.61)	1000	6 950 (1 070)[r]	6.94 (1.07)[r]	...	...	57 000	57	0.22
29	Chloral formamide	515-82-2	$C_3H_4Cl_3NO_2$	192.44	0.796	3330	6 820	2.05	...	...	3 700	124 to 6	...
30	Acetophenone (Hypnone)	98-86-2	C_8H_8O	120.15	1.581 (1.58)	111	824	7.42	3 030	∞	15 000	19.7	...
31	Sulfonal	115-24-2	$C_7H_{16}O_4S_2$	228.33	−0.431 [1.10][s] [0.980][u]	2000	106 000 [4 290][t] [5 510][v]	52.7 [2.14][t] [2.75][v]	2 000 (2 740)[l]	9 090	105 000	124 to 126	0.66 (0.18)
32	Trional	76-20-0	$C_8H_{18}O_4S_2$	242.36	0.098 [1.70][w] [1.52][y]	1670	37 000 [1 300][x] [1 890][z]	22.2 [0.78][x] [1.13][z]	3 100 (5 000)[l]	50 000	198 000	74 to 76	1.21 (0.72)

[a] Calculated using the CLOGP3.3 computer program [*8*]; values in parentheses are measured and were obtained from the STARLIST file supported in Medchem Software Release 3.33 [*8*] or from Ref *9*.

[b] Observed concentration producing complete narcosis in tadpoles, *R. temporaria,* from Overton [*2*].

[c] Predicted narcosis concentration based upon Eq 4.

[d] Excess toxicity (predicted narcosis concentration)/(observed narcosis concentration) (Eq 5).

[e] Olive oil.

[f] Water solubility prediction based upon CLOGP3.3 calculated log *P* values and literature melting points using Eq 3.

[g] Melting point data were taken from Refs *12* through *14*.

[h] Logarithm to the base 10 of the measured olive oil/water partition coefficient from Overton [*2*]. Values in parentheses are from Macy [*10*].

[i] Calculated based upon the structure 1,2-*O*-trichloroethylidene-*D*-glucofuranose assigned by Forsén [*21*] based upon proton nuclear magnetic resonance studies.

[j] From Barker [*19*].

[k] Calculated from a log *P* value of 1.6 from Ref *19*.

[l] From M. Windholtz [*12*].

[m] From Barker [*20*].

[n] Calculated from a log *P* value of 0.8 from Ref *20*.

[o] Calculated from a water solubility of 6700 mg/L from Overton [*2*] using Eq 2.

[p] Calculated from a log *P* value of −0.163.

[q] Calculated from a water solubility of 4440 mg/L using Eq 2.

[r] Calculated from a log *P* of 0.004.

[s] Calculated from a water solubility of 2000 mg/L from Overton [*2*] using Eq 2.

[t] Calculated from a log *P* value of 1.10.

[u] Calculated from a water solubility of 2740 mg/L using Eq 2.

[v] Calculated from a log *P* value of 0.980.

[w] Calculated from a water solubility of 3100 mg/L from Overton [*2*] using Eq 2.

[x] Calculated from a log *P* value of 1.70.

[y] Calculated from a water solubility of 5000 mg/L using Eq 2.

[z] Calculated from a log *P* value of 1.52.

TABLE 5—*Comparison of observed and predicted narcosis concentrations for various classes of simple nonelectrolytes.*

Compound No.	Chemical Name	Chemical Abstracts Registry No.	Molecular Formula	Molecular Weight	Log P[a]	Narcosis Concentration, mg/L		T_e[d]	Solubility (Overton), mg/L		Predicted Water Solubility, mg/L[f]	Melting Point, °C[g]	Log K[h]
						Observed[b]	Predicted[c]		Water	Oil[e]			
33	Carbon disulfide	75-15-0	CS_2	76.14	[2.61][i] [2.04][k]	40.0[j]	60.6[j] 199[l]	1.52[j] 4.97[l]	670 (2 940)[m]	∞	. . .	liquid	(2.08)
34	Ethyl mercaptan	75-08-1	C_2H_6S	62.13	1.175	500	997	1.99	10 000	∞	22 700	liquid	. . .
35	Pentane	109-66-0	C_5H_{12}	72.15	3.339 (3.39)	200	12.5	0.0626	500	∞	97	liquid	. . .
36	Amylene (2-methyl-2-butene)	513-35-9	C_5H_{10}	70.13	2.665	167	49.8	0.298	590	∞	540	liquid	. . .
37	Ethyl chloride	75-00-3	C_2H_5Cl	64.52	1.465 (1.43)	285	564	1.98	10 000	∞	11 000	liquid	. . .
38	Ethyl bromide	74-96-4	C_2H_5Br	108.98	1.605 (1.61)	285	711	2.49	10 000	∞	13 000	liquid	. . .
39	Ethyl iodide	75-03-6	C_2H_5I	155.98	1.995 (2.00)	167	450	2.70	2 500	∞	6 800	liquid	. . .
40	Ethylene dichloride	107-06-2	$C_2H_4Cl_2$	98.96	1.458 (1.48)	236	878	3.73	. . .	. . .	17 000	liquid	. . .
41	Tetrachloromethane	56-23-5	CCl_4	153.84	2.875 (2.83)	111	70.4	0.633	. . .	. . .	690	liquid	(2.66)
42	Nitromethane	75-52-5	CH_3NO_2	61.04	−0.284 (−0.35)	5 000	20 700	4.15	105 000	118 000	970 000	liquid	0.050 (−0.32)
43	Methyl cyanide (acetonitrile)	75-05-8	C_2H_3N	41.05	−0.394 (−0.34)	15 000	7 600	1.17	∞	25 000	870 000	liquid	. . .
44	Acetaldoxime	107-29-9	C_2H_5NO	59.07	−0.144 (−0.13)	6 900	15 000	2.17	. . .	. . .	290 000	44 to 46	. . .
45	Acetoxime	127-06-0	C_3H_7NO	73.09	−0.168	5 560	19 500	3.51	. . .	. . .	220 000	60	. . .
46	Ethyl nitrate	625-58-1	$C_2H_5NO_3$	91.08	1.385	667	942	1.41	10 000	. . .	19 000	liquid	. . .
47	Triethyl phosphate	78-40-0	$C_6H_{15}O_4P$	182.16	0.735 (0.80)	2 000	7 340	3.67	∞	. . .	210 000	liquid	−1.07
48	Pinacone 6H2O	76-09-5	$C_6H_{14}O_2 \cdot 6H_2O$	226.28	0.047	45 000	38 500	0.854	∞	. . .	690 000	45.4	. . .

49	Monochlorohydrin	96-24-2	$C_3H_7ClO_2$	110.54	−0.997	18 800	167 000	8.92	∞	100 000	1×10^7	liquid	. . .
50	Dichlorohydrin	96-23-1	$C_3H_6Cl_2O$	128.99	0.200	1 430	15 900	11.1	110 000	∞	590 000	liquid	. . .
51	Triacetin	102-76-1	$C_9H_{14}O_6$	218.20	0.368	5 000	19 000	3.79	. . .	. . .	640 000	liquid	. . .
52	Ethyl acetoacetate	141-97-9	$C_6H_{10}O_3$	130.14	0.265	2 500	14 000	5.62	77 000	290 000	500 000	liquid	0.042
53	Valeramide	626-97-1	$C_5H_{11}NO$	101.15	0.353	5 000	9 070	1.81	110 000	7 700	91 000	114 to 116	. . .
54	Phenylurea	64-10-8	$C_7H_8N_2O$	136.15	0.845 (0.83)	625	4 360	6.97	5 000	. . .	1 000	147	. . .
55	Phenylthiourea	103-85-5	$C_7H_8N_2S$	152.22	0.745 (0.73)	1 000	6 010	6.00	. . .	. . .	1 100	154	. . .
56	Methyl acetyl urea	623-59-6	$C_4H_8N_2O_2$	116.14	−0.628	20 000	8 100	4.06	. . .	. . .	11 000	180	. . .
57	Triethyl thiourea	. . .	$C_7H_{16}N_2S$	160.28	[1.34][n]	4 000	[1 640][o]	[0.41][o]	17 000	100 000	. . .	46	0.76
58	Azobenzene	103-33-3	$C_{12}H_{10}N_2$	182.22	3.850 (3.82)	3.33	10.8	3.25	50	VS	12	68	. . .
59	Turpentine oil	. . .	$C_{10}H_{16}$	136.24	4.122	66.7	4.58	0.069	. . .	. . .		liquid	. . .
60	Menthol	89-78-1	$C_{10}H_{20}O$	156.26	3.233	16.7	33.8	2.02	670	200 000	138	41 to 43	2.48
61	Camphor	76-22-2	$C_{10}H_{16}O$	152.23	2.117	200	340	1.71	. . .	. . .	12	179.75	. . .
62	Phthalide	87-41-2	$C_8H_6O_2$	134.13	1.887	666	485	0.728	6 700	22 000	1 160	72 to 74	0.52
63	Coumarin	91-64-5	$C_9H_6O_2$	146.14	1.412 (1.39)	83.3	1 430	17.1	2 500	22 000 to 25 000	5 030	68 to 70	0.95 to 1.0 (1.21)

[a] Calculated using the CLOGP3.3 computer program [*8*]; values in parentheses are measured and were obtained from the STARLIST file supported in Medchem Software Release 3.33 [*8*] or from Ref *9*.
[b] Observed concentration producing complete narcosis in tadpoles, *Rana temporaria*, from Overton [*2*].
[c] Predicted narcosis concentration based upon Eq 4.
[d] Excess toxicity (predicted narcosis concentration)/(observed narcosis concentration) (Eq 5).
[e] Olive oil.
[f] Water solubility prediction based upon CLOGP3.3 calculated log P values and literature melting points using Eq 3.
[g] Melting point data were taken from Refs *12* through *14*.
[h] Logarithm to the base 10 of the measured olive oil/water partition coefficient from Overton [*2*]. Values in parentheses are from Macy [*10*].
[i] Calculated from a water solubility of 670 mg/L from Overton [*2*] using Eq 2.
[j] Calculated from a log P value at 2.61.
[k] Calculated from a water solubility of 2940 mg/L using Eq 2.
[l] Calculated from a log P value of 2.04.
[m] From Windholtz [*12*].
[n] Calculated from a water solubility of 17 000 mg/L from Overton [*2*] using Eq 2.
[o] Calculated from a log P value of 1.34.

alcohol toxicants within the tadpole bloodstream. This interpretation accounts for the comparatively poor correlation ($r^2 = 0.828$) between log P and the 96-h fathead minnow LC_{50} found by Veith et al. [*24*] for esters. Kamlet et al. [*25*] reported that the addition of a QSAR parameter, corresponding to the rate of base-catalyzed ester hydrolysis, yielded a greatly improved correlation for the 48-h LC_{50} of simple acetates to the golden orfe fish, consistent with Overton's interpretation.

Acetamide, succinimide, and urea are considerably more toxic than predicted ($T_e = 15$ to 62) (Table 6). Each of these compounds produces death without narcosis. All three compounds have very negative calculated log P values. Overton found the solubility of each in oil to be very low and observed that these substances either required very long time periods to permeate the gill membrane and enter the tadpole bloodstream or did not cross at all. On this basis, their lethality was attributed solely to osmotic effects at the very high aqueous concentrations required.

Phenol ($T_e = 1.6$) (Table 8) produced narcosis in tadpoles, but this effect was accompanied by spasms not produced by simple nonelectrolytes, as well as an effect upon the circulation leading to death in 36 h at concentrations of 1:4000. After 1 to 2 h, the circulation was slower and there was such a considerable drop in blood pressure that blood no longer flowed into all of the capillaries. This response continued for up to 36 h, with spontaneous jerky spasmodic movements. In more concentrated solutions (1:2000), Overton noted that narcosis disappeared almost immediately when the tadpoles were transferred to fresh water, but the blood circulation did not improve for some time. The cresols behaved similarly. *t*-Amylphenol and thymol, with higher partition coefficients, showed fewer side effects on spasmodic activity and on the blood circulation.

In contrast to phenol, phenol ethers elicited a purely narcotic effect. Anisole, for example, produced complete narcosis in 6 to 8 min at concentrations of 1:6000, which could be sustained beyond 20 h without producing death. In addition, no spasms or effects on the heart were noted for this compound.

Resorcinol and its two isomers, pyrocatechol and hydroquinone, produced nonnarcotic effects. Tadpoles were reported to continue moving in solutions of 1:400 for more than half an hour, after which the circulation stopped. In contrast, the dimethyl ethers of resorcinol and hydroquinone produced purely narcotic effects.

The two trihydroxybenzenes, phloroglucinol and pyrogallol, have T_e values of 7.1 and 17.6, respectively. The nephrotoxicity of catechols and related compounds has been related to their ease of metabolism to the corresponding quinoid derivatives, which can bind to critical cellular constituents via a Michael-type addition [*26*]. The greater T_e values of pyrocatechol and hydroquinone relative to resorcinol can be interpreted in terms of the inability of resorcinol (with *meta*-hydroxy groups) to undergo this quinoid oxidation. The T_e of 7.1 for phloroglucinol, which also cannot undergo quinoid oxidation, may reflect electrophilicity of the partially enolized tautomer, which should act as a Michael-type acceptor.

The excess toxicity of piperonal can be rationalized by electrophilic action via Schiff base covalent binding to amino groups. Turpentine oil in current commerce is mainly a mixture of three $C_{10}H_{16}$ isomeric terpenes, that is, α-terpene, β-terpene, and carene [*27*]. They have a common calculated log P of 4.12 and can be treated together for predicting toxicity. Since the turpentine tested by Overton is much less toxic than predicted ($T_e = 0.069$), it may have been of an entirely different composition.

Although coumarin is 17 times more toxic than predicted, Overton reported that on exposure to an 83 mg/L solution, tadpoles are completely narcotized in 10 min and that this condition could be maintained undisturbed for more than 6 h. Furthermore, when the aforementioned treated tadpoles were transferred to fresh water, they were again moving normally in just a few minutes. Solutions of 63 mg/L produced only partial narcosis which

TABLE 6—*Data for very hydrophilic compounds with low lipid solubility, producing lethality through osmotic effects without narcosis.*

Compound No.	Chemical Name	Chemical Abstracts Registry No.	Molecular Formula	Molecular Weight	Log P^a	Lethal Concentration, mg/L		T_e^d	Solubility (Overton), mg/L		Predicted Water Solubility, mg/L[f]	Melting Point, °C[g]	Log K^h
						Observed[b]	Predicted[c]		Water	Oil[e]			
64	Urea	57-13-6	CH_4N_2O	60.06	−2.110 (−2.11)	15 000	932 000	62.1	10^6	...	2×10^6	132.7	...
65	Acetamide	60-35-5	C_2H_5NO	59.07	−1.234 (−1.26)	10 000	147 000	15.0	$>2 \times 10^6$	5000	1×10^6	81	...
66	Succinimide	123-56-8	$C_4H_5NO_2$	99.09	−1.892	20 000	975 000	48.7	...	...	2×10^6	124 to 126	...
67	Ethylene glycol	107-21-1	$C_2H_6O_2$	62.07	−1.369 (−1.36)	40 000	204 100	5.11	∞	...	2×10^7	liquid	...

[a] Calculated using the CLOGP3.3 computer program [*8*]; values in parentheses are measured and were obtained from the STARLIST file supported in Medchem Software Release 3.33 [*8*] or from Ref *9*.
[b] Observed concentration producing complete narcosis in tadpoles, *Rana temporaria,* from Overton [*2*].
[c] Predicted narcosis concentration based upon Eq 4.
[d] Excess toxicity (predicted narcosis concentration)/(observed narcosis concentration) (Eq 5).
[e] Olive oil.
[f] Water solubility prediction based upon CLOGP3.3 calculated log P values and literature melting points using Eq 3.
[g] Melting point data were taken from Refs *12* through *14*.
[h] Logarithm to the base 10 of the measured olive oil/water partition coefficient from Overton [*2*]. Values in parentheses are from Macy [*10*].

TABLE 7—*Comparison of observed and predicted narcosis concentration for esters.*

Compound No.	Chemical Name	Chemical Abstracts Registry No.	Molecular Formula	Molecular Weight	Log P^a	Narcosis Concentration, mg/L		T_e^d	Solubility (Overton), mg/L		Predicted Water Solubility, mg/L[f]	Melting Point, °C[g]	Log K^h
						Observed[b]	Predicted[c]		Water	Oil[e]			
68	Methyl acetate	79-20-9	$C_3H_6O_2$	74.08	0.142 (0.18)	5 710	10 300	1.81	...	...	390 000	liquid	...
69	Ethyl formate	109-94-4	$C_3H_6O_2$	74.08	0.265	5 000	7 980	1.60	...	...	280 000	liquid	0.60
70	Ethyl acetate	141-78-6	$C_4H_8O_2$	88.10	0.671 (0.73)	2 500	4 060	1.62	66 000	∞	120 000	liquid	...
71	Ethyl propionate	105-37-3	$C_5H_{10}O_2$	102.13	1.200 (1.21)	1 110	1 560	1.40	20 000	∞	35 000	liquid	...
72	Propyl acetate	109-60-4	$C_5H_{10}O_2$	102.13	1.200	1 110	1 560	1.40	20 000	∞	35 000	liquid	...
73	Ethyl *n*-butyrate	105-54-4	$C_6H_{12}O_2$	116.16	1.729	500	585	1.17	5 300	∞	10 000	liquid	...
74	Ethyl *iso*-butyrate	97-62-1	$C_6H_{12}O_2$	116.16	1.509	667	926	1.39	7 100	∞	18 000	liquid	...
75	*n*-Butyl acetate	123-86-4	$C_6H_{12}O_2$	116.16	1.729	572	585	1.02	5 600	∞	10 000	liquid	...
76	*iso*-Butyl acetate	110-19-0	$C_6H_{12}O_2$	116.16	1.599	667	767	1.15	5 600	∞	14 000	liquid	...
77	Ethyl valerianate	539-82-2	$C_7H_{14}O_2$	130.19	2.258	250	217	0.867	2 000	∞	2 900	liquid	...
78	Amyl acetate	628-63-7	$C_7H_{14}O_2$	130.19	2.258	250	217	0.867	2 000	∞	2 900	liquid	...

79	Butyl valerianate	591-68-4	$C_9H_{18}O_2$	158.23	3.316	40	28.8	0.719	290	∞	230	liquid	. . .
80	Methyl urethane	598-55-0	$C_2H_5NO_2$	75.07	−0.704 (−0.66)	20 000	61 500	3.08	. . .	. . .	1×10^6	52 to 54	. . .
81	Ethyl urethane	51-79-6	$C_3H_7NO_2$	89.09	−0.175 (−0.15)	3 640	24 100	6.63	10^6	. . .	405 000	48 to 50	(−0.92)
82	Phenyl urethane	101-99-5	$C_9H_{11}NO_2$	165.19	1.095 (1.08)	1 390	3 130	2.25	50 000	270 000	25 000	52 to 53	0.73 (1.99)
83	Tartaric acid, ethyl ester	87-91-2	$C_8H_{14}O_2$	206.19	−1.137 (−0.29)	12 500	418 000 (71 000)	33.4 (5.68)	∞	14 000	3×10^7	liquid	. . .
84	Citric acid, ethyl ester	77-93-0	$C_{12}H_{20}O_7$	276.29	1.101	2 500	5 180	2.07	40 000	∞	120 000	liquid	. . .

[a] Calculated using the CLOGP3.3 computer program [*8*]; values in parentheses are measured and were obtained from the STARLIST file supported in Medchem Software Release 3.33 [*8*] or from Ref *9*.

[b] Observed concentration producing complete narcosis in tadpoles, *Rana temporaria,* from Overton [*2*].

[c] Predicted narcosis concentration based upon Eq 4.

[d] Excess toxicity (predicted narcosis concentration)/(observed narcosis concentration) (Eq 5).

[e] Olive oil.

[f] Water solubility prediction based upon CLOGP3.3 calculated log P values and literature melting points using Eq 3.

[g] Melting point data were taken from Refs *12* through *14*.

[h] Logarithm to the base 10 of the measured olive oil/water partition coefficient from Overton [*2*]. Values in parentheses are from Macy [*10*].

TABLE 8—*Comparison of observed narcosis or lethal concentration of phenols and their ethers with narcosis predictions.*

Compound No.	Chemical Name	Chemical Abstracts Registry No.	Molecular Formula	Molecular Weight	Log P^a	Narcosis or Lethal (L) Concentration, mg/L		T_e^d	Solubility (Overton), mg/L		Predicted Water Solubility, mg/L[f]	Melting Point, °C[g]	Log K^h
						Observed[b]	Predicted[c]		Water	Oil[e]			
85	Phenol	108-95-2	C_6H_6O	94.11	1.475	500	806	1.61	70 000	250 000	15 800	liquid	(0.78)
86	*o*-Cresol	95-48-7	C_7H_8O	108.13	2.124 (1.95)	133	238	1.80	25 000	...	2 790	30	...
87	*m*-Cresol	108-39-4	C_7H_8O	108.13	2.124 (1.96)	200	238	1.19	5 000	...	3 380	liquid	...
88	*p*-Cresol	106-44-5	C_7H_8O	108.13	2.124 (1.94)	200	238	1.19	10 000	...	2 200	36	...
89	*t*-Amylphenol[i]	14938-35-3	$C_{11}H_{16}O$	164.24	4.907	4.99	1.07	0.214	...	...	3.8	liquid	...
90	Thymol	89-83-8	$C_{10}H_{14}O$	150.21	3.701 (3.30)	8.33	12.2	1.46	830	10^6	29	49 to 51	...
91	α-Napthol	90-15-3	$C_{10}H_8O$	144.16	2.649 (2.98)	12.5	105.8	8.47	...	...	74	96	...
92	Anisole	100-66-3	C_7H_8O	108.13	2.061 (2.11)	167	272	1.63	...	...	4 000	liquid	...
93	Resorcinol	108-46-3	$C_6H_6O_2$	110.11	0.808 (0.80)	2 500	3 810	1.52	...	...	3 690	109 to 111	...
94	Pyrocatechol	120-80-9	$C_6H_6O_2$	110.11	0.808 (0.88)	833	3 810	4.57	...	...	4 650	105	...
95	Hydroquinone	123-31-9	$C_6H_6O_2$	110.11	0.808 (0.59)	833	3 810	4.57	...	...	360	170 to 171	...

96	Resorcinol dimethyl ether	151-10-0	$C_8H_{10}O_2$	138.17	2.150 (2.22)	62.4	288	4.62	1 050 to 1 100	. . .	4 040	liquid	. . .
97	Dimethyl hydroquinone	150-78-7	$C_8H_{10}O_2$	138.17	2.150	125	288	2.30	770	286 000	1 040	56 to 60	2.57 (2.15)
98	Guaiacol	90-05-1	$C_7H_8O_2$	124.13	1.294 (1.32)	334	1 550	4.65	20 000	$>2 \times 10^6$	28 500	27 to 29	>2.00
99	Eugenol	97-53-0	$C_{10}H_{12}O_2$	164.20	2.457	20.0	180	9.00	1 000	∞	2 170	liquid	. . .
100	Phloroglucinol	6099-90-7	$C_6H_6O_3$	126.11	0.141 (0.16)	2 500(L)[j]	17 600	7.05	1 700	400	373	218	−0.63
101	Pyrogallol	87-66-1	$C_6H_6O_3$	126.11	0.141	1 000(L)[j]	17 600	17.6	. . .	. . .	10 100	131 to 133	. . .
102	Vanillin	121-33-5	$C_8H_8O_3$	152.14	1.354 (1.21)	500	1 680	3.35	10 000	25 000 to 28 600	3 970	80 to 81	0.40 to 0.46 (0.41)
103	Piperonal	120-57-0	$C_8H_6O_3$	150.13	1.256 (1.05)	250	2 030	8.12	1 820	167 000	28 000	37	2.0 (1.47)

[a] Calculated using the CLOGP3.3 computer program [*8*]; values in parentheses are measured and were obtained from the STARLIST file supported in Medchem Software Release 3.33 [*8*] or from Ref *9*.
[b] Observed concentration producing complete narcosis in tadpoles, *Rana temporaria*, from Overton [*2*].
[c] Predicted narcosis concentration based upon Eq 4.
[d] Excess toxicity (predicted narcosis concentration)/(observed narcosis concentration) (Eq 5).
[e] Olive oil.
[f] Water solubility prediction based upon CLOGP3.3 calculated log *P* values and literature melting points using Eq 3.
[g] Melting point data were taken from Refs *12* through *14*.
[h] Logarithm to the base 10 of the measured olive oil/water partition coefficient from Overton [*2*]. Values in parentheses are from Macy [*10*].
[i] Overton does not indicate which amyl isomer he employed. The *t*-amyl isomer is the only one known in the literature eight years later [*13*].
[j] Lethality without narcosis.

TABLE 9—*Comparison of observed and predicted narcosis concentrations for some basic compounds and alkaloids.*

Compound No.	Chemical Name	Chemical Abstracts Registry No.	Molecular Formula	Molecular Weight	Log P^a	Narcosis Concentration, mg/L Observed[b]	Predicted[c]	T_e^d	Solubility (Overton), mg/L Water	Oil[e]	Predicted Water Solubility, mg/L[f]	Melting Point, °C[g]	Log K^h
104	Acetanilide	103-84-4	C_8H_9NO	135.16	1.161 (1.16)	667	2 230	3.35	5 000	8 300	1 550	113 to 115	0.22 (0.004)
105	Methacetin	55-66-1	$C_9H_{11}NO_2$	165.19	1.264 (1.03)	1330	2 200	1.65	2 000	4 000	750	130 to 132	0.30 (−0.20)
106	Phenacetin	62-44-2	$C_{10}H_{13}NO_2$	179.21	1.793 (1.58)	500	189	1.58	710	2 700	165	137 to 138	0.58 (0.43)
107	Diphenyl amine	122-39-4	$C_{12}H_{11}N$	169.22	3.620 (3.50)	6.24	16.3	2.61	100	...	37	53	...
108	Aniline	62-52-3	C_6H_7N	93.12	0.915 (0.90)	1000	2 570	2.57	33 000	∞	67 000	liquid	...
109	Dimethyl aniline	121-69-7	$C_8H_{11}N$	121.18	2.337 (2.31)	167	171	1.02	1 250	∞	2 190	liquid	...
110	Pyridine	110-86-1	C_5H_5N	79.10	0.665	2000	3 690	1.86	∞	∞	108 000	liquid	−0.40 (−0.02)
111	Quinoline	91-22-5	C_9H_7N	129.15	2.049 (2.03)	125	333	2.66	5 000	∞	4 900	liquid	...

112	Antipyrine	60-80-0	$C_{11}H_{12}N_2O$	188.22	0.574 (0.38)	2500[i]	10 600	4.24	$>10^6$	18 000	10 700	111 to 113	<-1.74
113	Coniline	458-88-8	$C_8H_{17}N$	127.22	2.132	41.8	276	6.60	11 000	∞	3 900	liquid	. . .
114	Nicotine	54-11-5	$C_{10}H_{14}N_2$	162.23	1.343 (1.17)	50.0	1 830	36.6	∞	250 000	38 300	liquid	. . .
115	Sparteine	90-39-1	$C_{15}H_{26}N_2$	234.37	2.147	83.3[j]	492	5.90	. . .	. . .	6 900	liquid	. . .
116	Morphine	57-27-2	$C_{17}H_{19}NO_3$	285.33	0.482 (0.76)	500[k]	19 500	39.4[k]	. . .	. . .	86.3	decomposes at 254	. . .
117	Strychnine	57-24-9	$C_{21}H_{22}N_2O_2$	334.40	−0.215 (1.93)	15.4	98 400 (1 110)	6390 (71.8)	. . .	. . .	152 (0.59)	268 to 290	. . .
118	Caffeine	58-08-2	$C_8H_{10}N_4O_2$	194.19	−0.240 (−0.07)	2250	60 200	26.8	. . .	. . .	708	238	. . .

[a] Calculated using the CLOGP3.3 computer program [*8*]; values in parentheses are measured and were obtained from the STARLIST file supported in Medchem Software Release 3.33 [*8*] or from Ref *9*.
[b] Observed concentration producing complete narcosis in tadpoles, *R. temporaria,* from Overton [*2*].
[c] Predicted narcosis concentration based upon Eq 4.
[d] Excess toxicity (predicted narcosis concentration)/(observed narcosis concentration) (Eq 5).
[e] Olive oil.
[f] Water solubility prediction based upon CLOGP3.3 calculated log P values and literature melting points using Eq 3.
[g] Melting point data were taken from Refs *12* through *14*.
[h] Logarithm to the base 10 of the measured olive oil/water partition coefficient from Overton [*2*]. Values in parentheses are from Macy [*10*].
[i] Tadpoles maintain movement for 24 h; sluggish after 36 h; death usually on the second or third day.
[j] Highly caustic effect on the epithelia and all other tadpole tissue; no narcosis.
[k] Saturated solution has almost no effect on tadpoles; only a small degree of agitation is produced after some time.

could be sustained for more than 30 h. The tadpoles recovered almost immediately on transfer to fresh water. Coumarin has several sites of chemical reactivity. Metabolic oxidation is known to occur preferentially at the 3, 7, and 8 positions [*28*]. Carbon atoms 2 and 4 are the only sites that are calculated to be electron deficient [*28*] and are known sites of electrophilic reactivity, susceptible to attack by nucleophiles [*29–30*]. The lactone carbonyl moiety (C-2) is known chemically to be attacked by water followed by ring opening. In contrast, C-4 can serve as a Michael-type acceptor site because of its conjugation with the lactone carbonyl. Michael-type reactions to coumarin at C-4 with bisulfite and cyanide nucleophiles have been reported in the chemical literature [*29–30*]. Roberts[4] has suggested that reversible covalent bond formation to nucleophilic moieties present at the narcosis site of action may lead to higher activity than predicted, based upon simple partitioning, terming this effect "enhanced narcosis."

Overton considered weakly basic amines to occupy a special toxicological category because they form a bridge between the true neutral narcotics and the stronger organic bases that exhibit a more specific mechanism of action. The former group includes aromatic amines such as aniline (Compound 108) (Table 9) and diphenylamine (Compound 107), and heteroaromatic compounds such as pyridine (Compound 110) and quinoline (Compound 111). He reported that diphenylamine and dimethylaniline, which are very weak bases, produce purely narcotic effects. In contrast, Overton found that for nicotine (Compound 114) and other alkaloids containing strongly basic amino groups, the paralyzing concentration varied significantly, depending upon the species and age of the tadpoles tested. Overton concluded that such substances act upon the ends of the motor nerves and directly upon the muscle fibers and nervous system. The range in calculated excess toxicity in comparing the range from weak to strong bases in Table 9 can be interpreted in terms of Overton's observations.

Acknowledgments

The author wishes to thank Dr. James H. Gilford, chief of the Environmental Effects Branch, Office of Toxic Substances, EPA, for his encouragement in carrying out this investigation; Dr. Gilman D. Veith, Director, EPA Environmental Research Laboratory (Duluth, MN), for making available the Cambridge Digital computer for CLOGP computations; Professor Corwin Hansch and Dr. Albert J. Leo, Pomona College (Claremont, CA), and Dr. David Roberts, Unilever, Port Sunlight Laboratory (U.K.), for helpful discussions. The use of the Modeling Laboratory (MLAB) computer program on the DEC10 computer at the National Institutes of Health under Interagency Agreement No. DW931218010 is gratefully acknowledged.

References

[*1*] Overton, E., "Ueber die Allgemeinen Osmotischen Eigenschaften der Zelle, ihre Vermutlichen Ursachen und ihre Bedeutung für die Physiologie," *Vierteljahrsschrift der Naturforschenden Gesellschaft in Zuerich,* Vol. 44, 1899, pp. 88–135.

[*2*] Overton, E., *Studien über die Narkose Zugleich ein Beitrag zur Allgemeinen Pharmakologie,* Gustav Fischer, Jena, Germany, 1901.

[*3*] Meyer, H., "Zur Theorie der Alkoholnarkose, Welche Eigenschaft der Anästhetica bedingt ihre narkotische Wirkung?" *Archiv fuer Experimentelle Pathologie und Pharmakologie (Naunyn-Schmiedebergs),* Vol. 42, Nos. 2–4, 1899, pp. 109–118.

[4] Roberts, D. W., Unilever, Port Sunlight Laboratory, United Kingdom, personal communication, 1 May 1987.

[4] Baum, F., "Theorie der Alkoholnarkose: Zweite Mittheilung—Ein Physikalisch-Chemischer Beitrag zur Theorie der Narcotica," *Archiv fuer Experimentelle Pathologie und Pharmakologie (Naunyn-Schmiedebergs),* Vol. 42, Nos. 2–4, pp. 119–137.
[5] Lipnick, R. L., "Charles Ernest Overton: Narcosis Studies and a Contribution to General Pharmacology," *Trends in Pharmacological Sciences,* Vol. 7, No. 5, 1986, pp. 161–164.
[6] Leo, A., Hansch, C., and Church, C., "Comparison of Parameters Currently Used in the Study of Structure-Activity Relationships," *Journal of Medicinal Chemistry,* Vol. 12, No. 5, 1969, pp. 766–771.
[7] Hansch, C. and Dunn, W. J. III, "Linear Relationships Between Lipophilic Character and Biological Activity of Drugs," *Journal of Pharmaceutical Sciences,* Vol. 61, No. 1, 1972, pp. 1–19.
[8] Leo, A. and Weininger, D., "Medchem Software Release 3.33," Medicinal Chemistry Project, Pomona College, Claremont, CA, 1985.
[9] Hansch, C. and Leo, A., *Substituent Constants for Correlation Analysis in Chemistry and Biology,* Wiley-Interscience, New York, 1979.
[10] Macy, R., "Partition Coefficients of Fifty Compounds Between Olive Oil and Water at 20°C," *Journal of Industrial Hygiene and Toxicology,* Vol. 30, 1948, pp. 140–143.
[11] Banerjee, S., Yalkowsky, S. H., and Valvani, S. S., "Water Solubility and Octanol/Water Partition Coefficients of Organics, Limitations of the Solubility-Partition Coefficient Correlations," *Environmental Science and Technology,* Vol. 14, No. 10, 1980, pp. 1227–1229.
[12] Windholtz, M., Ed., *The Merck Index,* 10th ed., Merck and Co., Inc., Rahway, NJ, 1983.
[13] *Beilstein's Handbuch der Organischen Chemie,* Hauptwerk, Germany, 1909.
[14] *Catalog Handbook of Fine Chemicals,* Aldrich Chemical Co., Milwaukee WI, 1986–1987.
[15] Könemann, H., "Quantitative Structure-Activity Relationships in Fish Toxicity Studies, Part 1: Relationship for 50 Industrial Pollutants," *Toxicology,* Vol. 19, No. 3, 1981, pp. 209–221.
[16] Veith, G. D., Call, D. J., and Brooke, L. T., "Structure-Toxicity Relationships for the Fathead Minnow, *Pimephales promelas:* Narcotic Industrial Chemicals," *Canadian Journal of Fisheries and Aquatic Sciences,* Vol. 40, No. 6, 1983, pp. 743–748.
[17] Lipnick, R. L., Watson, K. R., and Strausz, A. K., "A QSAR Study of the Acute Toxicity of Some Industrial Organic Chemicals to Goldfish: Narcosis, Electrophile, and Proelectrophile Mechanisms," *Xenobiotica,* Vol. 17, No. 8, 1987, pp. 1011–1025.
[18] Williams, R. T., *Detoxication Mechanisms,* 2nd ed., John Wiley and Sons, New York, 1959, pp. 96–97.
[19] Barker, J. L., "Activity of CNS Depressants Related to Hydrophobicity," *Nature* (London), Vol. 252, No. 5478, 1974, pp. 52–54.
[20] Barker, J. L., "CNS Depressants Effects on Post-Synaptic Pharmacology," *Brain Research,* Vol. 92, No. 1, 1975, pp. 35–55.
[21] Forsén, S., Lindberg, B., and Silvander, B.-G., "Trichloroethylidene Derivatives of *D*-Glucose," *Acta Chemica Scandinavica,* Vol. 19, No. 2, 1965, pp. 359–369.
[22] Balis, G. U. and Monroe, R. R., "The Pharmacology of Chloralose," *Psychopharmacologia,* Vol. 6, No. 1, 1964, pp. 1–30.
[23] Hanriot and Richet, C., "Les Chloraloses," *Archiv de Pharmacodynamie,* Vol. 3, No. 3, 1897, pp. 191–211.
[24] Veith, G. D., De Foe, D., and Knuth, M., "Structure-Activity Relationships for Screening Organic Compounds for Potential Ecotoxicity Effects," *Drug Metabolism Reviews,* Vol. 15, No. 7, 1984–1985, pp. 1295–1303.
[25] Kamlet, M. J., Doherty, R. M., Taft, R. W., Abraham, M. H., Veith, G. D., and Abraham, D. J., "Solubility Properties in Polymers and Biological Media: 8. An Analysis of the Factors that Influence Toxicities of Organic Nonelectrolytes to the Golden Orfe Fish *(Leuciscus idus melanotus),*" *Environmental Science and Technology,* Vol. 21, No. 2, 1987, pp. 149–155.
[26] Calder, I. C., Williams, P. J., Woods, R. A., Funder, C. C., Green, C. R., Ham, K. N., and Tange, J. D., "Nephrotoxicity and Molecular Structure," *Xenobiotica,* Vol. 5, No. 5, 1975, pp. 303–307.
[27] Gosselin, R. E., Hodge, H. C., Smith, R. P., and Gleason, M. N., *Clinical Toxicology of Commercial Products,* 4th ed., Williams and Wilkins, Baltimore, 1976.
[28] Feuer, G., "The Metabolism and Biological Actions of Coumarins," *Progress in Medical Chemistry,* Vol. 10, 1979, pp. 85–158.
[29] Shah, N. M., "The Chemistry of Coumarins," *Chemical Reviews,* Vol. 36, 1945, pp. 1–62.
[30] *Comprehensive Organic Chemistry,* D. Barton and W. D. Ollis, Eds., Vol. 4, *Heterocyclic Compounds,* P. G. Sommes, Ed., Pergamon, Oxford, 1979, pp. 648–649.

David W. Roberts[1]

Acute Lethal Toxicity Quantitative Structure-Activity Relationships for Electrophiles and Pro-Electrophiles: Mechanistic and Toxicokinetic Principles

REFERENCE: Roberts, D. W., "**Acute Lethal Toxicity Quantitative Structure-Activity Relationships for Electrophiles and Pro-Electrophiles: Mechanistic and Toxicokinetic Principles,**" *Aquatic Toxicology and Environmental Fate: Eleventh Volume, ASTM STP 1007*, G. W. Suter II and M. A. Lewis, Eds., American Society for Testing and Materials, Philadelphia, 1989, pp. 490–506.

ABSTRACT: Published data on acute lethal toxicity of aniline and nitrobenzene derivatives to fish can be correlated with an electrophilic reactivity descriptor based on substituent constants (proposed to relate to electrophilicity of the corresponding C-nitroso compounds, the aniline and nitrobenzene derivatives being regarded as pro-electrophiles). Likewise, published data on acute lethal toxicity of reactive halogen compounds can be correlated with an electrophilic reactivity parameter based on measured rate constants for reaction with a reference nucleophile, 4-nitrobenzylpyridine. In both cases, incorporation of a lipophilicity descriptor does not improve the quantitative structure-activity relationship (QSAR). A toxicokinetic model is proposed, whereby the electrophilic reaction leading to the toxic effect is assumed to take place in an aqueous biological medium separated from the external aqueous environment by a lipid barrier. Based on consideration of whether environment-lipid partition, lipid-aqueous biological medium partition, or reaction with biological nucleophiles is likely to be the rate-determining step, outlying behavior of some of the compounds can be rationalized, and the QSARs can, with justification, be improved by elimination of these compounds.

KEY WORDS; aquatic toxicity, structure-activity relationships, aniline derivatives, nitrobenzene derivative, reactive halogen compounds, electrophiles, pro-electrophiles, aquatic toxicology

Quantitative structure-activity relationships (QSARs) are an increasingly important tool in ecotoxicology, finding uses, for example, in setting priorities for toxicity testing and in reducing the need for range-finder tests. The more a QSAR is based on an understanding of the physico-chemical aspects of the mode of toxic action (that is, the underlying toxicokinetics/toxicodynamics and the nature of binding processes and forces), the more confidence one can have in it. Conversely, a QSAR developed empirically can provide useful evidence as to the chemistry underlying the mode of toxic action, if it is based on descriptors that have clear physico-chemical significance.

In acute lethal toxicity studies on several aquatic species, it has been demonstrated that many organic compounds, covering a diverse range of structures, fit a correlation between

[1] Scientist, Unilever Research Port Sunlight Laboratory, Bebington, Wirral, Merseyside L63 3JW, England.

toxicity as quantified by the log [1/median lethal concentration (LC_{50})] values and hydrophobicity as quantified by log P values, P being the octanol/water partition coefficient [*1–3*]. Thus, Könemann [*1*] obtained, for a range of aliphatic and aromatic compounds, the equation

$$\log\left(\frac{1}{LC_{50}}\right) = 0.871 \log P + 1.13 \quad (n = 50; r = 0.988; s = 0.237) \tag{1}$$

where n equals the number, r equals the correlation coefficient, and s equals the standard error for 14-day LC_{50} test data on guppies *(Poecilla reticulata).*

In this equation, as in all others in this paper, LC_{50} is expressed in SI units, that is, in kilogram moles per cubic metre (equivalent to moles per litre).

Equations analogous to Eq 1, with similar values for the slopes and intercepts, have been found to apply to fathead minnows *(Pimephales promelas)* [*2*] and to several other aquatic species in tests of various durations [*3*]. Compounds which conform to these equations are thought to act by a general nonspecific narcosis mechanism [*4*]. It has been widely considered that narcosis is caused by membrane perturbation resulting from dissolution of the narcotic agent in the lipid bilayer, but recent studies in anesthesia suggest an alternative possibility, that the narcotic agent produces its effect as a result of nonspecific binding to membrane-associated receptor protein [*5*].

Certain functional groups, when present in a compound, can lead to higher toxicity than that predicted by Eq 1, and several QSAR studies for such compounds have been published. Examples of toxicity-enhancing groups are phenolic-OH, allylic and propargylic -OH, aromatic $-NH_2$ and $-NO_2$, and halogen (when labile towards displacement by nucleophiles).

Log ($1/LC_{50}$) values for phenols can be correlated with their log P values, but the intercepts are larger than and the slopes different from the values in Eq 1. For example, for the guppy [*6*]

$$\log\left(\frac{1}{LC_{50}}\right) = 0.63 \log P + 2.52 \quad (n = 17; r = 0.982; s = 0.16) \tag{2}$$

and for fathead minnow (Roberts [*7*] analysis, of data by Hall et al. [*8*])

$$\log\left(\frac{1}{LC_{50}}\right) = 0.63 \log P + 2.42 \quad (n = 16; r = 0.983; s = 0.16) \tag{3}$$

There is a substantial body of evidence to suggest that phenols of the types covered by Eqs 2 and 3 act by a narcosis mechanism, but with a polar contribution to binding of the toxicant to the membrane. This mode of action is known as polar narcosis, or the narcosis II mechanism [*9,10*].

Allylic and propargyllic alcohols are believed to act by a pro-electrophile mechanism whereby they are metabolically oxidized to α,β-unsaturated aldehydes or ketones, which act as Michael acceptors for nucleophilic groups in biological macromolecules [*11*]. No QSAR describing their toxicity has yet been developed.

The remainder of this paper is concerned with the toxicity-enhancing effects of aromatic-

NH_2 and $-NO_2$ groups and of labile halogen substituents. A QSAR analysis of data for aromatic-NH_2 and $-NO_2$ compounds has recently been published [*7*] and is summarized only briefly here.

Aromatic Amino and Nitro Compounds [*7*]

In a reanalysis of guppy toxicity data for chloro- and alkyl-substituted anilines [*12*] and fathead minnow toxicity data for variously substituted mono-, di-, and tri-NO_2-substituted benzenes [*8*], it was found that both sets of log ($1/LC_{50}$) values could be correlated with a descriptor $\Sigma\sigma^-$, which is the sum of σ^- constants for all substituents other than the $-NH_2$ group (substituted anilines) or the most electron-deficient $-NO_2$ group (nitro compounds). A composite QSAR covering both data sets was obtained

$$\log\left(\frac{1}{LC_{50}}\right) = 1.17\ \Sigma\sigma^- + 3.63 \quad (n = 32;\ r = 0.955;\ s = 0.24;\ F = 310) \tag{4}$$

This correlation, illustrated as a graphical plot in Fig. 1, is based on toxicity data for 16 substituted anilines (guppy [*12*]), 15 aromatic nitro compounds (fathead minnow [*8*]), and 3,5-dichloronitrobenzene (guppy [*13*]). Aniline [*12*] is excluded from the data set used to derive Eq 4. It is a negative outlier from Eq 4, having a negative residual more than three times the s value for Eq 4. This outlying behavior may, as discussed later, be connected with aniline's having a lower log P value than any of the compounds used to derive Eq 4. Nevertheless, aniline is not sufficiently anomalous to significantly modify Eq 4 when it is added to the data set used for regression

$$\log\left(\frac{1}{LC_{50}}\right) = 1.21\Sigma\sigma^- + 3.58 \quad (n = 33;\ r = 0.947;\ s = 0.27;\ F = 270) \tag{5}$$

Toxicities of NH_2-substituted and OH-substituted nitrobenzenes fit Eq 4, except that some such compounds are largely ionized and tend to be negative outliers, and in some

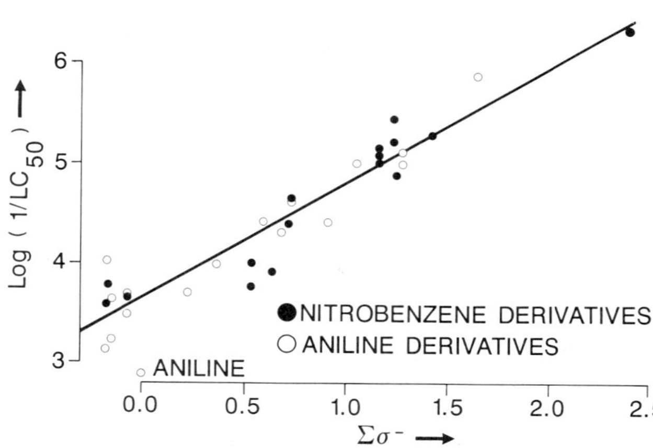

FIG. 1—*Nitrobenzene and aniline derivatives—graphical representation of Eq 4.*

cases, the corresponding C-nitroso compounds can have tautomeric Michael acceptor quinonoid forms and tend to be positive outliers. This is discussed in detail in Ref *7*.

The $\Sigma\sigma^-$ descriptor is based on σ^- constants, which are related to the more well-known σ constants of Hammett [*14*]. The σ constant for a given substituent is a measure of that substituent's influence, by electronic effects transmitted through the sigma- and pi-bond system (that is, inductive and resonance effects, respectively) on the electron density at a specified carbon atom in the benzene ring.

The σ constants are defined in terms of pK_a values of substituted benzoic acids. Thus, σ (meta-chlorine) = pK_a (meta-chlorobenzoic acid) − pK_a (benzoic acid). Many reaction rates and equilibrium constants can be correlated with σ constants, such as hydrolysis rates of substituted benzoic acid esters and equilibrium constants for cyanohydrin formation from substituted benzaldehydes and hydrogen cyanide [*15*]. Hammett σ constants are often used in drug design QSARs to model electronic effects of substituents on receptor binding [*16*].

In cases where the reaction center is (or becomes in the course of the reaction) conjugated with the benzene ring, resonance effects have increased importance, and the original σ constants of Hammett are found to show inadequacies. For cases where, as the reaction proceeds, there is an increase in electron density at a center conjugated with the benzene ring, σ^- constants are applicable [*14*]. Ionization reactions of phenols and anilinium ions are examples of such cases. The σ^- value for a given substituent can be measured by determining its effect on the pK_a value for phenol or anilinium ion. For many substituents, σ^- values do not differ significantly from σ values, but for substituents which are electron attracting through the pi-system, the differences can be large. For example [*14*]

$$\sigma(\text{para-NO}_2) = 0.78$$

and

$$\sigma^-(\text{para-NO}_2) = 1.23$$

The fact that the toxicity of aromatic amino and nitro compounds is correlated, in a QSAR common to both types of compounds, by the $\Sigma\sigma^-$ descriptor suggests that both types of compounds are metabolized to a third type of compound whose toxic action involves an increase in electron density at a center conjugated with the aromatic ring. The simplest rationale for the composite QSAR (Eq 4) is a pro-electrophile mechanism whereby $-NH_2$ groups are oxidized, and $-NO_2$ groups are reduced to C-nitroso groups, whose electrophilic reactivity [*17*] is quantified by the $\Sigma\sigma^-$ descriptor

Metabolism:

$$\text{Ar}-\text{NH}_2 \rightarrow \text{Ar}-\text{N}=\text{O} \leftarrow \text{Ar}-\text{NO}_2$$

Toxicity producing reactions:

(*i*) $$\text{Ar}-\text{N}=\text{O} + \text{biological nucleophile (NuH)} \rightarrow \underset{\text{I}}{\text{ArN(O}^-\text{)Nu}}$$

(*ii*) $$\text{ArN(O}^-\text{)Nu} \rightarrow \underset{\text{II}}{\text{ArN(OH)Nu}}$$

If NuH is an $-NH_2$ group, the likely product is ArN=N-, from the dehydration of II [*17*]. The rate of the toxicity producing reaction sequence will be determined by the stability of the intermediate I. Stability of I will be increased by delocalization of the nitrogen lone pair, thereby reducing the unfavorable electronic repulsion between this lone pair and the electrons on the negatively charged oxygen atom. This can rationalize the significance of the $\Sigma\sigma^-$ descriptor in Eq 4, $\Sigma\sigma^-$ being appropriate to model delocalization of the nitrogen lone pair in I.

If the nitroaromatics alone were being considered, it could be argued that the $\Sigma\sigma^-$ descriptor relates to the rate of metabolism, since this, being a reductive step, must involve an increase in electron density at the -NO_2 group. However, this argument is not applicable to the aminoaromatics, and the fact that both types of compound conform to the same QSAR (Eq 4) is not consistent with metabolic activation being the rate-determining step.

C-nitroso compounds are not the only candidates for the role of ultimate toxicants. Bailey and Spanggord [*18*], who originally reported the toxicity data on the nitro compounds used to derive Eq 4, suggest the nitroxy radicals, derived from one-electron reduction of the corresponding C-nitroso compounds, as the species responsible for the observed toxicity of nitrobenzene derivatives, and they argue that the degree of toxicity is determined by the ease of one-electron reduction of the nitroso compounds.

$$\mathrm{Ar{-}NO_2 \rightarrow Ar{-}N{=}O \rightarrow Ar{-}NH{-}O\cdot}$$

For the nitro compounds, this is a plausible alternative to the idea of C-nitroso compounds as ultimate toxicants. $\Sigma\sigma^-$ should be a good descriptor for the one-electron reduction of C-nitroso compounds. However, if the Bailey and Spanggord mechanism is extended to aniline derivatives, then for these compounds, the degree of toxicity should be determined by the ease of one-electron oxidation of the corresponding hydroxylamines

$$\mathrm{Ar{-}NH_2 \rightarrow Ar{-}NHOH \rightarrow Ar{-}NH{-}O\cdot}$$

Since the rate-determining steps in production of the toxic effect by the Bailey and Spanggord mechanism are different, nitro compounds and aniline derivatives would not be expected to fit the same QSAR. Thus, the Bailey and Spanggord mechanism is not consistent with the present findings of a QSAR common to nitro and aniline compounds. However, a variant of it can be postulated. This is that a hydroxylamine derivative is the reactive common metabolite. In the field of mammalian carcinogenicity, there is a substantial body of evidence implicating hydroxylamine derivatives as electrophilic metabolites of carcinogenic aromatic amines and nitro compounds. The derivatives have the -OH group acetylated or sulfated, and sometimes the NH group is acetylated [*19*]

$\mathrm{Ar{-}NH{-}OH}$	$\mathrm{ArN(H\ or\ COCH_3)O(COCH_3\ or\ SO_3^-)}$
hydroxylamine	hydroxylamine derivative

In the context of fish toxicity, the derivatives $ArNHOCOCH_3$ or $ArNHOSO_3^-$ seem the most plausible alternatives to the C-nitroso compounds as the ultimate toxic metabolites. S_N2 attack at nitrogen by biological nucleophiles can be envisaged

$$\mathrm{Nu(H) + Ar{-}NH{-}O{-}X \rightarrow ArNH{-}Nu + XO^-}$$

Reactivity in such reactions would be increased by delocalization of the nitrogen lone pair into the Ar group, and this would be appropriately represented by the $\Sigma\sigma^-$ descriptor.

On the basis of the data available, it is not possible to decide between the hypotheses of the C-nitroso compounds and the hydroxylamine derivatives as the ultimate toxicants. The C-nitroso hypothesis seems simpler and involves fewer assumptions.

Electrophilic Halogen Compounds

Hermens, Busser, Leeuwangh, and Musch [*13*] present QSAR equations for the toxicity of 15 reactive halogen compounds to the guppy. The enhanced toxicity (as compared with the predictions of the narcosis QSAR, Eq 1) of such compounds is attributed to their ability to react as alkylating agents with biological nucleophiles

$$\mathrm{Nu} + \mathrm{R{-}Hal} \rightarrow \mathrm{Nu{-}R} + \mathrm{Hal}^-$$

The data are shown in Table 1. Data on two further compounds, 1-chloro-2-butene and 3-chloro-1-butene, were omitted by Hermens et al. from their QSAR analysis because these compounds rapidly hydrolyzed.

The QSAR equations are based on a chemical reactivity parameter, k, the pseudo first-order rate constant for reaction with the nucleophile 4-nitrobenzylpyridine (NBP) under standard conditions. A plot of log ($1/LC_{50}$) against log k (Fig. 2) shows a more or less linear portion where log ($1/LC_{50}$) rises with increasing log k, followed by a plateau region where log ($1/LC_{50}$) does not change significantly with a further increase in log k. The data, including the plateau region, are fitted by the equation

$$\log\left(\frac{1}{LC_{50}}\right) = -1.30 \log (1604 + k^{-1}) + 10.35 \quad (6)$$
$$(n = 15;\ r = 0.938;\ s = 0.44)$$

Of the equations based on one descriptor, this was found by Hermens et al. to give the best fit.

One point included in the regression is much further off line than the others. This is chloroacetone, whose toxicity is badly overestimated by Eq 6 [log ($1/LC_{50}$): calculated, 6.02; found, 5.12].

Hermens et al. found that modification of this QSAR by inclusion of a log P term improves the fit somewhat

$$\log\left(\frac{1}{LC_{50}}\right) = 0.224 \log P - 1.32 \log (2484 + k^{-1}) + 10.05 \quad (7)$$
$$(n = 15;\ r = 0.956;\ s = 0.39)$$

However, examination of the residuals (Table 1) reveals that the improvement results from bringing the chloroacetone point more into line, the fit for the other compounds being slightly worse. This is shown by a comparison of the standard deviations of the residuals for all compounds except chloroacetone: for Eq 6 it is 0.34 and for Eq 7 it is 0.35.

It therefore seems reasonable, as an alternative to using the log P term, to exclude chloroacetone from the regression. Treatment of chloroacetone as an outlier may be justified in view of its low log P value in comparison with those of the other compounds *(vide infra).*

TABLE 1—*Reactive halogen compounds.*

Compound	Log k [*12*]	Log P [*12*]	Log $(1/LC_{50})$ Observed [*12*]	Residual [Log $(1/LC_{50})$ Observed Minus Log $(1/LC_{50})$ Calculated] Based on				
				Equation 6	Equation 7	Equation 8	Equation 9	Equation 10
2,3-Dichloro-1-propene	−4.30	1.99	4.99	+0.27	+0.20	+0.22	+0.25	+0.47
α,α'-Dichloro-*m*-xylene	−3.26	2.87	6.16	+0.40	+0.21	+0.29	+0.05	+0.03
2,2′-Dichlorodiethyl ether	−4.90	1.81	3.46	−0.51	−0.51	−0.51	−0.50	−0.11
2,4-α-Trichlorotoluene	−3.14	3.87	5.92	+0.09	−0.34	−0.03	+0.03	−0.01
Hexachlorobutadiene	−2.03	4.63	6.20	+0.05	−0.47	−0.08	−0.0	−0.13
Allyl chloride	−4.02	1.53	4.80	−0.24	+0.20	−0.31	−0.27	−0.13
Benzyl chloride	−3.55	2.48	5.51	0.08	−0.16	−0.11	−0.05	−0.03
1,2-dichloroethane	−5.86	1.76	2.97	+0.24	+0.23	+0.30	+0.08	+0.96[a]
1-Chloro-2,4-dinitrobenzene	−2.71	2.20	6.19	+0.16	+0.19	+0.05	+0.12	+0.03
1,3-Dichloropropane	−5.10	1.71	3.13	−0.58	−0.60	−0.57	−0.56	−0.12
1,5-Dichloropentane	−5.23	2.77	4.08	+0.53	+0.26	+0.55	+0.55	+1.04[a]
Chloroacetone	−2.73	0.45	5.12	−0.90	−0.45	−1.03[a]	−0.95[a]	−1.04[a]
4-Nitrobenzyl bromide	−1.32	2.41	6.30	+0.12	+0.15	−0.01	+0.07	−0.06
1,4-Dichloro-2-butene, *trans*	−3.34	1.94	6.16	+0.46	+0.48	+0.36	+0.12	+0.11
1,3-Dichloropropene, *cis/trans*	−3.69	1.41	5.34	−0.05	+0.05	−0.14	−0.09	−0.04

[a] Corresponding data point not included in regression.

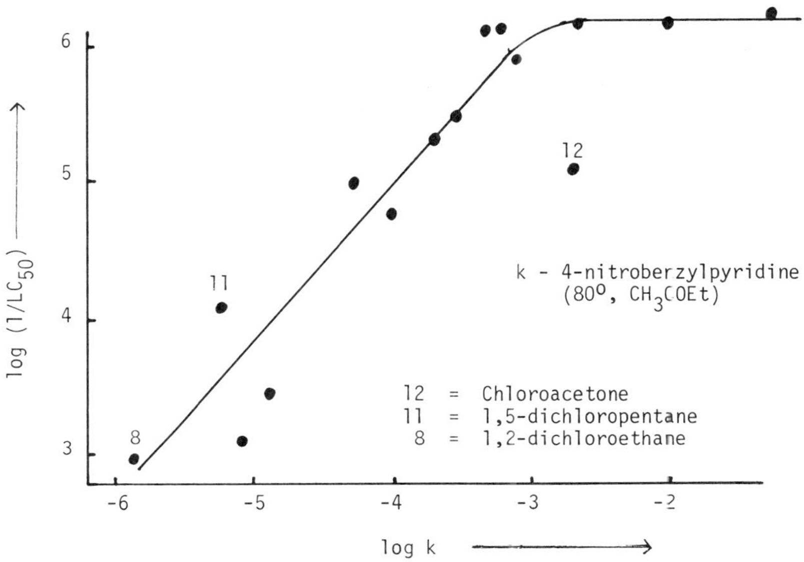

FIG. 2—*Electrophilic halogen compounds—log (1/LC$_{50}$) against log* k.

With chloroacetone omitted, Eq 6 is modified to

$$\log\left(\frac{1}{LC_{50}}\right) = -1.37 \log (1604 + k^{-1}) + 10.71 \tag{8}$$
$$(n = 14;\ r = 0.964;\ s = 0.34;\ F = 155)$$

A slight improvement is obtained if the log (1/LC$_{50}$) values for α,α'-dichloro-*m*-xylene and *trans*-1,4-dichloro-2-butene are adjusted by subtraction of log 2

$$\log\left(\frac{1}{LC_{50}}\right) = -1.34 \log (1604 + k^{-1}) + 10.53 \tag{9}$$
$$(n = 14;\ r = 0.968;\ s = 0.31;\ F = 180)$$

The justification for making these amendments to the log (1/LC$_{50}$) values is that these compounds have two reactive halogens which cannot both be displaced by the same nucleophilic group because of their wide separation. Five other compounds have two reactive sites (2,2′-dichlorodiethyl ether, hexachlorobutadiene, 1,2-dichloroethane, 1,3-dichloropropane, and 1,5-dichloropentane), but these are able to react twice with the same nucleophilic group, provided this group carries an active hydrogen atom (compare it with the cyclization of ω-aminoalkyl bromides [*20*]). These points are illustrated, taking the nucleophile as an -NH$_2$ group, in Fig. 3.

Whether this argument is sufficient justification remains to be established by studies on further bifunctional electrophiles. In any event, irrespective of whether Eq 8 or 9 is the more valid, the overall conclusion is that the log (1/LC$_{50}$) data for these electrophiles, with the exception of one relatively hydrophilic negative outlier, are well correlated by a reactivity parameter alone, without requiring a lipophilicity parameter. In this respect, there is

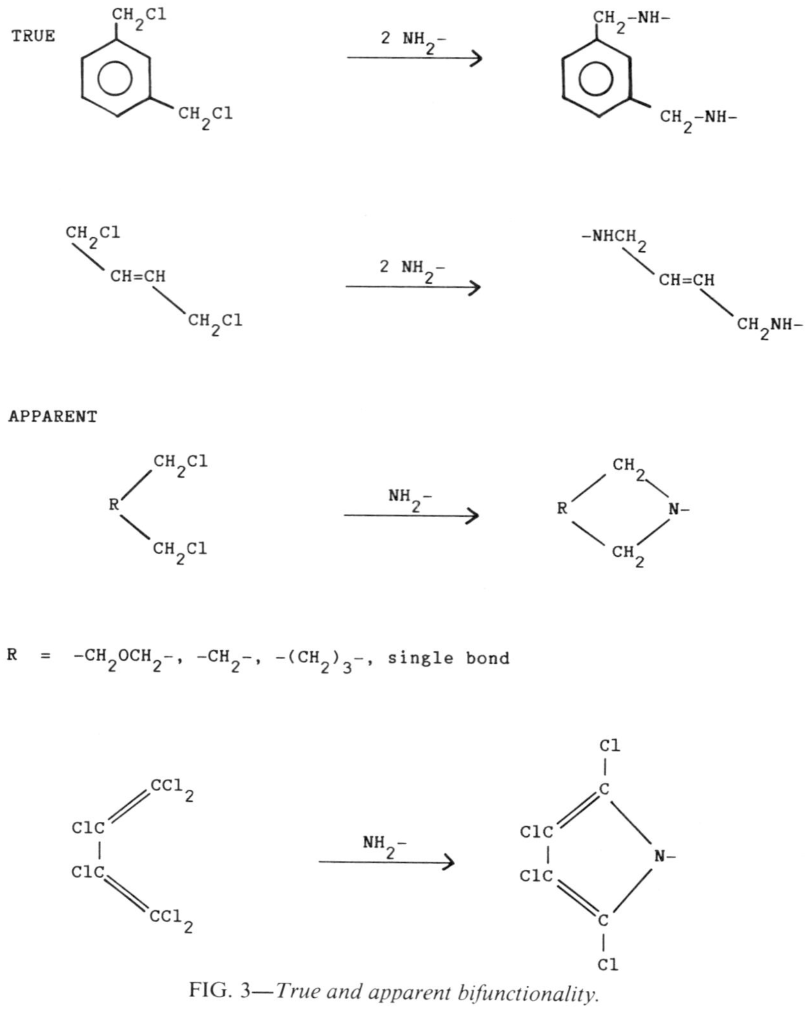

FIG. 3—*True and apparent bifunctionality.*

a close analogy with the findings presented above for aromatic nitro and amino compounds, which may be regarded as pro-electrophiles. Other types of pro-electrophiles (allylic and propargylic alcohols [*11*] and 4-aminophenol [*21*]) have been recognized as being more toxic than predicted by the narcosis QSAR, and it will be interesting to determine whether their toxicity values can also be correlated with a reactivity descriptor alone.

As discussed in the next section, 1,5-dichloropentane and 1,2-dichloroethane can justifiably be excluded from the QSAR. This improves Eq 9 significantly

$$\log\left(\frac{1}{LC_{50}}\right) = -1.64 \log(1604 + k^{-1}) + 11.62 \qquad (10)$$
$$(n = 12; r = 0.988; s = 0.17; F = 399)$$

Two of the compounds classified in this section as reactive halogen compounds contain nitro groups. It is interesting to examine how well their toxicity is predicted by Eq 10 (treating them as reactive halogen compounds) and by Eq 5 (treating them as nitro compounds).

	log $(1/LC_{50})$		
		Calculated from Equation	
Compound	Observed	Equation 5 (NO_2/NH_2 Compounds)	Equation 10 (Reactive Halogen Compounds)
1-Chloro-2,4-dinitrobenzene	6.19	5.26	6.17
4-Nitrobenzyl bromide	6.30	3.79	6.35

Clearly, these compounds behave as reactive halogen compounds rather than as nitro compounds.

Of the two rapidly hydrolyzed compounds which Hermens et al. left out from the QSAR analysis, one (1-chloro-2-butene) is much more reactive than the other (3-chloro-1-butene) toward NBP, but the log $(1/LC_{50})$ values are very similar and in poor agreement with the predictions of Eq 10, as shown below.

	1-Chloro-2-butene	3-Chloro-1-butene
Log k (NBP)	−3.70	−5.20
Log $(1/LC_{50})$ observed	4.15	4.18
Log $(1/LC_{50})$ calculated (Eq 10)	3.08	5.35

The first-order hydrolysis rate constants reported by Hermens et al. are very similar, the half lives (at 25°C in 1:1 acetone water) being 0.35 and 0.37 h for 1-chloro-2-butene and 3-chloro-1-butene, respectively. In the absence of a nucleophile (other than water), these two compounds can react via the S_N1 mechanism, involving ionization of the C—Cl bond to give, from either compound, the 3-methylallyl cation $(CH_2 \overset{...}{=} CH \overset{...}{=} CH-CH_3)^+$. On reaction with water, this cation can be hydroxylated in either the 1- or the 3-position. Thus, either of the two chlorine compounds can give the same mixture of 1-hydroxy-2-butene and 3-hydroxy-1-butene. These are both allylic alcohols of the type recognized by Lipnick [*11*] as the pro-electrophiles capable of behaving as precursors of Michael acceptors and likely to exhibit enhanced toxicity (in comparison with the predictions of Eq 1). Thus, Lipnick's pro-electrophile model can rationalize the behavior of 1-chloro-2-butene and 3-chloro-1-butene in the toxicity tests of Hermens et al.

An Outline Model for Aquatic Toxicity of Electrophiles and Pro-Electrophiles

The toxic action of electrophiles and pro-electrophiles is assumed to involve alkylation of biological nucleophiles, the pro-electrophiles being first activated by metabolism to electrophiles. The LC_{50} is the concentration in the external aqueous medium that will give rise to sufficient biological alkylation to kill half of the test population in the time scale of the experiment. The term alkylation is used in a broad sense to mean the formation of a covalent bond between a nucleophilic center and any grouping.

First, assume that the biological nucleophiles are situated in a lipid environment (for example, a membrane) which is in contact with the external aqueous medium. If the alkylation reaction is rate determining, then the overall rate of alkylation, dA/dt, is given by

$$\frac{dA}{dt} = KP_{L/W}C_W$$

where k is the pseudo first-order rate constant (the biological nucleophiles are assumed to be in excess over the electrophile), $P_{L/W}$ is the lipid-water partition coefficient, and C_W is the concentration of electrophile in water. Integrating (taking C_W as constant) over the time scale, t_e, of the experiment gives

$$A = kP_{L/W}C_W t_e$$

If A_{50} is defined as the extent of alkylation required (assumed to be independent of the structure of the alkylating agent) to kill half of the test population in the time scale, then the corresponding C_W value is the LC_{50}, and we have

$$LC_{50} = \frac{A_{50}}{t_e(kP_{L/W})}$$

that is

$$\log\left(\frac{1}{LC_{50}}\right) = \log k + \log P_{L/W} - \log\left(\frac{A_{50}}{t_e}\right)$$

Since $\log (A_{50}/t_e)$ may be taken as constant, and $\log P_{L/W}$ can be related to $\log P$ by a Collander relationship [*22*]

$$\log P_{L/W} = a \log P + \text{constant}$$

we have

$$\log\left(\frac{1}{LC_{50}}\right) = \log k + a \log P + \text{constant}$$

Thus, a QSAR for electrophiles would be expected to need both $\log k$ (a measured or calculated value, which can be taken as proportional to $\log k$ for the *in vivo* reaction) and $\log P$ as descriptors. Now, consider the case in which the alkylation reaction is faster than the rate of transfer of electrophile from the external aqueous environment to the lipid environment. This implies a relatively hydrophilic electrophile, for which the logarithm of the rate constant for transfer to the lipid phase will be proportional to $\log P$ [*23*]. It is easy to show, for this case, that a QSAR for electrophiles would be expected to need $\log P$, but not $\log k$, as the descriptor. For pro-electrophiles there is a third possibility, that metabolism to electrophilic species is rate determining. In this case, neither $\log P$ nor $\log k$, nor a combination of both, would be expected to correlate with $\log (1/LC_{50})$.

The QSARs for reactive halogen compounds and for aromatic nitro and amino compounds do not conform to any of these cases, being based solely on reactivity descriptors. This suggests that the initial assumption, that the alkylation reaction takes place in a lipid environment in contact with the external aqueous medium, is incorrect.

Now assume, instead, that the biological nucleophiles are situated in an aqueous biological medium, separated by a lipid barrier (for example, a membrane) from the external aqueous environment. As will now be demonstrated, this assumption enables the form of the QSARs to be rationalized. There are three steps in the alkylation of these nucleophiles by an electrophile: transfer from the external aqueous environment to the lipid phase, transfer from the lipid phase to the aqueous biological medium, and reaction with the biological nucleophiles. If the alkylation reaction is rate determining, then, by definition, phase equilibration is more rapid than the alkylation reaction, and the electrophile concentration in the aqueous biological medium may be taken as being proportional to C_W, the concentration in the external aqueous environment. The integrated alkylation rate equation, taking C_W as constant, is

$$A = kC_w t_e$$

from which

$$\log\left(\frac{1}{\mathrm{LC}_{50}}\right) = \log k - \log\left(\frac{A_{50}}{t_e}\right)$$

Thus, the QSAR should be a linear relationship between log (1/LC$_{50}$) and a logarithmic reactivity descriptor, independent of log P, as found by Hermens et al. [*13*] for reactive halogen compounds with log k values up to about -3, and as found for the nitro and amino aromatic compounds [*7*] discussed earlier.

Next, consider the case where transfer from the lipid phase to the biological aqueous phase is rate determining. This would apply for fast-reacting lipophilic electrophiles. The rate of alkylation is given by

$$\frac{dA}{dt} = k_{L-A}C_L$$

where k_{L-A} is the rate constant for transfer from the lipid to the aqueous phase, and C_L is the concentration of electrophile in the lipid phase. C_L can be substituted by $P_{L/W} \cdot C_W$ and, for lipophilic compounds, k_{L-A} is inversely proportional to $P_{L/W}$ [*23*]. It follows that

$$\frac{dA}{dt} \propto C_W$$

from which

$$\log\left(\frac{1}{\mathrm{LC}_{50}}\right) = \text{constant}$$

Thus, the plateau region observed by Hermens et al. is rationalized.

Next, consider the case where transfer from the external aqueous medium to the lipid phase is rate determining. This would apply for fast-reacting hydrophilic electrophiles, such as chloroacetone. It is easily shown that, in this case, log (1/LC$_{50}$) should be linearly correlated with log P. Thus, the outlying behavior of chloroacetone in the QSARs based on k (Eqs 6, 8, 9, and 10) and its improved fit when a log P term is included in the QSAR (Eq 7) is rationalized.

The above arguments also apply to pro-electrophiles, for which there is also the possibility that metabolic activation could be rate determining. For the nitro and amino aromatics considered previously, there is no evidence for metabolic activation or transfer from the lipid phase to the biological aqueous medium being rate determining. However, the outlying behavior of aniline may be rationalized in terms of transfer from the external aqueous medium to the lipid phase being rate determining.

The overall conclusion of this section is that the toxic action of electrophiles and pro-electrophiles involves alkylation of biological nucleophiles in an aqueous biological medium. The model is illustrated in Fig. 4. This mechanism will be competitive with the general narcosis mechanism which is believed to apply for chemically inert compounds, and for which the site of action is the lipid membrane. Some electrophilic compounds may have a higher toxicity by the narcosis mechanism than by alkylation in the aqueous biological medium. Thus, by comparison of Eqs 1 and 9, it may be estimated that reactive halogen compounds will kill by the narcosis mechanism if

$$0.871 \log P + 1.34 \log (1604 + k^{-1}) > 9.4$$

This can only be a rough estimate, since the narcotic toxicity of an electrophile is likely to be somewhat higher than predicted by Eq 1 as a result of alkylation of membrane-bound receptor protein (compare the enhanced upper respiratory tract irritancy of electrophiles with that of chemically inert compounds) [*24*]. Of the 15 reactive halogen compounds shown in Table 1, only 1,5-dichloropentane obeys the above inequality, having a 0.871 log P + 1.34 log (1604 + k^{-1}) value of 9.42. This is the justification for removing 1,5-dichloropentane from the QSAR of Eq 9. The resulting equation is

$$\log\left(\frac{1}{LC_{50}}\right) = -1.42 \log (1604 + k^{-1}) + 10.82$$
$$(n = 13;\ r = 0.978;\ s = 0.26;\ f = 0.239)$$

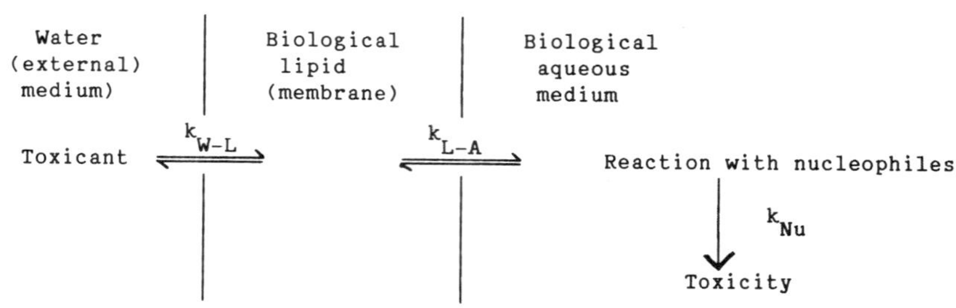

1. k_{Nu} rate determining - no log P dependence.
2. k_{W-L} rate determining (highly hydrophilic compounds, e.g. aniline) - dependence on log P.
3. k_{L-A} rate determining (highly reactive, hydrophobic compounds) - no dependence.
4. Metabolism (pro-electrophiles) rate determining - ?

FIG. 4—*Model for toxicity of pro-electrophiles.*

When compared with Eq 1, it gives

$$0.871 \log P + 1.42 \log (1604 + k^{-1}) > 9.7$$

as the revised criterion in order for the narcosis mechanism to apply. 1,5-Dichloropentane meets this revised criterion, having a $0.871 \log P + 1.42 \log (1604 + k^{-1})$ value of 9.84. However, 1,2-dichloroethane also meets this revised criterion, having a $0.871 \log P + 1.42 \log (1604 + k^{-1})$ value of 9.86. Therefore, 1,2-dichloroethane must also be removed from the QSAR of Eq 9. The resulting Eq 10 gives

$$0.871 \log P + 1.64 \log (1604 + k^{-1}) > 10.5$$

as the new criterion for the narcosis mechanism. Other than 1,5-dichloropentane and 1,2-dichloroethane, none of the compounds of Table 1 meets this criterion.

The problem of deciding whether or not a given electrophile will obey Eq 10 can now be approached by means of a map of log P versus log k territory, as shown in Fig. 5. The precise position of the lower boundary cannot at present be defined and is shown only schematically in Fig. 5.

In Fig. 5 the upper hatched zone represents the positive outliers from Eq 10. Narcosis toxicity (Eq 1) is greater than the aqueous biological medium electrophile toxicity (Eq 10). The half-shaded points in this zone correspond to 1,2-dichloroethane and 1,5-dichloropentane.

The central zone of Fig. 5 represents the toxicity by electrophilic reaction in an aqueous biological medium. Equation 10 applies. The broken line marks the approximate boundary between linear dependence and independence of toxicity with log k. Shaded points in this zone correspond to compounds listed in Table 1, excluding 1,2-dichloroethane, 1,5-dichloropentane, and chloroacetone.

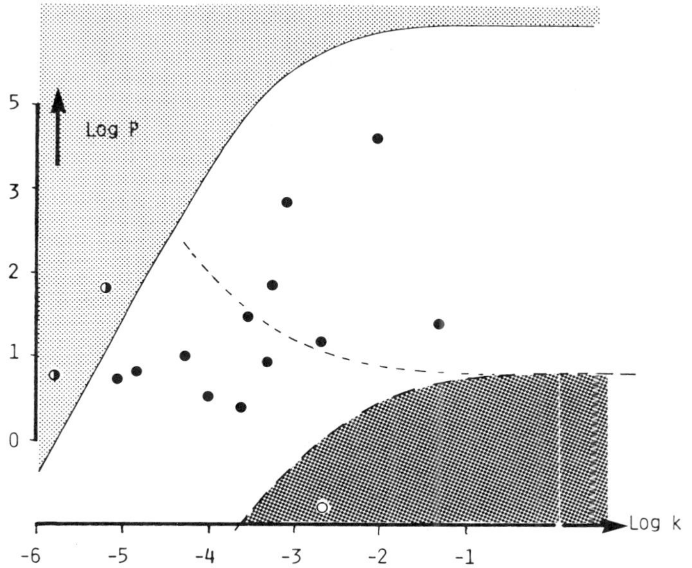

FIG. 5—*Mapping of lipophilicity-electrophilicity territory. See the text for an explanation.*

The lower hatched zone represents the negative outliers from Eq 10. Toxicity is by electrophilic reaction in an aqueous biological medium, but environment-lipid mass transfer is rate limiting. The unshaded point in this zone corresponds to chloroacetone.

The upper boundary shown in Fig. 5 is defined by the equation

$$0.871 \log P + 1.64 \log (1604 + k^{-1}) = 10.5$$

Since it seems unlikely that nucleophiles will be entirely absent from the lipid environment, which is the site of action for narcosis, it is possible that compounds close to the upper boundary may be positive outliers from both the narcosis QSAR (Eq 1) and the electrophile QSAR (Eq 10), their toxic activity resulting from alkylation of nucleophiles in the lipid environment. For such compounds, a two-descriptor QSAR based on both log P and log k should apply.

For the nitrobenzenes and aminobenzenes, competition between the pro-electrophile mode of action and narcosis is to be expected, but in these cases, the narcosis mechanism is likely to be of Type II (polar narcosis). Correlations between log ($1/LC_{50}$) and log P, which can be interpreted in terms of polar narcosis, have in fact been reported for some aminobenzenes [*12*] and mononitrobenzenes [*25,26*], for which $\Sigma\sigma^-$ is not large; it is therefore possible that some of the compounds included in the regression for Eq 4 act by the polar narcosis mechanism.

Conclusions

The QSAR for aromatic nitro and amino compounds is based on mixed data for two species, the fathead minnow and the guppy. This fact, together with the close resemblances between the separate species QSARs for phenols and between separate and mixed species QSARs for chemically inert compounds [*7*], indicates that from the viewpoint of acute toxicity, the fathead minnow and guppy can be treated as if they were one species.

The acute lethal toxicity data on reactive halogen compounds can be correlated with an electrophilicity descriptor, and the data are independent of log P. The same applies to aromatic nitro and amino compounds whose toxicity can be interpreted in terms of a pro-electrophile mechanism. The nondependence on log P can be rationalized in terms of a model whereby the reaction with biological nucleophiles takes place in an aqueous biological medium separated by a lipid barrier from the external aqueous environment. This model provides a rationale for negative outlying behavior of highly hydrophilic compounds and for the existence of a plateau region in the QSAR for reactive halogen compounds.

The original QSAR for reactive halogen compounds (Eq 6) was based on 15 compounds—insufficient, it has been argued [*27*] for the QSAR to be applicable predictively. Here the number of compounds has been reduced to twelve. Nevertheless, the present analysis increases predictive capability for this class of compounds. The degree of confidence one can have in a QSAR depends on the following points:

(*a*) how well the training set data are correlated and the statistical significance of the correlation,

(*b*) the number of outliers that can be rationalized and, inversely, the number of outliers that cannot be rationalized, and

(*c*) the availability or unavailability of criteria for deciding whether a given compound is expected to fit the QSAR.

For a one-descriptor bilinear QSAR like Eq 10, twelve data points with the statistical indexes of Eq 10 are ample to satisfy Point (*a*) above. As for Point (*b*), two outliers from the original QSAR (Eq 6) have now been rationalized, and the removal here of further compounds as outliers has been rationalized on toxicokinetic and mechanistic grounds. Most important, for Point (*c*), the boundaries of applicability of the QSAR, although still not completely defined, can now be recognized.

References

[*1*] Könemann, H., "Quantitative Structure-Activity Relationships in Fish Toxicity Studies: Part 1—Relationships for 50 Industrial Pollutants, *Toxicology,* Vol. 19, 1981, pp. 209–221.

[*2*] Veith, G. D., Call, D. J., and Brooke, L. T., "Structure Toxicity Relationships for the Fathead Minnow, *Pimephales promelas:* Narcotic Industrial Chemical," *Canadian Journal of Fisheries and Aquatic Sciences,* Vol. 40, 1983, pp. 743–748.

[*3*] Sloof, W., Canton, J. H., and Hermens, J. "Comparison of the Susceptibility of 22 Freshwater Species to 15 Chemical Compounds: I.—(Sub)acute Toxicity Tests," *Aquatic Toxicology,* Vol. 4, 1983, pp. 113–128.

[*4*] Hermens, J., Könemann, H., Leeuwangh, P., and Musch, A., "Quantitative Structure-Activity Relationships in Aquatic Toxicity Studies of Chemicals and Complex Mixtures of Chemicals," *Environmental Toxicology and Chemistry,* Vol. 4, 1985, pp. 273–279.

[*5*] Franks, N. P. and Lieb, W. R., "The Firefly Throws Light on Anaesthesia," *Chemistry in Britain,* Vol. 21, 1985, pp. 919–921.

[*6*] Saarikoski, J. and Viluksela, M., "Relation Between Physicochemical Properties of Phenols and Their Toxicity and Accumulation in Fish," *Ecotoxicology and Environmental Safety,* Vol. 6, 1982, pp. 501–512.

[*7*] Roberts, D. W. "An Analysis of Published Data on Fish Toxicity of Nitrobenzene and Aniline Derivatives," *QSAR in Environmental Toxicology—II,* K. L. E. Kaiser, Ed., D. Reidel Publishing, Dordrecht, The Netherlands, 1987.

[*8*] Hall, L. H., Kier, L. B., and Phipps, G., "Structure-Activity Relationship Studies on the Toxicities of Benzene Derivatives: 1—An Additivity Model," *Environmental Toxicology and Chemistry,* Vol. 3, 1984, pp. 355–365.

[*9*] Veith, G. D. and Broderius, S. J., *Structure-Toxicity Model for Type (II) Narcotic Chemistry in QSAR in Environmental Toxicology—II,* K. L. E. Kaiser, Ed., D. Reidel Publishing Co., Dordrecht, Holland, 1987.

[*10*] Schultz, T. W., Holcombe, G. W., and Phipps, G. L., "Relationships of Quantitative Structure-Activity to Comparative Toxicity of Selected Phenols in the *Pimephales promelas* and *Tetrahymena pyriformis* Test Systems," *Ecotoxicology and Environmental Safety,* Vol. 12, 1986, pp. 146–153.

[*11*] Lipnick, R. L., Johnson, D. E., Gilford, J. H., Bickings, C. K., and Newsome, L. D., "Comparison of Fish Toxicity Screening Data for 55 Alcohols with the Quantitative Structure-Activity Relationship Predictions of Minimum Toxicity for Non-Reactive Non-Electrolyte Organic Compounds," *Environmental Toxicology and Chemistry,* Vol. 4, 1985, pp. 281–296.

[*12*] Hermens, J., Leeuwangh, P., and Musch, A., "Quantitative Structure-Activity Relationships and Mixture Toxicity Studies of Chloro- and Alkylanilines at an Acute Lethal Toxicity Level to the Guppy *(Poecilla reticulata),*" *Ecotoxicology and Environmental Safety,* Vol. 8, 1984, pp. 388–394.

[*13*] Hermens, J., Busser, F., Leeuwangh, P., and Musch, A., "Quantitative Correlation Studies Between the Acute Lethal Toxicity of 15 Organic Halides to the Guppy *(Poecilla reticulata)* and Chemical Reactivity Towards 4-Nitrobenzylpyridine," *Toxicology and Environmental Chemistry,* Vol. 9, 1985, pp. 219–236.

[*14*] Hine, J., *Structural Effects on Equilibria in Organic Chemistry,* Wiley, New York, 1975.

[*15*] Gould, E. S., *Mechanism and Structure in Organic Chemistry,* Holt, Rinehart, and Winston, New York, 1975.

[*16*] Seydel, J. K. and Schafer, K. J., "Quantitative Structure-Pharmacokinetic Relationships and Drug Design," *Pharmacology and Therapeutics,* Vol. 15, 1982, pp. 131–182.

[*17*] Smith, P. A. S., *The Chemistry of Open-Chain Organic Nitrogen Compounds,* Vol. II, W. A. Benjamin, Inc., New York, NY, 1966.

[*18*] Bailey, H. C. and Spanggord, R. J., "The Relationship Between the Toxicity and Structure of

Nitro Aromatic Compounds," *Aquatic Toxicology and Hazard Assessment: Sixth Symposium, ASTM STP 802,* American Society for Testing and Materials, Philadelphia, 1984.

[*19*] Timbrell, J. A., *Principles of Biochemical Toxicity,* Taylor and Francis, London, 1982.

[*20*] Freundlich, H. and Kroepelin, H., "Kinetics of Transformation of Halogen Alkylamines into Heterocyclic Compounds (IV)," *Zeitschrift fuer Physikalische Chemie,* Vol. 122, 1926, pp. 39–48.

[*21*] Lipnick, R. L., Bickings, C. K., Johnson, D. E., and Eastmond, D. A., "Comparison of QSAR Predictions with Fish Toxicity Screening Data for 110 Phenols," *Aquatic Toxicology and Hazard Assessment (Eighth Symposium), ASTM STP 891,* American Society for Testing and Materials, Philadelphia, 1986, pp. 153–176.

[*22*] Leo, A., Hansch, C., and Elkins, D., "Partition Coefficients and Their Uses," *Chemical Reviews,* Vol. 71, 1971, pp. 525–616.

[*23*] van de Waterbeemd, H., "The Theoretical Basis for Relationships Between Drug Transport and Partition Coefficients," *Quantitative Approaches to Drug Design,* J. C. Dearden, Ed., Elsevier Science, Amsterdam, 1983, pp. 183–192.

[*24*] Roberts, D. W., "QSAR for Upper-Respiratory Tract Irritation," *Chemico-Biological Interactions,* Vol. 57, 1986, pp. 325–345.

[*25*] Schultz, T. W., Cajina-Quezada, M., Chang, M., Lin, D. T., and Jain, R., "Structure-Toxicity Relationships of Para-Position Alkyl- and Halogen- Substituted Monoaromatic Compounds," this publication, pp. 410–423.

[*26*] Deneer, J. W., Hermens, J. L. M., Sinnige, T. L., and Seinen, W., "Quantitative Structure-Activity Relationships for the Toxicity and Bioconcentration Factor of Nitrobenzene Derivatives Towards the Guppy *(Poecilia reticulata),*" *Aquatic Toxicology,* Vol. 10, 1987, pp. 115–129.

[27] Hermens, J. L. M., "Quantitative Structure-Activity Relationships in Aquatic Toxicology," *Pesticide Science,* Vol. 17, 1986, pp. 287–296.

John D. Walker[1] *and Robert H. Brink*[1]

New Cost-Effective, Computerized Approaches to Selecting Chemicals for Priority Testing Consideration

REFERENCE: Walker, J. D. and Brink, R. H. **"New Cost-Effective, Computerized Approaches to Selecting Chemicals for Priority Testing Consideration,"** *Aquatic Toxicology and Environmental Fate: Eleventh Volume, ASTM STP 1007,* G. W. Suter II and M. A. Lewis, Eds., American Society for Testing and Materials, Philadelphia, 1989, pp. 507–536.

ABSTRACT: The Interagency Testing Committee (ITC) was established in 1976 under Section 4(e) of the Toxic Substances Control Act (TSCA). The ITC consists of representatives from 15 member and liaison agencies. The committee recommends chemicals to the U.S. Environmental Protection Agency (EPA) administrator for priority testing consideration and potential rule-making under Section 4(a) of TSCA.

Between 1977 and 1983 the ITC conducted five scoring exercises to select existing chemicals for detailed review. These five scoring exercises focused on chemicals of regulatory concern, high-production chemicals, and chemicals recommended by other organizations. In 1986, in preparation for a sixth scoring exercise, we developed new approaches to selecting chemicals for review by the ITC. Expert panels were organized to identify substructures indicative of chemicals with potential to cause adverse health effects or adverse ecological effects. The expert ecological effects panel identified 109 organic chemical substructures that might cause adverse effects to any of eight organism types. These substructures were used to develop a questionnaire which was sent to 86 environmental chemists and toxicologists.

Results from this questionnaire were used to reduce the list of substructures identified by the expert panel to 63 substructures with potential to cause adverse ecological effects in one to six organism types. These chemical substructures were used to search a data base of approximately 18 100 discrete organic chemicals. These searches identified about 5200 chemicals which might produce adverse ecological effects. The data base containing 18 100 chemicals was also searched for about 5300 chemicals that were in eleven computer-searchable, environmental-monitoring data bases. These searches identified about 2300 chemicals. The 2300 Chemical Abstracts Service (CAS) numbers for chemicals with potential for environmental exposure were compared with the 5200 CAS numbers for chemicals that might produce adverse ecological effects in order to identify matches.

Over 100 chemicals were identified that had both the potential to cause adverse ecological effects and the potential for environmental exposure. A scoring system was developed to rank these chemicals. After obtaining available information on domestic production, many of these chemicals were screened by panels of toxicologists, chemists, ecologists, and engineers to select chemicals of environmental concern for detailed review by the ITC.

KEY WORDS: chemical selection, scoring, ecotoxicology, computerized selection process, Toxic Substances Control Act (TSCA), aquatic toxicology, ecological effects, environmental exposure

[1] U.S. EPA statutory member and executive secretary, respectively, Interagency Testing Committee, Office of Toxic Substances, U.S. Environmental Protection Agency, Washington, DC 20460.

Mission and Organization of the Interagency Testing Committee

The Toxic Substances Control Act (TSCA) (U.S. Public Law 94-469) established the Interagency Testing Committee (ITC) to recommend chemicals for priority testing consideration to the U.S. Environmental Protection Agency (EPA). The only existing chemicals not considered by the ITC (because they are not subject to regulation under TSCA) are pesticides, tobacco, nuclear materials, firearms, foods, food additives, drugs, and cosmetics. The list of chemicals recommended to the EPA is known as the "Priority List." In establishing the Priority List, the ITC must give priority attention to those substances and mixtures which are known to cause or contribute to, or which are suspected of causing or contributing to cancer, gene mutations, or birth defects. At least every six months the ITC must submit a report to the EPA Administrator providing reasons for revising the Priority List. The committee selects chemicals for detailed review and subsequent recommendation or deferral after conducting extensive screening and scoring exercises or evaluating lists of chemicals submitted by member and liaison agencies, other federal agencies, state agencies, and others.

The ITC consists of statutory members from eight agencies specified in TSCA and representatives from seven liaison agencies invited to participate in ITC activities (Table 1).

Previous Scoring Exercises

The chemical screening, scoring, and selection methods used by the ITC in the first through fifth scoring exercises (1977, 1979, 1980, 1982, and 1983) were devised in 1977 and were used for the first through fifth scoring exercises without significant modifications [*1,2*]. The general approach is summarized in Fig. 1 and described in detail in a Scoring Task Force Report prepared for the ITC [*3*].

The initial list contained chemicals with high annual production (for example, greater than 10^6 kg/year), analogues of known carcinogens, and chemicals of concern to federal or state agencies and others. The initial list for the fifth scoring exercise contained 3370 chem-

TABLE 1—*Structure of the ITC.*

Statutory Member Agencies
Council on Environmental Quality
Department of Commerce
Environmental Protection Agency
National Cancer Institute
National Institute of Environmental Health Sciences
National Institute for Occupational Safety and Health
National Science Foundation
Occupational Safety and Health Administration
Liaison Agencies
Consumer Product Safety Commission
Department of Agriculture
Department of Defense
Department of the Interior
Food and Drug Administration
National Library of Medicine
National Toxicology Program

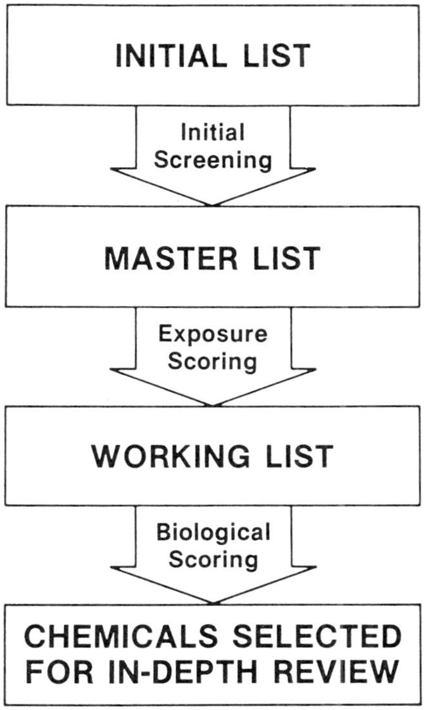

FIG. 1—*Chemical selection approaches for the first through the fifth ITC scoring exercises.*

ical substances and mixtures. Initial screening removed chemicals that were previously scored, already regulated, well understood, under review, considered to be innocuous, or considered to be poorly characterized mixtures (Fig. 1). In the fifth scoring exercise, the initial screening was conducted by a team of organic chemists and toxicologists who reduced the initial list of 3370 chemicals to a master list of 471 chemicals.

Key elements of the previous scoring exercises were sequential exposure and biological scoring and manual searching and selecting of chemicals. A panel of experts assigned consensus scores to the eleven exposure factors listed in Table 2. Scores ranged from 0 to 3; for example, the scores for persistence were 0 (hours), 1 (days), 2 (months), or 3 (years). The consensus scores for the eleven exposure factors were combined into three exposure algorithms (Table 2). In the fifth scoring exercise, exposure scoring reduced the master list of 471 chemicals to a working list of 214 chemicals.

Four panels of experts were used to score the working list of 214 chemicals for the eight biological effects factors listed in Table 3. Scores could range from −3 (strong suspicion of an adverse effect at low exposures but no data) to +3 (data supporting an adverse effect at low exposures). The greater the score (either negative or positive) the greater the potential concern. A consensus score was assigned to each chemical on the working list for each factor. The working list was ranked by consensus scores, and the ranked list was submitted to the ITC. At the completion of the fifth scoring exercise, 82 chemical substances and mixtures were selected for in-depth review [*4*].

TABLE 2—*Exposure factors and algorithms used in the ITC's fifth scoring exercise.*

Exposure Factors

- (Q) Annual production
- (R_i) Fraction released in plant
- (N_w) Number of workers
- (R_e) Fraction released to the environment
- (N_g) Number of people exposed in general population
- (F) Frequency of general population exposure
- (I) Intensity of general population exposure
- (H) Persistence
- (P) Penetrability
- (L) Influence on the environment
- (B) Bioaccumulation potential

Exposure Algorithms

$$\frac{Q + R_i}{2} + P + N_w = t_{oe} \qquad \frac{t_{oe}}{9} = \text{occupational exposure index}$$

$$\frac{Q + R_e}{2} + F + \frac{I + P}{2} + N_g = t_{ge} \qquad \frac{t_{ge}}{12} = \text{general population exposure index}$$

$$\frac{Q + R_e}{2} + H + \frac{B + L}{2} = t_{ee} \qquad \frac{t_{ee}}{9} = \text{ecological/environmental exposure index}$$

TABLE 3—*Biological effects factors used in the fifth scoring exercise.*

Mutagenicity
Carcinogenicity
Teratogenicity
Reproductive effects
Acute toxicity
Subchronic toxicity
Bioaccumulation potential
Ecological effects

Recognizing a Need for Change

Methods used for the first through the fifth scoring exercises were developed to identify easy-to-recognize, high-production, high-exposure, and poorly tested chemicals. When this approach was first used, the 1977 TSCA inventory containing 61 434 chemicals, reaction mixtures, and polymers had just been completed. Classifying these chemicals by production volume, knowing that the same chemical could be produced by a number of manufacturers at varying production levels (99 793 chemicals), revealed that 27% had annual production volumes greater than 10^6 kg (Table 4). Initially it was easy to recognize high-production chemicals and large categories of chemicals that needed testing (Table 5). As the ITC and other organizations such as the National Toxicology Program and the Chemical Industry Institute for Toxicology began to select chemicals for testing, it became more difficult for the ITC to select chemicals that satisfied the statutory factors it should consider (Table 6).

TABLE 4—*Number of chemicals in the 1977 TSCA inventory classified by production volume.*

Production Volume, kg/year	Number of Chemicals
0 to 2200	22 956
2200 to 22 000	15 217
22 000 to 220 000	18 833
220 000 to 2.2×10^6	16 0[illegible]1
2.2×10^6 to 22×10^6	9 900
22×10^6 to 220×10^6	5 791
110×10^6 to 220×10^6	2 424
220×10^6 to 1100×10^6	4 953
1100×10^9 to 2.2×10^9	1 636
$>2.2 \times 10^9$	2 072

TABLE 5—*Easy-to-recognize high-production chemicals and large chemical categories recommended for testing by the ITC.*[a]

Acetonitrile	Halogenated alkyl epoxides
Acrylamide	Hexachloro-1,3-butadiene
Alkyl epoxides	Hexachlorocyclopentadiene
Alkyl phthalates	Hydroquinone
Anilines	Isophorone
Antimony substances	Mesityl oxide
Aryl phosphates	4,4′-Methylenediamine
Benzidine-based dyes	Methyl ethyl ketone
Chlorinated benzenes	Nitrobenzene
Chlorinated naphthalenes	Phenylenediamines
Chlorinated paraffins	Polychlorinated terphenyls
Chloromethane	Pyridine
o-Dianisidine-based dyes	Quinone
Dichloromethane	*o*-Tolidine-based dyes
1,2-Dichloropropane	Toluene
Cyclohexanone	1,1,1-Trichloroethane
Glycidols	Xylenes

[a] Recommended in October 1977, April 1978, October 1978, April 1979, November 1979, or April 1980.

TABLE 6—*Statutory factors which the ITC considers before recommending a chemical for priority testing consideration.*

Quantities manufactured
Quantities released to environment
Number of individuals exposed or duration of exposure
Extent of human exposure
Structural relationships to known toxicants
Toxicity data
Reliability of test data to predict hazard
Availability of testing facilities

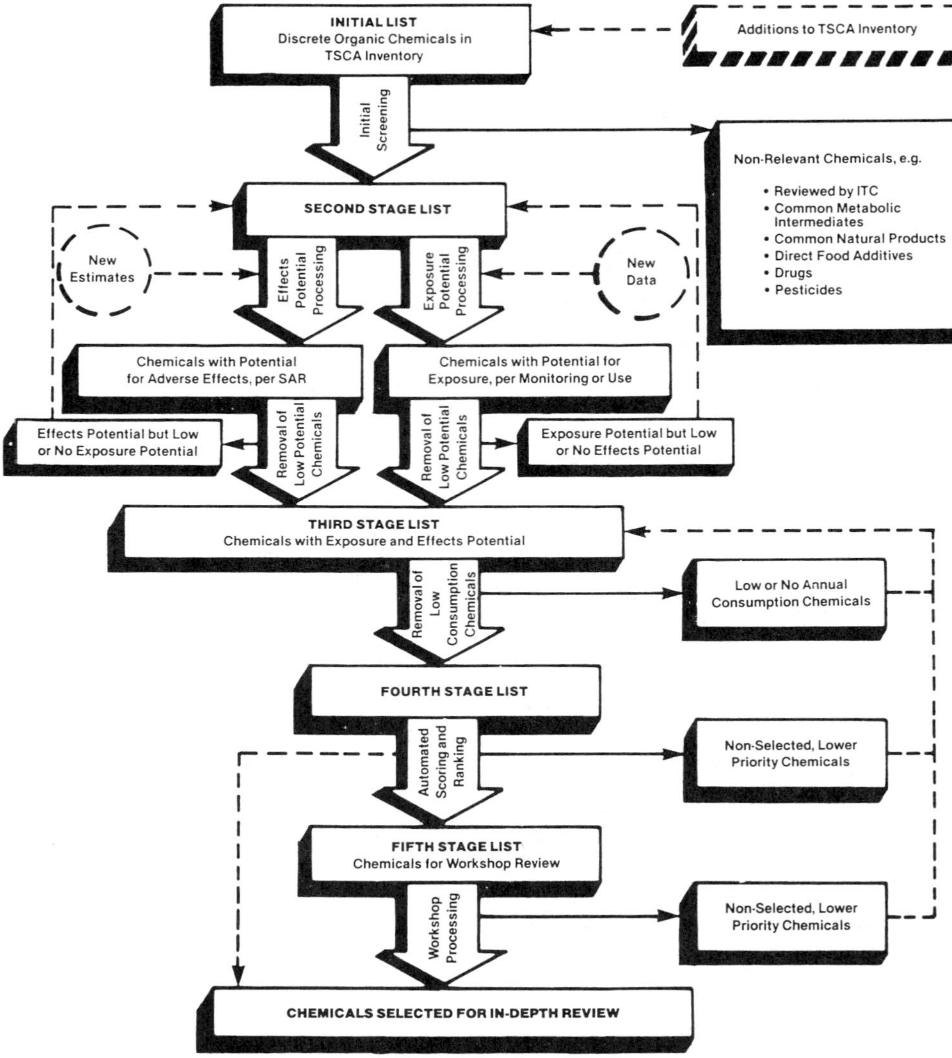

FIG. 2—*New chemical selection processes for the ITC's sixth scoring exercise.*

The need for a change was recognized. The new chemical selection processes illustrated in Fig. 2 and described below were developed to satisfy this need.

New Chemical Selection Processes

Several factors influenced our decisions for revising the ITC's chemical selection processes (Table 7). A key factor was the opportunity to develop cost-effective, computerized processes which simultaneously integrated effects and exposure information. Revision of the process began by using a new initial list from which to select chemicals. The initial list was a computer file obtained from Drs. Gil Veith and Robert Hunter, who recognized that the 1977 TSCA inventory could be organized into several large categories: (1) discrete

TABLE 7—*Factors influencing decisions for revising the ITC chemical selection processes.*

Opportunity to develop cost-effective, computerized chemical selection processes
Need to simultaneously integrate effects and exposure information
Desire to utilize structure-activity relationships for predicting physical-chemical properties, persistence, and potential adverse health and ecological effects of chemical substances
Opportunity to develop or utilize computer-searchable monitoring data bases

organics (20 to 30%), (2) inorganics and organometallics (10 to 20%), reaction mixtures (20 to 30%), and polymers (20 to 30%). The list contained the ninth Collective Index (CI) name, Chemical Abstracts Service (CAS) Registry number, and Simplified Molecular Input Line Entry System (SMILES) notation for about 20 000 discrete organic chemicals from the 1977 inventory (Fig. 2). The CAS numbers from the initial list were compared with the CAS numbers for the more than 188 000 searchable structures in the Structure and Nomenclature Search System (SANSS). SANSS is a component of the Chemical Information System (CIS), a commercial data base managed by Fein Marquart Co. in Baltimore, MD. Comparing CAS numbers from the initial list to CAS numbers for chemicals in SANSS identified 19 166 chemicals that had searchable structures. Initial screening was accomplished by removing readily identifiable, nonrelevant chemicals (Fig. 2).The list of chemicals removed from the initial list, the CAS numbers for about 18 133 chemicals in the second stage list, and details of the processes for selecting chemicals with potential to cause adverse human health effects are contained in Brink and Walker [5].

Ecological-Effects Potential Processing

An ecological effects panel, consisting of the six members listed in Table 8, met on 6 March 1986 and drafted a list of 109 chemical descriptors with potential to produce adverse ecological effects in eight target organisms (algae, aquatic invertebrates, birds, fish, mammals, microbes, plants, and terrestrial invertebrates). The descriptors consisted, for the most part, of substructural molecular features of certain classes of organic chemicals.

A questionnaire containing the 109 substructures was sent to 57 domestic and 29 foreign scientists (recommended by the panel) on 7 April 1986. They were asked to indicate, by check marks, if a chemical containing a given substructure had a high, moderate, or low potential to cause an adverse effect in a particular organism. They were also asked to indicate by check marks if they had no opinion on the substructure's toxicity, and they were invited to add comments or suggestions. Table 9 provides an example of one page from the questionnaire that was sent to the 86 scientists. About 44% of the questionnaires were completed and returned between 30 April and 30 July 1986.

Based on an analysis of the completed questionnaires, 63 substructures with potential to cause adverse effects in one to six organism types were used to search the second stage list (Table 10). None of the substructures had the potential to adversely effect all eight organ-

TABLE 8—*ITC ecological effects structure-activity relationships panel.*

Dr. James W. Gillett	Cornell University
Dr. Klaus L. E. Kaiser	Canada Centre for Inland Waters
Dr. Robert L. Lipnick	U.S. Environmental Protection Agency
Dr. Robert K. Ringer	Michigan State University
Dr. Gilman D. Veith	U.S. Environmental Protection Agency
Dr. John D. Walker, chairman	U.S. Environmental Protection Agency

TABLE 9—*Potential for substructures to cause adverse ecological effects.*

Chemical Class Requirements	Potential for Causing Adverse Effects	Potential Sensitive Species								
		Algae	Aquatic Invertebrates	Birds	Fish	Mammals	Microbes	Plants	Terrestrial Invertebrates	Comments
5-Nitrofuryl	high	____	____	____	____	____	____	____	____	
	moderate	____	____	____	____	____	____	____	____	
	low	____	____	____	____	____	____	____	____	
	no opinion	____	____	____	____	____	____	____	____	
Nitroaromatics	high	____	____	____	____	____	____	____	____	
[≥2 *N*s on	moderate	____	____	____	____	____	____	____	____	
the same	low	____	____	____	____	____	____	____	____	
ring; trisubstituted]	no opinion	____	____	____	____	____	____	____	____	
Nitro	high	____	____	____	____	____	____	____	____	
compounds	moderate	____	____	____	____	____	____	____	____	
aliphagic	low	____	____	____	____	____	____	____	____	
	no opinion	____	____	____	____	____	____	____	____	
Nitrohalides	high	____	____	____	____	____	____	____	____	
[aromatic]	moderate	____	____	____	____	____	____	____	____	
	low	____	____	____	____	____	____	____	____	
	no opinion	____	____	____	____	____	____	____	____	
Nitroquinolines	high	____	____	____	____	____	____	____	____	
	moderate	____	____	____	____	____	____	____	____	
	low	____	____	____	____	____	____	____	____	
	no opinion	____	____	____	____	____	____	____	____	
C-Nitroso	high	____	____	____	____	____	____	____	____	
compounds	moderate	____	____	____	____	____	____	____	____	
	low	____	____	____	____	____	____	____	____	
	no opinion	____	____	____	____	____	____	____	____	
N-Nitroso	high	____	____	____	____	____	____	____	____	
compounds	moderate	____	____	____	____	____	____	____	____	
	low	____	____	____	____	____	____	____	____	
	no opinion	____	____	____	____	____	____	____	____	

ism types; aromatic amines, aziridines, and azo compounds had the potential to adversely affect six organism types, and most substructures had the potential to adversely affect from one to three organism types (Table 10). Seven substructures were not contained in any second stage list chemicals; five were contained in only one second stage list chemical; most were contained in 1 to 40 second stage list chemicals (Table 11*a*). Based on the analysis of the questionnaire responses, chemicals containing about 30 substructures were most likely to find mammals or fish as potential target organisms (Table 11*b*). Chemicals with only half of these substructures were likely to find mammals and fish as potential target organisms (Table 10). The number of chemicals per substructure in the second stage list ranged from 0 to 1020. The total number of second stage list chemicals which contained the 63 substructures was about 5182 (Fig. 3). This number includes and has not been corrected for redundancies; for example, the same chemical could be scored as an aromatic amine and an azo dye. This suggests that at least 28% of the chemicals in the second stage list have a potential to cause adverse ecological effects (Fig. 2). Forty-six substructures were not used to search the second stage list because the questionnaire responses indicated that these substructures were likely to have low or uncertain potential to cause adverse ecological effects (Table 12).

Environmental Exposure Potential Processing

The process of selecting chemicals with potential for environmental exposure involved matching CAS numbers of 5318 chemicals from sources of environmental monitoring data with the CAS numbers in the second stage list (Fig. 3). Results from matching chemicals from different sources of environmental monitoring data with chemicals in the second stage list are summarized in Table 13. Air toxics is a data base developed from the papers by Hunt et al. and Pellizari et al. [*6,7*]. CESARS is an acronym for the Chemical Evaluation Search and Retrieval System; it is an on-line data base available from CIS, Inc. The food and feral animals data base was obtained from Oak Ridge National Laboratory; it is available commercially as a file from DIALOG called "Chemical Exposure." The Great Lakes chemical inventory is an EPA data base of about 1000 chemicals, including many pesticides, found in the Great Lakes region. Papers coauthored by R. Hites, including Sheldon and Hites, Nelson and Hites, Lopez-Avila and Hites, Elder et al., and Jungclaus et al. were used to develop a data base [*8–13*]. Other papers from which data bases were developed are listed below: Lonneman et al. wrote a paper on hydrocarbons in Florida's ambient air [*14*]; Lucas authored a paper on analysis of chemicals in drinking water and wastewater [*15*]; Pankow et al. examined trace organics in rainwater [*16*]; Brodzinsky et al. assessed volatile organics in the atmosphere [*17*]; and Eckel et al. studied the concentrations of chemicals in hazardous waste dump sites [*18*]. Approximately 44% of the chemicals from sources of environmental monitoring data could be matched with chemicals in the second stage list to generate a list of 2349 second stage list chemicals with environmental exposure potential (Figs. 2 and 3).

Scoring Chemicals with Potential for Environmental Exposure and Potential to Cause Adverse Ecological Effects

Chemicals from sources of environmental monitoring data and chemicals representing substructures with potential to cause adverse ecological effects were assigned codes and values (Tables 10 and 14). Typical environmental codes and values for four chemicals in the second stage list are provided in Table 15. Some chemicals received a score only for ecological effects, for example, CAS No. 60093; some received a score for ecological effects

TABLE 10—*Results of substructure searching of the second stage list.*

Name	Substructure	Code	Chemicals per Substructure	Organisms potentially susceptible to adverse ecological effects
Aliphatic aldehydes	RCHO	10	199	birds, fish, mammals
Anhydrides	-C(=O)-O-C(=O)-	02	39	mammals
Anilides	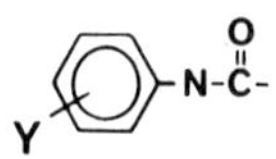	03	811	fish, microbes
Anthracenes	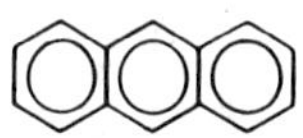	50	21	aquatic invertebrates, fish
Anthraquinones	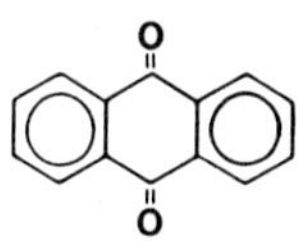	04	384	fish, microbes
Aromatic amines		13	760	algae, aquatic invertebrates, birds, fish, mammals, microbes

Aromatic nitriles	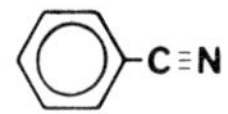	51	180	plants, soil invertebrates
Aromatic phosphorus	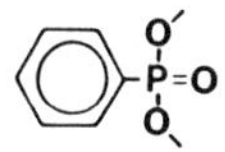	32	10	birds
Aryl ureas	R-N-C(=O)-N-R'	52	19	plants
Auxins	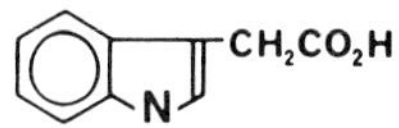	38	1	plants
Aziridines		05	22	algae, aquatic and soil invertebrates birds, fish, mammals
Azo compounds	-N=N-	15	1,020	algae, aquatic invertebrates, birds, fish, mammals, microbes
Azoxy compounds	R-N(→O)=N-R	16	16	mammals, microbes, soil invertebrates

TABLE 10—*(Continued).*

Name	Substructure	Code	Chemicals per Substructure	Organisms potentially susceptible to adverse ecological effects
1,8-Dihydroxyacridine	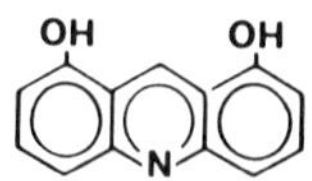	06	0	algae, aquatic and soil invertebrates, fish, microbes
Benzoylphenylureas	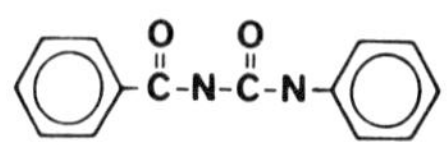	53	1	plants
Benzothiazoles	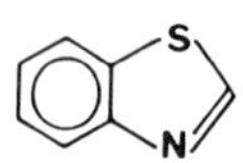	54	148	aquatic and soil invertebrates, fish, mammals
Benzyl chlorides	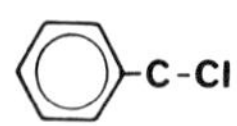	55	149	fish
Carbamates	R-N-C(=O)-O-R	07	148	birds, fish, aquatic and soil invertebrates
Carbazoles	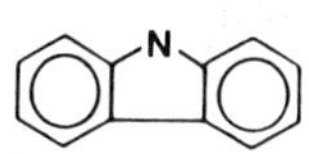	56	17	fish, mammals
Carbodiimides	-N=C=N-	39	10	fish, mammals, plants, soil invertebrates

Chlorinated benzoic acids	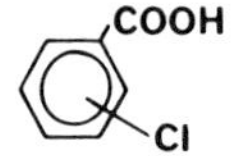	58	6	plants
Chlorophenyl silatranes	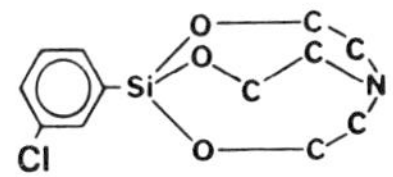	08	0	microbes, soil invertebrates
Chromenes	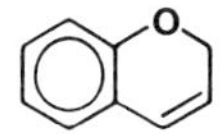	59	57	mammals
Cyanates	$R-O-C\equiv N$	09	0	mammals
Cyanohalides	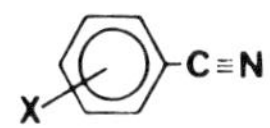	60	5	algae, aquatic and soil invertebrates, birds
Cyanohydrins	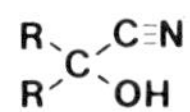	11	7	mammals
Cycloheximides	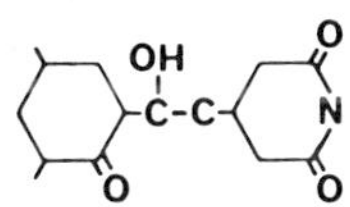	12	0	algae, aquatic invertebrates, microbes

TABLE 10—*(Continued).*

Name	Substructure	Code	Chemicals per Substructure	Organisms potentially susceptible to adverse ecological effects
Epoxides	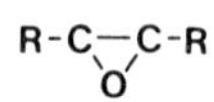	14	74	algae, fish, mammals
Formamidines, amidines	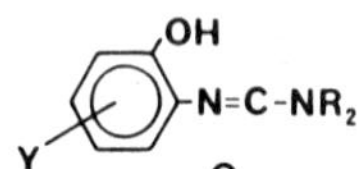	17	0	fish, mammals
Glycidyl amines	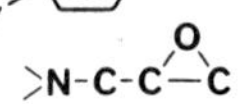	41	12	birds, fish, mammals, soil invertebrates
Haloalkyl ethers	X-CH_2-O-	18	7	fish
Haloalkyl sulfides	X-CH_2-S-	61	2	algae, aquatic intertebrates, fish
Hydrazines	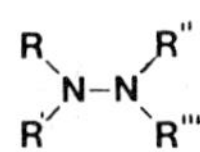	23	84	birds, fish, mammals
Isocyanates	R-N=C=O	24	86	mammals, microbes

Isothiocyanates	R-N=C=S	01	30	mammals, soil invertebrates
Imidazoles	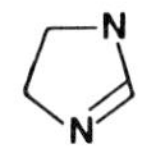	63	34	fish, mammals
Mustards	R-Het-C-C-X	19	12	fish
Nitroaromatics	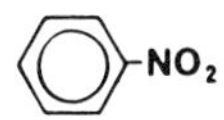	27	48	algae, fish, mammals, microbes
N-Nitroso compounds	R-N-N=O	25	20	aquatic and soil invertebrates, birds, fish, mammals
N-,S-heterocyclics	N-S-Het	20	0	mammals
Organophosphates	-O-P(=O)(-O-)-O-	29	187	aquatic and soil invertebrates, birds, fish, mammals

TABLE 10—*(Continued).*

Name	Substructure	Code	Chemicals per Substructure	Organisms potentially susceptible to adverse ecological effects
Peroxides	R-O-O-R	21	66	fish
Phosphines	H_3P	22	4	birds, mammals, soil invertebrates
Phosphonothioates	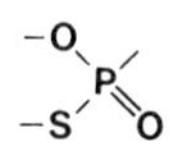	26	0	aquatic invertebrates, fish
Phosphoramidates	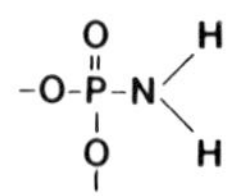	28	6	fish, mammals
Phosphorodithioates	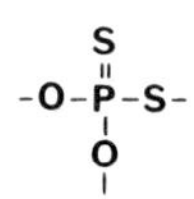	30	25	aquatic and soil invertebrates, birds, fish
Pyrethroids	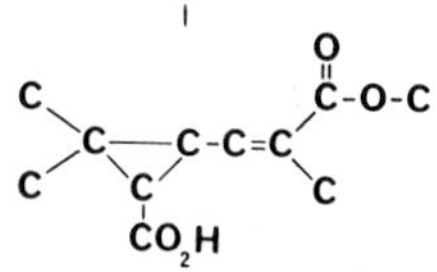	43	11	aquatic and soil invertebrates, fish

Pyridines (n≥2)	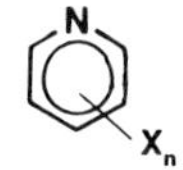	62	4	fish, mammals
Pyrrolidines	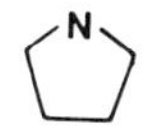	42	87	mammals, soil invertebrates
Quarternary amines	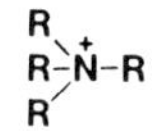	44	52	aquatic invertebrates, microbes
Quinones	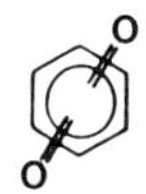	33	24	algae
Sulfanilamides	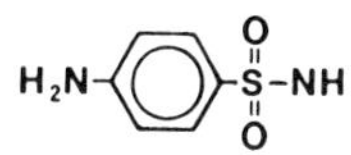	45	29	microbes
Sulfonates	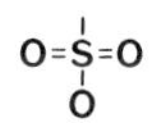	34	64	fish
Sulfonyl Halides	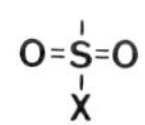	46	126	mammals, soil invertebrates

TABLE 10—*(Continued)*.

Name	Substructure	Code	Chemicals per Substructure	Organisms potentially susceptible to adverse ecological effects
Tetraalkyl pyrophosphates	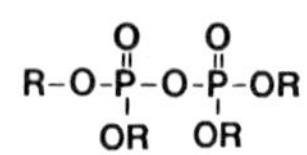	35	1	mammals, microbes
Thiadiazines	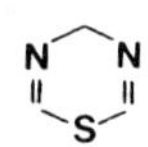	36	1	mammals
Thiocarbamates	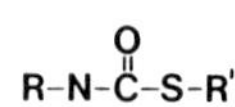	37	32	fish, mammals, microbes
Thiophosphates	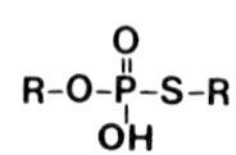	40	2	birds, fish, mammals
Thioxiranes	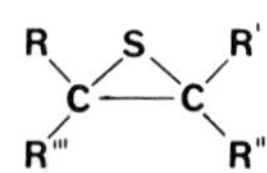	31	5	microbes

Thiram disulfides	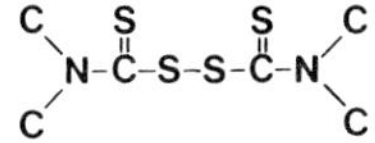	47	3	algae, aquatic invertebrates microbes, birds, mammals
Triazines	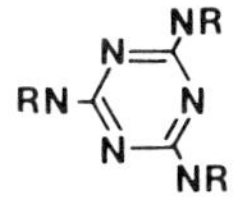	57	7	algae, plants
Triazoles	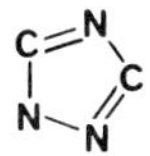	48	30	plants
Trichloromethyl thiophthalimides	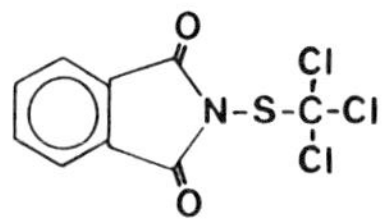	49	1	microbes

X = halide
Y = CH_3 or Cl
HET = heteroatom

TABLE 11*a*—*Characterization of substructures used to search the second stage list—number of substructures.*

Number of Substructures	Number of Chemicals in the Second Stage List
7	0
5	1
30	$>1 \leq 40$
17	$>40 \leq 200$
4	>200

TABLE 11*b*—*Characterization of substructures used to search the second stage list—potential target organisms.*

Number of Substructures	Potential Target Organisms
32	mammals
31	fish
18	microbes
18	soil invertebrates
16	aquatic invertebrates
13	birds
10	algae
8	plants

and environmental monitoring, for example, CAS Nos. 78853 and 91941; and some received a score only for environmental monitoring, for example, CAS No. 132650. Most chemicals that received scores for ecological effects had codes for one or more target organisms and one substructure; a few chemicals had codes for more than one chemical, for example, CAS No. 60093, which had an aromatic amine (Code 13) and an azo dye (Code 15) substructure (Tables 10 and 15). Chemicals in the second stage list that were coded to indicate potential environmental exposure or potential to cause adverse ecological effects were integrated to produce the third stage list by removing chemicals with effects potential, but low or no exposure potential, and by removing chemicals with exposure potential, but low or no effects potential (Fig. 2). Both of these lists of chemicals were recycled to the second stage list (Fig. 2). Integrating the 2349 chemicals with potential environmental exposure with the 5182 chemicals with potential to cause adverse ecological effects generated a list of over 100 chemicals that had both exposure and effects codes in the third stage list (Figs. 2 and 3). For the 63 substructures used to search the second stage list, 29 substructures (46%) had no chemicals with effects (E) or monitoring (M) codes (Table 16). For the remaining 34 substructures, the absolute number of chemicals with E and M codes per substructure and the percentage of E and M chemicals (E + M chemicals divided by the total chemicals per substructure) for each substructure are listed in Table 16. For most of the substructures, less than 10% of the chemicals per substructure were E + M chemicals; a few of the substructures had a moderate absolute number of E + M chemicals, and a few of the substructures had a relatively high percentage of E + M chemicals, for example, nitroaromatics, *N*-nitroso compounds and sulfanilimides (Table 16).

Since the number of chemicals in the third stage list was relatively small compared with the number of chemicals used to generate that list, we used some alternative procedures to

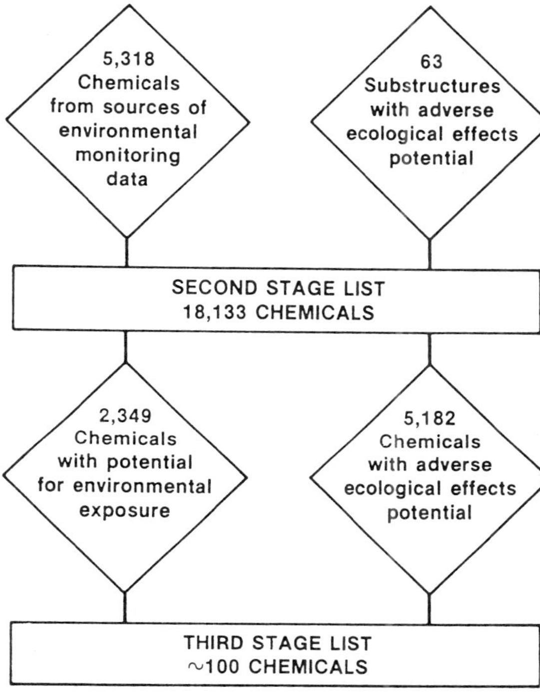

FIG. 3—*Chemicals from sources of environmental monitoring data and substructures with potential to cause adverse ecological effects used to select chemicals with environmental exposure and adverse effects potential.*

identify chemicals for the fourth and fifth stage lists. To generate CAS numbers of chemicals for the fourth stage list, data on annual production and importation volume were obtained for chemicals in the third stage list and selected chemicals in the second stage list (those with high scores for environmental exposure potential or potential to cause adverse ecological effects). Recent (1985 or later) annual production and importation volumes were obtained by comparing names of chemicals in the third stage list and selected chemicals in the second stage list with names of chemicals in the most recent report from the U.S. International Trade Commission [*19*]. These data were available for less than 50% of the chemicals; for the remaining chemicals, production data from the 1977 TSCA inventory had to be used. Computerized searching to obtain recent production volumes should be facilitated in the future because of the 1986 Inventory Update Rule, which requires industry to provide EPA with current production and importation information [*20*].

A minimum production-importation ceiling of 22 000 kg/year was used to select chemicals for the fourth stage list (Fig. 2). Nonselected chemicals were recycled to the third stage list. Details of the algorithms for scoring chemicals for potential adverse effects, potential exposure, and production volume have been described by Brink and Walker [*5*].

Chemicals from the fourth stage list with high exposure and effects scores (third list chemicals) and high production volumes and chemicals with high exposure or effects scores and high production volumes that had been poorly studied were selected for the fifth stage list. Chemicals not selected were recycled to the third stage list (Fig. 2). The ITC contractor developed an information profile (IP) for all of the chemicals in the fifth stage list. Each IP contained the following: (1) one page which indicated whether information

TABLE 12—*Substructures not used to search the second stage list.*

Name	Rationale for Not Searching
Acetals	L[a]
Allylic compounds	L or U[a]
Aliphatic amines	L or U
Beta-diketones	L
BHT derivatives	L or U
Brominated hydrocarbons	SNAD[b]
Cationic dyes	SNAD
CF_3-aromatics	L
Chlorinated hydrocarbons	SNAD
Coumarins	L or U
1,4-Dioxanes	U
Disulfides	SNAD
Dithiocarbamates	L or U
F-alkanes	L
Fluorene	U
Glycidyls	U
Beta-haloethyl amines	U
Haloformic esters	U
Hexyls	L
Hydroxamates	U
Alpha-hydroxyamines	U
Isocyanides	U
Ketenes	L or U
Lactones	L
Methylene dioxybenzenes	U
Nitric acid esters	L or U
Nitriles	U
5-Nitrofuryls	U
Nitroaliphatics	U
Nitrohalides	U
Nitroquinolenes	U
C-Nitroso compounds	L or U
Nonionic surfactants	SNAD
Phenanthrenes	L or U
Phenanthridenes	U
Phenols	L or U
Phosphoniums	U
Phosphoric esters	U
Propargylic esters	L or U
Pyrazines	L or U
Pyridazinones	U
Quinolenes	L or U
Sulfuric acid esters	L or U
Thiocyanates	U
Uracils	U
Urethanes	L or U

[a] L = low; U = uncertain and reflects the potential for causing adverse effects to the eight organisms listed in the questionnaire, as a result of analyzing the responses to that questionnaire.

[b] SNAD = structure not adequately defined.

TABLE 13—*Sources, codes, descriptions, and numbers of chemicals associated with environmental monitoring data used to indicate potential environmental exposure for chemicals in the second stage list.*

Source of Environmental Monitoring Data	Codes	Description	Number of TSCA Chemicals	Number of Matching Chemicals in Second Stage List
Air toxics	01	unpublished papers	28	27
CESARS	02	commercial data base	193	7
Food and feral animals	05	commercial data base	1558	436
Greak Lakes chemical inventory	06	chemicals in air, water, or sediment in Great Lakes region	163	110
Hites et al.	07	published papers	126	111
Lonneman et al.	09	published paper	47	46
Lucas	10	published paper	38	34
Pankow et al.	13	published paper	22	21
STORET data base	14	water quality data	2052	1097
Volatile organic chemicals	15	EPA report	107	96
Statistical data base	16	paper on hazardous waste	984	364

TABLE 14—*Environmental monitoring and ecological effects codes*[a] *and values.*

Category	Code	Value
Monitoring		
Animal tissue (food/feral)	MA	2
Drinking water	MD	1
Groundwater	MG	2
Soil	ML	2
Dump site	MP	1
Rural air	MR	1
Surface water	MS	2
Sediment	MT	2
Urban air	MU	1
Wastewater	MW	1
Effects		
Algae	EA	1
Aquatic invertebrates	EI	1
Birds	EB	2
Fish	EF	2
Mammals	EM	2
Microbes	EO	1
Plants	EP	2
Terrestrial invertebrates	ET	1

[a] Codes for chemical substructures with potential to cause adverse effects are listed in Table 10.

TABLE 15—*Environmental codes and scores for selected chemicals in the second stage list.*

CAS No.	Codes	Score
60-09-3	EA13, EB13, EB15, EF13, EI13, EM13	12
78-85-3	EB10, EF10, EM10, MS14	8
91-94-1	EA13, EB13, EF13, EI13, EM13, MA05, MA14, MP16, MS14, MT14, MW14	18
132-65-0	MA06, MA14, MP16, MS14, MT14	9

TABLE 16—*Chemicals with substructures that have potential to cause adverse ecological effects (E) or for which there are monitoring (M) data.*

Substructure Name	Number of E + M Chemicals per Substructure	Percentage of E + M Chemicals per Substructure
Aldehydes	13	6.5
Anhydrides	0	0
Anilides	0	0
Anthracenes	2	9.5
Anthraquinones	1	0.3
Aromatic amines	29	3.8
Aromatic nitriles	3	1.7
Aromatic phosphorus	0	0
Aryl ureas	2	10.5

TABLE 16—(*Continued*).

Substructure Name	Number of E + M Chemicals per Substructure	Percentage of E + M Chemicals per Substructure
Auxins	0	0
Aziridines	1	4.5
Azo dyes	2	0.2
Azoxy compounds	0	0
Benzacridines	0	0
Benzophenyl ureas	1	100
Benzothiazoles	4	2.7
Benzyl chlorides	5	3.4
Carbamates	9	6.1
Carbazoles	2	11.8
Carboimides	0	0
Chlorinated benzoic acids	3	50
Chlorophenyl silatranes	0	0
Chromenes	2	3.5
Cyanates	0	0
Cyanohalides	0	0
Cyanohydrins	0	0
Cycloheximides	0	0
Epoxides	0	0
Formamidines, amidines	0	0
Glycidyl amines	0	0
Haloalkyl ethers	1	14.3
Haloalkyl sulfides	0	0
Hydrazines	1	1.2
Isocyanates	3	3.5
Isothiocyanates	2	6.7
Imidazoles	0	0
Mustards	0	0
Nitroaromatics	45	94
N-Nitroso compounds	9	45
N-,*S*-Heterocyclics	0	0
Organophosphates	8	4.3
Peroxides	0	0
Phosphines	0	0
Phosphonodithionates	0	0
Phosphoramidates	0	0
Phosphordithionates	0	0
Pyrethroids	1	9.1
Pyridines (≥2 halogens)	4	100
Pyrrolidines	1	1.1
Quaternary amines	2	3.8
Quinones	2	8.3
Sulfanilmides	8	27.6
Sulfonates	2	3.1
Sulfonyl halides	0	0
Tetraalkyl pyrophosphates	0	0
Thiadiazines	1	100
Thiocarbamates	0	0
Thiophosphates	0	0
Thiooxiranes	1	20
Thiram disulfides	1	33.3
Triazines	0	0
Triazoles	2	28.6
Trichloromethyl thiophthalimides	0	0

TABLE 17—*Review panels for the ITC's Sixth Chemical Selection Workshop.*

Panel A		Panel B	
Name	Affiliation	Name	Affiliation
Dr. Paul Bickart	U.S. EPA	Dr. Peter Chapman	U.S EPA
Dr. Chester J. Cavallito	University of North Carolina	Dr. James W. Gillett	Cornell University
Dr. Theodore Mill	SRI International	Dr. Phillip H. Howard	Syracuse Research Corp.
Dr. Ronald Raineri	Microbiological Associates	Dr. David Roberts	Mitre Corp.
Dr. John D. Walker	U.S. EPA	Dr. James K. Selkirk	NIEHS
		Dr. Peter W. Woodman	Dynamac Corp.
Support group: Dr. Robert H. Brink, ITC Louis C. Borghi, Dynamac Corp. Willard W. Perry, Dynamac Corp.		Support group: Dr. Robert H. Brink, ITC Louis C. Borghi, Dynamac Corp. Willard W. Perry, Dynamac Corp.	

was or was not found as a result of searching over 30 information sources; (2) the structure from SANSS or SFILES; (3) synonyms from CHEMLINE; (4) one page on the physical-chemical properties, production volume, use, exposures, and "regulatory" status within EPA; (5) estimates of biodegradation, hydrolysis, or photolysis half-lives; and (6) copies of computer printouts from sources where information was found. The development of the IP is discussed in greater detail by Brink and Walker [*5*]. IPs on chemicals of potential environmental concern were distributed simultaneously to two panels at an ITC Chemical Selection Workshop (Table 17). Both panels reviewed chemicals of potential environmental concern and potential health concerns and then made consensus decisions as to which of the chemicals should be reviewed by the ITC. Based on consensus decisions, chemicals of environmental concern were selected for in-depth review [*21*]. Chemicals not selected for in-depth review were recycled to the third stage list (Fig. 2).

Assessing the Reliability of the Chemical Selection and Scoring Processes

In response to the promulgation of laws to protect human health and the environment, a number of chemical selection or scoring processes have been developed to rank or group chemicals for subsequent testing consideration [*1,2,22–27*]. Most of these processes were developed to evaluate a few hundred chemicals; many focus primarily on human health; few, if any, are computerized or have formally provided procedures for assessing the reliability for chemical selection or scoring. Our processes are computerized and were developed to evaluate several thousand chemicals. Procedures for assessing the reliability of our chemical selection and scoring processes are described below.

Our computerized chemical selection processes incorporate several feedback loops to assure that chemicals are reconsidered as new estimates or new data become available (Fig. 2). Nonetheless, there are several stages in the chemical selection processes that are assessed for reliability.

The selection of substructures that might produce adverse ecological effects was accomplished by analyzing responses to the questionnaire (Table 9). Decisions to categorize substructures as having the potential to cause adverse ecological effects were made when a majority of respondents who expressed an opinion about a substructure listed on the questionnaire thought the substructure's potential to cause adverse effects to any of the eight organism types was moderate or high (Table 10). If a majority of respondents indicated no opinion or thought the substructure's potential to cause adverse ecological effects to any of the eight organism types was low, then decisions were made to categorize these substructures accordingly and not to use them to search the second stage list (Table 12). The reliability of information obtained from conducting a survey could be tested by submitting the questionnaire to a new group of environmental toxicologists and chemists, if a sufficiently large and expert group could be identified, and if there were some reasonable assurances that most of the questionnaires would be completed and returned. The reliability of the information obtained from the survey approach could also be tested by employing new data-supported mathematical models that would use more precisely defined structure-activity relationships (SARs) and incorporate information on the mode or mechanism of action to predict the toxicity of the 109 substructures in the questionnaire. Depending on its cost-effectiveness, the QSAR approach will probably be used to assess the reliability of the substructure selection process.

Decisions to categorize generic substructures as having the potential to cause adverse ecological effects and to indicate which, if any, of eight organism types might be adversely affected are extremely difficult to make because of the uncertainties involved in the decision-making processes. The toxicity of a discrete chemical with a generic substructure will be influenced by the type, number, and position of the substituents. These may determine a discrete chemical's mode or mechanism of action, which may subsequently influence the type of organism most likely to be adversely affected. Undoubtedly there are discrete chemicals among the generic substructures listed in Table 10 which do not have potential to cause adverse ecological effects. Similarly, there are probably discrete chemicals among the 46 generic substructures listed in Table 12 which do have potential to produce adverse ecological effects. The reliability of generic substructure searching should be improved by more accurately characterizing the substructures used to search the second stage list. This should enhance the ability to identify discrete chemicals which have high potential to adversely affect any of the eight organism types.

The selection of chemicals with potential for environmental exposure was based on reports describing measured concentrations of chemicals. Assuming that the reported chemical was measured in the environment and identified correctly (as indicated by an analytical chemistry validation procedure) and was not present as a contaminant in or on solvents, glassware, tubing, and so forth, used to extract and quantify the chemical, it is reasonable to believe the chemical has potential for environmental exposure. Assessing the reliability of this potential is difficult but can be accomplished by assessing the manufacturing, processing, use, and disposal practices and estimating the probability that releases will occur. This assessment is conducted for all chemicals selected for in-depth review.

The selection of chemicals with potential for high consumption was based on production data obtained from the 1977 TSCA inventory or from the U.S. ITC [*20*]. Assessing the reliability of these data is accomplished by consulting the Chemical Update System which contains recent production volume data submitted in response to the 1986 Inventory Update Rule [*21*] or by contacting the manufacturers.

The selection of chemicals for detailed review will be based on an automated scoring and ranking system described by Brink and Walker [*5*]. The reliability of using assigned values for monitoring and effects information (Table 14) and incorporating these values into computerized algorithms for automatically scoring and ranking chemicals was assessed by loading several nonrelevant chemicals (previously recommended by the ITC for environmental effects and chemical fate testing and for which effects and fate data are available and indicate a concern) into the second stage list or file and running the computer program. Most of the chemicals were found in the fifth stage list or file.

The selection of chemicals for in-depth review was accomplished by interdisciplinary panels of experts during the ITC's Sixth Chemical Selection Workshop (Table 17). The reliability of this process was assessed by convening two panels to review the chemicals in opposite order (to assess review fatigue). Over 90% of the chemicals were selected by both panels. As part of the in-depth review, information is requested on the chemicals that are selected [*20*]. Information received is used to assess the reliability of the final chemical selection process and to determine whether the ITC should continue to evaluate the chemical and possibly recommend it to the EPA for priority testing consideration under Section 4(a) of TSCA.

Conclusions

The new computerized chemical selection processes will allow the ITC to cost-effectively screen, select, and score chemicals that may be potentially hazardous to humans and the environment. The processes are dynamic because they allow chemicals to be recycled and permit entry of new data or estimates. Advances in predictive toxicology, increases in searchable data bases on monitoring (and new monitoring information), improvements in chemical substructure searching, and reliable and current data on chemical production and import quantities will all enhance the processes.

Acknowledgments

We are grateful to Dr. Gilman D. Veith, Director, EPA-Duluth Environmental Research Laboratory and Dr. Robert Hunter, Center for Data Systems and Analysis, Montana State University, for providing us with the computer file we used for the initial list. We appreciate the efforts of Louis C. Borghi, Willard W. Perry, and Roberta M. Wedge of Dynamac Corp. We wish to thank Drs. Sanjivani Diwan, Martha Johnson, and Satish Bhalla of CRCS, Inc. for helping us code and file data for the new ITC chemical selection processes. We very much appreciate the efforts of James Bradshaw from EPA's Information Management Division for using SANSS to conduct substructure searching of chemicals in the second stage list. We appreciate receiving the completed questionnaire from Dr. William S. Ewell of the Eastman Kodak Co. on which answers to questions on 14 substructures were based on analysis of their ecological effects testing data. We thank Drs. Bill Adams, Gerald Brooks, Joel Coates, Ferenc Darvis, John Fletcher, Jeff Giddings, Bill Gingerich, Paul Gurba, Dave Hanson, Mike Harrass, Shane Hee, Gary Heinz, Joop Hermens, Phil Howard, Gene Kenaga, Hal Kibby, Albert Klein, W. Klein, Rainer Koch, Dennis Konasewich, Eric Lien, A. Malcolm, Fumio Matsumura, John Matthews, Sonny Mayer, Ikuo Moriguchi, Harry Painter, Fred Pfaender, Rich Purdy, Bob Ringer, Ed Schaefer, Dick Schoettger, John Street, Milon Tichy, Len Torstensson, Gil Veith, Ken Wallace, Gerry Walsh, Yoshitada Yoshioka, and Vlado Zitko for responding to the questionnaire. We are grateful to Patricia Harrigan of EPA's Exposure Evaluation Division (EED) for estimating physical-chemical properties and to Daryl Kaufman of EED for generating structures from SMILES for chemicals in the third stage list. We are also grateful to Drs. Asa Leifer and Robert S. Boethling of EED for estimating the photolysis, hydrolysis, and biodegradation half-lives of many chemicals in the third stage list. We thank Dr. Robert L. Lipnick and other members of EPA's Environmental Effects Branch for commenting on the potential ecotoxicity of chemicals in the third stage list. Administrative support for the ITC is provided by EPA; contractor support is provided by Dynamac Corp.

References

[*1*] Muir, W. R., "Initial Report of the TSCA Interagency Testing Committee to the Administrator, Environmental Protection Agency," EPA 560-10-78/001, U.S. Environmental Protection Agency, Washington, DC, 1978.

[*2*] "Technical Support to the TSCA Interagency Testing Committee," final report to the Toxic Substances Control Act Interagency Testing Committee under EPA Contract No. 68-01-6650, CRCS, Inc. and Dynamac Corp., Rockville, MD, 1985.

[*3*] "TSCA Interagency Testing Committee Scoring Task Force," final report to the Toxic Substances Control Act Interagency Testing Committee under EPA Contract No. 68-01-6650, CRCS, Inc. and Dynamac Corp., Rockville, MD, 1984.

[*4*] "Chemicals to be Reviewed by the Toxic Substances Act Interagency Testing Committee: Public Meeting and Request for Information," *Federal Register,* 1983, Vol. 48, p. 51519.

[*5*] Brink, R. H. and Walker, J. D. "TSCA Interagency Testing Committee Sixth Scoring Exercise: Interim Report," EPA Contract No. 68-02-4251, Dynamac Corp., Rockville, MD, 1987.

[*6*] Hunt, W. F., Jr., Faoro, R. B., Freas, W. P., and Nelson, K., "An Interim Data Base for State and Local Air Toxic Volatile Organic Chemical Measurements," presented at the 79th annual meeting of the Air Pollution Control Association, Minneapolis, MN, June 1986.

[*7*] Pellizarri, E. D., et al., *Environmental International,* Vol. 12, 1986, pp. 619–623.

[*8*] Sheldon, L. S. and Hites, R. A., *Environmental Science and Technology,* Vol. 12, No. 10, 1978, pp. 1188–1194.

[9] Sheldon, L. S. and Hites, R. A., *Environmental Science and Technology*, Vol. 13, No. 5, 1979, pp. 574–579.
[10] Nelson, C. R. and Hites, R. A., *Environmental Science and Technology*, Vol. 14, No. 9, 1980, pp. 1147–1150.
[11] Lopez-Avila, V. and Hites, R. A., *Environmental Science and Technology*, Vol. 14, No. 11, 1980, pp. 1382–1390.
[12] Elder, V. A., Proctor, B. L., and Hites, R. A., *Environmental Science and Technology*, Vol. 15, No. 10, 1981, pp. 1237–1243.
[13] Jungclaus, G. A., Lopez-Avila, V., and Hites, R. A., *Environmental Science and Technology*, Vol. 12, No. 10, 1978, pp. 88–96.
[14] Lonneman, W. A., Seita, R. L., and Bufalini, J. J., *Environmental Science and Technology*, Vol. 12, No. 4, 1978, pp. 459–463.
[15] Lucas, S. V., "GC/MS Analysis of Organics in Drinking Water Concentrates and Advanced Waste Treatment Concentrates: Volume I—Analysis Results for 17 Drinking Water, 16 Advanced Waste Treatment, and 3 Process Blank Concentrates," EPA-600/1-84-020a, U.S. Environmental Protection Agency, Office of Research and Development, Washington, DC, November 1984.
[16] Pankow, J. F., Isabelle, L. M., and Asher, W. E., *Environmental Science and Technology*, Vol. 18, No. 5, 1984, pp. 310–318.
[17] Brodzinsky, R. and Singh, H. B., "Volatile Organic Chemicals in the Atmosphere: An Assessment of Available Data," EPA project summary ESRL, EPA-600/53-83-027, U.S. Environmental Protection Agency, Washington, DC, June 1983.
[18] Eckel, W. P., Trees, D. P., and Kovell, S. P., *Proceedings*, National Conference on Hazardous Wastes and Environmental Emergencies, Superfund, Hazardous Materials Control Research Institute, Silver Spring, MD, 1986, pp. 250–257.
[19] "Synthetic Organic Chemicals, U.S. Production and Sales, 1985," Investigation No. (332-135), U.S. International Trade Commission, Publication No. 1892, U.S. Government Printing Office, Washington, DC, 1986.
[20] "Partial Updating of TSCA Inventory Data Base; Production and Site Reports," *Federal Register*, 1986, Vol. 51, p. 21438.
[21] "Toxic Substances Control Act Chemicals to be Reviewed by the Interagency Testing Committee; Public Meeting and Request for Information," *Federal Register*, 1987, Vol. 52, p. 10409.
[22] O'Bryan, T. and Ross, R., "Chemical Scoring System for Hazard and Exposure Assessment: Draft Report," U.S. Environmental Protection Agency, Washington, DC, 1986.
[23] Organization for Economic Cooperation and Development, "Chemicals on Which Data are Currently Inadequate: Selection Criteria for Health and Environmental Purposes, Volumes I and II," Umweltbundesamt, Berlin, Germany, 1984.
[24] Sampaolo, A. and Binetti, R., *Regulatory Toxicology and Pharmacology*, Vol. 6, 1986, pp. 129–154.
[25] Welch, J. L. and Ross, R. H., *Environmental Toxicology and Chemistry*, Vol. 1, No. 1, 1982, pp. 95–102.
[26] Ross, R. H. and Lu, P., "Chemical Scoring System Development," Contract No. W-7405-eng-26, Oak Ridge National Laboratory, Oak Ridge, TN, 1981.
[27] Hushon, J. M. and Kornreich, M. R., *Hazard Assessment of Chemicals: Current Developments*, Vol. 3, 1984, pp. 63–109.

Gerhardt F. Riedel[1]

Interspecific and Geographical Variation of the Chromium Sensitivity of Algae

REFERENCE: Riedel, G. F., **"Interspecific and Geographical Variation of the Chromium Sensitivity of Algae,"** *Aquatic Toxicology and Environmental Fate: Eleventh Volume, ASTM STP 1007,* G. W. Suter II and M. A. Lewis, Eds., American Society for Testing and Materials, Philadelphia, 1989, pp. 537–548.

ABSTRACT: The toxic effect of hexavalent chromium on the growth rate of ten species of estuarine phytoplankton was examined in a defined growth medium. The concentration causing 50% inhibition of growth rate varied by a factor of more than 50 among the species. Three possible causes of interspecific variation in the sensitivity were examined: detoxification by extracellular reduction, interspecific differences in growth rate, and interspecific differences in chromium (VI) uptake. Results indicate that none of these hypotheses explains the interspecific variation in chromium sensitivity.

The chromium (VI) sensitivity of each algal species was also tested at two different sulfate concentrations. In every case, the toxicity of chromium (VI) was greater in the medium with lower sulfate concentrations. This confirms and extends previous results showing that chromate sensitivity in algae is closely correlated to the sulfate concentration of the growth medium.

KEY WORDS: algae, chromium, phytoplankton, toxicity testing, aquatic toxicology

Chromium is among the most common industrial chemicals, widely employed for such varied activities as corrosion inhibition, leather tanning, and metallurgy. Some of this chromium is released into the environment [*1–4*], where it constitutes an environmental and human health threat because of its toxic and mutagenic properties.

Of the two forms found in the environment, trivalent and hexavalent, hexavalent chromium (chromate) is the form considered to be the greatest threat because of its high solubility, ready penetration of cell membranes, and strong oxidizing ability [*5,6*]. It has long been known that aquatic plants are particularly susceptible to chromate [*7,8*], and this has prompted several studies of its toxic effects on freshwater and marine plants [*9–17*]. These studies, using several species and a wide variety of methods for determining toxic effect, in varying and often poorly characterized growth medium, report a wide range for the toxic dose of chromate, from 1.6 to 5000 μg/L. Clearly, such a wide range of results is undesirable to government and industrial managers, who find it difficult to make good regulatory decisions based on conflicting evidence.

The purpose of this paper is to examine several possible reasons for the wide range of chromate toxicity reported for algae, including the water chemistry, particularly the relationship between chromate toxicity and sulfate concentration, previously observed for one species [*18*]; the means of reporting toxicity; and such properties intrinsic to algae as growth rate, chromate uptake rate, and possible chromate reduction and detoxification.

[1] Postdoctoral research associate, Academy of Natural Sciences, Benedict Estuarine Research Laboratory, Benedict, MD 20685.

This paper presents the results of chromate toxicity experiments performed on ten species of phytoplankton, from six different classes, using identical growth conditions and a single method for determining the toxicity, with each species cultured in two media of different salt composition.

Procedure

Experimental Organisms

The algae used in this study were *Synecchococcus bacillaris,* Clone SYN; *Synechococcus* sp., Clone UW405; *Chlorella capsulata,* Clone FLAE; *Chlorella* sp., Clone 017; *Isochrysis* sp., Clone TISO; *Pavlova lutheri,* Clone MONO; *Thalassiosira weisflogii,* Clone ACTIN; *Thalassiosira pseudonana,* Clone 3H; *Prorocentrum mariae-lebouriae,* Clone PROML; *Cryptomonas obovoida,* Clone UW354; and *Chroomonas* sp., Clone LB2007. Clones SYN, UW405, FLAE, 017, TISO, MONO, and UW354 were obtained from the Provasoli-Guillard Center for the Culture of Marine Phytoplankton, Boothbay, MA. Clone 3H was obtained from Dr. David Nelson, College of Oceanography, Oregon State University, Corvallis, OR; ACTIN from Dr. Dennis Hanisak, Harbor Branch Foundation, Fort Pierce, FL; PROML from Dr. Richard Rivkin, Johns Hopkins University, Shady Side, MD; and LB2007 from the Algal Culture Collection of the University of Texas, Austin, TX. These species are from six classes of algae including: Cyanophyceae, Chlorophyceae, Prymnesiophyceae, Bacillariophyceae, Dinophyceae, and Cryptophyceae. These experimental organisms were chosen because they were known to be tolerant to a wide range of salinities, each having been originally isolated from estuaries or coastal areas. It was hoped that this also would mean that they were tolerant to large variations in the salt composition of the medium, which indeed proved to be true.

Culturing

Experimental cultures were grown in an artificial seawater based on "standard ocean water" [*19*] diluted to the equivalent of three parts per thousand salinity. One inoculum culture of each species was begun in the standard medium (2.9 m*M* sulfate medium) and another in a similar medium prepared with one tenth of the normal concentration of sulfate at that salinity (0.29 m*M* sulfate medium). Details of the media preparation are published elsewhere [*18*]. Experimental culturing was carried out in acid-washed 50-mL Pyrex test tubes, containing 30 mL of medium. It was previously established that adsorption to culture vessels did not cause measurable loss of chromate from the medium. All cell culturing was carried out in a temperature-controlled incubator at 20 $\pm$ 1°C and 65-μeinsteins/m^2/s illumination using soft-white fluorescent lights, on a 12:12 h light/dark cycle. The inoculum cultures were grown in acid-washed glass 100-mL Erlenmeyer flasks under the same conditions. Experimental cultures were started with a small inoculum (less than 0.5 mL) of logarithmically growing cells from identical medium.

Cell Growth Measurements

Cell growth was measured by *in vivo* fluorescence using a Turner Designs fluorometer. The experimental cultures were measured once each day, at the same time, after gentle mixing. During the logarithmic growth phase, the *in vivo* fluorescence was approximately proportional to the cell density for any individual clone [*20*], although the fluorescence/cell density ratio varies considerably between different species, depending on the cell size and chlorophyll content. Growth rate was determined by regressing the log of *in vivo* fluorescence versus time, using results from the fourth to the seventh day after the inoculation. This time range was chosen for several reasons. The first three days were not included because, when some clones were placed in medium containing moderately inhibitory chromium concentrations, they continued to grow at their original growth rate for up to three days and, conversely, because there was often some initial lag phase after inoculation. Thus, waiting three days allowed full acclimation to the test medium. Results after the seventh day were not used to avoid possible growth rate decreases due to light and nutrient limitation in fast growing cultures, and to eliminate from consideration cultures which grew after extremely prolonged lag phases. Results of a chromate toxicity experiment for a single clone in both experimental media are shown in Fig. 1.

Chromate, as potassium dichromate, was added to the experimental media in a logarithmic series, 0, 1, 2.15, 4.64, 10, 21.5, 46.4 and 100 *μM* (later experiments used the series 0, 0.1, 0.215, 0.464, 1, 2.15, 4.64, 10 *μM* for the treatments with 0.29 m*M* sulfate). Hydrolysis reactions of hexavalent chromium are extremely rapid, and at the pH of this medium (8.1) virtually all the chromium is present as the chromate ion, CrO_4^{2-} [*21*].

To determine the concentration of chromate that inhibited the growth rate of the algae by 50% (EC_{50}), the growth rates of two replicate control cultures were averaged; the replicate cultures on either side of 50% of this value were found; and the concentration estimated to cause 50% inhibition was determined by regressing these growth rates versus the log of the concentration and finding the point on this line that yielded 50% of the mean control growth rate. In one case, the Clone 017 experiment, the highest chromate treatments failed to inhibit the growth rate by 50%, but did inhibit it as much as 40%, so the EC_{50} was estimated by extrapolating growth rates from the two pairs of concentrations with

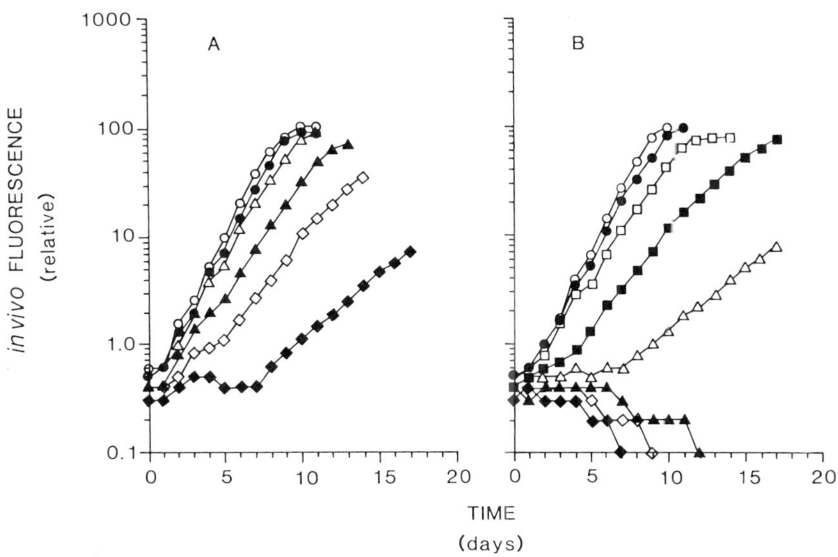

FIG. 1—*Results of a typical chromium toxicity bioassay in* (a) *2.9* mM *sulfate and* (b) *0.29* mM *sulfate medium. Open circles are control cultures, closed circles are 1.0* μM *chromate, open squares are 2.15* μM, *closed squares are 4.64* μM, *open triangles are 10* μM, *closed triangles are 21.5* μM, *open diamonds are 46.4* μM, *and closed diamonds are 100* μM. *Each point is the mean of two replicate cultures. The points for 2.15 and 4.64* μM *chromate have been left out of* (a) *for clarity. The actual clone is* Synechococcus *bacillaris (Clone SYN).*

the highest inhibition. Due to the convex shape of the response curve above the center, these values probably slightly overestimate the EC_{50} compared with those for the other clones.

Uptake of Chromium (VI)

To determine the amount of chromium taken up by the algae, 24-h uptake experiments in one medium (2.9 mM sulfate) at a single chromate concentration (100 μM) were carried out under the same growth conditions as the toxicity experiments. In a previous study [22], it was shown that for Clone 3H, chromate uptake is proportional to chromate concentration and inversely proportional to the sulfate concentration of the medium. Nuclepore filters (1.6 cm, 0.4 μM pore size) were cleaned by soaking them in 50% volume/volume nitric acid (Baker Analyzed Reagent grade) for 24 h, followed by five rinses with deionized water; dried in a 60°C drying oven for 48 h; and preweighed. Replicate cultures of cells were grown in 200 mL of medium in acid-washed 250-mL polycarbonate flasks, until the cells were dense but still growing in log phase. To start the experiment, the chromate was added, and 100 mL of the culture was immediately filtered to be used as a time-zero blank. After 24 h, 50-mL samples were filtered to determine chromate uptake. All samples were also rinsed with two 25-mL aliquots of medium containing no added chromium. The filters were again dried at 60°C and weighed to determine the weight of the phytoplankton by the weight difference. Dry weight was used to normalize cell chromium content rather than cell numbers because the various species of phytoplankton used differed greatly in cell size.

The filters and cells were wet digested with 1 mL of concentrated nitric acid, and made up to 5 mL with 1% volume/volume nitric acid. Chromium in the digests was determined by graphite furnace atomic absorption spectrophotometry, using a Perkin-Elmer Model 603 spectrophotometer with a HGA graphite furnace. A 50-μL sample volume was injected into the furnace, via a Model AS-40 autosampler. The graphite tubes employed were pyrolytically coated graphite, with a L'vov platform. Argon was used as the sheathing gas, at 100 mL/min, with "miniflow" during the atomization step. The absorbance signal was simultaneously corrected for background with a deuterium lamp. Standards were made up using commercially available atomic absorption standard (potassium dichromate) in 1% nitric acid. Calibration was carried out using a second-order polynomial fit to the standard curve. Samples were diluted by as much as a factor of ten to fit the working range of standards (10 to 500 nM).

Results and Discussion

One of the primary objectives of this study was to verify that all taxonomic classes of phytoplankton show the same dependence of chromate susceptibility on sulfate previously observed only in a single species of diatom [18]. This relationship most likely results from the inability of the cell sulfate active transport system to discriminate between chromate and sulfate required for growth because of their similarity in size, shape, and charge [22]. Thus, in media containing a higher ratio of chromate to sulfate, more chromate would be transported into the cell. However, given the wide range of susceptibility to chromate reported in the literature, it would appear that substantial differences in the metabolism of chromate must exist, and that this generalization may not be warranted. The growth rates from the same clone as in Fig. 1 are plotted versus the total chromium and the chromium/sulfate ratio in Figs. 2*a* and 2*b*. Note that when the growth rate is plotted against the ratio of chromate to sulfate, the two experimental media give nearly identical results. Table 1 lists the mean concentration of chromate toxic to all the phytoplankton clones in the two

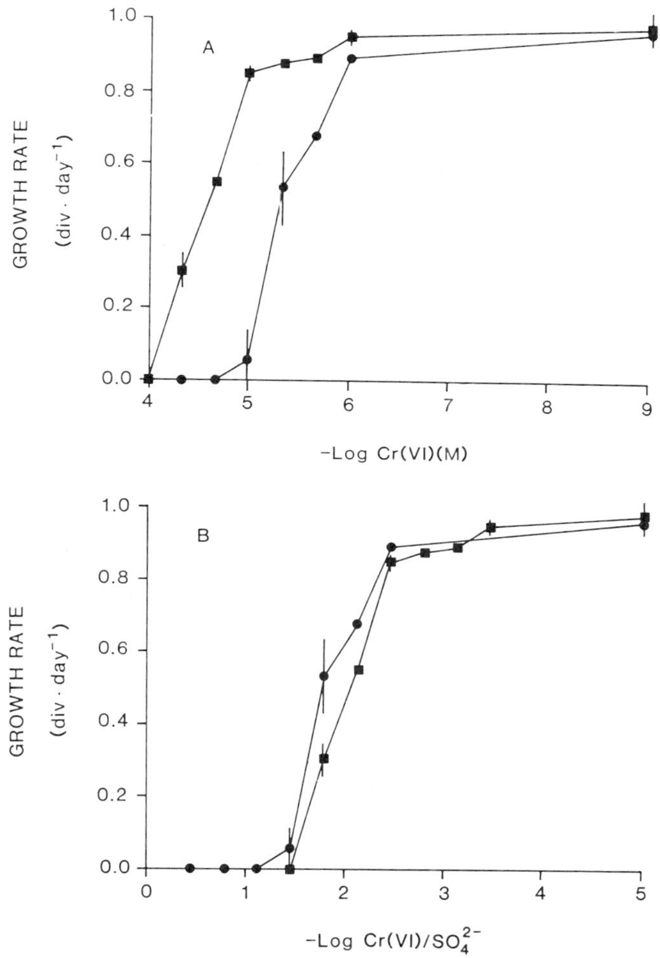

FIG. 2—*Growth rate* [± *1 standard deviation (SD)*] *for the same clone shown in Fig. 1:* (a) *Growth rate versus* −*log of total chromate concentration,* (b) *growth rate versus* −*log (chromate/sulfate) (molar). The squares are 2.9* mM *sulfate medium and circles are 0.29* mM *sulfate.*

media. As predicted by the hypothesis above, the toxicity of chromate is approximately ten times greater in media with 0.29 m*M* sulfate than in the 2.9 m*M* sulfate media in eight of eleven clones, and approximately five times greater in three clones. The range of algal phyla represented is sufficiently wide that the assumption that all phytoplankton, and probably all algae, are sensitive to chromate in inverse proportion to the sulfate concentration appears reasonable.

Another important goal of this study was to explain the variation of the values reported for chromate toxicity to phytoplankton. This goal is at least partially fulfilled by the finding that chromate toxicity is tightly coupled to sulfate concentrations. In seawater, the sulfate concentration is proportional to the salinity, so that sulfate concentrations in marine and estuarine water used in toxicity tests may be readily determined. In fresh water, sulfate

TABLE 1—*Results of the chromate toxicity bioassay in 2.9* mM *and 0.29* mM *sulfate medium for the phytoplankton.*

Taxon	EC_{50} in 2.9 m*M* Sulfate, μ*M*	EC_{50} in 0.29 m*M* Sulfate, μ*M*
Cyanophyceae		
SYN	26	4.9
UW405	13	2.3
Chlorophyceae		
FLAE	39	3.7
017	117	24
Prymnesiophyceae		
TISO	2.4	0.25
MONO	3.3	0.36
Bacillariophyceae		
3H[a]	6.6	0.66
ACTIN	3.7	0.39
Dinophyceae		
PROML	1.6	0.13
Cryptophyceae		
UW354	5.2	0.7
LB2007	23.5	2.5

[a] Results from a previous study [*18*] using the identical medium and temperature but a different light regime.

concentrations average about 100 μ*M* [*23*]; however, there is considerable variability in the composition of fresh waters, and variation by a factor of ten on either side of this value within a single watershed has been observed [*24*]. In such a watershed, we would expect to find similar variations in the chromate susceptibility of single clones of phytoplankton. Table 2 shows the chromate sensitivity of the various clones of algae normalized to two different sulfate concentrations, that of full-strength seawater (29 m*M* sulfate) and average fresh water (0.1 m*M* sulfate). It is interesting to note that the lowest of these susceptibilities is at a concentration of chromate only about a factor of ten higher than the reported average concentration of chromium in fresh water, 0.2 μg/L [*25*].

Despite the similar behavior with respect to sulfate, however, large differences in the sensitivity of different species of phytoplankton to chromate remain unexplained by the water chemistry. Overall, the different species range over a factor of about 70 in their sensitivity to chromate in identical media. The algae have been grouped by taxonomic class to show whether or not certain groups of algae tend to be more or less sensitive to chromate. Of the species examined, the two green algae were the least sensitive, in one case requiring sulfate to chromate ratios of less than about 30 to inhibit growth. Species of blue-green algae and cryptophytes were the next least sensitive. Blue-green algae have previously been noted to be unusually tolerant of chromate [*10,26*]. Finally, all the algae of the groups Prymesiophyceae, Bacillariophyceae, and Dinophyceae were very sensitive, with inhibition of growth occurring when the ratio of sulfate to chromate dropped below about 400 to 1800.

A note of caution must be added here. The algal clones chosen for this study are not meant to represent accurately the average chromate susceptibility of their class. The algal clones represented in culture collections are apt to be unusually tolerant of various stresses; those that are not frequently disappear from collections. These particular clones were further chosen to be tolerant to the variations of media salt content required by the design of these experiments. However, no consideration was given in advance to selection of cells particularly sensitive or insensitive to chromate, so that these results are likely to represent at least reasonably well the chromate sensitivity of the more robust species of their class.

Another cause of the extreme variation in the reported sensitivity of algae to chromium are differences in the methodology used to evaluate sensitivity. At one extreme, some workers report the least concentration of a toxicant that causes detectable decreases in growth rate or biomass as incipient inhibition [*16*]. This is justified by the argument that a species inhibited to any extent is at a competitive disadvantage relative to any species not so inhibited. The lowest concentration of chromate reported as toxic (1.6 μg/L chromium) was reported for the blue-green algae *Microcystis aeruginosa* using incipient inhibition as the toxic effect [*16*]. However, the actual determination of such an effect is very much a function of the design of the experiment, taking into account the sample size, replication, and other factors which are not directly properties of the organism in question. For example, in Fig. 1, the growth of the algae was somewhat retarded at all concentrations of chromate above zero with statistical significance at the 0.05 level. Thus, the lowest concentration tested, 1 μ*M*, causes incipient inhibition in at least one sense. However, I would argue that this difference is trivial.

At the other extreme, one could follow the toxicological practices on larger organisms and report as toxic those concentrations which kill the algae. This approach is almost never used with algae since the death of a unicellular organism is difficult to detect and growth of algae is easy to detect. However, this approach can be justified by acknowledging that pollutants frequently are released in pulses, which are later mixed and diluted away, so that an organism is only exposed to a high concentration for a short period. If it survives the short period in high concentrations, it will be able to resume growth after the dilution of the pollutant. This results in a very high concentration of toxicant being reported. In one study [*27*], cited in Ref *28*, it was noted that 184 μg/L chromate inhibited the growth

TABLE 2—*Extrapolation of chromate toxicity bioassay results to average freshwater and average seawater sulfate concentrations.*

Taxon	EC_{50} in Mean Fresh Water (0.1 m*M* Sulfate), μg/L	EC_{50} in Mean Seawater (29 m*M* Sulfate), μg/L
Cyanophyceae		
SYN	44	12 800
UW405	24	6 800
Chlorophyceae		
FLAE	71	20 500
017	210	61 900
Prymnesiophyceae		
TISO	4.7	1 300
MONO	5.7	1 700
Bacillariophyceae		
3H[a]	8.8	2 500
ACTIN	6.8	1 900
Dinophyceae		
PROML	3.1	900
Cryptophyceae		
UW354	11.0	3 200
LB2007	45.3	13 600

[a] Results from a previous study [*18*] using the identical medium and temperature but a different light regime.

rate of *Selenastrum capricornutum,* but more than 1000 μg/L were required to actually kill the cells. Similarly, in Fig. 1*a,* in 2.9 m*M* sulfate medium, no concentration of chromate up to 100 μ*M* (5200 μg/L) actually killed the cells. After a prolonged lag phase (seven days), *Synechococcus bacillaris* (Clone SYN) began to grow and after a very long time (not shown) achieved the same densities as the control cultures.

The middle ground between these two approaches, and that chosen for this paper, is the EC_{50} approach, the concentration of toxicant that reduces some property (usually growth rate or biomass) by 50%. This method has the advantage that it uses concentrations usually quite close to the concentration of substance that causes the greatest change in inhibition per unit change in concentration and, consequently, yields the greatest precision in estimation of toxic concentration; it can also be considered to represent a fixed relationship between the cells and the toxicant rather than depending on the precision of the experimental design. However, even with this method, differences in results can arise as a result of the various means used to measure inhibition. It is most often found that, for a given species, concentrations of toxicant can be found which inhibit the growth rate to some extent but which eventually allow the algae to grow to their maximum abundance allowed by the nutrient concentration of the medium. This is clearly shown in Fig. 1. Thus, EC_{50}s based on growth rate are usually lower than EC_{50}s based on final biomass yield by a factor of two or more [*18,29,30*].

Frequently, cell yield at some arbitrary time after the start of the experiment (four, five, or seven days, for example) is used as the end point for the EC_{50} determination [*9,10,27*]. Either a lag phase or reduced growth rate can contribute to the depression of cell yield by the toxicant, and depending on the contribution of each, the concentration of 50% depression varies somewhat with the length of the experiment. Unlike final yield, this method frequently gives toxic concentrations that are less than those based on growth rate alone. It has been argued, based on a study of many cases [*31*], that growth rate is superior to biomass as the response variable in that the results from different laboratories are more comparable.

Even when growth rate is used as the basis for an EC_{50}, the treatment of data may differ between studies. For example, in Fig. 1*a,* all the experimental cultures, including the control, show a slight lag in growth between inoculation and the first day. However, the 100 μ*M* chromate treatment algae remained stationary for seven days before starting slow, steady growth. In other studies, including some, but not all, of the species in this study, cells have been observed to continue to grow for as many as three days at maximal rates before slowing their growth rate in response to a toxic concentration of metal [*32*]. In this study, the growth rate was calculated between the fourth and seventh day. In other studies, the growth rate has been averaged over the whole experiment or over other lengths of time [*20*]. Some suggest that the area under the response curve (with or without log transformation) should be used to compare growth, thus including elements of both growth rate and biomass into a single measurement, without assuming a specific growth pattern [*33*].

Therefore, when comparing the results of algal toxicity tests from different sources, it is important to be aware that there are several sources of variability. First, media of different compositions can cause large differences for the same test organism, as seen in this study with chromate. Second, there may be large differences in the susceptibility of different algae to a toxic compound under identical conditions. Finally, given identical growth curves, there are a variety of different ways of treating the data, all of which have some virtue, but which can also cause significant differences in the concentration of a compound reported as toxic.

The chromate uptake experiment was prompted by the hypothesis that the difference in toxicity of chromate to algae might be related to the extent to which the different clones concentrated chromium from the medium. Thus, a species better able to discriminate between sulfate and chromate might transport less chromate into the cell, and be able to tolerate greater concentrations in the medium. If this were the case, high tolerance to chromate would be correlated with low concentrations of cellular chromium. The cellular chromium contents for the nine clones tested is plotted as a scattergram versus the EC_{50} in the same medium (Fig. 3). No significant relationship is observed. A value of approximately 5 μg/g dry weight was found for seven clones, including three of the most sensitive (3H, TISO, and MONO) and two of the least sensitive (FLAE and 017), while two sensitive clones (ACTIN and PROML) had much greater chromium contents. While these results are somewhat suggestive, the differences between the sensitivity of different clones cannot be explained by this hypothesis.

Another possible explanation for the varying sensitivities of clones could be differences in the growth rate—the hypothesis being that faster growing species might dilute out chromium more rapidly with new biomass and could thus tolerate higher concentrations of chromate. This hypothesis also requires that chromate be transported at a constant rate despite changes in cell growth rate, which has been observed for clone 3H [*22*]. This hypothesis was tested by regressing the mean growth rate in the control cultures of each clone in the 2.9 m*M* sulfate medium against the EC_{50} determined in the same medium. No significant relationship was found between these variables. A scatter plot of the data is shown in Fig. 4. The two clones with the highest and lowest growth rate (3H and PROML, respectively) were both relatively sensitive, with the remainder of the clones having similar growth rates (0.5 to 1.3 division per day).

A final possibility for the differing susceptibilities to chromate among the various algal clones is that some species might be able to reduce chromate to the less toxic Cr^{3+}. To determine whether any such reduction occurred in the experimental cultures, the chromate remaining in the water of the highest concentration of chromium that showed growth for each culture was measured using the *s*-diphenylcarbazide method [*34*], which is specific for chromium (VI). No loss of chromate was observed in any of these cultures.

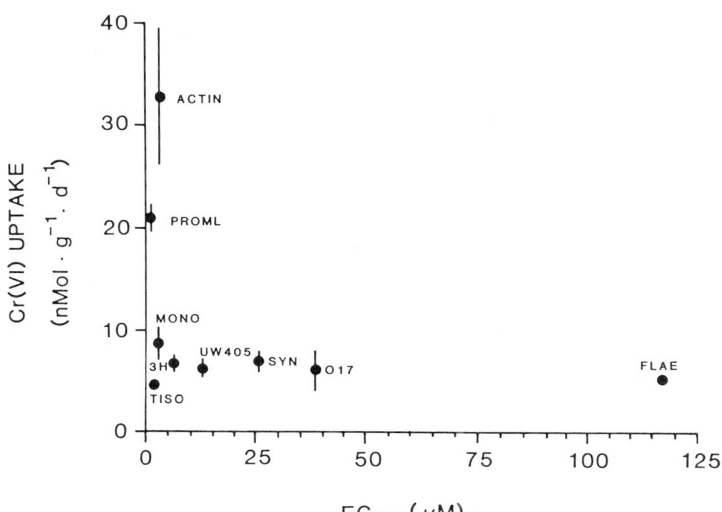

FIG. 3—*Scatter plot of chromium uptake from 100* μM *chromate in 2.9* mM *sulfate medium (± 1 SD) versus chromate EC_{50} in the same medium for nine algal clones.*

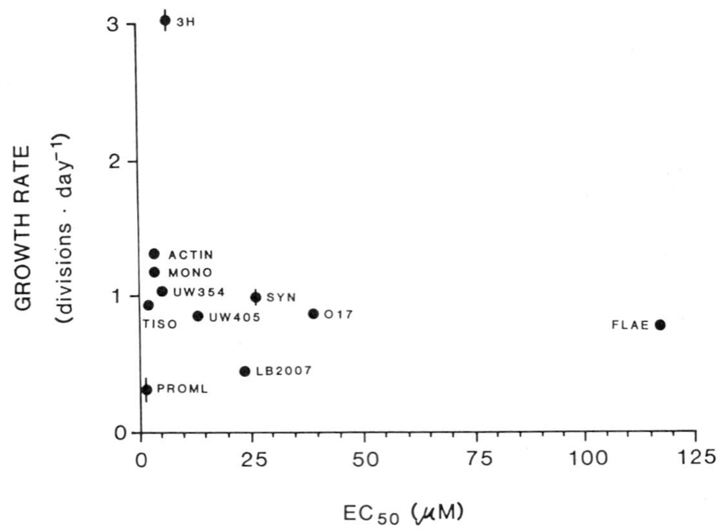

FIG. 4—*Scatter plot of control growth rate (±1 SD) in 2.9* mM *sulfate medium versus chromate EC_{50} in the same medium for eleven algal clones.*

Conclusions

One of the most important outcomes of this study is the evidence that the toxicity of chromate to all algae is a function of the sulfate ion concentration of the test medium. This has several important consequences. First, when performing laboratory studies on the effect of chromate on plants, the results will be comparable to those of other laboratories only if the sulfate concentration is known and taken into account. For this reason, it is advisable to perform studies in a defined medium when the source of the test medium is not an issue (for example, when determining some aspect of the physiology of chromate or when testing the sensitivity of a specific organism to chromate for general use). When the toxicity of a chromate bearing waste at a specific site is to be determined using water collected at that site, the concentration of sulfate in the site water should be determined, as well, for reference. Moreover, for similar reasons relating water chemistry to the toxicity of many compounds, a general analysis of the site water would be advisable for all toxicity tests.

The U.S. Environmental Protection Agency ambient water quality standards for chromium [*28*] ought to be reevaluated in light of the chromate/sulfate relationship as well. The agency currently acknowledges a correlation between salinity and chromate toxicity, which for algae arises through the relationship between chromate and sulfate, but fails to acknowledge that equal variation could occur because of variation in the sulfate concentrations of fresh water alone. A similar correlation between chromate toxicity and salinity is known for some marine invertebrates [*35,36*], but it is not known whether this arises via competition between chromate and sulfate. Before new standards are determined, this possibility ought to be investigated as well.

Another important result of this paper is that, under identical experimental conditions of light, temperature, and growth medium, and after identical means of analyzing the results, there are still large differences in the susceptibility of different species of algae.

Moreover, although only a few species of each class were tested, there is evidence that some classes, notably green and blue-green algae, may tend to be less sensitive to chromate than most other groups.

Finally, three hypotheses to help explain the differences between the chromate sensitivity of different clones of algae were tested: first, algae that take up chromate more readily might be more sensitive; second, algae with higher growth rates might be less sensitive because of the uncoupling of growth and chromate uptake; and third, resistant species might detoxify chromate by reduction. The data from these studies do not provide significant support for these hypotheses; therefore, apart from the general correlation between algal class and tolerance, there is as yet no simple way of predicting the toxicity of chromate to an untested algal population.

Acknowledgments

The author is grateful to Jim Sanders for his comments and criticisms on this manuscript and to The Academy of Natural Sciences for financial support and facilities for this study. He also wishes to thank Steve Cibik, JoAnne Bianchi, and Lorraine Currence for their help with the experiments.

References

[*1*] Cheremisinoff, P. N. and Habib, Y. H., *Water and Sewage Works,* Vol. 119, No. 7, 1971, pp. 73–86.

[*2*] Stratton, C. L. and Lee, G. F., *Journal of the Water Pollution Control Federation,* Vol. 47, No. 7, 1975, pp. 1901–1912.

[*3*] Elwood, J. W., Beauchamp, J. J., and Allen, C. P., *International Journal of Environmental Studies,* Vol. 14, No. 4, 1980, pp. 289–298.

[*4*] Dad, N. K., Rao, K. S., and Qureshi, S. A., *International Journal of Environmental Studies,* Vol. 15, No. 1, 1980, pp. 55–56.

[*5*] Mertz, W., *Physiological Reviews,* Vol. 49, No. 2, 1969, pp. 163–239.

[*6*] Towill, L. E., Shriner, C. R., Crury, J. S., Hammons, A. S., and Holleman, J. W., "Reviews of Environmental Effects of Pollutants: III-Chromium," Report 600/1/78/023, U.S. Environmental Protection Agency, Health Effects Laboratory, Cincinnati, OH, 1978.

[*7*] Richards, O. W., *Physiological Zoology,* Vol. 9, No. 2, 1936, pp. 246–253.

[*8*] Hervey, R. J., *Botanical Gazette,* Vol. 111, No. 1, 1949, pp. 1–11.

[*9*] Academy of Natural Sciences, "The Sensitivity of Aquatic Life to Certain Chemicals Commonly Found in Industrial Wastes," Philadelphia, PA, 1960.

[*10*] Patrick, R., Cairns, J., Jr., and Scheier, A., *Progressive Fish-Culturist,* Vol. 30, No. 3, 1968, pp. 137–140.

[*11*] Nollendorf, A., Pakalne, D., and Upitis, V., "Little Known Trace Elements in *Clorella* Culture," *Latvijas PSR Zinatnu Akaderuijas Vestis (Latvian S.S.R.),* Vol. 7, 1972 pp. 33–43.

[*12*] Garton, R. B. in *Water—1972,* G. F. Bennet, Ed., American Institute of Chemical Engineers, New York, 1973, pp. 284–292.

[*13*] Wium-Andersen, S., *Physiologia Plantarum,* Vol. 32, No. 4, 1974, pp. 308–310.

[*14*] Patrick, R., Booth, T., and Larson, R., "The Role of Trace Elements in the Management of Nuisance Growths," Report 66012-75-008, U.S. Environmental Protection Agency, Corvallis, OR.

[*15*] Mangi, J., Schmidt, K., Pankow, J., Gaines, L., and Turner, P., *Environmental Pollution,* Vol. 16, No. 4, 1978, pp. 285–291.

[*16*] Bringmann, G. and Kuhn, R., "Testing of Substances for Their Toxicity Threshold: Model Organisms *Microcystis (Diplocystis) aeruginosa* and *Scenedesmus quadicauda,*" *Internationale Vereinigung fur Theoretische und Angwandte Limnologie, Mittelungen,* Vol. 21, No. 1, June, 1978, pp. 275–284.

[*17*] Frey, B. E., Riedel, G. F., Bass, A. E., and Small, L. F., *Estuarine, Coastal, and Shelf Science,* Vol. 17, No. 2, 1983, pp. 181–187.

[*18*] Riedel, G. F., *Journal of Phycology,* Vol. 20, No. 4, 1984, pp. 496–500.

[*19*] "Methods of Chemical Analysis of Water and Wastes," Report 5501-0067, U.S. Environmental Protection Agency Quality Control Laboratory, Cincinnati, OH, 1971.
[*20*] Brand, L. E. and Guillard, R. R. L., *Journal of Plankton Research,* Vol. 3, No. 2, 1981, pp. 193–201.
[*21*] Stumm, W. and Morgan, J. J., *Aquatic Chemistry,* Wiley Interscience, New York, 1971.
[*22*] Riedel, G. F., *Aquatic Toxicology,* Vol. 7, No. 3, 1985, pp. 191–203.
[*23*] Wetzel, R. G., *Limnology,* Saunders, Philadelphia, 1975.
[*24*] Herlihy, A. T. and Mills, A. L., *Applied Environmental Microbiology,* Vol. 49, No. 1, 1985, pp. 179–187.
[*25*] Martin, J. M. and Meybeck, M., *Marine Chemistry,* Vol. 7, No. 3, 1979, pp. 173–206.
[*26*] Filip, D. S., Peters, T., and Adams, V. D., *Water Research,* Vol. 13, No. 3, 1979, pp. 305–343.
[*27*] Richter, J. E., Memorandum to C. E. Stephan, U.S. Environmental Protection Agency, Duluth, Minnesota, 3 June 1982.
[*28*] "Ambient Water Quality Criteria for Chromium—1984," Report 440/5-84-029, U.S. Environmental Protection Agency, Office of Water Regulations and Standards Criteria and Standards Division, Washington, DC, 1985.
[*29*] Miller, W. E., Greene, J. C., and Shiroyama, T., "The *Selenastrum capricornutum* Printz Algal Assay Test—Experimental Design, Application, and Data Interpretation Protocol," Report 600/9-78-108, U.S. Environmental Protection Agency, Corvallis, OR, 1978.
[*30*] Hanstveit, A. O. and Oldersma, H., "Evaluation of the Result of the Second ISO-Interlaboratory Study with an Algal Toxicity Test," Report ISO/TC 147/SC 5/WG5, Nederlands Normalistie-institut, Delft, The Netherlands, 1981.
[*31*] Nyholm, N., *Water Research,* Vol. 19, No. 3, 1985, pp. 273–279.
[*32*] Sunda, W. G. and Guillard, R. R. L., *Journal of Marine Research,* Vol. 34, No. 4, 1976, pp. 511–529.
[*33*] Nusch, E. A., "Evaluation of Growth Curves in Bioassays," ISO Report ISO/TC 147/SC5/WG5 N62, Nederlands Normalisatie-institut, Delft, The Netherlands, 1982.
[*34*] *Standard Methods for the Examination of Water and Wastewater,* 16th ed., American Public Health Association, Washington, DC, 1985, pp. 201–204.
[*35*] Olson, K. R. and Harrel, R. C., *Contributions in Marine Science,* Vol. 17, No. 10, 1973, pp. 9–13.
[*36*] Fales, R. R., *Bulletin of Environmental Contamination and Toxicology,* Vol. 20, No. 4, 1978, pp. 447–450.

New Methods for Aquatic Toxicology

George A. Blondin,[1] Lynda M. Knobeloch,[1] Harry W. Read,[1] and John M. Harkin[1]

An *In Vitro* Submitochondrial Bioassay for Predicting Acute Toxicity in Fish

REFERENCE: Blondin, G. A., Knobeloch, L. M., Read, H. W., and Harkin, J. M., "**An *In Vitro* Submitochondrial Bioassay for Predicting Acute Toxicity in Fish,**" *Aquatic Toxicology and Environmental Fate: Eleventh Volume, ASTM, STP 1007,* G. W. Suter II and M. A. Lewis, Eds., American Society for Testing and Materials, Philadelphia, 1989, pp. 551–563.

ABSTRACT: A new, rapid, and reliable bioassay for monitoring the acute toxicity of aquatic samples has been developed. The method is based on metabolic changes in preparations made from isolated animal mitochondria which affect the ratio of reduced to oxidized nicotinamide adenine dinucleotide (NADH/NAD) as measured in a spectrophotometer. Stable submitochondrial electron-transfer particles (SMP), made by sonic disruption of the cristae membranes of bovine heart mitochondria, afford the most practical biological test material because these vesicles can be stored indefinitely prior to use as frozen or freeze-dried preparations with full retention of biological activity. Under appropriate conditions, SMP can be programed to reduce exogenous NAD^+ by coupled reverse electron transfer (RET) or to oxidize NADH by the conventional electron transfer pathway (ETR). Exposure to toxicants inhibits one or both redox activities by an amount proportional to the toxicant concentration. Hence, bioassay procedures can be based on RET and ETR tests which are often complementary. Reliable data can be obtained in 2 to 18 min without interference caused by "hard water" conditions. The toxicities of 64 industrial chemicals, inorganics, and pesticides to SMP correlate closely with the results from tests with the fathead minnow (r^2 = 83%). The correlation was substantially poorer between fish toxicity and available data for the same toxicants from another short-term bioassay utilizing the Microtox procedure. In the new bioassay, the biological materials of the SMP test are handled like a chemical reagent. The test requires only small sample volumes, has good reproducibility, and promises to be relatively inexpensive, rapid, and simple. Bioassay variants and potential enhancements of the SMP test system are discussed.

KEY WORDS: mitochondria, submitochondrial particles, reverse electron transfer, respiration, nicotinamide adenine dinucleotides, toxicity bioassays, toxic chemicals, aquatic toxicology, hazard assessment

Biological monitoring of acute toxicity in aquatic environments using whole-organism bioassays is often impractical because the tests are too time-consuming and costly. Most fish or aquatic invertebrate toxicity tests cost in excess of $700 and require 48 to 96 h to complete. A wide range of *in vitro* bioassays that can be carried out in minutes and at far less expense have been proposed or developed in recent years in the search for an ideal prescreening test to precede whole-organism bioassays. Availability of suitable prescreening tests could reduce the necessity for routine use of whole-organism bioassays for testing many samples which do not evoke acute toxic response.

[1] Associate scientist, research assistant, research assistant, and scientist, respectively, Water Resources Center, University of Wisconsin, 1975 Willow Drive, Madison, WI 53706.

The utility of an *in vitro* bioassay in prescreening toxicity depends on its having a high correlation with the indices of toxicity toward diverse chemical toxicants afforded by whole-organism testing. A test based on toxicant-induced reduction of light emission from the luminescent bacterium *Photobacterium phosphoreum* [*1*] has been acclaimed as the bioassay most capable of adequately fulfilling this requirement [*2–4*]. The test has gained rapid popularity since its introduction in 1979 and subsequent marketing by Beckman Instruments, Inc. (Fullerton, CA) as the Microtox test.

Recent scrutiny of the Microtox test has led several investigators [*5–9*] to conclude that it alone cannot substitute for whole-organism hazard assessment. Over 100 toxicants have now been tested by the Microtox bioassay; the results reveal many outliers in regression correlations with fish and aquatic invertebrate toxicity test results. Anomalous results occurred mostly from tests with inorganics and hydrophobic chemicals. Although the relative deviations of the Microtox test toward many toxicants should diminish its utility as a prescreening test, its use as a short-term test is still recommended [*6,10*] for monitoring certain previously defined mixtures, for example, in process control of suitable industrial effluents. Consequently, no single *in vitro* bioassay now in use correlates well enough with toxicity data established by conventional toxicological measurements to justify its use as a general prescreening bioassay.

A new *in vitro* test based on the redox status of nicotinamide adenine dinucleotide (NAD) in the presence of electron-transporting submitochondrial particles (SMP) from bovine heart mitochondria has recently been introduced as a promising prescreening bioassay [*11*]. The reported versions of the SMP test exploit the phenomenon of reverse electron transfer (RET), discovered by Chance and Hollunger [*12*] in 1957. The RET reaction measures the rate of NAD^+ reduction to NADH by the NADH-coenzyme Q reductase enzyme complex of the mitochondrial inner membrane. The reaction requires energy which can be supplied by exergonic biochemical events in the adenosine triphosphatase (ATPase) or cytochrome oxidase complexes of the mitochondrion. The RET reaction is readily quantified by measuring the rate of production of NADH, which absorbs light strongly at 340 nm. Therefore, it can be used for rapid, simple, and sensitive determinations of acute toxicity by measuring inhibition of NADH production. Tests with 17 diverse inorganic and organic toxicants gave results [5-min median effective concentration (EC_{50}) values] which correlated well (r^2 = 82%) with the results of fish toxicity bioassays [*11*]. Regression of fish toxicities on Microtox toxicities for the same toxicants produced a poor correlation (r^2 = 29%).

This paper provides more extensive evidence of the suitability of SMP as a biomonitoring reagent for prescreening toxicity in environmental samples. SMP EC_{50} values were measured for a broad selection of 64 diverse chemical toxicants to evalute the assay as a predictor of fish median lethal concentration (LC_{50}) values. A variation of the original SMP test is also described which permits more sensitive detection of lipophilic toxicants by spectrophotometric measurement of the rate of NADH oxidation by the mitochondrial electron transfer chain. The new test version, designated ETR-NAD (conventional electron transfer—NAD), complements RET bioassays and enables SMP test results to maintain a high degree of correlation when compared statistically with the 96-h LC_{50} values for fish.

Methods and Procedures

Preparation of the SMP reagent and procedures for the RET bioassay driven by adenosine triphosphate (ATP) through ATPase (RET-ATP) or cytochrome oxidase (RET-COX) have been previously described [*11*]. The new ETR bioassay is carried out as follows: Aliquots (2.9 mL) of ETR-NAD incubation medium containing 0.25 M sucrose, 6.0 m*M* magnesium chloride ($MgCl_2$), and 20 m*M* potassium HEPES (hydroxyethylpiperazine ethane–

sulfonic acid) buffer (pH 7.4) are equilibrated at 25°C. A 5-min preincubation period to activate the particles is initiated by adding 0.05 mL of SMP suspension containing 0.1 mg biuret protein. The ETR reaction is started by adding 0.2 μmol (50 μL) of freshly prepared (daily) NADH solution, and the decrease in absorbance at 340 nm is monitored with time. The absorbance change after 18 min of incubation is a quantitative measure of NADH oxidation [*13*]. Shorter test times of 2 to 5 min can be employed by using two to six-fold higher protein concentrations or elevated temperatures (30 to 38°C).

The wide array of hydrophobic chemical toxicants examined in the present study necessitated expanding the range of organic solvents for dispersing toxicants in the incubation media. Absolute ethanol, dimethyl sulfoxide, or *N,N*-dimethylformamide were suitable solubilizing vehicles; EC_{50} values for these solvents (Table 1) establish their innocuous behavior in all three test systems. The concentration of standard toxicants in stock solutions and dilutions were not confirmed by analysis; this was deemed unnecessary because of the brevity of the testing period. EC_{50} values were calculated from nominal concentrations based on weighing the test chemicals. Toxicant solution additions were at or below 50 μL per 3.0 mL test volume.

Percent inhibition values for each toxicant concentration were averaged from three or four separate measurements. These were plotted against the logarithm of the toxicant concentration to obtain a dose-effect response, the linear portion of which was fitted by regression analysis. EC_{50} values were calculated from the equation for the fitted line. Several EC_{50} values were also derived by plotting the gamma function versus toxicant concentration on log-log axes—a standard practice for interpreting Microtox results [*14*]. Gamma was taken as the ratio of the activity lost to the activity remaining, and EC_{50} values were derived at gamma = 1. Both methods gave identical results.

SMP and Microtox EC_{50} values were compared statistically with 96-h LC_{50} values for fish. Microtox values were obtained from published reports appearing from 1979 through 1986. Fish LC_{50} values are predominantly from one species, the fathead minnow *(Pimephales promelas).* Because of the diverse nature of the toxicants examined in the present study, fathead minnow values could not be found for several toxicants. Following the recommendation of Kimerle et al. [*15*], values for bluegill *(Lepomis macrochirus)* or rainbow trout *(Salmo gairdneri)* were used in these instances. The di-2-ethylhexyl phthalate value is from a study utilizing *Daphnia magna,* because fish appear to be unusually insensitive to this toxicant. Several review compilations [*16–19*] or original papers traced through the *Journal of the Water Pollution Control Federation* annual reviews served as sources of fish LC_{50} values.

Results

Figure 1 shows SMP EC_{50} values obtained using RET and ETR bioassay protocols. Panel A graphically describes the relative responses for chemicals more sensitively detected by the ETR bioassay; the average EC_{50} in the ETR test is about 230 times lower than that in the RET test. Panel B shows the EC_{50} values for chemicals more sensitively detected in the RET bioassay, on average by a factor of about 490. Other chemicals (data not graphically shown) are detectable with equal sensitivities in the RET or ETR bioassay. For example, the EC_{50} responses (see Table 1) for alachlor,[2] diazinon, di-*n*-butyl phthalate, zinc, and silver differ on average by a factor of only 1.08 (range 1.03 to 1.17), depending on the test procedure utilized.

[2] The complete chemical name for the compounds discussed in this paper may be found in the Appendix.

TABLE 1—*Minimum SMP EC_{50}s for organic and inorganic chemicals compared with Microtox (Mct) EC_{50}s and fish LC_{50}s. All values in micrograms per litre.*

Code	Chemical/Toxicant	SMP Test	SMP EC_{50}	Mct EC_{50}	Fish LC_{50}
1	Acetone	RET-ATP[a]	26 000 000	22 000 000	8 120 000
2	Aflatoxin-B1	RET-ATP	10 000	23 000	N/A[b]
3	Alachlor	ETR-NAD[c]	55 000	N/A	5 000
4	Aroclor-1242 (PCB)	ETR-NAD	590	700	300
5	Aroclor-1248 (PCB)	ETR-NAD	890	N/A	690
6	Arsenic	RET-COX[d]	21 000	38 000	27 000
7	Bromoxynil	RET-ATP	1 000	N/A	600
8	Cadmium	RET-ATP	520	41 400	630
9	Carboxin	RET-ATP	480	N/A	84
10	Chlordane	ETR-NAD	1 260	600	115
11	Chlorobenzene	RET-ATP	85 000	16 000	38 000
12	Chromium-III	RET-ATP	33 000	15 300	33 200
13	Copper	RET-ATP	380	9 300	530
14	Cyanide	RET-COX	70	8 400	114
15	Cyhexatin	RET-ATP	34	10 000	7
16	*p,p'*-DDE	ETR-NAD	550	N/A	240
17	*p,p'*-DDT	ETR-NAD	112	7 000	12
18	Di-*n*-butyl phthalate	ETR-NAD	8 500	N/A	2 600
19	Di-2-ethylhexyl phthalate	ETR-NAD	360	N/A	2 000
20	Diazinon	ETR-NAD	39 000	5 750	6 050
21	Dimethylformamide	ETR-NAD	14 500 000	18 700 000	13 000 000
22	Dimethyl sulfoxide	RET-ATP	37 000 000	99 000 000	37 000 000
23	4,6-Dinitro-*o*-cresol (DNOC)	RET-ATP	440	6 400	360
24	Dinoseb	RET-ATP	60	N/A	67
25	Disulfoton	ETR-NAD	9 500	N/A	4 300
26	Ethanol	RET-ATP	27 000 000	45 000 000	13 500 000
27	Fenaminosulf	RET-ATP	90	N/A	60
28	Fentin hydroxide	RET-ATP	10	N/A	15
29	Guthion (azinphos-methyl)	ETR-NAD	18 000	N/A	1 900
30	Kelthane (dicofol)	ETR-NAD	575	450	510
31	Lead	RET-ATP	2 000	11 000	1 390
32	Lindane (gamma-HCH)	ETR-NAD	26 000	11 100	7 000
33	*m*-Cresol	RET-ATP	115 000	9 600	55 900
34	Mancozeb	RET-ATP	5 400	80	1 900
35	Mercury	RET-ATP	130	59	170
36	Nabam	RET-ATP	8 000	102 000	5 800
37	Nickel	RET-ATP	2 200	251 000	27 000
38	*o*-Cresol	RET-ATP	110 000	26 400	18 200
39	*o*-Dichlorobenzene	RET-ATP	21 000	10 250	9 500
40	*o*-Phenylphenol	RET-ATP	7 200	2 100	6 000
41	*p*-Cresol	RET-ATP	144 000	1 400	23 800
42	Pentachlorophenol (PCP)	RET-ATP	81	510	240
43	Permethrin	ETR-NAD	220	560	16
44	Phenol	ETR-NAD	133 000	33 000	34 000
45	Rotenone	ETR-NAD	2	N/A	7
46	Selenium	ETR-NAD	229 000	33 400	40 000
47	Silver	RET-ATP	140	20 620	11
48	Sodium dodecyl sulfate	RET-ATP	1 900	2 200	4 700
49	Triton X-100	RET-ATP	7 700	210	N/A
50	Zinc	RET-ATP	1 700	33 000	2 990
51	Zineb	RET-ATP	3 200	6 200	970
52	1,1,2-Trichloroethylene	RET-ATP	190 000	130 000	44 100
53	1,2,3-Trichlorobenzene	RET-ATP	5 300	3 150	3 380
54	1,2,4-Trichlorobenzene	RET-ATP	1 820	3 970	1 770
55	1,3,5-Trichlorobenzene	RET-ATP	62 000	14 100	99 700

TABLE 1—(*Continued*).

Code	Chemical/Toxicant	SMP Test	SMP EC_{50}	Mct EC_{50}	Fish LC_{50}
56	1-Propanol	RET-ATP	19 000 000	18 000 000	4 630 000
57	2,3,4,6-Tetrachlorophenol	RET-ATP	450	1 270	580
58	2,3,5,6-Tetrachlorophenol	RET-ATP	500	2 210	400
59	2,3,5-Trichlorophenol	RET-ATP	360	1 110	800
60	2,3,6-Trichlorophenol	RET-ATP	8 800	12 700	5 100
61	2,4,5-T	RET-ATP	38 000	118 000	10 400
62	2,4,5-Trichlorophenol	RET-ATP	600	1 270	610
63	2,4,6-Trichlorophenol	RET-ATP	10 000	7 200	9 160
64	2,4-Dinitrophenol	RET-COX	2 000	11 000	6 580

[a] RET-ATP = ATP-driven reverse electron transfer.
[b] N/A = toxicity value is not available.
[c] ETR-NAD = electron transfer from NADH to oxygen with production of NAD.
[d] RET-COX = cytochrome oxidase-driven reverse electron transfer.

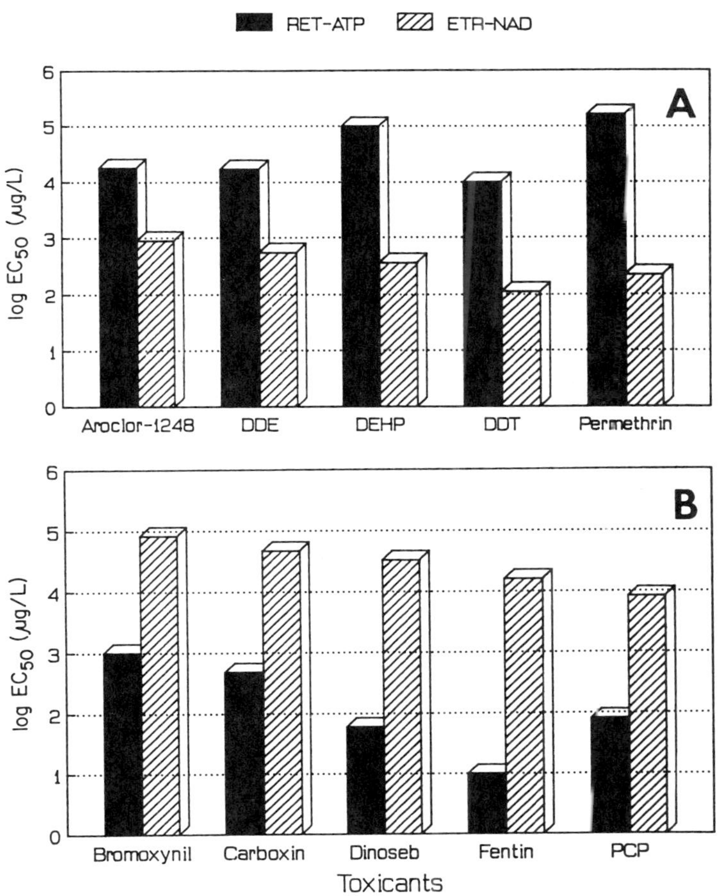

FIG. 1—*Comparison of SMP EC_{50} values for ten toxicants using the RET-ATP and ETR-NAD bioassays. PCP = pentachlorophenol; DEHP = di-2-ethylhexyl phthalate; Fentin = fentin hydroxide; and Aroclor-1248 = polychlorinated biphenyl mixture.*

Results of the SMP tests on the 64 individual chemicals are presented in Table 1. Microtox EC_{50} and fish LC_{50} values are included for comparison. SMP values represent those obtained using the most sensitive bioassay variant presently available, which is noted in Column 3 for each toxicant. A sequential number code is included in the table (Column 1) to facilitate identification of certain chemicals on the regression plots in Figs. 2 and 3 and for the identification of the grouped chemicals in Table 2.

A statistical comparison of SMP and fish toxicities is presented in Fig. 2. Values for 57 chemicals were used in a regression of fish LC_{50}s versus SMP EC_{50}s. Values for the five organic solvents (Entries 1, 21, 22, 26, and 56 in Table 1) were eliminated from the regressed data because these were examined only to document the low toxicity of prospective toxicant solvents. The detergent triton-X100 and the microbial toxin aflatoxin-B1 were also eliminated because fish LC_{50} values for these chemicals are unavailable.

If the fish and SMP data had displayed equal sensitivities to all the compounds tested, all data points would have fallen on a line with a slope of 1 and an intercept of 0.0 in the figure. The slope of the observed line was 0.85 and the intercept was 0.28. Thus, overall, fish may be slightly more sensitive to the compounds tested than are SMP. The correlation coefficient for the regressed line was 0.91, and r^2 was 83%; r^2 is a measure of the similarity in the degree and ordering of responses to the toxicants tested in both systems. The r^2 value obtained is almost identical to that reported earlier (82%) for a regression on only 17 data points [*11*]. Hence, the SMP test continues to be a good predictor of toxicity response in fish.

Also shown in Fig. 2 are the 95% confidence bands for the fitted solid line (inner dot-dot set) and for the prediction of fish toxicity from any single SMP measurement (employing

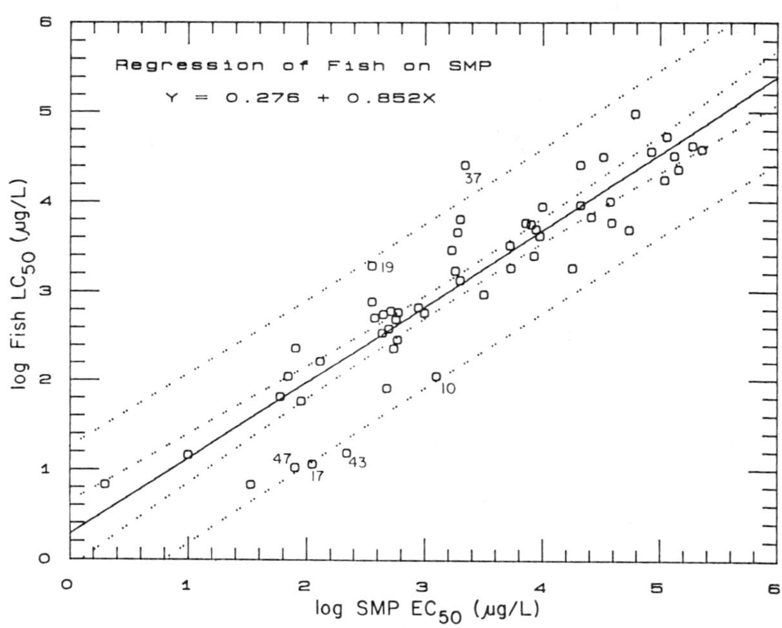

FIG. 2—*Logarithm of SMP EC_{50} (micrograms per litre) versus the log of fish LC_{50} (micrograms per litre) for 64 selected organic and inorganic chemicals. The inner broken lines represent the 95% confidence limits of the regressed line; the outer broken lines represent the 95% prediction limits. Labeled points refer to the chemicals in Table 1.*

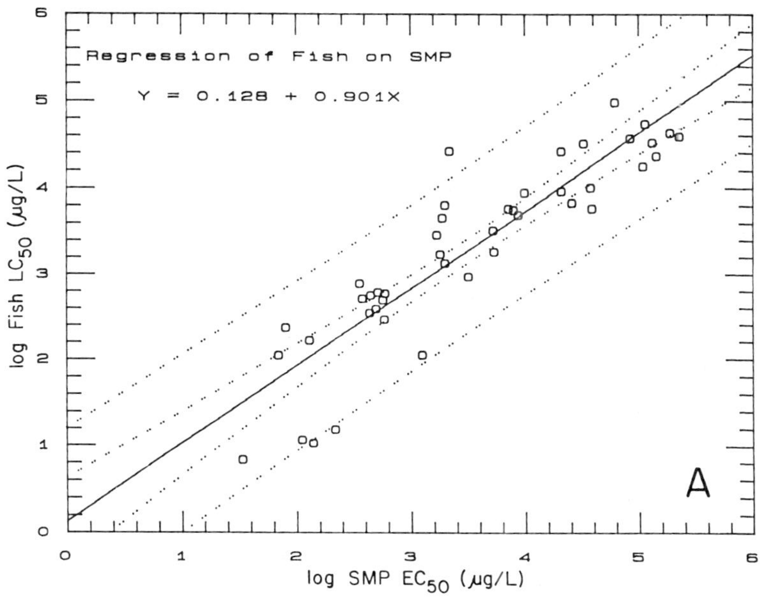

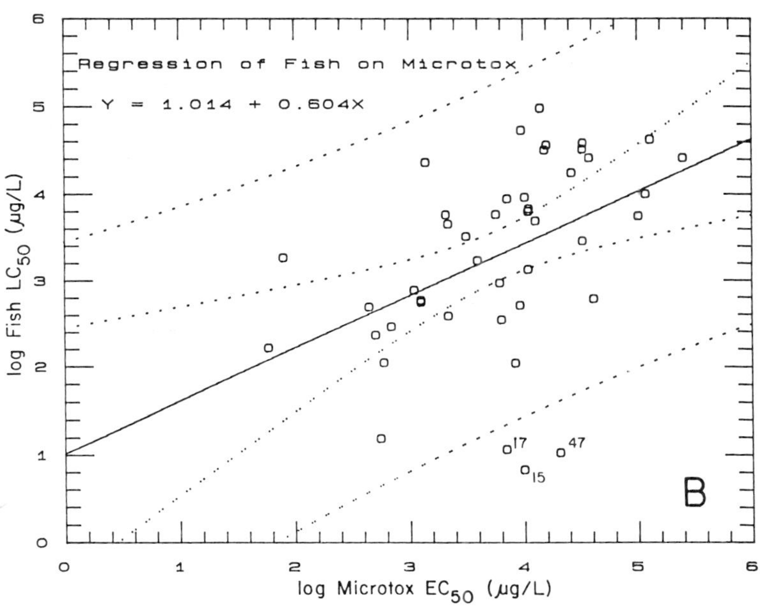

FIG. 3—*Logarithm of SMP* (A) *or Microtox* (B) *EC_{50} (micrograms per litre) versus the log of fish LC_{50} (micrograms per litre) for 44 selected organic and inorganic chemicals. The inner broken lines represent the 95% confidence limits of the regressed line; the outer broken lines represent the 95% prediction limits. Labeled points refer to the chemicals in Table 1.*

TABLE 2—*Correlation of fish versus SMP and Microtox toxicities for select groups of toxic chemicals.*

Group Definition	Data Set Size	r^2, %		Slope		Intercept	
		SMP	Microtox	SMP	Microtox	SMP	Microtox
Miscellaneous organics[a]	8	80.6	47.8	0.633	0.670	1.30	1.44
Inorganics[b]	11	77.9	20.8	0.940	0.605	0.19	0.70
Halogenated aromatics[c]	12	94.2	82.8	0.825	1.480	0.59	−1.80
Pesticides[d]	13	89.6	8.0	1.09	0.354	−0.84	1.49

NOTE—The numbers in each group refer to the code in Table 1.
[a] Miscellaneous organics = 4,33,38,41,44,48,52,64.
[b] Inorganics = 6,8,12,13,14,31,35,37,46,47,50.
[c] Halogenated aromatics = 11,39,42,53,54,55,57,58,59,60,62,63.
[d] Pesticides = 10,15,17,20,23,30,32,34,36,40,43,51,61.

the most sensitive result) of any toxicant (outer dot-dot set). Only six points lie on or slightly outside the 95% confidence level of the prediction. Chlordane, *p,p'*-DDT, permethrin, and silver (Points 10, 17, 43, and 47) occupy outlier positions on or below the lower prediction band, indicating only relatively marginal sensitivity of the SMP to these reagents in the bioassays used here. Di-2-ethylhexyl phthalate and nickel (Points 19 and 37) lie on or outside the prediction band on the high side of the regressed line, indicating a higher than predicted relative tolerance of the fish to these chemicals.

Figure 3 shows the results of statistical analysis applied to regressions of the logarithm of fish LC_{50} values on log SMP EC_{50} values (A in Fig. 3) and log Microtox EC_{50} values (B in Fig. 3) using fewer data (44 points) because Microtox values for several chemical species in Table 1 were not available from the literature. For these regressions, r^2 = 81.1% for fish versus SMP and 20.3% for fish versus Microtox toxicities. Estimates for the intercept and slope of the fitted line are shown in the respective panels. These estimates differ very little from our earlier [*11*] statistical comparisons of fish, SMP, and Microtox toxicities using a much smaller data base.

Table 2 presents a compilation of various statistical values that enable comparison of SMP and Microtox bioassay results with fish values for different categories of toxicants. Results of regressions for log fish LC_{50} values versus log SMP or log Microtox EC_{50} values are separately compiled for inorganics, halogenated aromatics (chlorobenzenes and chlorophenols), pesticides, and miscellaneous organics. The r^2 values for fish versus SMP correlations are quite similar (range 77.9 to 94.2%) for all subgroups, indicating a more uniform capacity of SMP tests to predict fish toxicities. However, r^2 values for fish versus Microtox comparisons differ markedly (range 8 to 82.8%); inorganics and pesticides yield the poorest r^2 values of only 20.8 and 8%.

Discussion

The SMP bioassay reagent has been used to evaluate toxicities for 64 diverse organic and inorganic chemicals. The results reveal that SMP tests afford more suitable predictions of fish toxicity than another popular, rapid *in vitro* bioassay (Microtox) which utilizes photoluminescent bacteria as an indicator of toxicity. Microtox test results were chosen as a basis of comparison because the speed and simplicity of the testing procedure parallels that of the mitochondrial bioassay. Moreover, the Microtox test has been touted as one of the best *in vitro* bioassay technologies available heretofore for rapid evaluation of toxicity [*2–5*].

Several investigators [*4,5,20,21*] have reported results of statistical analyses of fish versus Microtox toxicities for diverse organic chemicals. Their results showed correlations with an average r^2 value of 71.7% and a range of 65 to 82.8%. In the most extensive study, Curtis et al. [*5*] analyzed the relationship between the toxicity of 68 organic chemicals to fathead minnow and Microtox. The correlation for their composite had an r^2 value of 65%. Correlations we have found between fish and Microtox toxicities have much poorer r^2 values, that is to say, 29% in an earlier report [*11*] and 20.3% here.

Two features of the data analysis in Table 2 help clarify this discrepancy. First, fish versus Microtox data composites containing a substantial portion of toxicity values for pesticides and inorganics will likely give poorer correlations than composites from which such chemicals have been eliminated. The 44-chemical composite used here for fish versus Microtox statistical analysis contains data for 23 pesticides and inorganics which are equivalent to 52% of the total composite. Most earlier statistical comparisons of fish versus Microtox toxicity used composites that were devoid of pesticides and inorganics [*4,20,21*] or contained only a small percentage (5.9%) of these chemicals [*5*]. Hence, our estimates of the relationship between fish and Microtox toxicity are not directly comparable to those of other investigators. Curtis et al. [*5*] already alluded to this problem by indicating that preliminary studies with nine metals indicate that the relationship between fish and bacteria toxicity is much weaker in those compounds.

Second, had we restricted our statistical analysis to chemicals in the halogenated aromatic and miscellaneous organic chemical subgroups, our estimate of the relationship between fish and Microtox toxicity would likely have agreed better with the findings of earlier investigators. The chemicals in these two subgroups more closely approximate those employed by others for statistical analysis of the fish versus Microtox toxicity relationship, and the information contained in Table 2 shows better r^2 values (47.8% and 82.8%) in these instances.

To verify this assumption we combined the data shown here for miscellaneous organics with those for halogenated aromatics and supplemented the composite with toxicity values for the four pesticides (dinitro-*o*-cresol, Kelthane (dicofol), diazinon, and permethrin) used in the Curtis et al. [*5*] study. Regression analysis of this selective data set gave a fish versus Microtox correlation with an r^2 of 61.2%, closer to the value (65%) for the Curtis et al. [*5*] data composite. Hence, the poor correlation of fish versus Microtox toxicity for our 44-chemical data base seems to result from the greater diversity of toxins utilized in the present study compared with a more selective sampling by earlier investigators.

Despite the superiority of the SMP test system over the Microtox bioassay as a predictor of fish toxicity, the breadth of the 95% prediction limits in the two fish versus SMP regression plots (Figs. 2 and 3) suggests that present SMP EC_{50} values can be used only for order-of-magnitude estimates of fish toxicity. Attempts to decrease the spread of the prediction limits about the regression line are being made using additional SMP test variants that are likely to detect outliers with greater sensitivity. The new ETR-NAD test described above provides one example of a more sensitive bioassay (relative to RET tests) for detecting many hydrophobic toxicants.

Other tests that exploit such features as active ion transport, energized transhydrogenation, and prooxidant state effects are presently being explored to extend the SMP test system. This implies that the SMP bioassay may become a multiple testing system. The notion of multiple testing is not new. Current U.S. Environmental Protection Agency water quality criteria require tests for acute toxicity with at least eight species from eight different families. In view of the wide variations in the response of different aquatic species to many toxicants, this protocol greatly increases the likelihood that no environmental toxicant will go undetected. However, multiple species testing is extremely costly and time-consuming to implement.

Multiple testing using the SMP bioassay system is simple, rapid, and inexpensive. All three bioassays used to obtain the data shown in Table 1 use as the end-point the oxidation-reduction status of NAD^+. The bioassays differ only in their cofactor supplements. An appropriate choice of cofactors selects or inactivates different metabolic pathways in SMP to segregate unique sets of complex biochemical reactions that are interdependent and characteristically sensitive to different kinds of toxicants. This ease of manipulation of mitochondrial biochemistry in SMP makes the test system adaptable for the detection and quantitation of a wide range of toxic chemicals.

Information in Fig. 1 helps illustrate this versatility. RET-ATP responses to the chemicals in Panel B permit detection of these toxicants with far greater sensitivity than the ETR-NAD test. These results are readily rationalized in terms of what is known about ATP-driven reverse electron transfer [*22*] and knowledge of the mechanism of action of these toxins [*23*]. The RET reaction requires a source of energy and participation of the mitochondrial respiratory enzyme complexes I, II, and V. Bromoxynil, dinoseb, and pentachlorophenol are all uncouplers of oxidative phosphorylation, that is, chemicals which prohibit utilization of free energy for work performances such as reverse electron transfer. Carboxin and fentin are specific inhibitors of Complexes II and V, respectively. Hence, all chemicals described in Panel B are targeted toward biochemical systems that are obligatory participants in the RET-ATP reaction. In contrast, the ETR-NAD reaction does not require input of free energy and involves participation of only the mitochondrial respiratory enzyme complexes I, III, and IV; none is involved with the direct mechanism of action of these toxins. Consequently, only weak, nonspecific inhibition of the ETR-NAD reaction is caused by these toxicants.

The situation described in Panel A is quite different. Permethrin, DDT,[3] and DDE[3] are neurotoxins that modulate cation movements in nerve membranes. DDT behaves similarly in the mitochondrial membrane system [*24*]. Why these hydrophobic toxins are more sensitively detected by the ETR-NAD reaction can be explained in terms of two interdependent phenomena: (1) the paired moving charge hypothesis (PMC) of energy transduction [*25*], which requires compensatory movement of a positively charged species (for example, Na^+ or K^+) for electrical balance during electron transport in the various electron transfer complexes; and (2) the electron transfer events in Complex IV, which involve the largest free energy change of any mitochondrial electron transfer complex [*26*]. Following concepts developed by Szent-Gyorgi [*27*] and Calvin [*28*], the capacitance requirement (that is, insulation of the electron and cation channels by a lipid) for these ionic/electron transfer processes should be greatest in Complex IV. Complex IV is known to contain a high proportion of tightly bound lipid that is functionally associated with enzymatic phenomena catalyzed by the complex [*29*]. Consequently, in the ETR-NAD reaction, Complex IV should possess the highest affinity for hydrophobic toxicants, such as DDT or permethrin, which disturb the ion channels for compensatory movement of Na^+ and K^+. According to the Onsager principle, two transport processes in a confined system can interfere with each other. Hence, hydrophobics presumably elicit strong inhibition of electron transfer in Complex IV by selectively partitioning into the tightly bound lipid portion of the complex, where they interfere with compensatory cation movement.

The results shown in Fig. 1 indicate the ETR-NAD bioassay as a preferred test for *p,p′*-

---

[3] See Appendix for complete chemical name.

DDT and permethrin but do not validate our proposed mechanism of action of these reagents in submitochondrial particles. However, if the compensatory ion transport model is valid, the unexpectedly high sensitivity of the ETR-NAD test system to di-2-ethylhexyl phthalate and the polychlorinated biphenyl (PCB) mixture (Aroclor-1248) provides a new perspective on the mechanism of action of these toxicants, because neither has hitherto been recognized as modulators of ion transport.

Conclusion

Data obtained in a continuing evaluation of SMP as a rapid prescreening biomonitor of aquatic toxicity demonstrate that SMP is a versatile, sensitive reagent capable of assuming many different biochemical roles to quantify a broad spectrum of toxicants. The value of this new test system will likely extend beyond its simple use as a rapid, effective, and inexpensive method for toxicity detection and prescreening of environmental samples or industrial effluents. Additional bioassay protocols should provide a toxicity "fingerprint" of hazardous chemicals to impart a diagnostic feature to the test system. Moreover, information obtained in SMP tests should add to our understanding of the mechanisms of toxic activity of chemicals and facilitate some investigations into the fundamental properties of mitochondria.

Acknowledgments

Supported by the U.S. Geological Survey, Department of the Interior, under award number 14-08-0001-G1295 and by the College of Agriculture and Life Sciences, University of Wisconsin-Madison. We thank Eva Ruzicka and Anne Blondin for assistance in the preparation of mitochondria and submitochondrial particles, Enrique Valdivia for monitoring the biochemical and morphological quality of our biological preparations, and Raymond Hansen for many illuminating discussions which have led to a fuller understanding of the SMP test system.

APPENDIX

Chemical Equivalents of Generic Names Used in This Paper

Alachlor,
2-chloro-2′,6′-diethyl-*N*-(methoxymethyl)acetanilide

Aroclor-1242 (PCB), Aroclor 1248 (PCB)
polychlorinated biphenyl mixture containing 42 or 48% chlorine

bromoxynil,
3,5-dibromo-4-hydroxybenzonitrile

carboxin,
5,6-dihydro-2-methyl-1,4-oxathi-ine-3-carboxanilide

chlordane,
1,2,4,5,6,7,8,8-octachloro-2,3,3*a*,4,7,7*a*-hexahydro-4,7-methanoindene

cyhexatin,
tricyclohexyltin hydroxide

p,p'-DDE,
1,1-dichloro-2,2-bis(4-chlorophenyl)ethene

p,p'-DDT,
1,1,1-trichloro-2,2-bis(4-chlorophenyl)ethane

diazinon,
O,O-diethyl *O*-2-isopropyl-6-methylpyrimidin-4-yl phosphorothioate

DNOC,
4,6-dinitro-*o*-cresol

dinoseb,
2-*sec*-butyl-4,6-dinitrophenol

disulfoton,
O,O-diethyl *S*-2-ethylthioethyl phosphorodithioate

fenaminosulf,
sodium 4-dimethylaminobenzenediazosulfonate

Fentin hydroxide,
triphenyltin hydroxide

Guthion (azinphos-methyl),
S-(3,4-dihydro-4-oxobenzo[*d*][1,2,3]triazin-3-ylmethyl) *O,O*-dimethyl phosphorodithioate

Kelthane (dicofol),
2,2,2-trichloro-1,1-bis(4-chlorophenyl)ethanol

lindane (γ-HCH),
γ-1,2,3,4,5,6-hexachlorocyclohexane

Mancozeb,
manganese-zinc ethylenebis(dithiocarbamate)

nabam,
disodium ethylenebis(dithiocarbamate)

Permethrin,
3-phenoxybenzyl (1RS)-*cis/trans*-3-(2,2-dichlorovinyl)-2,2-dimethylcyclopropanecarboxylate

rotenone,
1,2,6,6*a*,12,12*a*-hexahydro-2-isopropenyl-8,9-dimethoxychromeno[3,4-*b*]furo[2,3-*h*]-chromen-6-one

zineb,
ethylenebis(dithiocarbamato) zinc

2,4,5-T,
2,4,5-trichlorophenoxyacetic acid

References

[*1*] Bulich, A. A. in *Aquatic Toxicology (Second Conference), ASTM STP 667,* American Society for Testing and Materials, Philadelphia, 1979, pp. 98–106.

[*2*] Bulich, A. A., Greene, M. W., and Isenberg, D. L. in *Aquatic Toxicology and Hazard Assessment*

(Fourth Conference), ASTM STP 737, Amercian Society for Testing and Materials, Philadelphia, 1981, pp. 338–347.

[*3*] Dutka, B. J. and Kwan, K. K., *Bulletin of Environmental Contamination and Toxicology,* Vol. 27, 1981, pp. 753–757.

[*4*] Lebsack, M. E., Anderson, A. D., DeGraeve, G. M., and Bergman, H. L. in *Aquatic Toxicology and Hazard Assessment (Fourth Conference), ASTM STP 737,* American Society for Testing and Materials, Philadelphia, 1981, pp. 348–356.

[*5*] Curtis, C., Lima, A., Lozano, S. J., and Veith, G. D. in *Aquatic Toxicology and Hazard Assessment (Fifth Conference), ASTM STP 766,* American Society for Testing and Materials, Philadelphia, 1982, pp. 170–178.

[*6*] McFeters, G. A., Bond, P. J., Olson, S. B., and Tchan, Y. T., *Water Research,* Vol. 17, No. 12, 1983, pp. 1757–1762.

[*7*] Grange, D. and Pescheux, F., *Bulletin Liaison Laboratoire Physique et Chimie,* Vol. 136, Mars-Avril, 1985, pp. 37–48.

[*8*] Hermens, J., Busser, F., Leeuwangh, P., and Musch, A., *Ecotoxicology and Environmental Safety,* Vol. 9, 1985, pp. 17–25.

[*9*] Ribo, J. M., Zaruk, B. M., Hunter, H., and Kaiser, K. L. E., *Journal of Great Lakes Research,* Vol. 11, No. 3, 1985, pp. 297–304.

[*10*] Qureshi, A. A., Flood, K. W., Thompson, S. R., Janhurst, S. M., Inniss, C. S., and Rokosh, D. A. in *Aquatic Toxicology and Hazard Assessment (Fifth Conference), ASTM STP 766,* American Society for Testing and Materials, Philadelphia, 1982, pp. 179–195.

[*11*] Blondin, G. A., Knobeloch, L. M., Read, H. W., and Harkin, J. M., *Bulletin of Environmental Contamination and Toxicology,* Vol. 38, 1987, pp. 467–474.

[*12*] Chance, B. and Hollunger, G., *Federation Proceedings,* Vol. 16, 1957, p. 163.

[*13*] Ernster, L. and Lee, C. P. in *Methods in Enzymology,* Vol. 10, R. W. Estabrook and M. E. Pullman, Eds., Academic Press, New York, 1967, pp. 738-744.

[*14*] Sheehan, K. C., Sellers, K. E., and Ram, N. M., "Establishment of a Microtox Laboratory and Presentation of Several Case Studies Using Microtox Data," Environmental Engineering Report No. 77-83-8, University of Massachusetts, Amherst, April 1984.

[*15*] Kimerle, R. A., Werner, A. F., and Adams, W. J. in *Aquatic Toxicology and Hazard Assessment (Seventh Symposium), ASTM STP 854,* American Society for Testing and Materials, Philadelphia, 1985, pp. 538–547.

[*16*] Johnson, W. W. and Finley, M. T., "Handbook of Acute Toxicity of Chemicals to Fish and Aquatic Invertebrates," Fish and Wildlife Service Resource Publication 137, U.S. Department of the Interior, Washington, DC, 1980.

[*17*] Brooke, L. T., Call, D. J., Geiger, D. L., and Northcott, C. E., "Acute Toxicities of Organic Chemicals to Fathead Minnows *(Pimephales promelas),*" Center for Lake Superior Environmental Studies, University of Wisconsin—Superior, Superior, WI, 1984.

[*18*] Mayer, F. L. and Ellersieck, M. R., "Manual of Acute Toxicity: Interpretation and Data Base for 410 Chemicals and 66 Species of Freshwater Animals," Fish and Wildlife Service Resource Publication 160, U.S. Department of the Interior, Washington, DC, 1986.

[*19*] Verschueren, K., *Handbook of Environmental Data on Organic Chemicals,* Van Nostrand, New York, 1983.

[*20*] Ribo, J. M. and Kaiser, K. L. E., *Chemosphere,* Vol. 12, No. 11/12, 1983, pp. 1421–1442.

[*21*] Indorato, A. M., Snyder, K. B., and Usinowicz, P. J. in *Toxicity Screening Procedures using Bacterial Systems,* Vol. 1, D. Liu and B. J. Dutka, Eds., Marcel Dekker, New York, 1983, pp. 37–53.

[*22*] Ernster, L. and Lee, C. P. in *Methods in Enzymology,* Vol. 10, R. W. Estabrook and M. E. Pullman, Eds., Academic Press, New York, 1967, pp. 729–738.

[*23*] Corbett, J. R., Wright, K., and Baillie, A. C., *The Biochemical Mode of Action of Pesticides,* second ed., Academic Press, New York, 1984, pp. 1–49.

[*24*] Chefurka, W., Zahradka, P., and Bajura, S. T., *Biochimica et Biophysica Acta,* Vol. 601, 1980, pp. 349–357.

[*25*] Blondin, G. A. and Green, D. E., *Chemical and Engineering News,* Vol. 53, 10 Nov. 1975, pp. 26–42.

[*26*] Tzagoloff, A., *Mitochondria,* Plenum, New York, 1982, p. 88.

[*27*] Szent-Gyorgyi, A., *Nature,* Vol. 148, No. 3745, August 1941, pp. 157–159.

[*28*] Calvin, M. in *Recent Progress in Photobiology,* E. J. Bowin, Ed., Blackwell, Oxford, 1964, pp. 225–258.

[*29*] Fry, M., Blondin, G. A., and Green, D. E., *Journal of Biological Chemistry,* Vol. 255, 1980, pp. 9967–9970.

R. A. Voyer,[1] *J. C. Sinnett,*[1] *and G. Modica*[1]

An Apparatus for Exposing Estuarine Aquatic Organisms to Toxicants in Constant and Fluctuating Salinity Regimes

REFERENCE: Voyer, R. A., Sinnett, J. C., and Modica, G., **"An Apparatus for Exposing Estuarine Aquatic Organisms to Toxicants in Constant and Fluctuating Salinity Regimes,"** *Aquatic Toxicology and Environmental Fate: Eleventh Volume, ASTM STP 1007,* G. W. Suter II and M. A. Lewis, Eds., American Society for Testing and Materials, Philadelphia, 1989, pp. 564–572.

ABSTRACT: A programmable control system for salinity has been developed and coupled with a flow-through toxicant exposure system. The resulting apparatus allows study of influences of constant and fluctuating salinity regimes on responses of estuarine organisms exposed to selected pollutants.

Constant and fluctuating salinity profiles available through use of the system (*a*) may be selected to range from fresh water to full-strength ambient seawater values; (*b*) are automatically repeatable, allowing long-term study; (*c*) are reproducible, thus permitting replication of treatment combinations over time and sequential testing of several species at a common salinity regime; and (*d*) can be quickly and easily modified or substituted for by another profile.

KEY WORDS: aquatic toxicology, exposure apparatus, constant salinity regime, fluctuating salinity regime, toxicants

A programmable control system for salinity has been developed and coupled with a flow-through toxicant exposure system. The resulting apparatus allows study of influences of constant and fluctuating salinity regimes on responses of estuarine organisms exposed to selected pollutants. The purpose of this report is to describe this apparatus and to demonstrate its effectiveness with operational results.

Levels of salinity in an estuary may undergo considerable change due to tidal influences, seasonal changes, and meteorological events. Salinity affects the physiology, behavior, and distribution of estuarine organisms. Because of its pervasive influence on aquatic life, salinity has been described as a "controlling factor" [*1*] as well as an "ecological master factor" [*2*]. Evidence indicates that salinity modifies chemical speciation of heavy metals entering an estuary [*3–5*] and, as a consequence, affects their distribution and availability to estuarine life. Although experimental evidence suggests that salinity may alter a number of response parameters ranging from survival to population growth [*6–8*], there are insufficient toxicity data from flow-through tests to permit development of salinity-toxicity relationships for saltwater water quality criteria.

Most data that are available to describe effects of salinity on various physiological and toxicological responses of aquatic life are based on experiments with fixed salinity regimes.

[1] Research aquatic biologist, electronics engineer, and biologist, respectively, U.S. Environmental Protection Agency, South Ferry Road, Narragansett, RI 02882.

Such data may offer only limited insight into potential influences of this fluctuating environmental variable. For example, experimentation with fixed salinities generally involves direct transfer of test organisms from holding to experimental conditions with no allowance for acclimation. The resulting osmotic gradient created, and the absence of opportunity for physiological adjustment, may cause an additional stress, accentuate the possible impact of an extreme salinity, or misrepresent an organism's tolerance to extreme salinities and its limits of distribution. Conversely, constant-salinity experiments may lead to overstatements of salinity effects by failing to account for physiological "damping," a process exhibited by various estuarine animals which serves to moderate internal osmotic changes relative to changes in the surrounding medium [*9–12*].

Experimental Apparatus and Procedures

A schematic of our apparatus is presented in Fig. 1. It is composed of a microcomputer, solenoid valves arranged in pairs, a series of mixing and exposure boxes, a toxicant delivery pump, and provisions for temperature regulation. The microcomputer described here is an additional application of the electronics included in a turbidistat previously described [*13*]. As in the earlier instance, the eight-bit microcomputer was assembled in-house from standard electronic components. Its operating instructions, along with selected constant data (for example, a sine table), are permanently incorporated into 4091 bytes of read-only memory (ROM). Variable parameters, such as salinity set points, are stored in 1024 bytes of read/write memory. Access to the microcomputer is gained through a keypad with an eight-digit display.

Our microprocessor provides salinity, calibration, and light cycle control functions (Table 1). In terms of the salinity function, the electronic system was set up to allow outputs to each of four pairs of solenoid valves. The salinity function controls opening and closing of each valve of a pair according to a time ratio. That is, every 30-s period is broken into two portions: saltwater delivery and freshwater delivery. One valve of each pair is open during the first portion and the other during the second portion. The ratio of these portions determines the ratio of salt to fresh water (for example, deionized tap water) delivered to the mixing boxes. Each of the four pair of valves may be programmed either to deliver a unique salinity profile or to replicate another. Flow rates, mixing volumes, and experimental volumes are rationalized so that water delivered to the test organisms is not subject to 30-s cycles but is easily varied according to tidal cycles if desired.

For a constant salinity regime, the ratio of saltwater delivery time to that of fresh water is held constant. The fluctuating salinity function, on the other hand, utilizes the sine table stored in ROM and allows simulation of an idealized tidal cycle. Lower and upper limits may be chosen to range between fresh water and full-strength ambient seawater. For a fluctuating regime, each valve of a couplet is energized for progressively shorter and then longer intervals. Proportions of fresh water and seawater are thus automatically changed with time to produce a sinusoidal pattern of salinity variation. Periodicity of fluctuations may be selected through choice of time constants in ROM to produce, for example, a 12-h cycle. High and low values of a selected cycle can be programmed to occur at either the same times each day or to shift as in a true tidal cycle. The theoretical salinity being delivered to an exposure box instantaneously is shown on the eight-digit display of the microprocessor unit. Comparison of this value with an actual measurement of salinity in an exposure box provides a convenient means of monitoring operational efficacy.

For the actual salinity delivered to be equal to a programmed set point, both valves of a pair must deliver the same volume of water per unit of time. To accomplish this, we inserted sections of capillary tubing between the solenoid valves and mixing boxes. Once

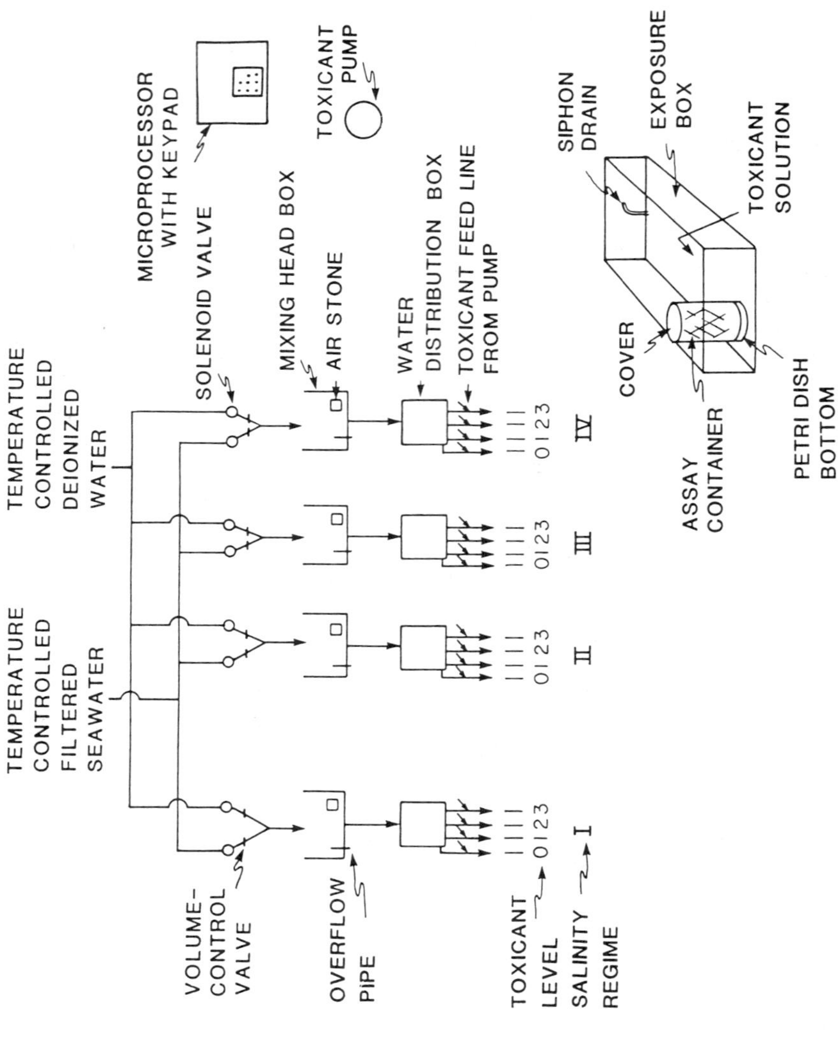

FIG. 1—*Schematic of dosing/salinity control apparatus.*

TABLE 1—*Summary of available control functions.*

Function	Choice of Levels	Pattern of Variation
Fixed salinity	any value between 0 and 32 parts per thousand	none
Fluctuating salinity	upper and lower limits of selected regime between 0 and 32 parts per thousand	sinusoidal (*a*) fixed time base—true tidal cycle (*b*) variable time base—for example, two high and two low parts per thousand in a 24-h period
Calibration	valves opened for 6-s interval	none
Light cycle		trapezoidal (*a*) julian light cycle (*b*) manual light cycle

volume equality had been established, it was easily checked by using the calibration function, which causes a pair of solenoid valves to open simultaneously on an automatic basis for precisely 6-s intervals. The volumes of water discharged are collected, and flow volumes are adjusted as necessary.

The light cycle function allows on-off control of laboratory lighting. The cycle begins with a gradual increase in light intensity simulating "sunrise," followed by an interval characterized by light of uniform intensity. The cycle ends with a gradual diminution and eventual extinction of light corresponding to "sunset." The photoperiod of an experiment can be selected manually or it can be set to follow automatically the Julian cycle stored in ROM.

Once a desired salinity profile and set points have been selected, the information is entered into the microcomputer. Each valve of a pair automatically opens and closes on a repeating and programed schedule. Volumes of fresh water and seawater discharged through any pair of valves are introduced into a mixing box (fiberglass-covered box of plywood) 15 by 20 by 20 cm, aerated, and blended. The resulting medium flows to an enclosed Plexiglas distribution box 10 by 11 by 13 cm and, subsequently, into a series of exposure boxes 36 by 27 by 13 cm [*14*]. In trials with fixed and fluctuating salinity regimes, the rates of flow of water to the exposure boxes were set at approximately 30 and 90 mL per min, respectively, using sections of capillary tubing.

In initial tests, a peristaltic pump metered toxicant solution directly into exposure boxes. This pump was later replaced with a Harvard[2] syringe pump modified to accommodate eight rather than the standard six syringes. Gentle aeration was applied to facilitate and ensure mixing of seawater and toxicant within an exposure box.

The performance of the system was evaluated by monitoring salinity and cadmium levels during 4 and 28-day trials. The salinity of each treatment was measured daily during the course of a trial with a temperature-compensated refractometer. Heated graphite atomization (HGA)/atomic absorption (AA) measurements of cadmium solutions were made with a Perkin-Elmer Model 2100 HGA unit coupled to a Perkin-Elmer Model 603 atomic

[2] Mention of trade names or commerical products does not constitute endorsement or recommendation of their use.

absorption instrument. The Model 603 was equipped with a D_2 arc background correction system, and the 2100 HGA unit was equipped with an autoinjector, Model AS-1. Transient HGA-AA signals were recorded with a Perkin-Elmer strip-chart recorder.

The cadmium content of "whole water" samples was determined by direct injection into the HGA unit [*15–16*]. Other than adding 100 μL of concentrated nitric acid (ultrapure) to 900 μL of the seawater samples, no chemical separation techniques were employed to concentrate cadmium from the seawater matrix. It was necessary, however, to matrix-match unknown samples with standards, since chemical interferences are not corrected by D_2 arc background correction. Accordingly, standard solutions of known concentrations of cadmium were prepared in identical fashion at each level of salinity to account for differences in the seawater matrices of the various test media. Three measurements of cadmium were made in the 4-day trials, and nine in the 28-day trial.

The levels of cadmium and salinity examined were selected based on use of a two-factor central composite design [*17*]. For the 4-day trials, various combinations of the two factors were examined in two incomplete blocks. Block 1 included the 2 by 2 factorial portion of the design, along with replicated center points. The star segment and additional center points constituted Block 2. For these tests, the cadmium concentrations ranged from 5 to 55 μg/L and the salinity from 6 to 32 parts per thousand. In the 28-day trial, only those combinations represented by Block 1 of the design were monitored. In this extended trial, the ranges of concentrations of cadmium and salinities evaluated were 2.2 to 7.8 μg/L and 13 to 21 parts per thousand, respectively.

Results

Information recorded shows that the actual test salinities closely approximated programed set points in both short- and long-term trials (Table 2). In each instance, for example, the observed mean salinity was within 1 g/kg of the intended level. The magnitude of calculated standard deviations associated with each signifies that observed salinities deviated little between measurements. In addition, the replicability of our apparatus is indicated by the consistency of repeated center points of the design within and between experimental blocks (30 μg of cadmium per litre, 19 parts per thousand salinity.

Repeated measurements of cadmium over the monitoring period (three in acute trials and nine during the 28-day trial) show that mean cadmium concentrations of test solutions corresponded to nominal levels. The largest deviation was 3.3 μg/L in the 4-day trial and 0.5 μg/L in the 28-day evaluation. As with salinity, the constancy of cadmium content among replicates (5 μg of cadmium per litre, 21 parts per thousand salinity) was excellent. Further review of Table 2 suggests that the syringe pump yielded more consistent cadmium solutions during the 28-day trial than the peristaltic pump did during the 4-day trial. In addition to being more reliable, the former pump was also easier to use and to maintain.

The overall pattern of salinity change depicted in Fig. 2 indicates the performance of the fluctuating salinity function. The three regimes illustrated were run concurrently. Each follows a 12-h cycle, and each was programmed with an upper limit of 32 parts per thousand salinity. The lower limits were arbitrarily set at 5, 12, and 19 parts per thousand. In each instance, the levels of salinity actually measured closely corresponded to theoretical values and followed the sinusoidal pattern of variation programed. The flow rates of water to the exposure boxes (90 mL/min) were adequate to preclude development of differences between programed and observed salinities in exposure boxes due to slow turnover rates in various chambers of the apparatus, that is, the mixing, distribution, and exposure boxes.

TABLE 2—*System performance comparing the nominal and measured mean concentrations [± standard derivation (SD)] of cadmium, and the respective intended and observed salinities.*[a]

Cadmium, μ/L				Salinity, parts per thousand			
Nominal	Measured	SD	N^b	Intended	Observed	SD	N^b
			4-Day Trial[c] *Block 1*				
12	12.3	0.4	3	10	9.8	0.4	5
48	50.3	1.0		10	9.8	0.4	
12	12.1	0.2		28	27.4	0.5	
48	47.6	0.6		28	27.4	0.5	
30	33.3	1.3		19	19.8	0.4	
30	29.5	1.1		19	19.8	0.4	
			Block 2				
5	5.1	0		19	20.0	0	
55	55.5	2.1		19	20.0	0	
30	31.9	0.4		6	5.8	0.5	
30	28.4	0.7		32	31.3	0.5	
30	31.4	0.5		19	20.0	0	
30	31.6	0.2		19	20.0	0	
			28-Day Trial[d] *Block 1*				
7.8	8.3	0.3	9	13	12.9	0.7	9
2.2	2.3	0.2		13	12.9	0.7	
7.8	7.9	0.3		29	28.2	0.4	
2.2	2.3	0.3		29	28.2	0.4	
5.0	5.3	0.3		21	20.3	0.5	
5.0	4.9	0.2		21	20.3	0.5	
5.0	4.9	0.3		21	20.9	0.9	
5.0	5.1	0.2		21	20.9	0.9	

[a] Levels of cadmium and salinity examined predicated by two-factor central composite design.
[b] N = number of measurements.
[c] Various combinations tested as two incomplete blocks in the 4-day trials.
[d] Factorial and replicated center points assessed in the 28-day trial.

Discussion

Salinity is one of the most difficult variables to control in conducting toxicity tests with estuarine species [*18*]. Operative results signify that our salinity-control system provides the requisite control. It also offers a variety of salinity regime options. For example, constant and fluctuating regimes may be studied separately or simultaneously. Salinity set points for a specified profile may be fixed at any whole value ranging from that of fresh water to that of full-strength seawater. In addition, a selected profile is automatically repeatable, allowing long-term studies. It is also reproducible, permitting replication of the experimental conditions over time, a feature which in turn allows selected species and various life stages of a single species to be tested sequentially at a chosen salinity regime. A selected profile may be easily and quickly modified or replaced with another simply through keyboard interaction with the microprocessor. Operational results, with actual values for salinity closely corresponding to programed levels, signify that the salinity control system is also reliable. Coupled with the dosing system described, this system provides a

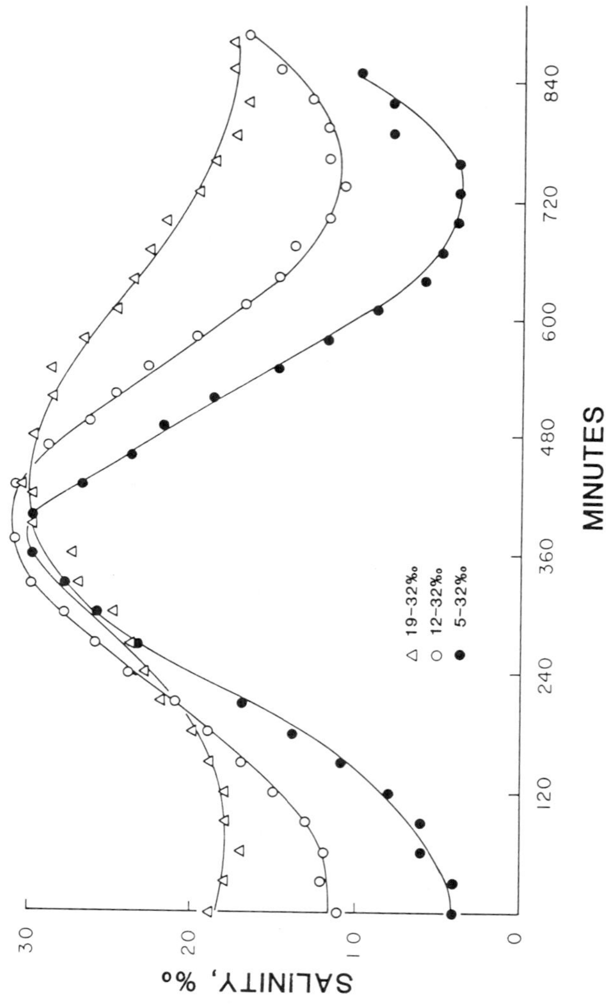

FIG. 2—*Three fluctuating salinity regimes showing a sinusoidal pattern of salinity change.*

useful tool for evaluating influences of salinity on toxicity responses of estuarine organisms.

Factorial experimental designs are frequently used in assessing effects of factor combinations on estuarine organisms. Unfortunately, as the number of variables studied is increased, the number of treatment combinations increases dramatically. Modified factorial designs [*19–20*] have been suggested as efficient and effective alternative designs for evaluating influences and interactions of multiple variables. Design requirements, however, impose contraints related to choice of factor levels. In order to utilize these designs in aquatic, flow-through toxicity testing, exposure systems must be sufficiently flexible to allow choice of a wide range of levels of test variables. Thus, from an experimental design perspective, our results illustrate the versatility of our apparatus, further demonstrating its utility.

Other salinity control systems reported [*18,21*] seemingly do not offer either the same ease of operation or range of salinity profiles as does our apparatus. On the other hand, an apparatus described by Davenport et al. [*21*] shares many of the operating features of our own. Both, for example, use solenoid valves in mixing fresh water and seawater. Whereas Davenport et al. used photoelectronic scanning of a cutout in the shape of the desired salinity profile, we employed time-based signals to control the opening and closing of these valves. Such signals can be easily and inexpensively generated with a microprocessor. Their use reduces the number of components to be adjusted and thus adds operational simplicity to our system. Moreover, time-based signals provide so precise a means of regulating opening and closing of solenoid valves, that no electronic feedback is needed to establish and maintain set-point salinity values. As a result, calibration of our system simply involves use of the calibration function as a means of checking that both valves of a pair are delivering the same volumes of water per unit of time.

Acknowledgments

It is a pleasure to recognize the very useful contributions of Dave Hansen, Roxanne Johnson, and Drs. Jerry Pesch, Hal Walker, and Jerry Zaroogian during preparation of this manuscript. Special thanks go to Ray Highland for his electronic wizardry.

References

[*1*] Fry, F. E. J., *Effects of the Environment on Animal Activity*, University of Toronto Press, Toronto, Ontario, Canada, 1947, p. 26.
[*2*] Kinne, O., *Netherlands Journal of Sea Research*, Vol. 3, No. 2, 1966, pp. 222–244.
[*3*] Helz, G. R., Huggett, R. J., and Hill, J. M., *Water Research*, Vol. 9, No. 1, 1975, pp. 631–636.
[*4*] Mantoura, R. F. C., Dickson, A., and Riley, J. P., *Estuarine and Coastal Marine Science*, Vol. 6, 1978, pp. 387–408.
[*5*] Frenet, M., *Water Research*, Vol. 15, No. 12, 1981, pp. 1343–1350.
[*6*] Nelson, D. A., Calabrese, A., and MacInnes, J. R., *Marine Biology*, Vol. 43, 1977, pp. 293–297.
[*7*] Voyer, R. A., Wentworth, C. E., Barry, E. P., and Hennikey, R. J., *Marine Biology*, Vol. 44, 1977, pp. 117–124.
[*8*] Roed, K. H., *Sarsia*, Vol. 64, 1979, pp. 245–252.
[*9*] Tucker, L. E., *Comparative Biochemistry and Physiology*, Vol. 36, 1970, pp. 301–319.
[*10*] Spaargaren, D. H., *Comparative Biochemistry and Physiology*, Vol. 47A, 1974, pp. 449–512.
[*11*] Stickle, W. B. and Ahokas, R., *Comparative Biochemistry and Physiology*, Vol. 47A, 1974, pp. 469–476.
[*12*] Shumway, S. E. and Davenport, J., *Journal of the Marine Biological Association of the United Kingdom*, Vol. 57, 1977, pp. 907–924.
[*13*] Sinnett, J. and Davis, W., *Journal of Experimental Marine Biology and Ecology*, Vol. 73, 1983, pp. 167–174.

[14] Voyer, R. A., Heltshe, J. F., and Krause, R. A., *Bulletin of Environmental Contamination and Toxicology,* Vol. 23, 1979, pp. 475–481.
[15] Ediger, R. D., Peterson, G. E., and Kerber, J. D., *Atomic Absorption Newsletter,* Vol. 13, 1974, pp. 61–64.
[16] Slavin, W., *Atomic Spectroscopy,* Vol. 1, No. 3, 1980, pp. 66.
[17] Cochran, W. G. and Cox, G. M., *Experimental Designs,* Wiley, New York, 1957, p. 317–334.
[18] Bahner, L. H. and Nimmo, D. R., *Transactions of the American Fisheries Society,* Vol. 104, 1975, pp. 388–389.
[19] Meyers, R. H., *Response Surface Methodology,* Allyn & Bacon, Boston, 1971.
[20] Alderdice, D. F. in *Marine Ecology,* O. Kinne, Ed., Wiley-Interscience, New York, 1972, Vol. 1, Part 3, pp. 1559–1722.
[21] Davenport, J., Gruffydd, L. D., and Beaumont, A. R., *Journal of the Marine Biology Association of the United Kingdom,* Vol. 55, 1975, pp. 391–409.

Guy R. Stehly[1] *and William L. Hayton*[2]

Errors in the Use of the Accelerated Bioconcentration Test

REFERENCE: Stehly, G. R. and Hayton, W. L., "**Errors in the Use of the Accelerated Bioconcentration Test,**" *Aquatic Toxicology and Environmental Fate: Eleventh Volume, ASTM STP 1007,* G. W. Suter II and M. A. Lewis, Eds., American Society for Testing and Materials, Philadelphia, 1989, pp. 573–584.

ABSTRACT: The accelerated bioconcentration test is commonly used in aquatic toxicology to calculate the bioconcentration factor (BCF) for waterborne xenobiotics. This method assumes that when aquatic organisms accumulate chemicals, they behave as a single compartment with the chemical coming to an equilibrium between the organism compartment and the water. Experiments have shown, however, that for many lipophilic substances, the aquatic organism must be modeled as two or more compartments to characterize the accumulation process adequately. We studied the errors associated with the estimation of the BCF when the accelerated test is applied to substances that require two compartments to represent the fish.

Body level/time data for xenobiotic uptake and depuration were simulated using the computer simulation program Minidynamo. The aquatic organism was modeled as two compartments; the amount of xenobiotic in the organism was a biexponential function of time; that is, for the uptake phase, $X = Ae^{-\alpha t} + Be^{-\beta t}$ where X is the amount in the organism, t is time, and A, B, α, and β are constants. These data sets were then fitted using NONLIN, an iterative nonlinear least-squares computer program, to the single-compartment model of the accelerated bioconcentration test to determine the apparent uptake and depuration rate constants, k_1 and k_2, and to estimate the apparent BCF from their ratio. The larger the ratio of the 0 time intercepts of the two components of the biexponential curve, A/B, the larger was the error in estimating BCF. Estimates for the uptake rate constant, k_1, were generally good but 10 to 30% smaller than the actual values, while the depuration rate constant, k_2, was poorly estimated, even for cases where BCF was accurately estimated. In extreme cases, the calculated BCF was less than 5% of the actual BCF. Testing for fit to multicompartmental models and knowledge of multicompartmental characteristics and the infusion input can aid in reducing errors involved in estimates of BCF, k_1, and k_2 by the accelerated bioconcentration test.

KEY WORDS: accelerated bioconcentration test, multicompartment model, uptake, distributive phase, terminal phase, biphasic elimination, aquatic toxicology

Organisms exposed to a xenobiotic of constant concentration in water will reach a steady state between the uptake and depuration processes. The bioconcentration factor (BCF) is defined as the ratio of the concentration of a chemical in an organism or one of its tissues, to the average water concentration at equilibrium or steady state [*1*]. The BCF of a compound in aquatic organisms is a major concern for risk assessment of chemicals that are present in surface waters. Concern for compounds with a high BCF has focused mainly on

[1] Aquatic toxicologist, Great Lakes Environmental Research Laboratory, National Oceanic and Atmospheric Administration, Ann Arbor, MI 48105-1593.

[2] Chair, Pharmacology/Toxicology Program, and professor, College of Pharmacy, Washington State University, Pullman, WA 99164.

the hazard associated with their consumption. Processes involved in bioconcentration are also important in consideration of accumulation of chemicals by aquatic organisms and subsequent acute or chronic toxicity [*2*]. If the organism is consumed by other animals, including man, toxicity to these organisms may occur, along with a biomagnification of the compound in the food chain [*3*]. Bioconcentration factors have been measured in a variety of aquatic organisms; discussion in this paper will center on BCF in fish, although it can readily apply to other aquatic species.

The accelerated bioconcentration test was developed as a time-saving method to determine BCF. This method assumes that uptake and depuration of compounds by aquatic organisms can be described kinetically by a model in which the organism and the water each act as a single compartment [*4,5*]

$$C_W \underset{k2}{\overset{k1}{\rightleftharpoons}} C_F \tag{1}$$

where C_F is the concentration of chemical in the organism (amount/body weight), C_W is the concentration of chemical in the water (amount/volume), and k_1 (volume of water/body weight/time) and k_2 (time^{-1}) are the rate constants for the inward and outward movement of chemical into the organism. If C_W is constant then

$$C_F = \left(\frac{k_1}{k_2}\right) C_W(1 - e^{-k2t}) \tag{2}$$

As time increases, the amount of compound in the organism reaches steady state and

$$\frac{C_F}{C_W} = \frac{k_1}{k_2} = \text{BCF} \tag{3}$$

For many compounds, aquatic species are thought to behave as multicompartmental systems; for example, di-2-ethylhexyl phthalate (DEHP) [*6–8*], 4-chlorophenyl ether [*9*], gentamicin sulfate [*10*] ^{95m}technetium [*11*], trifluralin [*12*], chlorobenzenes [*13*], and pesticides [*14*]. Compounds which are best modeled by multicompartment models have a biphasic elimination or depuration phase after an intravenous (iv) bolus. The initial more rapid elimination is a combination of elimination processes such as metabolism, back diffusion, and so forth, as well as loss to a peripheral compartment (that is, distribution) and is referred to as the α or distributive phase. When distribution ceases to be an important factor, the second phase of a biphasic elimination curve is observed and is referred to as the β, terminal, or postdistributive elimination phase. In other words, compounds characterized by biphasic elimination curves can be separated into distributive and terminal elimination phases. Most studies in aquatic toxicology measure whole body xenobiotic levels, and, therefore, loss of compound from a central compartment to a peripheral compartment is not observed directly, as in classical plasma concentration versus time studies, but is controlled by distribution in the sense that all the compound is initially in the central compartment where it is available for elimination. The compound then moves to the peripheral compartment where elimination will not occur until the chemical moves back to the central compartment. The use of single-compartment models for compounds for which the organism behaves as two compartments can result in errors in estimation of the absorption and elimination rate constants and the apparent volume of distribution [*15*]. Thus, the possibility exists for errors in the estimated rate constants for uptake and depur-

ation and the estimate for BCF if a single-compartment model is fitted to data for a compound that exhibits multicompartmental characteristics.

Simple models of BCF, such as empirical models relating log P or water solubility to BCF and the accelerated bioconcentration test, are necessary to make a risk assessment of the large number of chemicals in the aquatic environment and to quantify their ability to accumulate in fish [*16,17*]. While these methods save time, their limitations need to be understood.

The present research investigated errors involved in using the one-compartment BCF model to estimate the BCF of compounds for which the organism behaves as two compartments. Data sets based on the two-compartment model were generated by computer simulation, and the BCF for each set was calculated as the ratio k_1/k_2, which was estimated by a least-squares computer fit of the simulated data by the one-compartment model. Estimates for BCF and k_1 were generally good, but in some cases there were large errors; the estimates for k_2 were generally poor. The results illustrated the need to characterize the nature of the uptake kinetics before using the accelerated BCF method.

Methods

Compounds which behave kinetically as if the body were two compartments, a central compartment and a peripheral compartment, display a biphasic elimination after instantaneous input [Fig. 1 *(top)*] into the central compartment that is described by

$$X = Ae^{-\alpha t} + Be^{-\beta t} \tag{4}$$

where X is the amount of xenobiotic in the fish, t is time, and A, B, α, and β are constants. The values for A and B are obtained by resolving the plot of log amount versus time by the method of residuals into its two exponential components, as shown in Fig. 1 *(bottom)* [*2,18*], and are the zero time intercepts of the two components of the biexponential curve. The constants α and β can be calculated from the slopes of these lines, which are equal to $-\alpha/2.303$ and $-\beta/2.303$, respectively [Fig. 1 *(bottom)*].

Minidynamo, a computer program for simulating dynamic systems, was used to generate data sets for the two-compartment model shown in Fig. 2 for both an uptake and an elimination (depuration) phase [19]. For a two-compartment model (Fig. 2), the amounts of compound (amount/body weight) in the central, X_1, and peripheral, X_2, compartments, are defined by the differentials

$$\frac{dX_1}{dt} = k_1 C_W + k_{21} X_2 - (k_{12} + k_M + k_2) X_1 \tag{5}$$

and

$$\frac{dX_2}{dt} = k_{12} X_1 - k_{21} X_2 \tag{6}$$

where k_{12} and k_{21} (time^{-1}) are intercompartmental rate constants, k_M (time^{-1}) is a metabolism rate constant, k_2 (time^{-1}) is the depuration rate constant, and k_1 (volume water/body weight/time) is the uptake rate constant (Fig. 2). Equations 5 and 6 were integrated [*20*] to give

$$X_1 = \frac{k_1 k_{21} C_W}{\alpha\beta} - \frac{k_1 C_W \alpha - k_1 k_{21} C_W}{\alpha(\alpha - \beta)} e^{-\alpha t} + \frac{k_1 C_W \beta - k_1 k_{21} C_W}{\beta(\alpha - \beta)} e^{-\beta t} \tag{7}$$

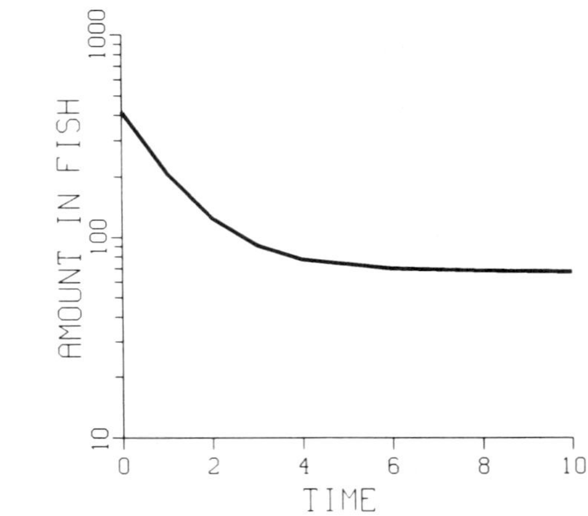

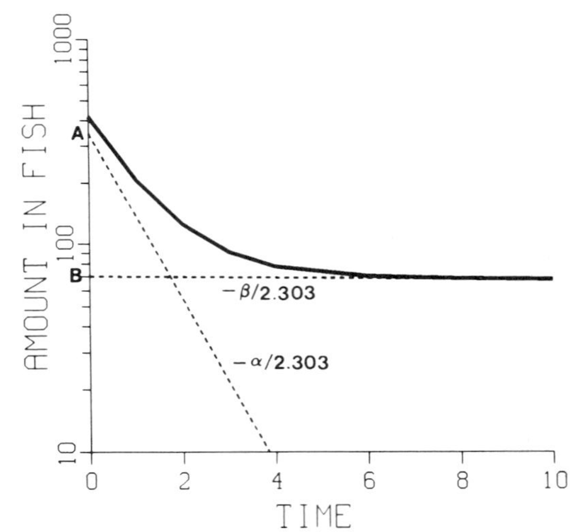

FIG. 1—(Top) *Biphasic elimination of a compound with two-compartment characteristics after instantaneous input, described by* $X = Ae^{-\alpha t} + Be^{-\beta t}$. (Bottom) *Resolution of biphasic curve into its two exponential components by the method of residuals to calculate* A *and* B *from the zero time intercept and* α *and* β *from the slope of the residual lines* ($-\alpha/2.303$ *and* $-\beta/2.303$).

and

$$X_2 = k_{12}k_1C_W\left(\frac{1}{\alpha\beta} + \frac{1}{\alpha(\alpha - \beta)}\,e^{-\alpha t} - \frac{1}{\beta(\alpha - \beta)}\,e^{-\beta t}\right) \tag{8}$$

where α and β are hybrid rate constants that are functions of the model rate constants [*18*]. At time $t = \infty$, when the fish accumulate compound to steady state, *SS*, the exponential terms $e^{-\alpha t}$ and $e^{-\beta t}$ go to 0 and

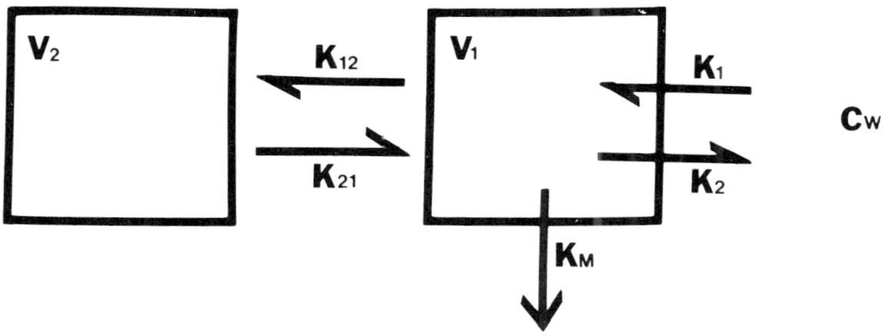

FIG. 2—*Two-compartment model used to simulate xenobiotic uptake and elimination.* k_1 *is the uptake rate constant (volume water/body weight/time);* k_2 *is the rate constant for transfer of compound out of the fish;* k_{12} *and* k_{21} *are the rate constants for intercompartmental transfer; and* k_M *is the rate constant for metabolism (time*$^{-1}$*).*

$$X_1^{SS} = \frac{k_1 k_{21} C_W}{\alpha\beta} = \frac{k_1 C_W}{k_M + k_2} \tag{9}$$

and

$$X_2^{SS} = \frac{k_1 k_{12} C_W}{\alpha\beta} = \frac{k_{12} k_1 C_W}{k_{21}(k_M + k_2)} \tag{10}$$

The total amount of compound in the fish at steady state, X^{SS}, is equal to the sum of the amount of compound in the two compartments

$$X^{SS} = X_1^{SS} + X_2^{SS} \tag{11}$$

The amount in the fish, X^{SS}, divided by the water concentration, C_W, is the BCF

$$\frac{X^{SS}}{C_W} = \frac{k_1(k_{21} + k_{12})}{k_{21}(k_M + k_2)} \tag{12}$$

Equation 12 indicates that the BCF for a two-compartment model includes intercompartmental rate constants and is complex compared with the equation derived for the one-compartment-model, accelerated bioconcentration test (Eq 3).

If the biexponential elimination curve is observed during the depuration phase, after uptake or infusion has stopped, it can be resolved into its exponential components by the method of residuals; the intercepts (R and S in Fig. 3) are the values extrapolated to the time at which depuration begins. They are not the same as the A and B intercepts [Fig. 1 *(bottom)*] but are calculated by [*18,21*]

$$R = \frac{k_1 C_W(\alpha - k_{21})}{(\alpha - \beta)\alpha}(1 - e^{-\alpha\tau}) \tag{13}$$

and

$$S = \frac{k_1 C_W(k_{21} - \beta)}{(\alpha - \beta)\beta}(1 - e^{-\beta\tau}) \tag{14}$$

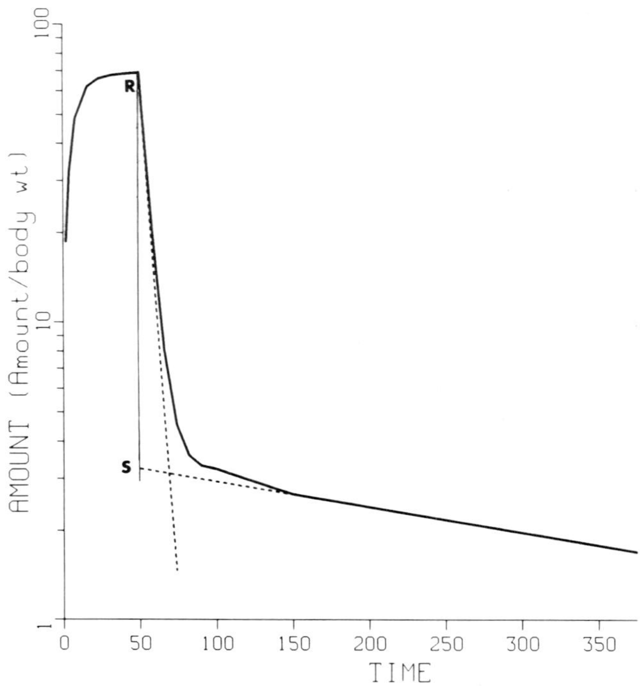

FIG. 3—*Resolution of biphasic elimination curve, postuptake or postinfusion, into its two exponential components by the method of residuals to calculate* R, S, α, *and* β. R *and* S *are the intercepts of the residuals at the start of the depuration phase, and* α *and* β *are calculated from the slope of the residuals* ($-\alpha/2.303$ *and* $-\beta/2.303$)*. The data are from Parameter Set 5, Table 1.*

where τ (time) is the duration of the uptake or infusion phase. When uptake or the infusion reaches steady state, τ is large, and $e^{-\alpha\tau}$ and $e^{-\beta\tau}$ equal 0. In this case

$$R = \frac{k_1 C_W(\alpha - k_{21})}{(\alpha - \beta)\alpha} \tag{15}$$

and

$$S = \frac{k_1 C_W(k_{21} - \beta)}{(\alpha - \beta)\beta} \tag{16}$$

The corresponding values for A and B that would be observed after an iv bolus administration may be calculated by

$$A = \frac{R X_0 \alpha}{k_1 C_W} \tag{17}$$

and

$$B = \frac{S X_0 \beta}{k_1 C_W} \tag{18}$$

where X_0 is the amount of compound in the organism after the uptake phase. For a two-compartment model, the ratio of the zero time intercepts A/B after intravenous injection is a measure of its multicompartmental characteristics. Multicompartmental characteristics are usually discernible for large A/B ratios during elimination phases but can appear monoexponential if the ratio is very large and the postdistributive phase occurs when body levels have dropped below detection levels. As A goes to 0 and the A/B ratio becomes very small, again the curves become monoexponential [18].

The ratio R/S in the pre-steady-state case can be calculated by

$$\frac{R}{S} = \frac{(\alpha - k_{21})\beta(1 - e^{-\alpha\tau})}{(k_{21} - \beta)\alpha(1 - e^{-\beta\tau})} \tag{19}$$

and after steady state is achieved

$$\frac{R}{S} = \frac{(\alpha - k_{21})\beta}{(k_{21} - \beta)\alpha} \tag{20}$$

After an iv bolus

$$\frac{A}{B} = \frac{\alpha - k_{21}}{k_{21} - \beta} \tag{21}$$

and this ratio is always larger than R/S since β is smaller than α by definition. Thus, depuration after steady state compared with that observed after an iv bolus results in the diminution of the α or distributive phase; the smaller β/α and the closer to steady state the depuration begins, the faster the distributive phase disappears. For uptake to non-steady-state situations $(1 - e^{-\alpha t})/(1 - e^{-\beta t})$ will be greater than or equal to one, but β/α always has a greater effect. The net result is that any period of uptake will result in a disappearance of some part of the distributive phase during a depuration experiment.

To test for characteristics that would result in errors in the estimation of BCF, each rate constant in the model was individually increased and decreased. The model parameters used are listed in Table 1. Estimates of BCF were calculated by fitting a one-compartment model (accelerated bioconcentration test) to the data sets for the uptake and depuration phases simultaneously, using NONLIN [22], an iterative nonlinear least squares computer program. The NONLIN program gave estimates for k_1 and k_2, which were used to calculate the estimated BCF using Eq 3. The actual BCF was calculated using Eq 12 and verified by simulations to steady state.

Results

The model-simulated and model-estimated values of k_1 and k_2 for the model parameter values listed in Table 1 are given in Table 2, and the model-simulated and model-estimated values of BCF with A/B and R/S ratios are given in Table 3. Models with large A/B ratios tended to give the poorest estimate of BCF by the accelerated bioconcentration test.

Discussion

The absorption of xenobiotics from the water by fish is similar to an infusion; that is, input occurs at a constant rate. The kinetics during and after a constant rate input have

TABLE 1—*Rate constants used for simulated two-compartment models.*

Model	V_1[a]	V_2[a]	k_1[b]	k_2[b]	k_{12}[b]	k_M[b]
1	50	1 000	10	0.05	0.1	0.001
2	500	1 000	10	0.05	0.1	0.001
3	500	500	10	0.05	0.1	0.001
4	1 000	50	10	0.05	0.1	0.001
5	100	50	10	0.05	0.001	0.1
6	500	1 000	10	0.05	0.5	0.000 1
7	50	1 000	10	0.05	0.001	0.1
11	50	5 000	10	0.05	0.1	0.001
12	50	1 000	10	0.05	0.1	0.1
131	50	1 000	10	0.05	0.01	0.01
132	50	1 000	10	0.05	1.0	0.001
141	50	1 000	100	0.05	0.1	0.001
142	50	1 000	1	0.05	0.1	0.001
151	50	1 000	10	0.5	0.1	0.001
152	50	1 000	10	0.005	0.1	0.001
211	500	1 000	10	0.05	0.01	0.001
212	500	1 000	10	0.05	1.0	0.001
221	500	1 000	10	0.05	0.1	0.1
222	500	1 000	10	0.05	0.1	0.000 01
231	500	1 000	100	0.05	0.1	0.001
232	500	1 000	1	0.05	0.1	0.001
241	500	1 000	10	0.001	0.1	0.001
242	500	1 000	10	0.1	0.1	0.001

[a] V_1 and V_2 are apparent volumes of distribution of the central and peripheral compartments, respectively.

[b] k_1, k_2, k_{12}, and k_M are the rate constants for uptake, depuration, central compartment to peripheral compartment, and metabolism, respectively (Fig. 2).

TABLE 2—*Simulated and estimated rate constants.*

Model	k_1 Simulated[a]	k_1 Estimated[a]	k_2 Simulated[b]	k_2 Estimated[c]
1	10	7.03	0.051	0.001 79
2	10	8.39	0.051	0.014 8
3	10	9.02	0.051	0.023 0
4	10	10.8	0.051	0.053 9
5	10	10.1	0.150	0.147
6	10	9.67	0.050 1	0.016 2
7	10	9.41	0.150	0.148
11	10	6.84	0.051	0.000 356
12	10	4.80	0.150	0.004 53
131	10	6.88	0.051	0.021 6
132	10	9.59	0.051	0.002 34
141	100	70.3	0.051	0.001 79
142	1	0.70	0.051	0.001 80
151	10	1.93	0.501	0.005 62
152	10	9.52	0.006	0.000 271
211	10	7.45	0.051	0.025 6
212	10	9.91	0.051	0.016 9
221	10	6.65	0.150	0.034 5
222	10	8.44	0.050	0.014 6
231	100	83.9	0.050 1	0.014 8
232	1	0.84	0.051	0.014 8
241	10	9.94	0.002	0.000 664
242	10	7.22	0.101	0.025 5

[a] k_1 simulated and estimated are the uptake rates simulated by Minidynamo and estimated by NONLIN, respectively.

[b] k_2 simulated is the sum of k_2 and k_M.

[c] k_2 estimated is the NONLIN calculated value.

TABLE 3—*Simulated and estimated BCFs and model values for* A/E *and* R/S.

Model	BCF Simulated[a]	BCF Estimated[a]	A/B[b]	R/S[c]
1	4 120	3 930	44.6	0.477
2	588	567	2.07	0.150
3	392	392	0.523	0.051 0
4	206	201	0.0525	0.001 21
5	100	68.8	113	0.510
6	599	597	2.28	0.046 5
7	2 000	68.2	4 560	0.510
11	19 800	19 200	227	0.503
12	1 400	1 060	122	1.43
131	4 120	319	7 400	5.06
132	4 120	4 100	22.0	0.464
141	41 200	39 200	44.6	0.477
142	412	392	37.5	0.401
151	419	343	710	4.91
152	35 000	34 900	22.3	0.054 7
211	588	280	64.6	4.30
212	600	586	2.14	0.023 2
221	200	187	9.90	1.00
222	600	580	3.73	0.268
231	5 880	5 670	3.77	0.274
232	58.8	56.7	3.77	0.274
241	15 000	15 000	2.05	0.008 96
242	297	283	6.40	0.617

[a] The BCFs simulated and estimated are the bioconcentration factors calculated by Minidynamo and NONLIN, respectively.

[b] A/B is the ratio of the 0 time intercepts of residuals for iv bolus [Fig. 1 *(bottom)*].

[c] R/S is the ratio of the intercepts of the residuals at the beginning of depuration, after an uptake (infusion) period (Fig. 3).

been well studied [*18*]; the distributive phase of compounds with multicompartmental characteristics is diminished in the postinfusion elimination phase (see the section on Methods). This characteristic is often utilized experimentally to study compounds that may be too insoluble or too toxic to administer as an iv bolus, and it may be used to study compounds with a terminal phase that is difficult to characterize because the blood level drops below the sensitivity of the analytical method. In these cases, the compound is infused and the kinetics are studied postinfusion.

The standard procedure for measurement of the BCF uses an analogous approach: a constant-rate uptake phase followed by an elimination phase (Fig. 4). A biphasic elimination phase in the uptake-elimination test for BCF indicates that the compound displays multicompartment model characteristics, which suggests that it should be analyzed by methods other than the simple BCF test [*5,23*]. The results of this study show that the biphasic elimination discerned after uptake of the compound from the water may be diminished or absent. It may therefore be difficult to ascertain when the simple BCF test is inappropriate.

The disappearance of the distributive phase after uptake to 10, 50, and >99% of steady state is shown in Figs. 5 through 7 for three models with A/B ratios of 2, 65, and 710; these model results are compared with a bolus injection of the amount in the fish after uptake to steady state. In all cases, the postuptake distributive phase has for practical purposes disappeared after >99% of steady-state uptake. In some cases in which uptake was a smaller percentage of steady state, the distributive phase was also eliminated. When the

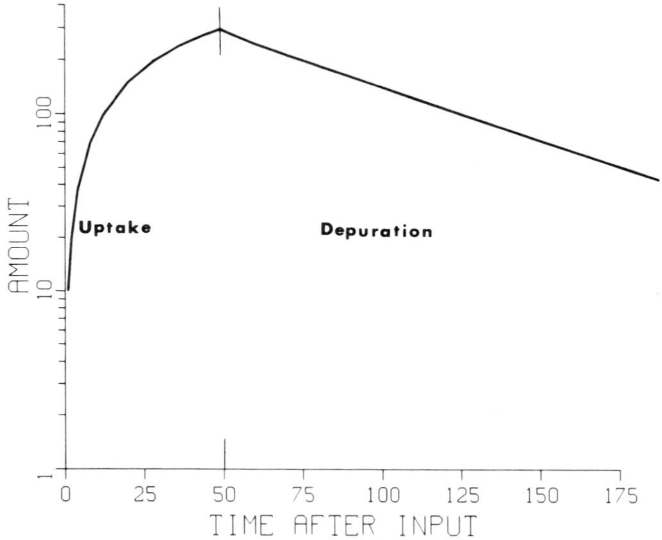

FIG. 4—*Typical uptake and elimination (depuration) phases for BCF test or infusion elimination experiment. The data are from Parameter Set 2, Table 1.*

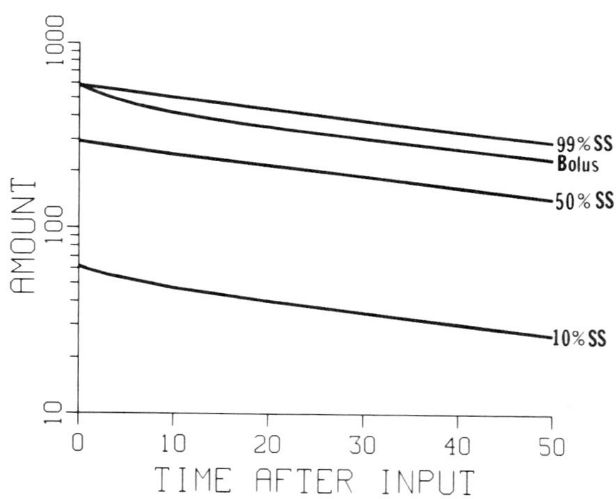

FIG. 5—*Elimination phase after uptake to greater than 99% of steady state (99% SS), 50% of steady state (50% SS), 10% of steady state (10% SS), and after iv bolus (Bolus) of a dose equal to the body amount at steady state for Parameter Set 2 (Table 1) with* A/B = *2.*

distributive phase was diminished, it probably could not be detected in the normal scatter of experimentally determined data; the elimination curve postuptake may therefore appear to display the characteristics of a compound leaving a single compartment. These results indicate that, in almost all cases, compounds that display multicompartmental characteristics will not have prominent distributive phases during the depuration that follows uptake of the compound. The disappearance of the distributive phase was observed for simulations of steady-state infusion compared with iv bolus administration for DEHP in rainbow trout [*6*].

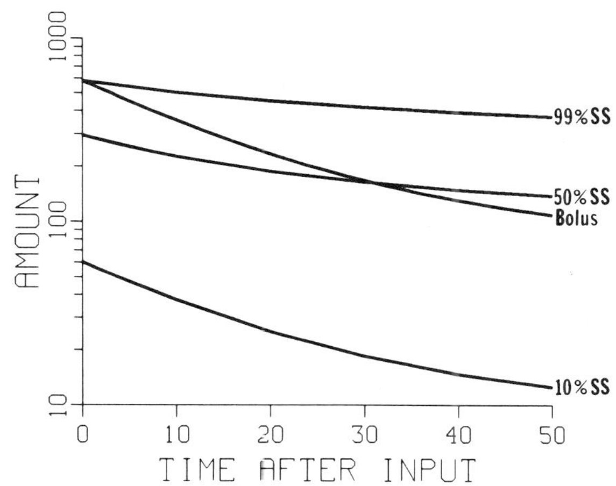

FIG. 6—*Elimination phase after uptake to greater than 99% of steady state (99% SS), 50% of steady state (50% SS), 10% of steady state (10% SS), and after iv bolus (Bolus) of a dose equal to the body amount at steady state for Parameter Set 211 (Table 1) with* A/B = *65.*

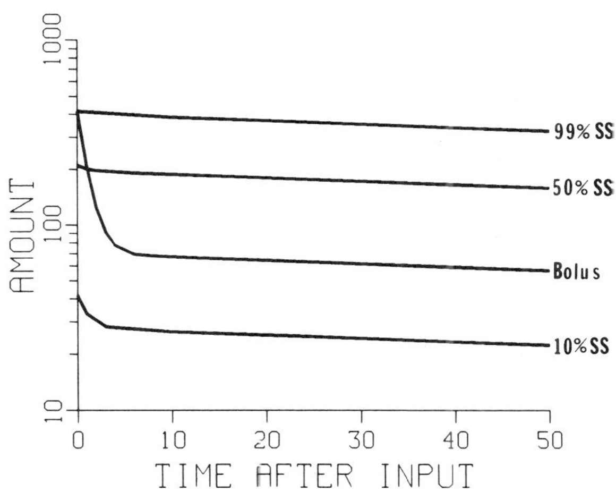

FIG. 7—*Elimination phase after uptake to greater than 99% of steady state (99% SS), 50% of steady state (50% SS), 10% of steady state (10% SS), and after iv bolus (Bolus) of a dose equal to the body amount at steady state for Parameter Set 151 (Table 1) with* A/B = *710.*

The accelerated bioconcentration test gave an accurate estimate of BCF in most cases, although it produced errors in some cases. Errors occurred when the ratio A/B was high (Table 3). The characteristic biphasic elimination during depuration was more visible if uptake was restricted to less than 50% of the steady state level before depuration began. The accelerated bioconcentration test can be used for its original purpose, to predict BCF; the use of k_2 from this test for correlation with or extrapolation to other factors [*2,24*] is unlikely to produce meaningful results, as there were generally large differences between the actual and estimated k_2 values (Table 2).

Constant water concentrations of xenobiotic are not necessary to measure BCF; writing an additional equation for the water concentration and fitting this equation simultaneously with the amount of compound in the fish [*7,8,12*] permits an accurate assessment of multicompartmental characteristics and provides a model that accurately predicts the disposition and BCF of the test compound.

Acknowledgments

The authors gratefully acknowledge A. Rashid for assistance with Minidynamo and the Great Lakes Environmental Research Laboratory, National Oceanic and Atmospheric Administration, in Ann Arbor, Michigan, for computer time. Financial support was provided by Research Grant ES 01995 from the National Institutes of Health, United States Public Health Service, and by Research Grant R812818-01-0 from the U.S. Environmental Protection Agency.

This paper is GLERL Contribution No. 590.

References

[*1*] Rand, G. M. and Petrocelli, S. R. in *Fundamentals of Aquatic Toxicology,* G. M. Rand and S. R. Petrocelli, Eds., Hemisphere Publishing, New York, 1985.

[*2*] Spacie, A. and Hamelink, J. L., *Environmental Toxicology and Chemistry,* Vol. 1, 1982, pp. 309–320.

[*3*] Spacie, A. and Hamelink, J. L. in *Fundamentals of Aquatic Toxicology,* G. M. Rand and S. R. Petrocelli, Eds., Hemisphere Publishing, New York, 1985.

[*4*] Branson, D. R., Blau, G. E., Alexander, H. C., and Neely, W. B., *Transactions of the American Fisheries Society,* Vol. 104, 1975, pp. 785–792.

[*5*] Hamelink, J. L. in *Aquatic Toxicology and Hazard Evaluation (First Symposium), ASTM STP 634,* American Society for Testing and Materials, Philadelphia, 1977, pp. 149–161.

[*6*] Barron, M. G., Tarr, B. D., and Hayton, W. L., *Toxicology and Applied Pharmacology,* Vol. 88, 1987, pp. 305–312.

[*7*] Karara, A. H. and Hayton, W. L., *Aquatic Toxicology,* Vol. 5, 1984, pp. 181–195.

[*8*] Tarr, B. D., Ph.D. thesis, Washington State University, Pullmann, WA, 1985.

[*9*] Chui, Y. C., Addison, R. F., and Law, F. C. P., *Aquatic Toxicology,* Vol. 8, 1986, pp. 41–49.

[*10*] Rolf, L. L., Setzer, M. D., and Walter, J. C., *Veterinary and Human Toxicology,* Vol. 28, No. S1, 1986, pp. 25–30.

[*11*] Blaylock, B. G., Frank, M. L., and DeAngelis D. L., *Health Physics,* Vol. 42, 1982, pp. 257–266.

[*12*] Hayton, W. L. and Archer, B. G., *Proceedings of the Western Pharmacological Society,* Vol. 24, 1981, pp. 147–149.

[*13*] Konemann, H. and van Leeuwen, K., *Chemosphere,* Vol. 9, 1980, pp. 3–19.

[*14*] Kreminski, S. F., Gilbert, J. T., and Ritts, J. A., *Archives of Environmental Contamination and Toxicology,* Vol. 5, 1977, pp. 157–166.

[*15*] Riegleman, S., Loo, J. C. K., and Rowland, M., *Journal of Pharmaceutical Sciences,* Vol. 57, 1968, pp. 117–123.

[*16*] Brungs, W. A. and Mount, D. I. in *Estimating the Hazard of Chemical Substances to Aquatic Life, ASTM STP 657,* J. Cairns, Jr., K. L. Dickson, and A. W. Maki, Eds., American Society for Testing and Materials, Philadelphia, 1978, pp. 30–31.

[*17*] Veith, G. D., Defoe, D. L., and Bergstedt, B. V., *Journal of the Fisheries Research Board of Canada,* Vol. 36, 1979, pp. 1040–1048.

[*18*] Gibaldi, M. and Perrier, D., *Pharmacokinetics,* Marcel Dekker, Inc., New York, 1982.

[*19*] Shafer, W. A., "Minidynamo User's Guide," Pugh-Roberts Associates, Cambridge, MA, 1978.

[*20*] Benet, L. Z., *Journal of Pharmaceutical Sciences,* Vol. 61, 1972, pp. 536–541.

[*21*] Gibaldi, M., *Journal of Pharmaceutical Sciences,* Vol. 58, 1969, pp. 1133–1135.

[*22*] Metzler, C. M., Elfring, G. L., and McEwen, A. J., *Biometrics,* Vol. 30, 1974, pp. 562–563.

[*23*] Esser, H. O., *Pesticides Science,* Vol. 17, 1986, pp. 265–276.

[*24*] Hawker, D. W. and Connell, D. W., *Ecotoxicology and Environmental Safety,* Vol. 11, 1986, pp. 184–197.

Summary

Summary

The primary role of the ASTM symposium series on aquatic toxicology is to provide a forum for discussion of recent developments that are important to those who perform environmental risk assessments and who work in ecotoxicology and environmental chemistry. The success of the eleventh symposium in fulfilling this role is shown in the 38 papers appearing in this publication. These papers, continuing the trend of previous meetings, include topics significant to the standard-writing process, such as chronic endpoints, statistical data analysis, and description of new methodologies. In addition, sections of this volume are devoted to three subjects that are becoming increasingly important to the scientific and regulatory communities. The number of papers in the section on effluent toxicity assessments reflects the increasing importance of the U.S. Environmental Protection Agency's national policy of control of toxic substances on the basis of water quality. The large section on current developments in understanding the fate of chemicals reflects the recent recognition of the importance of realistic exposure assessments. Likewise, the realization that not all species, chemicals, and conditions can be tested or measured has resulted in high interest in quantitative structure-activity relationships and similar models. Key aspects of these fields and brief summaries of the papers in this volume follow. It can be expected that the information presented in these papers will serve as a focus for future presentations at ASTM symposia in this series. This expectation reflects the dynamic and ever-challenging nature of aquatic toxicology, risk assessment, and the standard-writing process.

Fate of Chemicals in Water and Sediment

Understanding the fate of chemicals in the environment is obviously important in risk assessment and management. Two papers in this volume describe fate and risk assessment models. *Nocito et al.* present a risk framework for marine disposal of sludge. The procedure uses the risk information system of the U.S. Environmental Protection Agency (EPA) and represents a preliminary assessment of ecological and human health risks associated with municipal waste disposal. A model for predicting the effects of dredged material plumes in Long Island Sound is described by *Munns and Paul.* The model provides reliable estimates of suspended solid concentrations and specific contaminant levels in the ecosystem when compared with *in situ* measurements. The toxicity of percolate from decontaminated pesticide-contaminated soils is described by *Graves et al. Cowgill* discusses the chemical dynamics of the mud-water interface in aerobic and anaerobic eutrophic lakes. It was found that the elements in anaerobic muds of Linsley Pond (Connecticut) diffuse into the hypolimnion, whereas those in aerobic muds of Lake Hubeh (Israel) either diffuse in conjunction with the ferric hydroxide complex or are retained by organic matter.

Several papers describe the biodegradation potential of pure compounds and the fate of chemicals in estuarine and freshwater systems. *Adams et al.* measured the biodegradation of butyl benzyl phthalate (BBP) in water and sediment during a 30-day microcosm study. The data show that BBP is readily degradable in water and sediment. The half-life in water was 2.0 days or less. It was also found that BBP metabolites were 100 times less toxic to

Daphnia magna than the parent compound. The response of microbial communities in rivers and lakes to three detergent surfactants was studied by *Ventullo et al.* Exposure for 3 h to 23 days had no detrimental effect on bacterial heterotrophy, and chronic exposure resulted in an adaptive response, indicated by increased degradation of the chemicals after exposure. *Sanders and Abbe* found that silver uptake in estuarine and marine phytoplankton is rapid and inversely proportional to the salinity. Diatoms and dinoflagellates were considerably more sensitive to silver than other flagellated species. The authors state that integrated studies are needed to determine the health of estuarine and marine ecosystems. *Fulthorpe et al.* describe the influence of chemical concentration and the presence of mixtures of aromatic contaminants on the catabolic activities of bacteria. According to the authors, some species isolated from natural streams evolve unstable catabolic phenotypes.

Effluent Toxicity

The primary objective of the EPA's water-quality-based toxics control program is to prevent the entry of toxic effluents into the Nation's receiving waters. Nine papers address effluent assessment techniques and site-specific safety evaluations. An automated biological early warning system for detecting acutely toxic effluents is presented by *Gruber et al.* The computerized system measures the ventilatory responses of bluegill and was found to be useful and reliable in a validation study. A fractionation procedure for effluents using simple column chromatography techniques is described by *Doi and Grothe.* The fractionation format proved very useful in identifying toxic components in environmental samples. Two examples of the effects of effluents on marine ecosystems are presented. *Schimmel et al.* conducted an on-site evaluation of the toxicity of a pulp mill discharge using five test species. Sea urchins and an alga were the most sensitive test organisms. Unionized ammonia was the primary toxic constituent. *Frithsen et al.* conducted a four-month project to evaluate the ability of two test methods to detect the toxicity of sewage effluents in coastal areas. The methods were a single-species test (sea urchin) and outdoor mesocosms. The authors concluded that the single-species test was useful for indicating the magnitude and persistence of toxicity, and the mesocosms identified sensitive components and processes. *Westerman* evaluated the effects of point and nonpoint pollution sources in the Yellow Creek drainage in Kentucky. Sediment toxicity tests were useful for describing long-term impacts, while water column tests were more representative of short-term effects. Sediment toxicity testing was preferred, therefore, for assessing in-stream effects. *Goodfellow et al.* found *Ceriodaphnia dubia* to be more sensitive to a municipal wastewater effluent discharged into Baltimore Harbor than *Mysidopsis bahia* and Microtox. The utility of these data are discussed in the context of a toxicity reduction evaluation. *Pratt et al.* determined the effects of a municipal effluent containing textile dyes on bacterial communities, invertebrates, and fish. The bacterial microcosm tests were less sensitive than the single-species tests in detecting toxicity. *Finlayson and Wilson* determined the effects of acid mine waste on salmonids and monitored metal levels in the receiving water for potential toxic effects. They suggest that new criteria are needed for receiving water if effective protection is to be afforded. *Freeman et al.* investigated the effects of mitigating low pH levels on Atlantic salmon. A salt-enriched diet appeared to be helpful, as was neutralization with limestone.

Chronic Endpoints: Statistical Versus Biological Significance

The experience accrued in chronic toxicity testing has made it clear that application of hypothesis testing statistics to individual measured responses of test organisms results in endpoints that are inconsistent and, in some cases, that would not protect populations. There is no consistent relationship between statistical significance and biological significance, and equal changes in two responses, such as growth and mortality, do not have equal significance. Two suggested changes in the endpoints are (1) use of an integrated population response parameter rather than individual measured responses, and (2) use of concentrations that cause particular levels of effect rather than concentrations that differ from controls at some level of statistical significance. New insights concerning this problem are provided in this volume. *Stephan* summarizes current thinking concerning the analysis of chronic toxicity test results. The author discusses appropriate endpoints, terminology, and statistical data treatments. He stresses that these issues need to be resolved before additional chronic toxicity studies are developed. The utility of the instantaneous rate of increase, *r*, as an effect parameter is examined by *Barbour et al.* This parameter was found to be a valid measure of toxicity, based on a review of 100 chronic toxicity tests with *Ceriodaphnia.* However, it was not more sensitive than other chronic endpoints and is not cost-effective to calculate. *Barnthouse et al.* state that biological significance, for which there is no uniform definition, cannot be derived from the statistical significance of test results. Barnthouse and his co-workers show how simple population models developed for assessing effects of fishing and power-plant-induced mortality on fish populations can be used to summarize and interpret chronic toxicity test results and provide a basis for inferring biological significance.

Statistical Procedures

Advances in the statistical analysis of toxicity test results and the validation of various data analysis procedures are discussed in four papers. *Staples and Sebaugh* evaluate sensitivity and uncertainty analysis for use with chemical environmental exposure models. The utility and efficiency of the Latin hypercube sampling method is demonstrated using a sample data set. The effect of withdrawals during chronic toxicity tests is addressed by *Groggel et al.* A withdrawal is an observation that exists in the study prior to the measurement of the variable of interest, and only the value up to that point in time is noted. The authors note that results of studies with high withdrawal rates and unequal sample sizes are misleading. Two papers evaluate the available computer programs in their ability to conduct probit analyses. *Bromaghin and Engeman* evaluate the ASTM proposed standard practice for using probit analysis and compare it with a well-known SAS package. They conclude that the ASTM program is a subset of the SAS procedure and recommend that the ASTM program be expanded. *Roberts* has compared the analysis of 20 data sets with four software packages and found that most of the programs provide consistent results.

Microcosms

Numerous papers in previous ASTM publications in this series have described various microcosm and field testing procedures. The validation, reproducibility, and utility of these assessment techniques, however, have been rarely described. Two papers in this volume provide this information for one proposed test method called the standardized aquatic microcosm (SAM). SAM has been used in seven experiments by four laboratories, as reported by *Taub et al.* All the SAM experiments conducted with copper exhibited the same sequence of events, although the timing of the effects varied. Copper initially eliminated *Daphnia* and reduced algal populations, but the algae rapidly recovered. *Landis et al.* recommend the use of this microcosm based on their experience with three test compounds. The primary advantage of its use is the capability of this method to observe effects and fate simultaneously; the key disadvantage is the amount of labor required.

QSARs and Other Data Extrapolations

Many of the chemicals entering the environment lack comprehensive fate and effects data. Predictive equations to provide these data are obviously helpful to risk assessment managers. Several papers report QSARs (quantitative structure-activity relationships) for various compounds, species, and endpoints. A model to estimate LC_{50} values for *Daphnia magna* from rat oral LD_{50} values and molecular descriptors based on 190 chemicals is presented by *Enslein et al.* A model is also presented by *Niemi and Veith* for classification of chemicals as biodegradable or persistent under conditions used in conventional waste treatment. The authors indicate that the usefulness of this model would greatly increase if a BOD data base existed for chemicals with a wider variety of structures. *Schultz et al.* present models to predict the acute toxicity of various compounds to the ciliate *Tetrahymena pyriformis.* The effects of pyridines, cyanobenzenes, phenols, and benzyl alcohols, among others, are included. *Lipnick* describes the relationship of the toxicity of alcohols, saturated monoketones, and hydrocarbons to the tadpole, *Rana temporaria. Holcombe et al.* describe a method that uses acute toxicity data for the fathead minnow to predict effects on communities of organisms.

Standard QSARs do not reliably predict effects for some compounds. One such example is presented by *Kaiser and Gough* and is based on the acute response of *Photobacterium phosphoreum* to mono-substituted benzenes. *Roberts* reviews various QSARs reported in the literature for chemicals shown to be more toxic to fish than predicted. These predictive equations are based on electrophilicity parameters for which the chemical basis is discussed.

Walker and Brink describe a new procedure that the Interagency Testing Committee (ITC) is using to select chemicals for toxicity screening. This procedure utilizes expert panels of ecologists, engineers, chemists, and toxicologists to recommend target chemicals. These panels have identified 250 compounds having the potential for toxicity. *Riedel* presents a means of extrapolating the effects of hexavalent chromium on phytoplankton based on the sulfate concentration in fresh water and correlating these data with the salinity of marine and estuarine waters.

New Methods for Aquatic Toxicology

New test methods and improvements to established methods are constantly needed by the scientific and regulatory communities. Three papers in this volume describe recent developments in this area. *Stehly and Hayton* evaluate the ability of the commonly used accelerated bioconcentration test to predict BCF values for lipophilic compounds and fish. Differences between the calculated and estimated values are as great as 95%. The discrepancy in values could be reduced by considering the fit to multicompartmental models. A programmable microcomputer control system is described by *Voyer et al.* which permits the study of the effects of fluctuating salinity on responses of estuarine organisms to chemicals. The apparatus (eight-bit microcomputer) has been used in toxicity tests to confirm that cadmium toxicity increases with decreasing salinity. *Blondin et al.* present a bioassay based on metabolic changes in isolated animal mitochondria. Effects data on electron transfer and transport systems are obtained in 2 to 15 min. The correlation between fathead minnow acute lethality and mitochondria toxicity for 44 test compounds was greater than 85%, which is better than the correlation between fish toxicity and the commonly used Microtox procedure.

Michael A. Lewis

Procter and Gamble Co., Cincinnati, OH 45217; symposium chairman and editor.

Glenn W. Suter II

Oak Ridge National Laboratory, Oak Ridge, TN 37831; symposium chairman and editor.

Indexes

Author Index

Subject Index

N

O

P

T